Proceedings of the Conference on Categorical Algebra

This Conference was supported by the
United States Air Force Office of Scientific Research

Proceedings of the Conference on Categorical Algebra

La Jolla 1965

Edited by

S. Eilenberg · D. K. Harrison · S. MacLane · H. Röhrl

Springer-Verlag Berlin · Heidelberg · New York 1966

Professor Dr. S. EILENBERG
Columbia University, Hamilton Hall, New York 27, N. Y.

Professor Dr. D. K. HARRISON
Department of Mathematics, University of Oregon, Eugene, Ore.

Professor Dr. S. MACLANE
Department of Mathematics, The University of Chicago, Chicago, Ill.

Professor Dr. H. RÖHRL
Department of Mathematics, University of California at San Diego, La Jolla

ISBN-13: 978-3-642-99904-8 e-ISBN-13: 978-3-642-99902-4
DOI: 10.1007/978-3-642-99902-4

Softcover reprint of the hardcover 1st edition 1966

Title No. 1339

Preface

This volume contains the articles contributed to the Conference on Categorical Algebra, held June 7—12, 1965, at the San Diego campus of the University of California under the sponsorship of the United States Air Force Office of Scientific Research. Of the thirty-seven mathematicians, who were present seventeen presented their papers in the form of lectures. In addition, this volume contains papers contributed by other attending participants as well as by those who, after having planned to attend, were unable to do so.

The editors hope to have achieved a representative, if incomplete, coverage of the present activities in Categorical Algebra within the United States by bringing together this group of mathematicians and by soliciting the articles contained in this volume. They also hope that these Proceedings indicate the trend of research in Categorical Algebra in this country.

In conclusion, the editors wish to thank the participants and contributors to these Proceedings for their continuous cooperation and encouragement. Our thanks are also due to the Springer-Verlag for publishing these Proceedings in a surprisingly short time after receiving the manuscripts.

S. Eilenberg
D. K. Harrison
S. MacLane
H. Röhrl

Resolved

The members of the Conference on Categorical Algebra wish to express their deep appreciation to those who have made this conference possible:

To the University of California at San Diego and especially to Professors ECKART, GOLDBERG, STEWART and WARSCHAWSKI for holding the conference in this idyllic setting,

To the Air Force Office of Scientific Research, and especially to Dr. R. J. POHRER, for providing the support and encouragement necessary for the conference,

To Mrs. ILSE WARSCHAWSKI and Mrs. VIVIAN RÖHRL, for hospitable reception and entertainment,

To Mrs. BARI SACCOMAN, for daily assistance with many practical problems,

Finally and most especially to Professor HELMUT RÖHRL, for his manifest mastery of manifold arrangements.

Unanimously adopted
June 11, 1965

Contents

List of Participants

Auslander, B. L.
Auslander, M.
Bass, H.
Beck, J.
Buchsbaum, D.
Chase, S. U.
Dickson, S. E.
Diener, K. H.
Duskin, J. W.
Dyson, V. H.
Eilenberg, S.
Engeler, E.
Faber, R.
Freyd, P.
Give'on, Y.
Gray, J. W.
Harrison, D. K.
Heller, A.
Isbell, J. R.
Kan, D. M.
Kelly, G. M.
Knighten, C. M.
Knighten, R. L.
Lawvere, F. W.
Leicht, J. B.
Linton, F. E. J.
MacLane, S.
Morimoto, A.
Rinehart, G. S.
Röhrl, H.
Sonner, J.
Tierney, M.
Verdier, J. L.
Walker, C. L.
Walker, E. A.
Watts, C. E.
Wyler, O.

The Category of Categories as a Foundation for Mathematics*,**

By

F. William Lawvere

In the mathematical development of recent decades one sees clearly the rise of the conviction that the relevant properties of mathematical objects are those which can be stated in terms of their abstract structure rather than in terms of the elements which the objects were thought to be made of. The question thus naturally arises whether one can give a foundation for mathematics which expresses wholeheartedly this conviction concerning what mathematics is about, and in particular in which classes and membership in classes do not play any role. Here by "foundation" we mean a single system of first-order axioms in which all usual mathematical objects can be defined and all their usual properties proved. A foundation of the sort we have in mind would seemingly be much more natural and readily-useable than the classical one when developing such subjects as algebraic topology, functional analysis, model theory of general algebraic systems, etc. Clearly any such foundation would have to reckon with the Eilenberg-MacLane theory of categories and functors. The author believes, in fact, that the most reasonable way to arrive at a foundation meeting these requirements is simply to write down axioms descriptive of properties which the intuitively-conceived category of all categories has until an intuitively-adequate list is attained; that is essentially how the theory described below was arrived at. Various metatheorems should of course then be proved to help justify the feeling of adequacy. The system to be described is an improved version of the one sketched in Chapter 1 of the author's doctoral dissertation [Columbia, 1963].

By the *elementary theory of abstract categories* we mean the notions of formula and theorem defined as follows

0. For any letters x, y, u, A, B the following are formulas

$$\Delta_0(x) = A\,, \quad \Delta_1(x) = B\,, \quad \Gamma(x, y; u)\,, \quad x = y\,.$$

* Research partially supported by an NSF-NATO Postdoctoral Fellowship.
** Received September 8, 1965

These are to be read, respectively, "A is the domain of x", "B is the codomain of x", "u is the composition x followed by y", and "x equals y".

1. If Φ and Ψ are formulas, then

$$[\Phi] \text{ and } [\Psi]$$
$$[\Phi] \text{ or } [\Psi]$$
$$[\Phi] \Rightarrow [\Psi]$$
$$\text{not } [\Phi]$$

are also formulas.

2. If Φ is a formula and x is a letter, then

$$\forall x[\Phi], \quad \exists x[\Phi]$$

are also formulas. These are to be read, as usual, "for every x, Φ" and "there is an x such that Φ", respectively.

3. A string of marks is a *formula* of the elementary theory of abstract categories iff its being so follows from 0, 1, 2 above. Of course we immediately begin to make free use of various ways of abbreviating formulas. The notion of free and bound variables in a formula can now be defined; we mean by a *sentence* any formula with no free variables, i.e. in which every occurence of each letter x is within the scope of a quantifier $\forall x$ or $\exists x$.

The theorems of the elementary theory of abstract categories are all those sentences which can be derived by logical inference from the following axioms (it is understood that Δ_0, Δ_1 are unary function symbols)

Four bookkeeping axioms

$$\Delta_i(\Delta_j(x)) = \Delta_j(x), \quad i, j = 0, 1.$$

$$\Gamma(x, y; u) \text{ and } \Gamma(x, y; u') \Rightarrow u = u',$$
$$\exists u[\Gamma(x, y; u)] \Longleftrightarrow \Delta_1(x) = \Delta_0(y),$$
$$\Gamma(x, y; u) \Rightarrow \Delta_0(u) = \Delta_0(x) \text{ and } \Delta_1(u) = \Delta_1(y).$$

Identity axiom

$$\Gamma(\Delta_0(x), x; x) \text{ and } \Gamma(x, \Delta_1(x); x).$$

Associativity axiom

$$\Gamma(x, y; u) \text{ and } \Gamma(y, z; w) \text{ and}$$
$$\Gamma(x, w; f) \text{ and } \Gamma(u, z; g) \Rightarrow f = g.$$

Besides the usual means of abbreviating formulas, the following (as well as others) are special to the elementary theory of abstract categories:

$$A \xrightarrow{f} B \text{ means } \Delta_0(f) = A \text{ and } \Delta_1(f) = B,$$

$fg = h$ means $\Gamma(f, g; h)$,

$$\begin{array}{ccc} A & \xrightarrow{f} & B \\ & h \searrow & \downarrow g \\ & & C \end{array} \quad \textit{commutes} \text{ means } \begin{array}{l} \Delta_0(f) = \Delta_0(h) = A \text{ and} \\ \Delta_1(f) = \Delta_0(g) = B \text{ and} \\ \Delta_1(g) = \Delta_1(h) = C \text{ and} \\ \Gamma(f, g; h). \end{array}$$

(Notice that we write compositions in the order of the arrows from left to right.)

Commutative diagrams in general are regarded as abbreviated formulas, signifying the usual indicated systems of equations. For example, our statement above of the associativity axiom becomes transparent on contemplating the following commutative diagram, made up of four elementary triangles of the above sort.

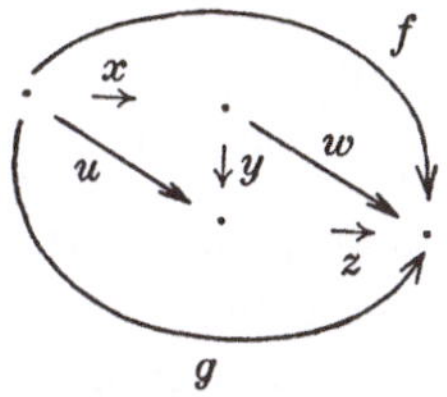

Further abbreviated formulas are

Obj(A) means a) $A = \Delta_0(A) = \Delta_1(A)$,

b) $\exists x[A = \Delta_0(x)]$ or $\exists y[A = \Delta_1(y)]$,

c) $\forall x \forall u[\Gamma(x, A; u) \Rightarrow x = u]$ and $\forall y \forall v[\Gamma(A, y; v) \Rightarrow y = v]$.

That is, the three formulas a, b, c express provably equivalent properties of A, and this common property is that of being an *object*. It is usually understood that a capital letter used as a variable (free or bound) is restricted to refer only to objects.

Mono(f) means $\forall x \forall y[xf = yf \Rightarrow x = y]$.

Epi(f) means $\forall x \forall y[fx = fy \Rightarrow x = y]$.

Endo(f) means $\Delta_0(f) = \Delta_1(f)$.

Iso(f) means $\exists g[fg = \Delta_0(f)$ and $gf = \Delta_1(f)]$.

$A \cong B$ means $\exists f[A \xrightarrow{f} B$ and Iso$(f)]$.

A is a *retract* of B means $\exists f \exists g[A \xrightarrow{f} B$ and $fg = A]$.

G is a *generator* means $\forall f \forall g[\Delta_0(f) = \Delta_0(g)$ and $\Delta_1(f) = \Delta_1(g)$ and $f \neq g \Rightarrow \exists x[\Delta_0(x) = G$ and $\Delta_1(x) = \Delta_0(f)$ and $xf \neq xg]]$.

In a similar way a great number of the usual categorical notions can be expressed as formulas in the elementary theory of abstract categories;

for example, Prod $(A, B; P, p, q)$, meaning that P with projections p, q is a product of A with B, the notions of coproduct, terminal object, coterminal object, equalizer, coequalizer, meet (pullback), and comeet (pushout) are all elementary. However the notions of infinite limits and colimits, or of an object being "finitely generated" are not always elementary from the point of view of a given category, although they do become elementary if the category is viewed as an object in the category of categories, as explained below.

By a category we of course understand (intuitively) any structure which is an interpretation of the elementary theory of abstract categories, and by a functor we understand (intuitively) any triple consisting of two categories and a rule T which assigns, to each morphism x of the first category, a unique morphism xT of the second category in such a way that always

$$\begin{array}{lll} \text{if} \quad \Delta_i(x) = A\,, & \text{then} \quad \Delta'_i(xT) = AT & \text{for} \quad i = 0, 1\,, \\ \text{if} \quad \Gamma(x, y; u)\,, & \text{then} \quad \Gamma'(xT, yT; uT)\,. & \end{array}$$

Here "morphism" is the usual name for the "elements" of a category, the primes denote the interpretations of Δ_0, Δ_1, Γ in the second category, and calling T a "rule" is not supposed to have any connotation of effectiveness, etc.

With the evident definitions of Δ_0, Δ_1, Γ, the world of all functors becomes itself a category. Our purpose for the remainder of this article will then be to indicate certain axioms which hold for this intuitively-conceived category; actually there will be two theories, *a basic theory and a stronger theory*.

Both the basic theory and the stronger theory have the same notion of formula, which is essentially that of the elementary theory of abstract categories except that two individual constants ∂_0, ∂_1 are adjoined. These are needed in order to enable us to distinguish in a fixed way between a category and its dual, and they are intended to denote the two constant endofunctors of the ordinal number 2, considered as the category pictured below

$$\underset{0}{\cdot} \rightarrow \underset{1}{\cdot}\,.$$

Formally 2 is defined by (any one of) the equations

$$\Delta_i(\partial_j) = 2\,, \quad i, j = 0, 1\,.$$

Of course, now that we are in the category of categories, the things denoted by capitals will be called categories rather than objects, and we shall speak of functors rather than morphisms.

The axioms of the basic theory are those of the elementary theory of abstract categories plus several more axioms.

First we assume the existence of the category with exactly one morphism.

$$\exists 1 \, \forall A \, \exists! x [A \xrightarrow{x} 1] .$$

A functor is called *constant* iff it factors through 1. We also find it a great notational convenience to assume the following "partial skeletal axiom":

$$\forall x [A \xrightarrow{x} A \quad \text{and} \quad \mathrm{Iso}(x) \Rightarrow x = A] \quad \text{and} \quad A \cong B \Rightarrow A = B .$$

That is, if the identity is the only endofunctor of A which is an automorphism, then A is the only category in its isomorphism class. For example, 1 is the unique terminal category. We now state axioms characterizing 2:

∂_0 and ∂_1 are constant .
$\Gamma(\partial_i, \partial_j; \partial_j)$, $\quad i, j = 0, 1$.
$\partial_0 \neq \partial_1$, $\quad \partial_i \neq 2$, $\quad i = 0, 1$.
$\forall x [2 \xrightarrow{x} 2 \Rightarrow x = \partial_0 \quad \text{or} \quad x = \partial_1 \quad \text{or} \quad x = 2]$.
2 is a generator.
If C is any generator, then 2 is a retract of C.

The intuitive validity of the last statement is easily seen with the help of the category E to be defined presently.

Proposition. If C is any generator with exactly three endofunctors, two of which are constant, and which is a retract of any other generator, then $C = 2$.

We remark that a simpler set of properties hoped by FREYD to characterize 2 [*Abelian Categories*, HARPER and ROW 1964] fails to do so since the following category also has exactly two objects and three endofunctors:

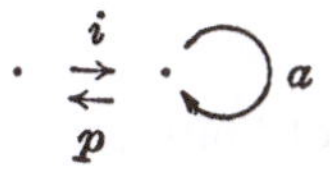

$$i p = \text{identity}, \quad a = p i .$$

The symbols τ, $\bar{\partial}_i$ will be used to denote the unique functors making the following diagram commutative.

$$\begin{array}{ccc} 2 & \xrightarrow{\partial_i} & 2 \\ & {}_{\tau}\searrow \quad \nearrow_{\bar{\partial}_i} & \\ & 1 & \end{array} \qquad i = 0, 1 .$$

Basic is the following

Definition. $x \in A$ means $2 \xrightarrow{x} A$.

This will be read "x is a morphism in A". While this notation has a strong intuitive appeal, it should not be thought to have much formal connection with class elementhood in the usual sense; for example, if two categories have at least one morphism in common, they are equal.

Definition. If $x \in A$, then

$$A \models \Delta_i(x) = a \quad \text{means} \quad \partial_i x = a \qquad i = 0,1\,.$$

These are read "a is the domain [respectively codomain] of x in A"; they clearly imply that $2 \xrightarrow{a} A$ also. Using this we can define $A \models \mathrm{Obj}\,(a)$ in the obvious way. We sometimes confuse an object $a \in A$ with the corresponding $1 \xrightarrow{\bar{a}} A$ such that $a = \tau\bar{a}$. Note that it is provable that

$$\partial_0 \xrightarrow{2} \partial_1 \quad \text{"in 2"}\,.$$

Axiom of Finite Roots. *There is a coterminal category* 0. *Any two categories have a product and a coproduct. Any two functors with a common domain category and a common codomain category have an equalizer and a coequalizer.*

It is well known that meets and comeets, etc., in particular inverse images and intersections can then be proved to exist. We also assume at this point the following axiom.

If $A \xrightarrow{i} A + B \xleftarrow{j} B$ is a coproduct diagram, then

$$x \in A + B \Rightarrow \exists y\,[x = y\,i] \quad \text{or} \quad \exists z\,[x = z\,j]\,.$$

Incidentally, if

$$1 \underset{\bar{\partial}_1}{\overset{\bar{\partial}_0}{\rightrightarrows}} 2 \to N$$

is a coequalizer diagram, then we call N the additive monoid of non-negative integers. This shows that the basic theory needs no explicit "axiom of infinity".

Another consequence of the axiom of finite roots is that the colimit of the following diagram exists:

$$\begin{array}{ccc} 1 & \xrightarrow{\bar{\partial}_0} & 2 \\ & \bar{\partial}_1 \times \bar{\partial}_0 & \\ 1 & \xrightarrow[\bar{\partial}_1]{} & 2 \end{array}$$

Denote this colimit by E and the two injections $2 \to E$ by φ and ψ.

Axiom. *The category E has exactly four morphisms, namely*

$$\varphi,\ \psi,\ \partial_0\psi = \partial_0\varphi\,,\quad \partial_1\psi = \partial_1\varphi\,.$$

This axiom may well be provable from the others; at any rate it allows us to picture the "inside" of E as follows:

$$\dot{0} \overset{\varphi}{\underset{\psi}{\rightrightarrows}} \dot{1}$$

Although we have characterized 2, we still have not assumed enough about it, for all axioms stated so far are valid in the category of directed graphs (in particular our remark about N is really only sensible in view of axioms still to be stated). We need another

Definition. 3, α, β will always mean the unique category and functors in the following *comeet* (pushout) diagram:

$$\begin{array}{ccc} 1 & \xrightarrow{\overline{\partial}_1} & 2 \\ {\scriptstyle \overline{\partial}_0}\downarrow & & \downarrow{\scriptstyle \alpha} \\ 2 & \xrightarrow[\beta]{} & 3 \end{array}$$

Axiom. 3 *has exactly one morphism* γ *besides the five implied by the definition (which are distinct); it satisfies*

$$\partial_0 \gamma = \partial_0 \alpha, \quad \partial_1 \gamma = \partial_1 \beta .$$

Thus 3 may be pictured internally:

$$\begin{array}{ccc} \overset{0}{\bullet} & \overset{\alpha}{\rightarrow} & \overset{1}{\bullet} \\ & \gamma\searrow & \downarrow\beta \\ & & \underset{2}{\bullet} \end{array}$$

Using 3, α, β, γ we can make the all-important

Definition. If $f \in A$, $g \in A$, $h \in A$, then

$$A \models \Gamma(f, g; h) \quad \text{means} \quad \exists t[\alpha t = f, \ \beta t = g, \ \gamma t = h].$$

This is read "h is the composition f followed by g in A". For example, $3 \models \Gamma(\alpha, \beta; \gamma)$ can be proved.

We can also now formally prove that every f in our world is a functor according to our earlier intuitive definition. Of course such a t as above must satisfy $3 \xrightarrow{t} A$; hence the letter t for "triangle".

We have defined $A \models \Delta_0$, $A \models \Delta_1$, $A \models \Gamma$. Now given any formula Φ of the elementary theory of abstract categories, we can make in Φ the following substitutions

$$\begin{array}{lll} A \models \Delta_i & \text{for} & \Delta_i \quad i = 0,1 \\ A \models \Gamma & \text{for} & \Gamma \\ \forall x[x \in A \Rightarrow \ ] & \text{for} & \forall x[\ ] \\ \exists x[x \in A \text{ and } \ ] & \text{for} & \exists x[\ ] \end{array}$$

and thus obtain a new formula $A \models \Phi$ in the basic theory which has one more free variable (namely A) than Φ has, and which expresses intuitively the statement that Φ is true in A (of the morphisms in A denoted by the free variables if there are any). If Φ is a sentence of the elementary theory of abstract categories and if $A \models \Phi$ holds, we say that A is a *model* for Φ or that Φ is *true in* A. Similar remarks hold for formulas of the basic theory if ∂_0, ∂_1 are thought of as variables in the modified formula. We have not yet described all of the basic theory; however its axioms will only be finite in number, so the conjunction of all of them (more precisely of their universally quantified closures) will be a single sentence. Thus

"there exists a model A of the basic theory having property Ψ"

(where Ψ is any other formula of the basic theory) is also a formula of the basic theory, which will be a sentence if Ψ has only one free variable A. Of course such a formula could not be proved in the basic theory; in fact, the passage from the basic theory to the stronger theory will involve just the addition of two formulas of the above type to the list of axioms.

At present, we have not yet assumed enough to insure that we are not talking about a category of objects with a non-associative partial multiplication. We now remedy this.

Definition. 4 is the category in the comeet diagram.

$$\begin{array}{rcc} & 1 \xrightarrow{\bar{\partial}_0} 2 \\ \bar{\partial}_1 \gamma \downarrow & \quad \downarrow \\ & 3 \rightarrow 4 \end{array}$$

Axiom. 4 *has exactly ten morphisms, satisfying the evident equations so that* 4 *may be pictured*:

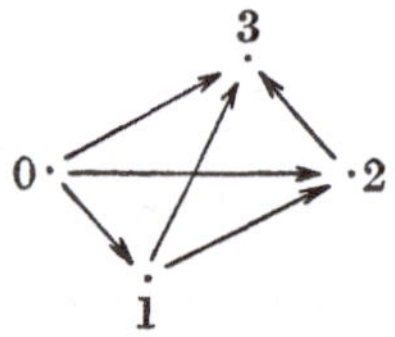

Evidently the above picture is closely related to the diagram used to describe the associativity axiom in our discussion of the elementary theory of abstract categories. (Let

$$0 \xrightarrow{x} 1 \xrightarrow{y} 2 \xrightarrow{z} 3 .$$

Then f, g are represented by the same arrow in the above picture.) In fact

Theorem Schema. *If Φ is any theorem of the elementary theory of abstract categories, then*

$$\forall A [A \models \Phi]$$

is a theorem of the basic theory of the category of all categories.

Thus every object in a world described by the basic theory is at least a category. The remaining five axioms of the basic theory are intended to help insure that the objects are no more than categories (i.e. have no further structure) by insuring that there are many functors, and also to help make sure that there are enough categories. The additional two axioms of the stronger theory will have also the latter aim.

One of the most important constructions of category theory is the formation of functor categories, and of course our intuition tells us that whenever two categories exist in our world, then so does the corresponding category of all natural transformations between the functors from the first category to the second. Thus

Axiom. Given two categories A, B, there is a category B^A and a functor (called evaluation)

$$A \times B^A \xrightarrow{e} B$$

such that for any C and for any $A \times C \xrightarrow{f} B$ there is exactly one $C \xrightarrow{h} B^A$ such that

$$f = (A \times h)\, e. \qquad \begin{array}{ccc} A \times C & & \\ A \downarrow \times \downarrow h & \searrow f & \\ A \times B^A & \xrightarrow{e} & B \end{array}$$

[The above statement of the exponentiation axiom is not quite precise since $A \times h$ is not meaningful until projections have been chosen for the two products. The intention is that the axiom as stated holds for any choice of projections (and of the product categories themselves) with the correct universal properties; this can easily be written out directly in the elementary language in about a page and a half.]

The exponentiation axiom implies that products distribute over sums, that the usual laws of exponents hold, and that for any three categories A, B, C there is a single "composition functor"

$$B^A \times C^B \xrightarrow{\circ} C^A .$$

To deduce the usual internal description, let $C = 1$ in the exponentiation axiom; from the fact that $A \times 1 \cong A$, it then follows immediately that the objects in B^A correspond exactly to the functors from A to B. To see that the morphisms in B^A correspond exactly to the "natural transformations" between such functors, set $C = 2$ in the axiom, use commutativity of products and the fact that a functor $A \to B^2$ must have commutative squares in B as values since 2×2 has an easily-deduced internal picture, and verify that the equations which must be satisfied at each stage of the just-indicated transformations imply that all these individual squares fit together as they should*. If $2 \xrightarrow{\varphi} B^A$ is a morphism in B^A, then the functors $A \to B$ corresponding to its domain and codomain in B^A are explicitly

$$A \cong A \times 1 \xrightarrow{A \times \partial_i} A \times 2 \xrightarrow{\bar{\varphi}} B \qquad i = 0,1$$

where $\bar{\varphi} = (A \times \varphi)\, e$, the e being of course the evaluation functor. If $A \xrightarrow{f} B$, we denote by 1_f the object of B^A corresponding to it. In particular 1_A is a distinguished object $2 \to A^A$ in A^A, and $1_{fg} = 1_f \circ 1_g$.

Definition. The category A is said to be *discrete* (or to be a *set*) iff A^τ is an isomorphism $A^2 \xrightarrow{\sim} A^1$. That is, every morphism in a set is an object.

Axiom. *For any category A there is a discrete category A_c with a functor $A \to A_c$ such that for any functor $A \to B$ from A to a discrete category there is exactly one functor making this diagram commute*

$$\begin{array}{ccc} A & \to & A_C \\ & \searrow & \downarrow \\ & & B \end{array} \quad .$$

This A_c is called the *set of components* of A.

Axiom. Dualize the proceeding axiom. Thus every category A has a maximal discrete subcategory $|A|$, called for obvious reasons the *set of objects* of A. The "absolute value" notation for the set of objects will be used consistently. By the *set of morphisms* of A we understand the discrete category $|A^2|$, since A^2 is of course a category whose objects correspond to morphisms in A. In particular

$$|N^2|$$

is called the *set of nonnegative integers*, where N is the monoid of nonnegative integers as previously defined.

We also state now the

Axiom of Choice.

$$A \xrightarrow{f} B,\ 0 \not\cong A,\ B \text{ discrete} \Rightarrow \exists g[fgf = f].$$

Now it follows easily from the definition that if B is discrete, so is B^C for any C. Also, since the usual formal proof that adjoints preserve

---

* B^2 is not to be confused with $B \times B$; ∂_0, ∂_1 induce a functor $B^2 \to B \times B$ which is in general faithful but not full.

limits holds equally well for metacategories and metafunctors (i.e. "subcategories" of the universe defined by formulas but which, like the full metacategory of all sets, cannot necessarily be represented by an actual category in the universe,) it follows that sets are closed under the formation of finite roots. In fact,

Metatheorem. *Define a relative interpretation of the elementary theory of abstract categories into the basic theory by relativizing all quantifiers to functors between discrete categories. Then in the induced theory, all theorems of the elementary theory of the category of sets* [LAWVERE, Proc. Nat. Ac. Sc. USA Dec. 1964] *are provable.*

Thus one could, by referring only to discrete categories, develop on the basis of the axioms we have so far assumed such subjects as number theory, calculus, linear operators in Hilbert space, etc. (such a program, of course, would not make the most efficient use of the functorial method.) In such a development, as well as in our work here, it is convenient to use the following metatheorem, which, because it is provable in the elementary theory of the category of sets, is by the above also available in the basic theory of the category of categories.

Predicative Subset Schema. *Suppose that Φ is any formula of the elementary theory of abstract categories whose free variables are $A_0, \ldots, A_{m-1}, x_0, \ldots, x_{n-1}$, and in which all bound variables are restricted to range only over morphisms whose domains and codomains are among the A_j, $j < m$.* Let σ_0, σ_1 *be any mappings from n to m, and think of x_i as ranging over morphisms $A_{i\sigma_0} \to A_{i\sigma_1}$ for $i < n$. Then (it is provable in the elementary theory of the category of sets that) there exists a subset y_Φ of*

$$\prod_{i<n} A_{i\sigma_1}^{A_{i\sigma_0}}$$

such that the members of y_Φ are exactly those elements $\langle x_0, \ldots, x_{n-1}\rangle$ of the above product for which

$$\Phi(A_0, \ldots, A_{m-1}, x_0, \ldots, x_{n-1})$$

is true.

In the above assertion the definitions of subset, member, and element are those given in the above-cited article. Note that as a special case one could have $A_{i\sigma_0} = 1$ and $A_{i\sigma_1} = A$ for all $i < n$, so that in particular all predicatively definable relations on a given set A exist.

The last axiom of the basic theory is intended to express that the full (and finite) metacategory determined by 1, 2, 3 is *adequate* in the universe in the sense of ISBELL [Rozprawy Matematyczne XXXVI]. Of course, no elementary axiom could really express this, but at least we can express adequacy relative to the metacategory of sets as we have defined that. Essentially the axiom states that when given a function from the set of morphisms of a category A into the set of morphisms of a category B, if it satisfies the intuitive definition of "functor", then we can find in the universe the corresponding actual functor from A itself to B itself.

Axiom. *If* $|A^2| \xrightarrow{\bar{f}} |B^2|$ *and*

$$\forall t[3 \xrightarrow{t} A \Rightarrow \exists!\, u[3 \xrightarrow{u} B \text{ and } |u^2| = |t^2|\bar{f}]]$$

then

$$\exists!\, f[A \xrightarrow{f} B \quad \text{and} \quad \bar{f} = |f^2|].$$

Having presented the axioms for the basic theory of the category of categories, we now ask what can be done with them. Besides the possibility of developing analysis which was previously alluded to, one can also define easily the full metacategories of ordered sets, groups, or algebraic theories [LAWVERE, Proc. Nat. Ac. Sc. USA, Nov. 1963] and study these to a considerable extent; however comfortably complete *categories* (i.e. objects in the universe) corresponding to these cannot be shown to exist without adding the two axioms of the stronger theory which will be discussed presently. For any category A, the category of semisimplicial objects from A can also be shown to exist in the basic theory, although a much less messy proof can be given in the stronger theory. The general theories of triplable categories, of fibered categories, and of closed categories (when the latter is phrased so as not to refer to the category of sets) can all be developed quite nicely within the basic theory, as can many other things. Thus before we state the stronger axioms, we will discuss some principles which can be proved using only the basic theory.

First we point out that of the several definitions of "adjoint functors", all except the one involving hom-functors can be easily stated in the basic theory. The following general adjoint functor theorem can then also be proved in the basic theory.

Theorem. *A functor* $A \xrightarrow{f} B$ *has an adjoint iff*

i) *f preserves all (inverse) limits which exist in A.*

ii) *For every object $b \in B$, the category (b, f) has a final subcategory C_b which is among those over which A has (inverse) limits.*

Here (b, f) is a special case of an operation defined below, and to say that $C_b \to (b, f)$ is *final* is meant in the following sense:

> $C \xrightarrow{u} C'$ is *final* iff for every g such that $\Delta_0(g) = C'$, if $\varprojlim (ug)$ exists, so does $\varprojlim (g)$ and
>
> $$\varprojlim (g) \cong \varprojlim (ug)$$
>
> in $\Delta_1(g)$.

All other adjoint functor theorems are refinements of the above intended to make condition (ii) easier to verify in special cases. Usually one employs a notion of "small" category (which will be available in our stronger theory) and assumes that A is complete (has limits over small categories); then one need only say in condition (ii) that C_b is small and final in (b, f) for each b. Sometimes it is useful also to assume that the hom-sets (defined below) of A and B are small. The existence of such C_b's in this context is easily seen to be equivalent to Freyd's Solution-Set Condition [*Abelian Categories*].

The following operation is very convenient, and easily seen to exist in the basic theory. Given two functors

$$A_i \xrightarrow{f_i} B \qquad i = 0,1$$

with a common co-domain, define the category (f_0, f_1) so that all three squares below are meets (pullbacks).

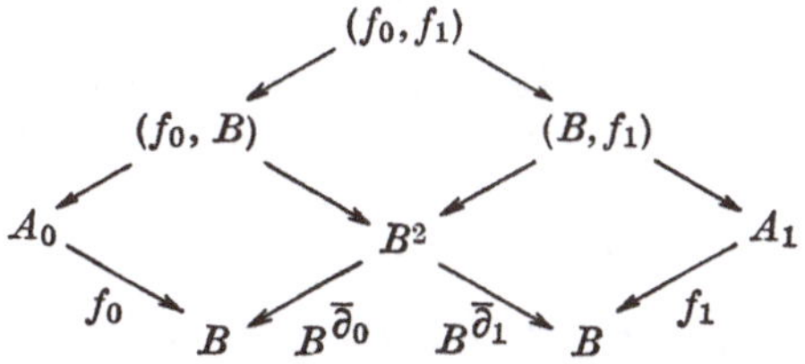

Note that (except for the canonical isomorphism $B^1 \cong B$) $B^{\bar{\partial}_i}$ represent the domain and codomain functors on B. There is a forgetful functor

$$(f_0, f_1) \to A_0 \times A_1$$

each object of (f_0, f_1) having the additional structure involving a morphism in B.

We consider some special cases of the $(\ ,\ )$ notation. If $A_0 = f_0 = B$, and if $A_1 = 1$, so that $f_1 = b$ is an object in B, then

$$(B, b)$$

is nothing but the category of "objects over b" as used for example by Beck in his triple cohomology [this volume]. If both A_0, A_1 are 1, then one can show that

$$(b_0, b_1)$$

is a discrete category, called the set of B-morphisms from b_0 to b_1, or simply a *hom-set*. (This does not mean, by the way, that a hom-functor exists for B. For one thing, we cannot show in the basic theory that a category of sets exists, and in any case no single category of sets could serve as the recipient of hom-functors for all categories B.) The third special case which we consider is that where $B = f_1 = A_1$ is a monoid (category with one object) with $A_0 = 1$ and $f_0 = e =$ the unique functor $1 \to B$. Then

$$(e, B)$$

is a category which contains the divisibility information about B. If the monoid B has cancellation, then (e, B) is a preorder (category in which every hom-set is 0 or 1) and in particular we define

$$\omega = (0, N)$$

the *well-ordered set of natural numbers.*

And of course still another case of the (,) notation was used in the statement of the adjoint functor theorem.

The following theorem has somewhat the same sort of use in constructing categories that the adequacy axiom has in constructing functors. Notice that the hypothesis of the theorem describes essentially a set equipped with a partial multiplication table which satisfies the axioms of the elementary theory of abstract categories.

Theorem. *Let A_1, A_2, A_3, A_4 be given discrete categories, and let $A_2 \xrightarrow{d_i} A_2$, $i = 0,1$ and $A_3 \xrightarrow{c} A_2$ be given functors, satisfying the following conditions:*

$$d_i d_j = d_i \qquad i, j = 0,1 .$$

There is given $A_1 \to A_2$ which is the equalizer of d_0 with d_1. A_3 is the meet of d_0 with d_1, with structural functors a_i:

$$\begin{array}{ccc} A_3 & \xrightarrow{a_0} & A_2 \\ {\scriptstyle a_1}\downarrow & & \downarrow{\scriptstyle d_1} \\ A_2 & \xrightarrow[d_0]{} & A_2 \end{array}$$

A_4 is the meet of a_0 with a_1, with structural functors b_i:

$$\begin{array}{ccc} A_4 & \xrightarrow{b_0} & A_3 \\ {\scriptstyle b_1}\downarrow & & \downarrow{\scriptstyle a_1} \\ A_3 & \xrightarrow[a_0]{} & A_2 \end{array}$$

The functor $A_3 \xrightarrow{c} A_2$ satisfies the "book-keeping" axioms:

$$c\, d_i = a_i\, d_i \qquad i = 0,1$$

as well as the identity axioms

$$e_i\, c = A_2 \qquad i = 0,1$$

and the associativity axiom

$$f_0\, c = f_1\, c$$

where $A_4 \xrightarrow{f_i} A_3$ and $A_2 \xrightarrow{e_i} A_3$ are constructed in the evident fashion.

Conclusion. *There is a category A together with isomorphisms*

$$A_i = |A^i| \qquad i = 1, 2, 3, 4$$

so that (in addition to obvious compatibility conditions)

$$d_i = |A^{\partial_i}| \qquad i = 0,1$$
$$c = |A^{\gamma}|.$$

Any two such categories are canonically isomorphic.

In the proof of the above theorem, A is constructed as a quotient of the free category $A_2 \times 2$, the two functors along which the coequalizer is taken being constructed with the help of the given d_0, d_1, c.

Corollary. *Every category A has a dual A^*.*

For let $A_2 = |A^2|$, $d_0 = |A^{\partial_1}|$, $d_1 = |A^{\partial_0}|$, etc.

Combining the above theorem with the Predicative Subset Schema for sets and with the adequacy axiom, one can derive

Predicative Functor-Construction Schema. *Let $\Phi(a, b)$ be a formula (possibly with parameters) such that all bound variables are suitably restricted, and suppose that*

$$\forall a \in A \, \exists ! b \in B [\Phi(a, b)]$$
$$\Phi(a, b) \Rightarrow \Phi(\partial_i a, \partial_i b) \qquad i = 0,1$$
$$\Phi(a_i, b_i) \quad i = 0, 1, 2 \quad \text{and} \quad A \models \Gamma(a_0, a_1; a_2)$$
$$\Rightarrow B \models \Gamma(b_0, b_1; b_2).$$

Then

$$\exists ! f [A \xrightarrow{f} B \quad \text{and} \quad \forall a \forall b [a f = b \Longleftrightarrow \Phi(a, b)]].$$

Often in applying this schema (for example in the proof of the general adjoint functor theorem stated earlier), the given formula Φ does not quite have the absolute uniqueness property stated in our above hypothesis, but has it only up to isomorphisms in B which are themselves unique with respect to some other parameters (projections, etc.) in the formula. However, with the help of the axiom of choice a functor f as desired can still be constructed; of course it will itself only be unique up to natural equivalence.

Another corollary of our theorem on construction of categories with given set of morphisms and given multiplication table is the following

Theorem. *Let $B_0, B_1, \ldots, B_{n-1}$ be a finite number of categories, assumed distinct. Then there exists a category with n objects which is "isomorphic" to the full metacategory of all functors between the B_i's.*

Here the formal significance of the word "isomorphic" can be guessed from the construction which proves the theorem, this construction beginning by setting

$$A_2 = \sum_{i,j<n} | B_j^{B_i} |$$

Definition. *For any category A, let $\{A\}$ denote the category with four objects obtained by applying the above theorem to the list of four categories*

$$1, 2, 3, A .$$

(We do not bother to give a definition of $\{A\}$ in the three exceptional cases where A is 1 or 2 or 3.)

The reason for this "singleton" notation is that, intuitively, if C is a category of categories (i.e. a model for the basic theory), then A is ("isomorphic" to) an object in C iff $\{A\}$ is in a smooth way a subcategory of C. This will be made more precise before we state the axioms of the stronger theory. It makes sense to say that a category C is a model for the basic theory since, because the basic theory is finitely axiomatized, the conjunction of all its axioms is a single sentence Φ of the basic theory. (Strictly, it makes sense only relative to a given pair $2 \rightrightarrows C$ of morphisms in C which can play the roles of ∂_0, ∂_1 in C). Now of course (assuming consistency) we cannot prove in the basic theory that there exist models for the basic theory. However, it is useful to know that we can prove in the basic theory that its models are (essentially) just as numerous as models for the elementary theory of the category of sets.

Theorem. *Suppose C is any model for the basic theory. Then the full subcategory determined by the discrete objects in C is a model for the elementary theory of the category of sets. (Such a subcategory exists as the equalizer of the identity functor C with endofunctor of C corresponding to the notion $|\ \ |$). On the other hand, if Q is any model for the elementary theory of the category of sets (which also has a finite number of axioms) then there is a full subcategory of the functor category*

$$Q^{\{4\}*}$$

which is also constructible as a finite left root and which is (except for a minor adjustment to account for the partial skeletal axiom) a model of the basic theory.

The subcategory of the functor category in question is that determined by those contravariant functors from $\{4\}$ to Q which take comeets (of these finite ordinals) into meets (of "sets"). Notice that the $a_0, a_1, b_0, b_1, c, d_0, d_1, e_0, e_1, f_0, f_1$ occurring in our theorem of the construction of categories from multiplication tables have exactly the form of the category $\{4\}^*$, with A_i corresponding to i. Of course, by adequacy any category of categories is represented fully by a category of diagrams of sets which have only three vertices, but 4 is needed here since the category of such diagrams would contain many objects with *non-associative* partial multiplications.

We now prepare to state the axioms of the stronger theory with some definitions.

Definition. *A $(\in)$ C means that C is a model of the basic theory and that there is a full, faithful, root-preserving functor*

$$\{A\} \to C$$

which also preserves the morphisms with the names ∂_0, ∂_1.

Proposition. *A functor $\{A\} \to C$ as described in the above definition is unique if it exists (up to natural isomorphism).*

Notice that if A $(\in)$ C, then there are two objects A_C, 2_C in C with the correct significance:

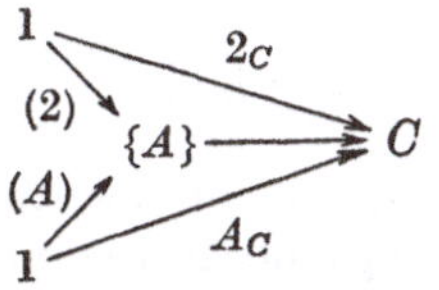

and that one has canonically

$$|A^2| \cong (2_C, A_C).$$

Thus, if A $(\in)$ C, B $(\in)$ C, and if $2 \xrightarrow{x} C$ is such that

$$C \models [\Delta_0(x) = A_C \quad \text{and} \quad \Delta_1(x) = B_C]$$

then there is an induced functor "$(2_C, x)$"

$$|A^2| \to |B^2|.$$

Definition. If $A \xrightarrow{f} B$, A $(\in)$ C, B $(\in)$ C, then

$$f\,(\in)\,C$$

shall mean that some $x \in C$ induces $|f^2|$.

Definition. *C is full in the universe means that C is a model of the basic theory and that whenever A $(\in)$ C, B $(\in)$ C, and*

$$A \xrightarrow{f} B$$

one has

$$f\,(\in)\,C.$$

Definition. *A is C-complete means that C is a model of the basic theory and for every functor f, if f $(\in)$ C then A^f has a co-adjoint. Dually, A is C-co-complete means that every such A^f has an adjoint.*

Actually only functors f with co-domain 1 have to do with limits as such, since they induce the diagonal functors

$$A \to A^D \quad D = \Delta_0(f).$$

However, it is well known that if A has enough co-limits, then lots of induced functors

$$A^{D'} \xrightarrow{A^f} A^D, \quad \text{where} \quad D \xrightarrow{f} D'$$

have adjoints, and since these arise very often (e.g. in algebra and sheaf theory) it seems more to the point to incorporate these directly into the definition of co-completeness.

Axiom. *For any A, and for any model C of the basic theory, there exists a smallest category*

$$C[A]$$

which is a model of the basic theory, which is C-complete and full in the universe, and for which

$$A\ (\in)\ C.$$

Here "smallest" means up to equivalence of categories. Actually the above axiom is not very strong since it does not give us the "inaccessible" category which we need (in fact we still cannot prove that there exist any models of the basic theory). Thus

Axiom. *There exists a category C_0 which is full in the universe (in particular is a model of the basic theory) and such that*

$$C_0 \text{ is } C_0\text{-complete.}$$

Further, any category satisfying these conditions is equivalent to C_0.

The last clause thus embodies the idea that only *one* inaccessible is needed for most mathematics; our world thus stops far short of the second Grothendieck universe if we assume the above axiom. Why not much category theory is lost thereby will be explained below. We could have of course assumed much stronger axioms. For example, by analogy with the work of the set-theorists BERNAYS and LEVY, we could alternately have assumed the following infinite set of axioms.

Strong Reflection Principle: Let Φ be any formula, with free variables $v_0, v_1, \ldots, v_{n-1}$. Then
$\Phi(v_0, v_1, \ldots, v_{n-1}) \Rightarrow \exists C$ [C is C-complete and full in the universe and $v_0, \ldots, v_{n-1}\ (\in)\ C$ and $C \models \Phi(\bar{v}_0, \ldots, \bar{v}_{n-1})$].
(Here $\bar{v}_i$ denotes the morphism in C corresponding to the functor v_i.)

However, we remain in this article with the finite list of axioms which we have presented, and call it the stronger theory (although as we have just pointed out there are much stronger ones still).

Definition. The full subcategory of C_0 determined by its discrete objects is denoted by S and called the *category of sets* (more precisely the category of small sets). A category A is *small* iff A ($\in$) C_0. A category is *complete* iff it is C_0-complete (i.e. has small limits). We write

$$\widetilde{C} = C_0[S].$$

Thus $\widetilde{C}$ is the smallest full, complete category of categories which contains the category of small sets as an object; $\widetilde{C}$ is itself an object in our world.

Actually most mathematics, including most category theory, can be done if we assume only the existence of S and $\widetilde{C}$ and the basic theory, provided we understand that structures are always to be small (i.e., modeled in S), at least whenever we collect structures into categories. For example, semantics functors for categories of small theories [Lawvere, Proc. Nat. Ac. Sc. USA, Nov. 1963 (Algebraic Theories) and Logic Colloquium Leicester 1965 (Elementary Theories)] all take their values in a part of $\widetilde{C}$, and all the usual examples of large fibered categories also involve only a "small" part of $\widetilde{C}$.

Theorem. *If A has small hom-sets then there is a hom-functor*

$$A^* \times A \to S.$$

We leave to the reader to make precise within our language what it means to be a hom-functor. Applying the exponentiation axiom and the usual argument for Yoneda's lemma, we obtain the usual.

Corollary. *If A has small hom-sets, then there is a full and faithful functor*

$$A \to S^{A^*}$$

which preserves any (inverse!) limits which may exist in A.

The above representation is of course the starting point of most investigations into the structure of categories (see for example the work of Freyd, Mitchell, Lawvere, Isbell, Linton, etc.), the aim being in general to cut down on the size of A^* and to say more about the image of the representation. This is also the basic method used in proving the following, which in order to make contact with previous work in foundations, we have phrased in the language of set theory, although it could also be phrased in the stronger theory itself (if the uniqueness of $\mathfrak{C}_0$ is dropped).

Metatheorem. Let θ_2 be the third stongly inaccessible ordinal (where the first is $\theta_0 = \omega$) and let $\mathfrak{C}$ be category whose morphisms are all functors (defined in the obvious set-theoretical way as triples) whose domain and codomain categories have their underling sets of rank less than θ_2. Then $\mathfrak{C}$ is a model for the stronger theory (in particular models for the stronger theory cannot contain an element anything like θ_2 itself). Conversely, given any category $\mathfrak{M}$ which is a model for the stronger theory, and all of whose hom-sets have cardinality less than θ_2, there is a functor

$$\mathfrak{M} \to \mathfrak{C}$$

which is an equivalence of categories if $\mathfrak{M}$ has products of size α for any $\alpha < \theta_2$. Actually, not θ_2 but $\theta_1 + \theta_1$ gives the smallest "natural" model for the stronger theory; this smallest natural model thus has cardinality $\beth_{\theta_1}(\theta_1)$ which is the number reached by starting at the first inaccessible beyond ω (namely θ_1) and iterating the power set operation θ_1 times.

We conclude by posing what seems to be a basic open problem in the foundations of category theory.

Problem. *Find a useful characterization of those complete categories A with small hom-sets such that every functor*

$$A \to S$$

which preserves (inverse) limits has an adjoint.

The aim of the problem is clearly to understand when one can ignore condition (ii) in the General Adjoint Functor Theorem. Such categories A do exist, by FREYD's Special Adjoint Functor Theorem [*Abelian Categories*] where (ii) is replaced by the assumption that A have a cogenerator and that A be well-powered. ISBELL's notion co-adequacy seems to be relevant to the problem. More particulary, one can ask which A's among some known class, say that of algebraic categories, have the property in question.

Footnote added 22 Oct. 1965: Professor ISBELL has since shown the author an example of functor (which can be constructed in our theory) from small groups to small sets which is left continuous but not representable by a small group.

Forschungsinstitut für Mathematik
Eidgenössische Technische Hochschule
Zürich

Fibred and Cofibred Categories*,**

By

John W. Gray

0. Introduction

Fibred categories were introduced by Grothendieck in [*SGA*] and [*BB190*]. As far as I know these are the only easily available references to the subject. Through sheer luck, during the final preparation of this paper I obtained a copy of handwritten notes [*BN*] of a seminar given by Chevalley at Berkeley in 1962 which treated these questions from a slightly different point of view. We discuss the "Chevalley condition" in 3.11.

I became interested in fibred categories as a tool in generalizing Čech cohomology (the results will appear in a subsequent paper). In particular I wanted to prove that if $P: \mathscr{E} \to \mathscr{B}$ is a fibration then for any $\mathscr{L}$, $P^{\mathscr{L}}: \mathscr{E}^{\mathscr{L}} \to \mathscr{B}^{\mathscr{L}}$ is also a fibration. A straightforward proof was rather complicated and indicated that there was more structure around that needed investigation. The idea behind this investigation is that the universal mapping property used in defining a fibration should be equivalent to the existence of a suitable adjoint functor. The point of this paper is that this adjointness can be phrased in a way that looks sufficiently like the notion of a fibration in topology to enable one to use techniques similar to topological ones to prove theorems and to make it possible to identify the correct notion of a cofibration. Clearly, a cofibration $J: \mathscr{A} \to \mathscr{B}$ ought to have the property that $\mathscr{L}^J: \mathscr{L}^{\mathscr{B}} \to \mathscr{L}^{\mathscr{A}}$ is a fibration for suitable categories $\mathscr{L}$. This is completely different from what is called a cofibration in [*SGA*], and hence we use the prefix "op" where [*SGA*] uses "co". As will be seen, if in a certain diagram, all categories are replaced by their opposites (duals) and all functors by the induced functors, then fibrations are turned into opfibrations; i.e., $P: \mathscr{E} \to \mathscr{B}$ is an opfibration if and only if $P^{\mathrm{op}}: \mathscr{E}^{\mathrm{op}} \to \mathscr{B}^{\mathrm{op}}$ is a fibration. On the other hand, if this diagram is replaced by *its* dual in the category

* This work has been partially supported by the National Science Foundation under Grant No. gP-3624.

** Received September 13, 1965.

of categories, then fibrations are turned into cofibrations. For this reason, we have tried wherever possible to give proofs of properties of fibrations within the category of categories in order to have the results automatically for cofibrations by duality. As is pointed out below this has not always been possible, presumably because of lack of technique on our part.

In this introductory paragraph we discuss terminology and prove a few propositions that will be needed later. In § 1 we review some of the results of [*SGA*] so as to have them easily available for reference. In § 2 we establish the equivalence of several defining properties of fibrations and show how the results we want follow almost trivially from these characterizations. In § 3 we treat the category of fibrations over a fixed base category $\mathscr{B}$ and show that it is a reflective subcategory of the category of all categories over $\mathscr{B}$. In § 4 we discuss the preservation of limits by fibrations and apply our results to categories of sheaves. In § 5 we discuss cofibrations and dualize the results on fibrations as far as possible, although a few proofs have so far eluded automatic dualization. We also discuss the relation of cofibrations with sieves (cribles) and obtain a definitive criterion for cofibrations with split cocleavages. I would like to thank ALEX HELLER and JON BECK for several helpful discussions. In particular, the notion of cofibration was developed in conversation with them.

We now turn to questions of notation and terminology.

0.1. Generalities

If $\mathscr{A}$ is a category, we denote the set of morphisms from A to B in $\mathscr{A}$ by $\mathscr{A}(A, B)$ or occasionally by $\mathrm{Hom}(A, B)$. We write

$$F : \mathscr{A} \to \mathscr{B} : A \to F(A) : f \to F(f)$$

for a functor from $\mathscr{A}$ to $\mathscr{B}$ which takes the object A (resp., morphism f) to $F(A)$ (resp., $F(f)$). Given functors

$$\mathscr{A} \xrightarrow{S} \mathscr{B} \underset{G}{\overset{F}{\rightrightarrows}} \mathscr{C} \xrightarrow{T} \mathscr{D}$$

and a natural transformation θ from F to G, we write $T\theta S$ or $T*\theta*S$ for the usual compositions of functors with natural transformations (cf. [*GOD*] or [*SVC*]). The identity morphism (resp., functor, natural transformation) of an object A (resp., category $\mathscr{A}$, functor F) is written i_A (resp., $I_{\mathscr{A}}$, 1_F). If $\mathscr{A}$ is a category then $\mathscr{A}^{\mathrm{op}}$ denotes the opposite (or dual) category and if $F : \mathscr{A} \to \mathscr{B}$ is a functor then $F^{\mathrm{op}} : \mathscr{A}^{\mathrm{op}} \to \mathscr{B}^{\mathrm{op}}$ denotes the induced functor.

0.2. Specific Categories and Functors

A. $*$ denotes the category consisting of one object $*$ and its identity morphism i_*.

B. $\mathscr{I}$ denotes the category $\gamma: 0 \to 1$ consisting of two objects and one non-identity morphism γ. The two functors from $*$ to $\mathscr{I}$ are denoted by

$$R: * \to \mathscr{I} : * \to 1\,,$$

$$D: * \to I\ : * \to 0\,.$$

C. $\mathscr{S}ets$ denotes the category of sets and functions.

D. $\mathscr{C}at$ denotes the category of categories and functors. We should really say a category $\mathscr{C}at$, but any one will do providing it satisfies LAWVERE's axioms [*FS*], which we hereby assume. In particular, $\mathscr{C}at$ has products, sums, pullbacks, pushouts, etc. We shall write $[\mathscr{A}, \mathscr{B}]$ or $\mathscr{B}^{\mathscr{A}}$ for the category whose objects are functors from $\mathscr{A}$ to $\mathscr{B}$ and whose morphisms are natural transformations between such functors. We shall always identify $* \times \mathscr{A}$ and $\mathscr{A}^*$ with $\mathscr{A}$. Hence there are functors

$$R \times \mathscr{A} : \mathscr{A} \to \mathscr{I} \times \mathscr{A}\,,$$

$$D \times \mathscr{A} : \mathscr{A} \to \mathscr{I} \times \mathscr{A}\,,$$

$$\mathscr{A}^R : \mathscr{A}^{\mathscr{I}} \to \mathscr{A}\,,$$

$$\mathscr{A}^D : \mathscr{A}^{\mathscr{I}} \to \mathscr{A}\,.$$

We shall make extensive use of the adjointness relation, written

$$[\mathscr{C}, \mathscr{B}^{\mathscr{A}}] \approx [\mathscr{C} \times \mathscr{A}, \mathscr{B}]$$

or $[\mathscr{C}, [\mathscr{A}, \mathscr{B}]] \approx [\mathscr{C} \times \mathscr{A}, \mathscr{B}]$ or $(\mathscr{B}^{\mathscr{A}})^{\mathscr{C}} \approx \mathscr{B}^{\mathscr{C} \times \mathscr{A}}$.

E. If $\mathscr{A}$ is a category and $A \in \mathscr{A}$, then $\mathscr{A}_A$ (resp., $\mathscr{A}^A$) denotes the category of objects over A (resp., under A); i.e., the objects of $\mathscr{A}_A$ (resp., $\mathscr{A}^A$) are morphisms $A' \to A$ (resp., $A \to A'$) in $\mathscr{A}$ and the morphisms are commutative triangles

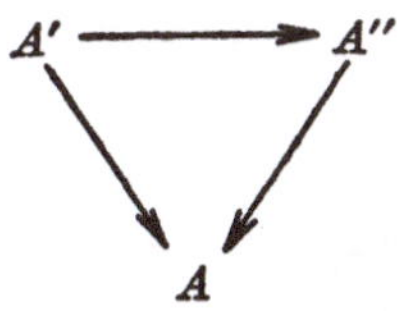

(resp.,

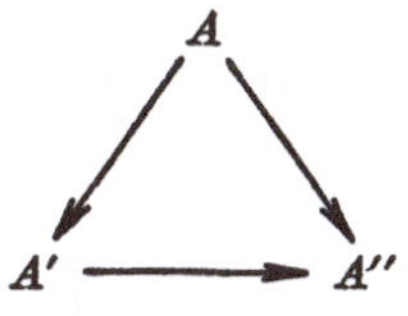

).

Note that $(\mathscr{A}^{\mathrm{op}})_A = (\mathscr{A}^A)^{\mathrm{op}}$. There is one ambiguous case. If $\mathscr{A} = \mathscr{C}at$ and $\mathscr{B} \in \mathscr{C}at$ then $\mathscr{C}at^{\mathscr{B}}$ will always denote categories under $\mathscr{B}$ and not the functor category which will always be denoted by $[\mathscr{B}, \mathscr{C}at]$.

This must be carefully distinguished from the notation for the *fibre* of a functor $P : \mathscr{E} \to \mathscr{B}$ over $B \in \mathscr{B}$, where, if $B \in \mathscr{B}$, then $\mathscr{E}_B = P^{-1}(B)$ denotes the subcategory of $\mathscr{E}$ consisting of all morphisms φ of $\mathscr{E}$ such that $P(\varphi) = i_B$. Intrinsically, $\mathscr{E}_B$ is the pullback in the adjoining diagram, where $B(*) = B$. Dually, if $J : \mathscr{A} \to \mathscr{B}$ then the *cofibre* of J, denoted by $\mathscr{B}/J(\mathscr{A})$ is the pushout in the adjoining diagram.

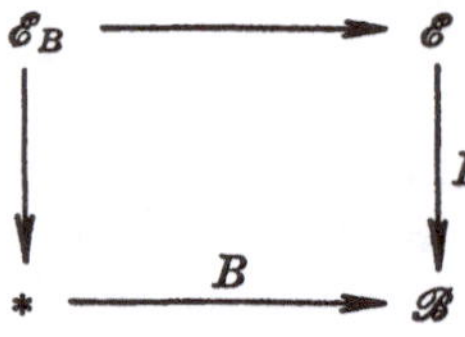

F. We shall use the following notation of Lawvere: Let $F_i : \mathscr{A}_i \to \mathscr{B}$, $i = 0,1$ be functors. In the diagram

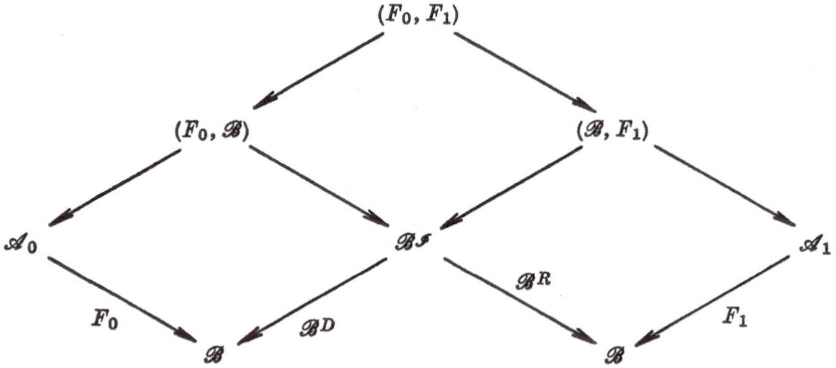

all squares are required to be pullbacks. Thus, in particular

$$\mathrm{ob}(\mathscr{B}, F_1) = \{(f : B \to F_1(A)) \mid B \in \mathscr{B}, A \in \mathscr{A}_1\}$$

and morphisms are commutative squares

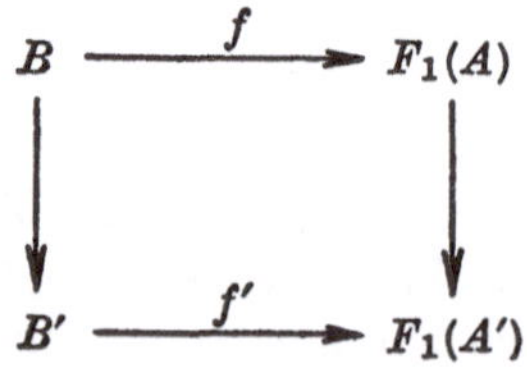

Dually, let $G_i : \mathscr{B} \to \mathscr{A}_i$, $i = 0,1$, be functors. In the diagram

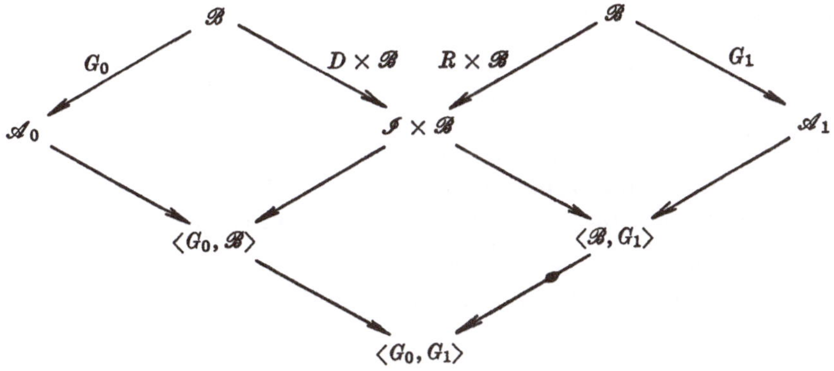

all squares are required to be pushouts. $\langle \mathscr{B}, G_1 \rangle$ can be represented as the disjoint union of $\mathscr{B}$ and $\mathscr{A}_1$ together with new hom sets so that

$$\langle \mathscr{B}, G_1 \rangle (B, A) = \mathscr{A}_1(G_1(B), A) \quad \text{for} \quad B \in \mathscr{B}, \quad A \in \mathscr{A}, \quad \text{while}$$
$$\langle \mathscr{B}, G_1 \rangle (A, B) = \Phi .$$

0.3. Adjointness

A. Let $S : \mathscr{A} \to \mathscr{B}$ and $T : \mathscr{B} \to \mathscr{A}$ be functors. If there exist natural transformations $\theta : ST \to I_{\mathscr{B}}$ and $\psi : I_{\mathscr{A}} \to TS$ such that

$$T\theta \circ \psi T = 1_T \quad \text{and} \quad \theta S \circ S\psi = 1_S$$

then S is called the left adjoint of T and T the right adjoint of S. This is equivalent to the existence of a natural equivalence,

$$\mathscr{B}(S(A), B) \approx \mathscr{A}(A, T(B)) .$$

(See, e.g., [SVC].)

B. We shall frequently be interested in adjoint pairs where T is the right adjoint, right inverse to S; i.e., θ is the identity natural transformation so $ST = I_{\mathscr{B}}$ and hence $\psi : I_{\mathscr{A}} \to TS$ is characterized by the property that $S\psi$ and ψT are identity natural transformations. We introduce the nonsense word *rari* for this and say that T is a rari of S, or $T =$ rari S. This last notation is alright since any two raris of S differ by a unique natural equivalence. In any case we shall usually require only that S have a rari. The various related notions in which one or more occurances of "right" are replaced by "left" will be denoted by *rali*, *lari*, and *lali*. Note that $T =$ rari S if and only if lali $T = S$, and $T =$ lari S if and only if rali $T = S$. For a number of simple examples which we shall use later, let $\nabla : 1 \to *$ be the unique such functor. Then $\nabla =$ lali R (0.2(C)) (or $R =$ rari ∇) and $\nabla =$ rali D. Hence for any category $\mathscr{A}$, $\nabla \times \mathscr{A} =$ lali $R \times \mathscr{A}$ and $\nabla \times \mathscr{A} =$ rali $D \times \mathscr{A}$. Similarly, if

$$\Delta_{\mathscr{A}} = \mathscr{A}^{\nabla} : \mathscr{A} \to \mathscr{A}^{\mathscr{I}} : A \to i_A : f \to (f, f)$$

then $\Delta_{\mathscr{A}} =$ rari $\mathscr{A}^R$ and $\Delta_{\mathscr{A}} =$ lari $\mathscr{A}^D$ (see [SVC], appendix A4). We shall frequently omit the subscript $\mathscr{A}$ from $\Delta_{\mathscr{A}}$.

C. If the natural transformation ψ: $I_{\mathscr{A}} \to TS$ is viewed as a functor $\overline{\psi}$: $\mathscr{A} \to \mathscr{A}^{\mathscr{I}}$ then raris have the following simple characterization.

Proposition. *Given* S: $\mathscr{A} \to \mathscr{B}$ *and* T: $\mathscr{B} \to \mathscr{A}$. *Then* $T =$ rari S *if and only if there is a functor* $\overline{\psi}$: $\mathscr{A} \to \mathscr{A}^{\mathscr{I}}$ *such that*

a) $\mathscr{A}^D \overline{\psi} = I_{\mathscr{A}}$

b) $\mathscr{A}^R \overline{\psi} = TS$; *i.e.*

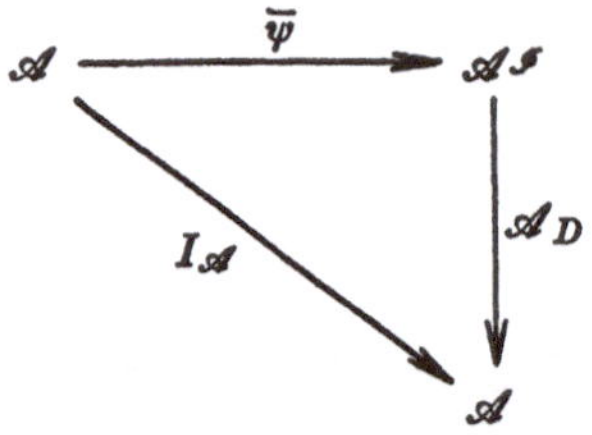

and

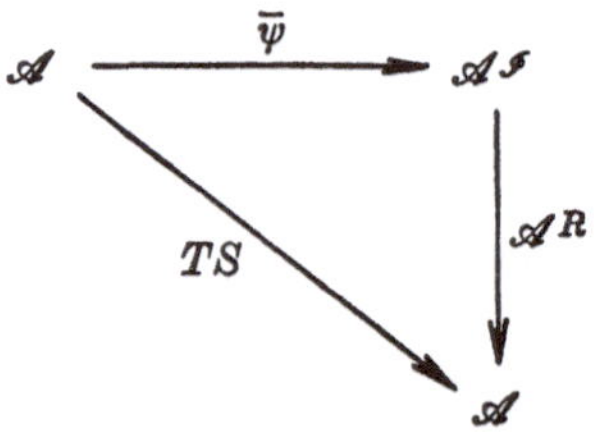

commute, and c) $S^{\mathscr{I}} \overline{\psi} = \Delta S$

d) $\overline{\psi} T = T^{\mathscr{I}} \Delta$; *i.e.*

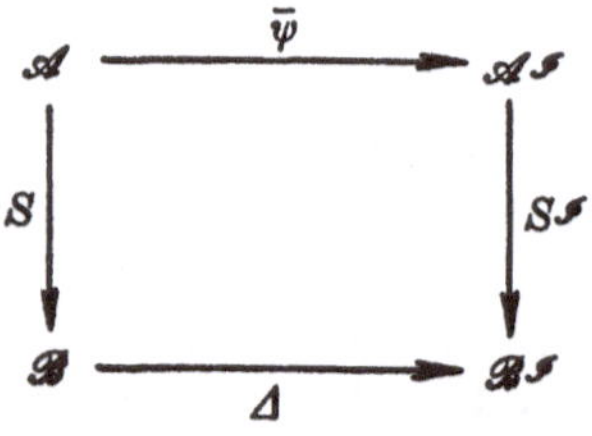

and

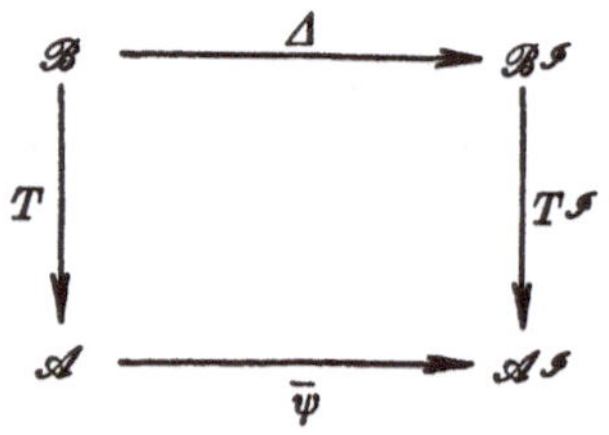

commute, and $ST = I_{\mathscr{B}}$.

Proof. This is simply a translation of the conditions given in 0.2, B. Note that adjoint functors in general can be characterized in a similar, but rather more complicated, way.

D. Proposition.

a) If the adjoining diagram is a pullback and if P has a rari, Q, then S has a rari, T, and the diagram with P and S replaced by Q and T still commutes.

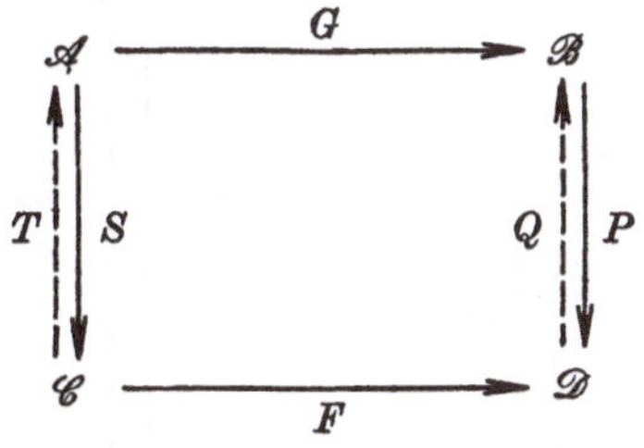

b) If the diagram is a pushout and S has a lali, T, then P has a lali, Q, and the diagram with P and S replaced by Q and T still commutes.

Proof. We give a proof for a) which dualizes to give a proof for b). A slightly simpler proof for a) can be given by representing the pullback $\mathscr{A}$ as $\mathscr{B} \times_{\mathscr{D}} \mathscr{C}$ but then this is of no use in proving b) where an explicit representation of the pushout $\mathscr{D}$ is much more complicated (see [FS]).

To prove a), observe that since $PQF = F = FI_{\mathscr{C}}$, there is a unique $T: \mathscr{C} \to \mathscr{A}$ such that $ST = I_{\mathscr{C}}$ and $GT = QF$

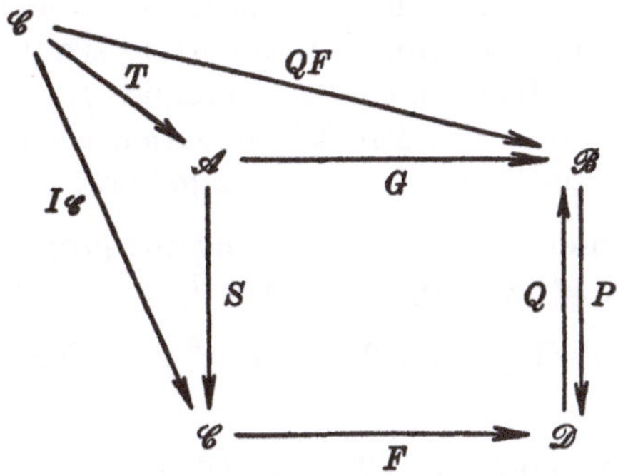

Thus T is a right inverse of S and the required diagram commutes. We use the preceding proposition to show that $T = \text{rari } S$.

Consider

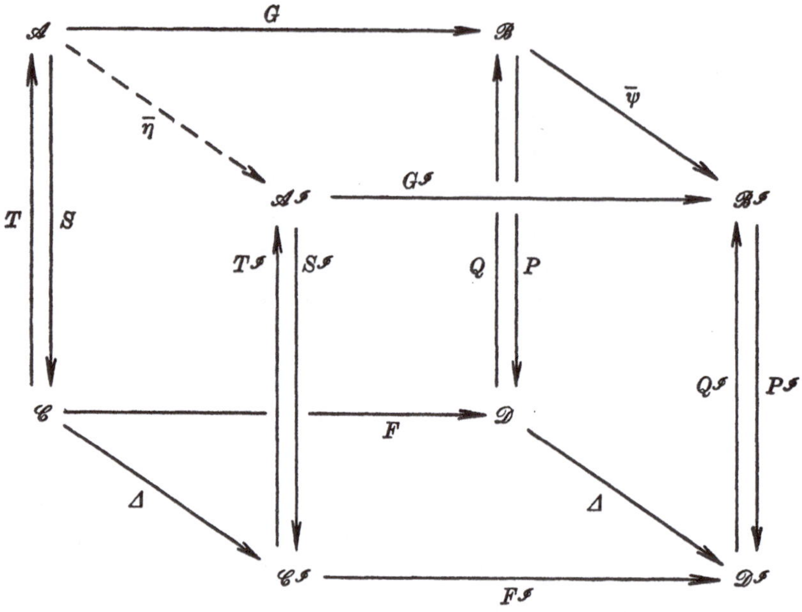

where $\bar{\psi}$ corresponds to the adjunction natural transformation $\psi\colon I_{\mathscr{B}} \to QP$. The bottom square, the back square, the two right hand squares and the two front squares all commute. Furthermore the front square with vertical arrows from top to bottom is a pullback since $(-)^{\mathscr{I}}$ has a left adjoint. Hence there is a unique $\bar{\eta}\colon \mathscr{A} \to \mathscr{A}^{\mathscr{I}}$ such that $G^{\mathscr{I}}\bar{\eta} = \bar{\psi}\, G$ and $S^{\mathscr{I}}\bar{\eta} = \Delta S$. This last equation is condition c) of those to be verified. For the others we argue as follows:

a) By the commutativity relations and the properties of $\bar{\psi}$, we have $G\mathscr{A}^{D}\bar{\eta} = G$ and $S\mathscr{A}^{D}\bar{\eta} = S$, so $\mathscr{A}^{D}\bar{\eta} = I_{\mathscr{A}}$.

b) Similarly, $G\mathscr{A}^{R}\bar{\eta} = GTS$ and $S\mathscr{A}^{R}\bar{\eta} = STS$ so $\mathscr{A}^{R}\bar{\eta} = TS$.

c) Finally,
$G^{\mathscr{I}}\,\bar{\eta}\,T = G^{\mathscr{I}}\,T^{\mathscr{I}}\Delta$ and $S^{\mathscr{I}}\,\bar{\eta}\,T = S^{\mathscr{I}}\,T^{\mathscr{I}}\Delta$ so, $\bar{\eta}\,T = T^{\mathscr{I}}\Delta$.

Therefore $T = \operatorname{rari} S$. We leave the dual diagram and calculations to the reader for his amusement.

E. We need one other property of adjoint functors. Consider a

diagram

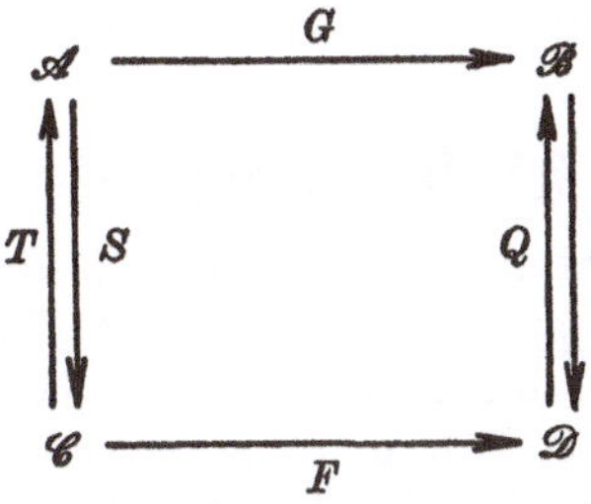

in which $PG = FS$ and where Q is a right adjoint of P and T a right adjoint of S. The adjunction natural transformations $\psi\colon I_{\mathscr{B}} \to QP$ and $\theta'\colon ST \to I_{\mathscr{C}}$ give rise to a *transfer* natural transformation

$$\eta = QF\theta' \circ \psi GT : GT \to QF$$

since $QPGT = QFST$. We shall make use of this in three increasingly special cases:

1. If $T =$ rari S then $\theta' =$ id, so this reduces to $\eta = \psi GT : GT \to GF$
2. If, furthermore, $Q =$ rari P, then $P\eta = 1_F$.
3. If, furthermore, the square with the vertical arrows from top to bottom is a pullback, then η is a natural equivalence; since, by the preceding result there is a $T^1 =$ rari S such that $\eta^1\colon GT^1 \to QF$ is the identity natural transformation, and T differs from T^1 by a natural equivalence.

Dually, if Q is a left adjoint of P and T is a left adjoint of S, then the adjunction natural transformations $\theta : QP \to I_{\mathscr{B}}$ and $\psi\colon I_{\mathscr{C}} \to ST$ give rise to a *transfer* natural transformation

$$\eta = \theta\, GT\, QF\psi' : QF \to GT .$$

a′) If $Q =$ lali P then $\theta =$ id so this reduces to

$$\eta = QF\psi' : QF \to GT .$$

b′) If, furthermore, $T =$ lali, S, then $\eta S = 1_G$.

c′) If, furthermore, the square with the vertical arrows from top to bottom is a pushout, then η is a natural equivalence.

F. The Categorical Yoneda Lemma. *Let $\mathscr{A}$ and $\mathscr{B}$ be categories and let $\tau\colon [\mathscr{A}, -] \to [\mathscr{B}, -]$ be a natural transformation between functors from $\mathscr{C}at$ to $\mathscr{C}at$. Then there is a unique $T\colon \mathscr{B} \to \mathscr{A}$ such that for any $\mathscr{C}$,*

$$\tau_{\mathscr{C}} = \mathscr{C}^T : [\mathscr{A}, \mathscr{C}] \to [\mathscr{B}, \mathscr{C}] .$$

Proof. This is only a special part of the general categorical Yoneda Lemma which is true because $\mathscr{C}at$ is an enriched category with limits. That the action of τ on functors is given by $\mathscr{C}^T$ for a unique T follows from the ordinary Yoneda Lemma. But this determines the action of τ on natural transformations since they can be identified with functors from $\mathscr{A}$ to $\mathscr{C}^{\mathscr{I}}$.

§ 1. Review of [SGA]

Let $P: \mathscr{E} \to \mathscr{B}$ be a functor and, for each $B \in \mathscr{E}$, let $\mathscr{E}_B = P^{-1}(B)$ denote the (possibly empty) fibre of P over B (0.2, E). This gives a category covering each object of B. We would like to have functors covering each morphism of B in either a contravariant or covariant "pseudo-functorial" way. To do this, let $J_B: \mathscr{E}_B \to \mathscr{E}$ denote the inclusion functor. A *cleavage* (resp., *opposite cleavage* or *opcleavage*) consists of functors $f^*: \mathscr{E}_B \to \mathscr{E}_{B'}$ (resp., $f_*: \mathscr{E}_{B'} \to \mathscr{E}_B$) for each morphism $f: B' \to B$ in $\mathscr{B}$, together with natural transformations $\theta_f: J_{B'} f^* \to J_B$ (resp., $\psi_f: J_{B'} \to J_B f_*$) satisfying two axioms.

Axiom 1.1. *$P(\theta_f) = f$ (resp., $P(\psi_f) = f$) and if $\varphi: E' \to E$ satisfies $P(\varphi) = f$ then there is a unique $\varphi': E' \to f^* E$ in $\mathscr{E}_{B'}$ (resp., $\varphi'': f_* E' \to E$ in $\mathscr{E}_B$) such that $(\theta_f)_E \circ \varphi' = \varphi$ (resp., $\varphi'' \circ (\psi_f)_{E'} = \varphi$).*

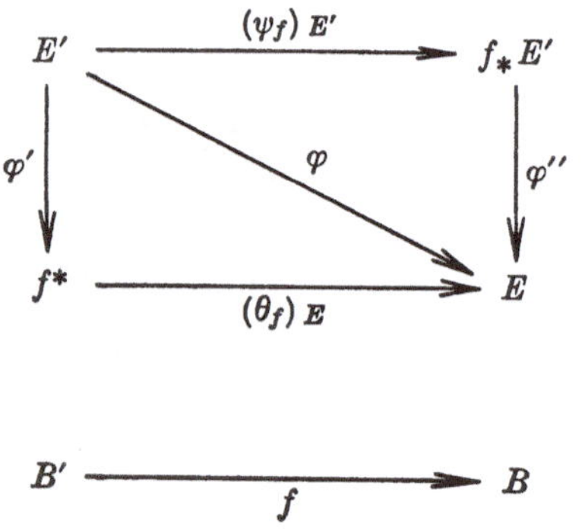

We shall usually omit the subscript E from $(\theta_f)_E$ (resp., $(\psi_f)_E$) when it is clear which component of the natural transformation is relevant. Another way to state axiom 1.1 is to say that for each $E \in \mathscr{E}_B$ (resp., $E' \in \mathscr{E}_{B'}$), f^*E represents the functor $\mathscr{E}(-, E)_f: \mathscr{E}_B \to \mathscr{S}ets$, (resp., $f_* E'$ represents the functor $\mathscr{E}(E', -)_f: \mathscr{E}_{B'} \to \mathscr{S}ets$), where the subscript (f) indicates morphisms φ with $P(\varphi) = f$. From this it is immediate that if there are functors f^* and f_* satisfying axiom 1.1 then f_* is the left adjoint of f^*, for in this case we would have

$$\mathscr{E}_B(f_* E', E) \approx \mathscr{E}(E', E)_f \approx \mathscr{E}_{B'}(E', f^* E).$$

To state the second axiom, consider the composition

$$B'' \xrightarrow{f} B' \xrightarrow{g} B$$

in $\mathscr{B}$. Then for each $E \in \mathscr{E}_B$ there is a uniquely determined morphism $(c_{f,g})\colon f^* g^* E \to (g f)^* E$ in $\mathscr{E}_{B''}$ such that

$$(\theta_{gf})_E \circ (c_{f,g})_E = (\theta_f)_{g^*(E)} \circ (\theta_g)_E = [(\theta_f^* g^*) \circ \theta_g]_E .$$

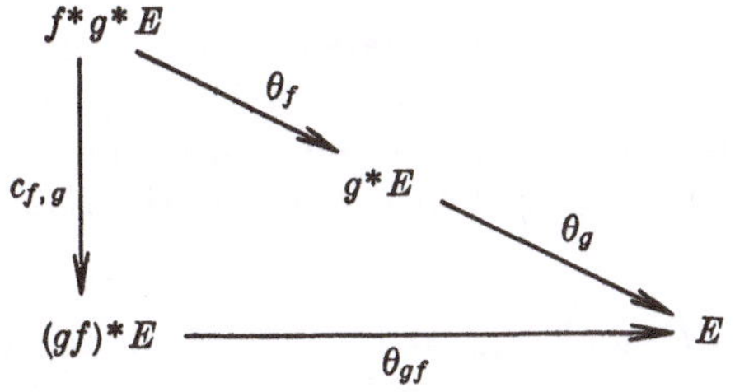

It is easily checked that these are the components of a natural transformation $c_{f,g}\colon f^* g^* \to (g f)^*$. In the opposite case one obtains a natural transformation

$$\tilde{c}_{f,g}\colon (g f)_* \to g_* f_* .$$

Axiom 1.2. *Each $c_{f,g}$ (resp., $\tilde{c}_{f,g}$) is a natural equivalence.*
These two axioms are equivalent to a single axiom.

Axiom 1.3. *$P(\theta_f) = f$ (resp., $P(\psi_f) = f$) and if $\varphi\colon E'' \to E$ satisfies $P(\varphi) = fg$ (resp., gf), for some g, then there is a unique*

$$\varphi'\colon E'' \to f^* E \ (\text{resp., } \varphi'\colon f_* E'' \to E)$$

such that

$$P(\varphi') = g \quad \text{and} \quad (\theta_f)_E \circ \varphi' = \varphi \quad (\text{resp., } \varphi' \circ (\psi_f)_{E''} = \varphi) .$$

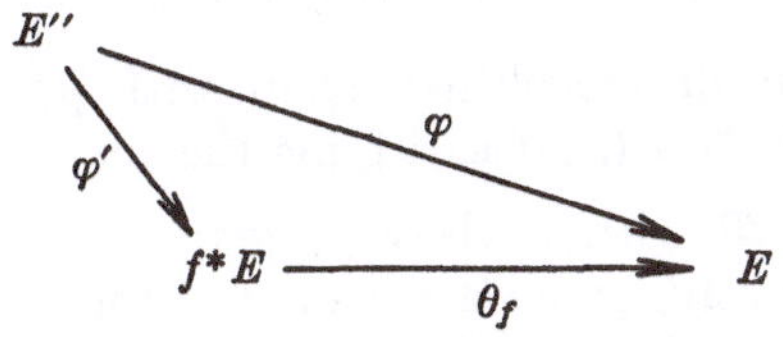

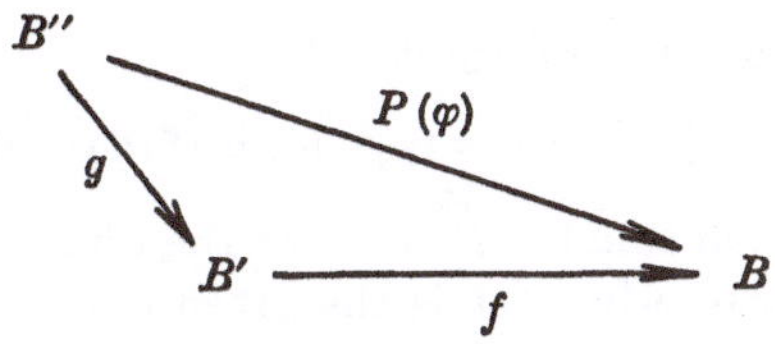

This axiom clearly implies the other two. Conversely, given those two axioms, then

$$\varphi' = (\theta_g)_{f^*E}\,(c_{g,f})^{-1}\,\varphi''$$

where φ'' is the unique factorization of φ through $(fg)^*\,E$.

It is easily checked that an opcleavage for $P\colon \mathscr{E} \to \mathscr{B}$ is equivalent to a cleavage for $P^{\mathrm{op}}\colon \mathscr{E}^{\mathrm{op}} \to \mathscr{B}^{\mathrm{op}}$, so from now on we shall just deal with cleavages with only brief remarks on the opposite situations.

A cleavage is called *normalized* if $(i_B)^* = I_{\mathscr{E}_B}$ for all B and it is called *split* (scindée) if each $c_{f,g}$ is the identity natural transformation. The $c_{f,g}$'s satisfy some other identities besides the defining ones; namely:

1.4.

i. $c_{f,\mathrm{id}} = \theta_{\mathrm{id}} * f^*$ and $c_{\mathrm{id},f} = f^* * \theta_{\mathrm{id}}$ or, in the case of a normal cleavage

i'. $c_{f,\mathrm{id}} = c_{\mathrm{id},f} = 1_{f^*}$

and

ii. $(c_{f,gh})_E \circ (c_{g,h})_{f^*E} = (c_{fg,h})_E \circ h^*(c_{f,g})_E$.

This last property says that identifying via the $c_{f,g}$'s does not lead to contradictions.

If, conversely, categories, functors and natural transformations are specified satisfying these data then a functor $P\colon \mathscr{E} \to \mathscr{B}$ can be reconstructed, as follows:

1.5. A (contravariant) *pseudo-functor* $\mathscr{B}^{\mathrm{op}} \to \mathscr{C}at$, consists of

a) a map $\mathrm{ob}\,\mathscr{B} \to \mathrm{ob}\,\mathscr{C}at\colon B \to \mathscr{E}_B$

b) for each B' and B a map

$$\mathscr{B}(B', B) \to \mathscr{C}at(\mathscr{E}_B, \mathscr{E}_{B'})\colon f \to f^*$$

c) for each pair of morphisms (f, g) with gf defined, a natural equivalence $c_{f,g}\colon g^*\,f^* \to (f\,g)^*$ satisfying the conditions given above.

A functor $P\colon \mathscr{E} \to \mathscr{B}$ with a cleavage gives rise to a pseudofunctor $\mathscr{B}^{\mathrm{op}} \to \mathscr{C}at$. Conversely, given a pseudo-functor $\mathscr{B}^{\mathrm{op}} \to \mathscr{C}at$, let $\mathscr{E}$ be the category such that $\mathrm{ob}\,\mathscr{E} = \coprod_{B \in \mathscr{B}} \mathrm{ob}\,\mathscr{E}_B$ and if $E \in \mathscr{E}_B$, $E' \in \mathscr{E}_{B'}$, then

$$\mathscr{E}(E', E) = \{(f, \varphi) \mid f\colon B' \to B \text{ in } \mathscr{B} \text{ and } \varphi\colon E' \to f^*E \text{ in } \mathscr{E}_{B'}\}.$$

Composition is given by the formula

$$(f, \varphi) \circ (g, \psi) = (f\,g, c_{f,g} \circ g^*(\varphi) \circ \psi).$$

The obvious projection functor $P\colon \mathscr{E} \to \mathscr{B}$ then has a canonical cleavage whose associated pseudo-functor is the given one. Clearly this cleavage is split if and only if the given pseudo-functor is a functor from $\mathscr{B}^{\mathrm{op}}$ to $\mathscr{C}at$.

We would like to extend this equivalence between the notion of functors with cleavages and the notion of pseudo-functors to an equivalence of categories. To do so, consider a commutative diagram

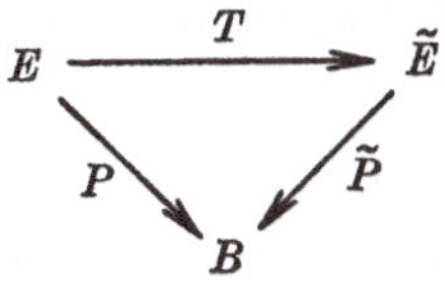

of functors, where P and $\tilde{P}$ have cleavages $\{f^*, \theta_f, c_{f,g}\}$ and $\{\tilde{f}^*, \tilde{\theta}_f, \tilde{c}_{f,g}\}$ respectively. Then

$$T \mid \mathscr{E}_B = T_B : \mathscr{E}_B \to \tilde{\mathscr{E}}_B$$

for all $B \in \mathscr{B}$ and if $f\colon B' \to B$ then there is a unique natural transformation

$$\eta_f : T_{B'} \circ f^* \to \tilde{f}^* \circ T_B$$

such that $P(\eta_f) = i_{B'}$ and $\tilde{\theta}_f \circ \eta_f = T(\theta_f)$. These transformations satisfy a complicated relation with the c's; namely,

1.6.

$$\tilde{c}_{f,g} \circ \tilde{g}^*(\eta_f) \circ (\eta_g)_{f^*E} = \eta_{fg} \circ T_{B''}[(c_{f,g})_E]$$

for $B'' \xrightarrow{g} B' \xrightarrow{f} B$ or

$$\tilde{c}_{f,g} \circ (\tilde{g}^* * \eta_f) \circ (\eta_g * f^*)) = \eta_{fg} \circ (T * c_{f,g}).$$

1.7. Three special cases are of interest.

a) If both cleavages are split, then

$$(\tilde{g}^* * \eta_f) \circ (\eta_g * f^*) = \eta_{fg}.$$

In a later paper we shall see that this is a desirable situation.

b) If η is the identity for all f; i.e., if

$$T_{B'} f^* = \tilde{f}^* T_B \quad \text{and} \quad T(\theta_f) = \tilde{\theta}_f, \quad \text{then} \quad \tilde{c}_{f,g} = T(c_{f,g}).$$

In this case T is called a *cleavage preserving* functor.

c) If η_f is a natural equivalence for all f, then T is called a *cartesian* functor.

1.8. Let $\mathscr{C}at_{\mathscr{B}}$ denote the category of categories over $\mathscr{B}$ (0.2, E). Then we can distinguish the following categories related to $\mathscr{C}at_{\mathscr{B}}$.

i. $\mathscr{C}\ell eav(\mathscr{B})$ (resp., $\mathscr{S}plit(\mathscr{B})$) denotes the category whose objects are functors $P\colon \mathscr{E} \to \mathscr{B}$ with given normal cleavages (resp., split cleavages) and such that the morphisms from P to $\tilde{P}$ are all morphisms between them, regarded as objects in $\mathcal{C}at_{\mathcal{B}}$; i.e., the forgetful functor $\mathcal{C}leav(\mathcal{B}) \to$ $\to \mathcal{C}at_{\mathcal{B}}$ is full.

ii. $\mathcal{C}leav_{\text{cart}}(\mathcal{B})$ (resp., $\mathcal{S}plit_{\text{cart}}(\mathcal{B})$) is the subcategory with the same objects but only cartesian morphisms allowed.

iii. $\mathcal{C}leav_0(\mathcal{B})$ (resp., $\mathcal{S}plit_0(\mathcal{B})$) is the subcategory with the same objects but only cleavage preserving morphisms allowed.

1.9. One has the following inclusions of subcategories

$$\begin{array}{ccccc} \mathcal{S}plit_0(\mathcal{B}) & \subset & \mathcal{S}plit_{\text{cart}}(\mathcal{B}) & \subset & \mathcal{S}plit(\mathcal{B}) \\ \cap & & \cap & & \cap \\ \mathcal{C}leav_0(\mathcal{B}) & \subset & \mathcal{C}leav_{\text{cart}}(\mathcal{B}) & \subset & \mathcal{C}leav(\mathcal{B}) \end{array}$$

Each of the categories corresponds to a suitable category of pseudofunctors where the maps between pseudofunctors are systems of functors $T_B\colon \mathcal{E}_B \to \tilde{\mathcal{E}}_B$ and natural transformations $\eta_f\colon T_{B'} f^* \to \tilde{f}^* T_B$ satisfying analogous identities. In particular, $\mathcal{S}plit_0(\mathcal{B})$ is equivalent to the functor category $[\mathcal{B}^{\text{op}}, \mathcal{C}at]$.

Examples 1.10. Perhaps the most familiar example of a functor with a cleavage and an opcleavage is the projection functor P from the category $\mathcal{M}$ of all modules to the category $\mathcal{R}$ of rings; i.e., the objects of $\mathcal{M}$ are pairs (R, M) where R is a ring and M is a (left) R-module, while the morphisms are pairs $(f, \varphi)\colon (R, M) \to (S, N)$ where $f\colon R \to S$ is a ring homomorphism and $\varphi\colon M \to N$ is a group homomorphism such that $\varphi(rm) = f(r)\,\varphi(m)$. P is given by

$$P\colon \mathcal{M} \to \mathcal{R}\colon (R, M) \to R\colon (f, \varphi) \to f.$$

If $f\colon R \to S$, then define

$$f^*\colon \mathcal{M}_S \to \mathcal{M}_R\colon (S, N) \to (R, {}_{[f]}N)$$

where ${}_{[f]}N$ is the group N regarded as an R-module via the action $rn = f(r)\,n$. Furthermore

$$\theta_f\colon J_R \circ f^* \to J_S$$

is given by $(\theta_f)_{(S,N)} = (f, i_N)$. This clearly defines a split cleavage. The opcleavage is given by

$$f_*\colon \mathcal{M}_R \to \mathcal{M}_S\colon (R, M) \to (S, S_{[f]} \otimes_R M)$$

and

$$(\psi_f)_{(R,M)} = (f, j_M) \quad \text{where}$$
$$j_M\colon M \to S \otimes_R M\colon m \to 1 \otimes m.$$

This opcleavage is not split, and in fact is not even normal, but it can be normalized by replacing $S \otimes_S S$ by S.

As a final review example, let $\mathscr{E}$ and $\mathscr{B}$ both be categories with a single object and such that every morphism is an equivalence, i.e., $\mathscr{E}$ and $\mathscr{B}$ are groups. If $P\colon \mathscr{E} \to \mathscr{B}$ is an onto functor — that is, a group epimorphism — then the cleavages of P are in 1-1 correspondence with the set-theoretic inverse functions $\gamma\colon \mathscr{B} \to \mathscr{E}$. A cleavage is normal if and only if the corresponding function satisfies $\gamma(1) = 1$ and it is split if and only if γ is also a group homomorphism. This accounts for the term split and shows that a functor which admits a cleavage need not admit a split cleavage.

§ 2. Fibred Categories

Definition 2.1. *Let $\mathscr{F}\mathit{ib}(\mathscr{B})$ denote the image of $\mathscr{C}\mathit{leav}_{\mathrm{cart}}(\mathscr{B})$ in $\mathscr{C}\mathit{at}_{\mathscr{B}}$ by the forgetful functor. $\mathscr{F}\mathit{ib}(\mathscr{B})$ is called the category of fibred categories over $\mathscr{B}$.*

Thus the objects of $\mathscr{F}\mathit{ib}(\mathscr{B})$ are functors $P\colon \mathscr{E} \to \mathscr{B}$ which admit some cleavage and the morphisms are functors over $\mathscr{B}$ which are cartesian for some choice of cleavages. In this section we shall give a number of equivalent intrinsic characterizations of $\mathscr{F}\mathit{ib}(\mathscr{B})$ together with a number of properties of fibrations which follow easily from these characterizations. We begin with the characterization in Grothendieck [*SGA*].

Definition 1.2. *Let $P\colon \mathscr{E} \to \mathscr{B}$ be a functor. A morphism φ in $\mathscr{E}$ which satisfies the same universal mapping property that $(\theta_f)_E$ satisfies in Axiom* 1.1 *(resp.,* 1.3*) is called a (resp., strong) cartesian morphism (over f). If for every $f\colon B' \to B$ in $\mathscr{B}$, each $E \in \mathscr{E}_B$ is the range of a (resp., strong) cartesian morphism then we shall say that there are enough (resp., strong) cartesian morphisms. Opcartesian morphisms are defined dually; i.e., $\varphi \in \mathscr{E}$ is opcartesian if and only if $\varphi \in E^{\mathrm{op}}$ is cartesian. We shall use θ_f as a generic notation for a cartesian morphism over f.*

Lemma 2.3. *[SGA] i. If*

$$\begin{array}{ccc} E' & \xrightarrow{\quad\theta_f\quad} & E \\ \sigma\big\downarrow & & \big\downarrow\tau \\ F' & \xrightarrow{\quad\theta'_f\quad} & F \end{array}$$

is a commutative diagram in $\mathscr{E}$ with σ and τ equivalences such that $P(\sigma)$ and $P(\tau)$ are identity morphisms, then θ_f is cartesian if and only if θ'_f is.

ii. If $\theta_f\colon E' \to E$ and $\theta'_f\colon E'' \to E$ are cartesian over f then there is a unique $\gamma\colon E' \to E''$ with $P(\gamma) = id$ and $\theta'_f\gamma = \theta_f$, and γ is an equivalence.

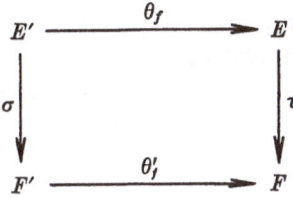

Definition 2.4. *[SGA] $P: E \to B$ is called a fibration if*

i. There are enough cartesian morphisms

ii. The composition of cartesian morphisms is cartesian.

Opfibrations are defined dually in terms of opcartesian morphisms.

Proposition 2.5. *[SGA] The following are equivalent.*

i. $P \in \mathscr{F}ib(\mathscr{B})$

ii. P is a fibration

iii. P is a fibration and every cartesian morphism is strong

iv. There are enough strong cartesian morphisms.

Proof. "i. implies ii." and "iii. implies iv." are obvious. "ii. implies iii." is a simple exercise and "iv. implies i." is shown by choosing a strong cartesian morphism for each $f: B' \to B$ in $\mathscr{B}$ and each $E \in \mathscr{E}_{\mathscr{B}}$. The domains of these morphisms determine functions f^*: ob $\mathscr{E}_B \to$ ob $\mathscr{E}_{B'}$ and the morphisms themselves are the components of the desired transformations. The universal mapping properties imply that there is a unique way to extend these functions to functors for which the transformations are natural. By Lemma 2.3 the cleavage may be chosen to be normal.

Definition 2.6. *[SGA] Let $P: \mathscr{E} \to \mathscr{B}$ and $\tilde{P}: \tilde{\mathscr{E}} \to B$ be fibrations and let $T: \tilde{\mathscr{E}} \to \mathscr{E}$ be a functor over $\mathscr{B}$ (i.e., $\tilde{P}T = P$). T is called cartesian if T preserves cartesian morphisms.*

Proposition 2.7. *[SGA] $T: \mathscr{E} \to \tilde{\mathscr{E}}$ over $\mathscr{B}$ is cartesian if and only if for any cleavages $\{f^*, \theta_f, c_{f,g}\}$ and $\{\tilde{f}^*, \tilde{\theta}_{f,g}, \tilde{c}_{f,g}\}$ of P and $\tilde{P}$ respectively, T is cartesian in the sense of* 1.7c.

Proof. We have $T(\theta_f) = \tilde{\theta}_f \circ \eta_f$. By Lemma 2.3, $T(\theta_f)$ is cartesian if and only if η_f is an equivalence.

To translate these characterizations of [*SGA*] into forms more reminiscent of topology, we consider a number of properties, all of which turn out to be equivalent.

Properties in $\mathscr{C}at^{\mathscr{B}}$ 2.8. *Let $P: \mathscr{E} \to \mathscr{B}$, $\tilde{P}: \tilde{\mathscr{E}} \to \mathscr{B}$ and $T: \mathscr{E} \to \tilde{\mathscr{E}}$ with $\tilde{P}T = P$.*

i. For each $E \in \mathscr{E}$, P induces a functor $P_E: \mathscr{E}_E \to \mathscr{B}_{P(E)}$ (see 0.2, E). Suppose for each P_E there is a $Q_E =$ rari P_E, and similarly a $\tilde{Q}_{\tilde{E}} =$ rari $\tilde{P}_{\tilde{E}}$

for $\tilde{E} \in \tilde{\mathscr{E}}$. If $T: E \to \tilde{E}$ satisfies $\tilde{P}T = P$ then $\tilde{P}_{T(E)}\, T_E = P_E$.

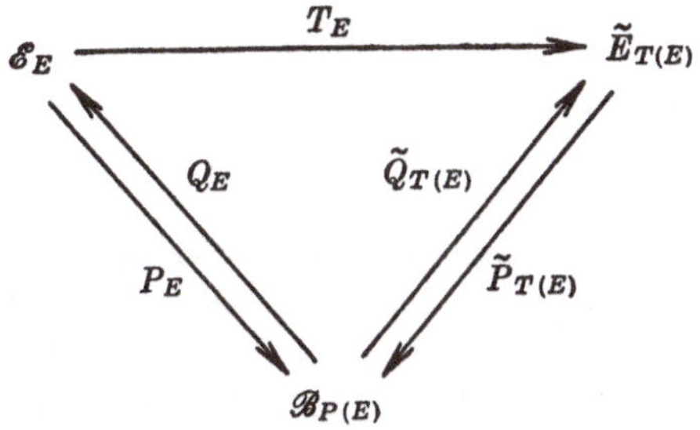

However, T need not commute with Q_E and $\tilde{Q}_{T(E)}$, but the adjunction natural transformation

$$\tilde{\psi}: I \to \tilde{Q}_{T(E)}\, \tilde{P}_{T(E)}$$

gives rise to a transfer natural transformation

$$\eta_E = \tilde{\psi}\, T_E\, Q_E : T_E\, Q_E \to Q_{T(E)}$$

(since $\tilde{Q}\tilde{P}TQ = \tilde{Q}PQ = \tilde{Q}$, omitting subscripts. See 0.3, E) with $P_{T(E)} * \eta_E = 1$, where 1 is the identity natural transformation of the identity functor of $\mathscr{B}_{P(E)}$. For the next few pages, T will be called 1 — *cartesian* if, for every $E \in \mathscr{E}$, η_E is a natural equivalence. {Each P_E having a rari is clearly equivalent to P being a fibration and equally clearly, 1 — cartesian is equivalent to cartesian. The entire treatment so far could be redeveloped rather economically in these terms. However, we prefer to complicate the notions so as to make later proofs simpler.}

ii. Let $G: * \to \mathscr{E}$ be given; i.e., $G(*) = E \in \text{ob}\, \mathscr{E}$. Consider

$$\mathscr{E}^R: [\mathscr{I}, \mathscr{E}] \to [*, \mathscr{E}]\ (0.2, D).$$

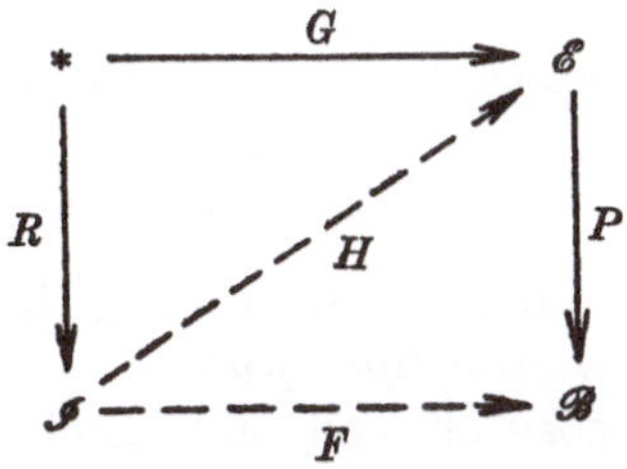

Let $[\mathscr{I}, \mathscr{E}]_G = (\mathscr{E}^R)^{-1}(G) = \{H: \mathscr{I} \to \mathscr{E} \mid HR = G\} \cup \{t: H \to H' \mid tR = 1_G,\}$

i.e., $[\mathscr{I}, \mathscr{E}]_G$ is the fibre of $\mathscr{E}^R$ over G.

Similarly, let

$$[\mathscr{I},\mathscr{B}]_{PG} = (\mathscr{B}^R)^{-1}(PG) = \{F:\mathscr{I}\to\mathscr{B} \mid FR = PG\} \cup \cup \{s: F\to F' \mid sR = 1_{PG}\}.$$

Then P induces a functor

$$P_G: [\mathscr{I},\mathscr{E})_G \to [\mathscr{I},\mathscr{B}]_{PG}.$$

If P_G has a right inverse, then any "directed path" $F:\mathscr{I}\to\mathscr{B}$ ending at $PG(*)$ can be "lifted" to a "directed path" $H:\mathscr{I}\to\mathscr{E}$ ending at $G(*)$. If this right inverse is a right adjoint to P_G, then there is a canonical, universal choice for such a lifting. Thus if each P_G has a rari, we shall say that *directed paths have cartesian liftings.*

Now consider

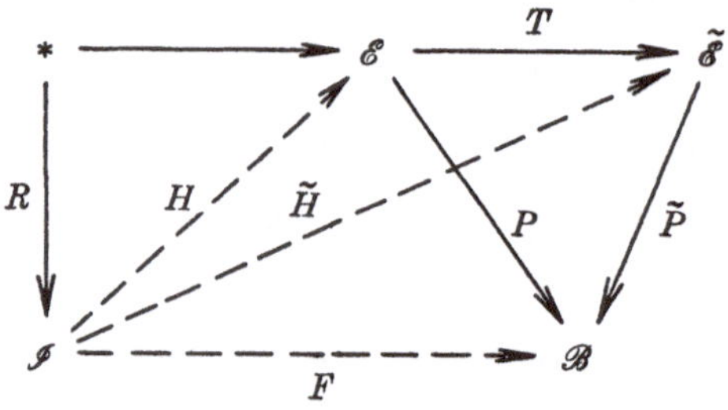

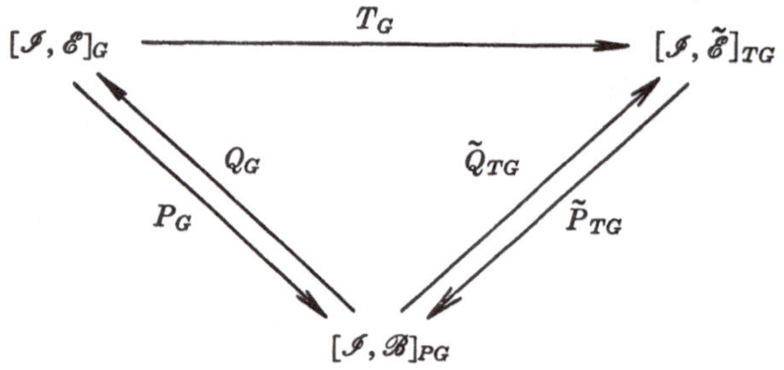

where $\tilde{P}T = P$ and where $Q_G = \text{rari } P_G$ and $\tilde{Q}_{TG} = \text{rari } \tilde{P}_{TG}$. As before we have the transfer

$$\eta_G = \overline{\psi}\, T_G Q_G : T_G Q_G \to Q_{TG}$$

with $P_{TG} * \eta_G = 1$. (See 0.3, E). If every η_G is a natural equivalence then T will be said to *preserve lifted paths.*

iii. Given any category $\mathscr{N}$ and any functor $G:\mathscr{N}\to\mathscr{E}$, then P induces a functor

$$P_G : [\mathscr{I}\times\mathscr{N},\mathscr{E}]_G \to [\mathscr{I}\times\mathscr{N},\mathscr{B}]_{PG}$$

(defined as above with $*$ replaced by $\mathscr{N}$). A functor $F\in[\mathscr{I}\times\mathscr{N},\mathscr{B}]_{PG}$

satisfies $F \circ (R \times \mathscr{N}) = P\,G$ and is the same as a natural transformation between functors from $\mathscr{N}$ to $\mathscr{B}$ ending at $P\,G$. We prefer to think of it as a "homotopy" ending at $P\,G$. If there is an $H \in [\mathscr{I} \times \mathscr{N}, \mathscr{E}]_G$ with $PH = F$, then F can be "lifted" to a "homotopy" ending at G. Thus, if, for every $\mathscr{N}$, every such P_G has a rari, then we shall say that *homotopies have cartesian liftings.*

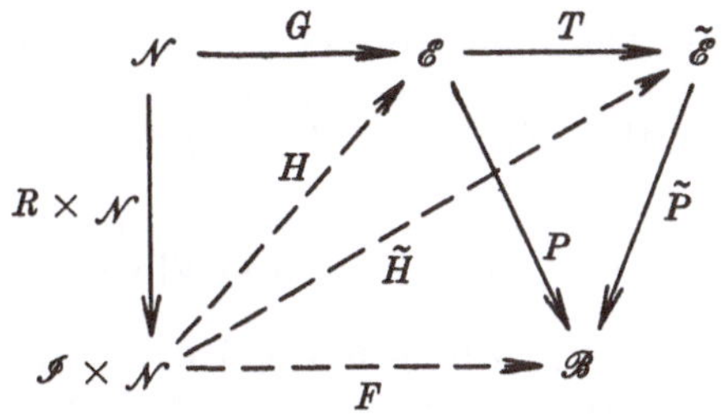

Now suppose $P\colon \mathscr{E} \to \mathscr{B}$ and $\tilde{P}\colon \tilde{\mathscr{E}} \to \mathscr{B}$ both satisfy this condition and $T\colon \mathscr{E} \to \tilde{\mathscr{E}}$ with $\tilde{P}T = P$. If $G\colon \mathscr{N} \to \mathscr{E}$, let $Q_G = \text{rari } P_G$ and $\tilde{Q}_{TG} = \text{rari } \tilde{P}_{TG}$. As before $\eta_G = \tilde{\psi}\, T_G Q_G$ satisfies $P_{TG} * \eta_G = 1$. If every such η_G is a natural equivalence then T will be said to *preserve lifted homotopies.*

iv. Given a functor $R\colon \mathscr{N} \to \mathscr{M}$ for which there exists $S = \text{lali } R$, and given $G\colon \mathscr{N} \to \mathscr{E}$, then P induces a functor

$$P_G\colon [\mathscr{M}, \mathscr{E}]_G \to [\mathscr{M}, \mathscr{B}]_{PG}$$

(defined as above, with $\mathscr{I} \times \mathscr{N}$ replaced by $\mathscr{M}$). If for every such $\mathscr{N} \to \mathscr{M}$, every such P_G has a rari then we shall say that *generalized homotopies have cartesian liftings.*

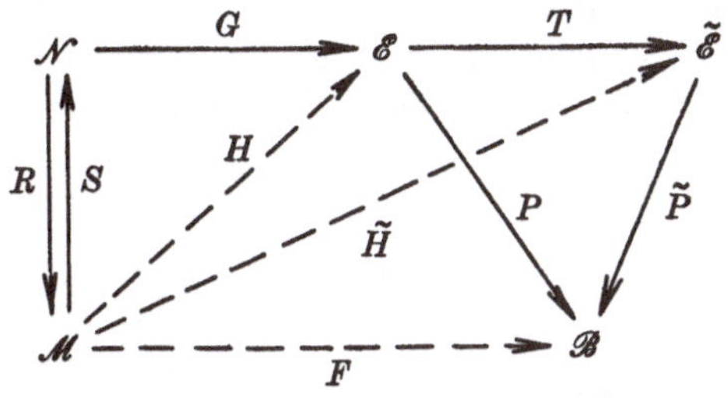

Let P and $\tilde{P}$ satisfy this condition and let $T\colon \mathscr{E} \to \tilde{\mathscr{E}}$ satisfy $\tilde{P}T = P$. If the transformations η_G defined as above are natural equivalences then T will be said to *preserve lifted generalized homotopies.*

v. In the same situation as iv., P induces a functor

$$P_{GS}\colon [\mathscr{M}, \mathscr{E}]_{GS} \to [\mathscr{M}, \mathscr{B}]_{PGS} \;\; (0.2, \text{E}),$$

where $[\mathcal{M}, \mathcal{E}]_{GS}$ is the category of functors over GS; i.e., functors $H: \mathcal{M} \to \mathcal{E}$ with an "augmentation" natural transformation $\eta: H \to GS$, and natural transformations $t: H \to H'$ such that $\eta' t = \eta$.

If $H: \mathcal{M} \to \mathcal{E}$ satisfies $HR = G$ (i.e., $H \in [\mathcal{M}, \mathcal{E}]_G$) then H has a canonical augmentation $H\psi: H \to HRS = GS$ where $\psi: I_{\mathcal{M}} \to RS$ is the adjunction natural transformation. Let $[\mathcal{M}, \mathcal{E}]'_{GS}$ denote the full subcategory of $[\mathcal{M}, \mathcal{E}]_{GS}$ determined by functors $H \in [\mathcal{M}, \mathcal{E}]_G$ with the canonical augmentation. Similarly, let $[\mathcal{M}, \mathcal{B}]'_{PGS}$ denote the full subcategory of $[\mathcal{M}, \mathcal{E}]_{GS}$ determined by functors $F \in [\mathcal{M}, \mathcal{B}]_{PG}$ with a canonical augmentation $F\psi: F \to FRS = PGS$. Clearly, P induces a functor

$$P'_{GS}: [\mathcal{M}, \mathcal{E}]'_{GS} \to [\mathcal{M}, \mathcal{B}]'_{PGS}.$$

If for every $\mathcal{N} \to \mathcal{M}$, each such P'_{GS} has a rari, we shall say that *canonically augmented functors have cartesian liftings.*

Let P and $\tilde{P}$ satisfy this condition and let $T: \mathcal{E} \to \tilde{\mathcal{E}}$ satisfy $\tilde{P}T = P$. Then P induces

$$T_{GS}: [\mathcal{M}, \mathcal{E}]'_{GS} \to [\mathcal{M}, \mathcal{E}]'_{TGS}$$

and as before there is a transfer natural transformation

$$\eta_{GS}: T_{GS}\, Q_{GS} \to \tilde{Q}_{TGS}$$

with $\tilde{P}_{TGS} * \eta_{GS} = 1$, where $Q_{GS} = \text{rari } P_{GS}$ and $\tilde{Q}_{TGS} = \text{rari } \tilde{P}_{TGS}$. If all such η_{GS} are natural equivalences then T will be said to *preserve liftings of canonically augmented functors.*

Lemma 2.9. *In the situation described above in* v.

i. If $\mu: H \to GS$ is a natural transformation then $\mu RS \cdot H\psi = \mu$. Hence if $HR = G$ and $\mu RS = 1_{GS}$ then $\mu = H\psi$.

ii. If H and H' satisfy $HR = H'R = G$ and if t is a natural transformation then $tR = 1_G$ if and only if $H'\psi \circ t = H\psi$.

Proof. i. Consider

$$\begin{array}{ccc} H & \xrightarrow{\ \mu\ } & GS \\ {\scriptstyle H\psi}\downarrow & & \downarrow{\scriptstyle GS\psi} \\ HRS & \xrightarrow[\ \mu RS\ ]{} & GSRS \end{array}$$

This commutes by the rules of functoral calculus [*GOD*] and since $S = \text{lali } R$, $S\psi = 1_S$ so $GS\psi = 1_{GS}$.

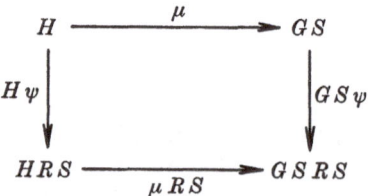

ii. Consider

$$\begin{array}{ccc} H & \xrightarrow{H\psi} & HRS = GS \\ \Big\downarrow{\scriptstyle t} & & \Big\downarrow{\scriptstyle tRS} \\ H' & \xrightarrow{H'\psi} & H'RS = GS \end{array}$$

If $tR = 1_G$ then $tRS = 1_{GS}$ so $H'\psi \circ t = H\psi$.

Conversely, consider

$$\begin{array}{ccc} HRS & \xrightarrow{H\psi RS} & HRSRS = GSRS \\ \Big\downarrow{\scriptstyle tRS} & & \Big\downarrow{\scriptstyle (1_{GS})RS} \\ H'RS & \xrightarrow[H'\psi RS]{} & H'RSRS = GSRS \end{array}$$

If $H'\psi \circ t = H\psi$ then this diagram commutes. But, since $S = \operatorname{lali} R$, we have $\psi R = 1$ so $H\psi RS = 1_{GS}$ and $H'\psi RS = 1_{GS}$ and hence $tRS = 1_{GS}$. Since S is onto, this implies that $tR = 1_G$.

Theorem 2.10. *The following are equivalent for a functor* $P\colon \mathscr{E} \to \mathscr{B}$:

o. P is a fibration

i. For each $E \in E$, $P_E\colon \mathscr{E}_E \to \mathscr{B}_{P(E)}$ has a rari

ii. Directed paths have cartesian liftings.

iii. Homotopies have cartesian liftings.

iv. Generalized homotopies have cartesian liftings.

v. Canonically augmented functors have cartesian liftings.

Proof. We dispose of the elementary implications first.

0. $\Leftrightarrow$ i. Let $B = P(E)$ and let $(f\colon B' \to B) \in \operatorname{ob} \mathscr{B}_B$,

$$(\varphi : E'' \to E) \in \operatorname{ob} \mathscr{E}_E, \quad (\theta_f\colon E' \to E) \in \operatorname{ob} \mathscr{E}_E$$

and

$$(P_E\,\varphi\colon PE'' \to B) \in \operatorname{ob} \mathscr{B}_B.$$

Then

$$\mathscr{B}_B(P_E(\varphi), f) \approx \mathscr{E}_E(\varphi, \theta_f)$$

with $P_E(\theta_f) = f$ (i.e., $\theta_{(-)} = \operatorname{rari} P_E$) if and only if θ_f is a strong cartesian morphism over f with range E.

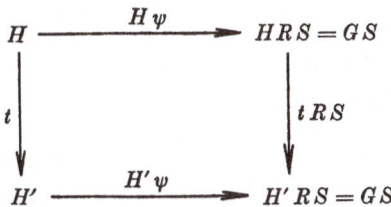

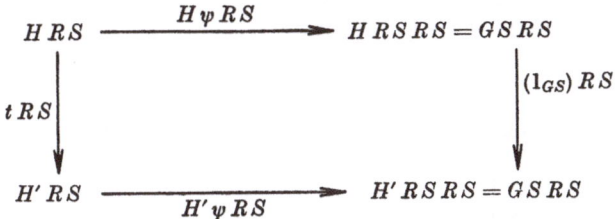

i. ⇔ ii. This is trivial since there are obvious categorical isomorphisms

$$[\mathcal{I}, \mathcal{E}]_G \approx \mathcal{E}_{G(*)} \quad \text{and} \quad [\mathcal{I}, \mathcal{B}]_{PG} \approx \mathcal{B}_{PG(*)}.$$

iii. ⇒ ii. Immediate since ii. is the special case of iii. in which $N = *$.

iv. ⇒ iii. Immediate, since $\nabla \times \mathcal{N} = \text{lali}\, R \times \mathcal{N}$ (0.2, C) so iii. is the special case of iv. in which $\mathcal{M} = \mathcal{I} \times \mathcal{N}$.

iv ⇔ v. Let $\Gamma_{\mathcal{E}}\colon [\mathcal{M}, \mathcal{E}]_G \to [\mathcal{M}, \mathcal{E}]_{GS}$ be the functor such that $\Gamma_{\mathcal{E}}(H) = (H\psi\colon H \to GS)$ and $\Gamma_{\mathcal{E}}(t) = t$. By Lemma 2.9 ii. $\Gamma_{\mathcal{E}}(t)$ is a morphism in $[\mathcal{M}, \mathcal{E}]_{GS}$ since $tR = 1_G$ and by the same Lemma,

$$\Gamma_{\mathcal{E}}\colon [\mathcal{M}, \mathcal{E}]_G \to [\mathcal{M}, \mathcal{E}]'_{GS}$$

is a full, faithful bijection.
Defining $\Gamma_{\mathcal{B}}$ analogously, the diagram

$$\begin{array}{ccc} [\mathcal{M}, \mathcal{E}]_G & \xrightarrow{\Gamma_{\mathcal{E}}} & [\mathcal{M}, \mathcal{E}]'_{GS} \\ \downarrow{\scriptstyle P_G} & & \downarrow{\scriptstyle P'_{GS}} \\ [\mathcal{M}, \mathcal{B}]_{PG} & \xrightarrow[\Gamma_{\mathcal{B}}]{} & [\mathcal{M}, \mathcal{B}]'_{PGS} \end{array}$$

obviously commutes and the existence of a rari for P_G is therefore equivalent to the existence of a rari for P'_{GS}.

0. ⇒ v. This is the only step that requires any work. Let $P\colon \mathcal{E} \to \mathcal{B}$ be a fibration. By Proposition 2.5 we may choose a normal cleavage $\{f^*, \theta_f, c_{f,g}\}$ of P. Consider $F\colon \mathcal{M} \to \mathcal{B}$ with $FR = PG$ and with its canonical augmentation $F\psi\colon F \to FRS = PGS$. For any $M \in \mathcal{M}$, define

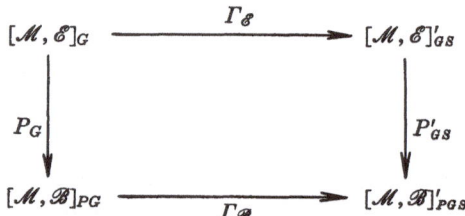

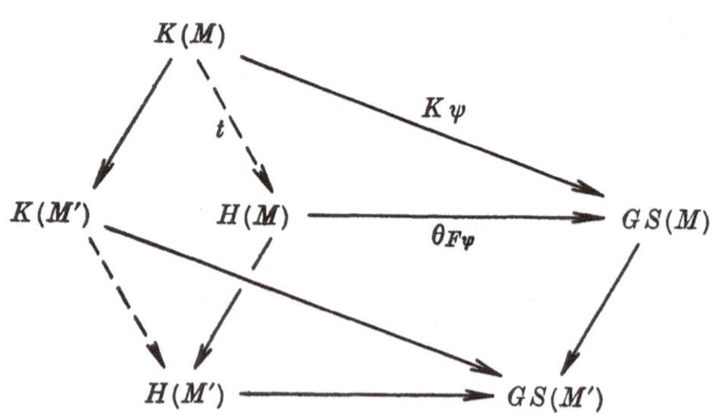

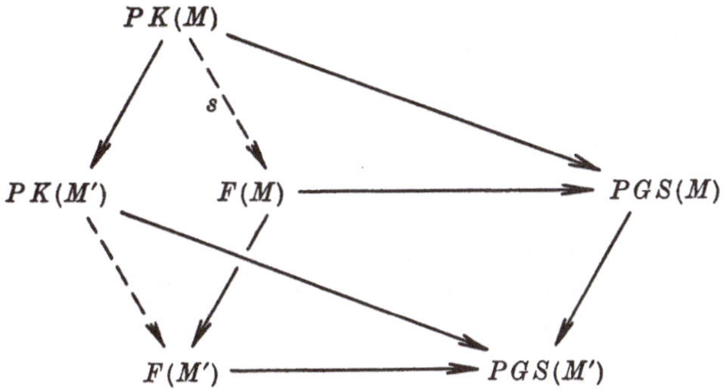

$H(M) = (F\psi_M)^* [GS(M)]$ and if $m\colon M \to M'$ in $\mathcal{M}$, define $H(m)$ to be the unique morphism in $\mathcal{E}$ such that

$$GS(m) \circ \theta_{F\psi_M} = \theta_{F\psi_{M'}} \circ H(m) \quad \text{and} \quad PH(m) = F(m) \quad \text{(by 1.3)}.$$

Then by construction $PH = F$ so H is a lifting of F. We must now verify a number of things.

a) $HR = G$. This is immediate since for any $N \in \mathcal{N}$, we have

$$H\,R(N) = (F\,\psi_{RN})^* [GSR(N)] = (F\,i_{RN})^* [G(N)] = G(N)$$

the second equality holding since $S = \text{lali}\ R$ and hence $\psi R = 1$, and the third equality holding since the cleavage is normal.

b) $(\theta_{F\psi}\colon H \to GS) \in [\mathcal{M}, \mathcal{E}]'_{GS}$; i.e., $\theta_{F\psi} = H\psi$. By Lemma 2.9, i. it is sufficient to verify that $(\theta_{F\psi})\,RS = 1_{GS}$. But

$$P(\theta_{F\psi})\,RS = F\,\psi\,RS = 1_{PGS}$$

since $\psi R = 1$ which gives the desired result because the cleavage is normal.

c) What we have accomplished so far can be expressed by saying that each homotopy of F ending at PG can be lifted to a homotopy of H ending at G. We must now verify that this lifting is cartesian; i.e., if $K\colon \mathcal{M} \to \mathcal{E}$ also satisfies $KR = G$ then

$$\text{Hom}_{\mathcal{E}}(K, H) \approx \text{Hom}_{\mathcal{B}}(PK, F)$$

the hom-sets being taken in either of the equivalent categories $[\mathcal{M}, \mathcal{E}]_G$ or $[\mathcal{M}, \mathcal{E}]'_{GS}$ and similarly for $\mathcal{B}$.

Let K have its canonical augmentation $K\psi$ and let $s\colon PK \to F$ be a natural transformation in $[\mathcal{M}, \mathcal{B}]'_{PGS}$; i.e., $F\psi \circ s = PK\psi$. Each component s_M has a unique lifting to a morphism

$$t_M\colon K(M) \to H(M)$$

in $\mathcal{E}$ such that $H\psi_M \circ t_M = K\psi_M$ and such that $P(t_M) = s_M$ (by 1.3). Finally, to verify that this transformation is natural, let $m\colon M \to M'$.

Then there is a unique map $r: K(M) \to H(M')$ such that $H\psi_{M'} \circ r = GS(m) \circ K\psi_M$ and such that

$$P(r) = F(m) \circ s_M = s_{M'} \circ PK(m).$$

Since $H(m) \circ t_M$ and $t_{M'} \circ K(m)$ both satisfy this condition on r, they are equal.

Remark. We outline briefly how the preceding theorem must be altered for opfibrations. Since an opfibration of $P: \mathscr{E} \to \mathscr{B}$ is the same as a fibration of $P^{\mathrm{op}}: \mathscr{E}^{\mathrm{op}} \to \mathscr{B}^{\mathrm{op}}$, opfibrations lead to diagrams

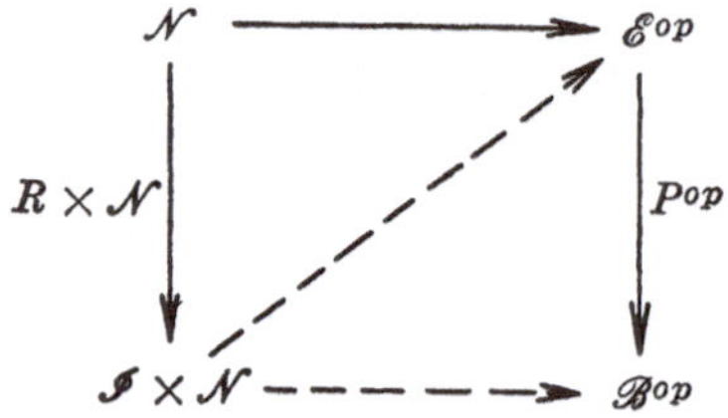

or

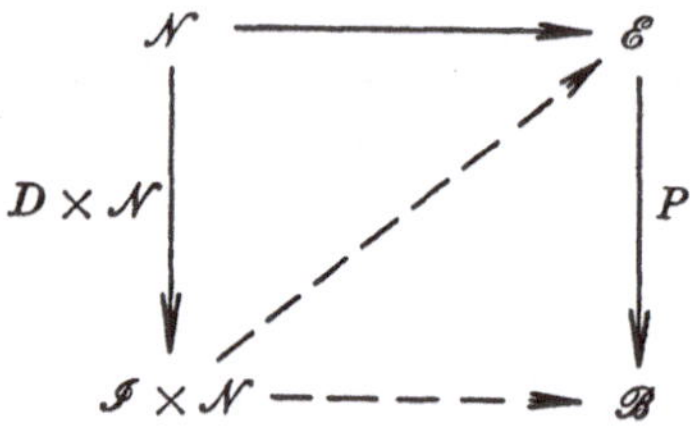

where the second is obtained from the first by replacing all categories by their opposites, then replacing $\mathscr{N}^{\mathrm{op}}$ by $\mathscr{N}$ since $\mathscr{N}$ was arbitrary, and observing that $\mathscr{I}^{\mathrm{op}} = \mathscr{I}$ but $R^{\mathrm{op}} = D$. The condition to be satisfied is that $P_G^{\mathrm{op}}: [\mathscr{I} \times \mathscr{N}, \mathscr{E}^{\mathrm{op}}]_G \to [\mathscr{I} \times \mathscr{N}, \mathscr{B}^{\mathrm{op}}]_{P^{\mathrm{op}}G}$ have a rari, or equivalently that

$$P_G: [\mathscr{I} \times \mathscr{N}, \mathscr{E}]_G \to [\mathscr{I} \times \mathscr{N}, \mathscr{B}]_{PG}$$

have a lari in the second diagram. Similarly, in the situation below with $S = \text{rali } D$, $P_G: [\mathscr{M}, \mathscr{E}]_G \to [\mathscr{M}, \mathscr{B}]_{PG}$ must have a lari. Finally, instead of augmented functors $H \to GS$, one considers "coaugmented" functors $GS \to H$.

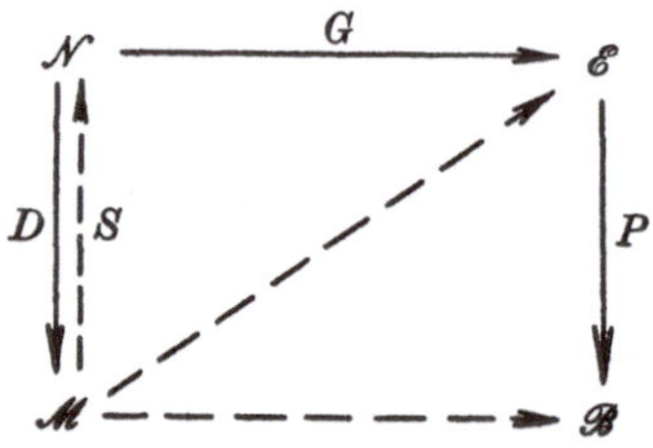

Theorem 2.10 characterizes the objects of $\mathscr{F}\!ib(\mathscr{B})$ and the next theorem similarly characterizes the morphisms of $\mathscr{F}\!ib(\mathscr{B})$.

Theorem 2.11. *Let $P: \mathscr{E} \to \mathscr{B}$ and $P: \tilde{\mathscr{E}} \to \mathscr{B}$ be fibrations and let $T: E \to \tilde{E}$ satisfy $\tilde{P}T = P$. Then the following are equivalent.*

o. *T is cartesian*
i. *T is 1-cartesian*
ii. *T preserves lifted paths*
iii. *T preserves lifted homotopies*
iv. *T preserves lifted generalized homotopies*
v. *T preserves liftings of canonically augmented functors.*

Proof. It is immediate that 0. ⇔ i. ⇔ ii. and iv. ⇔ v. as well as iv. ⇒ iii. ⇒ ii. It is therefore sufficient to show that 0. implies v.

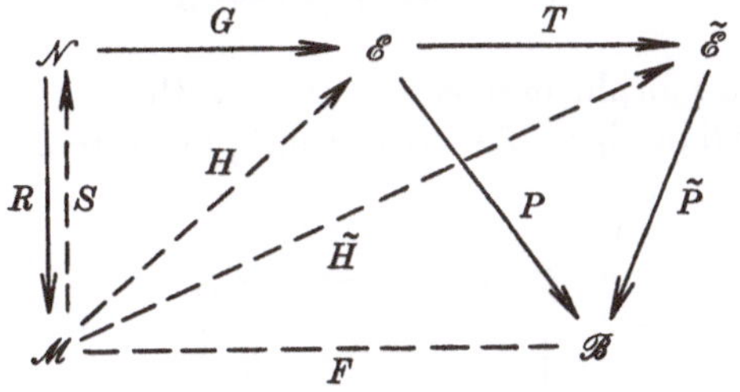

Suppose $T: \mathcal{E} \to \tilde{\mathcal{E}}$ is cartesian, and let $[f^*, \theta_f, c_{f,g}\}$ and $\{\tilde{f}^*, \tilde{\theta}_f, \tilde{c}_{f,g}\}$ be (normal) cleavages of P and $\tilde{P}$ respectively. If $F: \mathcal{M} \to \mathcal{B}$ satisfies $FR = PG$, let H and $\tilde{H}$ be cartesian liftings to $\mathcal{E}$ and $\tilde{\mathcal{E}}$ respectively; i. e.,

$$H(M) = (F\psi_M)^* [GS(M)] \quad \text{and} \quad \tilde{H}(M) = (F\psi_M)^* [TGS(M)].$$

Consider

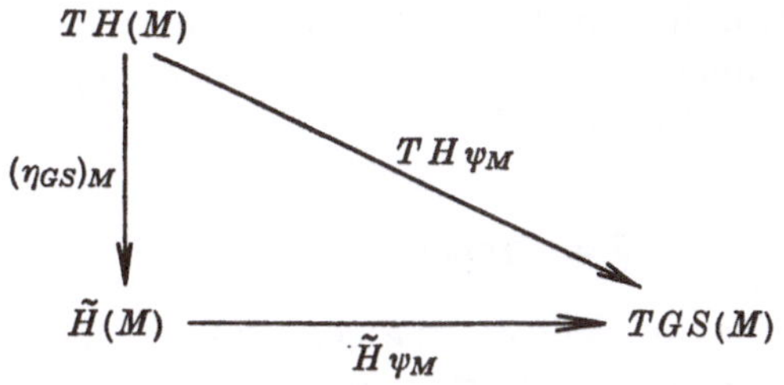

Since $H\psi_M = \theta_{F\psi_M}$ and $\tilde{H}\psi_M = \tilde{\theta}_{F\psi_M}$ it follows that $(\eta_{GS})_M = \eta_{F\psi_M}$ which is an isomorphism by hypothesis, and hence η_{GS} is a natural equivalence.

Example 2.12. *Let $\mathcal{B}$ be a category. Then*

i. $\mathcal{B}^R: \mathcal{B}^{\mathcal{I}} \to \mathcal{B}$ *is an opfibration and it is a fibration if and only if* $\mathcal{B}$ *has pullbacks.*

ii. $\mathcal{B}^D: \mathcal{B}^{\mathcal{I}} \to \mathcal{B}$ *is a fibration and it is an opfibration if and only if* $\mathcal{B}$ *has pushouts.*

To prove i., one checks immediately that if $f\colon B \to B'$ is a morphism in $\mathcal{B}$ and $(g\colon A \to B) \in \operatorname{ob} \mathcal{B}^{\mathcal{I}}$ then

$$\begin{array}{ccc} A & \xrightarrow{\ i_A\ } & A \\ {\scriptstyle g}\downarrow & & \downarrow{\scriptstyle fg} \\ B & \xrightarrow{\ f\ } & B' \end{array}$$

is an opcartesian morphism over f starting with g.

On the other hand if $f\colon B' \to B$ is a morphism in $\mathcal{B}$ and $(g\colon A \to B) \in \operatorname{ob} \mathcal{B}^{\mathcal{I}}$ then

$$\begin{array}{ccc} C & \longrightarrow & A \\ \downarrow & & \downarrow{\scriptstyle g} \\ B' & \xrightarrow[\ f\ ]{} & B \end{array}$$

is a cartesian morphism over f ending with g if and only if it is a pullback square. Pullback squares are sometimes called cartesian squares and this is presumably how the term "cartesian" gets into the study of fibred categories. Notice that our terminology forces us to call pushout squares "opcartesian" squares rather than "cocartesian" squares. For another way to prove i. and ii., see 5.5.

§ 3. Properties of $\mathcal{F}ib(\mathcal{B})$

In this section we shall use the characterizations of the preceding sections to derive some properties about fibrations as far as possible strictly within an axiomatic framework for the category of categories, taking iii., iv., or v. of 2.10 as the definition of a fibration. We state these properties for fibrations but they of course hold for opfibrations as well.

Proposition 3.1. *i. The composition of two fibrations is again a fibration.*

ii. The pullback of a fibration is a fibration.

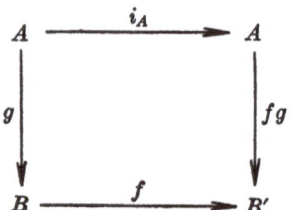

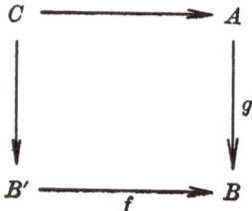

Proof. i. Let $P'\colon \mathcal{E}' \to \mathcal{E}$ and $P\colon \mathcal{E} \to \mathcal{B}$ be fibrations. In the situation illustrated, intuitively F is first lifted to H and then to H'.

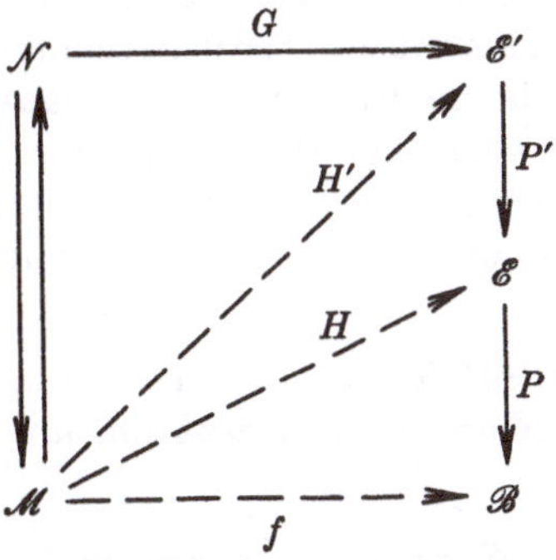

Formally, we have

$$[\mathcal{M}, \mathcal{E}']_G \xrightarrow{P_{G'}} [\mathcal{M}, \mathcal{E}]_{P'G} \xrightarrow{P_{P'G}} [\mathcal{M}, \mathcal{B}]_{PP'G}.$$

Since P'_G and $P_{P'G}$ both have raris, so does

$$P_{P'G}\, P'_G = (PP')_G$$

ii. Consider a diagram

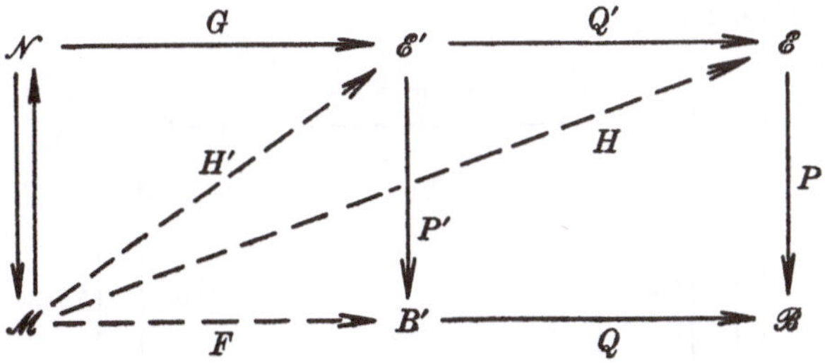

where the right hand square is a pullback and P is a fibration. Intuitively, QF can be lifted to an H with $PH = QF$ which therefore determines H' which is then shown to be the cartesian lifting of F. Formally, consider

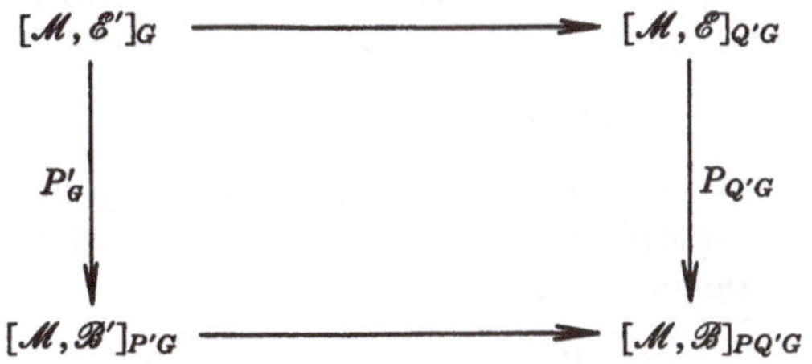

The corresponding diagram without the subscripts is a pullback since

$[\mathcal{M}, -]$ has a left adjoint and hence preserves pullbacks. The diagram with the subscripts is still a pullback because, for example, $[\mathcal{M}, \mathcal{E}]_G$ is a pullback (by 0.2, E), and because of the usual commutativity of pullbacks. We leave it to the reader to draw the appropriate diagram if he wishes. Finally, apply 0.3, D to conclude that if $P_{Q'G}$ has a rari then so does P'_G. Needless to say, a much simpler proof using cartesian morphisms and an explicit representation of $\mathcal{E}'$ can be given, but that proof does not dualize to cofibrations.

Definition 3.2. *Let $\mathcal{F}ib$ denote the subcaetgory of $[\mathcal{I}, \mathcal{C}at]$ whose objects are fibrations $P: \mathcal{E} \to \mathcal{B}$ and whose morphisms are commutative squares*

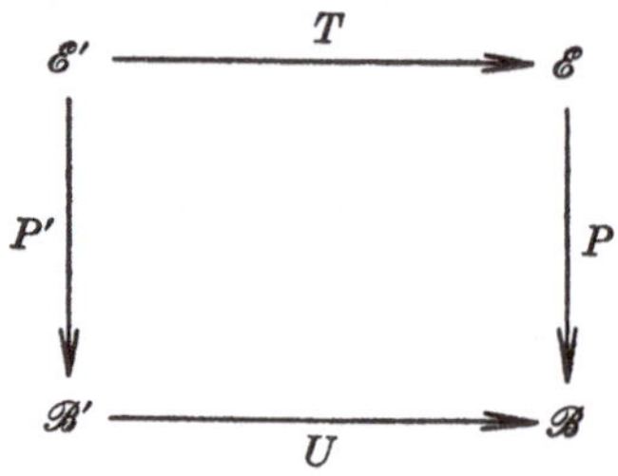

such that given any $R: \mathcal{N} \to \mathcal{M}$ with $S = lali\ R$ and any $G: \mathcal{N} \to \mathcal{E}'$, then, in the diagram

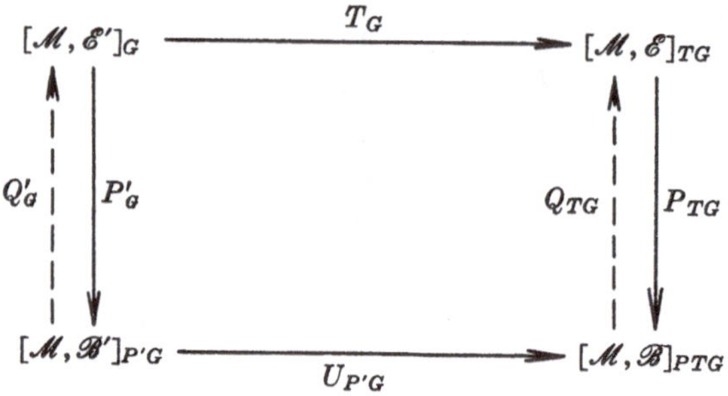

where $Q'_G = rari\ P'_G$ and $Q_{TG} = rari\ P_{TG}$, the induced transfer natural transformation (see 0.3, E*)*

$$\eta_G: T_G \circ Q'_G \to Q_{TG} \circ U_{P'G}$$

is a natural equivalence. In terms of cartesian morphisms this says that if φ' is a cartesian morphism in $\mathcal{E}'$ then $T(\varphi')$ is a cartesian morphism in $\mathcal{E}$, by a proof analogous to 2.11.

Let $\bar{R} = \mathcal{C}at^R \,|\, \mathcal{F}ib$, so

$$\bar{R}: \mathcal{F}ib \to \mathcal{C}at: (P: \mathcal{E} \to \mathcal{B}) \to \mathcal{B}: (T, U) \to U.$$

Using this terminology, 3.1, ii, can be restated as follows:

Proposition 3.3. *$\bar{R}: \mathscr{F}ib \to \mathscr{C}at$ is a fibration with fibres*

$$\mathscr{F}ib_{\mathscr{B}} = \mathscr{F}ib(\mathscr{B}).$$

Proof. The proof is an adaptation of Example 2.12. We want the cartesian morphisms to be pullback squares. Thus, given $U: \mathscr{B}' \to \mathscr{B}$ in $\mathscr{C}at$ and $(P: \mathscr{E} \to \mathscr{B}) \varepsilon$ ob $\mathscr{F}ib$, let the adjoining square be a pullback.

$$\begin{array}{ccc} \mathscr{E}' & \xrightarrow{\quad T \quad} & \mathscr{E} \\ {\scriptstyle P'}\downarrow & & \downarrow{\scriptstyle P} \\ \mathscr{B}' & \xrightarrow{\quad U \quad} & \mathscr{B} \end{array}$$

Then $P': \mathscr{E}' \to \mathscr{B}$ is a fibration by 3.1, ii. and it satisfies the required universal mapping property, providing (T, U) is a morphism in $\mathscr{F}ib$. But the relevant diagram 3.2 ii., with the vertical arrows from top to bottom, is a pullback diagram by the proof of 3.2, and hence by 0.3, E), η_G is a natural equivalence.

The objects of $\mathscr{F}ib_{\mathscr{B}}$ are clearly fibrations over $\mathscr{B}$ and the morphisms are pairs $(T, I_{\mathscr{B}})$ such that

$$\eta_G : T_G \circ Q'_G \to Q_{TG}$$

is a natural equivalence. Since this is one of the equivalent defining conditions (preservation of lifted generalized homotopies) of cartesian morphisms, it follows that $\mathscr{F}ib_{\mathscr{B}} = \mathscr{F}ib(\mathscr{B})$.

Proposition 3.4. *$\mathscr{F}ib$ is closed under the formation of products and sums in $[\mathscr{I}, \mathscr{C}at]$.*

Proof. It is easily checked that if $P_i: \mathscr{E}_i \to \mathscr{B}_i$ is a fibration for all i then so are $\prod P_i : \prod \mathscr{E}_i \to \prod \mathscr{B}_i$ and $\coprod P_i : \coprod \mathscr{E}_i \to \coprod B_i$ and the projections are morphisms in $\mathscr{F}ib$, since if S_i is the left adjoint of T_i then $\prod S_i$ is the left adjoint of $\prod T_i$, $\coprod S_i$ is the left adjoint of $\coprod T_i$, $[\mathscr{M}, -]$ preserves products and $[\mathscr{I}, -]$ preserves sums.

Remark. A similar argument shows that under suitable restrictions $\mathscr{F}ib$ is closed under left limits, the restrictions being that one considers only skeletal categories (i.e., isomorphisms are identities) or only inverse systems in which the η_G's are identities.

Proposition 3.5. *The categories $\mathscr{F}ib(\mathscr{B})$ have products which are pre-*

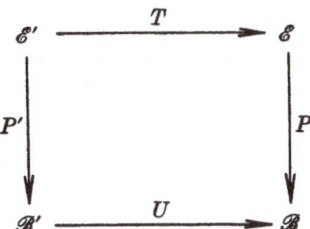

served by the functors F^* in any cleavage of $\bar{R}: \mathscr{F}ib \to \mathscr{C}at$, and by the inclusion functor $\mathscr{F}ib(\mathscr{B}) \to \mathscr{C}at_{\mathscr{B}}$.

Proof. Let $P: \mathscr{E} \to \mathscr{B}$ and $P': \mathscr{E}' \to \mathscr{B}$ be the fibrations and let the adjoining square be a pullback. Then

$$P' Q = P Q' : \mathscr{E}'' \to \mathscr{B}$$

is the product of P and P' in $\mathscr{C}at_{\mathscr{B}}$. Furthermore, Q and Q' are fibrations, by 3.1, ii. and hence $P'Q = PQ'$ is a fibration by 3.1, i.

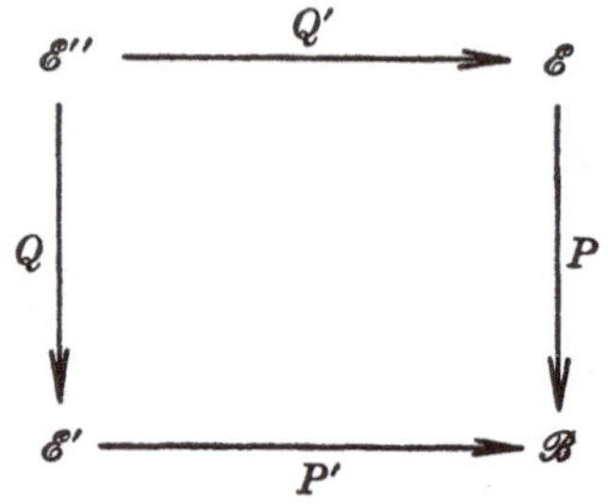

Therefore we need only check that Q and Q' are cartesian. But given any $R: \mathscr{N} \to \mathscr{M}$ with $S = \text{lali } R$ and any $G: \mathscr{N} \to \mathscr{E}''$, we have a diagram

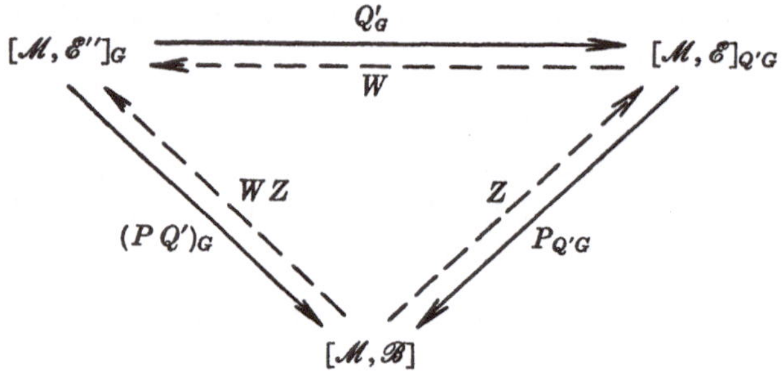

where $Z = \text{rari } P_{Q'G}$, $W = \text{rari } Q'_G$ and hence $WZ = \text{rari } (PQ')_G$. But then $Q'_G\, WZ = Z$ so the transfer natural transformation

$$\eta_G: Q'_G \circ (WZ) \to Z$$

is the identity. Hence Q' (and similarly Q) is cartesian. Finally, the commutativity of pullbacks shows that the functors F^* of a cleavage of $\bar{R}: \mathscr{F}ib \to \mathscr{C}at$ preserve this product.

Proposition 3.6. *$P: \mathscr{E} \to \mathscr{B}$ is a fibration if and only if for every (small) category $\mathscr{L}$,*

$$P^{\mathscr{L}}: [\mathscr{L}, \mathscr{E}] \to [\mathscr{L}, \mathscr{B}]$$

is a fibration.

Proof. The "if" part follows by taking $\mathscr{L}$ to be $*$. Conversely, let $P\colon \mathscr{E} \to \mathscr{B}$ be a fibration. Intuitively, the diagrams

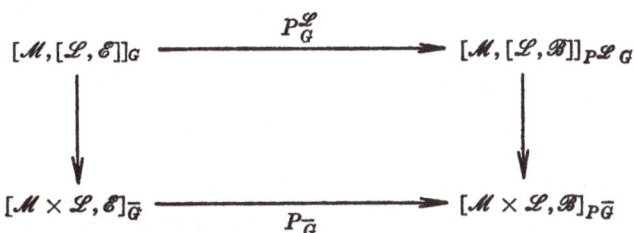

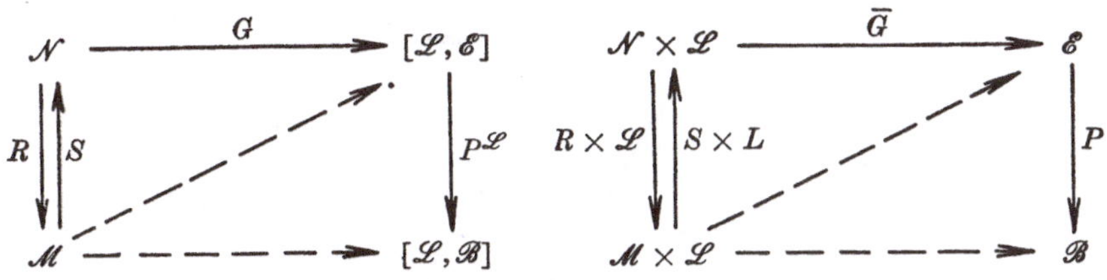

are equivalent by the adjointness between $[\mathscr{L}, -]$ and $- \times \mathscr{L}$. Hence since the right hand diagram can be filled properly, so can the left hand one.

Formally, consider the diagram

$$\begin{array}{ccc} [\mathscr{M}, [\mathscr{L}, \mathscr{E}]]_G & \xrightarrow{P_G^{\mathscr{L}}} & [\mathscr{M}, [\mathscr{L}, \mathscr{B}]]_{P^{\mathscr{L}} G} \\ \downarrow & & \downarrow \\ [\mathscr{M} \times \mathscr{L}, \mathscr{E}]_{\bar{G}} & \xrightarrow[P_{\bar{G}}]{} & [\mathscr{M} \times \mathscr{L}, \mathscr{B}]_{P\bar{G}} \end{array}$$

Without the subscript, the diagram commutes and the vertical morphisms are isomorphisms. A pullback argument, as above, shows that with the subscripts these properties still hold.

Hence $P_G^{\mathscr{L}}$ has a rari if and only if $P_{\bar{G}}$ does. But $P_{\bar{G}}$ has a rari since $S \times \mathscr{L} = \text{lali } R \times \mathscr{L}$.

Corollary 3.7. *If $P\colon \mathscr{E} \to \mathscr{B}$ is a fibration then a natural transformation in $[\mathscr{L}, \mathscr{E}]$ is cartesian if and only if all of its components are cartesian.*

Proof. Take $R\colon \mathscr{N} \to \mathscr{M}$ to be $R\colon * \to \mathscr{I}$. Then the cartesian natural transformations $H\colon \mathscr{I} \to [\mathscr{L}, \mathscr{E}]$ correspond to the cartesian lifted homotopies $\bar{H}\colon \mathscr{I} \times \mathscr{L} \to \mathscr{E}$. An examination of the proof of 2.10 shows that these can be chosen so that the components are cartesian morphisms. Since all choices are equivalent, this implies that a cartesian natural transformation has cartesian components. A similar examination shows that the only relevant factor for a lifted homotopy to be cartesian is that its components be cartesian.

Proposition 3.8. *Let $P\colon \mathscr{E} \to \mathscr{B}$ have a cleavage $\{f^*, \theta_f, c_{f,g}\}$ and let $\tilde{P}\colon \tilde{E} \to \mathscr{B}$ have a claevage $\{\tilde{f}^*, \tilde{\theta}_f, \tilde{c}_{f,g}\}$. If $T\colon \mathscr{E} \to \tilde{\mathscr{E}}$ is cleavage pre-*

serving then there is an $S = \text{lari}\ T$ *with* $PS = \tilde{P}$ *if and only if each* $T_B\colon \mathscr{E}_B \to \tilde{\mathscr{E}}_B$ *has a lari.*

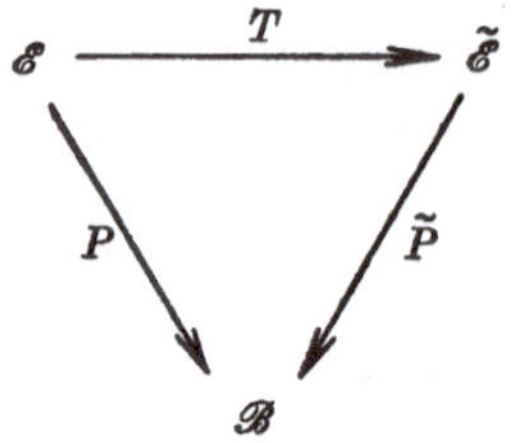

Proof. The "only if" is obvious. Conversely, suppose for each T_B, there is an $S_B = \text{lari}\ T_B$. Then under the adjointness relation

$$\begin{aligned}(S_{B'}\tilde{f}^*\tilde{E}, f^*S_B\tilde{E}) &\approx (\tilde{f}^*\tilde{E}, T_{B'}f^*S_B\tilde{E})\\ &\approx (\tilde{f}^*\tilde{E}, \tilde{f}^*T_BS_B\tilde{E}) \approx (\tilde{f}^*\tilde{E}, \tilde{f}^*\tilde{E})\end{aligned}$$

there is a unique $\eta_f\colon S_{B'}f^* \to f^*S_B$ with $T^{\eta_f} = 1_{\tilde{f}^*}$. One checks easily that these η_f's satisfy the required compatibility relations in 1.6 to determine a functor $S\colon \tilde{\mathscr{E}} \to \mathscr{E}$. By construction $PS = \tilde{P}$ and $TS = \text{id}$. Adjointness is easily checked directly.

We now wish to discuss some adjoint functors between $\mathscr{S}\!\mathit{plit}_0(\mathscr{B})$, $\mathscr{F}\!\mathit{ib}(\mathscr{B})$ and $\mathscr{C}\!\mathit{at}_{\mathscr{B}}$. Suppose $F\colon \mathscr{A} \to \mathscr{B}$ is a functor. Then, as in 0.2 F), we can consider

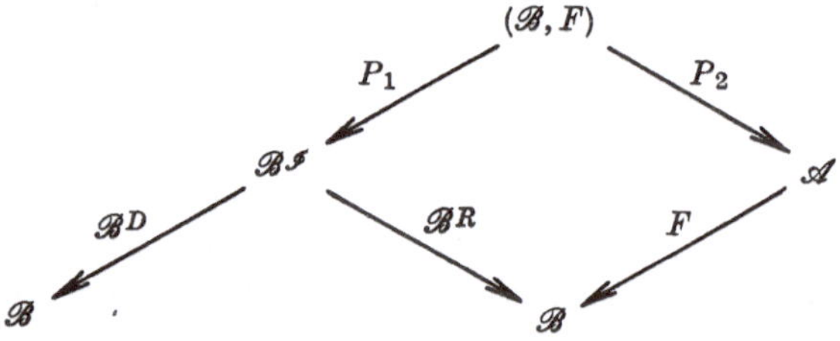

where the square is a pullback. Let

$$P_F = \mathscr{B}^D \circ P_1\colon (\mathscr{B}, F) \to \mathscr{B}$$

This clearly determines a functor

$$(\mathscr{B}, -)\colon \mathscr{C}\!\mathit{at}_{\mathscr{B}} \to \mathscr{C}\!\mathit{at}_{\mathscr{B}}\colon F \to P_F\colon T \to T_*$$

where, given $F\colon \mathscr{A} \to \mathscr{B}$, $F'\colon \mathscr{A}' \to \mathscr{B}$ and $T\colon \mathscr{A} \to \mathscr{A}'$ with $F'T = F$, $T_*\colon (\mathscr{B}, F) \to (\mathscr{B}, F')$ is the induced functor given by the pullback structure.

Let $\mathscr{C}\!\mathit{at}^h_{\mathscr{B}}$ (resp., $\mathscr{F}\!\mathit{ib}^h_{\mathscr{B}}$) denote the category with the same objects as $\mathscr{C}at$ (resp., $\mathscr{F}ib$) but where morphisms are equivalence classes of triangles

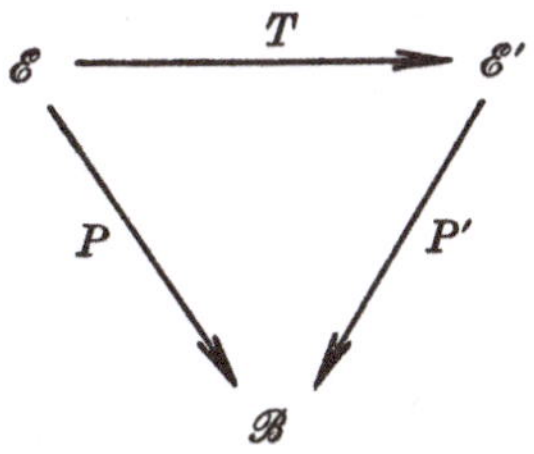

with $P'T = P$ (resp., and T cartesian) where two such, determined by T and T', are equivalent if and only if there is a natural equivalence $t: T \to T'$ with $P't = 1_P$.

Theorem 3.9. *i.* $(\mathscr{B}, -): \mathscr{C}at_{\mathscr{B}}^h \to \mathscr{F}ib_{\mathscr{B}}^h$ *is a reflection; i.e., it is left adjoint to the inclusion* $\mathscr{F}ib_{\mathscr{B}}^h \subset \mathscr{C}at_{\mathscr{B}}^h$.

ii. $(\mathscr{B}, -): \mathscr{C}at_{\mathscr{B}} \to \mathscr{S}plit_0(\mathscr{B})$ *is left adjoint to the forgetful functor* $\mathscr{S}plit_0(\mathscr{B}) \to \mathscr{C}at_{\mathscr{B}}$, *and hence*

ii'. $(\mathscr{B}, -): \mathscr{C}at_{\mathscr{B}} \to \mathscr{C}leav_0(\mathscr{B})$ *is left adjoint to the forgetful functor* $\mathscr{C}leav_0(\mathscr{B}) \to \mathscr{C}at_{\mathscr{B}}$.

iii. $(\mathscr{B}, -): \mathscr{C}leav(\mathscr{B}) \to \mathscr{S}plit_0(\mathscr{B})$ *is left adjoint to the inclusion* $\mathscr{S}plit_0(\mathscr{B}) \subset \mathscr{C}leav(\mathscr{B})$. *(The analogous adjoints for opfibrations and opcleavages are given by* $(-, \mathscr{B})$.)

Proof. We first check, as in Example 2.12 that if $F: \mathscr{A} \to \mathscr{B}$ then $P_F: (\mathscr{B}, F) \to \mathscr{B}$ is a fibration. We do not know a categorical proof for this but must use the explicit representation in 0.2, F). Given $f: B' \to B$ in $\mathscr{B}$ and given $(g: B \to F(A)) \in (\mathscr{B}, F)$ with $P(g) = B$ then

$$(\theta_f)_g = \left\{ \begin{array}{ccc} B' & \xrightarrow{\ f\ } & B \\ gf \downarrow & & \downarrow g \\ F(A) & \xrightarrow[F(i_A)]{} & F(A) \end{array} \right.$$

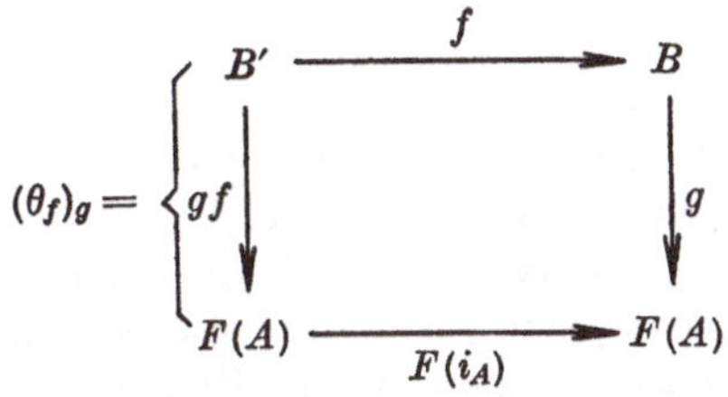

is easily checked to be a cartesian morphism over f ending at g. If $h: B'' \to B'$ then

$$(\theta_f)_g \circ (\theta_h)_{gf} = (\theta_{fh})_g$$

which implies that the canonical cleavage constructed from these cartesian morphisms as in 2.5 is split. Furthermore if $F: \mathscr{A} \to \mathscr{B}$, $F': \mathscr{A}' \to \mathscr{B}$

and $T: \mathscr{A} \to \mathscr{A}'$ satisfies $F'T = F$ then the induced functor is

$$T_*: (\mathscr{B}, F) \to (\mathscr{B}, F'): (g: B \to F(A)) \to$$
$$\to (g: B \to F'(T(A))): (f, h) \to (f, T(h)).$$

Clearly $T_*((\theta_f)_g) = (\theta_f)_g$ so T_* not only is cartesian but preserves the canonical split cleavages. Hence $(\mathscr{B}, -): \mathscr{C}at_{\mathscr{B}} \to \mathscr{F}ib_{\mathscr{B}}$, and, equally well, $(\mathscr{B}, -): \mathscr{C}at_{\mathscr{B}} \to \mathscr{S}plit_0(\mathscr{B})$. Furthermore, if T and T' differ by a natural equivalence $t: T \to T'$ with $F't = 1_F$ then T_* and T'_* differ by a natural equivalence t_* with $P_{F'}(t_*) = 1_{P_F}$, so

$$(\mathscr{B}, -): \mathscr{C}at^h_{\mathscr{B}} \to \mathscr{F}ib^h_{\mathscr{B}}.$$

Now, since $\Delta_{\mathscr{B}} = \text{rari } \mathscr{B}^R: \mathscr{B} \to \mathscr{B}^{\mathscr{I}}$ (by 0.3, B), there is a $Q_F = \text{rari}$ $P_2: \mathscr{A} \to (\mathscr{B}, F)$ (by 0,3, D) which satisfies $P_1 Q_F = \Delta F$ and hence

$$P_F Q_F = \mathscr{B}^D P_1 Q_F = \mathscr{B}^D \Delta F = F.$$

Q_F is clearly natural in F so we get a natural transformation

$$Q: I\mathscr{C}at_{\mathscr{B}} \to (\mathscr{B}, -).$$

(It does not in general seem to be true that F being a fibration implies that Q_F is cartesian.) Now we must check that Q_F satisfies the desired universal mapping property; i.e.,

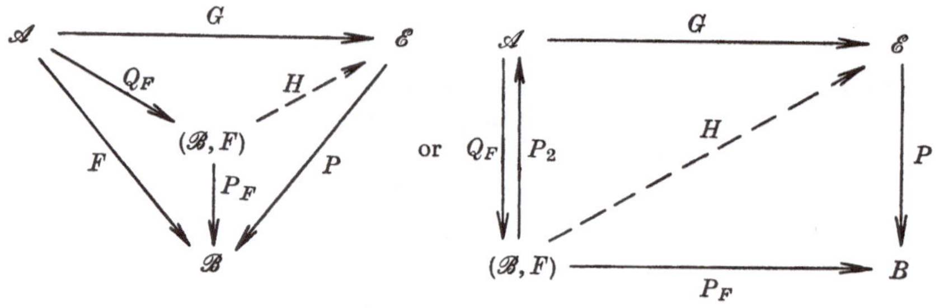

where the first diagram illustrates the universal mapping property for objects over $\mathscr{B}$ and the second shows that it is a special case of lifting a generalized homotopy, since $P_2 = \text{lali } Q_F$. Hence there is a "best possible" H making everything commute and we must show that "best possible" is the same as cartesian in this circumstance.

Suppose H is the "cartesian lifting" constructed as in 2.10 from some normal cleavage $\{f^*, \theta_f, c_{f,g}\}$ of P. A simple calculation shows that if

$$(f: B \to F(A)) \in (\mathscr{B}, F)$$

then $H(f) = f^* G(A)$, and if (g, h) is a morphism in $(\mathscr{B}, F)$ then $H(g, h)$ is the unique morphism such that

$$\begin{array}{ccc} H(f) & \xrightarrow{\theta_f} & G(A) \\ {\scriptstyle H(g,h)}\downarrow & & \downarrow{\scriptstyle G(h)} \\ H(f') & \xrightarrow[\theta_{f'}]{} & G(A') \end{array}$$

commutes with $PH(g, h) = g$.

Now cartesian morphisms in $(\mathscr{B}, F)$ are of the form $(g, 1)$ and it follows from 1.3 that

$$H(g, 1) = (\theta_g)_{f^*G(A)} \circ (c_{f,g})^{-1}$$

which is cartesian since it differs from a cartesian morphism by an equivalence, (2.3). In particular, for cartesian morphisms of the form

$$\begin{array}{ccc} B & \xrightarrow{g} & F(A) \\ {\scriptstyle g}\downarrow & & \downarrow{\scriptstyle i_{F(A)}} \\ F(A) & \xrightarrow[F(i_A)]{} & F(A) \end{array}$$

we have $H(g, 1) = (\theta_g)_{G(A)} : H(g) \to G(A)$.

Conversely, suppose $K: (\mathscr{B}, F) \to \mathscr{E}$ satisfies $K\,Q_F = G$, $PK = P_F$, and K is cartesian. Then, for cartesian morphisms $(g, 1)$ of the above special form, $K(g, 1) : K(g) \to G(A)$ is a cartesian morphism. Since $PK(g, 1) = PH(g, 1)$, it follows that K differs from H by a unique natural equivalence t with $Pt = 1_{P_F}$. Hence up to a natural equivalence, there is a unique cartesian H making the diagram commute.

Finally, observe that if $P: \mathscr{E} \to \mathscr{B}$ comes equipped with a fixed split cleavage $\{f^*, \theta_f, c_{f,g}\}$, then the above choice of H gives the unique functor which is cleavage preserving.

Corollary 3.10. *A functor $F: \mathscr{A} \to \mathscr{B}$ has a "best possible" factorization $F = P_F\,Q_F$ where Q_F has a lali and P_F is a fibration such that, if $F = P'_F Q'_F$ is another such factorization, then there is a cartesian functor T with* $T\,Q_F = Q'_F$ and $P'_F T = P_F$.

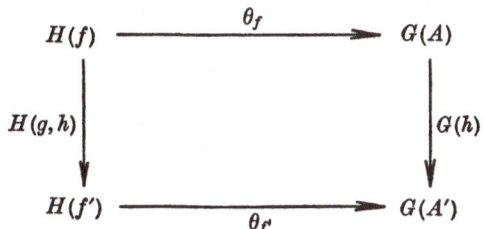

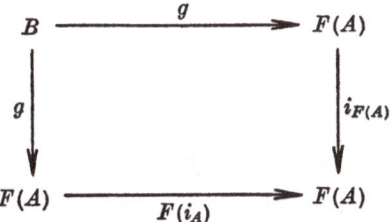

Remark. It is pointed out in $[GT]$, Ch. III, § 2, that if $F: \mathscr{A} \to \mathscr{B}$ is a left exact functor between abelian categories, then $(\mathscr{B}, F)$ is abelian, P_F is exact, as is P_2, and Q_F is left exact. $[GT]$ gives a characterization of $(\mathscr{B}, F)$ in this abelian case in terms of the existence of such functors with suitable properties.

Proposition 3.11. *(Chevalley criterion, [BN]) Let $P: \mathscr{E} \to \mathscr{B}$ be a functor and let $L: \mathscr{E}^I \to (\mathscr{B}, P)$ be the unique functor with $P_2 L = \mathscr{E}^R$ and $P_1 L = P^{\mathscr{I}}$. Then P is a fibration if and only if L has a rari.*

Proof. Consider

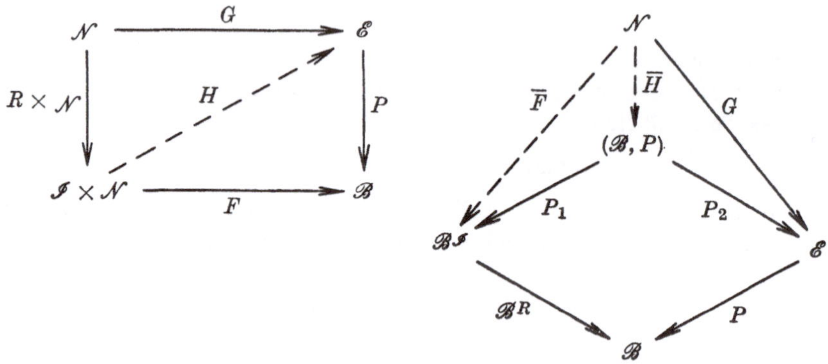

We have category isomorphisms

$$[\mathscr{I} \times \mathscr{N}, \mathscr{B}]_G \approx [\mathscr{N}, \mathscr{B}^{\mathscr{I}}]_{PG} \approx [\mathscr{N}, (\mathscr{B}, P)]_G$$

where $[\mathscr{N}, \mathscr{B}^{\mathscr{I}}]_{PG}$ consists of functors $\bar{F}: \mathscr{N} \to \mathscr{B}^{\mathscr{I}}$ with $\mathscr{B}^R \bar{F} = PG$ and natural transformations t with $\mathscr{B}^R t = 1_{PG}$ while $[\mathscr{N}, (\mathscr{B}, P)]_G$ consists of functors $\bar{H}: \mathscr{N} \to (\mathscr{B}, P)$ with $P_2 \bar{H} = G$ and natural transformations s with $P_2 s = 1_G$. Similarly, there is a category isomorphism

$$[\mathscr{I} \times \mathscr{N}, \mathscr{E}]_G \approx [\mathscr{N}, \mathscr{E}^{\mathscr{I}}]_G$$

where $[\mathscr{N}, \mathscr{E}^{\mathscr{I}}]_G$ consists of functors $\bar{H}: \mathscr{N} \to \mathscr{E}^{\mathscr{I}}$ such that $\mathscr{E}^R \bar{H} = G$ and natural transformations s such that $\mathscr{E}^R s = 1_G$. Furthermore, there is a commutative diagram

$$
\begin{array}{ccc}
[\mathscr{I} \times \mathscr{N}, \mathscr{E}]_G & \xrightarrow{\ \ P_G\ \ } & [\mathscr{I} \times \mathscr{N}, \mathscr{B}]_{PG} \\[2em]
\wr & & \wr \\[2em]
[\mathscr{N}, \mathscr{E}^{\mathscr{I}}]_G & \xrightarrow[\ \ L_G\ \ ]{} & [\mathscr{N}, (\mathscr{B}, P)]_G
\end{array}
$$

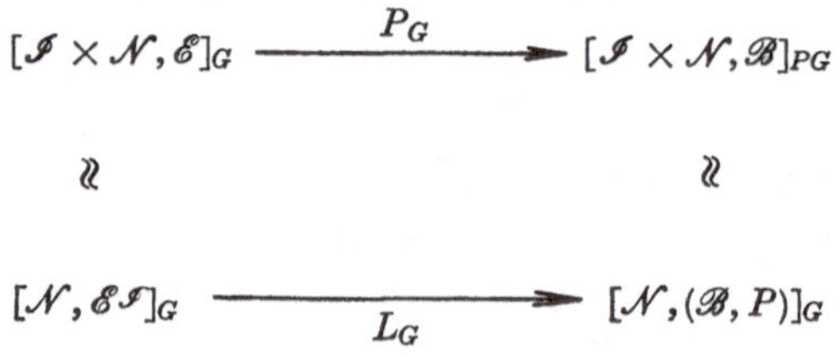

Hence P_G has a rari if and only if L_G has a rari.

Now, if L has a rari then so does every L_G ([SVG], Appendix A4) so P is a fibration. Conversely, suppose P is a fibration. Let $\{f^*, \theta_f, c_{f,g}\}$

be a cleavage of P and define

$$K: (\mathscr{B}, P) \to \mathscr{E}^{\mathscr{I}} : (g: B \to P(E)) \to ((\theta_g)_E : g^* E \to E).$$

It follows immediately from the definition of a fibration that $K = \text{rari}\ L$. Alternatively, observe that by 2.12, $\mathscr{B}^R: \mathscr{B}^{\mathscr{I}} \to \mathscr{B}$ has a canonical split opcleavage which gives rise to a split opcleavage of the pullback $P_2: (\mathscr{B}, F) \to \mathscr{E}$. Also $\mathscr{E}^R: \mathscr{E}^{\mathscr{I}} \to \mathscr{E}$ has a split opcleavage and in

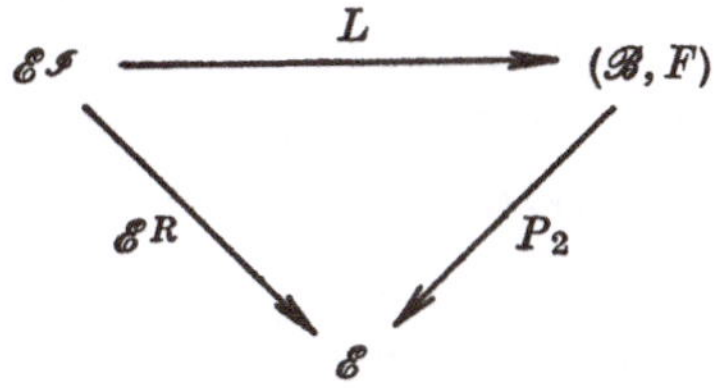

L preserves these opcleavages. Taking $N = *$ above shows that L restricted to each fibre has a rari. Hence by 3.7, L has a rari.

§ 4. Limits and Sheaves

We use the term *left limit* for (generalized) projective or inverse limit. If $F: \mathscr{A} \to \mathscr{B}$ is a functor then a sufficient condition guaranteeing that F carries sheaves with values in $\mathscr{A}$ into sheaves with values in $\mathscr{B}$ is that F preserves left limits. More precisely, it must preserve exactly those left limits which arise from the covering relations in the topology under consideration. The following results give conditions for a fibration to preserve various limits.

Proposition 4.1. *Let $P: \mathscr{E} \to \mathscr{B}$ be a fibration. Then the inclusion functors $J_B: \mathscr{E}_B \to \mathscr{E}$ preserve left limits (of a given type) for all $B \in \mathscr{B}$ if and only if the functors f^* in any cleavage of P preserve left limits (of the same type) for all morphisms f in B.*

Proof. Let $D: \mathscr{D} \to \mathscr{E}_B$, let $E_0 = \varprojlim D$ (in $\mathscr{E}_B$), let $f: B' \to B$ in $\mathscr{B}$ and let $E_0' = \varprojlim f^* D$ (in $\mathscr{E}_{B'}$).

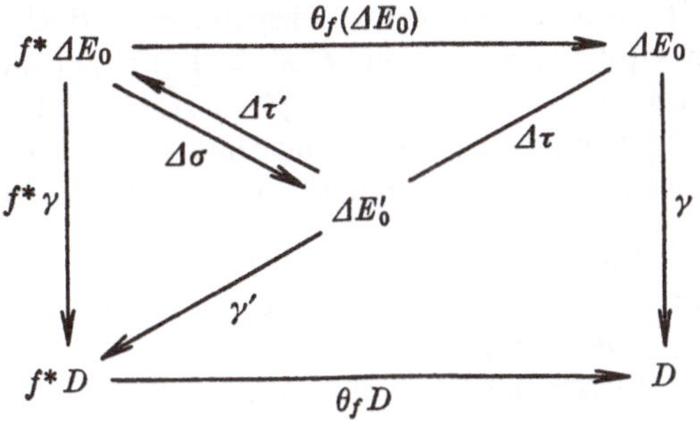

For any $E \in \mathscr{E}$, let $\Delta E: \mathscr{D} \to \mathscr{E}$ be the constant functor with value E and for any $\varphi: E \to E'$, let $\Delta\varphi: \Delta E \to \Delta E'$ be the constant natural transformation. Then there are adjunction natural transformations $\gamma: \Delta E_0 \to D$ and $\gamma': \Delta E_0' \to f^* D$. Furthermore there is a unique map $\sigma: f^* E_0 \to E_0'$ such that $\gamma' \circ \Delta\sigma = f^*\gamma$. Now, if J_B preserves left limits (of type $\mathscr{D}$) then there is a unique morphism $\tau: E_0' \to E_0$ such that $\gamma \circ \Delta\tau = (\theta_f D) \circ \gamma'$ and hence there is a unique morphism $\tau': E_0' \to f^* E_0$ such that $\theta_f(\Delta E_0) \circ \Delta\tau' = \Delta\tau$. Notice that then $\gamma': \Delta E_0' \to f^* D$ is the unique natural transformation such that $(\theta_f D) \circ \gamma' = \gamma \circ \Delta\tau$ since its components are uniquely determined by this relation; but

$$(\theta_f D)\, \gamma' \circ \Delta\sigma \circ \Delta\tau' = (\theta_f D)\, (f^* \gamma)\, \Delta\tau' = \gamma\, (\theta_f \Delta E_0)\, \Delta\tau' = \gamma \circ \Delta\tau\,.$$

Hence $\gamma' \circ \Delta\sigma \circ \Delta\tau' = \gamma'$, and therefore, by the uniqueness of factorization through γ', we have $\sigma \circ \tau' = i_{E_0'}$. On the other hand, since $E_0 = \varprojlim D$ (in $\mathscr{E}$), $\theta_f\, \Delta E_0$ is the unique natural transformation such that $\gamma\, (\theta_f \Delta E_0) = (\theta_f D)\, (f^* \gamma)$. But

$$\gamma\, (\theta_f \Delta E_0)\, \Delta\tau' \circ \Delta\sigma = \gamma \circ \Delta\tau \circ \Delta\sigma = (\theta_f D)\, \gamma' \circ \Delta\sigma = (\theta_f D)\, (f^* \gamma)\,.$$

Hence $(\theta_f \Delta E_0)\, \Delta\tau' \circ \Delta\sigma = \theta_f \Delta E_0$ and therefore by the uniqueness of factorizations through cartesian morphisms, we have $\tau'\sigma = i_{f^* E_0}$. Therefore f^* preserves left limits (of type $\mathscr{D}$). Conversely, a natural transformation $\delta: \Delta E \to D$ satisfies $P\delta = \Delta f$ for some morphism $f: B' \to B$ in

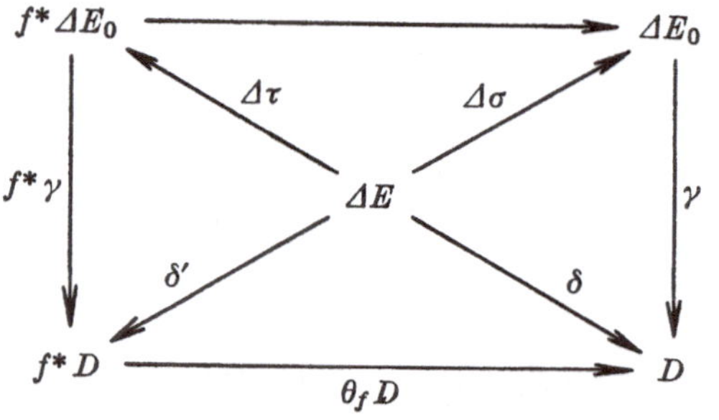

$\mathscr{B}$, and hence there is a unique factorization $\delta = (\theta_f D) \circ \delta'$. If f^* preserves left limits (of type $\mathscr{D}$), then $f^* E_0 = \varprojlim f^* D$ (in $\mathscr{E}_{B'}$) and hence there is a unique $\tau: E \to f^* E_0$ such that $f^* \gamma \circ \Delta\tau = \delta'$. Let $\sigma = (\theta_f)_{E_0} \circ \tau$. It is easily checked that $\gamma \circ \Delta\sigma = \delta$ and that σ is the unique morphism with this property. Hence $E_0 = \varprojlim D$ (in $\mathscr{E}$) so J_B preserves left limits.

Theorem 4.2. *Let $P: \mathscr{E} \to \mathscr{B}$ be a fibration and let $\mathscr{B}$ have left limits (of a given type). Then $\mathscr{E}$ has left limits (of the same type) which are preserved by P if and only if*

i. Each fibre $\mathscr{E}_B$ has left limits (of the given type).

ii. The inclusion functors $J_B\colon \mathscr{E}_B \to \mathscr{E}$ preserve left limits (of the given type).

(The analogous proposition for opfibrations is valid if "left" is replaced by "right".)

Proof. Let $\mathscr{D}$ be any small category (of the given type) and let $D\colon \mathscr{D} \to \mathscr{E}$. Let $F\colon \mathscr{I} \times \mathscr{D} \to \mathscr{B}$ be the functor which represents the natural transformation:

$$\varprojlim PD \to PD .$$

Then there is a best possible functor $H\colon \mathscr{I} \times \mathscr{D} \to \mathscr{E}$ with the following properties:

a) $PH = F$

b) $H \circ (R \times \mathscr{D}) = D$ so $H \mid 1 \times \mathscr{D} = D$

c) $H \mid 0 \times \mathscr{D} = H_0\colon \mathscr{D} \to \mathscr{E}_{\varprojlim PD}$

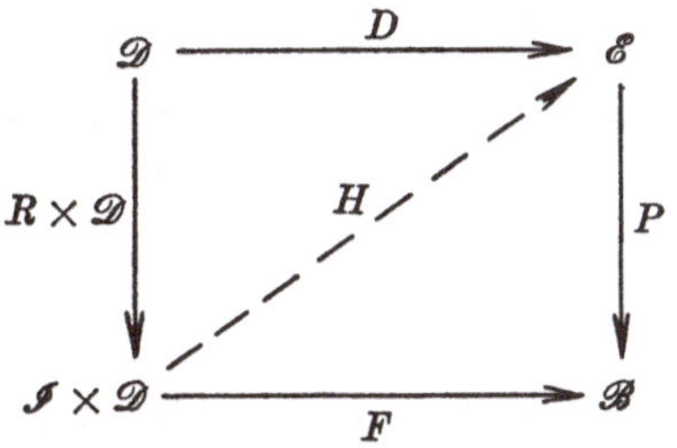

Using hypothesis i), $\mathscr{E}_{\varprojlim PD}$ has left limits, so we define $E_0 = \varprojlim H_0$ (in $\mathscr{E}_{\varprojlim PD}$). The problem is now to show that $E_0 = \varprojlim D$. The universal mapping property satisfied by H guarantees that for any $E \in \mathscr{E}$, any natural transformation from ΔE to D factors uniquely through H_0. If ii) is satisfied, then $E_0 = \varprojlim H_0$ in $\mathscr{E}$ so the natural transformation factors uniquely through E_0 and hence $E_0 = \varprojlim D$.

Conversely, if $\mathscr{E}$ has left limits which are preserved by P then for any $D\colon \mathscr{D} \to \mathscr{E}_B$, PD is the constant functor mapping $\mathscr{D}$ to B so $\varprojlim PD = B$. Hence $\varprojlim D \in \mathscr{E}_B$ which shows that $\mathscr{E}_B$ has left limits which are preserved by J_B.

Corollary 4.3. *If $P\colon \mathscr{E} \to \mathscr{B}$ is both a fibration and an opfibration and if $\mathscr{B}$ has left (resp., right) limits then $\mathscr{E}$ has left (resp., right) limits which are preserved by P if and only if all fibres $\mathscr{E}_B$ have left (resp., right) limits.*

Proof. If $P\colon \mathscr{E} \to \mathscr{B}$ is both a fibration and an opfibration then — by the discussion following 1.1 — for every morphism f in $\mathscr{B}$, we have that f^* is the left adjoint of f^*. Hence f^* preserves left limits and f^* preserves right limits, which by Proposition 3.1 is equivalent to condition

ii) of Theorem 3.2. Hence only condition i) is relevant to the existence of limits in $\mathscr{E}$.

We now turn to the question of the preservation of right limits by fibrations (or, equivalently, of left limits by opfibrations). The situation is not as satisfactory as it is with left limits where the theorem can be stated and proved without reference to cleavages, together with a proposition telling how to verify the condition in terms of cleavages. Here we do not seem to be able to avoid cleavages.

Proposition 4.4. *Let $P: \mathscr{E} \to \mathscr{B}$ be a fibration. The fibres have terminal objects which are preserved by the functors f^* of some cleavage if and only if P has a rari. Hence this condition implies that P preserves right limits.*

Proof. Recall that $T_B \in \mathscr{E}_B$ is terminal if and only if for each $E \in \mathscr{E}_B$, $\mathscr{E}_B(E, T_B)$ has exactly one element. Hence given any $E' \in \mathscr{E}$ and any $\varphi: E' \to T_B$, with its unique factorization $\varphi = \theta_f \circ \varphi'$ where

$$f = P(\varphi): B' \to B;$$

if $f^* T_B$ is terminal in $\mathscr{E}_{B'}$ then φ' is the unique morphism $E' \to f^* T_B$ in $\mathscr{E}_B$ so that φ is the only morphism from E' to T_B with $P(\varphi) = f$. In other words P establishes a bijection

$$\mathscr{E}(E', T_B) \approx \mathscr{B}(P(E'), B).$$

Thus the functor

$$Q: \mathscr{B} \to \mathscr{E}: B \to T_B : f \to (\theta_f)_{T_B}$$

is the rari of P.

Conversely, if $Q = \text{rari } P$, then for any $E \in \mathscr{E}_B$,

$$\mathscr{E}(E, Q(B)) \approx \mathscr{B}(B, B)$$

and there is exactly one $\varphi: E \to Q(B)$ with $P(\varphi) = i_B$. Hence $Q(B)$ is terminal in $\mathscr{E}_B$. Now suppose $f: B' \to B$. Since $PQ(f) = f$, there is a

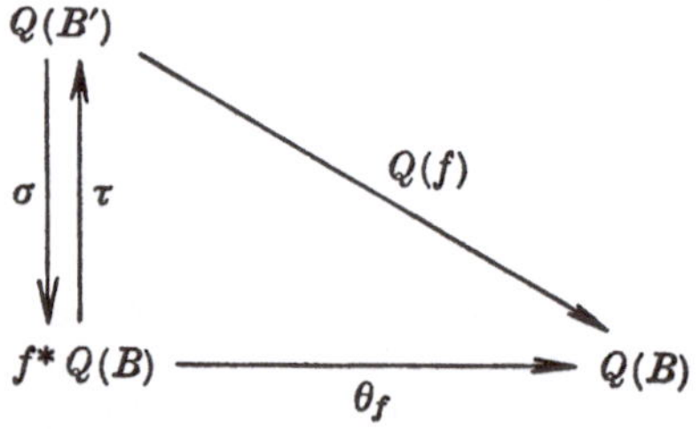

unique $\sigma: Q(B') \to f^*Q(B)$ in $\mathscr{E}_{B'}$ with $\theta_f \circ \sigma = Q(f)$ and, by adjointness, there is a unique $\tau: f^*Q(B) \to Q(B')$ in $\mathscr{E}_{B'}$ (i. e., with $P(T) = i_{B'}$) which, again by adjointness, satisfies $Q(f) \circ \tau = \theta_f$. But then $\sigma\tau = \text{id}$ by the uniqueness of factorizations through cartesian morphisms and

$\tau\sigma = \text{id}$ by adjointness. Hence f^* can be chosen so that $f^*(Q(B)) = Q(B')$

Finally, it is well known that functors that have right adjoints preserve right limits (see, e.g., [*SVC*])

Proposition 4.5. *If $P: \mathscr{E} \to \mathscr{B}$ is a fibration which preserves left limits, then a left limit of cartesian morphisms is cartesian. Dually, if $P: \mathscr{E} \to \mathscr{B}$ is an opfibration which preserves right limits, then a right limit of opcartesian morphisms is opcartesian.*

Proof. Let D and $D': \mathscr{D} \to \mathscr{E}$, let $E_0 = \varprojlim D$, $E_0' = \varprojlim D'$ and let γ and γ' be the indicated adjunction natural transformations. Let $\varphi: D \to D'$ be a natural transformation all of whose components are

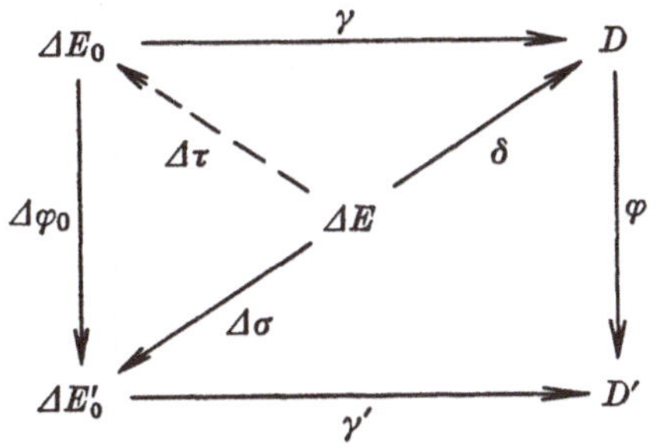

cartesian and let $\varphi_0 = \varprojlim \varphi: E_0 \to E_0'$. To show that φ_0 is cartesian, let $\sigma: E \to E_0'$ satisfy $P(\sigma) = P(\varphi_0)$. Then

$$P(\gamma' \circ \Delta\sigma) = P(\gamma' \circ \Delta\varphi_0) = P(\varphi \circ \gamma) = P(\varphi) \circ P(\gamma).$$

Since φ is a cartesian natural transformation in the functor category $[\mathscr{D}, \mathscr{E}]$ by 3.7, there is a unique $\delta: \Delta E \to D$ with $P\delta = P\gamma$ and $\varphi \circ \delta = \gamma' \circ \Delta\sigma$. Hence there is a unique $\tau: E \to E_0$ with $\gamma \circ \Delta\tau = \delta$. Since $P(E_0) = \varprojlim PD$, $P(\tau) = \text{id}$, so τ is the unique morphism with

$$\gamma' \circ \Delta\sigma = \varphi \circ \gamma \circ \Delta\tau = \gamma' \circ \Delta\varphi_0 \circ \Delta\tau.$$

Hence τ is the unique morphism with $\varphi_0 \circ \tau = \sigma$. Therefore φ_0 is cartesian. Note that we have proven that

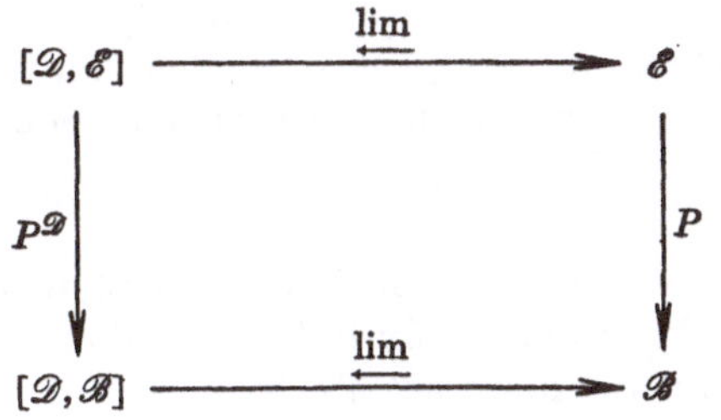

is a morphism in $\mathscr{F}ib$. We do not know a nice automatic proof of this fact that takes place in $\mathscr{C}at$.

We need one more result about fibrations which is related to the result on left limits and which we have placed in this section because it also lacks a categorical proof.

Proposition 4.6. *Let $T\colon \mathscr{E}_0 \to \mathscr{E}_1$ be a cartesian functor between fibrations $P_0\colon \mathscr{E}_0 \to \mathscr{B}$ and $P_1\colon \mathscr{E}_1 \to \mathscr{B}$. If $T\,|\,(\mathscr{E}_0)_B\colon (\mathscr{E}_0)_B \to (\mathscr{E}_1)_B$ is a fibration for all B and if the functors f^* in any cleavage of P_0 are cartesian then T is a fibration.*

Proof. Given $\varphi\colon E_1' \to E_1$ in $\mathscr{E}_1$ and $E_0 \in \mathscr{E}_0$ with $T(E_0) = E_1$, let $f = P_1(\varphi)\colon B' \to B$ and let $\theta_f\colon E_0' \to E_0$ be a cartesian morphism in

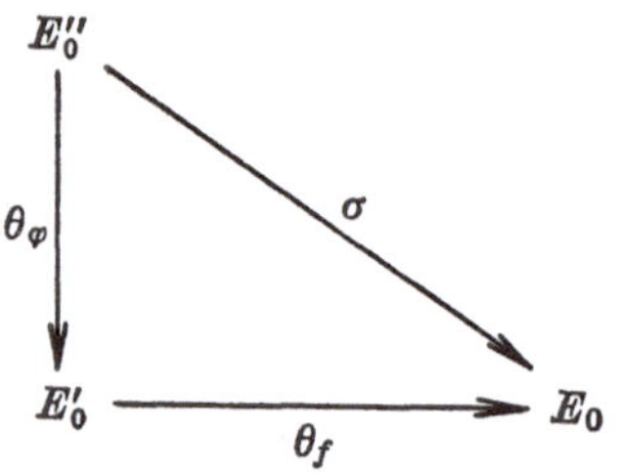

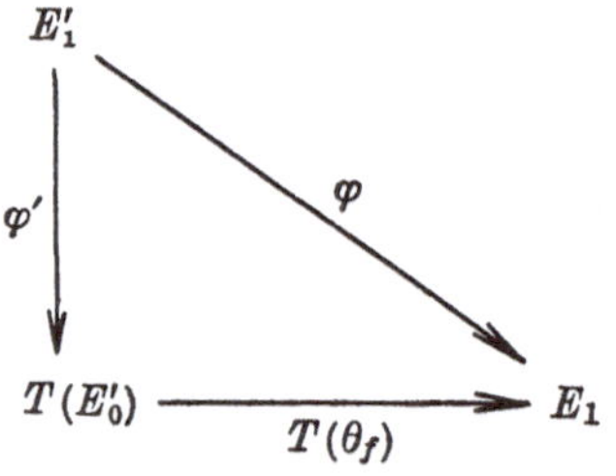

$\mathscr{E}_0$ with $P_0(\theta_f) = f$. Since T is cartesian, $T(\theta_f)$ is a cartesian morphism in $\mathscr{E}_1$, since $P_1 T(\theta_f) = f = P_1(\varphi)$ there is a unique $\varphi'\colon E_1' \to T(E_0')$ in $(\mathscr{E}_1)_{B'}$, such that $T(\theta_f) \circ \varphi' = \varphi$. Since $T\,|\,(\mathscr{E}_0)_{B'}$ is a fibration, there is a cartesian morphism $\theta_{\varphi'}$ in $(\mathscr{E}_0)_{B'}$ with $T(\theta_{\varphi'}) = \varphi'$. Let $\sigma = \theta_f \circ \theta_{\varphi'}$. It is easily checked that any morphism of this form is cartesian (in the sense of 2.2). Finally, our assumption about the functors f^* of a cleavage of P_0 guarantees that the composition of morphisms of this form is again of this form, and hence the composition of cartesian morphisms is cartesian. Therefore T is a fibration.

Example 4.7. In the case of the category of modules over the category of rings, the projection functor $P\colon \mathscr{M} \to \mathscr{R}$ is both a fibration and an opfibration. Hence $\mathscr{M}$ has limits of the same types that $\mathscr{R}$ does, which are preserved by P since the fibres have limits of all types. Furthermore, the fibres have zero objects, i. e., both initial and terminal, so the functor

$$Q : \mathscr{R} \to \mathscr{M} : R \to (R, 0) : f \to (f, 0)$$

is a right inverse to P which is both left and right adjoint to P. This same situation holds automatically if $P\colon \mathscr{E} \to \mathscr{B}$ is both a fibration and an opfibration and if the fibres are abelian.

We are now ready to discuss the relation between the category of sheaves with values in $\mathscr{E}$ and the category of sheaves with values in $\mathscr{B}$ when there is a fibration $P: \mathscr{E} \to \mathscr{B}$. We briefly recall the situation in general for sheaves with values in a "good" category $\mathscr{A}$. We consider only the case of ordinary topological spaces, although everything works perfectly well for arbitrary topologies in the sense of [GT] or [AS]. If X is a topological space, let $\mathscr{T}_X$ denote the category whose objects are the open sets of X and such that there is a unique $U \to V$ if and only if $U \supset V$. The category of presheaves on X with values in $\mathscr{A}$ is the functor category $[\mathscr{T}_X, \mathscr{A}]$. The category $\mathscr{S}(X, \mathscr{A})$ of sheaves on X with values in $\mathscr{A}$ is the full subcategory of $[\mathscr{T}_X, \mathscr{A}]$ determined by those functors which preserve (or in the general case, creat) certain left limits. (See [GT], [AS], or [SVC].) We call $\mathscr{A}$ a "good" category if $\mathscr{A}$ has left limits, directed right limits, and for every X, $\mathscr{S}(X, \mathscr{A})$ is a reflective subcategory of $[\mathscr{T}_X, \mathscr{A}]$; that is, given the inclusion functor

$$J_X : \mathscr{S}(X, \mathscr{A}) \to [\mathscr{T}_X, \mathscr{A}]$$

there is an $R_X = \text{lali}\, J_X$, which is constructed by a suitable combination of left and directed right limits. (See [GT], [AS], [HR] or [SVC].) Under these circumstances, given a continuous map $f: X \to Y$, there is a diagram

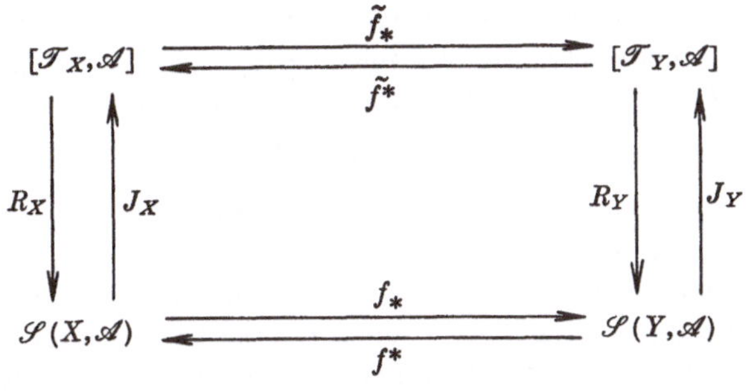

such that

i. $\tilde{f}_* \circ J_X = J_Y \circ f_*$ and there is a natural equivalence

$$f^* \circ R_Y \approx R_X \circ \tilde{f}^* \,.$$

ii. $\tilde{f}_*$ is the right adjoint of $\tilde{f}^*$ and f_* is the right adjoint of f^*.

iii. Given $g: Y \to Z$ then $(gf)_* = g_* f_*$ and there are compatible natural equivalences $c_{f,g} : (gf)^* \to f^* g^*$. (See, e.g. [$GT$], [$SGA$], [$EGA$], or [$SVC$].)

The only obstacle to constructing categories of presheaves and sheaves, both of which would be fibred and opfibred over the category $\mathscr{T}$ of topological spaces is that the adjointness relations in ii. above are in the wrong order. This can be remedied by replacing $\mathscr{T}$ by its opposite category (which is not done) or by replacing each $[\mathscr{T}_X, \mathscr{A}]$ and $\mathscr{S}(X, \mathscr{A})$ by its opposite category (which is done, e.g., in $[EGA]$.) Thus the pseudo functor

$$\mathscr{T} \to \mathscr{C}at : X \to [\mathscr{T}_X, \mathscr{A}]^{\mathrm{op}} : f \to (\tilde{f}^*)^{\mathrm{op}}$$

and the opposite pseudo functor

$$\mathscr{T} \to \mathscr{C}at : X \to [\mathscr{T}_X, \mathscr{A}]^{\mathrm{op}} : f \to (\tilde{f}_*)^{\mathrm{op}}$$

determine isomorphic categories, which we identify, denote by $\mathscr{P}(\mathscr{A})$ and call the category of presheaves with values in $\mathscr{A}$. The projection functor

$$\tilde{P} : \mathscr{P}(\mathscr{A}) \to \mathscr{T}$$

is both a fibration and an opfibration.

Similarly, one obtains the category $\mathscr{S}(\mathscr{A})$ of sheaves with values in $\mathscr{A}$ and a projection functor

$$P : \mathscr{S}(\mathscr{A}) \to \mathscr{T}$$

which is both a fibration and an opfibration. One recovers the standard description immediately; i.e., an object of $\mathscr{S}(\mathscr{A})$ is a pair (X, F) where $X \in \mathscr{T}$ and $F \in \mathscr{S}(X, \mathscr{A})$, while a morphism is a pair

$$(f, \varphi) : (X, F) \to (Y, G)$$

where $f : X \to Y$ is continuous and $\varphi : f^* G \to F$ (or equivalently, $\varphi : G \to f_* F$) is a sheaf morphism.

The relations in i. above imply that the inclusions $J_X : \mathscr{S}(X, \mathscr{A}) \to$ $\to [\mathscr{T}_X, \mathscr{A}]$ induce an inclusion functor $J : \mathscr{S}(\mathscr{A}) \to \mathscr{P}(\mathscr{A})$ which preserves the opcleavages but which is not cartesian as far as the cleavages are concerned. Similarly, the functors $R_X : [\mathscr{T}_X, \mathscr{A}] \to \mathscr{S}(X, \mathscr{A})$ induce a functor $R : \mathscr{P}(\mathscr{A}) \to \mathscr{S}(\mathscr{A})$ which is cartesian but not opcartesian. Because of the reversal of the categories we now have that $R = \mathrm{rali}\, J$ since $R \,|\, \mathscr{P}(\mathscr{A})_X = R_X^{\mathrm{op}}$ and $R_X = \mathrm{lali}\, J_X$.

Lemma 4.8. *Let $P : \mathscr{E} \to \mathscr{B}$ be a fibration. Suppose $J_e : \mathscr{E}' \to \mathscr{E}$ and $J_b : \mathscr{B}' \to \mathscr{B}$ are full inclusions such that there exist functors $R_e = \mathrm{rali}\, J_e$ and $R_b = \mathrm{rali}\, J_b$. Suppose further that $P' = P \,|\, \mathscr{E}' : \mathscr{E}' \to \mathscr{B}'$ and $P' R_e = R_b P$. Then P' is a fibration.*

(The analogous result opfibrations holds if "rali" is replaced by "lali".)

Proof. Consider

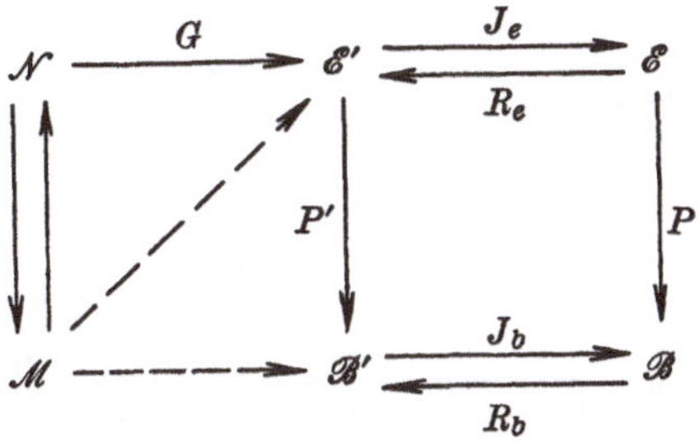

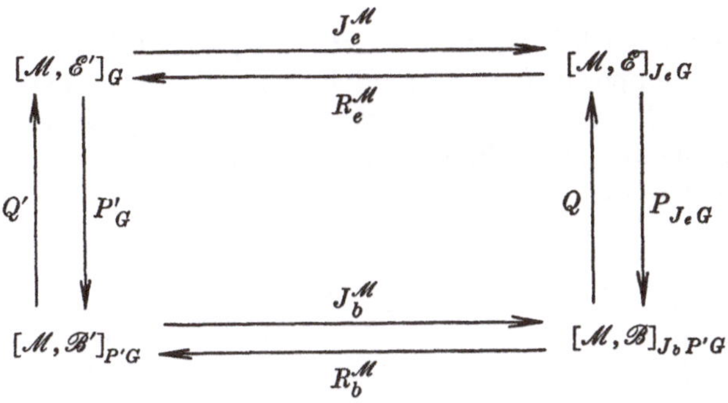

Observe that $R_e^{\mathscr{M}} = \text{rali}\, J_e^{\mathscr{M}}$ and $R_b^{\mathscr{M}} = \text{rali}\, J_b^{\mathscr{M}}$ (e.g., by [*SVC*] appendix A4). Hence if $Q = \text{rari}\, P_{J_e G}$ then $Q' = R_e^{\mathscr{M}} Q J_b^{\mathscr{M}} = \text{rari}\, P'_G$, since

i. $P'_G Q' = P'_G R_e^{\mathscr{M}} Q J_b^{\mathscr{M}} = R_b^{\mathscr{M}} P_{J_e G} Q J_b^{\mathscr{M}} = R_b^{\mathscr{M}} J_b^{\mathscr{M}} = \text{id}$.

ii. $$\begin{aligned}\operatorname{Hom}(Q'A, B) &= \operatorname{Hom}(R_e^{\mathscr{M}} Q J_b^{\mathscr{M}}(A), B)\\ &= \operatorname{Hom}(J_b^{\mathscr{M}}(A), P_{J_e G} J_e(B))\\ &= \operatorname{Hom}(J_b^{\mathscr{M}}(A), J_b^{\mathscr{M}} P'_G(B))\\ &= \operatorname{Hom}(A, P'_G(B))\end{aligned}$$

the last equality holding since J_b is a full inclusion.

Theorem 4.9. (cf. [*BN*]). *Let $P : \mathscr{E} \to \mathscr{B}$ be an opfibration between good categories which preserves left limits and directed right limits.*

i. For any topological space (or topology) X,

$$P^X : \mathscr{S}(X, \mathscr{E}) \to \mathscr{S}(X, \mathscr{B})$$

is an opfibration.

ii. $\bar{P} : \mathscr{P}(\mathscr{E}) \to \mathscr{P}(\mathscr{B})$ and $\bar{\bar{P}} : \mathscr{S}(\mathscr{E}) \to \mathscr{S}(\mathscr{B})$ are fibrations.

Proof. By 4.5, P preserves left limits if the fibres have initial objects which are preserved by the functors f_* in some opcleavage.

1. Consider

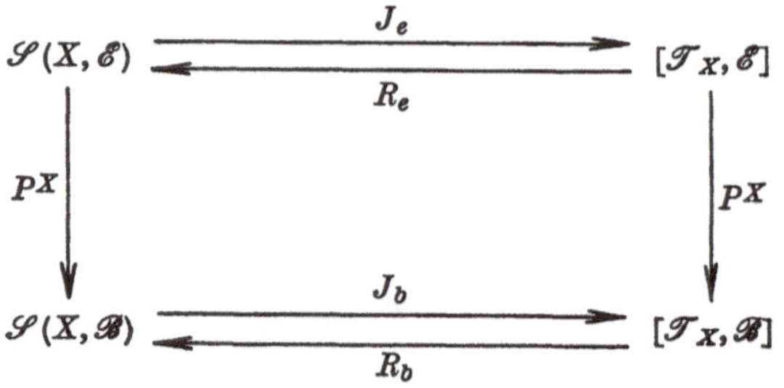

Since P preserves left limits, $P^X: \mathscr{S}(X, \mathscr{E}) \to \mathscr{S}(X, \mathscr{B})$. All the conditions of the opposite form of Lemma 4.8 are satisfied except possibly the condition $P^X R_e = R_b P^X$ which follows since P preserves the left limits and directed right limits used to construct R.

2. Now consider

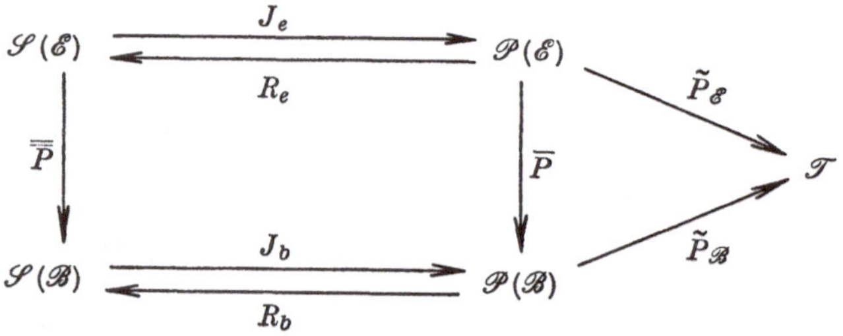

Again Lemma 4.8 implies that it is sufficient to show that $\bar{P}$ is a fibration. To use 4.6 on the right hand triangle observe first that $\bar{P}$ is cartesian since cartesian morphisms in $\mathscr{P}(\mathscr{E})$ are of the form $(f, i_{\tilde{f}^* F})$ and

$$\bar{P}(f, i_{\tilde{f}^* F}) = (f, i_{\bar{P}\tilde{f}^* F}) = (f, i_{\tilde{f}^* \bar{P} F})$$

where $\bar{P}$ commutes with $\tilde{f}^*$ since P preserves directed right limits. Second, by Corollary 3.7, a presheaf morphism in $[\mathscr{T}_X, \mathscr{A}]$ is opcartesian if and only if all of its components are. But $\tilde{f}^*$ is constructed by taking direct limits, which by 4.5 preserve opcartesian morphisms. Hence $(\tilde{f}^*)^{\mathrm{op}}: [\mathscr{T}_Y, \mathscr{E}]^{\mathrm{op}} \to [\mathscr{T}_X, \mathscr{E}]^{\mathrm{op}}$ preserves cartesian morphisms. 4.6 now implies that $\bar{P}$ is a fibration.

Theorem 4.10 (cf. [BN]). *Let $P: \mathscr{E} \to \mathscr{B}$ be a fibration which preserves left limits.*

i. For any topology X, $P^X: \mathscr{S}(X, \mathscr{E}) \to \mathscr{S}(X, \mathscr{B})$ is a fibration.
ii. $\bar{P}: \mathscr{P}(\mathscr{E}) \to \mathscr{P}(\mathscr{B})$ and $\bar{\bar{P}}: \mathscr{S}(\mathscr{E}) \to \mathscr{S}(\mathscr{B})$ are opfibrations.

Proof. i. The proof in this case is much simpler. It is sufficient to show that a cartesian morphism in $[\mathscr{T}_X, \mathscr{E}]$ that ends with a sheaf starts with a sheaf. But this follows immediately from 3.6 and 4.5.

ii. As in the preceding theorem, we want to use 4.6. In both cases

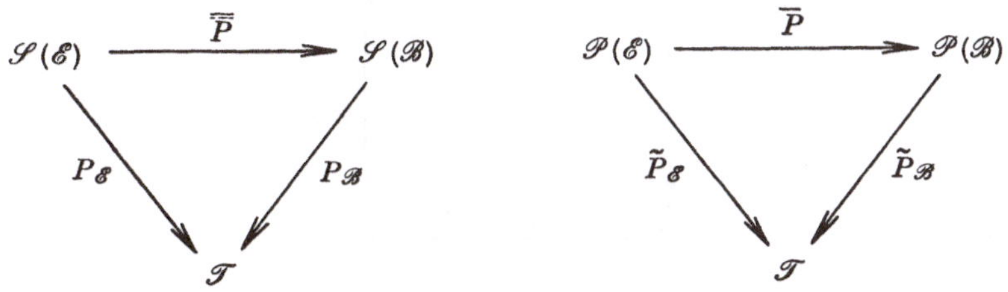

the projection functors are opfibrations and $\bar{P}$ and $\overline{\overline{P}}$ preserve the standard opcleavages; e.g., $\overline{\overline{P}} f_* = f_* \overline{\overline{P}}$. Furthermore, by 3.7 it is immediate that the functors $(f_*)^{\mathrm{op}}$ preserve opcartesian morphisms. Hence the result follows from 4.6.

Example 4.10. We mention one last relation between sheaves and fibrations. Let $i: A \to X$ be the inclusion of a closed subspace and let $U = X \sim A$ with $j: U \to X$ the inclusion. Then

$$i^* j_*: \mathscr{S}(A, \mathscr{A}) \to \mathscr{S}(U, \mathscr{A}).$$

It is shown in [GT] that there is an isomorphism

$$\mathscr{S}(X, \mathscr{A}) \approx (\mathscr{S}(U, \mathscr{A}), i^* j_*)$$

with $P_{i^* j_*} = i^*: \mathscr{S}(X, \mathscr{A}) \to \mathscr{S}(U, \mathscr{A})$. Thus i^* is not only a fibration, but it is the universal fibration with respect to compatible functors on $\mathscr{S}(A, \mathscr{A})$.

This suggests a relation between cohomology and fibrations — or at least universal fibrations — which we hope to explore further.

§ 5. Cofibrations

The first three properties characterizing fibrations in Theorem 2.10 have no obvious dualizations in $\mathscr{C}at$. However, the last three do, as follows.

Properties in $\mathscr{C}at^{\mathscr{A}}$ 5.1. Let

$$J: \mathscr{A} \to \mathscr{B},\ J': \mathscr{A} \to \mathscr{B}' \text{ and } T: \mathscr{B} \to \mathscr{B}' \text{ with } TJ = J'$$

i. Let $\mathscr{L}$ be a fixed category and let $G: \mathscr{B} \to \mathscr{L}$ be a functor. Then $(\mathscr{L}^R)^{\mathscr{B}}: [\mathscr{B}, \mathscr{L}^{\mathscr{I}}] \to [\mathscr{B}, \mathscr{L}]$ and we define

$$\begin{aligned}[\mathscr{B}, \mathscr{L}^{\mathscr{I}}]_G &= ((\mathscr{L}^R)^{\mathscr{B}})^{-1}(G)\\ &= \{H: \mathscr{B} \to \mathscr{L}^{\mathscr{I}} \mid \mathscr{L}^R H = G\} \cup \{t: H \to H' \mid L^R t = 1_G\}.\end{aligned}$$

Similarly $[\mathscr{A}, \mathscr{L}^{\mathscr{I}}]_{GJ} = ((\mathscr{L}^R)^{\mathscr{A}})^{-1}(GJ)$. Clearly J induces a functor

$$J_G : [\mathscr{B}, \mathscr{L}^{\mathscr{I}}]_G \to [\mathscr{A}, \mathscr{L}^{\mathscr{I}}]_{GJ}.$$

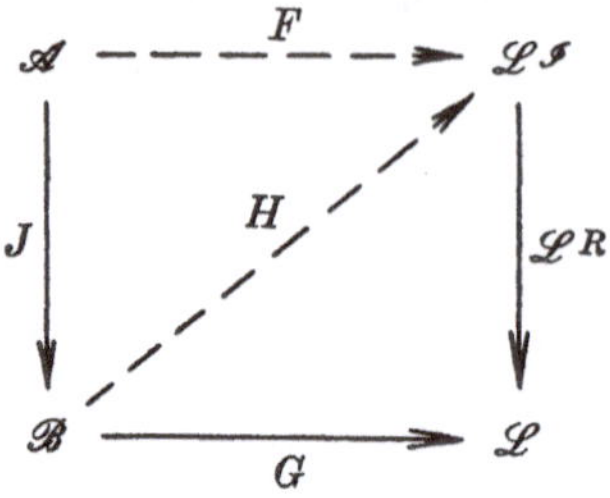

A functor $F \in [\mathscr{A}, \mathscr{L}^{\mathscr{I}}]_{GJ}$ is the same as a natural transformation or "homotopy" between functors from $\mathscr{A}$ to $\mathscr{L}$ ending at GJ. If J is an inclusion then GJ is just $G \mid \mathscr{A}$. If there is an $H \in [\mathscr{B}, \mathscr{L}^{\mathscr{I}}]_G$ with

$$J_G(H) = HJ = H \mid \mathscr{A} = F,$$

then F can be extended to a homotopy ending at G. Thus, for fixed $\mathscr{L}$, if every such J_G has a rari then we shall say that *homotopies have cocartesian extensions*. If this is satisfied then $J : \mathscr{A} \to \mathscr{B}$ is called an *$\mathscr{L}$-cofibration*. If Λ is a class of categories and if J is an $\mathscr{L}$-cofibration for all $L \in \Lambda$ then J is called a *Λ-cofibration*. If J is an $\mathscr{L}$-cofibration for all categories $\mathscr{L}$, then J is called a *cofibration*.

This condition is derived from 2.8 iii. by reversing all arrows and replacing $\mathscr{I} \times (-)$ by its adjoint $(-)^{\mathscr{I}}$. There is no corresponding restriction on the categories $\mathscr{N}$ in the fibration case since we showed that as soon as the fibration condition is satisfied for $*$ then it is satisfied for all $\mathscr{N}$.

Suppose now that $J : \mathscr{A} \to \mathscr{B}$ and $J' : \mathscr{A} \to \mathscr{B}'$ are $\mathscr{L}$-cofibrations and $T : \mathscr{B} \to \mathscr{B}'$ satisfies $TJ = J'$. Then given $G : \mathscr{B}' \to \mathscr{L}$, we get

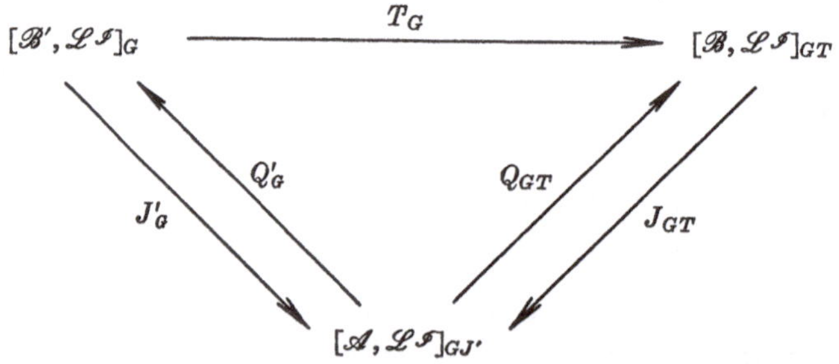

T_G commutes with J'_G and J_{GT} but not necessarily with $Q'_G = \text{rari } J'_G$ and $Q_{GT} = \text{rari } J_{GT}$. As in 0.3, E) there is a transfer natural transformation

$$\eta_G\colon T_G Q'_G \to Q_{GT}$$

with $J_{GT}\eta_G = 1$. If every such η_G is a natural equivalence then T is called *L-cocartesian* (with analogous definitions of *Λ-cocartesian* and *cocartesian.*)

ii. Given a functor $K\colon \mathscr{L} \to \mathscr{K}$ for which there exists $L = \text{rari}\, K$ and given $G\colon \mathscr{B} \to \mathscr{K}$, then J induces a functor

$$J_G\colon [\mathscr{B}, \mathscr{L}]_G \to [\mathscr{A}, \mathscr{L}]_{GJ}.$$

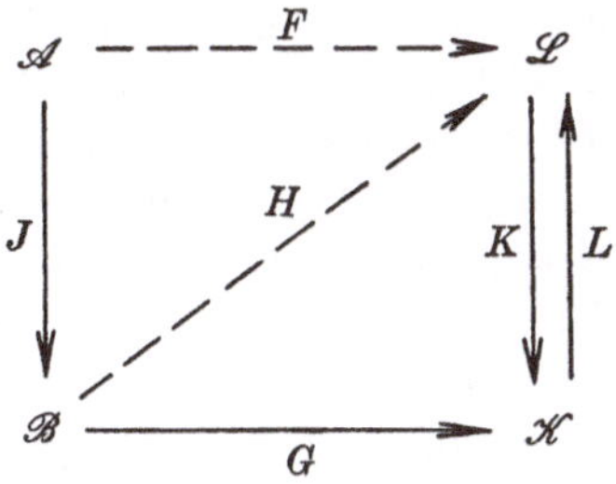

If, for fixed $K\colon \mathscr{L} \to \mathscr{K}$, every such J_G has a rari then we shall say that *generalized* $(\mathscr{L}, \mathscr{K})$*-homotopies have cocartesian extensions.* If this is satisfied by J then J is called a *strong* $(\mathscr{L}, \mathscr{K})$*-cofibration.* If this holds for all $\mathscr{L}$ and $\mathscr{K}$ in a class Λ then J is called a *strong Λ-cofibration* and if Λ is $\mathscr{C}at$ then J is called a *strong cofibration. Strong* $(\mathscr{L}, \mathscr{K})$*-cocartesian, strong Λ-cocartesian* and *strong cocartesian* functors are defined as above.

In particular, $\mathscr{L}$ has a terminal object if and only if $\mathscr{L} \to *$ has a rari, in which case one can consider strong $(L, *)$-cofibrations. If $G\colon \mathscr{B} \to *$ denotes the unique such functor then $[\mathscr{B}, \mathscr{L}]_G = [\mathscr{B}, \mathscr{L}]$ and $[\mathscr{A}, \mathscr{L}]_{GJ} = [\mathscr{A}, \mathscr{L}]$. Hence $J_G\colon [\mathscr{B}, \mathscr{L}] \to [\mathscr{A}, \mathscr{L}]$ has a rari if and only if every functor $F\colon \mathscr{A} \to \mathscr{L}$ has an "extension" $H\colon \mathscr{B} \to \mathscr{L}$ with $HJ = F$ such that for any $K\colon \mathscr{B} \to \mathscr{L}$, there is a natural bijection between the natural transformations from KJ to F and those from K to H.

iii. In the same situation as ii., J induces a functor

$$J_{LG}\colon [\mathscr{B}, \mathscr{L}]_{LG} \to [\mathscr{A}, \mathscr{L}]_{LGJ}$$

where as in 0.2, E), $[\mathscr{B}, \mathscr{L}]_{LG}$ is the category of functors over LG. As in 2.8, v), if $H \in [\mathscr{B}, \mathscr{L}]_G$ (i.e., $KH = G$) then the natural transformation $\psi\colon I_{\mathscr{L}} \to LK$ induces a canonical augmentation

$$\psi H\colon H \to LKH = LG.$$

Let $[\mathscr{B}, \mathscr{L}]'_{LG}$ denote the full subcategory of $[\mathscr{B}, \mathscr{L}]_{LG}$ determined by functors $H \in [\mathscr{B}, \mathscr{L}]_G$ with this canonical augmentation. Similarly, for

$[\mathscr{A}, \mathscr{L}]'_{LGJ}$. Then J induces

$$J'_{LG}\colon [\mathscr{B}, \mathscr{L}]'_G \to [\mathscr{A}, \mathscr{L}]'_{LGJ}.$$

If, for fixed $K\colon \mathscr{L} \to \mathscr{K}$, each J'_{LG} has a rari then we shall say that *canonically augmented* $(\mathscr{L}, \mathscr{K})$*-functors have cocartesian extensions.*

Functors $T\colon \mathscr{B} \to \mathscr{B}'$ are said to *preserve extensions of canonically augmented* $(\mathscr{L}, \mathscr{K})$*-functors* if the analogous natural transformations η_{LG} are natural equivalences. We adopt analogous terminology for extensions of *augmented* Λ*-functors* and of *augmented functors.*

Definition 5.2. *If* $J\colon \mathscr{A} \to \mathscr{B}$ *is a cofibration of one of the types described in* 5.1 *then a fixed choice of functors* $Q_G = \text{rari}\, J_G$ *for all relevant* G*'s is called a cocleavage of* J*. If* $\{Q_G\}$ *is a cocleavage of* J*,* $\{Q'_G\}$ *is a cocleavage of* J' *and if* $T\colon \mathscr{B} \to \mathscr{B}'$ *is cocartesian of the same type then* T *is called cocleavage preserving if* $T_G Q_G = Q'_{GT}$ *for all relevant* G*. The cocleavage is called normal if for all appropriate* $K\colon \mathscr{L} \to \mathscr{K}$ *with* $L = \text{rari}\, K$*, and* $G\colon \mathscr{B} \to K$*, we have*

$$Q_G(LGJ) = LG.$$

If $J\colon \mathscr{A} \to \mathscr{B}$ *is a cofibration then a cocleavage* $\{Q_G\}$ *is called split if for any* $E\colon \mathscr{L} \to \mathscr{M}$ *the diagram*

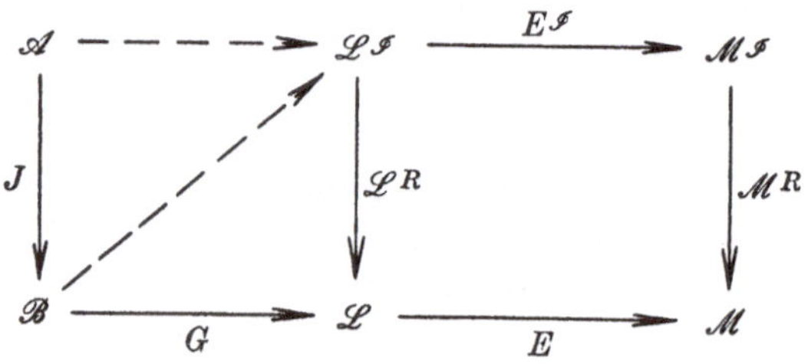

gives rise to a diagram

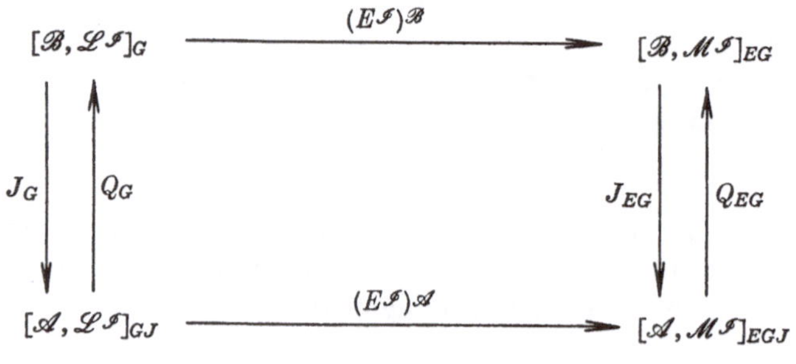

in which $(E^{\mathscr{I}})^B \circ Q_G = Q_{EG} \circ (E^{\mathscr{I}})^{\mathscr{A}}$.

Theorem 5.3. *Let* $J\colon \mathscr{A} \to \mathscr{B}$.

i. J is an $\mathcal{L}$-cofibration if and only if $\mathcal{L}^J: [\mathcal{B}, \mathcal{L}] \to [\mathcal{A}, \mathcal{L}]$ is a fibration. Furthermore there is a 1—1 correspondence between cocleavages of J and cleavages of $\mathcal{L}^J$.

ii. J is a strong $(\mathcal{L}, \mathcal{K})$-cofibration if and only if canonically augmented $(\mathcal{L}, \mathcal{K})$-functors have cocartesian extensions.

iii. If $\mathcal{L}$ has a terminal object then an $\mathcal{L}$-cofibration is a strong $(\mathcal{L}, *)$-cofibration.

iv. If Λ is a class of categories such that $\mathcal{L} \in \Lambda$ implies $\mathcal{L}^{\mathcal{I}} \in \Lambda$ then a strong Λ-cofibration is Λ-cofibration.

v. Strong cofibrations are cofibrations.

Remark. It seems likely that if J is an $\mathcal{L}$-cofibration and a $\mathcal{K}$-cofibration then J is a strong $(\mathcal{L}, \mathcal{K})$-cofibration. The proof would be dual to a proof that iii. implies v. in 2.10 without the use of cleavages. Unfortunately, we have so far been unable to carry out the details of such a proof. This would also of course imply that cofibrations are strong cofibrations. The special case of this given in iii. gives the curious result that the existence of cocartesian extensions of natural transformations (homotopies) implies that of functors.

Proof. i. Intuitively, the diagrams

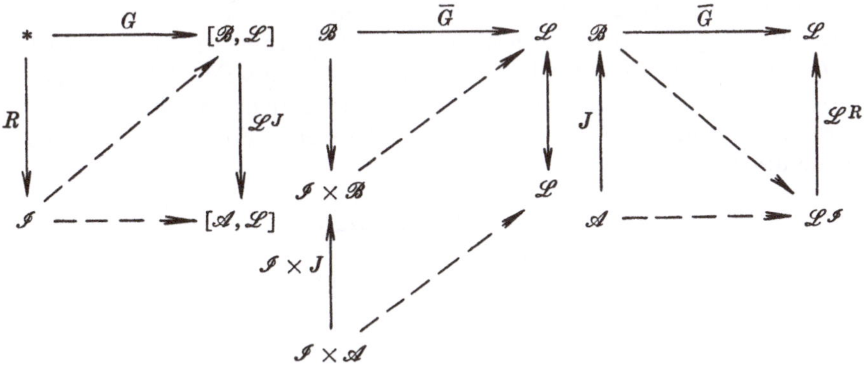

all express equivalent conditions. Formally, there is a commutative diagram.

$$\begin{array}{ccc} [\mathcal{I},[\mathcal{B},\mathcal{L}]]_B & \xrightarrow[(\mathcal{L}^J)_G]{} & [\mathcal{I},[\mathcal{A},\mathcal{L}]]_{\mathcal{L}^J G} \\ \wr & & \wr \\ [\mathcal{B},\mathcal{L}^{\mathcal{I}}]_{\bar G} & \xrightarrow[J_{\bar G}]{} & [\mathcal{A},\mathcal{L}^{\mathcal{I}}]_{\bar G J} \end{array}$$

where the vertical equivalences are isomorphisms. Hence $(\mathscr{L}^J)_G$ has a rari if and only if $J_{\bar{G}}$ does. Furthermore a choice of a rari of $(\mathscr{L}^J)_G$ for every $G: * \to [\mathscr{B}, \mathscr{L}]$ is equivalent to the choice of a cleavage of $\mathscr{L}^J$, by the construction in 2.5.

ii. The proof is dual to the proof in 2.10 that condition iv. there is equivalent to condition v. using a dual of 2.9.

iii. Let $\mathscr{L}$ have a terminal object 0 and let $J: \mathscr{A} \to \mathscr{B}$ be an $\mathscr{L}$-cofibration. Consider

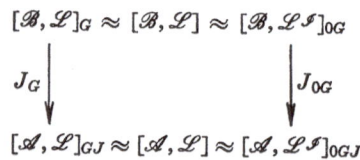

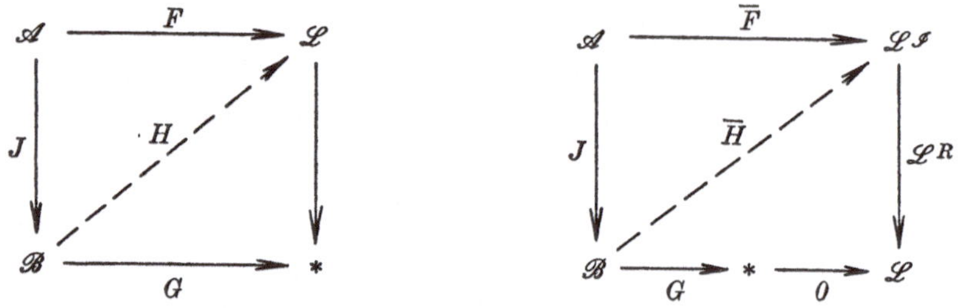

where for any $F: \mathscr{A} \to \mathscr{L}$, we let $\bar{F}: \mathscr{A} \to \mathscr{L}^{\mathscr{I}}$ be the functor representing the unique natural transformation $F \to \Delta 0$ ($\Delta 0: \mathscr{A} \to \mathscr{L}$ is the constant functor all of whose values are 0). The diagram at the right commutes, so $\bar{F}$ has an extension to $\bar{H}: \mathscr{B} \to \mathscr{L}^{\mathscr{I}}$ which by commutativity represents the unique natural transformation $H \to \Delta 0$ for some $H: \mathscr{B} \to \mathscr{L}$ with $HJ = F$. It is immediate that if $\bar{K}: \mathscr{B} \to \mathscr{L}^{\mathscr{I}}$ represents $K \to \Delta 0$ then natural transformations from $\bar{K}$ to $\bar{H}$ are equivalent to natural transformations from K to H, which implies the result. Formally, there is a commutative diagram

$$\begin{array}{ccccc} [\mathscr{B}, \mathscr{L}]_G & \approx & [\mathscr{B}, \mathscr{L}] & \approx & [\mathscr{B}, \mathscr{L}^{\mathscr{I}}]_{0G} \\ \downarrow{\scriptstyle J_G} & & & & \downarrow{\scriptstyle J_{0G}} \\ [\mathscr{A}, \mathscr{L}]_{GJ} & \approx & [\mathscr{A}, \mathscr{L}] & \approx & [\mathscr{A}, \mathscr{L}^{\mathscr{I}}]_{0GJ} \end{array}$$

Hence if J_{0G} has a rari, so does J_G.

iv. Trivial, since if J is a strong Λ-cofibration and $\mathscr{L} \in \Lambda$, then $\mathscr{L}^{\mathscr{I}} \in \Lambda$ and $\mathscr{L}^R: \mathscr{L}^{\mathscr{I}} \to \mathscr{L}$ has a rari (0.3, B) so J is a strong $(\mathscr{L}^{\mathscr{I}}, \mathscr{L})$-cofibration which is the same thing as an $\mathscr{L}$-cofibration.

v. Immediate from iv.

Theorem 5.4. *Let $J: \mathscr{A} \to \mathscr{B}$, $J': \mathscr{A} \to \mathscr{B}'$ and $T: \mathscr{B} \to \mathscr{B}'$ satisfy $TJ = J'$.*

i. If J and J' are $\mathscr{L}$-cofibrations then T is $\mathscr{L}$-cocartesian if and only if $\mathscr{L}^T: [\mathscr{B}', \mathscr{L}] \to [\mathscr{B}, \mathscr{L}]$ is cartesian. Furthermore, T preserves cocleavages *if and only if $\mathscr{L}^T$ preserves the corresponding cleavages.*

ii. If J and J' are strong $(\mathscr{K}, \mathscr{L})$-cofibrations then T is strong $(\mathscr{K}, \mathscr{L})$-cocartesian if and only if T preserves extensions of canonically augmented $(\mathscr{K}, \mathscr{L})$-functors.

iii. If $\mathscr{L}$ has a terminal object then an $\mathscr{L}$-cocartesian functor is a strong $(\mathscr{L}, *)$-cocartesian functor.

iv. If Λ is a class of categories such that $\mathscr{L} \in \Lambda$ implies $\mathscr{L}^{\mathscr{I}} \in \Lambda$ then a strong Λ-cocartesian functor is a Λ-cocartesian functor.

v. Strong cocartesian functors are cocartesian.

Proof. All of these results follow immediately from the excessive naturality of the constructions we have used.

Remark. The dual notions are called coopfibrations and coopcartesian functors. They are described by replacing $\mathscr{L}^R$ by $\mathscr{L}^D$, rari by lari and augmented by coaugmented. The dual forms of 5.3 and 5.4 are, of course, verified by duality.

Example 5.5. *Consider $R \times \mathscr{A}: \mathscr{A} \to \mathscr{I} \times \mathscr{A}$ and $D \times \mathscr{A}: \mathscr{A} \to \mathscr{I} \times \mathscr{A}$ as in 0.2, B).*

i. $R \times \mathscr{A}$ is a coopfibration with a canonical split coopcleavage and it is a Λ-cofibration for the class Λ of categories with pullbacks.

ii. $D \times \mathscr{A}$ is a cofibration with a canonical split cocleavage and it is a Λ-coopfibration for the class Λ of categories with pushouts.

We check i. directly. Consider

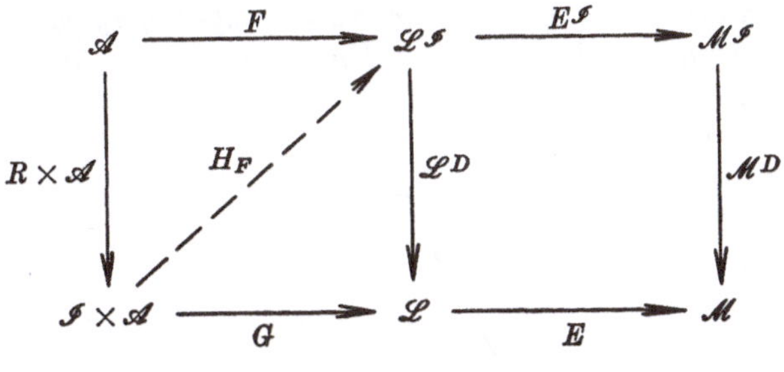

Given G and F with $\mathscr{L}^D \circ F = G \circ (R \times \mathscr{A})$, replace F by its adjoint $\bar{F}: \mathscr{I} \times \mathscr{A} \to \mathscr{L}$. Regarding $\bar{F}$ and G as natural transformations, commutativity says that the domain of $\bar{F}$ is the range of G. H_F is then defined as the commutative square

$$H_F = \left\{ G \begin{array}{c} \xrightarrow{\ 1\ } \\ \downarrow \qquad \downarrow \bar{F}G \\ \xrightarrow[\bar{F}]{} \end{array} \right.$$

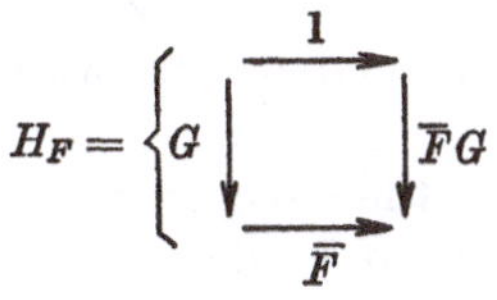

It is easily verified this determines a coopcleavage which is split since $H_{E\mathscr{I}_F} = E^{\mathscr{I}} \circ H_F$.

Note that this construction is also natural in $\mathscr{A}$; i.e., if $T: \mathscr{A} \to \mathscr{B}$, then in

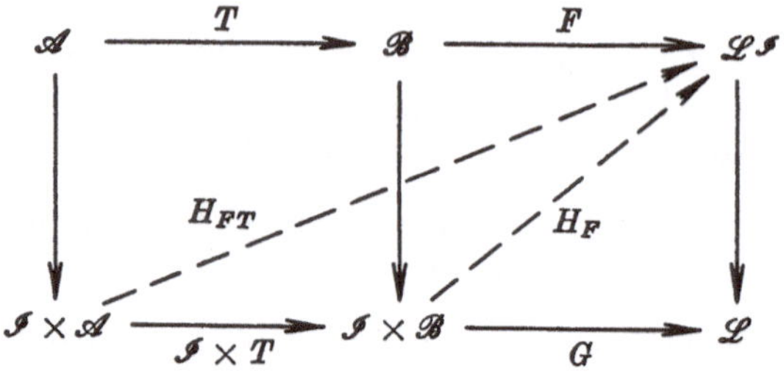

one has immediately that $H_{FT} = H_F \circ (\mathscr{I} \times T)$. The second part of i. follows by taking appropriate pullbacks.

Proposition 5.6. *i. The composition of two cofibrations of the same type is again a cofibration of the same type.*

ii. The pushout of a cofibration of a given type is a cofibration of the same type.

Proof. The proof is dual to the proof of 3.1. For $\mathscr{L}$-cofibrations it follows from 3.1 by 5.3, i.

Definition 5.7. *Let $\mathscr{C}ofib\,(\mathscr{A})$ denote the subcategory of $\mathscr{C}at^{\mathscr{A}}$ whose objects are cofibrations $J: \mathscr{A} \to \mathscr{B}$ and whose morphisms are commutative triangles*

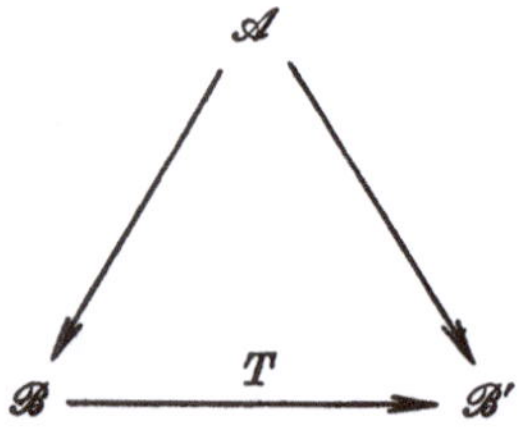

between cofibrations where T is cocartesian.

We use the obvious notation $\mathscr{L}$-$\mathscr{C}ofib\,(\mathscr{A})$, st- $(\mathscr{K}, \mathscr{L})$-$\mathscr{C}ofib\,(\mathscr{A})$, st-$\mathscr{C}ofib\,(\mathscr{A})$ for the various related categories of cofibrations of specific types.

Similarly $\mathscr{C}ofib$ denotes the subcategory of $[\mathscr{I}, \mathscr{C}at]$ whose objects are cofibrations $J: \mathscr{A} \to \mathscr{B}$ and whose morphisms are commutative squares

$$\begin{array}{ccc} \mathscr{A} & \xrightarrow{U} & \mathscr{A}' \\ {\scriptstyle J}\downarrow & & \downarrow{\scriptstyle J'} \\ \mathscr{B} & \xrightarrow{T} & \mathscr{B}' \end{array} \qquad \begin{array}{ccc} [\mathscr{B}', \mathscr{L}\mathscr{I}]_G & \xrightarrow{\quad T_G \quad} & [\mathscr{B}, \mathscr{L}\mathscr{I}]_{GT} \\ {\scriptstyle Q'_G}\uparrow\downarrow{\scriptstyle J'_G} & & {\scriptstyle Q_{GT}}\uparrow\downarrow{\scriptstyle J_{GT'}} \\ [\mathscr{A}', \mathscr{L}\mathscr{I}]_{GJ'} & \xrightarrow{\quad U_{GJ'} \quad} & [\mathscr{A}, \mathscr{L}\mathscr{I}]_{GTJ} \end{array}$$

such that, in the right hand diagram for any $\mathscr{L}$ and any $G: \mathscr{B}' \to \mathscr{L}$, if $Q'_G = \operatorname{rari} J'_G$ and $Q_{GT} = \operatorname{rari} J_{GT}$, then the induced transfer natural transformation (see 0.3, E)

$$\eta_G: T_G \circ Q'_G \to Q_{GT} \circ U_{GJ'}$$

is a natural equivalence. When necessary we also use the notation $\mathscr{L}$-$\mathscr{C}\!\mathit{ofib}$, st-$(\mathscr{K}, \mathscr{L})$-$\mathscr{C}\!\mathit{ofib}$ and st-$\mathscr{C}\!\mathit{ofib}$ for the related categories.

Let $\bar{D} = \mathscr{C}\!\mathit{at}^D | \mathscr{C}\!\mathit{ofib}$, so

$$\bar{D}: \mathscr{C}\!\mathit{ofib} \to \mathscr{C}\!\mathit{at}: (J: \mathscr{A} \to \mathscr{B}) \to \mathscr{A}: (U, T) \to U\,.$$

Using this terminology, 5.6, ii. can be restated as follows:

Proposition 5.8. *$\bar{D}: \mathscr{C}\!\mathit{ofib} \to \mathscr{C}\!\mathit{at}$ is an opfibration with fibres*

$$\mathscr{C}\!\mathit{ofib}_A = \mathscr{C}\!\mathit{ofib}(\mathscr{A})\,.$$

The same holds for the categories of cofibrations of various specific types.

Proof. Dual to 3.3.

Proposition 5.9. *$\mathscr{C}\!\mathit{ofib}$ is closed under the formation of sums in $[\mathscr{I}, \mathscr{C}\!\mathit{at}]$.*

Proof. Dual to the part of 3.4 concerning products. If we wanted to show that a product of cofibrations is a cofibration, we would need a category $\mathscr{I}'$ which serves as a test category for cofibrations and such that $[-, \mathscr{I}']$ preserves products. We know of no such $\mathscr{I}'$. However, again under suitable restrictions $\mathscr{C}\!\mathit{ofib}$ is closed under right limits. The same holds for the categories of cofibrations of various specific types.

Proposition 5.10. *The categories $\mathscr{C}\!\mathit{ofib}(\mathscr{A})$ have sums which are preserved by the functors F_* in any opcleavage of $\bar{D}: \mathscr{C}\!\mathit{ofib} \to \mathscr{C}\!\mathit{at}$ and by the inclusion functor $\mathscr{C}\!\mathit{ofib}(\mathscr{A}) \to \mathscr{C}\!\mathit{at}^{\mathscr{A}}$.*

Proof. Dual to 3.5, the sum of two cofibrations being their pushout. The same holds for the categories of cofibrations of various specific types.

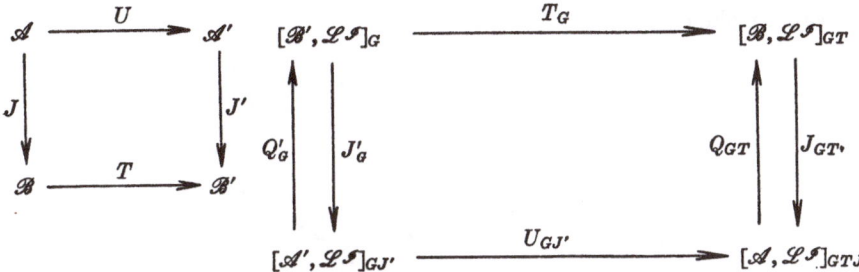

We now wish to discuss an adjoint functor from $\mathscr{C}at^{\mathscr{A}}$ to $\mathscr{C}ofib\,(\mathscr{A})$ and st-$\mathscr{C}ofib\,(\mathscr{A})$. Suppose $F\colon \mathscr{A} \to \mathscr{C}$ is a functor. Then, as in 0.2. F, we can consider

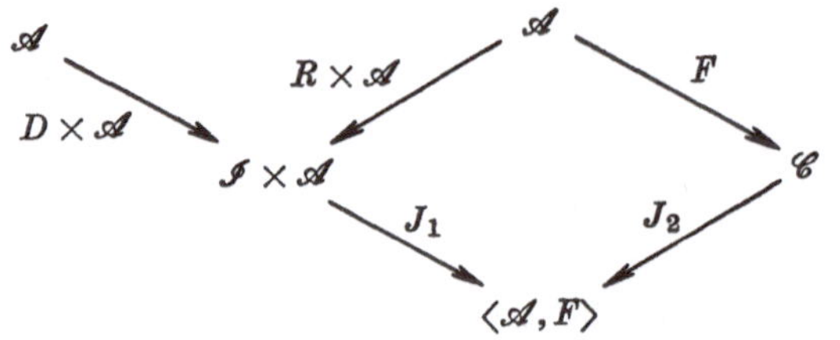

where the square is a pushout. Let

$$J_F = J_1 \circ (D \times \mathscr{A}) : \mathscr{A} \to \langle \mathscr{A}, F \rangle .$$

This determines a functor

$$\langle \mathscr{A}, - \rangle : \mathscr{C}at^{\mathscr{A}} \to \mathscr{C}at^{\mathscr{A}} : F \to J_F : T \to T_*$$

where given $F\colon \mathscr{A} \to \mathscr{C}$, $F'\colon \mathscr{A} \to \mathscr{C}'$ and $T\colon \mathscr{C} \to \mathscr{C}'$ with $TF = F'$, $T_*\colon \langle \mathscr{A}, F \rangle \to \langle \mathscr{A}, F' \rangle$ is the induced functor given by the pushout structure.

Let $(\mathscr{C}at^{\mathscr{A}})^h$ (resp., $\mathscr{C}ofib^h\,(\mathscr{A})$) denote the category with the same objects, but where morphisms are equivalence classes of morphisms in the appropriate category, where T and T' mapping J to J' are equivalent if there is a natural equivalence $t\colon T \to T'$ with $tJ = 1_{J'}$. We also let st-$\mathscr{C}ofib^h\,(\mathscr{A})$ denote the full subcategory of $(\mathscr{C}at^{\mathscr{A}})^h$ determined by the strong cofibrations. Finally $\mathscr{C}ocleav_0\,(\mathscr{A})$ denotes the category of cofibrations under $\mathscr{A}$ with given split cocleavages and cocleavage preserving functors.

Theorem 5.11. *i.* $\langle \mathscr{A}, - \rangle\colon (\mathscr{C}at^{\mathscr{A}})^h \to$ st-$\mathscr{C}ofib^h\,(\mathscr{A})$ *is a coreflection; i. e., it is right adjoint to the inclusion functor.*

ii. $\langle \mathscr{A}, - \rangle\colon (\mathscr{C}at^{\mathscr{A}})^h \to \mathscr{C}ofib^h\,(\mathscr{A})$ *is a coreflection.*

iii. $\langle \mathscr{A}, - \rangle\colon \mathscr{C}at^{\mathscr{A}} \to \mathscr{C}ocleav_0\,(\mathscr{A})$ *is right adjoint to the forgetful functor.*

Proof. To show that $J_F\colon \mathscr{A} \to \langle \mathscr{A}, F \rangle$ is a cofibration, one simply observes that the diagram

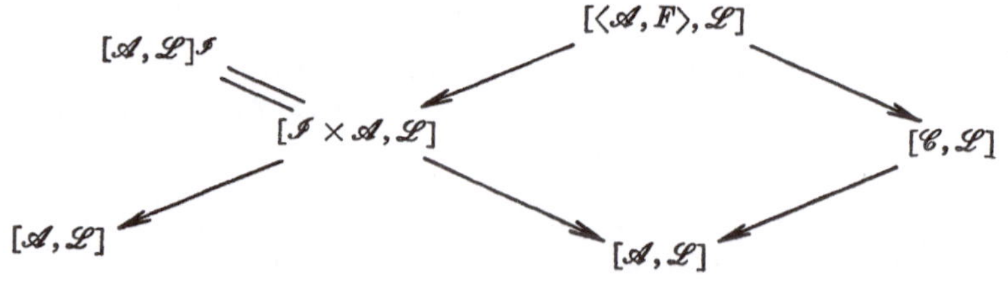

is a special case of the diagram determining universal fibrations; i.e.,

$$[\langle\mathcal{A}, F\rangle, \mathcal{L}] \approx ([\mathcal{A}, \mathcal{L}], \mathcal{L}^F) \quad \text{and} \quad [J_F, \mathcal{L}] \approx P_{L^F},$$

which is a fibration so, by 5.3, i., J_F is a cofibration.

To show that J_F is a strong cofibration, we have to use the explicit representation of $\langle\mathcal{A}, F\rangle$ given in 0.2, F as $\mathcal{A} \coprod \mathcal{C}$ together with

$$\langle\mathcal{A}, F\rangle (A, C) = \mathcal{C}(F(A), C) \quad \text{and} \quad \langle\mathcal{A}, F\rangle (C, A) = \Phi .$$

In this representation, $J_F: \mathcal{A} \to \langle\mathcal{A}, F\rangle$ is just the inclusion of $\mathcal{A}$ in $\mathcal{A} \coprod \mathcal{C}$. Let $K: \mathcal{L} \to \mathcal{K}$ be a functor with $L = \operatorname{rari} K$, let $G: \langle\mathcal{A}, F\rangle \to \mathcal{K}$

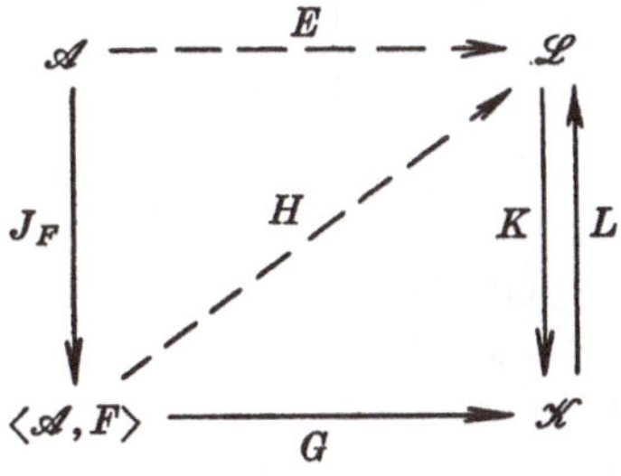

be fixed and suppose $E: \mathcal{A} \to \mathcal{L}$ satisfies $KE = GJ_F$. Define

$$H: \langle\mathcal{A}, F\rangle \to \mathcal{L} \quad \text{by} \quad H|\mathcal{A} = E, \quad H|\mathcal{C} = LG, \quad \text{and if}$$
$$\varphi: A \to C \quad \text{in} \quad \langle\mathcal{A}, F\rangle \quad \text{by}$$
$$H(\varphi) = LG(\varphi) \circ \psi_{E(A)} : E(A) \to LG(C);$$

where $\psi: I_{\mathcal{L}} \to LK$ is the adjunction natural transformation. Then clearly $KH = G$ and $HJ_F = E$. Now suppose $H': \langle\mathcal{A}, F\rangle \to \mathcal{L}$ also satisfies $KH' = G$ and $s: H' \circ J_F \to E$, with $Ks = 1_{GJ_F}$. We must show that there is a unique $t: H' \to H$ with $tJ_F = s$ and $Kt = 1_G$. The first requirement implies that $t|\mathcal{A} = s$ and the second, by a dual of 2.9 i., that

$$t|\mathcal{C} = \psi H'|\mathcal{C} : H'|\mathcal{C} \to LKH'|\mathcal{C} = LG|\mathcal{C} = H|\mathcal{C}.$$

Thus t is unique and the only problem is naturality for morphisms $\varphi: A \to C$ in $\langle\mathcal{A}, F\rangle$. Consider

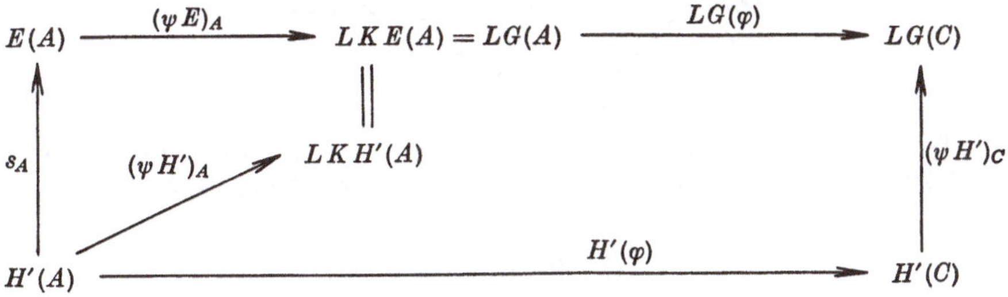

The right hand square commutes since $\psi : I \to LK$ is natural and the left hand square commutes by a dual of 2.9, ii. since $K s_A = 1_{GJ}$. We have in fact constructed a canonical normal cocleavage of J_F.

One checks easily that as far the cofibration structure is concerned this cocleavage is split since in the diagram

$$\begin{array}{ccccc}
\mathscr{A} & \longrightarrow & \mathscr{L}^{\mathscr{I}} & \xrightarrow{E^{\mathscr{I}}} & \mathscr{M}^{\mathscr{I}} \\
{\scriptstyle J_F}\downarrow & & {\scriptstyle \mathscr{L}^R}\downarrow\uparrow{\scriptstyle \Delta_{\mathscr{L}}} & & {\scriptstyle \mathscr{M}^R}\downarrow\uparrow{\scriptstyle \Delta_{\mathscr{M}}} \\
\langle \mathscr{A}, F \rangle & \xrightarrow[G]{} & \mathscr{L} & \xrightarrow[E]{} & \mathscr{M}
\end{array}$$

we have $E^{\mathscr{I}} \Delta_{\mathscr{L}} = \Delta_{\mathscr{M}} E$ and if $\psi_{\mathscr{L}} : I \to \Delta_{\mathscr{L}} \mathscr{L}^R$ denotes the adjunction natural transformation, then $E^{\mathscr{I}} \psi_{\mathscr{L}} = \psi_{\mathscr{M}} E^{\mathscr{I}}$.

Finally, if $F : \mathscr{A} \to \mathscr{C}$, $F' : \mathscr{A} \to \mathscr{C}'$ and $T : \mathscr{C} \to \mathscr{C}'$ satisfies $TF = F'$, then $T_* : \langle \mathscr{A}, F \rangle \to \langle \mathscr{A}, F' \rangle$ is the functor such that $T_* | \mathscr{A} = I_{\mathscr{A}}$, $T_* | \mathscr{C} = T$ and the diagram

$$\begin{array}{ccc}
\langle \mathscr{A}, F \rangle (A, C) & \xrightarrow{T_*} & \langle \mathscr{A}, F' \rangle (T_*(A), T_*(C)) \\
\| & & \| \\
\mathscr{C}(F(A), C) & \xrightarrow{T} & \mathscr{C}'(F'(A), T(C))
\end{array}$$

commutes. Hence in the diagram

$$\begin{array}{ccc}
[\langle \mathscr{A}, F' \rangle, \mathscr{L}]_G & \xrightarrow{T_G} & [\langle \mathscr{A}, F \rangle, \mathscr{L}]_{GT} \\
{\scriptstyle (J'_F)_G}\searrow\nwarrow{\scriptstyle Q'_G} & & {\scriptstyle Q_{GT}}\nearrow\swarrow{\scriptstyle (J_F)_{GT}} \\
& [\mathscr{A}, \mathscr{L}]_{GJ_F} &
\end{array}$$

if Q'_G and Q_{GT} are the canonical raris, then $T_G Q'_G = Q_{GT}$; i.e., T is cocleavage preserving. Thus $\langle \mathscr{A}, - \rangle : \mathscr{C}at^{\mathscr{A}} \to \mathscr{C}ocleav_0(\mathscr{A})$. Furthermore, it is immediate that

$$\langle \mathscr{A}, - \rangle : (\mathscr{C}at^{\mathscr{A}})^h \to st\text{-}\mathscr{C}ofib^h(\mathscr{A}).$$

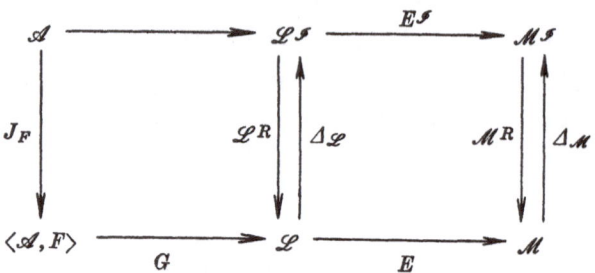

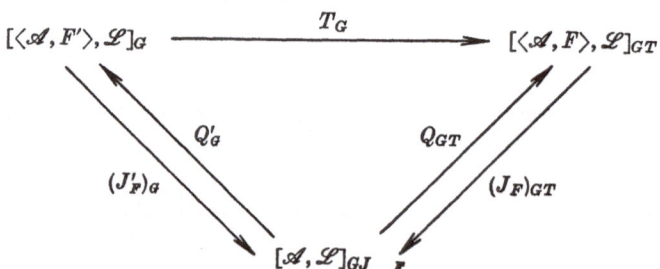

Now, since $\nabla \times \mathscr{A} = \text{lali}\, R \times \mathscr{A} : \mathscr{I} \times \mathscr{A} \to \mathscr{A}$, there is a $K_F = \text{lali}\, J_2$ (by 0.3, D) which satisfies $K_F J_1 = F \circ (\nabla \times \mathscr{A})$ and hence

$$K_F J_F = K_F J_1 (D \times \mathscr{A}) = F \circ (\nabla \times \mathscr{A}) \circ (D \times \mathscr{A}) = F\,.$$

K_F is clearly natural in F, so we get a natural transformation

$$K_{(-)} : \langle \mathscr{A}, - \rangle \to I_{\mathscr{C}at}\, \mathscr{A}\,.$$

We must check that K_F satisfies the desired universal mapping property, i.e.,

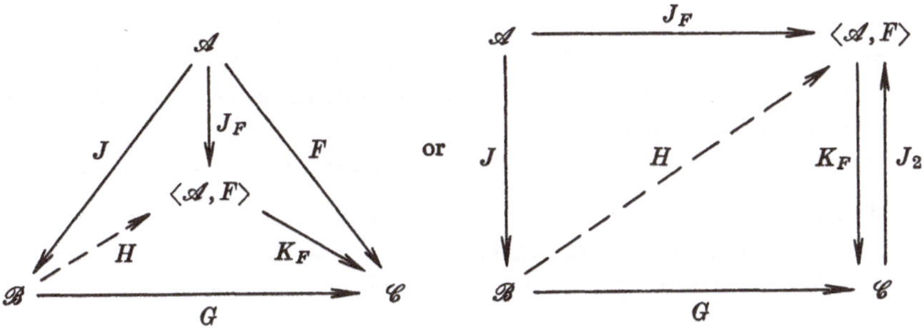

where the first diagram illustrates the universal mapping property for objects under $\mathscr{A}$ and the second shows that it is a special case of extending a generalized homotopy, since $J_2 = \text{rari}\, K_F$. Hence there is an H making everything commute which is unique up to a unique natural equivalence η with $K_F \eta = 1$. Notice that we do not assert that H is cocartesian. The reason is that, although it very likely is true that H is cocartesian, the proof of the analogous property in 3.9 is highly noncategorical and we have so far been unable to discover a dualizable proof.

If we restrict our attention to cofibrations then we can use duality. We have a diagram

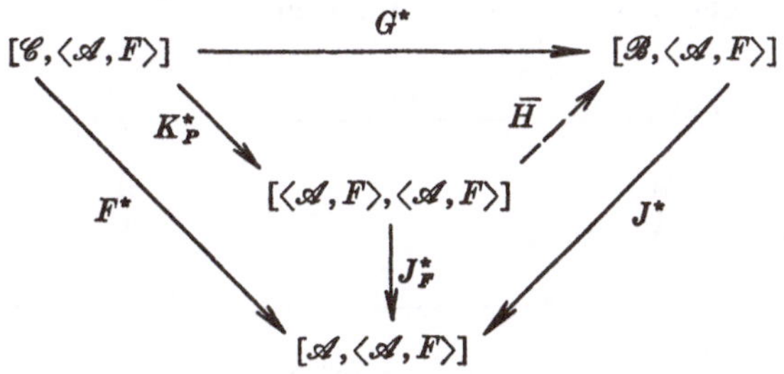

where J_F^* is the universal fibration corresponding to F^*. Hence, as in 3.9, there is a *cartesian* $\bar{H}$ which is unique up to an appropriate natural equivalence and which is in fact unique if it is required to be cleavage preserving for a given cleavage of J^*. Let

$$H = \bar{H}(I_{\langle \mathscr{A}, F\rangle}) : \mathscr{B} \to \langle \mathscr{A}, F\rangle .$$

If one has a split cocleavage of $J : \mathscr{A} \to \mathscr{B}$, it determines cleavages of $\mathscr{D}^J : [\mathscr{B}, \mathscr{D}] \to [\mathscr{A}, \mathscr{D}]$ for all $\mathscr{D}$ and hence functors $\bar{H}(\mathscr{D}) : [\langle \mathscr{A}, F\rangle, \mathscr{D}] \to [\mathscr{B}, \mathscr{D}]$ for all $\mathscr{D}$. It is easily checked that these are the components of a natural transformation

$$[\langle \mathscr{A}, F\rangle, -] \to [\mathscr{B}, -]$$

and thus by the categorical Yoneda Lemma (0.3, F), we have that $\bar{H} = H^* = \langle \mathscr{A}, F\rangle^H$. H satisfies $HJ = J_F$ and $K_F H = G$ (by choosing $\mathscr{D} = \mathscr{C}$). Finally, H is cocartesian by 5.4, and preserves cocleavages for a given cocleavage of J.

Finally, we wish to examine the dual of the Chevalley criterion, 3.11.

Proposition 5.12. *Let $J : \mathscr{A} \to \mathscr{B}$ be a functor and let $\langle \mathscr{A}, J\rangle \to \mathscr{I} \times \mathscr{B}$ be the unique functor such that $LJ_1 = \mathscr{I} \times J$ and $LJ_2 = R \times \mathscr{B}$. Then J is a cofibration with a split cocleavage if and only if L has a lali.*

Proof. Consider

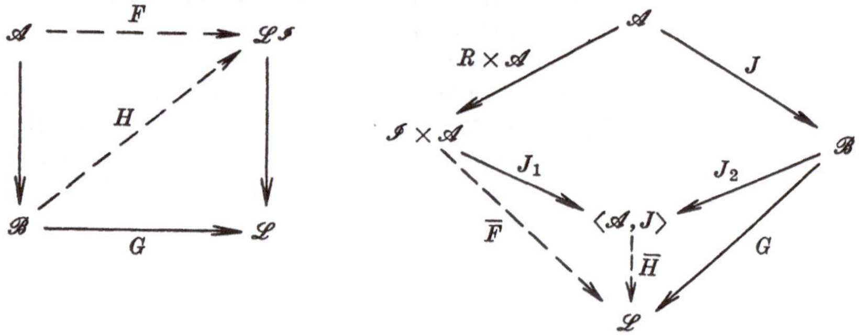

As in 3.11 we get a commutative diagram

$$\begin{array}{ccc} [\mathscr{B}, \mathscr{L}^{\mathscr{I}}]_G & \xrightarrow{\quad J_G \quad} & [\mathscr{A}, \mathscr{L}^{\mathscr{I}}]_{GJ} \\ \wr\wr & & \wr\wr \\ [\mathscr{I} \times \mathscr{B}, \mathscr{L}]_G & \xrightarrow{\quad \mathscr{L}^L \quad} & [\langle \mathscr{A}, J\rangle, \mathscr{L}]_G \end{array}$$

Hence J_G has a rari if and only if $\mathscr{L}^L$ has a rari. But if $K = \text{lali}\, L$, then $\mathscr{L}^K = \text{rari}\, \mathscr{L}^L$ ([SVG], Appendix A4). Hence if L has a lali then J is a cofibration and the cocleavage constructed this way is split.

The converse is rather involved. Observe first that $J_2 : \mathscr{B} \to \langle \mathscr{A}, J \rangle$ has a canonical split coopcleavage since it is a pushout of $\mathscr{A} \to \mathscr{I} \times \mathscr{A}$ which has a split coopcleavage (5.5). Furthermore the diagram

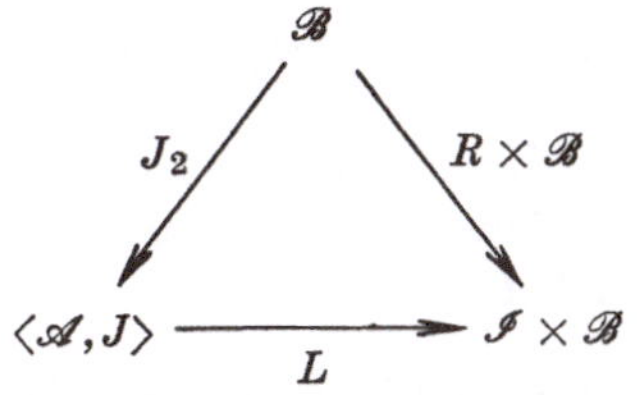

commutes, by construction. We claim that L preserves these coopcleavages, for, in the diagram

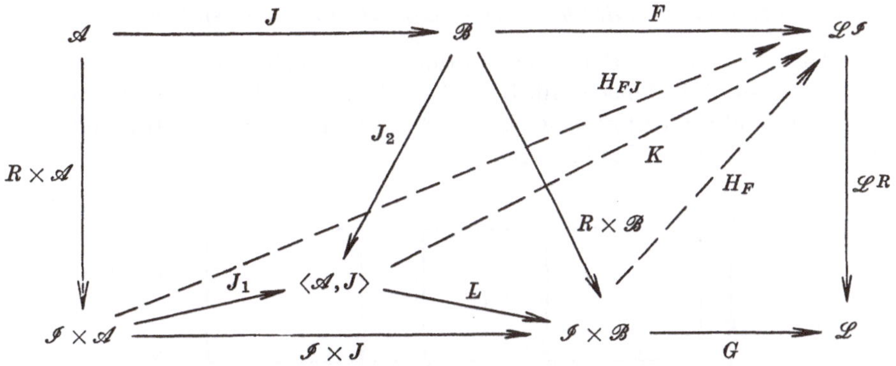

we have $H_{FJ} \circ (R \times \mathscr{A}) = FJ$, which induces a unique $K : \langle \mathscr{A}, J \rangle \to \mathscr{L}^{\mathscr{I}}$ (this is how the coopcleavage of J_2 is constructed). But, by 5.5,

$$H_F \circ (\mathscr{I} \times J) = H_{FJ} \quad \text{so} \quad H_F \circ L : \langle \mathscr{A}, J \rangle \to \mathscr{L}^{\mathscr{I}}$$

makes the same two triangles commute that K does and hence $H_F \circ L = K$.

Now, for any $\mathscr{L}$ consider

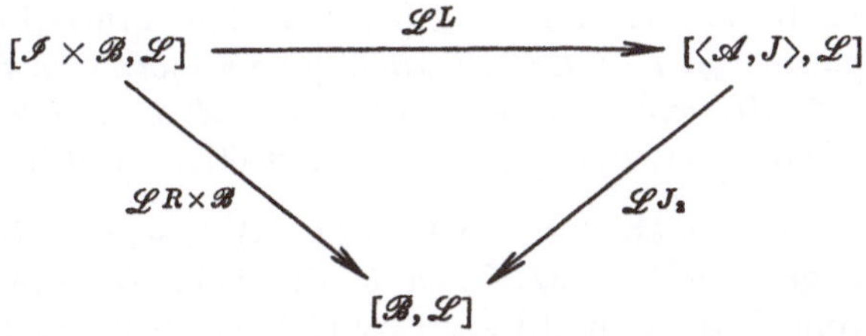

Then $\mathscr{L}^{R \times \mathscr{B}}$ and $\mathscr{L}^{J_2}$. have natural (in $\mathscr{L}$) opcleavages which are preserved by $\mathscr{L}^L$. If $J : \mathscr{A} \to \mathscr{B}$ is a cofibration with a split cocleavage, then, referring back to the diagram at the beginning of the proof, each J_G

has a rari and these raris are natural in which implies that the restrictions of $\mathscr{L}^L$ to the fibres of $\mathscr{L}^{R\times\mathscr{B}}$ have raris which are natural in $\mathscr{L}$, which implies by 3.7 that $\mathscr{L}^L$ has a rari $Q_{\mathscr{L}}$ which is natural in $\mathscr{L}$. That is, we have a natural transformation

$$Q:[\langle\mathscr{A},J\rangle,-]\to[\mathscr{I}\times\mathscr{B},-].$$

By the categorical Yoneda Lemma (0.3, F) Q is induced by $K:\mathscr{I}\times\mathscr{B}\to\langle\mathscr{A},J\rangle$; i.e., $Q_{\mathscr{L}}=\mathscr{L}^K$. One verifies easily that $\mathscr{L}^K=\text{rari }\mathscr{L}^L$ for all $\mathscr{L}$ if and only if $K=\text{lali }L$.

Corollary 5.13. *$J:\mathscr{A}\to\mathscr{B}$ is a cofibration with a split cocleavage if and only if*

i. J is injective; and

ii. Regarding $\mathscr{A}$ as a subcategory of $\mathscr{B}$, then the inclusion of $\mathscr{A}$ into the left ideal of $\mathscr{B}$ generated by $\mathscr{A}$ is an equivalence with a left inverse.

Proof. If $J:\mathscr{A}\to\mathscr{B}$ is a cofibration with a split cocleavage then $L:\langle\mathscr{A},J\rangle\to\mathscr{I}\times\mathscr{B}$ has a rali. But $\mathscr{A}$ can be regarded as a subcategory of $\langle\mathscr{A},J\rangle$ and then $L|\mathscr{A}=J:\mathscr{A}\to\{0\}\times\mathscr{B}$. Since L has a left inverse so does J and hence J is injective.

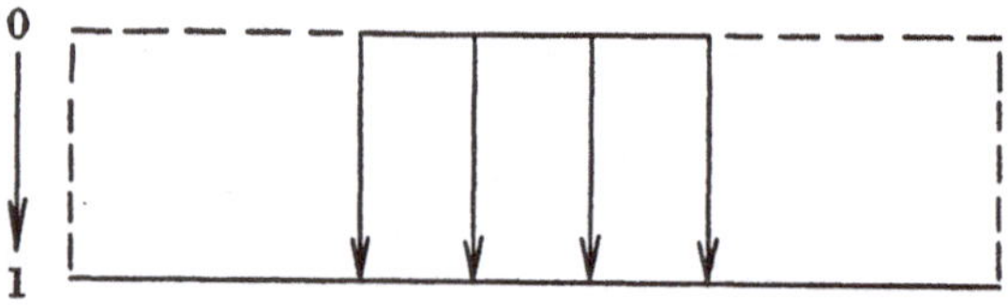

Now if J is injective, then $\langle\mathscr{A},J\rangle$ looks like the top hat illustrated, included in $\mathscr{I}\times\mathscr{B}$, the cylinder, by L. Note that in the illustration there are no maps from the bottom to the top. If $B\in\{0\}\times\mathscr{B}$ and $B\notin\{0\}\times\mathscr{A}$ then there are two cases in finding a reflection $K(B)$ of B in $\langle\mathscr{A},J\rangle$.

Case 1. For all $A\in\mathscr{A}$, $\mathscr{B}(B,A)=\Phi$; i.e., B is not in the left ideal of $\mathscr{B}$ generated by $\mathscr{A}$. Then $K(B)\notin\{0\}\times\mathscr{A}$ since otherwise the adjunction morphism $\psi_B:B\to LK(B)$ would give a morphism from B to something in $\mathscr{A}$. However, the description of $\langle\mathscr{A},J\rangle$ in 0.2 F shows that $K(B)=L(B)=(1,B)\in\{1\}\times\mathscr{B}$ is a reflection of B in this case.

Case 2. There exists an $A\in\mathscr{A}$ with $\mathscr{B}(B,A)\neq\Phi$; i.e., B is in the left idea of $\mathscr{B}$ generated by $\mathscr{A}$. Then $K(B)\notin\{1\}\times B$ since the non-trivial map from B to A in $\{0\}\times\mathscr{B}$ would have to factor through a morphism form the bottom to the top, which is forbidden. Hence the only possibility is that $K(B)\in\{0\}\times\mathscr{A}$ and this works exactly if $K(B)$ is a reflection of B in $\{0\}\times\mathscr{A}$ such that the adjunction natural transformation is a natural equivalence; i.e., if and only if the inclusion of $\mathscr{A}$ in the left ideal generated by $\mathscr{A}$ is an equivalence with a left inverse.

Examples 5.14. *The two extremes of* 5.13 *are*

i. $\mathscr{A} \subset \mathscr{B}$ *is a sieve (crible,* [*AS*]*); i.e.,* $\mathscr{A}$ *itself is a left ideal. Then* $J : \mathscr{A} \to \mathscr{B}$ *is a cofibration with a split cocleavage. For any* E, $J_F : \mathscr{A} \to \langle \mathscr{A}, F \rangle$ *is such an example.*

ii. The inclusion $J : \mathscr{A} \to \mathscr{B}$ *is an equivalence with a left inverse. Then the left ideal generated by* $\mathscr{A}$ *is* $\mathscr{B}$ *and* J *is a cofibration with a split cocleavage.*

As a last example to illustrate the difference between cofibrations and strong $(\mathscr{L}, *)$-cofibrations, observe that if $J : \mathscr{A} \to \mathscr{B}$ is a strong $(\mathscr{L}, *)$-cofibration then, as in the proof of 5.3 iii., $\mathscr{L}^J : [\mathscr{B}, \mathscr{L}] \to [\mathscr{A}, \mathscr{L}]$ has a rari. This is true for every $\mathscr{L}$ and the raris can be chosen to be natural in $\mathscr{L}$ (i.e., J is a strong Λ-cofibration with a split cocleavage, where Λ is the class of pairs $(L, *)$) if and only if J has a lali. For instance, the inclusion of the category of sheaves on a space X into presheaves on X satisfies this property.

References

[*GT*] Artin, M.: Grothendieck topologies. Seminar notes. Harvard University 1962.

[*BN*] Chevalley, C.: Categories and schemes (unpublished seminar notes). Berkeley: Univ. of Calif. 1962.

[*AS*] Giraud, J.: Analysis situs. Paris: Seminaire Bourbaki, n° 256, 1962—63.

[*GOD*] Godement, R.: Topologie algébrique et théorie des faisceaux. Paris: Hermann 1958.

[*SVC*] Gray, J. W.: Sheaves with values in a category. Topology **3**, 1—18 (1965),

[*SGA*] Grothendieck, A.: Catégories fibrées et descente. Seminaire de géométrie algébrique de l'Institut des Hautes Études Scientifiques. Paris 1961.

[*BB190*] — Technique de descente et théorémes d'existence en géométrie algébrique, I. Généralités. Descente par morphismes fidélement Plats. Paris: Seminaire Bourbaki, n° 190, 1959—60.

[*EGA*] — avec J. Diendonné: Éléments de géométrie algébrique, I. Paris: Publ. Inst. Hautes Études Sci. No. 4, 1960.

[*HR*] Heller, A., and K. A. Rowe: On the category of sheaves. Amer. J. Math. **84**, 205—216 (1962).

[*FS*] Lawvere, F. William: Functorial semantics of algebraic theories. Thesis. New York: Columbia University 1963.

Some Aspects of Equational Categories [1]

By

F. E. J. LINTON *

Introduction

The theory of equationally definable classes of algebras, initiated by BIRKHOFF in the early thirties, is, despite its power, elegance and simplicity, hampered in its usefulness by two defects. The first is its refusal to deal with infinitary operations; the second is the awkwardness inherent in the presentation of an equationally definable class in terms of operations and equations.

Quite recently, LAWVERE [*3*], by introducing the notion — closely akin to the clones (cf. [*1*], Ch. III, § 3, Exer. 3) of P. HALL — of an *algebraic theory*, rectified the second defect without, unfortunately, rectifying the first. Not long before LAWVERE's work, SŁOMINSKI [*5*] rectified the first defect without, however, rectifying the second. It is possible, as it turns out, that both defects can be rectified at once; the present paper will sketch the highlights of the resulting theory. More complete details must, for lack of time, appear elsewhere.

The following pages are divided into seven sections, entitled: 1. Equational theories and their models; 2. The adjointness of structure and semantics; 3. The characterization of varietal categories; 4. Proof of the characterization theorem; 5. Illustrations and applications; 6. Variations on the theme: the question of rank; triples versus theories; 7. When are epimorphisms onto?

1. Equational Theories and their Models

An *equational theory* is a product preserving covariant functor $T : \mathscr{S}^* \to \mathbb{T}$ from the dual $\mathscr{S}^*$ of the category $\mathscr{S}$ of sets and functions to a (large) category $\mathbb{T}$ whose class of objects is put by T in one-one

---

[1] This paper, written during the author's participation in the 1965 N.S.F. Summer Seminar in Homological Algebra at Bowdoin College, summarises the results of research engaged in with the support of the Air Force Office of Scientific Research (AFOSR 520—64) and the National Science Foundation (NSF GP-2432) while on leave from Wesleyan University at the University of Chicago.

* Received September 9, 1965.

correspondence with the objects of $\mathscr{S}^*$. It is convenient, then, to identify the classes of objects of $\mathscr{S}$ (or $\mathscr{S}^*$) and of $\mathbb{T}$. The equational theory T is *varietal* if the category $\mathbb{T}$ is locally small, that is, if each class $\mathbb{T}(n, k)$ of $\mathbb{T}$-morphisms from n to k is a set, whatever the sets (i.e., $\mathbb{T}$-objects) n, k.

A *morphism of theories* from T to $T' : \mathscr{S}^* \to \mathbb{T}'$ is any functor $j : \mathbb{T} \to \mathbb{T}'$ satisfying $j \cdot T \cong T'$. The resulting category of varietal theories will be denoted Th.

For each equational theory $T : \mathscr{S}^* \to \mathbb{T}$, we single out, from the category $\mathscr{S}^{\mathbb{T}}$ of all set valued functors on $\mathbb{T}$, the full subcategory $\mathscr{S}^T$ whose objects are those functors $X : \mathbb{T} \to \mathscr{S}$ for which the composite $X \cdot T : \mathscr{S}^* \to \mathscr{S}$ preserves products. Such a functor X is called a *model of T in $\mathscr{S}$*, or a *T-algebra*; any category $\mathscr{X}$ equivalent to the category $\mathscr{S}^T$ of T-algebras and T-homomorphisms is called an *equational category* (*varietal* if T is varietal).

Evaluation at the object $1 \in \mathbb{T}$ provides a functor $U_T : \mathscr{S}^T \to \mathscr{S}$, the *underlying set functor for T-algebras*, which makes $\mathscr{S}^T$ into a concrete category.

Proposition 1. *If the theory T is varietal, the underlying set functor for T-algebras has a left adjoint.*

Proof. For n (resp. k) an object or morphism of $\mathscr{S}$ (resp. $\mathbb{T}$), define

$$F_T(n)(k) = \mathbb{T}(T(n), k).$$

Then each $F_T(n)$ is a T-algebra[2] (called the *free T-algebra freely generated by the set n of free generators*), and F_T is a functor $\mathscr{S} \to \mathscr{S}^T$. That F_T is left adjoint to U_T is established by showing that the maps $\alpha_n : n \to U_T F_T(n)$, defined as the compositions

$$n \cong \mathscr{S}^*(n, 1) \xrightarrow{T} \mathbb{T}(T(n), T(1)) = F_T(n)(1) = U_T F_T(n),$$

serve as the front adjunction ("inclusion of the generators").

We shall say that a set valued functor $U : \mathscr{X} \to \mathscr{S}$ is *tractable* if, whatever the sets n and k, the class $\mathrm{n.t.}(U^n, U^k)$ of natural transformations from the functor U^n (given by $U^n(X) = (U(X))^n$) to the functor U^k is actually a set. Observe that the Yoneda Lemma guarantees the tractability[3] of any set valued functor U having a left adjoint F, since

$$\mathrm{n.t.}(U^n, U^k) \cong U^k(F(n)).$$

Corollary. *If the theory T is varietal, the underlying set functor for T-algebras is tractable.*

---

[2] Indeed, $F_T(n)$ preserves all products in T.

[3] See lemma 1, § 3, for a stronger result.

2. The Adjointness of Structure and Semantics

Every functor $U\colon \mathcal{X} \to \mathcal{S}$ yields an equational theory, varietal if U is tractable, by taking the category $\mathbb{T}_U$ whose morphisms are the k-tuples of n-ary operations on U. That is, $\mathbb{T}_U$ is the category whose class of objects is the class of all sets, whose morphisms are given by

$$\mathbb{T}_U(n, k) = \text{n. t.}\,(U^n, U^k)\,,$$

and whose composition rule is the usual composition of natural transformations. Because each function $f\colon k \to n$ between sets gives a natural transformation $U^f\colon U^n \to U^k$, we may define a contravariant functor $\exp_U\colon \mathcal{S} \to \mathbb{T}_U$ by

$$\exp_U(n) = n\,, \quad \exp_U(f) = U^f.$$

Product preserving when viewed as a covariant functor from $\mathcal{S}^*$ to $\mathbb{T}_U$, $\exp_U$ is the *equational theory of* the set valued functor U.

Taking as morphisms from one set valued functor $U\colon \mathcal{X} \to \mathcal{S}$ to another $U'\colon \mathcal{X}' \to \mathcal{S}$ those functors $F\colon \mathcal{X} \to \mathcal{X}'$ with $U' \cdot F \cong U$, and writing K for the category of tractable set valued functors, we have the following important result.

Proposition 2. *The passages*

$$U \rightsquigarrow \exp_U\colon K \to Th$$
$$T \rightsquigarrow U_T\colon Th \to K$$

are contravariantly functorial with respect to morphisms of set valued functors and of equational theories, and are adjoint on the right.

Proof. Given a K-morphism $F\colon U \to U'$, note that $U'^n \cdot F \cong U^n$, and define $\mathbb{T}_F\colon \mathbb{T}_{U'} \to \mathbb{T}_U$ by sending $t\colon U'^n \to U'^k$ to $\mathbb{T}_F(t) = t \cdot F = t_F\colon U^n \to U^k$. Given a Th-morphism $j\colon T \to T'$ and a T'-algebra X, observe that $X \cdot j$ is a T-algebra, and define $\mathcal{S}^j\colon \mathcal{S}^{T'} \to \mathcal{S}^T$ by $\mathcal{S}^j(X) = X \cdot j$. Writing

$$\begin{array}{ll} \mathfrak{S}(U) = \exp_U & (U \in K) \\ \mathfrak{S}(F) = \mathbb{T}_F & (F \text{ a } K\text{-morphism}) \\ \mathfrak{M}(T) = U_T & (T \in Th) \\ \mathfrak{M}(j) = \mathcal{S}^j & (j \text{ a } Th\text{-morphism}) \end{array}$$

we obtain functors $\mathfrak{S}\colon K \to Th$ and $\mathfrak{M}\colon Th \to K$, called *equational structure* and *equational semantics*, respectively, extending the terminology of LAWVERE.

For the adjointness assertion, one of the adjunctions is taken to be the natural *equivalence* id $\to \mathfrak{S} \cdot \mathfrak{M}$ provided by proposition 1 and the Yoneda Lemma:

$$\mathbb{T}(n, k) = F_T(n)\,(k) = U_T^k(F_T(n)) = \text{n. t.}\,(U_T^n, U_T^k) = \mathbb{T}_{U_T}(n, k)\,;$$

the other, $\Phi\colon \mathrm{id} \to \mathfrak{M} \cdot \mathfrak{S}$, sends the set valued functor $U\colon \mathscr{X} \to \mathscr{S}$ to the functor $\Phi_U\colon \mathscr{X} \to \mathscr{S}^{\exp_U}$ given by

$$\begin{array}{lll} \Phi_U(X)(n) = U^n(X) & (X \in \mathscr{X}, & n \in \mathbb{T}_U) \\ \Phi_U(X)(t) = t_X & (X \in \mathscr{X}, & t \text{ a } \mathbb{T}_U\text{-morphism}) \\ (\Phi_U(f))_n = U^n(f) & (f \text{ an } \mathscr{X}\text{-morphism}, & n \in \mathbb{T}_U). \end{array}$$

Corollary. *The semantics functor $\mathfrak{M}\colon Th \to K$ is both full and faithful.*

3. The Characterization of Varietal Categories

We outline here the form of LAWVERE's characterisation theorem for algebraic categories appropriate to the present more general context. Some preliminary comments are needed.

Let us remark first that each category $\mathscr{S}^T$ of T-algebras has all set-indexed inverse limits, and that U_T preserves them (and hence is faithful). What is more, the first isomorphism theorem is valid in $\mathscr{S}^T$, in a sense to be explained below, with respect to U_T. When T is varietal, moreover, $\mathscr{S}^T$ has all set-indexed direct limits, as may be proved by straightforward construction as a quotient of a free.

Next, we give two simple but useful lemmas regarding Φ_U.

Lemma 1. *Let $U\colon \mathscr{X} \to \mathscr{S}$ be a set valued functor with left adjoint F. Then $\exp_U$ is varietal and $\Phi_U \cdot F \cong F_{\exp_U}$.*

Proof.

$$\Phi_U(F(n))(k) = U^k(F(n)) \cong \text{n.t.}(U^n, U^k) = \mathbb{T}_U(n, k) = F_{\exp_U}(n)(k).$$

Lemma 2. *The functor $\Phi_U\colon \mathscr{X} \to \mathscr{S}^{\exp_U}$ is faithful if and only if the set valued functor $U\colon \mathscr{X} \to \mathscr{S}$ is.*

Proof. This is an immediate consequence of the faithfulness of the underlying set functor for $\exp_U$-algebras and the commutativity[4] of the diagram

$$\begin{array}{ccc} & & \mathscr{S}^{\exp_U} \\ & \overset{\Phi_U}{\nearrow} & \big\downarrow \\ \mathscr{X} & & \\ & \underset{U}{\searrow} & \\ & & \mathscr{S} \end{array}$$

Finally, in any category $\mathscr{X}$, consider three maps in the configuration

$$R \underset{p_1}{\overset{p_2}{\rightrightarrows}} X \overset{\pi}{\to} Q \quad (\pi \cdot p_1 = \pi \cdot p_2). \tag{3.1}$$

[4] Reflecting the fact that Φ_U is a K-morphism.

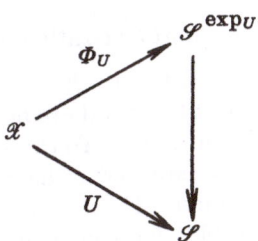

π is the *difference cokernel* of the pair (p_1, p_2) (loosely speaking, Q is the quotient of X by the relations R) if each map π' from X satisfying

$$\pi' \cdot p_1 = \pi' \cdot p_2$$

is of the form $\pi' = y \cdot \pi$ for a unique map y from Q. (p_1, p_2) is the *kernel pair* of the map π (loosely speaking, R is the congruence relation on X induced by π) if each pair (p_1', p_2') of maps to X satisfying

$$\pi \cdot p_1' = \pi \cdot p_2'$$

determines a unique map x to R with $p_i' = p_i \cdot x$ $(i = 1, 2)$. The configuration (3.1) is *exact* if both the above envisioned relations are valid. The *first isomorphism theorem* is said to hold in $\mathscr{X}$ with respect to a set valued functor $U: \mathscr{X} \to \mathscr{S}$ if

(FIT$_0$) $\mathscr{X}$ has all kernel pairs and difference cokernels;

(FIT$_1$) a necessary and sufficient condition for an $\mathscr{X}$-morphism π to be a difference cokernel is that $U(\pi)$ be, i.e., that $U(\pi)$ be onto;

(FIT$_2$) a pair (p_1, p_2) of $\mathscr{X}$-morphisms $p_i: R \to X$ is a kernel pair if and only if $(U p_1, U p_2)$ is, i.e., if and only if the function $U R \to U X \times U X$ induced by the pair of functions $U p_i$ $(i = 1, 2)$ is one-one and has an equivalence relation as image.

Remark. Although it follows from the conjunction of (FIT$_1$) and (FIT$_2$) that a necessary and sufficient condition for the exactness of (3.1) is the exactness of

$$U(R) \underset{U(p_1)}{\overset{U(p_2)}{\rightrightarrows}} U(X) \xrightarrow{U(\pi)} U(Q), \tag{3.2}$$

the converse is false, as JON BECK has kindly pointed out in his comments to an earlier draft of this paper.

Proposition 3. *An arbitrary category $\mathscr{X}$ is varietal if and only if the first isomorphism theorem holds in $\mathscr{X}$ with respect to some set valued functor U having a left adjoint F.*

4. Proof of the Characterization Theorem

It being clear from the discussion at the head of § 3 that the conditions stated in proposition 3 for $\mathscr{X}$ to be varietal are necessary, we confine our attention to the proof of their sufficiency. To this end, we agree, throughout this section, to work with a category $\mathscr{X}$, having all kernel pairs and difference cokernels, and a functor $U: \mathscr{X} \to \mathscr{S}$ having a left adjoint $F: \mathscr{S} \to \mathscr{X}$. Then U is tractable; for convenience, we write $\mathbb{T} = \mathbb{T}_U$, $T = \exp_U$, $|\ | = U_T$, and $\Phi = \Phi_U: \mathscr{X} \to \mathscr{S}^T$.

The proof of proposition 3 may then be summarised as follows.

Proposition 4. (FIT$_1$) *guarantees that Φ is full, faithful, and has a left adjoint.* (FIT$_2$), *in the presence of* (FIT$_1$), *ensures that the left adjoint to Φ is actually a two sided inverse to Φ, so that, under Φ, $\mathscr{X}$ is equivalent to $\mathscr{S}^T$.*

Proof. Assume (FIT_1). That Φ is faithful is a consequence of lemma 2, the faithfulness of U coming (cf. [2], Ch. II, Prop. 1.5) from the fact that U reflects epimorphisms (every difference cokernel always being epi).

All the rest of the proof depends on the fact that every object X of $\mathscr{X}$ has a presentation as a quotient of a free, in the sense that there is a set n and an $\mathscr{X}$-morphism $\pi: F(n) \to X$ which is a difference cokernel. This is proved by taking $n = UX$ and π the back adjunction; π is a difference cokernel, by (FIT_1), since $U(\pi): UFUX \to UX$ is onto.

To see that Φ is full, observe first, recalling lemma 1, that it is at least full between free $\mathscr{X}$-objects:

$$\mathscr{X}(F(k), F(n)) \cong U^k(F(n)) = |F_T(n)|^k \cong \mathscr{S}^T(F_T(k), T_T(n)).$$

Then, given $X, Y \in \mathscr{X}$ and $f: \Phi X \to \Phi Y$, represent X and Y as quotients of free $\mathscr{X}$-objects, say by maps

$$\pi: F(k) \to X, \quad \varrho: F(n) \to Y.$$

Using lemma 1 to identify $\Phi F(k)$, lift f over $\Phi(\pi)$ and $\Phi(\varrho)$ to an $\mathscr{S}^T$-morphism $\bar{f}: F_T(k) \to F_T(n)$; there is then a unique $\mathscr{X}$-morphism $\bar{\varphi}: F(k) \to F(n)$ with $\Phi(\bar{\varphi}) = \bar{f}$. It remains to check that $\varrho\bar{\varphi}p_1 = \varrho\bar{\varphi}p_2$ whenever $\pi \cdot p_1 = \pi \cdot p_2$; granting this, $\varrho\bar{\varphi}$ factors through π to give a unique $\mathscr{X}$-morphism $\varphi: X \to Y$ with $\varphi\pi = \varrho\bar{\varphi}$; for this φ, $\Phi(\varphi) = f$, as is readily checked at the level of the underlying sets.

Next, to construct the left adjoint to Φ, take a T-algebra $X \in \mathscr{S}^T$. Represent it as a quotient of a free T-algebra,

$$E \underset{p_1}{\overset{p_2}{\rightrightarrows}} F_T(n) \overset{\pi}{\to} X \quad (\pi = \mathrm{cok}(p_1, p_2), \quad (p_1, p_2) = \ker(\pi)),$$

and represent E in turn as a quotient of a free

$$F_T(k) \overset{p}{\to} E;$$

then $\pi = \mathrm{cok}(p_1 p, p_2 p)$. Now take $\mathscr{X}$-morphisms $\varrho_1, \varrho_2: F(k) \to F(n)$ with $\Phi(\varrho_i) = p_i \cdot p$ $(i = 1, 2)$, and form the difference cokernel (in $\mathscr{X}$)

$$t: F(n) \to \bar{X}$$

of the pair (ϱ_1, ϱ_2). It must then be checked that $\Phi(t) \cdot p_1 = \Phi(t) \cdot p_2$; granting that, there is a unique T-homomorphism $\alpha_X: X \to \Phi(\bar{X})$ making commutative the diagram

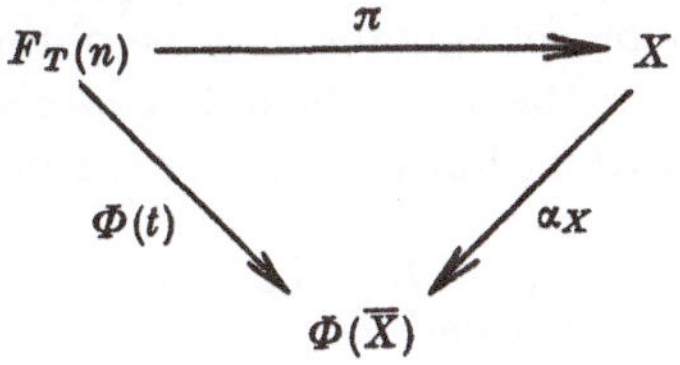

These maps α_X constitute the front adjunction making the passage $X \rightsquigarrow \bar{X}: \mathcal{S}^T \to \mathcal{X}$ a functor left adjoint to Φ.

Finally, assuming the validity of (FIT_2) in addition, Φ is an equivalence because each T-homomorphism α_X is an isomorphism, as can be seen by comparing the congruence relation of $U(t)$ with that of $|\pi|$ (they're the same).

This completes the proof.

5. Illustrations und Applications

It should be clear that both SLOMINSKI's equationally definable classes of algebras and LAWVERE's algebraic categories, taken with the standard underlying set functors, are varietal. There are, however, other varietal categories of interest, some of which we list here.

1. Compact Hausdorff spaces, with the usual underlying set functor, form a varietal category. The first isomorphism theorem is easily proved, using compactness; the adjoint to the underlying set functor is provided by the Stone-Čech compactification. (The first isomorphism theorem fails so badly for just plain topological spaces that, for them, the functor Φ *is* the underlying set functor.)

2. An equational category that is not varietal is provided by the category of complete boolean algebras (and complete homomorphisms). In fact, the obvious underlying set functor fails to be tractable, as is shown by GAIFMANN's proof[5] of the nonexistence of a free complete boolean algebra freely generated by a countable set.

3. Modifying example 2 slightly by taking the full subcategory of complete *atomic* boolean algebras does yield a varietal category (the adjoint to the underlying set functor assigns to the set n the boolean ring of subsets of the set of 2-valued functions on n).

4. Compact abelian groups form a varietal category. The first isomorphism theorem is not hard; the adjoint to the underlying set functor is obtained by passing from the set n to the character group $((G(n)^\wedge)_d)^\wedge$ of the character group made discrete $(G(n)^\wedge)_d$ of the free abelian group $G(n)$ generated by n.

5. More generally, the category of all compact groups is varietal — indeed, whatever the varietal theory T, the category of all compact T-algebras is varietal. Again, the first isomorphism theorem is clear; the adjoint to the underlying set functor may be obtained, for example, by use of the adjoint functor theorem (the only conceivable stumbling block is the solution set condition, but a counting argument disposes of that).

---

[5] GAIFMANN shows that n.t. $(U^{\aleph_0}, U)$ is a proper class, where U is the underlying set functor for complete boolean algebras.

These examples should suffice. We are tempted to point out that examples 3 and 4 arise as the duals of the algebraic categories of sets and discrete abelian groups, respectively, and to speculate when the dual of a varietal category need itself be varietal. Of greater interest, however, is the following very useful blanket assertion regarding the existence of adjoint functors between varietal categories.

Proposition 5. *Let* $U_i\colon \mathscr{X}_i \to \mathscr{S}\ (i = 1, 2)$ *be functors by virtue of which* $\mathscr{X}_i$ *is varietal, and suppose* $T\colon \mathscr{X}_1 \to \mathscr{X}_2$ *is a functor satisfying* $U_2 \cdot T \cong U_1$. *Then* T *has a left adjoint* $S\colon \mathscr{X}_2 \to \mathscr{X}_1$.

Proof. Writing F_i for the left adjoint to U_i, pass from $X \in \mathscr{X}_2$ to the quotient of $F_1(U_2(X))$ by the intersection of the congruence relations induced on $F_1(U_2(X))$ by those $\mathscr{X}_1$-morphisms $F_1(U_2(X)) \to X_1$ that arise as values of the transformation

$$\mathscr{X}_2(X, TX_1) \xrightarrow{U_2} \mathscr{S}(U_2X, U_2TX_1) \cong \mathscr{S}(U_2X, U_1X_1) \cong \mathscr{X}_1(F_1U_2X, X_1).$$

The details of the proof that this passage provides the desired adjoint are very similar to standard arguments.

Corollary. *Let* T *be a varietal theory. The forgetful functor from the category of compact* T*-algebras to* $\mathscr{S}^T$*, as well as that from compact* T*-algebras to compact spaces, has a left adjoint.*

In particular, one has compact abelian groups freely generated (in a manner of speaking) by compact spaces, compact groups freely generated by (or the universal compactifications of) discrete groups, and so on. The full import of proposition 5 awaits further exploitation.

6. Variations on the Theme: The Question of Rank; Triples Versus Theories

Call a cardinal number $\mathfrak{r}$ *regular* if each sum of fewer than $\mathfrak{r}$ cardinals each less than $\mathfrak{r}$ is itself less than $\mathfrak{r}$. The *rank* of a theory $T\colon \mathscr{S}^* \to \mathbb{T}$ is the least regular cardinal $\mathfrak{r} \geqq 2$ (if any; ∞ otherwise) for which it is the case that each $\mathbb{T}$-morphism $t\colon n \to 1$ can be represented as a composition

$$n \xrightarrow{T(f)} k \xrightarrow{t'} 1$$

with $f \in \mathscr{S}^*(n, k)$, $t' \in \mathbb{T}(k, 1)$, and $\operatorname{card}(k) < \mathfrak{r}$.

If $\mathbb{T}$ is any category whose objects are all sets, and if $\mathfrak{r}$ is a cardinal, ${}_{\mathfrak{r}}\mathbb{T}$ will denote the full subcategory of $\mathbb{T}$ whose objects have cardinality $< \mathfrak{r}$. Each equational theory $T\colon \mathscr{S}^* \to \mathbb{T}$ induces, by restriction, a product-preserving functor

$$_{\mathfrak{r}}T\colon {}_{\mathfrak{r}}\mathscr{S}^* \to {}_{\mathfrak{r}}\mathbb{T}, \tag{6.1}$$

the $\mathfrak{r}$-*truncation* of T.

Proposition 6. *For any varietal theory* $T\colon \mathscr{S}^* \to \mathbb{T}$ *and any regular cardinal* $\mathfrak{r} \geqq 2$, *the following three assertions are equivalent:*

1. $\operatorname{rank}(T) \leq \mathfrak{r}$;

2. *each functor* $X\colon {}_{\mathfrak{r}}\mathbb{T} \to \mathcal{S}$ *whose composition with* ${}_{\mathfrak{r}}T$ *preserves products has a unique extension to a* T*-algebra*;

3. $U_T F_T(n) = \bigcup_{\substack{\operatorname{card}(k)<\mathfrak{r} \\ f\in\mathcal{S}(k,n)}} U_T F_T(f)\,(U_T F_T(k))$.

Motivated by the salient features of (6.1), one makes the obvious modification in the definition of a varietal theory and comes up with a definition of $\mathfrak{r}$-*ary theory*. Calling a set valued functor U $\mathfrak{r}$-*tractable* if n.t.(U^n, U^k) is a set whenever the sets n and k have cardinality less than $\mathfrak{r}$, one can easily extend the definition of structure and semantics to provide a pair of contravariant functors, adjoint on the right, between the category $K_{\mathfrak{r}}$ of $\mathfrak{r}$-tractable set valued functors and the category $Th_{\mathfrak{r}}$ of $\mathfrak{r}$-ary theories. It then becomes possible to identify $Th_{\mathfrak{r}}$ with the full subcategory of Th whose objects are the varietal theories of rank $\leq \mathfrak{r}$. Bearing this in mind, we combine propositions 3 and 6 to obtain a sharper generalization of LAWVERE's characterization theorem.

Corollary. *The category* $\mathcal{X}$ *is equivalent to the category of all* T*-algebras, for some* $\mathfrak{r}$*-ary theory* T*, if and only if the first isomorphism theorem holds in* $\mathcal{X}$ *with respect to a set valued functor* U *having a left adjoint* F *of such a sort that each* $\mathcal{X}$*-morphism* $F(1) \to F(n)$ *has a factorization*

$$F(1) \to F(k) \xrightarrow{F(f)} F(n),$$

where $f \in \mathcal{S}(k, n)$ *and* $\operatorname{card}(k) < \mathfrak{r}$.

(LAWVERE's theorem is recovered from this corollary by taking $\mathfrak{r} = \aleph_0$.)

So much for the question of rank. Turning to another direction, we note that each varietal theory $T\colon \mathcal{S}^* \to \mathbb{T}$ has a left adjoint, namely the functor $\mathbb{T}(-, 1)$. If we now replace $\mathcal{S}$ by any category $\mathcal{A}$, we may speak of a *theory over* $\mathcal{A}$ as a functor $T\colon \mathcal{A}^* \to \mathbb{T}$ having a left adjoint and setting up a one-one correspondence between the objects of $\mathcal{A}$ and the objects of $\mathbb{T}$. In the same way can be justified the definition of a T-algebra as a functor $X\colon \mathbb{T} \to \mathcal{S}$ whose composition XT with T is a representable contravariant functor on $\mathcal{A}$.

It turns out that this notion of theory is entirely equivalent with BECK's notion (exposed elsewhere in these Proceedings) of triple, and that the algebras in the sense of theories are, modulo this equivalence, the same as BECK's algebras over triples. In particular, proposition 3 serves to characterise those categories that, in BECK's terminology, are triplable over $\mathcal{S}$. The obvious question, whether (or under what hypotheses on $\mathcal{A}$) this proposition is still valid when $\mathcal{S}$ is replaced by $\mathcal{A}$, remains unanswered. (Added in proof: BECK has just answered this question. Details will appear elsewhere.)

7. When are Epimorphisms onto?

This last section is devoted to the elucidation of a condition on a varietal category, necessary and sufficient for every epimorphism to be onto. In § 2, we have already vaguely referred to a natural transformation $U^n \to U^k$ as a k-tuple of n-ary operations on the set valued functor U. Here we shall deal with k-tuples of partial n-ary operations, that is to say, with natural transformations $V \to U^k$ (with V a subfunctor of U^n) of a special sort to be called *k-fold clusters of implicit n-ary operations*[6] (briefly, *implicit clusters*) *on* U. The main result is this:

Proposition 7. *Let T be a varietal theory and U_T: $\mathcal{S}^T \to \mathcal{S}$ the underlying set functor for T-algebras. $U_T(f)$ is onto for every epimorphism f in $\mathcal{S}^T$ if and only if every implicit cluster on U_T is explicit.*

It can be proved (cf. [*4*]) that the optimistic but precarious assumption of the existence of a non-trivial injective in the category of boolean σ-algebras forces every epimorphism of σ-algebras to be onto. It is to be hoped the technique of implicit clusters will permit a more satisfactory solution to the question, whether a σ-epimorphism must be onto.

Clearly, the proof of proposition 7 requires some definitions. A *k-fold cluster of implicit n-ary operations* on a set valued functor U: $\mathcal{X} \to \mathcal{S}$ is a natural transformation

$$\eta : V \to U^k, \tag{7.1}$$

with V a subfunctor of U^n, subject to two conditions whose description requires introduction of the class E of all pairs (e_1, e_2) of natural transformations e_i: $U^{n+k} \to U$ for which both compositions

$$V \xrightarrow{\text{inclu} \times \eta} U^n \times U^k \cong U^{n+k} \xrightarrow{e_i} U \qquad (i = 1, 2)$$

are equal. The conditions are that for each $X \in \mathcal{X}$ and $x \in U^n X$,

(IC_1) there is at most one $y \in U^k X$ with

$$(e_1)_X (x, y) = (e_2)_X (x, y) \quad \text{for all} \quad (e_1, e_2) \in E;$$

(IC_2) $x \in VX$ iff there is a y as envisioned in (IC_1).

(Necessarily, $\eta_X(x)$ is that element y. Moreover, the assumptions that V is a sub*functor* and that η is a *natural* transformation are superfluous; any collection of subsets VX and functions η_X satisfying (IC_1)and (IC_2) will in fact be natural. Finally, given a class of pairs of natural trans-

[6] To illustrate what is envisioned by this notion, let us say that inversion, in the category of semigroups (or rings) with unit, the identity element, in the category of possibly empty groups, and countable union, in the category of countably intersection-complete boolean rings (not necessarily having units), are examples of single implicit unary, nullary, and $\aleph_0$-ary operations, respectively.

formations, E, satisfying (IC_1), one obtains an implicit cluster whose subfunctor V is defined by (IC_2), and whose η is forced.)

An implicit cluster (7.1) is *explicit* if the following necessary condition (7.2) for the relation $x \in VX$ is also sufficient. The inclusion function $n \to n + k$ induces a projection $U^{n+k} \to U^n$, composition with which sends $\mathrm{n.t.}(U^n, U)$ to $\mathrm{n.t.}(U^{n+k}, U)$. Let E' be those pairs of natural transformations $U^n \to U$ sent in this way to pairs in E. The necessary condition that we have in mind is (that $x \in U^n X$ and)

$$(e_1')_X(x) = (e_2')_X(x) \quad \text{for all} \quad (e_1', e_2') \in E'. \tag{7.2}$$

Turning to the proof of proposition 7, let $\mathscr{X} = \mathscr{S}^T$ and $U = U_T$, T a varietal theory. The sets $\mathrm{n.t.}(U^{n+k}, U)$ and $\mathrm{n.t.}(U^n, U)$ can be reinterpreted as $UF(n + k)$ and $UF(n)$, respectively, and the sets of pairs E and E' are in fact congruence relations. The inclusion $F(n) \to F(n + k)$ happily drops through to a T-homomorphism

$$F(n)/_{E'} \to F(n + k)/_E \tag{7.3}$$

which is both monic and epic.

In the same setting, each epimorphism $f\colon A \to B$ gives rise to an implicit cluster by setting $n = f(A)$, $k = UB - n$, and taking E in $F(n + k) \times F(n + k)$ to be the kernel pair (i.e., congruence relation) of the projection $F(n + k) \to B$ afforded by the back adjunction. (IC_1) holds because f is epi, and the map (7.3) obtained from the induced implicit cluster is just the inclusion in B of the image of f (precisely f, if f is mono).

The proof of proposition 7 is clinched by the lemma below, which shows explicitly how non onto epimorphisms must always arise. The proof of the lemma can safely be omitted.

Lemma. *The map* (7.3) *is onto if and only if the implicit cluster* (7.1) *is explicit.*

References

[1] COHN, P. M.: Universal Algebra. New York: Harper & Row 1965.

[2] EILENBERG, S., and J. C. MOORE: Foundations of relative homological algebra. Mem. Amer. Math. Soc. 55 (1965).

[3] LAWVERE, F. W.: Functorial semantics of algebraic theories, dissertation. Columbia U., New York, 1963 (summary in Proc. Nat. Acad. Sci. 50, 869—872 (1963)).

[4] LINTON, F. E. J.: Injective boolean σ-algebras. Arch. der Math. (in press).

[5] SŁOMINSKI, J.: The theory of abstract algebras with infinitary operations. Rozprawy Mat. 18, Warzawa 1959.

Department of Mathematics
Wesleyan University
Middletown, Connecticut

Representations in Abelian Categories*

By

PETER FREYD

1. Full Embeddings into Abelian Categories

Consider the general but imprecise question: given a category how nicely can it be represented in an abelian category?

Without the "nicely" qualification the question evaporates. Any small category can be embedded in an abelian category and any small +'ive category can be fully embedded, namely via the representation functor $\mathscr{A} \to (\mathscr{A}^*, \mathscr{G})$, *(Abelian Categories 5. 35)*.

If $\mathscr{A}$ is an exact category, that is an $\oplus$'ive category together with a class of short exact sequences satisfying certain conditions, then there are very nice embeddings into abelian categories: besides being full they both preserve and reflect exactness (i.e. preserve non-exactness). *(Abelian Categories, 7. G)*.

In the other extreme we have the following

Proposition 1.1. *Let $\mathscr{A}$ be any small category, +' ive or not. There exists an abelian category $\mathscr{B}$ which is closed in the Eilenberg-Kelly sense and possesses a commutative associative tensor product $\mathscr{B} \otimes \mathscr{B} \to \mathscr{B}$. There exists a set $\bar{\mathscr{A}}$ of projective commutative associative co-algebras in $\mathscr{B}$, that is, a collection of projectives $\{A\}$ together with co-multiplications*

$$\{A \to A \otimes A\}.$$

The category of all non-zero co-algebra homomorphisms of the set $\bar{\mathscr{A}}$ is isomorphic to $\mathscr{A}$.

Proof. Let $\mathscr{B} = (\mathscr{A}^*, \mathscr{G})$. For $S, T \in (\mathscr{A}^*, \mathscr{G})$ define $S \otimes T : \mathscr{A} \to \mathscr{G}$ by $A \to SA \otimes TA$. Note that the functor $Z : \mathscr{A} \to \mathscr{G}$ which sends everything to the group of integers is a neutral object for $\otimes$. The right-adjoint of $\otimes$ exists by the special adjoint functor theorem but does not have a particularly nice description unless $\mathscr{A}$ is a partially ordered set.

For $A \in \mathscr{A}$ let $H_A : \mathscr{A}^* \to \mathscr{G}$ be defined by $H_A B \to \sum_{(B,A)} Z$ and define

* Received September 5, 1965

$\mathscr{A} \to (\mathscr{A}^*, \mathscr{G})$ by $A \to H_A$. Finally define $H_A \to H_A \otimes H_A$ to be the unique transformation which sends

$$1_A \in H_A(A) \quad \text{to} \quad 1_A \otimes 1_A \in H_A(A) \otimes H_A(A)\,. \blacksquare$$

This proposition is at best a curiosity. Because the representation it describes preserves none of the structure that the original category might have possessed it is almost certainly a useless representation. I think, however, that it points in the right direction. For a non-additive category $\mathscr{A}$ it is clearly impossible to *fully* embed $\mathscr{A}$ in an abelian category, hence we look for an abelian category in which we can ask for additional structure on the objects. I strongly suspect that co-algebras will be replaced by algebras for suitably restricted categories.

From now on all categories and functors shall be $+$'ive. Let $\boldsymbol{A}$ be the category of small $+$'ive categories and natural equivalence classes of $+$'ive functors. Let $\boldsymbol{B}$, be the full subcategory of abelian categories. We might ask if $\boldsymbol{B}$ is reflective in $\boldsymbol{A}$, that is, if for every $\mathscr{A} \in \boldsymbol{A}$ there exists $\mathscr{A} \to \mathscr{B}$, $\mathscr{B} \in \boldsymbol{B}$, such that for every $\mathscr{A} \to \mathscr{C}$, $\mathscr{C} \in \boldsymbol{B}$, there is a unique (up to natural equivalence by the definition of $\boldsymbol{A}$) completion to a commutative triangle

$$\begin{array}{ccc} \mathscr{A} & \to & \mathscr{B} \\ & \searrow \; \swarrow & \\ & \mathscr{C} & \end{array}$$

The answer is no — unless $\mathscr{A}$ is already an abelian category (in which case $\mathscr{B} = \mathscr{A}$) or if $\mathscr{A}$ "made amenable" (*Abelian Categories*, *5. G*) is abelian. Because of the peripheral interest of this statement I have delayed its proof until the periphery of the paper.

Suppose instead that we consider the subcategory of $\boldsymbol{B}_R$ defined as the category of abelian categories and (natural equivalence classes of) *right exact* functors.[1]

Proposition 1.2. *If $\mathscr{P}$ is the full subcategory of projectives in an abelian category $\mathscr{A}$ with enough projectives then $\mathscr{A}$ is the reflection of $\mathscr{P}$ in $\boldsymbol{B}_R$.*

Proof. Given any abelian $\mathscr{C}$ and $+$'ive $\mathscr{P} \to \mathscr{C}$ the right-exact extension $\mathscr{A} \to \mathscr{C}$ is constructed as if we were constructing a 0'th left derived functor. Namely for each $A \in \mathscr{A}$ we choose P', $P \in \mathscr{P}$ and exact $P' \to P \to A \to 0$ in $\mathscr{A}$ and then an exact sequence $F(P') \to F(P) \to$ $\to F(A) \to 0$ in $\mathscr{C}$. The rest of the proof can be gleaned from any source that describes the construction of 0'th left derived functors when there are enough projectives. One must notice only that the given functor's behavior on projectives is all that is used. $\blacksquare$

[1] In answer to questions directed by S. EILENBERG during the original exposition, "right exact" coincides with "cokernel preserving". I use the term "right exact" for the same reasons that I use the term "bounded linear transformation" instead of "continuous homomorphism" when applicable.

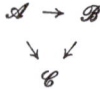

The hypotheses of the last proposition were too strong. $\mathscr{P}$ must be a full subcategory of projectives but not necessarily all projectives.

It clearly suffices to have a **resolving set** of projectives, that is a set $\mathscr{P}$ such that for all $A \in \mathscr{A}$ there exists $P \in \mathscr{P}$ and an epimorphism $P \to A$. Because of the additivity of the functors it suffices, indeed, for $\mathscr{P}$ to be a **nearly resolving set,** that is, a set $\mathscr{P}$ such that for all $A \in \mathscr{A}$ there exist $P_1, \ldots, P_n \in \mathscr{P}$ and an epimorphism $P_1 \oplus \cdots \oplus P_n \to A$. (Note that the next relaxation, namely that in which the finiteness of n is removed, leaves us with the definition of *generating set.*) Thus if $\mathscr{P}$ is the full subcategory of a nearly-resolving set of projectives in an abelian category $\mathscr{A}$ then $\mathscr{A}$ is the reflection of $\mathscr{P}$ in the category of abelian categories and (natural equivalence classes of) right exact functors. The converse is true:

Theorem 1.3. *If $\mathscr{P}$ is an +'ive category, $\mathscr{A}$ abelian, $T : \mathscr{P} \to \mathscr{A}$ a functor such that for every $S : \mathscr{P} \to \mathscr{B}$, $\mathscr{B}$ abelian there exists a right-exact $V : \mathscr{A} \to \mathscr{B}$ such that $VT \simeq S$ and V is unique up to natural equivalence then T is a full embedding and it represents $\mathscr{P}$ as the full subcategory of a nearly resolving set of projectives in $\mathscr{A}$.*

Proof. Consider the full embedding $\mathscr{P} \to (\mathscr{P}^*, \mathscr{G})$. The existence of a factorization $\mathscr{A} \to (\mathscr{P}^*, \mathscr{G})$ implies at least that $T : \mathscr{P} \to \mathscr{A}$ is an embedding.

Because $(\mathscr{A}, \mathscr{G})^*$ is a complete abelian category and $(\mathscr{P}^*, \mathscr{G})$ allows right-exact extensions of functors from $\mathscr{P}$ into such (*Abelian Categories,* 5.I) there exists right-exact $(\mathscr{P}^*, \mathscr{G}) \to (\mathscr{A}, \mathscr{G})^*$ such that

$$\begin{array}{ccc} \mathscr{P} & \to & (\mathscr{P}^*, \mathscr{G}) \\ \downarrow & & \downarrow \\ \mathscr{A} & \to & (\mathscr{A}, \mathscr{G})^* \end{array}$$

commutes. Now the representation functor $\mathscr{A} \to (\mathscr{A}, \mathscr{G})^*$ is right exact, so is $\mathscr{A} \to (\mathscr{P}^*, \mathscr{G}) \to (\mathscr{A}, \mathscr{G})^*$, and the two agree when preceeded by $\mathscr{P} \to \mathscr{A}$ hence are naturally equivalent. Thus $\mathscr{A} \to (\mathscr{P}^*, \mathscr{G})$ is an embedding. Because $\mathscr{P} \to (\mathscr{P}^*, \mathscr{G})$ is a full embedding it follows that $\mathscr{P} \to \mathscr{A}$ is a full embedding.

For the rest of the proof we thus assume that $\mathscr{P}$ is a full subcategory of $\mathscr{A}$.

The objects of $\mathscr{P}$ form a nearly-resolving set: Fix $B \in \mathscr{A}$ and define GA as the subset of (A, B) defined by: $f \in GA$ iff there exist $P_1, \ldots, P_n \in \mathscr{P}$ and a map $g : P_1 \oplus \cdots \oplus P_n \to B$, $\mathrm{Im}(f) \subset \mathrm{Im}(g)$. G is an additive subfunctor of $(-, B)$ and it is easily seen to carry cokernels into kernels, i.e. both G and $(-, B)$ are right exact functors from $\mathscr{A}$ to $\mathscr{G}^*$. Clearly $GP = (P, B)$ all $P \in \mathscr{P}$ and hence G is naturally equivalent to $(-, B)$.

It is true but not yet proven that the inclusion $G \subset (-, B)$ is onto. We know only that there is an equivalence $(-, B) \to G$.

Let $\eta : (-, B) \to G$ be a natural equivalence. $\eta_B(1_B) = \alpha \in GB \subset (B, B)$. Let $0 \to A \overset{k}{\to} B \overset{\alpha}{\to} B$ be exact and consider the commuting diagram

$$\begin{array}{ccc} (B, B) & \overset{(k, B)}{\rightarrow} & (A, B) \\ \eta_B \downarrow & & \downarrow \eta_A \\ GB & \rightarrow & GA \\ \downarrow & & \downarrow \\ (B, B) & \underset{(k, B)}{\rightarrow} & (A, B) \end{array}$$

starting with 1_B in the upper left corner and travelling counter-clockwise to (A, B) we obtain 0. The vertical arrows are one-to-one but travelling clockwise we find that $\eta_A(k) = 0$. Hence $k = 0$ and α is a monomorphism. By the definition of G there exists

$$P = P_1 \oplus \cdots \oplus P_n, \quad P \overset{\beta}{\to} B, \quad P_1, \ldots, P_n \in \mathscr{P}$$

such that $\operatorname{Im}(\alpha) \subset \operatorname{Im}(\beta)$.

Define $\gamma \in (P, B)$ to be such that $\eta_P(\gamma) = \beta$

$$\begin{array}{ccc} (B, B) & \overset{(\gamma, B)}{\rightarrow} & (P, B) \\ \eta_B \downarrow & & \downarrow \eta_P \\ GB & & GP \\ \downarrow & & \downarrow \\ (B, B) & \underset{(\gamma, B)}{\rightarrow} & (P, B) \end{array}$$

By starting with 1_B in the upper left corner of the commutative diagram we find that $P \overset{\gamma}{\to} B \overset{\alpha}{\to} B = G \overset{\beta}{\to} B$ and hence because α is a monomorphism, $\operatorname{Im}(\alpha) \subset \operatorname{Im}(\beta)$, it follows that $P \overset{\gamma}{\to} B$ is epimorphic. Thus $\mathscr{P}$ is a nearly resolving set.

Finally, the objects of $\mathscr{P}$ are projective. Let $P \in \mathscr{P}$ and $B \to P$ be an epimorphism. We wish to split $B \to P$. Choose $P_1, \ldots, P_n \in \mathscr{P}$ and an epimorphism $P_1 \oplus \cdots \oplus P_1 \to B$. It suffices to split the epimorphism $P_1 \oplus \cdots \oplus P_n \to B \to P$. The functor $(P, -) : \mathscr{P} \to \mathscr{G}$ extends to a right exact functor on $\mathscr{A}$ and hence $(P, P_1) \oplus \cdots \oplus (P, P_n) \to (P, P)$ is an epimorphism in $\mathscr{G}$ which implies that $P_1 \oplus \cdots \oplus P_n \to P$ splits. ∎

The remaining question is the characterization of those categories which may be represented as full subcategories of nearly-resolving sets of projectives. It is easier to first characterize those representable as full subcategories of resolving sets and then obtain as corollaries the characterization of those representable as a nearly-resolving and those as all projectives.

To that end we introduce the notions of **weak kernel** and **weak product.** The prefix "weak" should in general be used to remove the uniqueness conditions from definitions. Hence a weak kernel of a map $A \to B$ is a map such that $K \to A$

0) $K \to A \to B = 0$ and

1) For every $X \to A$ such that $X \to A \to B = 0$ there exists (possibly many) $X \to K$ such that $X \to K \to A = X \to A$.

Weak kernels are not unique in any sense. In a category with kernels the weak kernels of a map are precisely the split extensions of its kernel.

If $\mathscr{P}$ is the full subcategory of a resolving set of projectives in an abelian category $\mathscr{A}$ then every map in $\mathscr{P}$ has a weak kernel: given $P \to P'' \in \mathscr{P}$ let $0 \to K \to P \to P''$ and $P' \to K \to 0$ be exact in $\mathscr{A}$, $P' \in \mathscr{P}$. Then $P' \to P$ is a weak kernel of $P \to P''$ in $\mathscr{P}$. (Note that we must use the projectivity of the objects in $\mathscr{P}$.) In general, $P' \to P$ is a weak kernel of $P \to P''$ in $\mathscr{P}$ iff $P' \to P \to P''$ is exact in $\mathscr{A}$.

While here we may also observe that whereas $\mathscr{P}$ has weak kernels it can not have kernels unless the global dimension of $\mathscr{A}$ is two or less.

Given a pair of objects A_1, A_2 we say that A is a **weak product** of A_1, A_2 if there exist maps $A \xrightarrow{p_1} A_1$, $A \xrightarrow{p_2} A_2$ such that for every pair $X \xrightarrow{f_1} A_1$, $X \xrightarrow{f_2} A_2$ there exists a map (perhaps many) $X \xrightarrow{f} A$ such that $X \xrightarrow{f} A \xrightarrow{p_i} A_i = X \xrightarrow{f_i} A_i$, $i = 1, 2$. As one should expect the presence of weak kernels and weak products in an additive category implies the presence of weak finite limits.

If a category is +'ive, has weak products, and if every map factors as an epimorphism followed by a monomorphism then it has products as follows: let A be a weak product of A_1, A_2 and $A \xrightarrow{p_1} A_1$, $A \xrightarrow{p_2} A_2$ the maps which show it to be a weak product. Let $A_1 \xrightarrow{u_1} A$, $A_2 \xrightarrow{u_2} A$ be maps such that $p_j u_i = \delta_{ij}$. Then $u_1 p_1 + u_2 p_2$ is an idempotent. If $A \to P$ is epic and $P \to A$ monic and $A \to P \to A = u_1 p_1 + u_2 p_2$ then P is the product of A_1, A_2. For if $X \to P \to A \xrightarrow{p_i} A_i = 0$ $i = 1, 2$ then became $X \to P = X \to P \to A \to P \to A$ it follows that $X \to P = 0$.

Returning to P, the full subcategory of a resolving set of projectives in an abelian category $\mathscr{A}$ it is clear that $\mathscr{P}$ has weak products.

Theorem 1.4. *An +'ive category $\mathscr{P}$ is representable as the full subcategory of a resolving set of projectives in an abelian category iff*

1. *every map has a weak kernel,*

2. *every pair of objects has a weak product.*

Preproof. One of the pleasant aspects of this theorem is the first order nature of the conditions which imply a second order result, namely the existence of another category $\mathscr{A}$ and a functor $\mathscr{P} \to \mathscr{A}$ satisfying certain further conditions. Moreover the essential uniqueness of $\mathscr{A}$ (it is a reflection) forces upon us its construction. Assume the truth of the theorem and suppose that $\mathscr{P} \to \mathscr{A}$ has been found. For convenience let $\mathscr{P} \to \mathscr{A}$ be an inclusion $\mathscr{P} \subset \mathscr{A}$. Consider the category $\mathscr{P}^{\to}$ whose objects are $\mathscr{P}$-maps and whose maps are $\mathscr{P}$-squares. Define $\mathscr{P}^{\to} \to \mathscr{A}$ by $(P' \to P) \to \mathrm{Cok}(P' \to P)$. The fact that the objects of $\mathscr{P}$ form a resolving set says that the image of $\mathscr{P}^{\to} \to \mathscr{A}$ meets every isomorphism class of objects and hence may be assumed to meet all the objects. The fact that the objects of $\mathscr{P}$ are projective in $\mathscr{A}$ says that $\mathscr{P}^{\to} \to \mathscr{A}$ is full. Hence if we describe the kernel of $\mathscr{P}^{\to} \to \mathscr{A}$ we shall know $\mathscr{A}$. Given two objects $(P_1' \to P_1)$, $(P_2' \to P_2) \in \mathscr{P}^{\to}$ and a map between them:

$$\begin{array}{ccc} (P_1' & \to & P_1) \\ \downarrow & & \downarrow \\ (P_2' & \to & P_2) \end{array}$$

when is the induced map $\mathrm{Cok}(P_1' \to P_1) \to \mathrm{Cok}(P_2' \to P_2) = 0$? Precisely when there exists $P_1 \to P_2'$ such that

$$\begin{array}{ccc} & & P_1 \\ & \swarrow & \downarrow \\ P_2' & \to & P_2 \end{array}$$

commutes. Hence if we define the two-sided ideal $\mathfrak{A} \subset \mathscr{P}^{\to}$ to consist of all such maps then $\mathscr{A} \simeq \mathscr{P}^{\to}/\mathfrak{A}$. The embedding $\mathscr{P} \to \mathscr{P}^{\to}$ defined by $P \to (0 \to P)$ remains an embedding when followed by $\mathscr{P}^{\to} \to \mathscr{P}^{\to}/\mathfrak{A}$ and is the desired full inclusion. The description of $\mathscr{P}^{\to}$ and $\mathfrak{A}$ and hence of $\mathscr{P}^{\to}/\mathfrak{A}$ makes no reference to $\mathscr{A}$.

There is, indeed, an elementary proof that $\mathscr{P}^{\to}/\mathfrak{A}$ is abelian. The non-elementary proof that follows (it uses functor categories) seems much easier. An elementary proof can be found in section 4.

But keep in mind that once we know that there is such an abelian category than the know that $\mathscr{P}^{\to}/\mathfrak{A}$ is abelian.

Proof. Let $\mathscr{P}$ be an $+$'ive category with weak kernels and weak products. The representation functor $\mathscr{P} \to (\mathscr{P}^*, \mathscr{G})$ allows us to consider $\mathscr{P}$ to be the full subcategory of a generating set of projectives in an abelian category which we shall call $\mathscr{B}$.

We define $\mathscr{A}$ to be the full subcategory of $\mathscr{B}$ generated by the objects $\{A \in \mathscr{B} / \exists\, P, P', \in \mathscr{P},\ P' \to P \to A \to 0 \text{ exact}\}$. We shall show that $\mathscr{A}$ is an exact subcategory of $\mathscr{B}$, i.e. that it is closed under the formation of kernels, cokernels and finite direct sums. It is then automatically abelian and the objects of $\mathscr{P}$ remain projective in $\mathscr{A}$ and quite tautologically form a resolving set for $\mathscr{A}$.

Lemma 1.4.1. *$\mathscr{A}$ is closed under the formation of cokernels.*

Proof of Lemma. Let $A \to B \in \mathscr{A}$ and choose $P_1, P_1, P_2', P_2 \in \mathscr{P}$, $F \in \mathscr{B}$ and exact commutative

$$\begin{array}{ccccccc} & & P_1 & \to & A & \to & 0 \\ & & \downarrow & & \downarrow & & \\ P_2' & \to & P_2 & \to & B & \to & 0 \\ & & & & \downarrow & & \\ & & & & F & & \\ & & & & \downarrow & & \\ & & & & 0 & & \end{array}$$

We wish to show that $F \in \mathscr{A}$. Let $P \in \mathscr{P}$ be a weak product of P_1, P_2' and $P \to P_2$ the sum of the maps $P \xrightarrow{p_1} P_1 \to P_2$ and $P \xrightarrow{p_2} P_2' \to P_2$. Then $P \to P_2 \to F \to 0$ is exact. (We are inside an abelian category and one is free to chase elements if one must.)

Lemma 1.4.2. *For $P_1, P_2 \in \mathscr{P}$ it is the case that $P_1 \oplus P_2 \in \mathscr{A}$.*

Proof of Lemma. Let P be a weak product of P_1, P_2 with maps $P \xrightarrow{p_1} P_1$, $P \xrightarrow{p_2} P_2$, $P_1 \xrightarrow{u_1} P$, $P_2 \xrightarrow{u_2} P$ where $p_i u_j = \delta_{ij}$. Let $\bar{P}$ be the cokernel of the idempotent $1 - (u_1 p_1 + u_2 p_2)$. $\bar{P} \simeq P_1 \oplus P_2$. (The easiest verification rests on the observation that $P = P_1 \oplus P_2 \oplus X$ in $\mathscr{B}$ and that $1 - (u_1 p_1 + u_2 p_2) = u_3 p_3$.

Lemma 1.4.3. *$\mathscr{A}$ is closed under the formation of direct sums.*

Proof of Lemma. Given $A_1, A_2 \in \mathscr{A}$ choose exact $P_1' \to P_1 \to A_1 \to 0$, $P_2' \to P_2 \to A_2 \to 0$ with $P_1', P_1, P_2', P_2 \in \mathscr{P}$. We then obtain an exact sequence $P_1' \oplus P_2' \to P_1 \oplus P_2 \to A_1 \oplus A_2 \to 0$. By Lemma 1.42 both $P_1' \oplus P_2'$ and $P_1 \oplus P_2$ are in $\mathscr{A}$ and hence by Lemma 1.41,

$$A_1 \oplus A_2 \in \mathscr{A}.$$

Lemma 1.4.4. *If $P_1 \to P_2$ is a weak kernel of $P_2 \to P_3$ in $\mathscr{P}$ then $P_1 \to P_2 \to P_3$ is exact in $\mathscr{B}$.*

Proof of Lemma. For all $P \in \mathscr{P}$ the sequence of groups $(P, P_1) \to (P, P_2) \to (P, P_3)$ is exact. The functor $\Pi_{\mathscr{P}}(P, -)$ is an exact embedding ($\mathscr{P}$ is a generating set of projectives), and when applied to $P_1 \to P_2 \to P_3$ yields an exact sequence of groups, hence $P_1 \to P_2 \to P_3$ is exact in $\mathscr{B}$. (Embeddings preserve non-exactness.)

Lemma 1.4.5. *$\mathscr{A}$ is closed under the formation of kernels.*

Proof of Lemma. It suffices to show that for all $A \to B \in \mathscr{A}$ there exists $P \in \mathscr{P}$ and exact $P \to A \to B$ because it is then possible by iteration to find $P' \in \mathscr{P}$ and exact $P' \to P \to A \to B$ and

$$\mathrm{Ker}(A \to B) \simeq \mathrm{Cok}(P' \to P).$$

Accordingly let $P_1, P_2', P_2 \in \mathscr{P}$ and

$$\begin{array}{ccc} & P_1 \to A \to 0 & \\ & \bar{x} \downarrow \quad\;\; \downarrow x & \\ P_2' \xrightarrow[b]{} & P_2 \to B \to 0 & \end{array}$$

be exact and commutative. Let $P \to P_1 \oplus P_2'$ be an epimorphism, $P \in \mathscr{P}$, and $P_0 \to P$ a weak kernel of $P \to P_1 \oplus P_2' \xrightarrow{\bar{x}p_1 - bp_2} P_2$. Then not only is $P_0 \to P \to P_2$ exact but so is $P_0 \to P_1 \oplus P_2' \to P_2$. That is, P_0 maps epimorphically onto the pullback of

$$\begin{array}{cc} & P_1 \\ & \downarrow \\ P'_2 \to & P_2 \,. \end{array}$$

It follows that for $P_0 \to A = P_0 \to P_1 \oplus P_2' \xrightarrow{p_1} P_1 \to A$ it is the case that $P_0 \to A \to B$ is exact. (Choose elements if you must.) ▮

Recall that an **amenable** category is an $\oplus$'ive category in which **idempotents split,** that is for $A \xrightarrow{e} A$, $e^2 = e$ there exist $A \to B$, $B \to A$ such that

$$A \to B \to A = e$$
$$B \to A \to B = 1$$

(Note that if idempotents always split then the existence of weak products implies ordinary products because $u_1p_1 + u_2p_2$ is an idempotent and its image has already been observed to be the product.) ▮

Corollary 1.5. *A category is representable as the full subcategory of all projectives of an abelian category (with enough projectives) iff it is an amenable category in which every map has a weak kernel.*

Proof. The presence of weak kernels and products in a category $\mathscr{P}$ allows us to represent $\mathscr{P}$ as the full subcategory of a resolving set of projective in an abelian category $\mathscr{A}$. Now let $B \in \mathscr{A}$ be projective, and $P \to B$ an epimorphism, $P \in \mathscr{P}$.

Because B is projective we obtain $B \to P$ such that $B \to P \to B = 1$. The map $P \to B \to P$ is idempotent and hence there exists $P' \in \mathscr{P}$ such that $P' \simeq \mathrm{Im}(P \to B \to P)$ and $P' \simeq B$. To make the statement of the theorem strictly correct remove from $\mathscr{A}$ all projectives not in $\mathscr{P}$. What is left is equivalent to $\mathscr{A}$ and still abelian. ∎

Corollary 1.6. *An +'ive category $\mathscr{P}$ is representable as the full subcategory of a nearly resolving set of projectives in an abelian category iff for every finite sequence $P_1 \xrightarrow{x_1} P''$, $P_2 \xrightarrow{x_2} P''$, ..., $P_n \xrightarrow{x_n} P''$ there exists a finite sequence of objects P_1', P_2', ..., P_m' and maps*

$$\{P_i' \xrightarrow{y_{ji}} P_j\}\begin{matrix} i = 1, \ldots, m \\ j = 1, \ldots, n \end{matrix}$$

such that

0) $\sum\limits_i x_j y_{ji} = 0$ *each* j.

1) *For each object* X *and sequence* $\{X \xrightarrow{z_j} P_j\}$ *such that* $\sum\limits_i x_j z_j = 0$ *there exists* $X \xrightarrow{u_i} P_i'$ *such that* $\sum\limits_i y_{ji} u_i = x_j$, *each* j.

Proof. Consider the representation $\mathscr{P} \to (\mathscr{P}^*, \mathscr{G})$. Define $\mathscr{P}^\oplus$ to be the full subcategory of $(\mathscr{P}^*, \mathscr{G})$ generated by the finite direct sums of representable functors. $\mathscr{P}^\oplus$ automatically has products. The conditions on $\mathscr{P}$ imply that $\mathscr{P}^\oplus$ has weak kernels thusly.

Let $P \to P'' \in \mathscr{P}^\oplus$ where $P = P_1 \oplus \cdots \oplus P_n$, $P'' = P_1'' \oplus \cdots \oplus P_m''$. The conditions directly assert that $\mathscr{P}^\oplus$ contains a weak kernel

$$K_1 \to P \quad \text{of } P \to P'' \xrightarrow{p_1} P_1''$$

and then a weak kernel

$$K_2 \to K_1 \quad \text{of } K_1 \to P \to P'' \xrightarrow{p_2} P_2''$$

and so on — $K_{i+1} \to K_i$ is a weak kernel of

$$K_i \to K_{i-1} \to \cdots \to K_1 \to P \to P'' \xrightarrow{p_{i+1}} P_{i+1}''.$$

Finally then $K_n \to K_{n-1} \to \cdots \to K_1 \to P$ is a weak kernel of $P \to P''$.

$\mathscr{P}^\oplus$ is thus a resolving set of projectives in an abelian category $\mathscr{A}$ and $\mathscr{P}$ is a nearly resolving set. ∎

Note that if $\mathscr{P}$ has a single object, in other words that it is a ring, then the conditions of the theorem say that it is a coherent ring. For the logically minded it may be noted that whereas the conditions of Theorem 1.4 and Corollary 1.5 were elementary (e.g. the existence of weak kernels in an abstract category reads like $\mathsf{A}_w \exists_x \mathsf{A}_y \mathsf{E}_z [(wx = 0) \wedge (wy = 0 \Rightarrow xz = y)]$). The conditions of Cor. 1.6 are not elementary nor can they be replaced with elementary conditions. The latter can be seen as follows: Let R_j be the ring $Z[x_1, \ldots x_j]/(2x_1, \ldots, 2x_j)$ and let R be a non-principal ultra-product of the R_j's. Whereas each R_j is coherent, (indeed Noetherian), R is not. The annihilator of $2R$ can not be finitely generated.

A **Frobenius** category has been defined by Heller to be an abelian category with enough projectives and injectives, and in which projectives coincide with injectives. If $\mathcal{F}$ is a Frobenius category and $\mathcal{P}$ the full subcategory of a set of projectives which is not only a resolving set but co-resolving set then $\mathcal{P}$ has weak kernels, weak products and weak cokernels. Moreover, every map in $\mathcal{P}$ is a weak kernel and a weak cokernel This can most easily be seen by first noticing that $P_1 \to P_2 \to P_3 \in \mathcal{P}$ is exact in $\mathcal{F}$ iff $P_1 \to P_2$ is a weak kernel of $P_2 \to P_3$ (as defined in $\mathcal{P}$) iff $P_2 \to P_3$ is a weak cokernel of $P_1 \to P_2$.

Theorem 1.7. *A category is representable as a doubly resolving set of projectives-injectives in a Frobenius category iff it has weak kernels, weak products and every map is a weak kernel and a weak cokernel.*

Proof. Let $\mathcal{P}$ be as described and let $\mathcal{F}$ be an abelian category which represents $\mathcal{P}$ as the full subcategory of a resolving set of projectives (using theorem 1.4).

Given $A \in \mathcal{F}$ we wish to find $P \in \mathcal{P}$ and a monomorphism $A \to P$. Let $P_1, P_2 \in \mathcal{P}$ and $P_1 \to P_2 \to A \to 0$ be exact. Let $P_2 \to P_3$ be such that $P_1 \to P_2$ is a weak kernel of $P_2 \to P_3$. Then $P_1 \to P_2 \to P_3$ is exact and we obtain a monomorphism $A \to P_3$. Hence $\mathcal{P}$ is a co-resolving set.

Let $P \in \mathcal{P}$ and $P \to A$ a monomorphism. We wish to split $P \to A$. By the last paragraph it suffices to assume that $A \in \mathcal{P}$. Let $P' \to P$ be such that $P \to A$ is a weak cokernel (in $\mathcal{P}$) of $P' \to P$. Because $P \to A$ is a monomorphism (in $\mathcal{F}$) it must be the case that $P' \to P$ is the zero-map. Hence because $P' \to P \to P = 0$ and $P \to A$ is a weak cokernel of $P' \to P$ there exists $A \to P$ such that $P \to A \to P = 1$. ∎

Given a Frobenius category $\mathcal{F}$ and the full subcategory $\mathcal{P}$ of a doubly resolving set of projectives-injectives we have at least two ways of describing $\mathcal{F}$ in terms of $\mathcal{P}$, one from the fact that $\mathcal{P}$ is a resolving set of projectives; one from the fact that it is a resolving set of injectives. There is a third way. Every object in $\mathcal{F}$ is an image of a map in $\mathcal{P}$. The functor $\mathcal{P}^{\to} \to \mathcal{F} : (A \to B) \to Im(A \to B)$ has a representative image. Because the objects in $\mathcal{P}$ are both projective and injective, $\mathcal{P}^{\to} \to \mathcal{F}$ is full. Its kernel is the set of all squares.

$$\begin{matrix} (A & \to & B) \\ \downarrow & & \downarrow \\ (C & \to & D) \end{matrix}$$

such that $A \to D = 0$ and hence $\mathcal{F}$ may be described as $\mathcal{P}^{\to}$ mod such an ideal.

In our paper *Stable Homotopy* we shall give an elementary proof that the category $\mathcal{F}$ so constructed from a category $\mathcal{P}$ as described is abelian.

Note that an additive functor $\mathscr{P} \to \mathscr{A}$, $\mathscr{A}$ abelian, has two canonical extensions to $\mathscr{F}$ one left-exact, one right-exact. There is a unique transformation from the right-exact extension to the left-exact extension which is the identity transformation on $\mathscr{P}$ and the image of this transformation is a third canonical extension: it is described up to natural equivalence by the fact that it preserves monomorphisms and epimorphisms, which property is equivalent to preserving images. Hence $\mathscr{P} \to \mathscr{A}$ has a cokernel preserving extension, an image preserving extension, and a kernel preserving extension.

2. Ample Classes

Let $\mathscr{A}$ be an $\oplus$'ive category, $\boldsymbol{P}$ a class of objects closed under $\oplus$. We shall say that a map $A \to B \in \mathscr{A}$ is a **$\boldsymbol{P}$-fibration** if for every $P \in \boldsymbol{P}$ and $P \to B$ there exists a commutative triangle

$$\begin{array}{ccc} & P & \\ \swarrow & & \searrow \\ A & \to & B \end{array}$$

Given $A \in \mathscr{A}$ we say that $P \to A$ is a **$\boldsymbol{P}$-cover** if $P \in \boldsymbol{P}$ and $P \to A$ is a $\boldsymbol{P}$-fibration. We say that $\boldsymbol{P}$ is **ample** if every object in $\mathscr{A}$ has a $\boldsymbol{P}$-cover.

We shall use $\mathscr{A}/\boldsymbol{P}$ to denote the quotient category which kills the two sided ideal generated by $\{1_P / P \in \boldsymbol{P}\}$ and when the context warrants it, we shall use $\boldsymbol{P}$ to denote the full subcategory of $\mathscr{A}$ generated by $\boldsymbol{P}$.

Lemma 2.1. (The lifting lemma)

If $B \xrightarrow{f} C$ is a $\boldsymbol{P}$-fibration, then for any $g: A \to C$, $h: A \to B$ such that

$$\begin{array}{ccc} & A & \\ h \swarrow & & \searrow g \\ B & \underset{b}{\to} & C \end{array}$$

commutes in $\mathscr{A}/\boldsymbol{P}$ there exists $h' \in \mathscr{A}(A, B)$ *such that* $h' \equiv h$

in $\mathscr{A}/\boldsymbol{P}$ and

$$\begin{array}{ccc} & A & \\ h' \swarrow & & \searrow g \\ B & \underset{b}{\to} & C \end{array}$$

commutes in $\mathscr{A}$.

Proof. Let $fh - g = ab$ where $a \in \mathscr{A}(P, C)$, $b \in \mathscr{A}(A, P)$, $P \in \boldsymbol{P}$. Let

$$\begin{array}{ccc} & P & \\ d \swarrow & & \searrow a \\ B & \underset{b}{\to} & C \end{array}$$

commute. Define $h' = h - db$. Then $h' = h$ in $\mathscr{A}/\boldsymbol{P}$ and $fh' = fh - fdb = \\ = fh - ab = fh - (fh - g) = g$ in $\mathscr{A}$. ∎

Lemma 2.2. *If* $\boldsymbol{P}$ *is ample and* $\mathscr{A}$ *has weak-kernels then* $\mathscr{A}/\boldsymbol{P}$ *has weak kernels.*

Proof. Given $A \to B \in \mathscr{A}$. Let $P \to B$ be *a* $\boldsymbol{P}$-cover, and K a weak kernel of $A \oplus P \to B$ (which it may be noted is a $\boldsymbol{P}$-fibration). Then $K \to A \oplus P \xrightarrow{p_1} A$ is a weak kernel of $A \to B$ in $\mathscr{A}/\boldsymbol{P}$. ▮

Lemma 2.3. *If* $\boldsymbol{P}$ *is ample and* $\mathscr{A}$ *has weak kernels then the full subcategory of* $\boldsymbol{P}$ *has weak-kernels.*

Proof. Given $P_1 \to P_2 \in \boldsymbol{P}$ let $K \to P_1$ be a weak kernel in $\mathscr{A}$ and let $P_0 \to K$ be a $\boldsymbol{P}$-cover. Then $P_0 \to P_1$ is a weak kernel of $P_1 \to P_2$ in $\boldsymbol{P}$. ▮

Let $\boldsymbol{P}$ be ample in $\mathscr{A}$, and let $\mathscr{A}$ have weak kernels. Let $\mathscr{B}$ be the abelian category which represents $\boldsymbol{P}$ as a resolving set of projectives. Define $\mathscr{A} \xrightarrow{T} \mathscr{B}$ as follows: given $A \in \mathscr{A}$ let $P_1 \to A$ be a $\boldsymbol{P}$-cover, $K \to P_1$ a weak kernel of $P_1 \to A$, $P_2 \to K$ a $\boldsymbol{P}$-cover and TA a cokernel in $\mathscr{B}$ of $P_2 \to P_1$. T carries $\boldsymbol{P}$-fibrations into epimorphisms and conversely, i.e. if Tf is an epimorphism, then f is a $\boldsymbol{P}$-fibration. If $\mathscr{A}$ has cokernels, we can define $\mathscr{B} \xrightarrow{S} \mathscr{A}$ in the usual manner, and S is the left-adjoint of T, hence T preserves left-limits. If $\mathscr{A}$ is abelian and all $\boldsymbol{P}$-fibrations are epimorphisms (i.e. $\boldsymbol{P}$ generates $\mathscr{A}$), then T is full. Using $\boldsymbol{P}$ to define a relative homological structure on $\mathscr{A}$ (relative epies are fibrations) then $\mathscr{B}$ is the category described by the writer in *Abelian Categories*, in particular, given a relatively connected pair of functors from $\mathscr{A}$ there is a cannonical absolutely connected pair of extensions on $\mathscr{B}$, and the construction of ordinary satellites of functors on $\mathscr{B}$ restricts to relative satellites on $\mathscr{A}$.

Lemma 2.4. *If* $\boldsymbol{P}_1$ *and* $\boldsymbol{P}_2$ *are ample in* $\mathscr{A}$ *then* $\boldsymbol{P}_1 \oplus \boldsymbol{P}_2 = \{P_1 \oplus P_2 | P_i \in \boldsymbol{P}_i\}$ *is ample.*

Proof. If $P_1 \to A$ is a $\boldsymbol{P}_1$-cover and $P_2 \to A$ is a $\boldsymbol{P}_2$ cover then $P_1 \oplus P_2 \to A$ is a $\boldsymbol{P}_1 \oplus \boldsymbol{P}_2$ cover. ▮

Lemma 2.5. $\{(A \xrightarrow{1} A) | A \in \mathscr{A}\}$ *and* $\{(0 \to A) | A \in \mathscr{A}\}$ *are ample in* $\mathscr{A}^{\to}$. ▮

Lemma 2.6. $\{(A \to 0) | A \in \mathscr{A}\}$ *is ample in* $\mathscr{A}^{\to}$ *iff* $\mathscr{A}$ *has weak kernels.* ▮

Given $\mathscr{A}$ with weak kernels it may be noticed from the preproof of 1.4 that the abelian category which represents $\mathscr{A}$ as the full subcategory of a resolving sat of projectives is $\mathscr{A}^{\to}/\{A \to 0\} \oplus \{A \xrightarrow{1} A\}$.

Lemma 2.7. *Let* $\boldsymbol{P}$ *be an ample class in* $\mathscr{A}$, $\mathscr{F}_{\boldsymbol{P}}(\mathscr{A})$ *the full subcategory* *of* $\mathscr{A}^{\to}$ *generated by the* **P***-fibrations. Then* $(\mathscr{A}/\mathbf{P})^{\to}$ *is equivalent to* $\mathscr{F}_{\mathbf{P}}(\mathscr{A})/\mathscr{F}_{\mathbf{P}}(\mathbf{P})$ *where* $\mathscr{F}_{\mathbf{P}}(\mathbf{P}) = \{P \to 0 \mid P \in \mathbf{P}\} \oplus \{P \xrightarrow{1} P \mid P \in \mathbf{P}\}$.

Proof. Consider the functor $\mathscr{F}_{\mathbf{P}}(\mathscr{A}) \to \mathscr{A}^{\to} \to (\mathscr{A}/\mathbf{P})^{\to}$. Its image is representative: given $(A \to B) \in \mathscr{A}^{\to}$ let $P \to B$ be a **P**-cover and observe that $(A \oplus P \to B) \in \mathscr{F}_{\mathbf{P}}(\mathscr{A})$ and that $(A \to B)$ is isomorphic in $(\mathscr{A}/\mathbf{P})^{\to}$ to $(A \oplus P \to B)$.

It is a full functor: Given $(A \to B)$, $(A' \to B') \in \mathscr{F}_{\mathbf{P}}(\mathscr{A})$ and a square in $\mathscr{A}^{\to}$ that commutes in $\mathscr{A}/\mathbf{P}$

$$\begin{array}{ccc} A & \to & B \\ \downarrow & & \downarrow \\ A' & \to & B' \end{array}$$

we can use the lifting lemma to replace $A \to A'$ with a map which makes the square commute in $\mathscr{A}$. That is, given a map in $(\mathscr{A}/\mathbf{P})^{\to}$ from $(A \to B)$ to $(A' \to B')$ we can find an ancestor in $\mathscr{F}_{\mathbf{P}}(\mathscr{A})$.

The kernel of $\mathscr{F}_{\mathbf{P}}(\mathscr{A}) \to (\mathscr{A}/\mathbf{P})^{\to}$ consists of squares

$$\begin{array}{ccc} (A & \to & B) \\ f\downarrow & & \downarrow g \\ (A' & \to & B') \end{array}$$

such that both f and g lie in the two sided ideal generated by $\mathbf{P}$. Suppose $g = B \to P \to B'$. Because $A' \to B'$ is a $\mathbf{P}$-fibration we can find a commutative diagram of the form

$$\begin{array}{ccc} A & \to & B \\ \downarrow & & \downarrow \\ P & \xrightarrow{1} & P \\ \downarrow & & \downarrow \\ A' & \to & B'. \end{array}$$

Hence the given square is equivalent mod $\{P \xrightarrow{1} P \mid P \in \mathbf{P}\}$ to a square of the form

$$\begin{array}{ccc} A & \to & B \\ f'\downarrow & & \downarrow 0 \\ A' & \to & B' \end{array}$$

where f' is still in the two sided ideal of $\mathbf{P}$. Clearly then, if $f' = A \to P' \to A'$ we can factor the original square through $(P' \oplus P \to P)$. ∎

Proposition 2.8. *Let* $\mathscr{A}$ *be an* $\oplus$*'ive category with weak kernels,* $\mathbf{P}$ *an ample class in* $\mathscr{A}$, $\mathscr{F}_{\mathbf{P}}(\mathscr{A})$ *the full subcategory of* $\mathscr{A}^{\to}$ *of* **P**-fibrations, $\boldsymbol{S}(\mathscr{A})$ *the class of splitting epimorphisms. Then* $\mathscr{F}_{\mathbf{P}}(\mathscr{A})/\boldsymbol{S}(\mathscr{A})$ *is an abelian category and the functor* $\mathscr{A}/\mathbf{P} \to \mathscr{F}_{\mathbf{P}}(\mathscr{A})/\boldsymbol{S}(\mathscr{A})$ *which sends* A *to* $(P \to A) \in \mathscr{F}_{\mathbf{P}}(\mathscr{A})$, $P \in \mathbf{P}$ *represents* $\mathscr{A}/\mathbf{P}$ *as the full subcategory of a resolving set of projectives in* $\mathscr{F}_{\mathbf{P}}(\mathscr{A})/\boldsymbol{S}(\mathscr{A})$.

Proof. $\mathscr{A}/\mathbf{P}$ has weak kernels by 2.2 and hence $(\mathscr{A}/\mathbf{P})^{\rightarrow}/\mathbf{S}(\mathscr{A})$ is an abelian category by the preproof of 1.4.

By the last proposition $(\mathscr{A}/\mathbf{P})^{\rightarrow}$ is equivalent to $\mathscr{F}_{\mathbf{P}}(\mathscr{A})/\mathscr{F}_{\mathbf{P}}(\mathbf{P})$. Note that $\mathscr{F}_{\mathbf{P}}(\mathbf{P}) \subset \mathbf{S}(\mathscr{A})$ and hence $(\mathscr{A}/\mathbf{P})^{\rightarrow}/\mathbf{S}(\mathscr{A})$ is equivalent to $\mathscr{F}_{\mathbf{P}}(\mathscr{A})/\mathbf{S}(\mathscr{A})$.

The canonical embedding $\mathscr{A}/\mathbf{P} \to (\mathscr{A}/\mathbf{P})^{\rightarrow}/\mathbf{S}(\mathscr{A})$ sends A to $(0 \to A) \in \mathscr{A}/\mathbf{P}$. The equivalence $(\mathscr{A}/\mathbf{P})^{\rightarrow} \to \mathscr{F}_{\mathbf{P}}(\mathscr{A})/\mathscr{F}_{\mathbf{P}}(\mathbf{P})$ sends $(0 \to A)$ to $(P \to A) \in \mathscr{F}_{\mathbf{P}}(\mathscr{A})$ where $P \in \mathbf{P}$. ▮

Corollary 2.9. *If $\mathscr{A}$ is an abelian category and $\mathbf{P}$ a resolving set of projectives, Shex $(\mathscr{A})$ the category of short exact sequences (i.e. Objects (Shex $\mathscr{A}$) $= \{0 \to A' \to A \to A'' \to 0 \mid$ exact in $\mathscr{A}\}$) and $\mathbf{S}(\mathscr{A})$ the class of split short exact sequences, then Shex $\mathscr{A}/\mathbf{S}(\mathscr{A})$ is abelian and the functor $\mathscr{A}/\mathbf{P} \to$ Shex $\mathscr{A}/\mathbf{S}(\mathscr{A})$ which sends $A \in \mathscr{A}$ to $(0 \to K \to P \to A \to 0) \in$ Shex $\mathscr{A}$, $P \in \mathbf{P}$, represents $\mathscr{A}/\mathbf{P}$ as the full subcategory of a resolving set of projectives in Shex $\mathscr{A}/\mathbf{S}(\mathscr{A})$.*

Proof. One needs only to observe that $\mathbf{P}$ is an ample class, that $\mathscr{F}_{\mathbf{P}}(\mathscr{A})$ is the full subcategory of epimorphisms in $\mathscr{A}^{\rightarrow}$, that $\mathscr{F}_{\mathbf{P}}(\mathscr{A})$ is equivalent to *Shex* $(\mathscr{A})$ and that the class of splitting epimorphisms in $\mathscr{F}_{\mathbf{P}}(\mathscr{A})$ corresponds to the class of splitting sequences in *Shex* $\mathscr{A}$. ▮

Let $\mathscr{A}$ be an $\oplus'$ ive category with kernels, $\mathbf{P}$ an ample class in $\mathscr{A}$. Define $\Omega: \mathscr{A}/\mathbf{P} \to \mathscr{A}/\mathbf{P}$ by $\Omega(A) = Ker\,(P \to A)$ where $P \to A$ is a $\mathbf{P}$-cover.

Consider a map $A \to B \in \mathscr{A}$ and the diagram

$$\begin{array}{ccccc} D & \to & C & \to & P_0 \\ \downarrow & & \downarrow & & \downarrow \\ P_1 & \to & A & \to & B \end{array}$$

where $P_0 \to B$ and $P_1 \to A$ are $\mathbf{P}$-covers, and both squares are pullbacks. Then $C \to A$ is a weak kernel of $A \to B$ in $\mathscr{A}/\mathbf{P}$, and $D \to C$ is a weak kernel of $C \to A$ in $\mathscr{A}/\mathbf{P}$. But $D = \Omega(B)$ in $\mathscr{A}/\mathbf{P}$.

If we go a step further to the pullback.

$$\begin{array}{ccc} E & \to & P_2 \\ \downarrow & & \downarrow \\ D & \to & C \end{array}$$

where $P_2 \to C$ is a $\mathbf{P}$-cover we see that $E \to D$ is a map such that

$$\begin{array}{ccccc} E & \to & P_0 \oplus P_2 & \to & A \\ \downarrow & & \downarrow & & \downarrow \\ D & \to & P_0 \oplus P_1 & \to & B \end{array}$$

commutes and hence $E \to D = \Omega(A \to B)$ in $\mathscr{A}/\mathbf{P}$.

Thus we obtain an infinite sequence $\cdots \to \Omega^2(B) \to \Omega\, C \to \Omega A \to \Omega B \to C \to A \to B$ in which each map is a weak kernel of its neighbor.

In particular $\Omega : \mathscr{A}/\boldsymbol{P} \to \mathscr{A}/\boldsymbol{P}$ carries weak kernels to weak kernels and its unique right exact extention to $\Omega : \mathscr{F}_{\boldsymbol{P}}(\mathscr{A})/\boldsymbol{S}(\mathscr{A}) \to \mathscr{F}_{\boldsymbol{P}}(\mathscr{A})/\boldsymbol{S}(\mathscr{A})$ in exact.

Given $(0 \to A' \to A \to A'' \to 0) \in \mathit{Shex}\, \mathscr{A}/\boldsymbol{S}(\mathscr{A})$ we may compute $\Omega(A' \to A \to A'')$ as the top row of the exact commutative diagram:

$$\begin{array}{ccccccccc}
 & & 0 & & 0 & & 0 & & \\
 & & \downarrow & & \downarrow & & \downarrow & & \\
0 & \to & K' & \to & K & \to & K'' & \to & 0 \\
 & & \downarrow & & \downarrow & & \downarrow & & \\
0 & \to & P' & \to & P' + P'' & \to & P'' & \to & 0 \\
 & & \downarrow & & \downarrow & & \downarrow & & \\
0 & \to & A' & \to & A & \to & A'' & \to & 0 \\
 & & \downarrow & & \downarrow & & \downarrow & & \\
 & & 0 & & 0 & & 0 & &
\end{array}$$

where P', P'' are projective.

Proposition 2.10. *Let $\mathscr{A}$ be an abelian category with enough projectives. If we denote $\mathit{Shex}\, \mathscr{A}/\boldsymbol{S}\mathscr{A}$ by $\mathscr{E}xt\, \mathscr{A}$ then* $\mathscr{E}xt_{\mathscr{A}}(A, B) \simeq (\mathscr{E}xt\, \mathscr{A})\ (\Omega'' A, B)$ $n > 0$. ▮

We have the curious situation that the presence of either enough projectives or injectives insures that $\mathit{Shex}\, \mathscr{A}/\boldsymbol{S}\mathscr{A}$ is abelian (clearly the two hypotheses are dual and the conclusion is self dual). Indeed it is the case that $\mathit{Shex}\, \mathscr{A}/\boldsymbol{S}\mathscr{A}$ is abelian for any abelian $\mathscr{A}$. The proof appears in the next section. For the moment consider the opposite situation: $\mathscr{A}$ has enough projectives and injectives. Then we can pass to either the "projective-homotopy" category $\mathscr{A}/\boldsymbol{P}$ or the "injective-homotopy" category $\mathscr{A}/\boldsymbol{E}$. From the general theorems we know that $\mathscr{A}/\boldsymbol{P}$ can be represented as the full subcategory of a resolving set of projectives in an abelian category $\mathscr{B}$, and that $\mathscr{A}/\boldsymbol{E}$ can be represented as the full subcategory of a resolving set of injectives in an abelian category $\mathscr{B}_2$. From 2.9 we find that $\mathscr{B}_1 = \mathscr{B}_2$.

There are two functors $\mathscr{A} \to \mathit{Shex}\, \mathscr{A}/\boldsymbol{S}\mathscr{A}$.
Let $PA = (0 \to K \to P \to A \to 0)$, P projective.
Let $EA = (0 \to A \to E \to F \to 0)$, E injective.
Let Ω be the exact endomorphism described above, let S be the exact endomorphism defined dually. Then

$$\mathrm{Ext}^n(A, B) = (\Omega^n PA, PB) \simeq (EA, S^n EB) \qquad n > 0.$$

Both P and E may be extended to $\mathscr{A}^{\to}$.

$P(A \to B) = (0 \to K \to A \oplus P \to B \to 0)$ where P is projective and $P \to B$ an epimorphism. $E(A \to B) = (0 \to A \to B \oplus E \to F \to 0)$

where E is injective and $A \to E$ is a monomorphism. We are then in a position to define Ext of a pair of maps, but such is the territory of Eckmann and Hilton.

One final observation for this section. Consider an $\oplus$'ive category $\mathscr{P}$ with weak kernels and weak cokernels. We can represent $\mathscr{P}$ as the full subcategory of a resolving set of projectives in an abelian category $\mathscr{A}$. We can then pass to the projective-homotopy category $\mathscr{A}/\mathscr{P}$. Now $\mathscr{A} = \mathscr{P}^{\to}/\{P \to 0\} \oplus \{P \to P\}$ and the values of $\mathscr{P} \to \mathscr{A}$ look like $\{0 \to P\}$. Thus $\mathscr{A}/\mathscr{P} = \mathscr{A}^{\to}/\{P \to 0\} \oplus \{P \to P\} \oplus \{0 \to P\}$ and we would have arrived at the same category if we had represented $\mathscr{P}$ as the full subcategory of a resolving set of injectives in an abelian category and then passed to the injective-homotopy category.

3. Recognizing Abelian Categories

First a theorem which allows us to recognize that a category is abelian, then two applications.

Proposition 3.1. *If a category $\mathscr{A}$ with a zero-object has the property that for every $A \to B \in \mathscr{A}$ there exists a factorization $A \to I \to B$ such that $A \to I$ is a cokernel and $I \to B$ is a kernel then every map in $\mathscr{A}$ has a kernel and a cokernel and every monomorphism is a kernel and every epimorphism is a cokernel.*

Hence if $\mathscr{A}$ also has finite sums and products it is abelian. Or if $\mathscr{A}$ is also $+$'ive and has weak products it is abelian.

Proof. The hypothesis is self-dual. It suffices to prove

First: Every map has a kernel.

Given $A \to B \in \mathscr{A}$ let $A \to I \to B$ be a factorization as described. $A \to I$ is a cokernel, say of $C \to A$. Let $C \to K \to A$ be a factorization of $C \to A$ as described. $K \to A$ is a kernel, say of $A \to D$. Now $C \to K \to A \to D = 0$ hence there exists commutative $\begin{array}{ccc} A & \to & I \\ & \searrow & \downarrow \\ & & D \end{array}$. Suppose that $X \to A \to B = 0$. Because $I \to B$ is a monomorphism it follows that $X \to A \to I = 0$ and hence that $X \to A \to D = 0$ and there exists a unique $X \to K$ such that $\begin{array}{ccc} & X & \\ \swarrow & & \searrow \\ K & \to & A \end{array}$ commutes. Finally, in order to show that $K \to A$ is a kernel of $A \to B$ it suffices to show that $K \to A \to B = 0$. Clearly $C \to K \to A \to B = 0$, but $C \to K$ is an epimorphism and hence $K \to A \to B = 0$.

Second: Every monomorphism is a kernel.

Let $A \to B$ be a monomorphism, and $A \to I \to B$ a factorization as described. $A \to I$ is a cokernel, say of $C \to A$. Then $C \to A \to B = 0$ hence $C \to A = 0$ and $A \to I$ is an isomorphism. Thus $A \to B$ is just as much a kernel as $I \to B$. ▮

Now for the elementary proof of theorem 1.4.

Theorem 3.2. *Let $\mathscr{P}$ be an +'ive category with weak kernels and weak products. Then*

$$\mathscr{A} = \mathscr{P}^{\to}/\{P \to 0\} \oplus \{P \xrightarrow{1} P\}$$

is abelian and the functor $\mathscr{P} \to \mathscr{A} : P \to (0 \to P)$ represents $\mathscr{P}$ as the full subcategory of a resolving set of projectives.

Proof. We recall that $\begin{matrix}(A' \to A)\\ \downarrow \qquad \downarrow\\ (B' \to B)\end{matrix} \equiv 0\,(\text{in } \mathscr{A})$

iff there exists $\begin{matrix} & A & \\ \swarrow & & \searrow \\ B' & \to & B\end{matrix}$.

Given a map $\begin{matrix}(A' \to A)\\ \downarrow \qquad \downarrow\\ (B' \to B)\end{matrix}$ in $\mathscr{P}^{\to}$ let $\begin{matrix}P \to A\\ \downarrow \quad \downarrow\\ B' \to B\end{matrix}$

be a weak pullback in $\mathscr{P}$. There

exists $\begin{matrix}(A' \to A)\\ \downarrow \qquad \downarrow\\ (P \to A)\end{matrix}$ such that $\begin{matrix}(A' \to A)\\ \downarrow \qquad \downarrow 1\\ (P \to A)\\ \downarrow \qquad \downarrow\\ (B' \to B)\end{matrix} = \begin{matrix}(A' \to A)\\ \downarrow \qquad \downarrow\\ (B' \to B)\end{matrix}$

Guided by the last theorem we proceed to show:

Lemma 3.2.1

$\begin{matrix}(A' \to A)\\ \downarrow \qquad \downarrow 1\\ (P \to A)\end{matrix}$ *represents a cokernel in $\mathscr{A}$.*

Lemma 3.2.2

$\begin{matrix}(P \to A)\\ \downarrow \qquad \downarrow\\ (B' \to B)\end{matrix}$ *represents a kernel in $\mathscr{A}$.*

Proof of Lemma 3.2.1.

(3.2.1.1)

$$\begin{array}{cc}(A' \to & A)\\ \downarrow & \downarrow 1\\ (P \to & A)\end{array} \quad \text{represents an epimorphism in } \mathscr{A}$$

Suppose $\begin{array}{cc}(A' \to & A)\\ \downarrow & \downarrow\\ (P \to & A)\\ \downarrow & \downarrow\\ (X' \to & X)\end{array} \equiv 0.$ There exists $\begin{array}{cc} & A\\ \swarrow & \downarrow\\ X' \to & X\end{array}$ and hence $\begin{array}{cc}(P \to & A)\\ \downarrow & \downarrow\\ (X' \to & X)\end{array} \equiv 0$

(3.2.1.2.)

$\begin{array}{cc}(A' \to & A)\\ \downarrow & \downarrow\\ (P \to & A)\end{array}$ *represents a weak cokernel of* $\begin{array}{cc}(O \to & P)\\ \downarrow & \downarrow\\ (A' \to & A)\end{array}$ *because:*

(3.2.1.2.1.)

$$\begin{array}{cc}(O \to & P)\\ \downarrow & \downarrow\\ (A' \to & A)\\ \downarrow & \downarrow\\ (P \to & A)\end{array} \equiv 0 \quad \text{since there exists} \quad \begin{array}{cc} & P\\ \swarrow & \downarrow\\ P \to & A\end{array}$$

(3.2.1.2.2.) Suppose $\begin{array}{cc}(O \to & P)\\ \downarrow & \downarrow\\ (A' \to & A)\\ \downarrow & \downarrow\\ (X' \to & X)\end{array} \equiv 0.$ There exists $\begin{array}{cc} & P\\ \swarrow & \downarrow\\ X' \to & X\end{array}$

and hence there exists $\begin{array}{cc}(P \to & A).\\ \downarrow & \downarrow\\ (X' \to & X)\end{array}$ Finally $\begin{array}{cc}(A' \to & A)\\ \downarrow & \downarrow\\ (P \to & A)\\ \downarrow & \downarrow\\ (X' \to & X)\end{array} \equiv \begin{array}{cc}(A' \to & A)\\ \downarrow & \downarrow\\ (X' \to & X)\end{array}$

because the right hand legs are equal.

Proof of Lemma 3.2.2.

Let $\begin{array}{ccc} & \bar{b} & \\ P & \to & A\\ \bar{a}\downarrow & & \downarrow a\\ B' & \underset{b}{\to} & B\end{array}$ be a weak pullback in $\mathscr{P}$.

(3.2.2.1.)

$$\begin{array}{c}(P \to A)\\ \downarrow \quad \downarrow\\ (B' \to B)\end{array}$$ *represents a monomorphism in* $\mathscr{A}$.

Given $\begin{array}{c}(X' \to X)\\ \downarrow \quad \downarrow\\ (P \to A)\\ \downarrow \quad \downarrow\\ (B' \to B)\end{array} \equiv 0$ there exists $\begin{array}{c}X\\ \swarrow \ \downarrow\\ B' \to B\end{array}$ and hence there exist,

$\begin{array}{c}X\\ \swarrow \ \downarrow\\ P \to A\end{array}$ and $\begin{array}{c}(X' \to X)\\ \downarrow \quad \downarrow\\ (P \to A)\end{array} \equiv 0.$

(3.2.2.2.) Let S be a weak product of A and B', $A \xrightarrow{u_1} S$, $S \xrightarrow{p_1} A$, $B' \xrightarrow{u_2} S$, $S \xrightarrow{p_2} B'$ maps which show it to be a weak product.

$\begin{array}{ccc}(P & \xrightarrow{\bar{b}} & A)\\ \bar{a}\downarrow & & \downarrow a\\ (B' & \xrightarrow[b]{} & B)\end{array}$ *represents a weak kernel of* $\begin{array}{ccc}(B' & \xrightarrow{\quad b \quad} & B)\\ u_2\downarrow & & \downarrow'\\ (S & \xrightarrow[ap_1 + bp_2]{} & B)\end{array}$ *because*:

(3.2.2.2.1.)

$\begin{array}{c}(P \to A)\\ \downarrow \quad \downarrow\\ (B' \to B)\\ \downarrow \quad \downarrow\\ (S \to B)\end{array} \equiv 0$ because $\begin{array}{c}A\\ u_1\swarrow \ \downarrow\\ S \to B\end{array}$ commutes.

(3.2.2.2.2.)

Suppose $\begin{array}{ccc}(X' & \xrightarrow{x} & X)\\ y'\downarrow & & \downarrow y\\ (B' & \to & B)\\ \downarrow & & \downarrow\\ (S & \to & B)\end{array} \equiv 0.$ There exists $\begin{array}{c}X\\ f\swarrow \ \downarrow\\ S \to B\end{array}$

Then $\begin{array}{ccc}X' & \to & X\\ & & \downarrow p_1 f\\ y' - p_2 f x \Big\downarrow & & A\\ & & \downarrow a\\ B' & \xrightarrow[b]{} & B\end{array}$ commutes and there exists $\begin{array}{ccc}X' & \to & X\\ \downarrow & & \downarrow p_1 f\\ P & \to & A\end{array}$

Thus $\begin{matrix}(X' \to X)\\ \downarrow \quad \downarrow p_1 f\\ (P \to A)\\ \downarrow \quad \downarrow a\\ (B' \to B)\end{matrix} \equiv \begin{matrix}(X' \to X)\\ \downarrow \quad \downarrow y\\ (B' \to B)\end{matrix}$ because $\begin{matrix} & X \\ p_2 f \swarrow & \downarrow y - a p_1 f \\ B' \to B & \end{matrix}$ commutes.

From the last theorem we now know that $\mathscr{A}$ has kernels, cokernels, and that every monomorphism is a kernel and every epimorphism is a cokernel. Given an object $(P' \to P) \in \mathscr{A}$ lemma 3.21 tells us that

$$\begin{matrix}(O \to P)\\ \downarrow \quad \downarrow\\ (P' \to P)\end{matrix}$$

represents an epimorphism, hence the values of $\mathscr{P} \to \mathscr{A}$ form a resolving set. To show that they are projective it suffices to split every epimorphic

map of the form $\begin{matrix}(O \to P)\\ \downarrow \quad \downarrow\\ (O \to P'')\end{matrix}$ in $\mathscr{A}$.

Let $\begin{matrix}(P' \to P)\\ \downarrow \quad \downarrow\\ (O \to P'')\end{matrix}$ be a weak pullback in $\mathscr{P}$. $\begin{matrix}(O \to P)\\ \downarrow \quad \downarrow\\ (P' \to P)\\ \downarrow \quad \downarrow\\ (O \to P'')\end{matrix}$

is a canonical factorization hence $\begin{matrix}P' \to P\\ \downarrow \quad \downarrow\\ O \to P''\end{matrix}$ represents an isomorphism

in $\mathscr{A}$. Thus there exists $\begin{matrix}(O \to P'')\\ \downarrow \quad \downarrow\\ (P' \to P)\end{matrix}$ and necessarily $P'' \to P \to P'' = 1$.

Finally $\mathscr{A}$ has direct sums. We shall think of $\mathscr{P}$ as a full subcategory of $\mathscr{A}$. First given $P_1, P_2 \in \mathscr{P}$ let S be a weak produkt, $P_1 \xrightarrow{u_1} S$, $S \xrightarrow{p_1} P_1$, $P_2 \xrightarrow{u_2} S$, $S \xrightarrow{p_2} P_2$ maps which show it to be a weak product. The cokernel of $S \xrightarrow{1 - u_1 p_1 - u_2 p_2} S$ in $\mathscr{A}$ is the direct sum of P_1, P_2.

Given $A_1, A_2 \in \mathscr{A}$ let $P_1' \to P_1 \to A_1 \to 0$, $P_2' \to P_2 \to A_2 \to 0$ be exact. Then $A_1 \oplus A_2 = \operatorname{Cok}(P_1' \oplus P_2' \to P_1 \oplus P_2)$. ▮

Theorem 3.3. *Let $\mathscr{A}$ be an abelian category, Shex $\mathscr{A}$ the category whose objects are short exact sequences of the form $0 \to A' \to A \to A'' \to 0$, $\mathbf{S}\mathscr{A}$ the class of short splitting sequences. Then Shex $\mathscr{A}/\mathbf{S}\mathscr{A}$ is abelian.*

Lemma 3.3.1

$$\begin{array}{ccccccccc} 0 & \to & A' & \xrightarrow{x} & A & \xrightarrow{x''} & A'' & \to & 0 \\ & & \downarrow a' & & \downarrow a & & \downarrow a'' & & \\ 0 & \to & B' & \xrightarrow[y]{} & B & \xrightarrow[y'']{} & B'' & \to & 0 \end{array} \equiv 0 \text{ iff there exists } \begin{array}{cc} A' & \to A \\ \downarrow & \swarrow \\ B' & \end{array}$$

Proof of Lemma

Suppose we are given
$$\begin{array}{ccccccccc} 0 & \to & A' & \xrightarrow{x} & A & \to & A'' & \to & 0 \\ & & \downarrow a' & & \downarrow a & & \downarrow & & \\ 0 & \to & C' & \to & C' \otimes C'' & \to & C'' & \to & 0 \\ & & \downarrow c' & & \downarrow & & \downarrow & & \\ 0 & \to & B' & \to & B & \to & B'' & \to & 0 \end{array}$$

Then $\begin{array}{cc} A' & \xrightarrow{x} A \\ c'a\downarrow & \swarrow c'p_1 a \\ B' & \end{array}$ commutes.

Conversely suppose $\begin{array}{cc} A' & \xrightarrow{x} A \\ a'\downarrow & \swarrow f \\ B' & \end{array}$ commutes.

Then there exists $\begin{array}{cc} A & \to A'' \\ a - yf\downarrow & \swarrow g \\ B & \end{array}$

and

$$\begin{array}{ccccc} A' & \xrightarrow{x} & A & \xrightarrow{x''} & A'' \\ \downarrow a' & & \downarrow a & & \downarrow a'' \\ B' & \xrightarrow[y]{} & B & \xrightarrow[y'']{} & B'' \end{array} = \begin{array}{ccccc} A' & \xrightarrow{x} & A & \xrightarrow{x''} & A'' \\ \downarrow a' & & \downarrow u_1 f + u_2 x'' & & \downarrow 1 \\ B' & \xrightarrow{u_1} & B' \otimes A'' & \xrightarrow{p_1} & A'' \\ \downarrow 1 & & \downarrow y p_1 + g p_2 & & \downarrow a'' \\ B' & \xrightarrow[y]{} & B & \xrightarrow[y'']{} & B'' \end{array} \quad \text{q. e. d.}$$

The naive construction of direct sums in *Shex* $\mathscr{A}/S\mathscr{A}$ works.

Given $\begin{array}{ccccccccc} 0 & \to & A' & \to & A & \to & A'' & \to & 0 \\ & & \downarrow & & \downarrow & & \downarrow & & \\ 0 & \to & B' & \to & B & \to & B'' & \to & 0 \end{array}$ let $\begin{array}{ccc} E & \to & A'' \\ \downarrow & & \downarrow \\ B & \to & B'' \end{array}$

be a pullback and consider the exact diagram
$$\begin{array}{ccccccccc} 0 & \to & A' & \to & A & \to & A'' & \to & 0 \\ & & \downarrow & & \downarrow & & \downarrow 1 & & \\ 0 & \to & B' & \to & \dot{E} & \to & A'' & \to & 0 \\ & & 1\downarrow & & \downarrow & & \downarrow & & \\ 0 & \to & B' & \to & B & \to & B'' & \to & 0 \end{array}$$

The upper left square is a pushout. Indeed the lower right square is a pullback iff the upper left is a pushout. In the light of theorem 3.1 it suffices to prove:

Lemma 3.3.2.

If $\begin{array}{ccc} E & \to & A'' \\ \downarrow & & \downarrow \\ B & \to & B'' \end{array}$ *is a pullback in* $\mathscr{A}$ *then* $\begin{array}{ccccccccc} 0 & \to & B' & \to & E & \to & A'' & \to & 0 \\ & & \downarrow & & \downarrow & & \downarrow & & \\ 0 & \to & B' & \to & B & \to & B'' & \to & 0 \end{array}$

represents a kernel in $Shex\,\mathscr{A}/S\mathscr{A}$.

Proof of Lemma. (3.3.2.1.)

$\begin{array}{ccccc} B' & \to & E & \to & A'' \\ \downarrow & & \downarrow & & \downarrow \\ B' & \to & B & \to & B'' \end{array}$ represents a monomorphism in $Shex\,\mathscr{A}/S\mathscr{A}$

because if $\begin{array}{ccccc} X' & \to & X & \to & X'' \\ \downarrow & & \downarrow & & \downarrow \\ B' & \to & E & \to & A'' \\ \downarrow & & \downarrow & & \downarrow \\ B' & \to & B & \to & B'' \end{array} \equiv 0$ then by lemma 3.3.1

there exists $\begin{array}{ccc} X' & \to & X \\ \downarrow & \swarrow & \\ B' & & \end{array}$ and hence (again by lemma 3. 3.1) $\begin{array}{ccccc} X' & \to & X & \to & X'' \\ \downarrow & & \downarrow & & \downarrow \\ B' & \to & E & \to & A \end{array} \equiv 0$

(3.3.2.2.)

$\begin{array}{ccccc} B' & \to & E & \to & A'' \\ \downarrow & & \downarrow & & \downarrow \\ B' & \to & B & \to & B'' \end{array}$ *represents a weak kernel of* $\begin{array}{ccccccccc} 0 & \to & B' & \to & B & \to & B'' & \to & 0 \\ & & \downarrow & & \downarrow & & \downarrow & & \\ 0 & \to & E & \to & B \otimes A'' & \to & B'' & \to & 0 \end{array}$

because:

(3.3.2.2.1.)

$\begin{array}{ccccc} B' & \to & E & \to & B'' \\ \downarrow & & \downarrow & & \downarrow \\ B' & \to & B & \to & A'' \\ \downarrow & & \downarrow & & \downarrow \\ E & \to & B \otimes A'' & \to & B'' \end{array} \equiv 0$ because $\begin{array}{ccc} B' & \to & E \\ \downarrow & \swarrow & \\ E & & \end{array}$ commutes.

(3.3.2.2.2.)

Given $\begin{array}{ccccccccc} 0 & \to & X' & \to & X & \to & X'' & \to & 0 \\ & & {\scriptstyle x'}\downarrow & & \downarrow & & \downarrow & & \\ 0 & \to & B' & \to & B & \to & B'' & \to & 0 \\ & & {\scriptstyle h'}\downarrow & & \downarrow & & \downarrow & & \\ 0 & \to & E & \to & B \otimes A'' & \to & B'' & \to & 0 \end{array} \equiv 0$

there exists (lemma 3.3.1) $\begin{array}{ccc} X' & \to & X \\ h'x' \downarrow & \swarrow f & \\ E & & \end{array}$. Hence $\begin{array}{ccc} X' & \to & X \\ x' \downarrow & & \downarrow \\ B' & \underset{b}{\to} & E \end{array}$ commutes

and we obtain
$$\begin{array}{ccccc} X' & \to & X & \to & X'' \\ x' \downarrow & & \downarrow f & & \downarrow \\ B' & \to & E & \to & B'' \end{array}$$

Finally

$$\begin{array}{ccccc} X' & \to & X & \to & X'' \\ x' \downarrow & & \downarrow f & & \downarrow \\ B' & \to & E & \to & B'' \\ \downarrow & & \downarrow & & \downarrow \\ B' & \to & B & \to & B'' \end{array} \equiv \begin{array}{ccccc} X' & \to & X & \to & X'' \\ x' \downarrow & & x \downarrow & & x'' \downarrow \\ B' & \to & B & \to & B'' \end{array}$$

because the left hand legs are equal (lemma 3.3.1 once again). ∎

4. Abelian Reflections of Categories with Exactness Conditions

A **category-with-exactness-conditions** is a category $\mathscr{A}$ together with a class E of pairs of maps $\{A_i \to B_i \to C_i\}$. Given two categories $\mathscr{A}, \mathscr{A}'$, exactness conditions E on $\mathscr{A}$, E' on $\mathscr{A}'$ we define a functor $T; \mathscr{A} \to \mathscr{A}'$ to be an **exact** functor if

$$A \to B \to C \in E \Rightarrow TA \to TB \to TC \in E'.$$

Let $\boldsymbol{E}$ be the category of small categories-with-exactness-conditions and natural equivalence classes of exact functors. Define $\boldsymbol{B}$ to be the full subcategory of $\boldsymbol{E}$ of abelian categories where the set of exactness conditions on $\mathscr{B} \in \boldsymbol{B}$ is understood to be the set of exact sequences as usually defined.

Note that the category $\boldsymbol{A}$ of $\oplus$'ive categories and natural equivalence classes of additive functors appears as a full subcategory of $\boldsymbol{E}$. For if we are given an $\oplus$'ive $\mathscr{A}$ we may define $E = \{A \overset{u_1}{\to} A \oplus B \overset{p_1}{\to} B\}$. And of course the category of all small categories and all functors appears as a full subcategory of $\boldsymbol{E}$, simply by understanding the exactness conditions to be vacuous.

Theorem 4.1. *$\boldsymbol{B}$ is a reflective subcategory of $\boldsymbol{E}$.*

That is, given any small category $\mathscr{A}$ with exactness conditions E there exists an abelian category $\mathscr{B}$ and an exact functor $\mathscr{A} \to \mathscr{B}$ such that for every abelian $\mathscr{C}$ and exact $\mathscr{A} \to \mathscr{C}$ there exists, uniquely up to natural *equivalence, exact* $\mathscr{B} \to \mathscr{C}$ *such that* $\begin{array}{ccc} \mathscr{A} & \to & \mathscr{B} \\ & \searrow \ \swarrow & \\ & \mathscr{C} & \end{array}$ *commutes.*

Proof. The general adjoint functor theorems do not help here. $\boldsymbol{B}$ is not closed under the formation of difference kernels.

Given $\mathscr{A}$ with exactness conditions E we may construct a weak reflection in the conventional manner: let $\boldsymbol{S}$ be a representative set of abelian categories of cardinality bounded by $\max\{\aleph_0, \text{cardinality of } \mathscr{A}\}$. Define

$$\hat{\mathscr{A}} = \prod_{\mathscr{B} \in \boldsymbol{S}} \prod_{\boldsymbol{E}(\mathscr{A}, \mathscr{B})} \mathscr{B} \quad \text{and} \quad \mathscr{A} \to \hat{\mathscr{A}}$$

to be the exact functor whose $\langle \mathscr{B}, T\rangle$'th coordinate is $\mathscr{A} \xrightarrow{T} \mathscr{B}$. $\hat{\mathscr{A}}$ is rather easily seen to be a weak reflection, i.e. given any $\mathscr{C} \in \boldsymbol{B}$ and exact $\mathscr{A} \to \mathscr{C}$ we can find $\mathscr{B} \in \boldsymbol{S}$, exact $\mathscr{A} \xrightarrow{T} \mathscr{B}$, $\mathscr{B} \to \mathscr{C}$ such that $\begin{array}{ccc} \mathscr{A} & \to & \mathscr{B} \\ & \searrow \ \swarrow & \\ & \mathscr{C} & \end{array}$ Hence the $\langle \mathscr{B}, T\rangle$'th projection $\hat{\mathscr{A}} \to \mathscr{B}$ yields $\begin{array}{ccc} & & \hat{\mathscr{A}} \\ & \nearrow & \downarrow \\ \mathscr{A} & \to & \mathscr{B} \\ & \searrow & \downarrow \\ & & \mathscr{C} \end{array}$

The reflection of $\mathscr{A}$ will be an exact subcategory of $\hat{\mathscr{A}}$. In a paper in this collection by R. Faber and the writer entittled *Fill-in Theorems* a construction is given for „minimal'' exact subcategories. Let $\mathscr{A}_0$ be the image of $\mathscr{A} \to \hat{\mathscr{A}}$ (we may assume that $\mathscr{A} \to \hat{\mathscr{A}}$ is one-to-one on objects). Let $\mathscr{A}_4$ be the result of that construction, starting with $\mathscr{A}_0$.

$\mathscr{A}_4$ is still a weak reflection. We need only show that for abelian $\mathscr{C}$ and exact functors $T, T' : \mathscr{A}_4 \to \mathscr{C}$ that a natural equivalence

$$\eta_0 : T \,|\, \mathscr{A}_0 \to T' \,|\, \mathscr{A}_0$$

may be extended to $\eta_4 : T \to T'$.

To do so we must recall the construction. Let $\mathscr{A}_1$ be the subcategory of $\hat{\mathscr{A}}$ consisting of all finite sums of maps in $\mathscr{A}_0$ (same objects). It must only be checked that η_0 remains natural when considered as a transformation $\eta_1 : T \,|\, \mathscr{A}_1 \to T' \,|\, \mathscr{A}_1$. It checks easily.

Let $\mathscr{A}_2$ be the subcategory of $\hat{\mathscr{A}}$ consisting of all finite direct sums of $\mathscr{A}_1$-objects and matrices of $\mathscr{A}_1$-maps. There is clearly only one possible extension of η_1 to $\eta_2 : T \,|\, \mathscr{A}_2 \to T' \,|\, \mathscr{A}_2$ and it clearly works.

Let $\mathscr{A}_3$ be the subcategory of $\hat{\mathscr{A}}$ obtained by "adjoining the homology" of $\mathscr{A}_2$-pairs, i.e. given $A' \to A$, $A \to A'' \in \mathscr{A}_2$ define

$$H = \operatorname{Im}[\operatorname{Ker}(A \to A'') \to \operatorname{Cok}(A' \to A)]\,.$$

The maps of $\mathscr{A}_3$ arise as follows: given
$$\begin{array}{ccccc}\bar{A}' & \to & \bar{A} & \to & \bar{A}' \\ \downarrow & & \downarrow & & \downarrow \\ A' & \to & A & \to & A''\end{array}$$

in $\mathscr{A}_2$ there exists a canonical diagram
$$\begin{array}{ccc}\operatorname{Ker}(\bar{A} \to \bar{A}'') & \to & \operatorname{Cok}(\bar{A}' \to \bar{A}) \\ \downarrow & & \downarrow \\ \operatorname{Ker}(A \to A'') & \to & \operatorname{Cok}(A' \to A)\end{array}$$

and hence a canonical map $\bar{H} \to H$. Again there is only one possible extension of η_2 to $\eta_3 : T|\mathscr{A}_3 \to T'|\mathscr{A}_3$. It works. And it remains an equivalence (via the five lemma).

Finally, $\mathscr{A}_4$ is obtained by adjoining the inverses of maps in $\mathscr{A}_3$ that are isomorphisms in $\hat{\mathscr{A}}$. The only thing to check is that η_3 remains natural when considered as a transformation $\eta_4 : T|\mathscr{A}_4 \to T'|\mathscr{A}_4$. And this checks easily.

The fact that $\mathscr{A}_4$ is an exact subcategory of $\hat{\mathscr{A}}$ is proved in the above advertised paper.

5. The Periphery

Let A be the category of small $\oplus$'ive categories and natural equivalence classes of additive functors (otherwise known as cosacanecoaf). We have shown that the subcategory of abelain categories and right exact functors allows reflections only of categories with weak kernels (1.4), dually that the subcategory of abelian categories and left exact functors allows reflections only of categories with weak cokernels. We have shown that the subcategory of abelian categories and exact functors is reflective (4.1). An obvious case remains: the subcategory of abelian categories and additive functors.

Proposition 5.1. *If $\mathscr{A}$ is a small $+$'ive category, $\mathscr{B}$ abelian, and $\mathscr{A} \to \mathscr{B}$ is such that for every abelian $\mathscr{C}$ and functor $\mathscr{A} \to \mathscr{C}$ there exists commutative*
$$\begin{array}{ccc}\mathscr{A} & \to & \mathscr{B} \\ & \searrow \swarrow & \\ & \mathscr{C} & \end{array}$$

where $\mathscr{B} \to \mathscr{C}$ is unique up to natural equivalence, then $\mathscr{A} \to \mathscr{B}$ is a full embedding and every object in $\mathscr{B}$ is a direct summand of a finite sum of objects coming from $\mathscr{A}$. In particular if $\mathscr{A}$ is amenable then $\mathscr{A} \to \mathscr{B}$ is an equivalence.

Proof. $\mathscr{A} \to \mathscr{B}$ is an embedding because there exists at least one embedding of $\mathscr{A}$ into an abelian category, namely the representation functor $\mathscr{A} \to (\mathscr{A}^*, \mathscr{G})$. The category $(\mathscr{B}^*, \mathscr{G})$ is a complete abelian category and hence there exists commutative

$$\begin{array}{ccc} \mathscr{A} & \to & (\mathscr{A}^*, \mathscr{G}) \\ \downarrow & & \downarrow \\ \mathscr{B} & \to & (\mathscr{B}^*, \mathscr{G}) \end{array}$$

Thus the functor $\mathscr{B} \to (\mathscr{A}^*, \mathscr{G})$ which extends $\mathscr{A} \to (\mathscr{A}^*, \mathscr{G})$ when followed by $(\mathscr{A}^*, \mathscr{G}) \to (\mathscr{B}^*, \mathscr{G})$ must be equivalent to the embedding $\mathscr{B} \to (\mathscr{B}^*, \mathscr{G})$, and hence $\mathscr{B} \to (\mathscr{A}^*, \mathscr{G})$ is an embedding. Because $\mathscr{A} \to \to \mathscr{B} \to (\mathscr{A}^*, \mathscr{G})$ is full it follows that $\mathscr{A} \to \mathscr{B}$ is full.

Henceforth we consider $\mathscr{A}$ to be a subcategory of $\mathscr{B}$. Fix $B \in \mathscr{B}$. For $B' \in \mathscr{B}$ define $KB' \subset (B', B)$ by $f \in KB'$ iff $\exists\, A_1, \ldots, A_n \in \mathscr{A}$ and maps $B' \xrightarrow{g} A_1 \oplus \cdots \oplus A_n \xrightarrow{h} B = f$.

KB' is easily seen to describe an additive subfunctor of $(-, B)$. Consider the exact sequence $0 \to K \to (-, B) \to F \to 0$. For $A \in \mathscr{A}$, $FA = 0$. By the uniqueness condition on $\mathscr{B} \to \mathscr{G}$ in commutative triangles

$$\begin{array}{ccc} \mathscr{A} & \to & \mathscr{B} \\ & \searrow \swarrow & \\ & \mathscr{G} & \end{array}$$

it then follows that $F = 0$ everywhere and hence that $FB = 0$ and $1_B \in K_B$ which is precisely the conclusion we have been trying to prove.

Department of Mathematics
University of Pennsylvania
Philadelphia, Pennsylvania

Stable Homotopy*

By

PETER FREYD

1. Basic Definitions and Theorems

Let $\mathscr{T}$ be a good category of topological spaces with base points. That is, $\mathscr{T}$ shall be any one of the following categories (we'll specialize later):

The category of finite simplicial complexes and simplicial maps.

The category of finite cell complexes and cellular maps.

The category of finite dimensional CW-complexes and cellular maps.

The category of all CW-complexes and cellular maps.

The spheres live in each of these categories, and the definition of homotopic maps can be stated in each. Thus for each we obtain a quotient category $\mathscr{H}$, where $\mathscr{H}(X, Y)$ is the set of homotopy classes of maps in $\mathscr{T}(X, Y)$. For spheres S^n and S^m, $\mathscr{H}(S^n, S^m)$ is the same regardless of the particular chosen category. In the simplicial setting we can most easily see that $\mathscr{H}(S^n, S^m)$ is trivial for $n < m$.

For $X, Y \in \mathscr{T}$ we construct the **wedge**, $X \vee Y$ as

$$X \times \{y_0\} \cup \{x_0\} \times Y$$

where x_0 and y_0 are the base points. $X \vee Y$ is the sum of X and Y in $\mathscr{T}$ and $\mathscr{H}$.

The **smash** $X \wedge Y$ is constructed as $X \times Y / X \vee Y$, that is, the cartesian product with the wedge collapsed to the base point. Smashing is a functor $\mathscr{T} \times \mathscr{T} \to \mathscr{T}$ and it is compatable with homotopy, hence becomes a functor $\mathscr{H} \times \mathscr{H} \to \mathscr{H}$.

The **suspension**, SX is defined as the smash, $S^1 \wedge X$ where S^1 is the 1-sphere. Suspending is a functor $\mathscr{T} \to \mathscr{T}$ and becomes a functor mod homotopy $\mathscr{H} \to \mathscr{H}$.

The suspension of a sphere is a sphere.

For any integer n we define the **stable homotopy**,

$$\mathscr{S}_n(X, Y) = \lim_{j \to \infty} \mathscr{H}(S^{n+j} X, S^j Y).$$

Even for negative n the limit makes sense (start with $j = -n$).

* Received September 5, 1965.

The original motivation for considering $\mathcal{S}_n(S^1, S^1) = \mathcal{S}_0(S^{n+1}, S^1)$ was that $\mathcal{S}_n(S^1, S^1) = \mathcal{H}(S^{k+n}, S^k)$ for $k > n$ and $\mathcal{S}_n$ was deemed to be a "first approximation" of $\mathcal{H}$. And so it is. The orders of the sets $\mathcal{S}_n(S^1, S^1)$ have since appeared to be worthy of study on their own — they have had a tendency to appear in otherwise explicit formulae. For example, Kervaire and Milnor have expressed the order of the set of distinct differential structures on a sphere with formulae involving the Bernoulli numbers, a function that has $\{1, 2\}$ as values, and the order of $\mathcal{S}_n(S^1, S^1)$.

We shall here construct an abelian category in which stable homotopy lives very comfortably and examine some of the consequences. Because this appears in a volume officially devoted to algebra (though *categorical* algebra and hence all of mathematics that can be shown to be trivial)[1] a few concepts and facts in topology shall be briefly reviewed.

The **cone** of X, Cone X is defined as $I \wedge X$ where I is the unit interval with base point chosen as $0 \in [0, 1]$. The usual map $X \to \text{Cone } X$ may be written $\{0, 1\} \wedge X \to I \wedge X$.

$X \xrightarrow{f} Y$ is nullhomotopic iff there exists $\text{Cone } X \xrightarrow{H} Y$ such that $X \to \text{Cone } X \xrightarrow{H} Y = X \xrightarrow{f} Y$.

Given any $X \xrightarrow{f} Y \in \mathcal{T}$ define the **mapping cone** of f, Cone(f) to be the pushout (in $\mathcal{T}$)

$$\begin{array}{ccc} X & \rightarrow & Y \\ \downarrow & & \downarrow \\ \text{Cone } X & \rightarrow & \text{Cone}(f) \end{array}$$

Cone(f) is a **weak cokernel** of f in $\mathcal{H}$. That is, $X \to Y \to \text{Cone}(f)$ is nullhomotopic and, for any $Y \xrightarrow{g} Z$ such that $X \to Y \to Z$ is nullhomotopic, there exists (possibly many) $\text{Cone}(f) \to Z$ such that

$$\begin{array}{ccc} Y & \rightarrow & \text{Cone}(f) \\ & \searrow & \swarrow \\ & Z & \end{array}$$

commutes.

Puppe made the fundamental observation that for any $X \xrightarrow{f} Y$ there exists a diagram

$$\begin{array}{ccccccccccccc} X & \xrightarrow{f} & Y & \xrightarrow{f'} & \text{Cone}(f) & \xrightarrow{f''} & \text{Cone}(f') & \xrightarrow{f'''} & \text{Cone}(f'') & \xrightarrow{f''''} & \text{Cone}(f''') & \rightarrow & \cdots \\ 1\downarrow & & 1\downarrow & & 1\downarrow & & \downarrow & & \downarrow & & \downarrow & & \\ X & \xrightarrow[f]{} & Y & \xrightarrow[f']{} & \text{Cone}(f) & \xrightarrow[g]{} & SX & \xrightarrow[Sf]{} & SY & \xrightarrow[Sf']{} & S\,\text{Cone}(f) & \xrightarrow[Sg]{} & \cdots \end{array}$$

[1] Perhaps the purpose of categorical algebra is to show that which is trivial is trivially trivial.

in which each square either commutes or anti-commutes in $\mathscr{H}$ and each vertical map is an isomorphism in $\mathscr{H}$. The notion of *anti-commute* may be founded on the definition:

$$SX \xrightarrow{-1} SX = S^1 \wedge X \xrightarrow{(-1)\wedge 1} S^1 \wedge X .$$

Hence the sequence

$$X \xrightarrow{f} Y \to \text{Cone}\, f \to SX \to SY \to S\,\text{Cone}\,(f) \to S^2 X \to \cdots$$

has the property that each map is a weak cokernel in $\mathscr{H}$ of the map on its left.

There exist functors $\{H_n : \mathscr{T} \to \mathscr{G}\}_n$, the **homology functors,** with the following properties:

H_n respects homotopy

(Exactness) $H_n X \to H_n Y \to H_n \text{Cone}\,(f)$ *is exact.*

(Connecting) $H_n S$ is naturally equivalent to H_{n-1}.

(Coefficient) $H_n S^j = \begin{cases} \mathbf{Z} & \text{if } n = j \\ 0 & \text{if } n \neq j. \end{cases}$

(Hurewicz) *The natural transformation*

$$\mathscr{H}(S^n, X) \to \mathscr{G}(H_n S^n, H_n X) \to H_n X$$

is such that for $n > 1$ *if* $\mathscr{H}(S^j, X) = 0$ *all* $j < n$ *then* $\mathscr{H}(S^n, X) \to H_n(X)$ *is an isomorphism.*

For any X, Y the mapping cone of $X \xrightarrow{0} SY$ is $SX \vee SY$ and the exactness property yields

$$H_j(SY \vee SX) \cong H_j(SY) \oplus H_j(SX).$$

The connecting property can be used to remove the S's.

If X is a finite complex of dimension n we may inductively assume that for its $(n-1)$-skeleton X^{n-1}, $\sum_j H_j(X^{n-1})$ is finitely generated. The mapping cone of the inclusion $X^{n-1} \to X^n$ is homotopic to a wedge of n-spheres and hence the exactness and the coefficient properties reveal $\sum_j H_j(X)$ as an extension of finitely generated groups, hence finitely generated.

For any X, SX is connected i.e. $\mathscr{H}(S^1, X) = 0$. For connected X, SX is simply connected, i.e. $\mathscr{H}(S^1, SX) = 0$ (this can be proven geometrically rather easily). If X is n-connected i.e. if $\mathscr{H}(S^j, X) = 0$, $j \leqq n$, then from Hurewicz we obtain that SX is $(n+1)$-connected and from purely geometric considerations we can see that

$$\dim SX = \dim X + 1 .$$

The Freudenthal Theorem 1.1. *If* $\dim X = n$ *and* Y *is* j*-connected and* $n \geqq 2j$ *then* $\mathscr{H}(X, Y) \to \mathscr{H}(SX, SY)$ *is an isomorphism.* ▮

Corollary 1.2. *For finite dimensional X the sequence*

$$\mathscr{H}(X,Y) \to \mathscr{H}(SX,SY) \to \mathscr{H}(S^2X,S^2Y) \to \cdots$$

eventually stabilizes to isomorphisms. In particular the stable set $\mathscr{S}_0(X,Y)$ is isomorphic to $\mathscr{H}(S^nX,S^nY)$ for $n > \dim X$. ▮

The Serre Theorem 1.3. *$\mathscr{H}(S^n,S^m)$ is finite except when $n=m$ and when $m=2q$, $n=2q+1-1$.* ▮

Freudenthal and Serre combine to give

Corollary 1.4. *$\mathscr{S}_0(S^n,S^m)$ is finite except when $n=m$, and*

$$\mathscr{S}_0(S^n,S^n) = Z\,. ▮$$

This corollary is the most important ingredient in all that follows. We shall use it and its consequence not only in almost every theorem but in almost every line.

The Freudenthal theorem will be directly used for stable homotopy in only one other question. Its chief use is to provide a use for stable homotopy.

Note that the Hurewicz property suffices to prove that for j-connected X, $j>1$, $\mathscr{S}_0(S^{j+1},X) \cong \mathscr{H}(S^{j+1},X)$.

Proposition 1.5. *If X is a cell or simplicial complex and $\mathscr{S}_0(S^n,X)=0$ all n, then $X=0$ in $\mathscr{H}$.*

Proof. We wish to show that $X \to \operatorname{Cone} X$ is a retraction. By the remark above $\mathscr{S}_0(S^n,X)=0$ all n implies $\mathscr{H}(S^n,X)=0$ all n. The identity map $X \xrightarrow{1} X$ is extended piece by piece over the skeletons of $\operatorname{Cone}(X)$, by extending piece by piece over each cell, each little extension provided for by the hypothesis. ▮

Corollary 1.6. *If X is a simply connected cell or simplicial complex and $H_nX=0$ all n then $X=0$ in $\mathscr{H}$.* ▮

2. The Stable Categories

We first define a graded category $G\mathscr{S}$. Its objects are the objects of $\mathscr{T}$, its maps of degree n are triples $\langle X,Y,f\rangle$, where X and Y are objects of $\mathscr{T}$ and $f \in \mathscr{S}_n(X,Y) = \lim_{j\to\infty}(S^{n+j}X, S^jY)$.

Let $\mathscr{S}$ be the category whose objects are pairs $\langle X,n\rangle$, $X \in \mathscr{T}$, $n \in Z$, and define $\mathscr{S}(\langle X,n\rangle, \langle Y,m\rangle) = G\mathscr{S}_{n-m}(X,Y)$.

Define $S': \mathscr{S} \to \mathscr{S}$ by $\langle X,n\rangle \mapsto \langle X,n+1\rangle$. Clearly S' is an automorphism.

If $S: \mathscr{S} \to \mathscr{S}$ is defined by $\langle X,n\rangle \mapsto \langle SX,n\rangle$ then S and S' are naturally equivalent, and when we pass to the category of categories and natural equivalence classes of functions (COCANECOF), S itself is seen to be an automorphism.

The passage from $\mathscr{H}$ to $\mathscr{S}$ can be described as that which is necessary to make suspensions an automorphism. To wit:

Proposition 2.1. *The functor $\mathscr{H} \to \mathscr{S} : X \mapsto \langle X, 0\rangle$ yields a commutative diagram*

$$\begin{array}{ccc} \mathscr{H} & \xrightarrow{S} & \mathscr{H} \\ \downarrow & & \downarrow \\ \mathscr{S} & \xrightarrow[S]{} & \mathscr{S} \end{array}$$

and for every category $\mathscr{A}$ with automorphism $S : \mathscr{A} \to \mathscr{A}$ and functor $\mathscr{H} \to \mathscr{A}$ such that

$$\begin{array}{ccc} \mathscr{H} & \xrightarrow{S} & \mathscr{H} \\ \downarrow & & \downarrow \\ \mathscr{A} & \xrightarrow[S]{} & \mathscr{A} \end{array}$$

commutes there exists a unique $\mathscr{S} \to \mathscr{A}$ such that

$$\begin{array}{ccc} \mathscr{S} & \xrightarrow{S} & \mathscr{S} \\ \downarrow & & \downarrow \\ \mathscr{A} & \xrightarrow[S]{} & \mathscr{A} \end{array}$$

and

$$\begin{array}{ccc} \mathscr{H} & \to & \mathscr{S} \\ & \searrow & \downarrow \\ & & \mathscr{A} \end{array}$$

commute. (All statements are modulo the identification of functors by natural equivalence.) ▮

Given a space $X \in \mathscr{T}$ we can refer to its **desuspension**, $S^{-1}X$ in $\mathscr{S}$. If $\mathscr{T}$ is chosen large enough then $S^{-1}X$ is isomorphic in $\mathscr{S}$ to a space Y. We are most interested, however, in specializing to the case that $\mathscr{T}$ is the category of finite complexes. Whereas we have adjoined many abstract entities it shall be noticed that for any discussion involving only a finite number of objects in $\mathscr{S}$ the automorphism S may be applied a finite number of times to move all the objects into the image of $\mathscr{T} \to \mathscr{S}$.

Proposition 2.2. *$Y \to \mathrm{Cone}(f)$ is a weak cokernel of $X \xrightarrow{f} Y$ in $\mathscr{S}$ and $X \xrightarrow{f} Y$ is a weak kernel of $Y \to \mathrm{Cone}(f)$.*

Proof. We have already observed that $X \to Y \to \text{Cone}(f)$ is a null-homotopic. Suppose $g \in \mathcal{S}(Y, X)$ is such that $X \to Y \to Z = 0$ in $\mathcal{S}$. g is represented in $\mathcal{T}$ by $S^n Y \xrightarrow{\bar{g}} S^n Z$ some n (in particular $S^n Z$ is in the image of $\mathcal{T} \to \mathcal{S}$). And for some $m > 0$,

$$S^{m+n} X \xrightarrow{S^{m+n} f} S^{m+n} Y \xrightarrow{S^m \bar{g}} S^{m+n} Z$$

is nullhomotopic. Because $S^{m+n}\,\text{Cone}\, f$ is a weak cokernel in $\mathcal{H}$ of $S^{m+n} f$ there exists $h : S^{m+n}\,\text{Cone}(f) \to S^{m+n} Z$ such that

$$\begin{array}{ccc} S^{m+n} Y & \to & S^{m+n}\,\text{Cone}(f) \\ & \searrow & \swarrow h \\ S^m(\bar{g}) & & S^{m+n} Z \end{array}$$

commutes. h represents a map in $\mathcal{S}(\text{Cone}(f), Z)$.

For the other side, suppose $Z \xrightarrow{g} Y \in \mathcal{S}$ is such that $Z \xrightarrow{g} Y \to \text{Cone}(f) = 0$ in $\mathcal{S}$. Let $\bar{g} : S^n Z \to S^n Y$ in $\mathcal{T}$ represent g. For some $m > 0$,

$$S^{n+m} Z \xrightarrow{S^m \bar{g}} S^{n+m} Y \to S^{n+m}\,\text{Cone}(f)$$

is nullhomotopic.

We obtain

$$\begin{array}{ccccccccc} S^{m+n} Z & \xrightarrow{S^m \bar{g}} & S^{m+n} Y & \to & \text{Cone}(S^m \bar{g}) & \to & S^{m+n+1} Z & \xrightarrow{S^{m+1} \bar{g}} & S^{m+n+1} Y \\ 1\downarrow & & & & \downarrow & & h\downarrow & & 1\downarrow \\ S^{m+n} Z & \longrightarrow & S^{m+n} Y & \to & S^{m+n}\,\text{Cone}(f) & \to & S^{m+n+1} X & \to & S^{m+n+1} Y \end{array}$$

commuting (or anti-commuting) in $\mathcal{H}$. The fact that $1_{S^{m+n+1} Y}$ can reappear in the right-hand side is geometric. Thus $S^{m+1}(\bar{g})$ factors back through $S^{m+n+1}(f)$. But S is an automorphism, hence $\bar{g}$ factors through $S^n(f)$, hence $S^{-n}(\bar{g}) = g$ factors through f. ∎

Given $X \xrightarrow{f} Y \in \mathcal{S}$ let n be large enough so that $S^n X \to S^n Y$ is in the image of $\mathcal{T} \to \mathcal{S}$ and define $\text{Cone}(f) = S^{-n}\,\text{Cone}\, S^n f$. Because S is an automorphism and weak kernels and cokernels are categorically describable, $Y \to \text{Cone}(f)$ is a weak cokernel of $X \to Y$ and $X \to Y$ is a weak kernel of $Y \to \text{Cone}(f)$. Note that in the sequence

$$X \xrightarrow{f} Y \to \text{Cone}(f) \to SX \xrightarrow{Sf} SY \to S\,\text{Cone}(f) \to \cdots$$

every map is a weak cokernel of a predecessor and a weak kernel of its successor.

Proposition 2.3. *Every map in $\mathcal{S}$ has a weak kernel, a weak cokernel and is a weak kernel and a weak cokernel.*

Proof. All mentioned facts lie in the sequence

$$S^{-1}\,\mathrm{Cone}(f) \to X \xrightarrow{f} Y \to \mathrm{Cone}(f)\cdot \quad \blacksquare$$

$\mathscr{S}$ is an additive category and the wedge operation yields the direct sum. A number of things should be verified, almost all of them functorially. Very briefly: if the category of spaces $\mathscr{T}$ is chosen large enough then $S : \mathscr{T} \to \mathscr{T}$ has a right adjoint $\Omega : \mathscr{T} \to \mathscr{T}$ (ΩX is the space of maps from S^1 to X). S and Ω remain adjoint when viewed as functors $\mathscr{H} \to \mathscr{H}$.

The functor S has a co-group structure in $\mathscr{H}$, $S \xrightarrow{\delta} S \vee S$. Formally therefore Ω has a group structure $\Omega \times \Omega \xrightarrow{\sigma} \Omega$. $(S^2 X, Y)$ has, a priore, three group structures. One from $S\delta_X$, one from δ_{SX}, one from the isomorphism $(S^2 X, Y) \cong (SX, \Omega Y)$ and σ_Y. Because the two structures on $(SX, \Omega Y)$, one from δ_X the other from σ_Y are, a la ECKMANN-HILTON, the same (and abelian), and because the isomorphism $(S^2 X, Y) \cong (SX, \Omega Y)$ carries δ_{SX} multiplication into σ_Y multiplication by formal adjoint functor considerations, and finally because the δ_X multiplication on $(SX, \Omega Y)$ must be carried back to $S\delta_X$ multiplication on $(S^2 X, Y)$ again by formal reasons it is seen that the three group structures on $(S^2 X, Y)$ are the same and abelian. In particular $(SX, SY) \to (S^2 X, S^2 Y)$ is a group homomorphism. Thus $\mathscr{S}(X, Y)$ has a natural group structure and $\mathscr{S}$ becomes an additive category.

3. The Abelian Category for Stable Homotopy

Let $\mathscr{S}$ be any $\oplus$'ive category with weak kernels and weak cokernels and in which every map is a weak kernel and a weak cokernel. Let $\mathscr{F}$ be the category whose objects are maps in $\mathscr{S}$: $(A' \xrightarrow{f} A)$ is an object in $\mathscr{F}$ for $f \in \mathscr{S}(A', A)$. Given two objects $(A' \to A)$, $(B' \to B)$ the maps from one to the next are represented by squares

$$\begin{array}{c}(A' \to A)\\ \downarrow \quad \downarrow\\ (B' \to B)\end{array}$$

subject to the following identification:

$$\begin{array}{c}(A' \to A)\\ f'\downarrow \quad \downarrow f\\ (B' \to B)\end{array} \equiv \begin{array}{c}(A' \to A)\\ g'\downarrow \quad \downarrow g\\ (B' \to B)\end{array}$$

iff $A' \to A \xrightarrow{f} B = A' \to A \xrightarrow{g} B$
(iff $A' \xrightarrow{f'} B' \to B = A' \xrightarrow{g'} B' \to B$).

Theorem 3.1. *$\mathscr{F}$ is a Frobenius Category, that is an Abelian Category with enough projectives and injectives and in which projectives and injectives coincide.*

The functor $\mathscr{S} \to \mathscr{F}$, $A \mapsto (A' \xrightarrow{1} A)$ is a full embedding. Its values are projective and injective and every object in $\mathscr{F}$ may be resolved both on the left and right by values of $\mathscr{S} \to \mathscr{F}$.

For every Abelian Category $\mathscr{A}$ and additive functor $\mathscr{S} \to \mathscr{A}$ there exist functors $R, M, L: \mathscr{F} \to \mathscr{A}$ each of which yields a commutative diagram:

$$\begin{array}{ccc} \mathscr{S} & \to & \mathscr{F} \\ & \searrow \ \swarrow & \\ & \mathscr{A} & \end{array}$$

R is right-exact, L is left-exact, and M preserves images. R, M, L and are each unique up to natural equivalence.

$X \to Y \to Z$ in $\mathscr{S}$ is sent to an exact sequence in $\mathscr{F}$ iff $Y \to Z$ is a weak cokernel of $X \to Y$ in $\mathscr{S}$ iff $X \to Y$ is a weak kernel of $Y \to Z$ in $\mathscr{S}$.

Proof. In another paper in this collection we give an overly sophisticated proof of the above theorem. Here we tackle it with elementary methods. Note that the assumptions on $\mathscr{S}$ and the definition of $\mathscr{F}$ are self-dual. We shall automatically use the dual of any proven lemma.

Lemma 3.1.1. *Given squares* $\begin{array}{ccc} (A' & \to & A) \\ f' \downarrow & & \downarrow f \\ (B' & \to & B) \end{array}$ *and* $\begin{array}{ccc} (A' & \to & A) \\ g' \downarrow & & \downarrow g \\ (B' & \to & B) \end{array}$ *if either $f' = g'$ of $f = g$ then the squares represent the same map in $\mathscr{F}$.*

Lemma 3.1.2. $\begin{array}{ccc} (A' & \to & A) \\ \downarrow & & \downarrow 1 \\ (B' & \to & A) \end{array}$ *represents a monomorphism in $\mathscr{F}$.*

Proof of lemma. Let $\begin{array}{ccc} (X' & \to & X) \\ \downarrow & & \downarrow \\ (A' & \to & A) \\ \downarrow & & \downarrow 1 \\ (B' & \to & A) \end{array}$ represent the zero-map. Then $X' \to X \to A = 0$ and $\begin{array}{ccc} (X' & \to & X) \\ \downarrow & & \downarrow \\ (A' & \to & A) \end{array}$ represents the zero map. q.e.d.

Given any square $\begin{array}{ccc}(A' & \to & A)\\ \downarrow & & \downarrow\\ (B' & \to & B)\end{array}$ we may factor the map it represents as

$$\begin{array}{ccc}(A' & \to & A)\\ {\scriptstyle 1\cdot}\downarrow & & \downarrow\\ (A' & \to & B)\\ \downarrow & & \downarrow{\scriptstyle 1}\\ (B' & \to & B)\end{array}$$

that is, by an epimorphism followed by a monomorphism.

Lemma 3.1.3. *Given* $\begin{array}{ccc}(A' & \to & A)\\ \downarrow & & \downarrow\\ (B' & \to & B)\end{array}$ *let $K \to A'$ be a weak kernel of $A' \to B$.*

Then $\begin{array}{ccc}(K & \to & A)\\ \downarrow & & \downarrow\\ (A' & \to & A)\end{array}$ represents a kernel in $\mathscr{F}$ of $\begin{array}{ccc}(A' & - & A)\\ \downarrow & & \downarrow\\ (B' & - & B)\end{array}$

Proof of lemma. Clearly $\begin{array}{ccc}(K & \to & A')\\ \downarrow & & \downarrow\\ (A' & \to & A)\\ \downarrow & & \downarrow\\ (B' & \to & B)\end{array} \equiv 0$. Suppose that $\begin{array}{ccc}(X' & \to & X)\\ \downarrow & & \downarrow\\ (A' & \to & A)\\ \downarrow & & \downarrow\\ (B' & \to & B)\end{array} \equiv 0$

Then $X' \to A' \to B = 0$ and there exists $\begin{array}{ccc} & X' & \\ \swarrow & & \searrow\\ K & \to & A'\end{array}$ and $\begin{array}{ccc}(X' & \to & X)\\ \downarrow & & \downarrow\\ (K & \to & A)\end{array}$ is easily seen to factor $\begin{array}{ccc}(X' & \to & X)\\ \downarrow & & \downarrow\\ (A' & \to & A)\end{array}$ through $\begin{array}{ccc}(K & \to & A')\\ \downarrow & & \downarrow\\ (A' & \to & A)\end{array}$. Lemma 2 said that $\begin{array}{ccc}(K & \to & A)\\ \downarrow & & \downarrow\\ (A' & \to & A)\end{array}$ represents a monomorphism and hence the uniqueness condition is automatic.

Lemma 3.1.4. *Any square of the form* $\begin{array}{ccc}(A' & \to & A)\\ \downarrow & & \downarrow{\scriptstyle 1}\\ (B' & \to & A)\end{array}$ *represents a kernel in $\mathscr{F}$.*

Proof of lemma. By assumptions on $\mathscr{S}$, $A' \to A$ is a weak kernel, say of $A \to F$. We shall show that $\begin{array}{c}(A' \to A)\\ \downarrow \quad \downarrow \\ (B' \to A)\end{array}$ is a kernel of $\begin{array}{c}(B' \to A)\\ \downarrow \quad \downarrow \\ (B' \to F)\end{array}$ in $\mathscr{F}$.

The composition of the two maps is easily seen to represent a zero map.

Suppose $\begin{array}{c}(X' \to X)\\ \downarrow \quad \downarrow \\ (B' \to A)\\ \downarrow \quad \downarrow \\ (B' \to F)\end{array} \equiv 0$ then $X' \to A \to F = 0$ and there exists $\begin{array}{c} X' \\ \swarrow \quad \searrow \\ A' \to A \end{array}$

and hence $\begin{array}{c}(X' \to X)\\ \downarrow \quad \downarrow \\ (A' \to A)\\ \downarrow \quad \downarrow' \\ (B' \to A)\end{array}$ represents the same map as the given $\begin{array}{c}(X' \to X)\\ \downarrow \quad \downarrow \\ (B' \to A)\end{array}$ because their right hand legs are the same (Lemma 3.1.1). **End of lemma proof.**

Now let $\begin{array}{c}(A' \to A)\\ \downarrow \quad \downarrow \\ (B' \to B)\end{array}$ be a monomorphism. We wish to show that it is a kernel. We may factor it: $\begin{array}{c}(A' \to A)\\ \downarrow \quad \downarrow \\ (A' \to B)\\ \downarrow \quad \downarrow \\ (B' \to B)\end{array}$ By the dual of lemma 3.1.4, $\begin{array}{c}(A' \to A)\\ {\scriptstyle 1}\downarrow \quad \downarrow \\ (A' \to B)\end{array}$ represents a cokernel. But it is a monomorphism. Hence it is an isomorphism! Thus $\begin{array}{c}(A' \to A)\\ \downarrow \quad \downarrow \\ (B' \to B)\end{array}$ is a kernel of whatever $\begin{array}{c}(A' \to B)\\ \downarrow \quad \downarrow {\scriptstyle 1} \\ (B' \to B)\end{array}$ is a kernel, and lemma 3.1.4 said the latter is a kernel.

We will know therefore that $\mathscr{F}$ is abelian once we know that it has direct sums. But $(A' \to A) \oplus (B' \to B) \cong (A' \oplus A \to B' \oplus B)$ without pain and $\mathscr{F}$ is abelian.

Lemma 3.1.5. *$(B \to B)$ is projective, and dually, injective, in $\mathcal{F}$.*

Proof. Let $\begin{matrix}(A' \to A)\\ \downarrow \quad \downarrow\\ (B \to B)\end{matrix}$ be an epimorphism. We wish to split it.

$A' \to B$ is an $\mathcal{S}$-weak kernel of $B \to F$, some $F \in \mathcal{S}$. Hence

$$\begin{matrix}(A' \to A) & \\ \downarrow \quad \downarrow & \\ (B \to B) & \equiv 0\\ \downarrow \quad \downarrow & \\ (B \to F) & \end{matrix}$$

and $B \to B \to F = 0$. Thus there exists $\begin{matrix} & B & \\ \swarrow & & \searrow \\ A' & \to & B\end{matrix}$ and $\begin{matrix}(B \to B)\\ \downarrow \quad \downarrow\\ (A' \to A)\end{matrix}$. The composition $\begin{matrix}(B \to B)\\ \downarrow \quad \downarrow\\ (A' \to A)\\ \downarrow \quad \downarrow\\ (B \to B)\end{matrix}$ is easily seen to be the identity because its left-hand leg is the identity (Lemma 3.1.1). **End of lemma proof.**

For any $(A' \to A)$ we have an epimorphism $\begin{matrix}(A' \xrightarrow{1} A')\\ \downarrow \quad \downarrow\\ (A' \to A)\end{matrix}$ and $(A' \to A)$ may be resolved by values of $\mathcal{S} \to \mathcal{F}$. $\mathcal{S} \to \mathcal{F}$ is clearly a full embedding.

Given $\mathcal{S} \xrightarrow{T} \mathcal{A}$, $\mathcal{A}$ abelian, define $R(A' \to A) = \operatorname{Cok}(T(K) \to T(A'))$ where $K \to A'$ is a weak kernel of $A' \to A$.
$M(A' \to A) = Im(TA' \to TA)$
$L(A' \to A) = Ker\ (TA \to TF)$ where $A \to F$ is a weak kernel of $A' \to A$. ∎

A more geometric description is possible for $\mathcal{F}$ where $\mathcal{T}$ is a category of CW or simplicial complexes.

Theorem 3.2. *Let $\mathcal{T}$ be a category of CW complexes. Let $\overline{\mathcal{F}}$ be the category whose objects are triples $\langle X', X, m\rangle$, $X' \subset X \in \mathcal{T}$, $m \in Z$. The maps from $\langle X'\ X, m\rangle$ to $\langle Y', Y, n\rangle$ are represented by pairs $\langle f, j\rangle\ j \geq m, n$, $f{:}S^{n+j}X \to S^{m+j}Y \in \mathcal{T}$ such that $f(S^{n+j}X') \subset f(S^{m+j}Y')$ subject to the identification $\langle f, j\rangle \equiv \langle f,' j'\rangle$ iff $\exists\, k \geq j, j'$ and a homotopy between the* *maps*

$S^{n+k}X' \to S^{n+k}X \xrightarrow{S^{n+k-j}f} S^{n+k}Y$ *and* $S^{n+k}X' \to S^{n+k}X \xrightarrow{S^{n+k-j}f'} S^{m+k}Y$.

Then $\overline{\mathscr{F}}$ *is an abelian category. The functor* $\mathscr{S} \to \overline{\mathscr{F}}: X \to \langle X, X, 0\rangle$ *is a full embedding, and its values are both projective and injective. Every object of* $\overline{\mathscr{F}}$ *may be resolved on both the right and left by values of* $\mathscr{S} \to \mathscr{F}$. *The* $\overline{\mathscr{F}}$ *of this theorem is equivalent to the* $\mathscr{F}$ *of the last theorem.*

Proof. Let $\overline{\mathscr{F}} \to \mathscr{F}$ be the obvious functor $\langle X', X, m\rangle \mapsto (\langle X', m\rangle \to \langle X, m\rangle)$. It is rather easily seen to be an embedding.

$\overline{\mathscr{F}} \to \mathscr{F}$ is full. Suppose $X' \subset X$, $Y' \subset Y$ are given as are integers m, n. Consider then a map represented by an $\mathscr{S}$-square

$$\begin{array}{ccc}(\langle X', m\rangle & \to & \langle X, m\rangle) \\ f'\downarrow & & \downarrow f \\ (\langle Y', n\rangle & \to & \langle Y, n\rangle)\end{array}$$

where f' is represented by $\bar{f}' \in \mathscr{T}(S^{m+j}X, S^{n+j}Y')$ and f by

$$\bar{f} \in \mathscr{T}(S^{m+k}X, S^{n+k}Y).$$

The square commutes in $\mathscr{S}$, hence there exists $l \geq j, k$ such that $S^{m+l}X' \xrightarrow{S^{l-j}\bar{f}'} S^{n+l}Y' \to S^{n+l}Y$ is homotopic to

$$S^{m+l}X' \to S^{m+l}X \xrightarrow{S^{l-k}\bar{f}} S^{n+l}Y.$$

The homotopy extension theorem implies the existence of a map $h: S^{n+l}X \to S^{m+l}Y$ homotopic to $S^{l-k}\bar{f}$ such that $h \mid S^{m+l}X' = S^{l-j}\bar{f}'$. Hence the given map is equivalent to that which comes from $\langle h, l\rangle \in \overline{\mathscr{F}}$.

It thus suffices to show that every object in $\mathscr{F}$ is isomorphic to a value of $\overline{\mathscr{F}} \to \mathscr{F}$.

Let $(\langle X, m\rangle \xrightarrow{f} \langle Y, n\rangle) \in \mathscr{F}$ where f is represented by

$$\bar{f} \in \mathscr{T}(S^{m+j}X \to S^{n+j}Y).$$

The object

$$(\langle X, m\rangle \xrightarrow{f} \langle Y, n\rangle)$$

is isomorphic to

$$(\langle S^{m+j}X, -j\rangle \xrightarrow{\bar{f}} \langle S^{n+j}Y, -j\rangle).$$

Hence every object in $\mathscr{F}$ is isomorphic to an object of the form $(\langle X, m\rangle \xrightarrow{f} \langle Y, m\rangle)$ where f is represented in $\mathscr{T}(X, Y)$.

For such an object let $\begin{array}{ccc} X & \xrightarrow{f} & Y \\ u_0\downarrow & & \downarrow \\ I \wedge X & \to & P\end{array}$ be a pushout in $\mathscr{T}$. $X \xrightarrow{u_0} I \wedge X \to P$ is homotopic to the injection $X \xrightarrow{u_1} I \wedge X \to P$ and $Y \to P$ is a homotopy equivalence. Hence every object in $\mathscr{F}$ is isomorphic to one of the form

$(\langle X', m\rangle \xrightarrow{f} \langle X, m\rangle)$ where $X' \subset X$ and f is the inclusion, i.e. a value of $\overline{\mathscr{F}} \to \mathscr{F}$. (The above P is the mapping cylinder of f). ∎

4. Topological Review

With aid of the abelian category $\mathscr{F}$ we prove a number of standard theorems for the stable homotopy category $\mathscr{S}$. Most of them are true for the homotopy category $\mathscr{H}$ if we restrict attention to simply connected spaces, their proofs, however, often being much longer.

Let $\mathscr{G}^{\mathbf{Z}}$ be the category of graded abelian groups and degree zero maps. In other words, $\mathscr{G}^{\mathbf{Z}}$ is the cartesian power of $\mathscr{G}$ indexed over $\mathbf{Z}$. For $X \in \mathscr{H}$ let $H(X)$ be the total homology, i.e. the graded group whose n'th component is $H_n(X)$. Let $S: \mathscr{G}^{\mathbf{Z}} \to \mathscr{G}^{\mathbf{Z}}$ be the shifting automorphism. Then the connecting property on the H_n's gives a commutative diagram

$$\begin{array}{ccc} \mathscr{H} & \xrightarrow{S} & \mathscr{H} \\ H\downarrow & & \downarrow H \\ \mathscr{G}^{\mathbf{Z}} & \xrightarrow[S]{} & \mathscr{G}^{\mathbf{Z}} \end{array}$$

and by proposition *2.1* there exists a unique $H: \mathscr{S} \to \mathscr{G}^{\mathbf{Z}}$ still compatable with shifts and such that $\begin{array}{ccc} \mathscr{H} & \to & \mathscr{S} \\ & \searrow \ \swarrow & \\ & \mathscr{G}^{\mathbf{Z}} & \end{array}$ commutes.

We define $H_n: \mathscr{S} \to \mathscr{G}$ by $H_n: \mathscr{S} \xrightarrow{H} \mathscr{G}^{\mathbf{Z}} \xrightarrow{p_n} \mathscr{G}$. Every one of the properties listed for $H_n: \mathscr{H} \to \mathscr{G}$ holds for $H_n: \mathscr{S} \to \mathscr{G}$. (Note that $H_n\langle X, m\rangle = H_{n+m} X$.) H_n carries finite wedges into direct sums because $X \xrightarrow{0} SY \to SY \vee SX \to SX \xrightarrow{0} S^2 Y$ is a mapping cone sequence. Hence H_n is an additive functor and by Theorem 3.1 H_n extends to a left-exact functor on $\mathscr{F}$. It is in fact exact. Indeed the three canonical extension coincide. We have need of the general lemma:

Lemma 4.1. *If $\mathscr{A}$ is an abelian category $T: \mathscr{S} \to \mathscr{A}$ a functor that carries mapping cone sequences into exact sequences then there is an exact $\mathscr{F} \to \mathscr{A}$, unique up to natural equivalence, such that* $\begin{array}{ccc} \mathscr{S} & \to & \mathscr{F} \\ & \searrow & \downarrow \\ & & \mathscr{A} \end{array}$ *commutes.*

Proof. Let $T: \mathscr{F} \to \mathscr{A}$ be a left-exact extension. We shall show that it preserves epimorphism. Given exact $A \to B \to 0$ in $\mathscr{F}$ let $X \in \mathscr{S}$ and $X \to A \to 0$ be exact. It clearly suffices to show that $TX \to TB \to 0$ is exact.

Let $Y \in \mathcal{S}$ and $B \to Y$ mono, and let $Y \to Z$ be the mapping cone of $X \to Y$. Then
$$\begin{array}{ccccccc} & & TX & \to & TY & \to & TZ \\ & & \downarrow & & \downarrow & & \downarrow \\ 0 & \to & TB & \to & TY & \to & TZ \end{array}$$
is an exact diagram. It follows that $TX \to TB$ is epi. ▮

The converse, that if T extends to an exact functor then it carries mapping cone sequences into exact sequences, follows easily from Theorem 3.1. Such functors, $\mathcal{S} \to \mathcal{A}$ should be called exact. In fact they are called half-exact, which seems just to be a mistake. (For $X \subset Y$, Y/X is $\mathcal{H}$-isomorphic to the mapping cone of $X \to Y$. $0 \to X \to Y \to Y/X \to 0$ is exact in $\mathcal{T}$. Hence the mistake.) Almost certainly the term "half-exact" will be replaced by "exact" in this context. We shall avoid both terms, except of course for functors between abelian categories.

Theorem (the Stable Whitehead Theorem) 4.2. *Let X, Y be CW-complexes, $f: X \to Y$ a map such that $H_n(f)$ is an isomorphism, all n (or such that $\mathcal{S}(S,^n f)$ is an isomorphism, all n). Then f is an isomorphism in $\mathcal{S}$.*

Proof. Let $X \to Y \to Z \to SX \to SY$ be a mapping cone sequence. $H_n(Z) = 0$, all n ($\mathcal{S}(S^n, Z) = 0$, all n). Hence $Z = 0$ by Proposition 1.5 (1.6) and $0 \to SX \to SY \to 0$ is exact in $\mathcal{F}$. ▮

Corollary 4.3. *If X is a CW-complex such that $H_j(X) = 0$ all $j \neq n$, $H_n(X) = Z \oplus \cdots \oplus Z$ then X is isomorphic (in $\mathcal{S}$) to $S^n \vee \ldots \vee S^n$.*

Proof. $H_n(X) \cong \mathcal{S}(S^n, X)$. Let $\alpha_1, \ldots, \alpha_m$ be a basis for (S^n, X). Then for $S^n \vee \ldots \vee S^n \xrightarrow{\alpha} X$ it is clear that $H_j(\alpha)$ is an isomorphism, all j. ▮

For positive integers a, n, Cone$(S^a \xrightarrow{n} S^a)$ is an example of a **Moore Space.** From the mapping cone sequence $S^a \xrightarrow{n} S^a \to C_n^a \to S^{a+1} \xrightarrow{n} S^{a+1}$ we easily see that $H_j(C_n^a) = 0$ $j \neq a$, $H_a(C_n^a) = Z_n$.

Corollary 4.4. *For n a prime power, C_n^a is wedge-indecomposable, that is it can not be decomposed as the wedge of two non-trivial spaces.*

Proof. If $C_n^a = A_1 \vee A_2$ then $H_a(C_n^a) = H_a(A_1) \oplus H_a(A_2)$. But surely then for $i = 1$ or 2, $H_a(A_i) = 0$ hence $H_j(A_1) = 0$ all j and $A_i = 0$. ▮

(We shall not need, but should record the fact that for a CW-complex X such that $H_j(X) = 0$ all $j \neq a$ $H_a(X) = Z_n$, $n > 0$ then there exists an isomorphism $C_n^a \to X$).

Two lemmas which are most expediently proved simultaneously:

Lemma 4.5. *Let X be a CW-complex. X is $\mathcal{S}$-isomorphic to a finite simplicial complex iff X is $\mathcal{S}$-isomorphic to finite cell complex iff $H(X) = \sum_n H_n(X)$ is finitely generated.*

Lemma (the 1st stable Dold lemma) 4.6. *Let $\mathscr{A}$ be an abelian category, $T: \mathscr{S} \to \mathscr{A}$ a functor that carries mapping cone sequences into exact sequences. Let $\mathbf{C}$ be a class of objects in $\mathscr{A}$ closed under the formation of kernels, cokernels, and exact extensions* ($A_1, A_2, A_4, A_5 \in \mathbf{C}$ and $A_1 \to A_2 \to A_3 \to A_4 \to A_5$ *exact implies* $A_3 \in \mathbf{C}$).

If $TS^n \in \mathbf{C}$ all n, then for all $X \in \mathscr{S}$ such that $HX = \sum_n H_n X$ is a finitely generated abelian group (e.g. if X is a finite complex) then $TX \in \mathbf{C}$.

Proof. Let $\boldsymbol{A}$ be the class of non-trivial spaces such that $X \in \boldsymbol{A}$ iff

$H(X)$ is finitely generated
$H_j(X) = 0$ all $j > 0$
$H_0(X)$ is free (perhaps trivial).

We shall show for all $X \in \boldsymbol{A}$ that $T(S^n X) \in \mathbf{C}$ all n and that X is $\mathscr{S}$-equivalent to a finite simplicial complex. For any X, $H(X)$ finitely generated implies that $H_j(X) = 0$ large j, and there exists n such that $S^n X \in \boldsymbol{A}$. Thus the theorem will be proved.

For $X \in \boldsymbol{A}$ let $c(X)$ be the smallest integer j, such that $H_j(X) \neq 0$. Of course $c(X) \leq 0$. The induction is based on $-c(X)$. For $-c(X) = 0$ we have the situation of proposition 4.3 and X is a wedge of 0-spheres. Hence $TX \in \mathbf{C}$ and X is $\mathscr{S}$-equivalent to a finite simplicial complex.

For $-c(X) > 0$, note first that $(S^c, X) \cong H_c(X)$ is finitely generated. Let W be a wedge of c-spheres and $W \to X$ a map such that $(S^c, W) \to (S^c, X)$ is onto. Let $W \to X \to Y \to SW \to SY$ be a mapping cone sequence. Then for $j > 0$ $0 = H_j X \to H_j Y \to H_j SW = 0$ and $H_j Y = 0$, For $j = 0$, $0 \to H_0(X) \to H_0 Y \to H_0 SW \to 0$ is exact and $H_0 Y$ is an extension of free groups, hence free. Thus $Y \in \boldsymbol{A}$.

For $j < c$, $0 \to H_j Y \to 0$ is exact and for $j = c$, $H_c W \to H_c X$ is onto, $H_c SW = 0$ and hence $H_c Y = 0$. Thus $c(Y) > c(X)$, $-c(Y) < -c(X)$, the induction holds: we may assume that $T(S^n Y) \in \mathbf{C}$, all n and that Y is a finite complex.

For arbitrary n, $S^{n-1} Y \to S^n W \to S^n X \to S^n Y \to S^{n+1} W$ is a mapping cone sequence, and hence

$$T(S^{n-1} Y) \to T(S^n W) \to T(S^n X) \to T(S^n Y) \to T(S^{n+1} W)$$

is exact. Clearly $T(S^n W) \in \mathbf{C}$, the induction tells us that $T(S^{n-1} Y) \in \mathbf{C}$, thus $T(S^n X) \in \mathbf{C}$. From the same sequence with $n = 1$ SX is revealed as the mapping cone of finite complexes. ∎

Corollary 4.7. *$\mathscr{S}(X, Y)$ is finitely generated for finite complexes.*

Proof. Let $\mathbf{C}$ be the class of finitely generated abelian groups in $\mathscr{G}$.

First, for each n, $(S^n, -): \mathscr{S} \to \mathscr{G}$ carries mapping cone sequences into exact sequences, $(S^n, S^a) \in \mathbf{C}$ all a, (Serre) and hence by the 1st Dold lemma, $(S^n, Y) \in \mathbf{C}$, all finite complexes Y.

Next, $(-, Y): \mathscr{S} \to \mathscr{G}$ carries mapping cone sequence into exact sequences and we have just shown that $(S^n, Y) \in \boldsymbol{C}$ all n, hence $(X, Y) \in \boldsymbol{C}$ any finite complex X. ∎

The not so special case of this corollary, namely that the ring of endomorphisms $End(X)$ is finitely generated as an abelian group for finite complex X, plays a ubiquitous role in much that follows. (It is not-so-special because (X, Y) is a subgroup of $End(X \vee Y)$.)

Corollary (the Stable Co-Whitehead Theorem) 4.8. *For finite complex Z, if $(Z, S^n) = 0$ all n then $Z = 0$ (in $\mathscr{S}$).*

For finite complexes X, Y and map $f: X \to Y$ if (f, S^n) is an isomorphism all n then f is an isomorphism (in $\mathscr{S}$).

Proof. Let $\boldsymbol{C} = \{0\} \subset \mathscr{G}$. Then $(Z, -): \mathscr{S} \to \mathscr{G}$ carries mapping cone sequences into exact sequences. Hence if $(Z, S^n) \in \boldsymbol{C}$ all n, then $(Z, Z) \in \boldsymbol{C}$ and $Z = 0$.

If $X \xrightarrow{f} Y$ is such that (f, S^n) is an isomorphism all n, then consider the mapping cone sequence $X \to Y \to Z \to SX \to SY$. It follows that $(Z, S^n) = 0$ all n, and $0 \to X \to Y \to 0$ is exact in $\mathscr{F}$. ∎

Lemma (the 2nd stable Dold lemma) 4.9. *Let $\mathscr{A}$ be abelian, $T_1, T_2: \mathscr{S} \to \mathscr{A}$ functors which carry mapping cone sequences into exact sequences. If $\eta: T_1 \to T_2$ is a natural transformation such that η_{S^n} is an isomorphism, all n, then η_X is an isomorphism for all finite complexes X.*

Proof. Let $\mathscr{A}^{\to}$ be the category whose objects are maps in $\mathscr{A}$ and whose maps are squares in $\mathscr{A}$. Let $T: S \to \mathscr{A}^{\to}: X \mapsto (T, X \xrightarrow{\eta} T_2 X)$. Let $\boldsymbol{C}$ be the class of objects in $\mathscr{A}^{\to}$ such that $(A_1 \to A_2) \in \boldsymbol{C}$ iff $A_1 \to A_2$ is an isomorphism. The classical five lemma is equivalent to the fact that $\boldsymbol{C}$ is closed under the neccessary operations. Hence if η_{S^n} is an isomorphism then $TS^n \in \boldsymbol{C}$ and the first Dold lemma implies that $TX \in \boldsymbol{C}$ any finite complex X. ∎

Let Q denote the rational numbers.

Corollary 4.10. *For finite complexes X*

$$\mathscr{S}(S^n, X) \otimes Q \cong H_n(X) \otimes Q.$$

Proof. Both $\mathscr{S} \xrightarrow{(S^n, -)} \mathscr{G} \xrightarrow{-\otimes Q} \mathscr{G}$ and $\mathscr{S} \xrightarrow{H_n} \mathscr{G} \xrightarrow{-\otimes Q} \mathscr{G}$ carry mapping cone sequences into exact sequences (because $-\otimes Q$ is exact). And both (S^n, S^a) and $H_n S^a$ are finite for $n \neq a$ (Serre again), and $(S^n, S^n) \to H_n(S^n)$ is an isomorphism. Hence $(S^n, S^a) \otimes Q \to H_n S^a \otimes Q$ is an isomorphism all a and the second Dold lemma applies.

(Indeed, X can be any CW-complex. Both $(S^n, -)$ and H_n carry infinite wedges into sums, $- \otimes Q$ carries sums into sums. The $(n+1)$-

skeleton of X can be used instead of X and the proof of the Dold lemmas can be modified to cover this case.) ∎

Let G be the graded group whose n'th component, G_n is $\mathcal{S}(S^n, S^0)$. We introduce a multiplication on G as follows: given $\alpha \in G_a$, $\beta \in G^b$ then $\beta\alpha = \beta \cdot S^b\alpha \in G_{a+b}$ and G becomes a graded ring.

Lemma 4.(11). $\beta\alpha = \beta \wedge \alpha$. i.e.

$$S^{b+a} \xrightarrow{S^b\alpha} S^b \xrightarrow{\beta} S^0 = S^b \wedge S^a \xrightarrow{1 \wedge \alpha} S^b \wedge S^0 \xrightarrow{\beta \wedge 1} S^0 \wedge S^0 . \quad ∎$$

Consider the endomorphism on $S^1 \wedge S^1$ which twists the components. It clearly is an involution and $S^1 \wedge S^1 = S^2$, hence the endomorphism in question is either $+1$ or -1. Direct geometric inspection reveals that it is -1. For any permutation π on $\{1, \dots n\}$ the induced endomorphism on the n-fold smash $S^1 \dots \wedge S^1$ is thus equal to the sign of π. In particular the map from $S^a \wedge S^b$ to $S^b \wedge S^a$ that reverses the two components is $(-1)^{ab}$. Hence $\beta \wedge \alpha = (-1)^{ab} \beta \wedge \alpha$ and:

Proposition 4.12. *G is commutative in the graded ring sense.* ∎

G might be commutative in the ungraded sense. At this writing all known examples for $\alpha \in G_a$, $\beta \in G_b$, a and b odd are such that $\beta\alpha$ is of order two and $\alpha\beta = \beta\alpha$.

There exist contravariant functors from $\mathcal{S}$ to $\mathcal{G}$ with all the properties listed for the homology functors. They are called, of course, the cohomology functors. The Hopf-classification theorem is used to dualize the Hurewicz theorem: $\mathcal{S}(X, S^n) \to \mathcal{S}(H^n(S^n), H^n(X)) \to H^n(X)$ is an isomorphism if $\mathcal{S}(X, S^j) = 0$ all $j > n$. Whenever we prove a theorem for $\mathcal{S}$ using homology and $\mathcal{S}(S^n, -)$ it is the case that the dual theorem can be proved by using cohomology and $\mathcal{S}(-, S^n)$. Twice we shall wish to do just this.

Such a duality principal strongly suggests a full duality on $\mathcal{S}$ itself. We shall never need it but feel obliged to state it:

Theorem (Spanier-Whitehead duality) 4.123. *There exists a contravariant functor $D : \mathcal{S} \to \mathcal{S}$ such that D^2 is naturally equivalent to the identity. $D(S^n) = S^{-n}$, D carries mapping cone sequences into mapping cone sequences and $DH_n = H^n$ all.*

Moreover $\mathcal{S}(X \wedge Y, Z)$ is naturally equivalent to $\mathcal{S}(X, DY \wedge Z)$. ∎

The proof is based not just on a knowledge of homology and cohomology sepearately, but on the operations involving both.

If we write $X \otimes Y$ for $X \vee Y$ and $\mathcal{H}om(Y, Z)$ for $DY \wedge Z$ we obtain

$$\mathcal{S}(X \otimes Y, Z) \cong \mathcal{S}(X, \mathcal{H}om(Y, Z))$$

$$S^0 \otimes Y \cong Y$$

$$\mathcal{S}(S^0, \mathcal{H}om(Y, Z)) \cong \mathcal{S}(Y, Z)$$

and $\mathcal{S}$ is revealed as a closed category in the Eilenberg-Kelly sense.

The unique contravariant exact functor $D:\mathscr{F}\to\mathscr{F}$ that extends $D:\mathscr{S}\to\mathscr{S}$ is still such that D^2 is naturally equivalent to the identity. If we extend $X\wedge -:\mathscr{S}\to\mathscr{F}$ to a right exact functor $X\otimes -:\mathscr{F}\to\mathscr{F}$ and then extend $\mathscr{S}\to(\mathscr{F},\mathscr{F}):X\to X\otimes -$ to a right-exact functor we obtain $\otimes:\mathscr{F}\times\mathscr{F}\to\mathscr{F}$, right exact in both variables.

We extend $(DX)\wedge -:\mathscr{S}\to\mathscr{F}$ to a left exact functor

$$\mathscr{H}om(X\wedge -):\mathscr{F}\to\mathscr{F}\quad\text{and}$$
$$\mathscr{S}^*\to(\mathscr{F},\mathscr{F}):X\to\mathscr{H}om(X,-)$$

to a left exact functor we obtain $\mathscr{H}om:\mathscr{F}^*\times\mathscr{F}\to\mathscr{F}$ left-exact in both variables and $\mathscr{F}$ becomes a closed category in the Eilenberg-Kelly sense.

5. All Projectives in $\mathscr{F}$ come from $\mathscr{S}$

We shall consider $\mathscr{S}$ to be a subcategory of $\mathscr{F}$. We wish to show that it is the full subcategory of projectives.

Let $P\in\mathscr{F}$ be projective. Because the objects in $\mathscr{S}$ resolve any object in $\mathscr{F}$ we obtain $X\in\mathscr{S}$ and an epimorphism $X\to P$. Because P is projective we obtain $P\to X\to P=1$, and hence $X\to P\to X=e$ is an idempotent. If e splits in $\mathscr{S}$, i.e. if there exists $Y\in\mathscr{S}$ and maps $X\to Y\to X=e$, $Y\to X\to Y=1$ then $P\cong Y$.

In a paper entitled *Splitting Homotopy Idempotents*, in this collection, we show that if a category $\mathscr{S}$ has countable co-powers and if retracts in $\mathscr{S}$ have complements, then idempotenst split in $\mathscr{S}$.

Retracts do have complements in $\mathscr{S}$. Suppose $A\to B\to A=1$, $A,B\in\mathscr{S}$. Let $B\to C$ be the mapping cone of $A\to B$. Then $A\to B\to\to C\to SA\to SB$ is exact in $\mathscr{F}$. But $SA\to SB$ is a monomorphism (it splits) and $C\to SA=0$. Thus $0\to A\to B\to C\to 0$ is exact and $B=A\oplus C$.

Proposition 5.1. *If $\mathscr{S}$ is the stable category obtained from finite dimensional CW-complexes then $\mathscr{S}$ has countable co-powers.*

Corollary 5.2. *Idempotents split in $\mathscr{S}$ and $\mathscr{S}$ is the full category of projectives in $\mathscr{F}$.*

Proof. Let X be a CW-complex of dimension n and let $\vee X$ be the wedge of a countable collection of copies of X. Then

$$\mathscr{S}(\vee X, Y)\cong\mathscr{H}(S^{n+1}\vee X, S^{n+1}Y)$$

by the Freudenthal theorem and
$\mathscr{H}(S^{n+1}\vee X, S^{n+1}Y)\cong\mathscr{H}(\vee S^{n+1}X.\, S^{n+1}Y)\cong\Pi\mathscr{H}(S^{n+1}X, S^{n+1}Y)$.

The corollary now follows from the above mentioned work on splitting idempotents. To partially translate that work to this setting let $X \xrightarrow{e} X$ be an idempotent and let $\vee X \to Y$ be the mapping cone of

$$\begin{array}{ccccccc} X & \vee & X & \vee & X & \vee & \cdots \\ \downarrow & \searrow & \downarrow & \searrow & \downarrow & \searrow & \\ X & \vee & X & \vee & X & \vee & \cdots \end{array}$$

where the vertical maps are equal to $1 - e$ and the diagonal maps equal to e. Then Y splits e. ∎

Theorem 5.3. *If $\mathscr{S}$ is the stable category obtained from finite cell complexes, then idempotents split in $\mathscr{S}$ and hence $\mathscr{S}$ is the full subcategory of projectives in $\mathscr{F}$.*

Proof. Given a finite complex X and an idempotent $X \xrightarrow{e} X$ the last proposition asserted the existence of a finite dimensional CW-complex Y and maps $X \to Y$, $Y \to X$ such that $X \to Y \to X = e$, $Y \to X \to Y = 1$. The construction above did not yield a finite complex. But for each n, $H_n Y \to H_n X \to H_n Y = 1$ and hence $\Sigma H_n Y$ is finitely generated. Thus by Lemma 4.5, Y is isomorphic (in $\mathscr{S}$) to a finite complex.

A note on infinite sums and wedges:

We used the fact that for a sequence of CW complexes $\{X_i\}$ of bounded dimension the infinite wedge of the X_i's is a categorical sum in $\mathscr{S}$. When the dimensions are unbounded the wedge fails as a sum in $\mathscr{S}$, indeed the natural map $\mathscr{S}(\vee X_i, Y) \to \Pi \mathscr{S}(X_i, Y)$ need be neither one-one nor onto. The counter-examples rest on the fact that the Freudenthal theorem is not much improvable, that is, for each n there exist spaces (finite complexes, if you will) X, Y and a map $X \xrightarrow{f} Y \in \mathscr{T}$ such that $S^n(f) \neq 0 (\mathscr{H})$ but $S^j(f) = 0$ large j, and there exist spaces X, Y such that $\mathscr{H}(S^n X, S^n Y) \to \mathscr{S}(X, Y)$. is not onto.

To see that $\mathscr{S} \vee (X_i, Y) \to \Pi \mathscr{S}(X_i, Y)$ need not be an injection choose for each i, a map $f_i \colon X_i \to Y_i$ such that $S^i(f_i) \neq 0$ but $S^j(f_i) = 0$ large j. Then for $f \colon \vee X_i \to \vee Y_i$ where $f u_i = f_i$ it is the case that $\mathscr{S}^i(f) \neq 0$ all i but $f u_i$ is stably trivial all i.

To see that $\mathscr{S}(\vee X_i, Y) \to \Pi \mathscr{S}(X_i, Y)$ need not be onto choose for each i spaces X_i, Y_i and an element $f_i \in \mathscr{S}(X_i, Y_i)$ not in the image of $\mathscr{H}(S^i X, S^i Y) \to \mathscr{S}(X, Y)$. Then for $f \in \Pi \mathscr{S}(X_i, Y_i)$ such that $p_i f = f_i$, it is the case that f is not in the image of

$$\mathscr{S}(\vee X_i, \vee Y_i) \to \Pi \mathscr{S}(X_i, \vee Y_i).$$

There is a foreboding consequence, foreboding because it is rather common practice to represent $\mathscr{S}$ in more complete categories (of spectra, et al.)

Proposition 5.4. *If $\mathscr{S}$ is the stable category containg at least the countable CW-complexes and if $\mathscr{S} \xrightarrow{T} \mathscr{A}$ carries countable wedges into sums, then T is not an embedding. And if for finite complexes X,*

$$\mathscr{S}(X, Y) \to \mathscr{A}(TX, TY)$$

is an injection, then T is not full.

Proof. Just consider the commutative diagram

$$\begin{array}{ccc} \mathscr{S}(\vee X_i, Y) & \longrightarrow & \Pi\, \mathscr{S}(X_i, Y) \\ \downarrow & & \downarrow \\ \mathscr{A}(T(\vee X_i), TY) \to (\mathscr{A}(\Sigma\, T Y_i, T Y) & \to & \Pi\,(T X_i, T X). \end{array}$$ ∎

6. Wedge Decomposition

From now on $\mathscr{S}$ shall be the stable category obtained from finite cell complexes (or equivalently, simplicial complexes), with the exception of the very last section of the paper. We have just seen that idempotents split in $\mathscr{S}$. Hence a space X is **wedge indecomposable,** i.e. can not be expressed as the wedge of two non-trivial spaces iff the ring of endomorphisms of X has no proper idempotents (all statements are in the stable sense).

Given any $A \in \mathscr{F}$ we may factor $A \cong A_1 \oplus \cdots \oplus A_n$ where each A_i is indecomposable. To do so it is only necessary to see that there is an upper bound on the number n such that there exists an isomorphism $A \cong A_1 \oplus \cdots \oplus A_n$, $A_i \neq 0$. But clearly if we express the abelian group of endomorphisms of $A_1 \oplus \cdots \oplus A_n$ as a direct sum of indecomposable cyclic groups we obtain at least n such cyclic groups. Hence n is bounded by the number of indecomposable cyclic groups needed to express the abelian group of endomorphisms of A.

If $A \in \mathscr{S} \subset \mathscr{F}$ and $A = A_1 \oplus \cdots \oplus A_n$ then each $A_1 \in \mathscr{S}$ because idempotents split in $\mathscr{S}$ and $A \cong A_1 \vee \ldots \vee A_n$.

It is natural to ask whether wedge decomposition is unique. The answer is no. Before we exhibit a counter example it will be usefull to have:

Proposition 6.1. *Let $C = \mathrm{Cone}(S_n \xrightarrow{\alpha} S^0)$, $n > 0$ and let $m = \mathrm{ord}(\alpha)$. Define $J = \{\langle a, b\rangle \mid a, b \in Z,\ a \equiv b \,(\mathrm{mod}\ m)\}$. J is a ring under coordinatewise addition and multiplication.*

There exists a ring homomorphism from End(C) *onto J, the kernel of which has trivial multiplication.*

Proof. Let $S^n \xrightarrow{\alpha} S^0 \to C \to S^{n+1} \to S^1$ be a mapping cone sequence. Define $f\colon \mathrm{End}(C) \to \mathrm{End}(S^0) \times \mathrm{End}(S^{n+1})$ to be the additive correspondence $\langle x, a, b\rangle \in f$ if there exists commutative

$$\begin{array}{ccccc} S^0 & \to & C & \to & S^{n+1} \\ a\downarrow & & \downarrow x & & \downarrow b \\ S^0 & \to & C & \to & S^{n+1} \end{array}$$

f is everywhere defined because for all $x \in \operatorname{End}(C)$,

$$S^0 \to C \xrightarrow{x} C \to S^{n+1} = 0.$$

(Keep in mind that the mapping cone sequence is exact in $\mathscr{F}$, and its objects are both projective and injective.)

f is well defined because if

$$\begin{array}{ccccc} S^0 & \to & C & \to & S^{n+1} \\ a\downarrow & & 0\downarrow & & b\downarrow \\ S^0 & \to & C & \to & S^{n+1} \end{array}$$

commutes there exists $\begin{array}{ccc} & & S^0 \\ & \swarrow & \downarrow a \\ S^n & \underset{\alpha}{\to} & S^0 \end{array}$ but $(S^0, S^n) = 0$ and $a = 0$. Similarly $b = 0$. Hence $f\colon \operatorname{End}(C) \to \operatorname{End}(S^0) \times \operatorname{End}(S^{n+1})$ is a function, and rather clearly a ring homomorphism.

The image of f is seen to be the set of pairs $\{\langle a, b\rangle \mid a \equiv b \,(\mathrm{mod}\, m)\}$, as follows: Note first that there exists x such that

$$\begin{array}{ccccc} S^0 & \to & C & \to & S^{n+1} \\ m\downarrow & & \downarrow x & & \downarrow 0 \\ S^0 & \to & C & \to & S^{n+1} \end{array}$$

commutes because

$$\begin{array}{ccc} S^n & \xrightarrow{\alpha} & S^0 \\ & 0\searrow & \downarrow m \\ & & S^0 \end{array}$$

and hence there exists $\begin{array}{ccc} S^0 & \to & C \\ m\downarrow & \swarrow & \\ S^0 & & \end{array}$. Let $C \xrightarrow{x} C = C \to S^0 \to C$.

$f(1) = \langle 1, 1\rangle$ and thus if $a \equiv b \,(\mathrm{mod}\, m)$ let $cm = b - a$ and then $f(b - cx) = \langle a, b\rangle$.

For the other inclusion suppose first that $\langle a, 0\rangle = f(x)$. Then there exists $S^0 \to C \to S^0 = a$ (f is well defined) and hence

$$\begin{array}{ccc} S^0 & \to & C \\ a\downarrow & \swarrow & \\ S^0 & & \end{array} \quad \text{and} \quad \begin{array}{ccc} S^n & \xrightarrow{\alpha} & S^0 \\ & 0\searrow & \downarrow a \\ & & S^0 \end{array}$$

and $m|a$. Now suppose that $f(x) = \langle a, b\rangle$ then $f(x - b) = \langle a - b, 0\rangle$ and $m|(a - b)$.

To find $\mathrm{Ker}(f)$ suppose that

$$\begin{array}{ccccc} S^0 & \to & C & \to & S^{n+1} \\ 0\downarrow & & x\downarrow & & \downarrow 0 \\ S^0 & \to & C & \to & S^{n+1} \end{array}$$

There exists $\begin{array}{ccc} C & \to & S^{n+1} \\ x\downarrow & \swarrow & \\ C & & \end{array}$ and $S^{n+1} \to C \to S^{n+1} = 0$ (f is well defined).

Thus there exists

$$\begin{array}{ccc} & C & \to S^{n+1} \\ \swarrow & \downarrow & \swarrow \\ S^0 \to & C & \end{array}$$

i.e. The image of the natural map $\mathcal{S}(S^{n+1}, S^0) \to \mathrm{End}(C)$ is $\mathrm{Ker}(f)$ and the product of any two elements in $\mathrm{Ker}(f)$ is zero.

(More simple minded diagram chasing yields an exact sequence

$$(S^1, S^0) \xrightarrow{(\alpha, S^0)} (S^{n+1}, S^0) \to \mathrm{End}(C) \to Z \times Z \to Z/mZ \to 0.)$$

Corollary 6.2. *For* $\alpha \neq 0$, $n > 0$ $\mathrm{Cone}(S^n \xrightarrow{\alpha} S^0)$ *is wedge-indecomposable.*

Proof. We show that $\mathrm{Cone}(\alpha)$ has no proper idempotents. Let e be an idempotent. Then $f(e) \in J$ is an idempotent. J has no proper idempotents, hence $f(e) = 1$ or 0 and either $1 - e$ or e is in $\mathrm{Ker}(f)$. But $(\mathrm{Ker}(f))^2 = 0$ hence $e = 1, 0$. ∎

Curiosity 6.3. *Let* $\alpha \in (S^m, S^0)$ *and* $\beta \in (S^n, S^0)$ *be of relatively prime order. Then there exists a space* X *and an isomorphism*

$$S^0 \vee X \cong \mathrm{Cone}(\alpha) \vee \mathrm{Cone}(\beta)$$

where. $\mathrm{Cone}(\alpha)$, $\mathrm{Cone}(\beta)$, *and* S^0 *are wedge indecomposable and neither* $\mathrm{Cone}(\alpha)$ *nor* $\mathrm{Cone}(\beta)$ *is a sphere. Hence the existence of relatively prime orders among the maps between spheres implies that wedge-decomposition is not unique.*

Proof. Let $a = \mathrm{ord}(\alpha)$, $b = \mathrm{ord}(\beta)$. There exist

$$\begin{array}{ccc} S^0 & \xrightarrow{f} & \mathrm{Cone}(\alpha) \\ & a\searrow & \downarrow g \\ & & S^0 \end{array} \quad \text{and} \quad \begin{array}{ccc} S^0 & \xrightarrow{f'} & \mathrm{Cone}(\beta) \\ & b\searrow & \downarrow g' \\ & & S^0 \end{array}$$

Let $c, d \in Z$ be such that $ca + db = 1$. Then

$$S^0 \xrightarrow{cu_1 f + du_2 f'} \mathrm{Cone}(\alpha) \vee \mathrm{Cone}(\beta) \xrightarrow{agp_1 + bg'p_2} S^0 = 1$$

and take X to be the mapping cone of $S^0 \to \mathrm{Cone}(\alpha) \vee \mathrm{Cone}(\beta)$. ▮

Unique decomposition does not hold. We do not have a counterexample for the failure of cancellation.* But we do have:

Proposition 6.4. *A retract of a wedge of spheres is a wedge of spheres.*

Proof. Let e be an idempotent on a wedge of spheres W. We give an inductive proof, the induction based on the number of dimensions that occur in the wedge W. If only one dimension appears, then e is an idempotent matrix of integers and there exists an automorphism f such that $f^{-1} e f$ is diagonal, all coordinates being either 0 or 1.

$$\mathrm{Im}(e) \cong \mathrm{Im}(f^{-1} e f)$$

and the image of such a diagonal matrix is clearly a wedge of spheres.

If more than one dimension appears we may express W as $W_1 \oplus W_2$, each W_i a wedge of spheres, but the dimensions appearing in W_1 less than those in W_2. Hence $(W_1, W_2) = 0$.

If we express e in matrix notation.

$$e = \begin{pmatrix} e_{11}, & e_{12} \\ 0\ , & e_{22} \end{pmatrix}$$

where $e_{ij} \in (W_j, W_i)$ then

$$\begin{pmatrix} e_{11}, & e_{12} \\ 0\ , & e_{22} \end{pmatrix} = \begin{pmatrix} e_{11}^2, & e_{11} e_{12} + e_{12} e_{22} \\ 0\ , & e_{22}^2 \end{pmatrix},$$

e_{11} and e_{22} are idempotents. Let

$$f = \begin{pmatrix} 1, & -e_{12} \\ 0, & 1 \end{pmatrix}.$$

f is an automorphism,

$$\mathrm{Im}(e) \cong \mathrm{Im}(f e f) = \mathrm{Im}\begin{pmatrix} e_{11} & 0 \\ 0 & e_{22} \end{pmatrix} \cong \mathrm{Im}(e_{11}) \oplus \mathrm{Im}(e_{22})$$

and the induction yields the desired result. ▮

Define a **torsion** space $X \in \mathscr{S}$ to be any space whose identity map is of finite order. X is torsion iff $\mathrm{End}(X)$ is torsion iff $\mathrm{End}(X)$ is finite. Let $\mathscr{T}or$ be the full subcategory of $\mathscr{S}$ of torsion spaces.

Lemma 6.5. *If* $0 \to A \to X \to B \to 0$ *is exact and* $m 1_A = 0$, $n 1_B = 0$ *then* $mn 1_X = 0$.

* Added in Proof: There is an example for the failure of cancellation.

Proof. Consider

$$\begin{array}{ccccccccc}
0 & \to & A & \to & X & \to & B & \to & 0 \\
 & & n\downarrow & \swarrow n & \downarrow & & \downarrow 0 & & \\
0 & \to & A & \to & X & \to & B & \to & 0 \\
 & & 0\downarrow & & m\downarrow & \swarrow & \downarrow m & & \\
0 & \to & A & \to & X & \to & B & \to & 0
\end{array}$$

Lemma 6.6. *$\mathscr{T}or$ is closed under the formation of mapping cones.*

Proof. Let $X \xrightarrow{f} Y \in \mathscr{T}or$, $0 \to \mathrm{Cok}(f) \to \mathrm{Cone}(f) \to S\ \mathrm{Ker}(f) \to 0$ is exact, $m1_Y = 0 \Rightarrow m1_{\mathrm{Cok}(f)} = 0$ and $n \cdot 1_X = 0 \Rightarrow n \cdot 1_{\mathrm{Ker}(f)} = 0$. ▌

Lemma 6.7. *For any $X \in \mathscr{S}$, positive integer n, $\mathrm{Cone}(n1_X) \in \mathscr{T}or$.*

Proof.

$$0 \to \mathrm{Cok}(n) \to \mathrm{Cone}(n) \to S\,\mathrm{Ker}(n) = 0$$

and $n \cdot 1_{\mathrm{Ker}(u)} = 0$, $n \cdot 1_{\mathrm{Cok}(u)} = 0$. ▌

Proposition 6.8. *$\mathscr{T}or$ both generates and cogenerates $\mathscr{S}$.*

Proof. Let $X \xrightarrow{f} Y$ be a non-zero map in $\mathscr{S}$. Because (X, Y) is a finitely generated abelian group, there exists $n > 0$ such that

$$f \notin \mathrm{Im}\left((X, Y) \xrightarrow{n} (X, Y)\right).$$

Hence if $\begin{array}{ccc} X & & \\ f\downarrow & \searrow 0 & \\ Y & \to & \mathrm{Cone}(n\,1_Y) \end{array}$ then $\begin{array}{ccc} & & X \\ & \swarrow & \downarrow f \\ Y & \xrightarrow[n]{} & Y \end{array}$

which contradicts the assumption on n. $\mathrm{Cone}(n \cdot 1_Y) \in \mathscr{T}or$ by the last lemma and hence $\mathscr{T}or$ cogenerates. Dually $\mathscr{T}or$ generates. ▌

When we restrict our attention to $\mathscr{T}or$ wedge-decomposition becomes unique. This is, however, a complete formality. First a few preparation lemmas, well-known, but sufficiently elementary to be included.

Lemma 6.9. *If R is a finite ring without proper idempotents, then R is local.*

Proof. Let R have n elements. Given any sequence in R, $x_1, x_2, \ldots$ there must appear a repetition among their products: $x_1 x_2 \cdots x_j = x_1 \cdots x_j \cdots x_{n+1}$ $(j \leqq n)$ hence $(x_0 \cdots x_j)\,(1 - x_{j+1} \cdots x_m) = 0$.

If the x_1's are the same we conclude that there exist $j, k > 0$, $x^j = x^{j+k}$ hence $x^j = x^{j+ak}$ any $a > 0$ and thus there exists $l \geq 0$ such that $x^j = x^{2j+l}$, so $(x^{j+l})^2 = x^{2j+2l} = x^{j+l}$. Hence for any $x \in R$ some power of x is idempotent and hence x is either nilpotent or a unit.

Now let x, y be non-units. For any sequence $z_1, z_2, \ldots$ where $z_i = x$ or y each i we have $j \leq n$ and $z_1 \ldots z_j(1 - z_{j+1} \ldots z_{n+1}) = 0$. But $z_{j+1} \ldots z_{n+1}$ is a non-unit, hence nilpotent, and $1 - z_{j+1} \ldots z_{n+1}$ is a unit. Thus $z_1 \ldots z_j = 0$ and a fortiore $z_1 \ldots z_n = 0$. Finally then $(x + y)^n = 0$ and the sum of non-units is a non-unit. ▮

Lemma 6.10. *Let $\mathscr{A}$ be an additive category, A an object in $\mathscr{A}$ such that* End(A) *is finite and contains no proper idempotents. If*

$$A \to X \oplus Y \to A = 1$$

then either

$$A \to X \oplus Y \xrightarrow{u_1 p_1} X \oplus Y \to A \quad \text{or} \quad A \to X \oplus Y \xrightarrow{u_2 p_2} X \oplus Y \to A$$

is an automorphism.

Proof. Immediately from the last lemma. ▮

Lemma 6.11. *Let $\mathscr{A}$ be an additive category, A an object in $\mathscr{A}$ such that* End(A) *is finite and has no proper idempotents. Then for X, $Y \in \mathscr{A}$, $A \oplus X \cong A \oplus Y$ implies $X \cong Y$.*

Proof. Let $A \oplus X \xrightarrow{f} A \oplus Y$ be an isomorphism. If $p_1 f u_1$ is a unit, then standard matrix type manipulatons deliver an isomorphism from X to Y. (For those easily amused, the isomorphism in question is expressible as

$$p_2[f - 2 f u_1 (p_1 f u_1)^{-1} p_1 f + f u_1 (p_1 f u_1)^{-1} p_1 f u_1 (p_1 f u_1)^{-1} p_1 f] u_2 .)$$

If $p_1 f u_1$ is not a unit then $p_1 f^{-1} u_2 p_2 f u_1$ is a unit because

$$p_1 f^{-1} (u_2 p_2 + u_1 p_1) f u_1 = 1 \quad \text{and} \quad p_1 f^{-1} u_1 p_1 f u_1$$

is not a unit in End(A) and it follows that $p_1 f^{-1} u_2 p_2 f u_1$ is a unit (lemma 6.9). Let $h p_2 f u_1 = 1_A$, $h : Y \to A$. Then $(1 + u_1 h p_2) \in$ Aut$(A \oplus Y)$ and $(1 + \mathrm{w}_1 h p_2) f = g$ is an isomorphism. But

$$p_1 g u_1 = p_1 (1 + u_1 h p_2) f u_1 = p_1 f u_1 + h p_2 f u_1 = 1 + p_1 f u_1$$

which is a unit because $p_1 f u_1$ is nilpotent. Hence we apply the last paragraph to g in place of f. ▮

Corollary 6.12. *Let $\mathscr{A}$ be an additive category in which idempotents split, A an object in $\mathscr{A}$ such that* End(A) *is finite. Then $A \oplus X = A \oplus Y$ implies $X \cong Y$.*

Proof. Decompose A as a finite sum of objects without proper idempotents and apply the above lemma the necessary finite number of times. ▮

We shall not need the next corollary, but it is an amusing example of the usefulness of abstract categories.

Corollary 6.13. *Let $\mathscr{A}$ be any additive category whatever, A an object in $\mathscr{A}$ such that* End(A) *is finite. Then $A \oplus X \cong A \oplus Y$ implies $X \cong Y$.*

Proof. For every additive $\mathscr{A}$ there exists a full embedding $\mathscr{A} \to \hat{\mathscr{A}}$ such that idempotents split in $\hat{\mathscr{A}}$. (See *Abelian Categories* or use the functor category $(\mathscr{A}^*, \mathscr{G})$.)

$A \oplus X \cong A \oplus Y$ in $\hat{\mathscr{A}}$, the above corollary applies, hence $X \cong Y$ in $\hat{\mathscr{A}}$. But $\mathscr{A} \to \hat{\mathscr{A}}$ is full and $X \cong Y$ in $\mathscr{A}$. ▮

Proposition 6.14. *Let $\mathscr{A}$ be an additive category in which idempotents split, and in which* End(X) *is finitely generated for each $X \in \mathscr{A}$. Then for each X there exists a decomposition $X = A_1 \oplus \cdots \oplus A_n \oplus B$ where* End(A_i) *is finite each i and without proper idempotents, and* End(B) *has no torsion proper idempotents. Any two such decompositions are essentially unique.*

Proof. Let e, be a non-zero torsion idempotent in End(X) and express $X = C_1 \oplus X_1$, where $C_1 = \mathrm{Im}(e_1)$. Let e_2 be a non-zero torsion idempotent in End(X_1) and express $X_1 = C_2 \oplus X_2$, $C_2 = \mathrm{Im}(e_2)$. Continuing in this manner we must eventually obtain X without torsion proper idempotents (m is bounded above by the number of indecomposable cyclic groups needed to express End(X)). We express $C_1 \oplus \cdots \oplus C_n$ as a sum of indecomposables $A_1 \oplus \cdots \oplus A_n$ and let $B = Xm$.

For the uniqueness let $A_1 \oplus \cdots \oplus A_n \oplus B \xrightarrow{f} A_1' \oplus \cdots \oplus A_m' \oplus B'$ be an isomorphism where End(A_i) is finite and without idempotents each i, and End(B), End(B') are without torsion proper idempotents.

$$p_1 f^{-1}(u_{B'} p_{B'} + u_m p_m + \cdots + u_1 p_1) f u_1 = 1$$

and hence because End(A_1) is local there exists j, such that $p_1 f^{-1}(u_j p_j) f u_1$ is a unit. (Note that $p_1 f^{-1} p_B u_B f u_1$ can not be a unit, otherwise B' would have a torsion idempotent.) Because A_j' is indecomposable, $p_j f u_1$ is an isomorphism $A_1 \to A_j'$. Hence by lemma 6.11, $A_2 \oplus \cdots \oplus A_n \oplus B \cong A_1' \oplus \cdots \oplus A_{j-1}' \oplus A_{j+1}' \oplus \cdots \oplus A_m' \oplus B'$ and we may iterate this procedure to obtain the fact that $n = m$ and a permutation π, and isomorphisms $A_i \to A_{\pi(i)}'$, and $B \to B'$. ▮

If the order of 1_X for $X \in \mathscr{F}$ is finite but not a prime power, then X has a proper idempotent. (Let $ord(1_X) = ab$, $ax + by = 1$. Then $(ax)^2 = ax(1 - by) = ax$.) Hence X splits as a sum of objects each of primary exponent. For each prime p we define $\mathscr{T}or_p$ to be the full subcategory of spaces X such that p^j, $1_X = 0$ large j. We obtain functors $\mathscr{T}or \to \mathscr{T}or_p$ which exhibit $\mathscr{T}or$ to be isomorphic to the weak product of the $\mathscr{T}or_p$'s (i.e. the full subcategory of $\Pi_p \mathscr{T}or_p$ of objects almost all of whose coordinate objects are zero. Such is in fact a categorical sum, namely in the category of $\oplus$'ive categories).

7. Characterization of Mapping Cones

The abelian category $\mathcal{F}$ has much structure comming from its topological origin; such as the automorphism $S:\mathcal{F} \to \mathcal{F}$. In the proof of 4.6 it was observed that by starting with the spheres and adjoining mapping cones all spaces are obtained. It is thus a priori desirable to characterize mapping cones within $\mathcal{F}$ without reference to geometry.

We have not yet seen a way of doing so in general. However, if attention is restricted to torsion spaces, the situation improves. First:

Proposition 7.1. *If $X \xrightarrow{f} Y \in \mathcal{S}$ and either the kernel or cokernel of f is torsion (e.g. if either X or Y is torsion) then there is a unique projective extension $0 \to \operatorname{Cok}(f) \to P \to S\operatorname{Ker}(f) \to 0$ and $P \cong \operatorname{Cone}(f)$.*

Proof. Suppose $\operatorname{Cok}(f)$ is torsion. Because we have the exact sequence $0 \to \operatorname{Cok}(f) \to \operatorname{Cone}(f) \to S\operatorname{Ker}(f) \to 0$ the Schanuel lemma says that $\operatorname{Cok}(f) \oplus \operatorname{Cone}(f) \cong \operatorname{Cok}(f) \oplus P$. By lemma 6.12, $\operatorname{Cok}(f)$ cancels and $\operatorname{Cone}(f) \cong P$.

We may recall the Schanuel lemma: let $0 \to A \to P \to B \to 0$ and $0 \to A' \to P' \to B \to 0$ be exact, P, P' projective. Let $\begin{matrix} Q \to P \\ \downarrow \quad \downarrow \\ P' \to B \end{matrix}$ be a pullback. Then $Q \to P$ and $Q \to P'$ are epimorphic, thus $Q \cong P \oplus \operatorname{Ker}(Q \to P) \cong P' \oplus \operatorname{Ker}(Q \to P')$. But $\operatorname{Ker}(Q \to P) = A'$, $\operatorname{Ker}(Q \to P') = A$ and $P \oplus A' \cong P \oplus A$. ▮ *

We have already observed that the torsion spaces both generate and cogenerate $\mathcal{S}$ (6.8). The last proposition gives added interest to

Proposition 7.2. *For every space $X \in \mathcal{S}$ there exists a torsion space T and maps $S^{n_1} \xrightarrow{\alpha_1} T$, $S^{n_2} \xrightarrow{\alpha_2} T, \ldots, S^{n_j} \xrightarrow{\alpha_j} T$ such that*

$$X \cong \operatorname{Cone}(S^{n_1} \vee \ldots \vee S^{n_j} \xrightarrow{\alpha} T).$$

Proof. We shall here prove the dual. Spanier-Whitehead duality can then deliver the proposition as stated, or one may use cohomology and cohomotopy where we use homology and homotopy.

Given X, note first that because $(S^n, X) \otimes Q \cong (H_n X) \otimes Q$ (4.(10)) the rank of the free part of $\Sigma_n (S^n, X)$ is finite. Let $S^{n_1} \xrightarrow{\alpha_1} X, \ldots, S^{n_j} \xrightarrow{\alpha_j} X$ be a basis for the free part of $\Sigma_n (S^n, X)$. Let $W = S^{n_1} \vee \ldots \vee S^{n_j}$ and $W \to X$ be the map whose coordinate maps are the α_j's. Then $(S^n, W) \otimes Q \to (S^n, X) \otimes Q$ is an isomorphism. Hence for $T = \operatorname{Cone}(W \to X)$, $(S^n, T) \otimes Q = 0$ and $(H_n T) \otimes Q = 0$. Thus $\Sigma_n H_n T$ is a finite group.

* Added in Proof: There is an example which shows that the hypothesis on $X \xrightarrow{f} Y$ can not be dropped.

Let a be the exponent of $\Sigma_n H_n T$. Then $H_n(T \xrightarrow{a+1} T) = 1$ all n and by the stable Whitehead theorem (4.2) $T \xrightarrow{\alpha+1} T$ is an isomorphism. Thus $\mathrm{End}(T) \xrightarrow{a+1} \mathrm{End}(T)$ is onto and because $\mathrm{End}(T)$ is finitely generated it must be finite. (Note that the exponent of T could be larger than a.) X is the mapping cone of $S^{-1}T \to W$. ∎

Recall that E is said to be an *injective envelope* of A if E is injective and there exists a monomorphism $A \to E$ such that for all B and $E \to B$, $0 \to (A, E) \to (A, B)$ exact implies $0 \to E \to B$ exact. If A has an injective envelope E then for any other injective E' and exact $0 \to A \to E'$ we obtain $\begin{array}{ccc} A & \to & E \\ \downarrow & & \downarrow \\ A & \to & E' \end{array}$ where $E \to E'$ is monomorphic. Hence $E' = E \oplus E''$. There can be no injectives contained in E that also contain the image of A and injective envelopes are unique up to (non-unique) isomorphism. Two embeddings of A into its injective envelope can be related by an automorphism on the envelope.

Many objects in $\mathscr{F}$ do not have injective envelopes. For example $\mathrm{Im}(S^n \xrightarrow{\alpha} S^0)$, $(n \neq 0, \alpha \neq 0)$ does not. If it did, then it would appear as a direct summand of S^0. But S^0 has no proper idempotents hence S^0 would have to be the injective envelope. But for b relatively prime to the order of α, $\mathrm{Ker}(b) \cap \mathrm{Im}(\alpha) = 0$ and $\mathrm{Im}(\alpha) \to S^0 \to S^0/\mathrm{Ker}(b)$ is monomorphic.

We say that an object $A \in \mathscr{F}$ is **reduced** if it has no injective subobjects. Every object can be expressed as a direct sum of a reduced object and an injective object, and if the object is torsion, then this decomposition is unique.

Let $\mathscr{F}\mathscr{T}or$ be the abelian category obtained from $\mathscr{T}or$. That is, $\mathscr{F}\mathscr{T}or$ is the full subcategroy of objects that appear as images of maps between torsion spaces.

Proposition 7.3. *If Y is a torsion space, $A \subset Y$ a subobject such that no proper summand of Y contains A, then Y is the injective envelope of A.*

Proof. Suppose $B \subset Y$, $A \cap B = 0$. Then there exists

$$\begin{array}{ccc} A \oplus B & \to & Y \\ {\scriptstyle u_2 p_2}\downarrow & & \downarrow{\scriptstyle e} \\ A \oplus B & \to & Y \end{array}$$

Every power of e works just as well, and hence from the proof of lemma 6.9, we may assume that e is idempotent. But $A \subset \mathrm{Ker}(e)$ and $\mathrm{Ker}(e) \oplus \mathrm{Ker}(1-e) = Y$. Hence $e = 0$ and $B = 0$. ∎

The Piece-de-resistance:

Theorem 7.4. *For torsion spaces X, Y and map $X \xrightarrow{f} Y$, let* Ker $(f) =$ $R \oplus E$ *where R is reduced, (that is, has no injective subobjects) and E injective. Let G be injective envelope of* Cok (f). *Then* Cone $(f) = G \oplus SE$.

Proof. Because E is injective, $X \cong X' \oplus E$, and there exists $X' \xrightarrow{f'} Y$ such that $f = X' \oplus E \xrightarrow{p_1} X' \xrightarrow{f'} Y$. $\mathrm{Cok}(f') = \mathrm{Cok}(f)$, $\mathrm{Ker}(f') = R$. If we show that $G \cong \mathrm{Cone}(f')$, i.e. that $\mathrm{Cone}(f')$ is the injective envelope of $\mathrm{Cok}(f')$, then surely there exists $0 \to \mathrm{Cok}(f) \to G \oplus SE \to SR \oplus SE \to 0$ and proposition 7.1 implies the desired answer.

But surely $\mathrm{Cone}(f')$ contains no proper summand which allows $\mathrm{Cok}(f') \to \mathrm{Cone}(f')$ for if it did, i.e. if $\mathrm{Cone}(f') = A_1 \oplus A_2$ and $\mathrm{Cok}(f') \to \mathrm{Cone}(f')$ factored through $A_1 \to A_1 \oplus A_2$, then $A_2 \to A_1 \oplus A_2 \to S\,\mathrm{Ker}(f')$ would be a monomorphism and $S\,\mathrm{Ker}(f')$ could not be reduced.

Thus by proposition 7.3, $\mathrm{Cone}(f')$ is the injective envelope of $\mathrm{Cok}(f')$. ∎

For those who object to the appearance of the suspension functor in the above characterization of mapping cones, we have:

Corollary 7.5. *If $X \xrightarrow{f} Y \in \mathscr{T}or$ and* $\mathrm{Ker}(f)$, $\mathrm{Im}(f)$ *and* $\mathrm{Cok}(f)$ *are all* reduced *(i.e. have no injective subobjects) then* $\mathrm{Cone}(f)$ *is the injective envelope of* $\mathrm{Cok}(f)$, *and SX is the injective envelope of* $\mathrm{Cone}(f)$. ∎

Proposition 7.6. *For any $X \in \mathscr{T}or$ there exists $Y \in \mathscr{T}or$ and $X \to Y$ such that* $\mathrm{Ker}(f)$, $\mathrm{Im}(f)$ *and* $\mathrm{Cok}(f)$ *are reduced. Hence SX is identifiable in $\mathscr{F}$ with category predicates alone.*

Proof. Let $X \cong X_1 \oplus \cdots \oplus X_j$ where each X_i is indecomposable. By proposition 4.8 there exists $g_i\colon X_i \to S^{a_i} \neq 0$. Let n_i be a prime power such that $g_i \notin \mathrm{Im}[(X_1, S^{a_i}) \xrightarrow{n_i} (X_1, S^{a_i})]$. Define $f_i\colon X_i \to \mathrm{Cone}\,(n_i) = X_i \to S^{a_i} \to \mathrm{Cone}\,(n_i)$. $f_i \neq 0$. $\mathrm{Cone}(n_i)$ was shown to be indecomposable in 4.4.

Call $\mathrm{Cone}(n_i) = Y_i$. Because X_i and Y_i are indecomposable, and $f_i \neq 0$. It follows that $\mathrm{Ker}(f_i)$, $\mathrm{Im}(f_i)$ and $\mathrm{Cok}(f_i)$ are reduced unless f_i is an isomorphism.

We can avoid that last possibility as follows: If $X_i = \mathrm{Cone}(n_i)$, let $S^{a_i} \xrightarrow{n_i} S^{a_i} \to X \to S^{a_i+1}$ be a mapping cone sequence, and let $Y_i = \mathrm{Cone}(S^{a_i+1} \xrightarrow{m} S^{a_i+1})$ and $f_i = X_i \to S^{a_i+1} \to Y_i$ where m is a prime power that does not divide $X_i \to S^{a_i+1}$ in (X_i, S^{a_i+1}). Then $H_{a_i}(X_i) \neq 0$ and $H_{a_i}(Y_i) = 0$ hence f_i is not an isomorphism.

Define $Y = Y_1 \oplus \cdots \oplus Y_n$, and $f\colon X \to Y$ the map whose i'th coordinate is f_i. Then $\mathrm{Ker}(f) = \mathrm{Ker}(f_1) \oplus \cdots \oplus \mathrm{Ker}(f_n)$, each summand being reduced it follows that $\mathrm{Ker}(f)$ is reduced, and similarly $\mathrm{Im}(f)$ and $\mathrm{Cok}(f)$ are reduced. ∎

8. Interpretations and Computations of Ext

Let A, B be objects in $\mathscr{F}$, $B = \operatorname{Im}(X \to Y)$, $X, Y \in \mathscr{S}$. Let $X \to Y \to Z$ be a mapping cone sequence. We obtain in $\mathscr{F}$ an injective resolution for B:

$$0 \to B \to Y \to Z \to SX \to SY \to SZ \to \cdots$$

from which it is emediately apparent that $\operatorname{Ext}^{n+3}(A, B) \cong \operatorname{Ext}^n(A, S\,B)$ all $n > 0$.

Consider the doubly infinite exact sequence

$$\cdots \to S^{-1}\,Y \to S^{-1}Z \to X \to Y \to Z \to SX \to \cdots$$

We define for any integer n, $E^n(A, B)$ to be the nth component of the homology of the complex

$$\to (A, S^{-1}\,Y) \to (A, S^{-1}Z) \to (A, X) \to (A, Y) \to \cdots$$

indexed so as to insure $\operatorname{Ext}^n(A, B) = E^n(A, B)$ for $n > 0$.

The E's form an exactly connected sequence of functors. $E^{n+1}(A, -)$ is the right satellite of $E^n(A, -)$ and $E^n(A, -)$ is the left satellite of $E^{n+1}(A, -)$. Each E is a ballanced functor and $\{E^n(-, B)\}$ is an exactly connected sequence of functors each of which is a right satellite of its left neighbor and a left satellite of its right neighbor.

The natural transformation $(A, B) \to E^0(A, B)$ is epimorphic and its kernel is the set of maps from A to B which factor through an injective or equivalenly a projective.

Given $A = \operatorname{Im}(X' \to Y')$, $B = \operatorname{Im}(X \to Y)$, $X', Y', X, Y \in \mathscr{S}$ we wish to express $E^n(A, B)$ without reference to the mapping cones. The task divides into three cases depending on $n \bmod 3$. Two of the tree cases, namely $n \equiv 0$, $n \equiv 2$ have explicit solutions. The remaining case, $n \equiv 1$, does not. It is this case, however, that gives us a new interpretation of the Toda bracked operation.

We shall define a number of additive relations between certain abelian groups. Let us fix our notation: for an additive relation $R \subset G \times H$, G, H abelian groups, and subgroups $G_1 \subset G$, $H_1 \subset H$ we shall write $G_1 \underset{R}{|} H_1$ to indicate that the following four conditions hold:

i) for all $x \in G_1$, $\langle x, y\rangle \in R \Rightarrow y \in H_1$

ii) for all $x \in G_1$, there exists $y \in H$ such that $\langle x, y\rangle \in R$

iii) for all $y \in H_1$, $\langle x, y\rangle \in R \Rightarrow x \in G_1$

iv) for all $y \in H_1$, there exists $x \in G$ such that $\langle x, y\rangle \in R$.

Given $G_2 \subset G_1 \subset G$, $H_2 \subset H_1 \subset H$ such that $G_1 \underset{R}{|} H_1$ and $G_2 \underset{R}{|} H_2$ there exist exact sequences:

$$0 \to G_2 \to G_1 \to H/H_2 \to H/H_1 \to 0$$
$$0 \to G_1/G_2 \to H_1/H_2 \to 0 .$$

Case $n \equiv 0$. *Consider* $E^0(A, B)$. *Let* $X \to Y \to Z$ *be a mapping cone sequence. Define*

$$R \subset (A, Y) \times (X', Y) \text{ by } \langle f, g\rangle \in R \text{ iff}$$

$$\begin{array}{ccc} X' & \to & A \\ g\downarrow & \swarrow f & \\ Y & & \end{array} \quad \text{commutes.}$$

Lemma 8.1. $\operatorname{Ker}[(A, Y) \to (A, Z)] \underset{R}{|} \operatorname{Im}[(X', X) \to (X', Y)] \cap$
$\cap \operatorname{Im}[(Y, Y) \to (X', Y)]$.

Proof. We consider the four conditions listed above that define the use of "$\underset{R}{|}$".

i. Let $f \in \operatorname{Ker}[(A, Y) \to (A, Z)]$, $\langle f, g\rangle \in R$. We must show (ia) $g \in \operatorname{Im}[(X', X) \to (X', Y)]$ and (ib) $g \in \operatorname{Im}[(Y', Y) \to (X', Y)]$.

ia. $X \to Y \to Z$ is exact in $\mathscr{F}$, $X' \xrightarrow{g} Y \to Z = 0$, X' is projective, hence there exists

$$\begin{array}{ccc} & X' & \\ \swarrow & & \searrow g \\ X & \to & Y \end{array}$$

ib. $A \to Y'$ is mono, Y is injective, hence there exists $\begin{array}{ccc} A & \to & Y' \\ f\searrow & & \swarrow \\ & Y & \end{array}$ and

$$\begin{array}{ccccc} X' & \to & A & \to & Y' \\ & g\searrow & f\downarrow & \swarrow & \\ & & Y & & \end{array}$$

ii. Given $f \in \operatorname{Ker}[(A, Y) \to (A, Z)]$ define $X' \xrightarrow{g} Y = X' \to A \xrightarrow{f} Y$. Clearly $\langle f, g\rangle \in R$.

iii. Let $g \in \operatorname{Im}[(X', X) \to (X', Y)] \cap \operatorname{Im}[(Y', Y) \to (X', Y)]$ and $\langle f, g\rangle \in R$. We must show that $f \in \operatorname{Ker}[(A, Y) \to (A, Z)]$. The condition on g says that there exists

$$\begin{array}{ccc} X' & \to & Y' \\ h_1\downarrow & g\searrow & \downarrow h_2 \\ X & \to & Y \end{array}$$

and hence $X' \xrightarrow{g} Y \to Z = X' \xrightarrow{h_1} X \to Y \to Z = 0$. Now $X' \to A \xrightarrow{f} Y$ $= X' \xrightarrow{g} Y$ hence $X' \to A \xrightarrow{f} Y \to Z = 0$. $X' \to A$ is epi, thus $A \xrightarrow{f} Y \to Z = 0$ and $f \in \operatorname{Ker}[(A, Y) \to (A, Z)]$.

iv. Let g be as above and define $A \xrightarrow{f} Y = A \to Y' \xrightarrow{h_2} Y$. Then $\langle f, g \rangle \in R$. ▮

Lemma 8.2. $\operatorname{Im}[(A, X) \to (A, Y)] \underset{R}{|} \operatorname{Im}[(Y', X) \to (X', Y)]$.

Proof. i. Let $f \in \operatorname{Im}[(A, X) \to (A, Y)]$, $\langle f, g \rangle \in R$. There exists

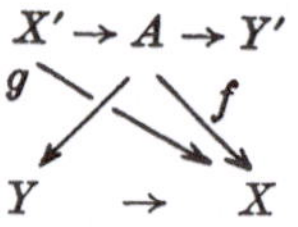

$A \to Y'$ is mono, $A \to X$ is injecture hence there exists $\begin{array}{c} A \to Y' \\ \searrow \swarrow \\ X \end{array}$ and $g \in \operatorname{Im}[(Y', X) \to (X', Y)]$.

ii. Let $f \in \operatorname{Im}[(A, X) \to (A, Y)]$. Define $X' \xrightarrow{g} Y = X' \to A \to Y$. $\langle f, g \rangle \in R$.

iii. Let $g \in \operatorname{Im}[(Y', X) \to (X', Y)]$, $\langle f, g \rangle \in R$. There exist

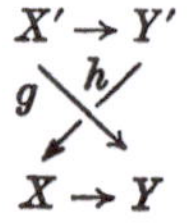 and $\begin{array}{c} X' \to A \\ {\scriptstyle g}\searrow \ \downarrow{\scriptstyle f} \\ Y \end{array}$

Define $A \xrightarrow{h'} X = A \to Y' \xrightarrow{h} X$. Then $\begin{array}{c} A \\ {\scriptstyle h'}\swarrow \ \searrow{\scriptstyle f} \\ X \to Y \end{array}$ commutes and

$$f \in \operatorname{Im}[(A\ X) \to (A, Y)].$$

(We have re-used the fact that $X' \to A$ is epi.)

iv. Let $g \in \operatorname{Im}[(Y', X) \to (X', Y)]$. From the diagram above define $A \xrightarrow{f} Y = A \to Y' \xrightarrow{h} X \to Y$. Then $\langle f, g \rangle \in R$. ▮

Lemma 8.3. $\operatorname{Im}[(A, X) \to (A, Y)] \subset \operatorname{Ker}[(A, Y) \to (A, Z)]$.

Proof. $X \to Y \to Z = 0$. ▮

Lemma 8.4. $\operatorname{Im}[(Y', X) \to (X', Y)] \subset \operatorname{Im}[(X', X) \to (X', Y)] \cap \operatorname{Im}[(Y', Y) \to (X', Y)]$. ▮

We now have the situation:

$$\begin{array}{l}\operatorname{Ker}[(A,Y)\to(A,Z)] \underset{R}{\Big|} \operatorname{Im}[(X',X)\to(X',Y)]\cap\operatorname{Im}[(Y',Y)\to(X',)]\\ \quad\ \cup \qquad\qquad\qquad\qquad\qquad\qquad\qquad \cup\\ \operatorname{Im}[(A,X)\to(A,Y)] \underset{R}{\Big|} \operatorname{Im}[(Y',X)\to(X',Y)].\end{array}$$

The quotient of the left hand subgroups of (A, Y) is, by definition $E^0(A, B)$. Hence:

Proposition 8.5.

$$E^0(A,B)=\frac{\operatorname{Im}[(X',X)\to(X',Y)]\cap\operatorname{Im}[(Y',Y)\to(X',Y)]}{\operatorname{Im}[(Y',X)\to(X',Y)]}. \ \blacksquare$$

In the special case when X', Y', X, Y are spheres we may use the commutativity of the graded ring $G=\{(S^n, S^0)\}$ to simplify the expression. Because $E^n(A, B)$ is less trivial for small (i.e. negative) values of n, it is convenient to write E_n for E^{-n}.

Proposition 8.6. *Given* $S^a \xrightarrow{\alpha} S^0$, $S^b \xrightarrow{\beta} S^0$,

$$E_{3n}(\operatorname{Im}(\alpha), \operatorname{Im}(\beta)) = E^{-3n}(\operatorname{Im}(\alpha), \operatorname{Im}(\beta)) = \frac{\beta G_{n+a-b}\cap\alpha G_n}{\alpha\beta G_{n-b}}. \ \blacksquare$$

Case $n \equiv 2$.

We shall consider $E^{-1}(A, B)$.

Define

$$R\subset(A,X)\times(Y',X) \quad \text{by} \quad \langle f,g\rangle\in R \quad \text{iff}$$

$$\begin{array}{ccc} A & \to & Y' \\ f\downarrow & \swarrow g & \\ X & & \end{array}$$

commutes.

Lemma 8.7. $(A, X) \underset{R}{\big|} (Y', X)$.

Proof. $A \to Y'$ is mono, X is injective. $\blacksquare$

Lemma 8.8. $\operatorname{Ker}[(A,X)\to(A,Y)] \underset{R}{\big|} \operatorname{Ker}[(Y',X)\to(X',Y)]$.

Proof. i. Let

$$f\in\operatorname{Ker}[(A,X)\to(A,Y)], \ \langle f,g\rangle\in R.$$

There exists

$$\begin{array}{ccc} A & \to & Y' \\ f\downarrow & \times & \searrow 0 \ \ g\swarrow \\ X & \to & Y \end{array}$$

Thus

$$X'\to Y'\xrightarrow{g}X\to Y = X'\to A\xrightarrow{f}X\to Y=0$$

and

$$g \in \operatorname{Ker}[(Y', X) \to (X', Y)].$$

ii. Follows from last lemma.

iii. Let $g \in \operatorname{Ker}[(Y', X) \to (X', Y)]$, $\langle f, g\rangle \in R$.

$$\begin{array}{ccc} X' & \to & Y' \\ & \searrow \!\!\!\! \swarrow^{g} & \\ & \searrow_{0} & \\ X & \to & Y \end{array}$$

commutes.

$$X' \to A \xrightarrow{f} X \to Y = X' \to A \to Y' \xrightarrow{g} X \to Y = 0.$$

$X' \to A$ is epi, hence $f \in \operatorname{Ker}[(A, X) \to (A, Y)]$.

iv. Follows from last lemma. ∎

Recall that $S^{-1}Z \to X \to Y$ is exact.

Lemma 8.9. $\operatorname{Im}[(A, S^{-1}Z) \to (A, X)] \,|_{R}\, \operatorname{Ker}[(Y', X) \to (X', X)] +$
$$+ \operatorname{Ker}[(Y', X) \to (Y', Y)].$$

Proof. i. Let $f \in \operatorname{Im}[(A, S^{-1}Z) \to (A, X)]$, $\langle f, g\rangle \in R$. There exists

$$\begin{array}{ccc} & A & \to Y' \\ {}^{h}\swarrow & f\downarrow & \swarrow g \\ S^{-1}Z & \to X & \end{array}$$

$A \to Y'$ is mono, $S^{-1}Z$ is injective, hence there exists

$$\begin{array}{cc} A & \to Y' \\ \downarrow & \swarrow h' \\ S^{-1}Z & \end{array}$$

Define $Y' \xrightarrow{g_1} X = Y' \xrightarrow{h'} S^{-1}Z \to X$. Clearly $g_1 \in \operatorname{Ker}[(Y', X) \to (Y', Y)]$. Moreover $g - g_1$ is such that

$$\begin{aligned} A \to Y' \xrightarrow{g - g_1} X &= (A \to Y' \xrightarrow{g} X) - (A \to Y' \xrightarrow{h'} S^{-1}Z \to X) \\ &= (A \xrightarrow{f} X) - (A \xrightarrow{f} X) = 0. \end{aligned}$$

A fortiori, $g - g_1 \in \operatorname{Ker}[(Y', X) \to (X', X)]$.

ii. Follows from lemma above.

iii. Let $g_1 \in \operatorname{Ker}[(Y', X) \to (X', X)]$, $g_2 \in \operatorname{Ker}[(Y', X) \to (Y', Y)]$, $\langle f, g_1 + g_2\rangle \in R$. We wish to show that $f \in \operatorname{Im}[(A, S^{-1}Z) \to (A, X)]$.

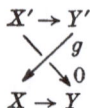

$S^{-1}Z \to X \to Y$ is exact, Y' is projective, hence there exists

$$\begin{array}{ccc} & & Y' \\ & {}^{h}\swarrow & \downarrow g_2 \\ S^{-1}Z & \to & X \end{array}$$

Define $A \xrightarrow{h'} S^{-1}Z = A \to Y' \xrightarrow{h} S^{-1}Z$. Then

$$\begin{aligned} X' \to A \xrightarrow{f} X &= X' \to A \to Y' \xrightarrow{g_1+g_2} X = (X' \to Y' \xrightarrow{g_1} X) + (X' \to Y \xrightarrow{g_2} X) \\ &= 0 + (X' \to Y \xrightarrow{h} S^{-1}Z \to X) = X' \to A \to Y \xrightarrow{h} S^{-1}Z \to X \\ &= X' \to A \xrightarrow{h'} S^{-1}Z \to X. \end{aligned}$$

$X' \to A$ is epi, hence

$$\begin{array}{ccc} & & A \\ & {}^{h}\swarrow & \downarrow f' \\ Y & \to & Z \end{array}$$

$$A \xrightarrow{f} X = A \xrightarrow{h'} S^{-1}Z \to X \quad \text{and} \quad f \in \operatorname{Im}[(A, S^{-1}Z) \to (A, X)].$$

iv. Follows from above lemma. ∎

Lemma 8.10. $\operatorname{Im}[(A, S^{-1}Z) \to (A, X)] \subset \operatorname{Ker}[(A, X) \to (A, Y)]$.
$\operatorname{Ker}[(Y', X) \to (X', X)] + \operatorname{Ker}[(Y', X) \to (Y', Y)] \subset$
$\subset \operatorname{Ker}[(Y', X) \to (X', Y)]$. ∎

Because $\dfrac{\operatorname{Ker}[(A, X) \to (A, Y)]}{\operatorname{Im}[(A, S^{-1}Z) \to (A, X)]}$ is, by definition, $E^{-1}(A, B)$ we obtain

Proposition 8.11.

$$E^{-1}(A, B) = \frac{\operatorname{Ker}[(Y', X) \to (X', Y)]}{\operatorname{Ker}[(Y', X) \to (X', X)] + \operatorname{Ker}[(Y', X) \to (Y', Y)]}. \ ∎$$

In the case that X', Y', X, Y are spheres we obtain:

Proposition 8.12. *Let* $A = \operatorname{Im}[S^a \xrightarrow{\alpha} S^0]$, $B = \operatorname{Im}[S^b \xrightarrow{\beta} S^0]$.

$$E_{3n+1}(\operatorname{Im}(\alpha), \operatorname{Im}(\beta)) = E^{-3n-1}(\operatorname{Im}(\alpha), \operatorname{Im}(\beta)) = \frac{\operatorname{Ann}_n(\alpha\beta)}{\operatorname{Ann}_n(\alpha) + \operatorname{Ann}_n(\beta)},$$

where $\operatorname{Ann}_n(\gamma) = \{\delta \in G_n \mid \gamma\delta = 0\}$. ∎

Case $n \equiv 1$. *We shall consider* $\operatorname{Ext}^1(A, B)$.

Define $R_1 \subset (A, Z) \times (X', Y)$ *by* $\langle f, g\rangle \in R_1$ *iff* $\begin{array}{ccc} X' & \to & A \\ g\downarrow & & \downarrow f \\ Y & \to & Z \end{array}$ *commutes.*

Lemma 8.13. $\operatorname{Dom} R_1 = R_1^{-1}(X', Y) = \operatorname{Ker}[(A, Z) \to (A, SX)]$.

Proof. 1. Given $\langle f, g\rangle \in R_1$ consider

$$X' \to A \xrightarrow{f} Z \to SX = X' \xrightarrow{g} Y \to Z \to SX = 0 .$$

$X' \to A$ is epi, hence $A \xrightarrow{f} Z \to SX = 0$ and $f \in \mathrm{Ker}[(A, Z) \to (A, SX)]$.

2. Given $f \in \mathrm{Ker}[(A, Z) \to (A, SX)]$ use the fact that X' is projective and that $Y \to Z \to SX$ is exact to obtain $g \in (X', Y)$ such that $\langle f, g\rangle \in R$.

Lemma 8.14.

$$\mathrm{Im}[(A, Y) \to (A, Z)] \underset{R_1}{|} \mathrm{Im}[(X', X) \to (X', Y)] + \mathrm{Im}[(Y', Y) \to (X', Y)] .$$

Proof. i. Let $f \in \mathrm{Im}[(A, Y) \to (A, Z)]$, $\langle f, g\rangle \in R_1$. There exists

$$\begin{array}{ccc} & A & \\ h\swarrow & & \searrow f \\ Y & \to & Z \end{array}$$

Define $X' \xrightarrow{g_1} Y = X' \to A \xrightarrow{h} Y$. Because $A \to Y'$ is mono and Y injective, there exists

$$\begin{array}{ccccc} X' & \to & A & \to & Y' \\ & g_1\searrow & \downarrow h & \swarrow & \\ & & Y & & \end{array}$$

and $g_1 \in \mathrm{Im}[(Y', Y) \to (X', Y)]$.

We wish to show that $g - g_1 \in \mathrm{Im}[(X', X) \to (X', Y)]$. But

$$\begin{aligned} X' \xrightarrow{g-g_1} Y \to Z &= (X' \xrightarrow{g} Y \to Z) - (X' \xrightarrow{g_1} Y \to Z) \\ &= (X' \to A \xrightarrow{f} Z) - (X' \to A \xrightarrow{h} Y \to Z) = 0 . \end{aligned}$$

Because X' is projective and $X \to Y \to Z$ exact there exists

$$\begin{array}{ccc} & X' & \\ \swarrow & & \searrow g - g_1 \\ X & \to & Y \end{array}$$

and $g - g_1 \in \mathrm{Im}[(X', X) \to (X', Y)]$.

ii. Given $f \in \mathrm{Im}[(A, Y) \to (A, Z)]$ let $\begin{array}{ccc} & A & \\ h\swarrow & & \searrow f \\ Y & \to & Z \end{array}$

commute and define $X' \xrightarrow{g} Y = X' \to A \xrightarrow{h} Y$. Then $\langle f, g\rangle \in R_1$.

iii. Let

$$g_1 \in \mathrm{Im}[(X', X) \to (X', Y)], \quad g_2 \in \mathrm{Im}[(Y', Y) \to (X', Y)]$$

and $\langle f_1 g_1 + g_2\rangle \in R_1$.

There exist $\begin{array}{ccc} & X' & \\ {\scriptstyle h_1}\swarrow & & \searrow{\scriptstyle g_1} \\ X & \rightarrow & Y \end{array}$ and $\begin{array}{ccc} X' & \rightarrow & Y' \\ & {\scriptstyle g_2}\searrow \quad \swarrow{\scriptstyle h_2} & \\ & Y & \end{array}$

Define $A \xrightarrow{h_3} Y = A \rightarrow Y' \xrightarrow{h_2} Y$. Then

$$\begin{aligned} X' \rightarrow A \xrightarrow{f} Z &= X' \xrightarrow{g_1+g_2} Y \rightarrow Z = (X' \xrightarrow{g_1} Y \rightarrow Z) + (X' \xrightarrow{g_2} Y \rightarrow Z) \\ &= (X' \xrightarrow{h_1} X \rightarrow Y \rightarrow Z) + (X' \rightarrow Y' \xrightarrow{h_2} Y \rightarrow Z) \\ &= 0 + (X' \rightarrow A \rightarrow Y' \xrightarrow{h_2} Y \rightarrow Z) = X' \rightarrow A \xrightarrow{h_3} Y \rightarrow Z. \end{aligned}$$

$X' \rightarrow A$ is epi hence $A \xrightarrow{f} Z = A \xrightarrow{h_3} Y \rightarrow Z$ and $f \in \operatorname{Im}[(A, Y) \rightarrow (A, Z)]$.

iv. Given $g_1 \in \operatorname{Im}[(X', X) \rightarrow (X', Y)]$ we obtain

$$\begin{array}{ccc} X' & \rightarrow & A \\ {\scriptstyle g_1}\downarrow & & \downarrow{\scriptstyle 0} \\ Y & \rightarrow & Z \end{array}$$

and $\langle 0, g_1\rangle \in R_1$ · Given $g_2 \in \operatorname{Im}[(Y', Y) \rightarrow (X', Y)]$ let

$$\begin{array}{ccc} X' & \rightarrow & Y' \\ & {\scriptstyle g_2}\searrow \quad \swarrow{\scriptstyle h} & \\ & Y & \end{array}$$

commute. We obtain

$$\begin{array}{ccc} X' & \rightarrow & A \\ & & \downarrow \\ {\scriptstyle g_2}\downarrow & & Y' \\ & & \downarrow{\scriptstyle h} \\ Y & \rightarrow & Z \end{array}$$

and $\langle A \rightarrow Y' \rightarrow Z, g_2\rangle \in R_1$. ∎

For any relations $R \subset G \times H$ and subgroups $G_1 \subset G$, $H_1 \subset H$ such that $G_1 \underset{R}{|} H_1$ we obtain a monomorphism

$$\frac{\operatorname{Dom} R}{G_1} \rightarrow \frac{H}{H_1}.$$

In this case the left hand side is, by definition, $\operatorname{Ext}'(A, B)$ and we obtain

Proposition 8.15. $\operatorname{Ext}'(A, B)$ *is a subgroup of*

$$\frac{(X', Y)}{\operatorname{Im}[(X', X) \rightarrow (X', Y)] + \operatorname{Im}[(Y', Y) \rightarrow (X', Y)]}. \quad ∎$$

Now consider the relation

$$R_2 \subset (Y', SX) \times (A, Z)$$

defined by $\langle f, g\rangle \in R_2$ iff there exists

$$\begin{array}{ccc} A & \to & Y' \\ g\downarrow & \swarrow & \downarrow f \\ Z & \to & SX \end{array}$$

Lemma 8.16. $\operatorname{Ker}[(Y', SX) \to (Y', SY]) \cap \operatorname{Ker}[(Y', SX) \to$
$\to (X', SX)] \underset{R_2}{|} \operatorname{Ker}[(A, Z) \to (A, SX)]$.

Proof. i. Let

$$f \in \operatorname{Ker}[(Y', SX) \to (Y', SY)] \cap \operatorname{Ker}[(Y', SX) \to (X', SX)],$$

$\langle f, g\rangle \in R_2$. There exists

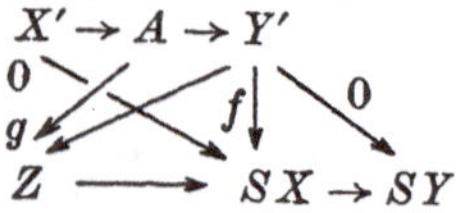

Because $X' \to A$ is epi, it follows that

$$A \xrightarrow{g} Z \to SX = 0 \quad \text{and} \quad g \in \operatorname{Ker}[(A, Z) \to (A, SX)].$$

ii. Let

$$f \in \operatorname{Ker}[(Y', SX) \to (Y', SY)] \cap \operatorname{Ker}[(Y', SX) \to (X', SX)].$$

Because Y' is projective and $Z \to SX \to SY = 0$ there exists

$$\begin{array}{ccc} & Y' & \\ h\swarrow & & \searrow f \\ Z & \to & SX \end{array}$$

Define $A \xrightarrow{g} Z = A \to Y' \xrightarrow{h} Z$. Then $\langle f, g\rangle \in R_2$.

iii. Let $g \in \operatorname{Ker}[(A, Z) \to (A, SX)]$ and $\langle f, g\rangle \in R_2$. There exists

$$\begin{array}{ccc} A & \to & Y' \\ g\downarrow & \overset{h}{\swarrow} & \downarrow f \\ Z & \to & SX \end{array}$$

Hence $Y' \xrightarrow{f} SX \to SY = Y' \xrightarrow{h} Z \to SX \to SY = 0$ and

$$f \in \operatorname{Ker}[(Y', SX) \to (Y', SY)].$$

Moreover,

$$X' \to Y' \xrightarrow{f} SX = X' \to A \xrightarrow{g} Z \to SX = 0$$

and $f \in \operatorname{Ker}[(Y', SX) \to (X', SX)]$.

iv. Let $g \in \operatorname{Ker}[(A, Z) \to (A, SX)]$. Because $A \to Y'$ is mono and Z

injective we obtain

$$\begin{array}{ccc} A & \to & Y' \\ g\updownarrow & \overset{h}{\swarrow} & \downarrow f \\ Z & \to & SX \end{array}$$

and $\langle f, g\rangle \in R_2$. ∎

Lemma 8.17. *Defect* $R_2 = R_2(0) \subset \mathrm{Im}[(A, Y) \to (A, Z)]$.

Proof. If $\langle 0, g\rangle \in R_2$ we have

$$\begin{array}{ccc} A & \to & Y' \\ g\downarrow & \overset{h}{\swarrow} & \downarrow 0 \\ Z & \to & SX \end{array}$$

Y' is projective, $Y \to Z \to SX$ exact, hence there exists $\begin{array}{ccc} & Y' & \\ h'\swarrow & & \searrow h \\ Y & \to & Z \end{array}$

Thus $A \xrightarrow{g} Z = A \to Y' \xrightarrow{h'} Y \to Z$ and $g \in \mathrm{Im}[(A, Y) \to (A, Z)]$. ∎

Given any relation $R \subset G \times H$ and subgroups $G_1 \subset G$, $H_1 \subset H$ such that $G_1 \underset{R}{|} H_1$ we obtain an epimorphism $G_1 \to H_1/\mathrm{Defect}\ R$.

Proposition 8.18. *There is an epimorphism*
$\mathrm{Ker}[(Y', SX) \to (Y', SY)] \cap \mathrm{Ker}[Y', SX) \to (X', SX)] \to \mathrm{Ext}^1(A, B)$. ∎

We may compose the relations R_1 and R_2.

$$R_1 R_2 \subset (Y', SX) \times (X', Y).$$

$\langle f, g\rangle \in R_1 R_2$ iff there exists

$$\begin{array}{ccccc} X' & \to & Y' & & \\ g\downarrow & & \downarrow & \searrow f & \\ Y & \to & Z & \to & SX \end{array} \tag{8.19}$$

$R_1 R_2$ is the Toda bracket operation. The image of $R_1 R_2$ is

$$\mathrm{Ext}'(\mathrm{Im}[X' \to Y'], \mathrm{Im}[X \to Y]).$$

For any $A \in \mathscr{F}$, $\{E^n(A, A)\}$ is a graded ring. If we specialize to $A = \mathrm{Im}[S^0 \xrightarrow{p} S^0]$ then $E_{3n}(A, A) = E^{-3n}(A, A) = \dfrac{pG_n}{p^2 G_n}$, $E_{3n+1}(A, A) = \dfrac{\mathrm{Ann}_n(p)}{\mathrm{Ann}_n(p^2)}$.
And in this case we obtain an explicit formula for $E_{3n+2}(A, A)$. We need

Theorem 8.20. *Let $S' \xrightarrow{p} S^0 \to C_p \to S' \xrightarrow{p} S'$ be a mapping cone sequence. Then*

$$p \cdot 1_{C_p} = 0 \text{ if } p \text{ is odd}$$

$$2 \cdot 1_{C_2} = C_2 \to S^1 \xrightarrow{\eta} S^0 \to C_2 \neq 0, \ \eta \text{ being the Hopf map.}$$

Proof. $\begin{array}{ccccccc} & S^0 & \to & C_p & \to & S^1 & \xrightarrow{p} S^1 \\ 1\swarrow & \downarrow p & & \downarrow p & & \downarrow p & \swarrow 1 \\ S^0 & \xrightarrow[p]{} & S^0 & \to & C_p & \to & S^1 \end{array}$ commutes, hence $\begin{array}{ccc} S^0 & \to & C_p \\ 0\searrow & \downarrow p & \searrow 0 \\ & C_p & \to S^1 \end{array}$

commutes.

There exists $\begin{array}{ccc} & & C_p \\ & g\swarrow & \downarrow p \\ S^0 & \to & C_p \end{array}$ and hence $\begin{array}{cccc} & & S^0 & \to C_p \\ & h\swarrow & & \swarrow g \\ S^0 & \xrightarrow[p]{} & S^0 & \end{array}$

$$S^0 \xrightarrow{p} S^0 \xrightarrow{h} S^0 \xrightarrow{p} S^0 = S^0 \xrightarrow{p} S^0 \to C_p \xrightarrow{g} S^0 = 0$$

and thus $h = 0$. There exists $\begin{array}{ccc} C_p & \to & S^1 \\ g\downarrow & \swarrow \alpha & \\ S^0 & & \end{array}$ and

$$C_p \xrightarrow{p} C_p = C_p \to S^1 \xrightarrow{\alpha} S^0 \to C_p.$$

Now $(S^1, S^0) \cong Z_2$ and hence $2\alpha = 0$ and $2p \cdot 1_{C_p} = 0$. For odd p therefore we know that $p^2 \cdot 1_{C_p}$ (6.5), and thus $p \cdot 1_{C_p} = 0$.

Specializing to the case $p = 2$ we recognize SC_2 as the projective plane and appeal to geometry to obtain the fact that $2 \cdot 1_{C_2} \neq 0$. Thus $\alpha \neq 0$ and if we call the non-zero element of (S^1, S^0), the Hopf map, η, we obtain the desired result. ▮

Theorem 8.21. $E_{3n+2}(\mathrm{Im}(p \cdot 1_{S^0}), \mathrm{Im}(p \cdot 1_{S^0}.)) = \dfrac{G_n}{\{\alpha \in G_n \mid \eta\alpha \in pG_{n-1}\}}$, *where η is the Hopf map.*

Proof. Consider the function $t\colon \mathrm{Ker}[(S^n, S^0) \xrightarrow{p} (S^n, S^0)] \to \dfrac{(S^n, S^{-1})}{p(S^n, S^{-1})}$ defined by $t(x) = \bar{y}$ if there exists

$$\begin{array}{ccccc} & & S^n & \xrightarrow{p} & S^n \\ & y\downarrow & & z\downarrow & \searrow x \\ S^{-1} & \to & C_p & \to & S^0 \end{array}$$

We showed above that $E_{3n+2} = \mathrm{Im}(t)$. (8.19)

If $\bar{y} = 0$, i.e. $y \in p(S^n, S^{-1})$ then there exists $\begin{array}{ccc} & & S^n \\ & \swarrow & \downarrow y \\ S^{-1} & \xrightarrow[p]{} & S^{-1} \end{array}$ and

$$S^n \xrightarrow{y} S^{-1} \to C_p = 0.$$

Hence $t(x) = 0$ iff $pz = 0$. Now using the last proposition $pz = S^n \xrightarrow{z} C_p \xrightarrow{p} C_p = S^n \xrightarrow{z} C_p \to S^0 \xrightarrow{\eta} S^{-1} \to C_p = S^n \xrightarrow{x} S^0 \xrightarrow{\eta} S^{-1} \to C_p$.
Thus $t(x) = 0$ iff $S^n \xrightarrow{x} S^0 \xrightarrow{\eta} S^{-1} \in p(S^n, S^{-1})$. Thus

$$\operatorname{Im}(t) = (S^n, S^0)/\{x \mid \eta x \in p(S^n, S^{-1})\}. \blacksquare$$

The following proposition contains more information than would a simple summary of the above. The additional information (i.e. when $j > 1$) is proved similarly but more easily:

Theorem 8.43. *For p a prime number*

$$E_{3n+a}(\operatorname{Im}(p^j \cdot 1_{S^0}), \operatorname{Im}(p \cdot 1_{S^0})) = \begin{cases} p^j G_n / p^{j+1} G_n & \text{if} \quad a = 0 \\ \operatorname{ann}_n(p^{j+1})/\operatorname{ann}_n(p^j) & \text{if} \quad a = 1 \\ G_n/\{\alpha \mid \eta\alpha \in 2\, G_{n-1}\} & \text{if} \quad a = 2,\ p = 2,\ j = 1 \\ 0 & \text{if} \quad a = 2,\ p^j > 2 \end{cases}$$

9. The Generating Hypothesis

In this section we explore some of the consequences of a single hypothesis,
The generating hypothesis:

The spheres generate the stable category of finite complexes.

A direct translation:

If X, Y are finite complexes, $X \xrightarrow{f} Y$ a map such that

$$\mathcal{S}(S^n, X) \to \mathcal{S}(S^n, Y) = 0$$

all n, then $f = 0$ in $\mathcal{S}$.

Because of Spanier-Whitehead duality the generating hypothesis is equivalent to:

The spheres co-generate $\mathcal{S}$, i.e. if X and Y are finite complexes and $f: X \to Y$ is such that $\mathcal{S}(Y, S^n) \to \mathcal{S}(X, S^n)$ all n, then $f = 0$ in $\mathcal{S}$.

Let $\mathcal{M}$ be the category of graded modules over the graded ring $G = \{\mathcal{S}(S^n, S^0)\}$. The maps of $\mathcal{M}$ are of degree 0. $\mathcal{M}$ is easily seen to be an abelian category. Define $\pi: \mathcal{F} \to \mathcal{M}$ by $\pi(A) = \{\mathcal{F}(S^n, A)\}$. We shall use $\pi_n X$ to denote the n'th component of πX. π is easily seen to be exact (the spheres are projective).

The generating hypothesis can be rerestated as: $\pi: \mathscr{F} \to \mathscr{M}$ is an embedding. For an ideal $\mathfrak{A} \subset G$ we define $\operatorname{ann}(\mathfrak{A}) = \{\beta \mid \beta\mathfrak{A} = 0\}$.

Proposition (G. Whitehead) 9.1. *The generating hypothesis implies that for each* $\alpha \in G$, $\operatorname{ann}(\operatorname{ann}(\alpha)) = (\alpha)$.

Proof. As always $(\alpha) \subset \operatorname{ann}(\operatorname{ann}(\alpha))$.

For the reverse inclusion let $\alpha \in G_n$ and $0 \to C \to S^0 \xrightarrow{\alpha} S^{-n}$ be exact and $\beta \in \operatorname{ann}(\operatorname{ann}(\alpha))$, $\beta \in G_m$. Observe that $\pi(C \to S^0 \xrightarrow{\beta} S^{-m}) = 0$ because for each $S^q \to C$ it is the case that $S^q \to C \to S^0 \in \operatorname{ann}(\alpha)$ hence $S^q \to C \to S^0 \xrightarrow{\beta} S^{-m} = 0$.

The generating hypothesis then says that $C \to S^0 \xrightarrow{\beta} S^{-m} = 0$ and hence there exists a commutative triangle
$$\begin{array}{ccc} S^0 & \xrightarrow{\alpha} & S^{-n} \\ & {\scriptstyle\beta}\searrow \quad \swarrow{\scriptstyle\gamma} & \\ & S^{-m} & \end{array}$$
and $\beta \in (\alpha)$. ∎

Corollary (Milnor-Kervaire) 9.2. *For each prime p and positive integer j there exists $\alpha \in G$ such that* $p^j\alpha = 0$, $p^{j-1}\alpha \neq 0$.

Proof based on Generating Hypothesis. If $\operatorname{ann}(p^j) = \operatorname{ann}(p^{j-1})$ then by the last proposition $(p^j) = (p^{j-1})$ which is surely not the case. Hence $\operatorname{ann}(p^j) \neq \operatorname{ann}(p^{j-1})$. But clearly $\operatorname{ann}(p^{j-1}) \subset \operatorname{ann}(pj)$ and there exists $\alpha \in \operatorname{ann}(p^j) - \operatorname{ann}(p^{j-1})$. ∎

Proposition 9.3. *The generating hypothesis implies that for* $\alpha, \beta \in G$, $(\alpha) \cap (\beta) = 0$ *iff the orders of α and β are relatively prime.*

Proof. $\operatorname{Im}(\alpha) \cap \operatorname{Im}(\beta) = 0 \Leftrightarrow \exists$
$$\begin{array}{ccc} \operatorname{Im}(\alpha) \oplus \operatorname{Im}(\beta) & \to & S^0 \\ {\scriptstyle u_1 p_1}\downarrow & & \downarrow j \\ \operatorname{Im}(\alpha) \oplus \operatorname{Im}(\beta) & \to & S^0 \end{array}$$
$\Leftrightarrow \exists_j$ such that $j\alpha = \alpha$, $j\beta = 0 \Leftrightarrow \exists_j$ such that $j \equiv 1 \pmod{\operatorname{ord}(\alpha)}$, $j \equiv 0 \pmod{\operatorname{ord}(\beta)} \Leftrightarrow \operatorname{ord}(\alpha)$ and $\operatorname{ord}(\beta)$ are relatively prime.

Now if $\operatorname{Im}(\alpha) \cap \operatorname{Im}(\beta) \neq 0$ then the generating hypothesis implies the existence of $\gamma \in G$, such that $0 \neq \operatorname{Im}(\gamma) \subset \operatorname{Im}(\alpha) \cap \operatorname{Im}(\beta)$ and $\gamma \in (\alpha) \cap (\beta)$.

Proposition 9.4. *The generating hypothesis implies that for each $\alpha \neq 0$ there exist* $\beta \in G$, $\deg \beta > 0$, $\alpha\beta \neq 0$.

Proof. If $\alpha \in G_0$ the existence of β comes from the Kervaire-Milnor result.

If $\alpha \in G_n$, $n > 0$ let p be a prime divisor of ord(α) and choose j large enough such that $p^j|\text{ord}(\beta) \Rightarrow \deg \beta \geq n$.

By the Kervaire-Milnor result there does exist $m > n$, $\beta \in Gm$, ord$\beta = p^j$. By the last proposition $(\alpha) \cap (\beta) \neq 0$. Hence there exists $\gamma \in G$ such that $0 \neq \gamma\alpha \in (\beta)$. But $m > n$ implies that $\gamma \notin G_0$. ∎

Proposition (G. Whitehead) 9.5. *The generating hypothesis implies that for $X \in \mathcal{S}$ if πX is finitely generated (as a graded module over G) then X is a wedge of spheres.*

Proof. Let $S^{a_1} \xrightarrow{\alpha_1} X, \ldots, S^{a_n} \xrightarrow{\alpha_n} X$ generate πX. Then for

$$B = S^{a_1} \vee \cdots \vee S^{a_n}$$

it is the case that $\pi(B \xrightarrow{\alpha} X)$ is epi and hence by the generating hypothesis, $B \xrightarrow{\alpha} X$ is epi. Thus X is a direct summand of a wedge of spheres and by prop. 6.4, X is a wedge of spheres.

Proposition 9.6. *The generating hypothesis implies that the ring G is totally non-coherent, i.e. for every exact $0 \to K \to S^{a_1} \vee \cdots \vee S^{a_n} \xrightarrow{\alpha} S^0$, α not zero or epimorphic, then K is not finitely generated.*

Proof. By the Shanuel lemma it suffices to assume that $\pi(\text{Im}(\alpha))$ can not be generated by fewer than n elements. Let

$$S^{a_1} \vee \cdots \vee S^{a_n} \xrightarrow{\alpha} S^0 \to C \to S^{a_1+1} \vee \cdots \vee S^{a_n+1} \to S^1$$

be a mapping cone sequence. If πK were finitely generated then πC would be finitely generated and by the last proposition $C = S^{b_1} \vee \cdots \vee S^{b_m}$ Now by the minimality of n it follows that $S^{a_i} \xrightarrow{\alpha_i} S^0 \neq 0$ all i and hence that $a_i \geq 0$ all i. If $b_i > 0$ then $S^0 \to S^{b_1} \vee \cdots \vee S^{b_m} \xrightarrow{p_i} S^{b_i} = 0$ and $S^{b_i} \xrightarrow{u_i} S^{b_1} \vee \cdots \vee S^{b_m} \to S^{a_1+1} \vee \cdots \vee S^{a_n+1}$ would be monomorphic and S^{b_i} would appear as a direct summand of Ker$S(\alpha)$ which contradicts the minimality of n. Hence $b_i \leq 0$ all i. Thus $S^{b_1} \vee \cdots \vee S^{b_m} \to S^{a_1+1} \vee \cdots \vee S^{a_n+1} = 0$ and $S^{a_1+1} \vee \cdots \vee S^{a_n+1} \to S^0$ is monomorphic. But there are no injective proper subobjects of S^0 because there are no proper idempotents of S^0. ∎

Proposition 9.7. *The generating hypothesis implies that $\pi: \mathcal{S} \to \mathcal{M}$ is a full embedding.*

Proof. Let $X, Y \in \mathcal{S}, f: \pi X \to \pi Y \in \mathcal{M}$. We wish to find $\bar{f}: X \to Y \in \mathcal{S}$ such that $\pi\bar{f} = f$.

For each n, chose $S^n \xrightarrow{\alpha_{n,1}} X, \ldots, S^n \xrightarrow{\alpha_{n,m_n}} X$ a generating set of $\pi_n X$, and let $W_n = \sum_{j \leq n} \sum_{i \leq m_j} S^j \xrightarrow{\alpha} X$ be the map whose $\langle j, i\rangle$'th component is

$\alpha_{j,i}$. Define $W_n \xrightarrow{h_n} Y$ to be the map whose $\langle j, i\rangle$'th component is $f\alpha_{j,i}$. Let $0 \to K_n \to W_n \to X$ be exact. $\pi(K_n \to W_n \xrightarrow{h_n} Y) = 0$ because for $S^q \to K_n$ we may express $S^q \to K_n \to W_n$ as $\sum n_{\langle j,i\rangle} \beta_{\langle j,i\rangle}$ and since $S^q \to W_n \to X = 0$ it is the case that $\sum \alpha_{\langle j,i\rangle} \beta_{\langle j,i\rangle} = 0$ hence

$$\sum f(\alpha_{\langle j,i\rangle}) \beta_{\langle j,i\rangle} = 0 \text{ i. e. } h_n \sum \mu_{\langle j,i\rangle} \beta_{\langle j,i\rangle} = 0 \text{ and } S^q \to W_n \xrightarrow{h_n} Y = 0.$$

The generating hypothesis thus implies that $K_n \to W_n \xrightarrow{h_n} Y = 0$ and the injectiveness of Y implies the existence of $\begin{matrix} W_n \to X \\ h_n \searrow \quad \swarrow f_n \\ Y \end{matrix}$. If d is large enough so that $H_m X = 0$, $m \geq d$ then $\sum_m H_m \operatorname{Cok}(W_a \to X)$ is torsion. Since for $m > d$ there exists $\begin{matrix} W_d \to W_n \\ \searrow \quad \swarrow \\ X \end{matrix}$ it follows that $W_d \to X \xrightarrow{f_m} Y = h_d$, and that there exists $\begin{matrix} X \to \operatorname{Cok}(W_d \to X) \\ f_d - f_m \searrow \quad \swarrow \\ Y \end{matrix}$. But $\mathscr{F}(\operatorname{Cok}(W_d \to X), Y)$ is finite and thus the sequence $\{f_m\}$ has a constant cofinal subsequence, i.e. a convergent subsequence, the value of which will serve as $\bar{f}$, $\pi(\bar{f}) = f$. ∎

Theorem 9.8. *The generating hypothesis is equivalent to the fullness of $\pi: \mathscr{F} \to \mathscr{M}$.*

Proof. $\Rightarrow$ Let $A, B \in \mathscr{F}$, $X' \to X$, $Y \to Y'' \in \mathscr{S}$, $X' \to X \to A \to 0$, $0 \to B \to Y \to Y''$ exact. Let f: $\pi A \to \pi B \in \mathscr{M}$ and use the last proposition to find $g: X \to Y \in \mathscr{F}$ such that $\pi g = \pi X \to \pi A \xrightarrow{f} \pi B \to \pi Y$. Because $\pi X' \to \pi X \xrightarrow{\pi g} \pi Y = 0$ the generating hypothesis say that $X' \to X \xrightarrow{g} Y = 0$ and there exists $A \xrightarrow{h} Y$ such that $\pi h = \pi A \xrightarrow{f} \pi B \to \pi Y$. Because $\pi A \xrightarrow{\pi h} \pi Y \to \pi Y'' = 0$ we obtain $A \xrightarrow{h} Y \to Y'' = 0$ and there exists $\bar{f}$: $A \to B \in \mathscr{F}$ such that $\pi(\bar{f}) = f$.
$\Leftarrow$ Let $X \xrightarrow{f} Y \in \mathscr{S}$ be such that $\pi f = 0$, and let $X \xrightarrow{f} Y \xrightarrow{g} F \to 0$ be exact. We wish to show that $0 \to Y \to F$ is exact. Because πg is an isomorphism and π is full, it follows that there exists $h: F \to Y$ such that $\pi(h\, g) = 1$. The stable Whitehead theorem (prop 4.2) implies that h is an automorphism and hence that g is a monomorphism. ∎

Theorem 9.9. *The generating hypothesis is true iff for each $X \in \mathscr{S}$ $|\operatorname{End}(X)| < \infty$ it is the case that πX is injective.*

Proof. $\Rightarrow$

Let $X \in \mathscr{S}$ be such that $|\mathrm{End}(X)| < \infty$ and suppose A is a finitely generated submodule of πS^n, and that $A \to \pi X \in \mathscr{M}$. Let W be a wedge of spheres and $W \to S^n$ a map such that $\pi[\mathrm{Im}(W \to S^n)] = A$. By proposition 9.7 there exists $W \xrightarrow{g} X$ such that $\pi g = \pi W \to A \xrightarrow{f} \pi X$ and as in the proof of proposition 9.7

$$\pi K \to \pi W \to \pi X = 0 \text{ where } K = \mathrm{Ker}(W \to S^n).$$

Hence there exists $\begin{array}{ccc} W & \to & S^n \\ g\searrow & & \swarrow \bar f \\ & X & \end{array}$ (X is injective in $\mathscr{F}$). Clearly then $\begin{array}{ccc} A & \to & \pi(S^n) \\ f\searrow & & \swarrow \pi\bar f \\ & \pi X & \end{array}$ commutes.

Now let A be an arbitrary submodule of πS^n. Let $A_1 \subset A_2 \subset \ldots$ be finitely generated submodules such that $\cup A_1 = A$. Given $f\colon A \to \pi X$ let $f_n\colon \pi S^n \to \pi X$ be such that $f_n | A_n = f | A$. The sequence $\{f_n\}$ has a constant cofinal subsequence because $\mathscr{M}(\pi S^n, \pi X) = \mathscr{F}(S^n, X)$ is finite.

Hence there exists $\begin{array}{ccc} A & \to & \pi S^n \\ f\searrow & & \swarrow \bar f \\ & \pi X & \end{array}$.

Suppose $\pi X \subset B$ and n is such that $\pi_n X \neq B_n$. Let $\beta \in B_n - \pi_n X$ and $\begin{array}{ccc} A & \to & S^n \\ f\downarrow & & \downarrow\beta \\ \pi X & \to & B \end{array}$ a pullback. Let $\begin{array}{ccc} A & \to & \pi S^n \\ f\searrow & & \swarrow \bar f \\ & \pi X & \end{array}$ commute and $\bar f(1_{S^n}) = \alpha \in \pi X$.

Then the submodule of B generated by $(\beta - \alpha)$ is a nontrivial submodule that meets πX triviality. Hence πX has no essential extensions.

$\mathscr{M}$ is a Grothendieck category. Theorem 6.13 *Abelian Categories* now implies that πX is injective.

Suppose πX is injective. Then

$$\mathscr{M}(\pi(-), \pi X)\colon \mathscr{F} \to \mathscr{G}$$

is exact, as is $\mathscr{F}(-, X)\colon \mathscr{F} \to \mathscr{G}$. The transformation

$$\mathscr{F}(-, X) \xrightarrow{\pi} \mathscr{M}(\pi(-), \pi X)$$

is an equivalence on spheres and hence by the 2nd Dold-lemma 4.7, it is an equivalence everywhere. Thus given $Y \xrightarrow{f} Z \neq 0$, $\pi f = 0$ we let X be a torsion space $Y \xrightarrow{f} Z \xrightarrow{g} X \neq 0$ (proposition 6.8 said that the torsion spaces cogenerate) and obtain a contradiction from the isomorphism $\mathscr{F}(Y, X) \xrightarrow{\pi_Y} \mathscr{M}(\pi Y, \pi X)$. ▮ A proposition having no dependence on the generating hypothesis, but relevant to this last proposition.

Proposition 9.(10). *πS^0 is not injective in $\mathcal{M}$.*

Proof. Let $G_+ = \{\pi_n S^0\}_{n=1}^{\infty}$. We shall define a map $G_+ \to \pi S^0$ which can not be extended.

Let l_n be the exponent of the group $\pi_1 S^0 \oplus \cdots \oplus \pi_n S^0$. Define $a_1 = 0$, $a_2 = l_1$, $a_3 = l_1 + l_2, \ldots, a_n = l_1 + \cdots + l_{n-1} \ldots$

Define $G_+ \xrightarrow{h} G_+$ by $h|\pi_n S^0 = a_n$. Now for $n > m$, $h/\pi_n S^0 = a_m$ because $a_n \equiv a_m (\bmod\, l_n)$. Hence given

$$\alpha \in \pi_n S^0, \quad \beta \in \pi_m S^0, \quad h(\alpha) = a_{n+m}\,\alpha, \quad h(\alpha\beta) = a_{n+m}\,\beta\alpha = h(\alpha)\,\beta$$

and $h \in \mathcal{M}$. Suppose $\bar{h} : \pi S^0 \to \pi S^0$, $\bar{h}|G_+ = h$. Let $a = \bar{h}(1_{S^0})$. Then $h|G_n = \underset{a}{\mathrm{a}}$ and $a \equiv a_n (\bmod \exp(\pi_n S^0))$. But be the Milnor-Kervaire result, $\{\exp(\pi_n S^0)\}$ is unbounded and hence $a = a_n$, surely a contradiction. ∎

Let $\mathcal{T}or$ be the full subcategory of $\mathcal{S}$ of torsion spaces ($X \in \mathcal{T}or$ $|\mathrm{End}(X)| < \infty$). $\mathcal{F}\,\mathcal{T}or$ the abelianization of $\mathcal{T}or$. $\mathcal{F}\,\mathcal{T}or$ is a full subcategory of $\mathcal{F}$ but should not be confused with the full subcategory of torsion objects of $\mathcal{F}$. Note that the image of the Hopf map $S^1 \xrightarrow{\eta} S^0$ is torsion but not an object of $\mathcal{F}\,\mathcal{T}or$, that is, it is not isomorphic to the image of a map between torsion spaces.

Theorem 9.(11). *The generating hypothesis implies that $\mathcal{F}\,\mathcal{T}or$ is equivalent to the minimal category of $\mathcal{M}$ which contains the injective envelopes of $\{P/n\,P/P$ finity generated projective, $n > 0\}$ and is closed under the formation of cokernels of maps, and injective envelopes.*

Proof. The generating hypothesis easily implies that $\pi : \mathcal{F} \to \mathcal{M}$ carries essential extension into essential extensions, and thus by Proposition 9.9 it carries torsion injective envelopes into injective envelopes. $\pi(\mathcal{F}\,\mathcal{T}or)$ contains $\pi\,\mathrm{Cone}(S^j \xrightarrow{n} S^j)$ which are now seen to be the injective envelopes of $\pi S^j/n\pi S^j$ by Proposition 7.4. πS^j is a finitely generated projective.

The repletion of $\pi(\mathcal{F}\,\mathcal{T}or)$ is readily closed under the operations of cokernel and injective envelope. And if $P \in \mathcal{M}$ is a finitely generated projective, then it is a retract of $\pi S^{j_1} \oplus \cdots \oplus \pi S^{j_k}$ hence is itself of the form $\pi S^{n_1} \oplus \cdots \oplus \pi S^{n_k}$ (Proposition 6.4). Thus

$$P/m\,P = \pi\,\mathrm{Cok}(S^{n_1} \vee \cdots \vee S^{n_k} \xrightarrow{m} S^{n_1} \vee \cdots \vee S^{n_k})$$

and the injective envelope of $P/m\,P$ is

$$\pi\,\mathrm{Cone}(S^{n_1} \vee \cdots \vee S^{n_k} \xrightarrow{m} S^{n_1} \vee \cdots \vee S^{n_k})\,.$$

We need only show that no proper subcategory of $\mathscr{F}\mathscr{T}or$ which contains all such cones is closed under the operation of cokernel, and injective envelope. The proof is an imitation of the proof of 4.5 with $\{\mathrm{Cone}(S^n/aS^n)\}_{a,n}$ playing the role of $\{S^n\}_n$. ∎

The repletion of $\pi(\mathscr{F}\mathscr{T}or)$ is describable using only categorical predicates in $\mathscr{M}$. ($P \in \mathscr{M}$ is a finitely generated projective iff $(P, -)$ preserves all right-limits). Among other things, we may conclude that it is subcategory invariant under all automorphisms. More philosophically, it is describable in entirely algebraic terms, and thus, modulo the generating hypothesis, $\mathscr{F}\mathscr{T}or$ and $\mathscr{T}or$ itself have been algebraically represented.

The exactness of π delivers a natural transformation $\mathrm{Ext}_{\mathscr{F}}(A, B) \to \to \mathrm{Ext}_{\mathscr{M}}(\pi A, \pi B)$, most easily seen by letting $0 \to A \to E \to B \to 0$ represent an element in $\mathrm{Ext}_{\mathscr{F}}(A, B)$ and sending it to the element in $\mathrm{Ext}_{\mathscr{M}}(\pi A, \pi B)$ represented by $0 \to \pi A \to \pi E \to \pi B \to 0$.

The fullness of π would imply that $\mathrm{Ext}_{\mathscr{F}}(A, B) \to \mathrm{Ext}(\pi A, \pi B)$ is mono, for if $\pi A \to \pi E$ splits, then so does $A \to E$. Moreover:

Theorem 9.(12). *The generating hypothesis implies that* $\mathrm{Ext}_{\mathscr{F}}(A, B) \to \to \mathrm{Ext}(\pi A, \pi B)$ *is an isomorphism, i.e., that given* $0 \to \pi A \to E \to \to \pi B \to 0$ *exact in* $\mathscr{M}$ *there exists* $C \in \mathscr{F}$ *and an isomorphism* $E \to \pi C$.

Proof. We shall let $\mathrm{Im}(\pi)$ be the repletion of the image of π, that is, $\mathrm{Im}(\pi)$ is closed under isomorphic copy. We wish to show that $\mathrm{Im}(\pi)$ is closed under exact extensions.

Lemma 9.(12) 1. *Let* $E_0 \subset E \in \mathscr{M}$ *be such that* $E/E_0 \in \mathrm{Im}(\pi)$, *and let* $\alpha \in E - E_0$. *Then for* $E_1 \subset E$ *generated by* E_0 *and* α *every map* $E_0 \to \pi X$, $X \in \mathscr{S}$ *extends to* E_1 *and* $E/E_1 \in \mathrm{Im}(\pi)$.

Proof of Lemma. Let $\deg \alpha = n$ and $\pi S^n \to E$ send 1_{S^n} to α. Then $\mathrm{Im}(E_0 \oplus \pi S^n \to E) = E_1$. Let $B \in \mathscr{F}$ and $0 \to E_0 \to E \to \pi B \to 0$ be exact. Because π is full there exists $g: S^n \to B$ such that $\pi g = \pi S^n \to \to E \to \pi B$. Let $0 \to K \to S^n \to B$ be exact. Then

$$\begin{array}{ccc} \pi K & \to & \pi S^n \\ \downarrow & & \downarrow \\ E_0 & \to & E \end{array}$$

is a pushout. Let $h: K \to X$ be such that $\pi h = \pi K \to E_0 \to \pi X$. Because X is injective in $\mathscr{F}$ there exists an extension of h to S^n and hence an extension of $E_0 \to \pi X$ to E_1.

E/E_1 is isomorphic to the cokernel of $E_0 \oplus \pi S^n \to E$ which is isomorphic to the cokernel of $\pi S^n \to \pi B$ which is isomorphic to

$$\pi\,\mathrm{Cok}(S^n \to B).$$

Lemma 9.(12) 2. *Let E be a countable module in $\mathcal{M}$, $E_0 \subset E$ such that $E/E_0 \in \operatorname{Im}(\pi)$. Then every map $E_0 \to \pi X$, $X \in \mathcal{S}$ extends to E.*

Proof of Lemma. There exists a sequence $E_0 \subset E_1 \subset \cdots$ such that for each n, E_{n+1} is generated by E_n and a single element of E, $E/E_n \in \operatorname{Im}(\pi)$, and $\cup E_n = E$. The last lemma thus allows us to extend any map $E_0 \to \pi X$ compatibly through each E_n and the "union" of such extensions is an extension to all of E.

Now for the theorem itself: let $A, B \in \mathcal{F}$ and $0 \to \pi A \to E \to \pi B \to 0$ be exact. Let $X \in \mathcal{S}$, $A \to X$ a monomorphism. E is certainly countable and the last lemma yields a commutative

$$\begin{array}{ccc} \pi A & \to & E \\ & \searrow & \downarrow \\ & & \pi X \end{array}$$

The map $E \to \pi X \oplus \pi B$ is a monomorphism.

Let $0 \to A \to X \to C \to 0$ be exact in $\mathcal{F}$. Then
$$\begin{array}{ccc} E & \to & \pi B \\ \downarrow & & \downarrow \\ \pi X & \to & \pi C \end{array}$$
is a pushout and hence $E \to \pi X \oplus \pi B \to \pi C \to 0$ is exact. Thus

$$E \cong \pi(\operatorname{Ker}(X \oplus B \to C)). \ ▮$$

Theorem 9.(13). *The generating hypothesis implies that $\mathcal{F}$ is equivalent to the minimal subcategory of $\mathcal{M}$ which contains the finitely generated projectives and is closed under the formation of kernels, cokernels, and exact extensions.* ▮

10. A Reduction of the Generating Hypothesis

Let N be the class of spaces which appear as

$$\operatorname{Cone}(S^{a_1} \vee \cdots \vee S^{a_n} \to S^a)$$

and let N^* be the class of spaces which appear as

$$\operatorname{Cone}(S^a \to S^{a_1} \vee \cdots \vee S^{a_n}).$$

In terms of cell complexes, $X \in N$ iff X is the result of attaching a finite number of cells simultaneously to a sphere, $Y \in N^*$ iff Y is the result of attaching a single cell to a finite wedge of spheres.

Theorem 10.1. *The generating hypothesis is true iff for every $X \in N$, $Y \in N^*$ and $f \in (X, Y)$ it is the case that $\pi f = 0 \Rightarrow f = 0$.*

Restatement 10.2. *If there is any counter-example to the generating hypothesis there is a counter example of the form $X \to Y$ $X \in N$, $Y \in N^*$.*

Lemma 10.3. *Given* $Y \in \mathscr{S}$ *such that for all* $N \in \mathbf{N}$ *it is the case that then* $\mathscr{S}(N, Y) \to \mathscr{M}(\pi N, \pi Y)$ *is mono for all* $X \in \mathscr{S}$ *then for all* $X \in \mathscr{S}$, $\mathscr{S}(X, Y) \to \mathscr{M}(\pi X, \pi Y)$ *is mono.*

Proof of Lemma. Fix an integer n. Define $\boldsymbol{A}$ to be the class of non-trivial spaces such that $X \in \boldsymbol{A}$ iff $H_j(X) = 0$ all $j > n$, and $H_n X$ is free (perhaps trivial). For $X \in \boldsymbol{A}$ let $c(X)$ be the dimension of the first non-vanishing homology group of X. Clearly $c(X) \leqq n$. If $c(X) = n$ then X is a wedge of spheres and $(X, Y) \to (\pi X, \pi Y)$ is mono.

Define for $X \in \boldsymbol{A}$, $g(X)$ to be the minimal number of generators needed for $H_{c(X)} X$. Assuming the hypothesis of the lemma we now give an inductive proof for the injectivity of $(X, Y) \to (\pi X, \pi Y)$, $X \in \boldsymbol{A}$. The induction is based on the function

$$F : \boldsymbol{A} \to Z \times Z : X \to \langle n - c(X), g(X) \rangle ,$$

where $Z \times Z$ is ordered lexicagraphically. Since F has only non-negative values, Im F is well-ordered. Furthermore gX is never zero, and we have already taken care of the case when $n - c(X) = 0$.

Sub-Lemma 10.31. *If* $X \in \boldsymbol{A}$, *and* $n - c(X) \neq 0$ *then there is a mapping cone sequence* $S^c \to X \to A \to S^{c+1} \to SX$ *such that* $A \in \boldsymbol{A}$ *and*

$$FA < FX .$$

Proof of Sub-Lemma. Let $\alpha_1, \ldots, \alpha_g$ be a set of generators for $H_c X = (S^c, X)$ where $c = c(X)$, $g = g(X)$. Let $S^c \xrightarrow{\alpha_1} X \to A \to S^{c+1} \to SX$ be a mapping cone sequence. Because $c < n$, we have for $j > n$

$$0 = H_j X \to H_j A \to H_j S^{c+1} = 0$$

exact, hence that $H_j A = 0$ and for $j = n$ we have

$$0 = H_n(S^c) \to H_n X \to H_n A \to H_n S^{c+1}$$

and $H_n A$ is an extension of free groups hence free. Thus $A \in \boldsymbol{A}$. For $j < c$ we have $0 = H_j X \to H_j A \to H_j S^{c+1}$ exact and hence $c(A) \geqq c(X)$. At $j = c$ we have $Z = H_c S^c \to H_c X \to H_c A \to H_c S^{c+1} = 0$ exact. The image under $H_c X \to H_c A$ of $\alpha_2 \ldots a_g$ generates $H_c A$. Thus if $g(X) > 1$ then $c(A) = c(X)$, $g(A) < g(X)$ and $FA < FX$. If $g(X) = 1$ then $H_c A = 0$ and $c(A) > c(X)$ and $FA < FX$. **End of Proof of Sublemma.**

Suppose $S^c \to X \to A \to S^{c+1} \to SX$ is as described in the sublemma, and that $(A, Y) \to (\pi A, \pi Y)$ is injective. Let $f : X \to Y$ be such that $\pi(f) = 0$. We shall, modulo the hypothesis of the lemma find $h : A \to Y$ such that

$$\begin{array}{ccc} X & \to & A \\ & f\searrow & \downarrow h \\ & & Y \end{array}$$

commutes and such that $\pi(h) = 0$. Hence from the inductive hypothesis, $f = 0$.

For each m we let W_m be a wedge of spheres and $W_m \to X$ a map such for all $j \leq m$ that $\pi_j W_m \to \pi_j X$ is onto, and $K_m \to W_m$ the kernel of $W_m \to A \to S^{c+1}$. Then $\begin{array}{ccc} K_m & \to & W_m \\ \downarrow & & \downarrow \\ \operatorname{Im}(X \to A) & \to & A \end{array}$ is a pullback. Let N be the desuspension of Cone $(W_j \to S^{c+1})$ and let $N \to W_j \to S^{c+1}$ be exact.

We obtain $\begin{array}{ccc} N & \to & W_m \\ \downarrow & & \downarrow \\ X & \to & A \end{array}$ where $N \to X \oplus W_m \to A$ is exact.

$N \in \boldsymbol{N}$ and hence the hypothesis of the lemma says that $N \to X \to Y = 0$. Because Y is injective there exists

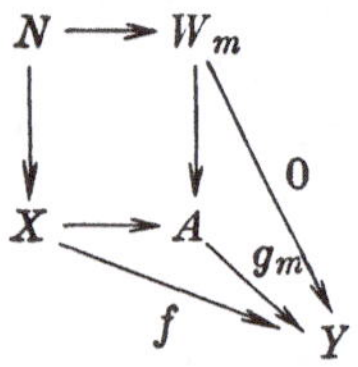

Cok $(W_n \to A) = F$ is torsion (Proposition 7.2). Because for all $j > n$, $W_n \to A \xrightarrow{g_j - g_n} Y = 0$ it follows that $g_j - g_n$ factors through F, and the sequence $\{g_j\}$ has a constant cofinal subsequence ((F, Y) is finite). Thus if we let g be the value of that cofinal sequence, $W_j \to A \xrightarrow{g} Y = 0$ all j. By inductive hypothesis it follows that $g = 0$. But $\begin{array}{ccc} W & \to & A \\ & f\searrow \;\; \swarrow g & \\ & Y & \end{array}$ commutes, hence $f = 0$. **End of proof of Lemma.**

Suppose there is a counter example $X \to Y$ to the generating hypothesis. From the lemma we can find $N \in \boldsymbol{N}$, $N \to Y$ such that $N \xrightarrow{f} Y$ is a counter-example. Either $(f, S^j) = 0$ all j, or not. If not, then there exists a counter-example $N \to Y \to S^j$ and surely $S^j \in \boldsymbol{N}^*$. In the other case we apply Spanier-Whitehead duality to obtain $Y^* \xrightarrow{f^*} N^*$ where $\pi(f^*) = 0$, hence f^* is a counter-example. Applying the lemma once again, we obtain $A \in \boldsymbol{N}$ and a counter-example $A \to N^*$. ∎

The proof generalizes: let $\boldsymbol{C}$ be a class of objects which contains the spheres, and is closed under wedging, suspending, and desuspending. If there is a counter-example to the statement that $\boldsymbol{C}$ generates $\mathscr{S}$ then there is a counter example of the form Cone $(A \to S^a) \to Y$ where $A \in \boldsymbol{C}$.

11. How a Category man finds Counter-Examples

We exhibit the existence of a finite dimensional, but not finite, complex X a stably non-trivial map $X \to S^a$ such that for each $S^b \to X$, it is the case that $S^b \to X \to S^a$ is stably trivial.

Let $\mathscr{F}_{CW}$ be the frobenious category whose projectives are finite dimensional countable CW-complexes. $\mathscr{F}_{CW}$ is not complete but we can ask whether it enjoys the Grothendieck property whenever applicable. That is, given $A \in \mathscr{F}_{CW}$ and an ascending sequence of subobjects $B_1 \subset B_2 \subset \cdots$ with a least upper bound $\cup B_i$ is it the case that for any $C \subset A$, $[C \cap B_i = 0 \text{ all } i] \Rightarrow [C \cap \cup B_i = 0]$?

Consider the ascending sequence $\{\text{Ker}(2^k)\}$ of subobjects of S^3. They do have a least upper bound, namely the image of $\vee C_{2^n} \to S^3$ where C_{2^n} is the mapping cone of $S^2 \xrightarrow{2^n} S^2$. (Infinite wedges of spaces of bounded dimension are categorical sums in $\mathscr{F}_{CW}$.) Let K_{2^∞} denote $\cup \text{Ker}(2^n)$. $\text{Ker}(3) \cap \text{Ker}(2^n) = 0$ each n, because there exist a, b such that $a\,3 + b\,2^n = 1$ and $a3 + b\,2^n$ kills $\text{Ker}(3) \cap \text{Ker}(2^n)$.

But $\text{Ker}(3) \cap \text{K}_{2^\infty} \neq 0$; for otherwise the injectiveness of S^3 would yield a commutative diagram

$$\begin{array}{ccc} \text{Ker}(3) \oplus \text{K}_{2^\infty} & \to & S^3 \\ {\scriptstyle u_1 p_1}\downarrow & & \downarrow{\scriptstyle a} \\ \text{Ker}(3) \oplus K_{2^\infty} & \to & S^3 . \end{array}$$

a is an integer, hence $aK_{2^\infty} = 0$, hence $aK_{2^n} = 0$ and $a \equiv 0 \pmod{2^n}$ all n, hence $a = 0$. But $a\,\text{Ker}(3) = 0$ a contradiction.

Thus $\mathscr{F}_{CW}$ is not a Grothendieck category.

The spheres are *small-projectives* in $\mathscr{F}_{CW}$. That is $(S^a, -)$ carries infinite wedges of bounded dimension into infinite sums, i.e. every map $S^a \to \vee X_i$ factors through a finite sub-wedge (S^a is compact).

In particular, for every $S^a \to K_{2^\infty}$ there exists n such that

$$\text{Im}(S^a \to K_{2^\infty}) \subset \text{Ker}(2^n).$$

Hence if the spheres generated $\mathscr{F}_{CW}$ then there would exist non-trivial $S^a \to \text{Ker}(3) \cap K_{2^\infty}$, but from the last sentence

$$\text{Im}(S^a \to \text{Ker}(3) \cap K_{2^\infty}) \subset \text{Ker}(3) \cap K(2^n),$$

a contradiction. Thus the spheres do not generate $\mathscr{F}_{CW}$.

One very pleasant aspect of this counterexample is that it shows that all the finite complexes together fail to generate $\mathscr{F}_{CW}$. We used only the compactness of the spheres. This is a pleasant aspect because the generating hypothesis and the fact that there exists non-trivial $X \xrightarrow{f} S^a$ such that $\pi f = 0$ implies that for all finite complexes B, $(B, f) = 0$.

Added in Proof: Let $S^3 \xrightarrow{\nu} S^0$ be a map of order *8*. *Cone(ν)* is not isomophic to *Cone(3ν)*. There exists an exact sequence

$$S^3 \xrightarrow{\nu} S^0 \rightarrow \mathit{Cone}\ (3\nu) \rightarrow S^4 \xrightarrow{\nu} S^1 .$$

There exist isomorphisms

$$S^0 \vee \mathit{Cone}\ (\nu) \simeq S^0 \vee \mathit{Cone}\ (3\nu)$$

$$S^4 \vee \mathit{Cone}\ (\nu) \simeq S^4 \vee \mathit{Cone}\ (3\nu)$$

$$\mathit{Cone}\ (\nu) \vee \mathit{Cone}\ (\nu) \simeq \mathit{Cone}\ (3\nu) \vee \mathit{Cone}\ (3\nu).$$

Department of Mathematics
University of Pennsylvania
Philadelphia, Pennsylvania

Splitting Homotopy Idempotents*

By

PETER FREYD

Let $\mathscr{A}$ be an abelian category with enough projectives and let $\mathscr{H}$ be the quotient category of $\mathscr{A}$ obtained by identifying with zero all maps which factor through projectives. ($\mathscr{H}$ is the Eckmann-Hilton homotopy category.) Do idempotents split in $\mathscr{H}$? That is, given $A \xrightarrow{e} A \in \mathscr{H}$, $e^2 \equiv e$ does there exist $B \in \mathscr{H}$ and maps $A \to B$, $B \to A$ such that $A \to B \to A \equiv e$, $B \to A \to B \equiv 1$?

The above question is equivalent to a question asked by M. AUSLANDER: Given a direct summand T of the functor Ext $(-, A)$: $\mathscr{A} \to \mathscr{G}$, does there exist $B \in \mathscr{A}$ such that $T \simeq$ Ext $(-, B)$? A positive answer leads to a characterization of Ext functors (see AUSLANDER's paper in this collection).

The equivalence of the two questions is easily seen by observing that the functors $\mathscr{H}(-, A)$: $\mathscr{A} \to \mathscr{G}$ and Ext$(-, A)$: $\mathscr{A} \to \mathscr{G}$ are left and right satellites of each other, hence

$$\mathscr{H}(A, B) \simeq (\mathscr{H}(-, A), \mathscr{H}(-, B)) \simeq (\mathrm{Ext}(-, A), \mathrm{Ext}(-, B))$$

and $\mathscr{H}$ is isomorphic to the full subcategory of $(\mathscr{A}^*, \mathscr{G})$ generated by the Ext functors.

In the case that $\mathscr{A}$ is complete the question has a positive answer: idempotents do split in $\mathscr{H}$. Indeed only countable co-powers are needed in $\mathscr{A}$. This is a corollary of a more general theorem, as is the fact that idempotents split in the stable homotopy category as discussed in our paper in this collection entitled *Stable Homotopy*.

It is useful to introduce a condition on categories weaker than that idempotents split. Given an additive category $\mathscr{B}$ we shall say that *Retracts have complements* in $\mathscr{B}$ if for every retraction $A_1 \xrightarrow{u_1} B \xrightarrow{p_1} A_1 = 1$ there exists $A_2 \in \mathscr{B}$ and maps $A_2 \xrightarrow{u_2} B \xrightarrow{p_2} A_2$ such that u_1, p_1, u_2, p_2 form a direct sum system: $p_1 u_1 = 1$, $p_2 u_2 = 1$, $p_1 u_2 = 0$, $p_2 u_1 = 0$, $u_1 p_1 + u_2 p_2 = 1$. (Equivalently, an idempotent e splits if $1 - e$ splits.)

To say that $\mathscr{B}$ has countable co-powers means that for each $A \in \mathscr{B}$ the countably repeated categorical sum $\Sigma_{\aleph_0} A$ exists in $\mathscr{B}$. To say that

* Received September 5, 1965.

$\mathscr{B}$ has countable powers means that $\Pi_{\aleph_0} A$ exists in $\mathscr{B}$. In the proposition below the parenthetically stated disjunction is simply the dual theorem.

Proposition. *Let $\mathscr{B}$ be an additive category with countable co-powers (or countable powers) and in which retracts have complements. Idempotents split in $\mathscr{B}$.*

Proof. Let $A \xrightarrow{e} A$ be an idempotent. Define
$B = \Sigma_{N_0} A = A + A + A + \cdots$

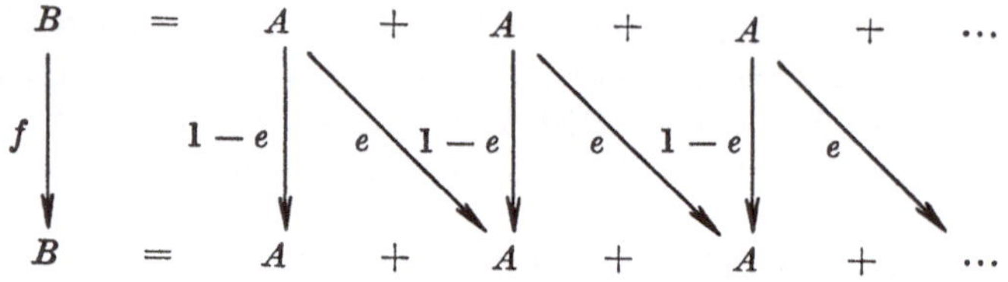

(That is, $f\, u_i = u_i\,(1-e) + u_{i+1}\, e,\ i = 1, 2, \ldots.$)

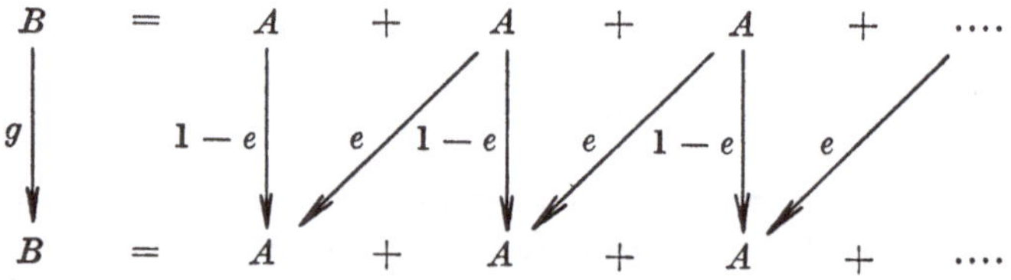

$$\left(g\, u_i = \begin{cases} u_1(1-e) & \text{if} \quad i = 1 \\ u_i(1-e) + u_{i-1} e & \text{if} \quad i = 2, 3, 4, \ldots. \end{cases} \right)$$

It follows that $B \xrightarrow{f} B \xrightarrow{g} B = 1$. Hence the condition that retracts have complements implies the existence of $C \xrightarrow{a} B \xrightarrow{b} C = 1$ such that

$$bf = 0, \quad ga = 0, \quad ab + fg = 1.$$

Consider

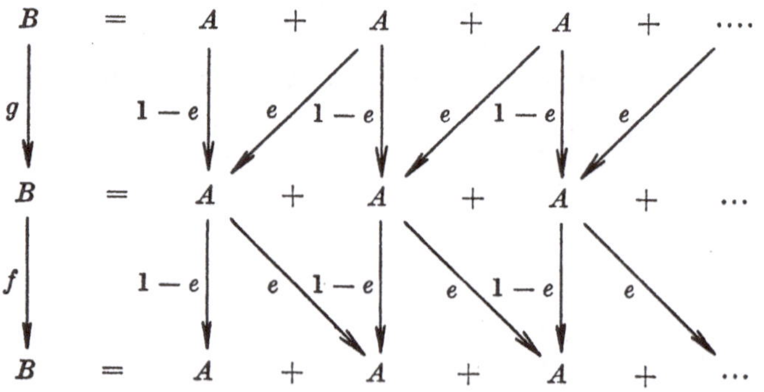

from which we see that

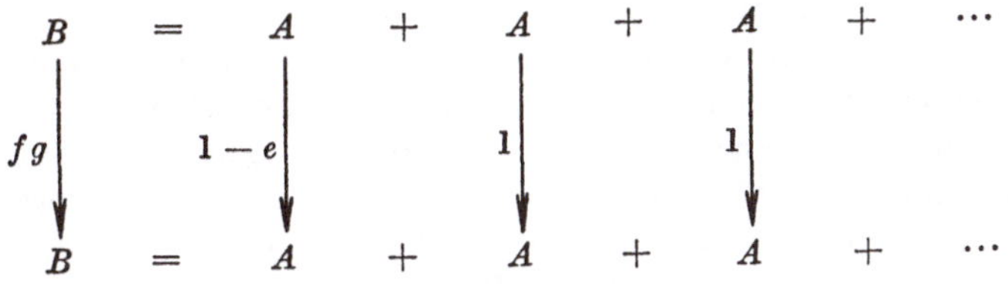

and that $ab = 1 - fg = B \xrightarrow{p_1} A \xrightarrow{e} A \xrightarrow{u_1} B$.
Consider

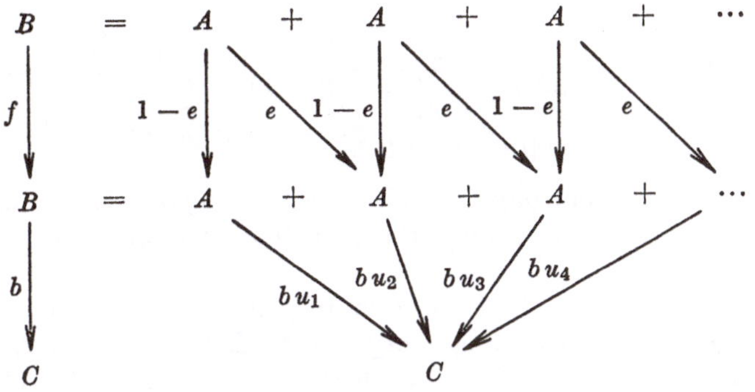

Now $bf = 0$ hence

$$b\,u_i(1-e) + b\,u_{i+1}e = 0 \qquad i = 1, 2, \ldots.$$

From which we obtain

$$[b\,u_i(1-e) + b\,u_{i+1}e]\,(1-e) = b\,u_i(1-e) = 0 \quad i = 1, 2, \ldots$$
$$[b\,u_i(1-e) + b\,u_{i+1}e]\,e = b\,u_{i+1}e = 0 \qquad i = 1, 2, \ldots$$

and for $i = 2, 3, \ldots,$ $b\,u_i = b\,u_i[(1-e) + e] = 0$

hence $$B \xrightarrow{b} C = B \xrightarrow{p_1} A \xrightarrow{u_1} B \xrightarrow{b} C.$$

Let $$A \to C = A \xrightarrow{u_1} B \xrightarrow{b} C$$

and $$C \to A = C \xrightarrow{a} B \xrightarrow{p_1} A.$$

Then
$$\begin{aligned} A \to C \to A &= A \xrightarrow{u_1} B \xrightarrow{b} C \xrightarrow{a} B \xrightarrow{p_1} A = \\ &A \xrightarrow{u_1} B \xrightarrow{1-fg} B \xrightarrow{p_1} A = \\ &A \xrightarrow{u_1} B \xrightarrow{p_1} A \xrightarrow{e} A \xrightarrow{u_1} B \xrightarrow{p_1} A = A \xrightarrow{e} A, \end{aligned}$$
$$\begin{aligned} C \to A \to C &= C \xrightarrow{a} B \xrightarrow{p_1} A \xrightarrow{u_1} B \xrightarrow{b} C = \\ &C \xrightarrow{a} B \xrightarrow{\qquad b \qquad} C = C \xrightarrow{1} C. \end{aligned}$$

Hence C splits e.

The above proposition suffices for the stable homotopy category of finite dimensional CW-complexes.

Returning to the original question, if the abelian category $\mathscr{A}$ has countable co-powers, then so does the homotopy category $\mathscr{H}$ obtained from $\mathscr{A}$ by killing the projectives. There are, indeed, two reasons for this fact, and the two reasons give rise to two different generalizations of the fact. One reason is as follows: For every $B \in \mathscr{A}$ there exists a projective P and an epimorphism $P \to B$ such that for all X

$$\mathscr{H}(X, B) \simeq \frac{(X, B)}{\operatorname{Im}[(X, P) \to (X, B)]}.$$

Hence

$$\mathscr{H}(\Sigma A, B) = \frac{(\Sigma A, B)}{\operatorname{Im}[(\Sigma A, P) \to (\Sigma A, B)]} = \frac{\Pi(A, B)}{\operatorname{Im}[\Pi(A, P) \to \Pi(A, B)]} =$$
$$= \Pi \frac{(A, B)}{\operatorname{Im}[(A, P) \to (A, B)]} = \Pi \mathscr{H}(A, B).$$

In the language of our paper in this collection *Representations in abelian categories*, we have used only the fact that the projectives form an "ample class".

On the other hand, if $\mathscr{A}$ has countable sums (not just co-powers) we may verify that the obvious epimorphism $\mathscr{H}(\Sigma A, B) \to \Pi(A, B)$ is mono as follows: given $f \in \mathscr{A}(\Sigma A, B)$ such that $f u_1 \equiv 0(\mathscr{H})$ all i, choose P_i projective each i and maps $A \xrightarrow{g_i} P_i \xrightarrow{h_i} A = f u_i$. Then $\Sigma A \xrightarrow{g} \Sigma P_i \xrightarrow{h} B = f$ and $f \equiv 0(\mathscr{H})$. We have used only the fact that the class of projectives is closed under countable sums.

Finally then we shall know that idempotents split in $\mathscr{H}$ once we have shown that retracts have complements in $\mathscr{H}$. Given $A_1 \xrightarrow{u_1} B$, $B \xrightarrow{p_1} A_1$ in $\mathscr{A}$ such that $p_1 u_1 \equiv 1(\mathscr{H})$ we let P be a projective and $A_1 \xrightarrow{b} P$, $P \xrightarrow{g} A_1$ maps such that $p_1 u_1 + g f = 1$ in $\mathscr{A}$. Then A_1 is seen to be a retract of $B \oplus P$ in $\mathscr{A}$; its complements A_2 will serve in $\mathscr{H}$ (remembering that B and $B \oplus P$ are isomorphic in $\mathscr{H}$). Hence the proposition applies and idempotents split in $\mathscr{H}$.

The generalizations: Let $\mathscr{A}$ be any additive category with countable co-powers (countable powers) and in which retracts have complements. Let $\boldsymbol{P}$ be any class of objects in $\mathscr{A}$ either ample or closed under countable sum (either co-ample or closed under countable products). Then idempotents split in $\mathscr{A}/\boldsymbol{P}$, the quotient category obtained by killing the objects in $\boldsymbol{P}$.

Department of Mathematics
University of Pennsylvania
Philadelphia, Pennsylvania

Fill-in Theorems*

By

Richard Faber and Peter Freyd

1. Introduction

In this paper, we study existential diagramatic theorems, i.e., theorems such as the connecting homomorphism lemma, in which certain exactness and commutativity conditions in a diagram imply the existence of an auxiliary map or „fill-in" in the diagram, together with certain properties for the enlarged diagram.

It is well known that if an existential theorem is true in the category of abelian groups, and if the fill-in is given by a canonical „diagram-chase" algorithm (as defined in [*1*], p. 14, and generalized herein), then the theorem is true in every abelian category. Our main result is the converse: If an existential theorem is true in every abelian category, then there is a diagram-chase algorithm for the fill-in (this means that the fill-in may always be constructed in a prescribed way from the maps in the diagram). It will be shown also that the fill-ins for such theorems have certain naturality properties with respect to exact functors and translations of diagrams.

2. Preliminaries

Let $x = x_1, x_2, \ldots, x_n$ be a collection of objects and morphisms that constitute a diagram in some abelian category (we assume this collection includes the domain and range of each member which is a morphism). Let $H(x_1, x_2, \ldots, x_n)$ be the statement that certain specified parts of this diagram are commutative or exact. We shall call such a statement *diagramatic*. Obviously, such statements, unless self-contradictory, may be made in any abelian category. Note that in terms of exactness and commutativity we may express a rather wide class of properties of diagrams, e.g., that an object is the zero object, or that an object is a (finite) limit of some part of the diagram.

$x_1, x_2, \ldots, x_n$ may be considered variables, some of which may have objects as values, the rest, morphisms. The specification of which may

* Received August 29, 1965.

be objects and which may be morphisms shall be called the *shape* of the diagram.

We shall use the term *diagramatic theorem* to denote a theorem of the following type.

Model 1. *If* $H(x_1, x_2, \ldots, x_n)$ *is true, then* $H'(x_1, x_2, \ldots, x_n)$ *is true.*

Here H and H' are two diagramatic statements defined for diagrams of the same shape. Such a theorem states that certain exactness and commutativity conditions in a diagram imply certain additional exactness and commutativity conditions in the same diagram. The truth of such a theorem depends, in general, upon the category or categories to which we assume x may belong.

A direct consequence of the Embedding Theorem for small abelian categories ([*2*], p. 140, [*3*]) is the

First Metatheorem 2.1. *Every diagramatic theorem true in the category of abelian groups is true in any abelian category.*

At first glance, this result sheds no light on whether the following lemma, true in any category of modules, is true in arbitrary abelian categories.

Lemma 2.2. *(connecting homomorphism). If, in* (2.3), *all columns and the two middle rows are exact, then there exists a map* $N \to G$ *such that* $L \to M \to N \to G \to H \to I$ *is exact.*

(2.3)

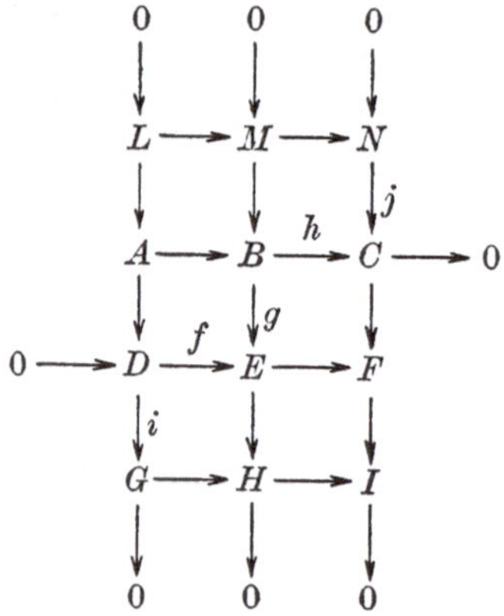

Such a theorem, in which is asserted the existence of a new morphism between objects in the diagram, and exactness/commutativity conditions for the completed diagram is called an *existential* or *fill-in theorem*. The general form is

Model 2. *If* $H(x_1, x_2, \ldots, x_n)$ *is true, then there exists a morphism* $x^*: x_\alpha \to x_\beta$ *such that* $C(x_1, x_2, \ldots, x_n, x^*)$ *is true.*

Here, H and C are diagramatic statements, but C is defined for diagrams with an additional morphism in the specified place. It is implicit here that x_α and x_β are object variables.

x^* is called the *fill-in* or *solution* morphism. In general, it is not unique (e.g., in the above lemma, the negative of any solution is a solution), but in many familiar cases, it is defined by a canonical constructive process referred to as "diagram-chasing", which we shall define precisely in section 4.

Full Embedding Theorem 2.4. *Every small abelian category enjoys an exact full embedding into a category of modules ([4]).*

There immediately follows the

Full Metatheorem 2.5. *Any fill-in theorem true in every category of modules is true in any abelian category.*

3. Naturality of Fill-ins

The foregoing metatheorems obviate the necessity for verifying categorically the myriads of classical diagram lemmas that frequently arise in the proof of more substantial results. One need prove such lemmas only for modules or abelian groups by „element-ary" arguments. On the other hand, a number of interesting questions on fill-in theorems are left unanswered, among them the functorial nature or naturality of fill-ins.

Let us enlarge upon our terminology for a moment. Suppose we have a fill-in theorem of the type exemplified by Model 2. We shall define an *occurrence* of H to be a pair $(x, \mathscr{A})$, where $\mathscr{A}$ is an abelian category, and $x = x_1, x_2, \ldots, x_n$ is a diagram in $\mathscr{A}$ on which H is defined and such that $H(x_1, x_2, \ldots, x_n)$ is true. We shall sometimes refer to $(x, \mathscr{A})$, or simply to x, as an occurrence of H in $\mathscr{A}$. (Note that a diagram x may be the first coordinate of many occurrences.) If the conclusion of our fill-in theorem is true for an occurrence $(x, \mathscr{A})$, then the solution x^* shall be called a solution for $(x, \mathscr{A})$, or, if ambiguity is precluded, a solution for x.

If $(x, \mathscr{A})$ and $(y, \mathscr{A})$ are occurrences of H in the same category, then by a *translation* from x to y, we shall mean a morphism of diagrams, i.e., a set of maps in $\mathscr{A}$ between corresponding object variables and commuting with the morphism variables.

If our fill-in theorem is always true in a particular abelian category $\mathscr{A}$, we would like to know that there is a canonical collection of solutions — one for each occurrence of H in $\mathscr{A}$ — with the property that if τ is a translation between occurrences x and y in $\mathscr{A}$, then τ is actually a translation between the filled-in diagrams $x_1, x_2, \ldots, x_n, x^*$ and $y_1, y_2, \ldots, y_n, y^*$, where x^* and y^* are the canonical solutions for x and y, respectively. Such a collection of solutions is called *translation-respecting*.

More generally, suppose our theorem is true in every abelian category. If $(x, \mathscr{A})$ is an occurrence of H, and $T: \mathscr{A} \to \mathscr{B}$ is an exact functor between abelian categories, then $(Tx, \mathscr{B})$ is an occurrence of H. (Although up to this point, we could have allowed diagramatic statements to include statements of non-exactness and non-commutativity, we must restrict ourselves, in what follows, to positive statements.) We can now ask if there is a canonical collection of solutions — one for each occurrence of H in *each* abelian category — such that if x^* is the canonical solution for $(x, \mathscr{A})$, then Tx^* is the canonical solution for $(Tx, \mathscr{B})$, for any occurrence $(x, \mathscr{A})$, any abelian category $\mathscr{B}$, and any exact functor $T: \mathscr{A} \to \mathscr{B}$. Such a collection is called *natural*. Shortly, we shall give necessary and sufficient conditions for the existence of natural collections.

4. Solutions by Diagram-chasing

If (2.3) is a diagram in $\mathscr{G}$, the category of abelian groups, then

$$(4.1) \qquad i f^{-1} g h^{-1} j$$

is a homomorphism from N to G. This is the usual definition of the connecting homomorphism. This might prompt us to define a *diagram-chase solution* of a fill-in theorem in $\mathscr{G}$ to be a map which satisfies the conclusion of the theorem and which is expressible as the relation obtained by composing certain maps *of the diagram* and/or their inverses. In fact, it will turn out that this definition is too narrow. For example, it is insufficient for the following theorem:

Splitting Lemma 4.2. *If* $0 \to A \xrightarrow{i} B \xrightarrow{j} C \to 0$ *is exact, and* $f: B \to A$ *satisfies* $fi = 1_A$, *then there exists a map* $g: C \to B$ *such that* $jg = 1_C$.

In this case, the fill-in map is $(1_B - if)\, j^{-1}$, but $1_B - if$ is not in the diagram.

A definition which will suit our purposes is the following.

Definition 4.3. *A diagram-chase solution (DCS) for an occurrence of a fill-in theorem is a morphism which satisfies the conclusion of the theorem and which is expressible as the relation obtained by composing certain maps and/or their inverses, where these maps are maps between finite sums of objects* **in the diagram** *and are induced by maps* **in the diagram** *(i.e., these maps may be represented by matrices whose entries are maps of the diagram).*

Note that the various maps and inverses which are composed to give a *DCS*, as well as the order of composition, are in general not unique.

Definition 4.4. *By the term diagram-chase algorithm (DCA) we shall mean a specified selection of maps and inverses, and the order of their composition, to yield a DCS.*

For example, (4.1) defines a *DCA* for the connecting homomorphism lemma. The above definitions apply to arbitrary abelian categories, since relations may be defined categorically (cf. [*5*], p. 51, [*2*], p. 103, and [*6*]). We briefly review this topic below.

A *relation* w between two objects C and D in an abelian category is defined to be a subobject of $C \oplus D$. We define a composition of relations that coincides with the usual definition when the relations occur in $\mathcal{G}$, and associate with each morphism g a relation g', so that composition of morphisms and composition of the associated relations are compatible. For a morphism $f\colon C \to D$, we define also a relation, called $f^{-1} \subseteqq D \oplus C$, which coincides with the usual f^{-1} when f is a group homomorphism.

For $f\colon B \to C$ and $g\colon B \to D$ in an arbitrary abelian category, the relation $g' f^{-1}$ is defined to be the image of $\binom{f}{g}\colon B \to C \oplus D$. In particular, if $C = B$, and $f =$ the identity, this defines the relation g' associated with g; and if $D = B$ and $g =$ the identity, this defines the relation f^{-1}. In fact, any relation is of the form $g' f^{-1}$, since any monomorphism into a direct sum has the form $\binom{f}{g}$ and represents its own image.

Now consider two relations $h' m^{-1}$ and $n' k^{-1}$ in an arbitrary abelian category. Assume these relations are associated with the morphisms depicted in the diagram below.

$$\begin{array}{ccccc} B & \xrightarrow{f} & C & \xrightarrow{m} & A \\ \downarrow{\scriptstyle g} & & \downarrow{\scriptstyle h} & & \\ D & \xrightarrow{k} & E & & \\ \downarrow{\scriptstyle n} & & & & \\ F & & & & \end{array}$$

Let the square be a pullback; then the composition $(n' k^{-1})\,(h' m^{-1})$ is the relation $(ng)'\,(mf)^{-1}$.

Proposition 4.5. *Let C and D be objects in an abelian category $\mathscr{A}$. By taking pullbacks, any relation from C to D in $\mathscr{A}$ may be reduced to the form $g' f^{-1}$, where $f\colon X \to C$ and $g\colon X \to D$ are some maps in $\mathscr{A}$. The subobject $\operatorname{im}(\binom{f}{g}\colon X \to C \oplus D)$ is independent of the choice of pullbacks, or on the order in which they are constructed.*

The criterion for a relation to be a map is the following:

Proposition 4.6. *Let $w\colon X \to C \oplus D$ be a (monomorphism representing a) relation. Then w has the form f', for some morphism $f\colon C \to D$ if and only if $p_1 w$ is an isomorphism. This f is unique, and is given by $p_2 w (p_1 w)^{-1}$ independently of the choice of representing monomorphism.*

(As usual, p_1, p_2 denote the projections from $C \oplus D$ onto its first and second summands, respectively.)

5. The Universal Metatheorem

We are now prepared to state and prove our main theorem.

Universal metatheorem 5.1. *Let T be a fill-in theorem with hypothesis, conclusion, and variables according to the notation of Model 2. Then the following statements are equivalent.*

1. T has solutions in $\mathcal{G}$ that are given by a fixed DCA.

2. T has a natural collection of solutions.

3. For each abelian category $\mathcal{A}$, T has a translation-respecting collection of solutions in $\mathcal{A}$.

4. T has a translation-respecting collection of solutions in $\mathcal{G}$.

5. T is true in any abelian category $\mathcal{A}$.

We shall show $1 \Rightarrow 2 \Rightarrow 3 \Rightarrow 4 \Rightarrow 5 \Rightarrow 1$, *but shall treat the more straightforward implications first.* $3 \Rightarrow 4$ *is automatic.*

Proof that $1 \Rightarrow 2$. In this argument, we make essential use of the following fact: equality of relations may be expressed as a condition of exactness and is therefore respected by exact functors ([*0*], p. 77, 78).

Let $\mathcal{A}$ be a small abelian category. Let $F: \mathcal{A} \to \mathcal{G}$ be an exact embedding, as guaranteed by the Embedding Theorem. By definitions 4.3 and 4.4, a *DCA* is a recipe for the concatenation of certain maps and inverses of maps, where these maps are matrices of maps in a diagram. According to our results on relations in general categories, we may apply such a recipe to any occurrence x of H in any abelian category $\mathcal{A}$ and obtain a relation form x_α to x_β. The maps occurring in the application of the recipe to $(x, \mathcal{A})$ are sent by F into the maps occurring in the application of the recipe to $(Fx, \mathcal{G})$.

By proposition 4.5, our relation from x_α to x_β may be reduced to the form $g' f^{-1}$ by taking a sequence of pullbacks in $\mathcal{A}$. Because exact functors preserve pullbacks, $F(g)\, F(f)^{-1}$ is the *DCS* for the diagram Fx in $\mathcal{G}$. If $\mathrm{im}\,(\binom{f}{g}) \subseteqq x_\alpha \oplus x_\beta$ is represented by the monomorphism w in $\mathcal{A}$, then $\mathrm{im}\,(\binom{Ff}{Fg}) \subseteqq Fx_\alpha \oplus Fx_\beta$ is represented by Fw in $\mathcal{G}$. By proposition 4.6, $Fp_1 Fw = F(p_1 w)$ is an isomorphism in $\mathcal{G}$, and since F is an embedding, $p_1 w$ is an isomorphism in $\mathcal{A}$. Therefore $g' f^{-1}$ is a map x^* which is sent by F into the *DCS* for Fx and so is a solution for the occurrence $(x, \mathcal{A})$. ($g' f^{-1}$ is actually $p_2 w (p_1 w)^{-1}$.)

Now, if $T: \mathcal{A} \to \mathcal{B}$ sends the occurrence $(x, \mathcal{A})$ to the occurrence $(y, \mathcal{B})$, then it sends the recipe for x^* to the recipe for y^*, and so $Tx^* = y^*$.

Note that we did not require the Full Embedding Theorem above. We do not need a *full* group-valued embedding provided we have a *DCA*.

Proof that $2 \Rightarrow 3$. Assume T has a natural collection of solutions, and let $\mathcal{A}$ be an abelian category. Let $(\to, \mathcal{A})$ denote the category of $\mathcal{A}$-

morphisms and translations of $\mathscr{A}$-morphisms: an object of $(\rightarrow, \mathscr{A})$ is a map $f: A \rightarrow B$ in $\mathscr{A}$, while a morphism between two objects $f: A \rightarrow B$ and $f': A' \rightarrow B'$ is a pair (α, β) of maps in $\mathscr{A}$ such that $\alpha: A \rightarrow A'$, $\beta: B \rightarrow B'$, and $f'\alpha = \beta f$. There are two exact functors $T_1, T_2: (\rightarrow, \mathscr{A}) \rightarrow \mathscr{A}$ obtained by assigning to an object of $(\rightarrow, \mathscr{A})$ its domain (resp., range), and to a morphism of $(\rightarrow, \mathscr{A})$ its first (resp., second) coordinate.

A translation $\tau: x \rightarrow y$ of occurrences in $\mathscr{A}$ may be considered a single occurrence w in $(\rightarrow, \mathscr{A})$. By 2, the natural solutions for x and y are the images under T_1 and T_2 of the natural solutions for w in $(\rightarrow, \mathscr{A})$. Since the filled-in diagram in $(\rightarrow, \mathscr{A})$ is really a translation of the two filled-in diagrams in $\mathscr{A}$, the result follows.

Proof that 4 ⇒ 5. We shall deduce from 4 that T is true in any category of modules. 5 will then follow from the Full Metatheorem. Let $(x, \mathscr{G}_R)$ be an occurrence of H, where $\mathscr{G}_R$ is the category of right modules over a unitary ring R. By applying the forgetful functor (which is an exact embedding), we may view x as an occurrence in $\mathscr{G}$. For each $r \in R$, we may construct a diagram morphism τ_r from x to itself in $\mathscr{G}$, by letting the map between each pair of corresponding objects be right multiplication by r. Since the maps of x are R-homomorphisms, τ_r is an honest translation. By 4, τ_r is actually a translation of the filled-in diagram $x_1, x_2, \ldots, x_n, x^*$, for each $r \in R$. This means that x^* is an R-homomorphism and a solution for $(x, \mathscr{G}_R)$.

Proof that 5 ⇒ 1. For any occurrence x in $\mathscr{G}$, there is a countable abelian subcategory $\mathscr{G}_x$ which contains x and whose inclusion functor $\mathscr{G}_x \rightarrow \mathscr{G}$ is exact (an "exact subcategory"). Hence, $H(x_1, x_2, \ldots, x_n)$ is true in $\mathscr{G}_x$, and if $x^*: x_\alpha \rightarrow x_\beta$ in $\mathscr{G}_x$ is such that $C(x_1, x_2, \ldots, x_n, x^*)$ is true in $\mathscr{G}_x$, then $C(x_1, x_2, \ldots, x_n, x^*)$ is true in $\mathscr{G}$.

Now consider all possible occurrences $(x, \mathscr{G}_x)$, where $\mathscr{G}_x$ is a countable exact abelian subcategory of $\mathscr{G}$. Let $\mathscr{B} = \times\, \mathscr{G}_x$ be the cartesian product category of all second coordinates of such occurrences. Let z be the diagram in $\mathscr{B}$ whose xth component is the diagram x. By 5, choose a solution z^* for $(z, \mathscr{B})$. For each $(x, \mathscr{G}_x)$, the exact projection functor $P_x: \mathscr{B} \rightarrow \mathscr{G}_x$ gives a solution $P_x z^*$ for $(x, \mathscr{G}_x)$. A priori, for a given occurrence x in $\mathscr{G}$, $P_x z^*$ depends on the choice of containing subcategory $\mathscr{G}_x$.

However, in the following section, we shall show that z^* may be chosen to be a *DCS*. In this case, it is easily seen that for any two occurrences $(x, \mathscr{G}_x)$ and $(y, \mathscr{G}_y)$, $x = y$ implies that $P_x z^* = P_y z^*$. Hence, the projection functors yield well-defined unique solutions for all occurrences of H in $\mathscr{G}$. The desired *DCA* is then furnished by the selection of a *DCA* for z^*.

6. The „Homology“ Construction

To show z^* is a *DCS*, we shall construct an exact abelian subcategory $\mathscr{C} \subseteqq \mathscr{B}$ which contains z (and hence also z^*), but which has so few maps that any map in $\mathscr{C}$ between objects in z is forced to have the form (described in Definition 4.3) of a *DCS*. This construction proceeds essentially in two steps. In the first, we adjoin all the objects we need for $\mathscr{C}$; while in the second, we adjoin the remaining maps we need (but add no new objects).

Let $\mathscr{B}_0 =$ the additive subcategory generated by z in $\mathscr{B}$ (adjoin identity maps, the zero object, finite direct sums of objects in z, matrices, etc.).

Step 1 (adjoin homology). For each short sequence $A = A_1 \xrightarrow{f} A_2 \xrightarrow{g} A_3$ in $\mathscr{B}_0$, adjoin to $\mathscr{B}_0$ an object $H(A)$ of $\mathscr{B}$ representing the "homology" of $A = (\ker(g) + \operatorname{im}(f))/\operatorname{im}(f)$, i.e., the image of the composed map $\ker(g) \to A_2 \to \operatorname{cok}(f)$. (We are not assuming $gf = 0$.) We identify an object X of $\mathscr{B}_0$ with the homology of the sequence $0 \to X \to 0$. For each pair of sequences $A = A_1 \to A_2 \to A_3$ and $B = B_1 \to B_2 \to B_3$, adjoin all maps in $\mathscr{B}$ from $H(A)$ to $H(B)$ which are induced by translations.

$$\begin{array}{ccccc} A_1 & \longrightarrow & A_2 & \longrightarrow & A_3 \\ \downarrow & & \downarrow & & \downarrow \\ B_1 & \longrightarrow & B_2 & \longrightarrow & B_3 \end{array}$$

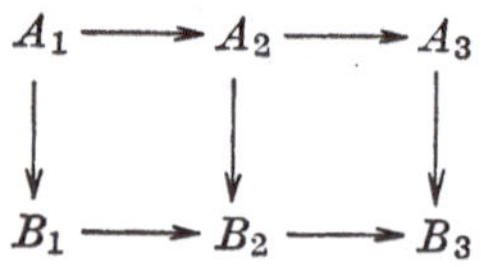

By assuming $\mathscr{B}$ to be replete ([2], p. 74), we may choose homology objects in such a way that no two sequences are assigned the same object. With the obvious composition, we obtain an additive category $\mathscr{B}_1$ contained in $\mathscr{B}$ and containing $\mathscr{B}_0$ as a full subcategory.

Lemma 6.1. *Each map $U \to V$ of $\mathscr{B}_1$ can be embedded in a sequence $X \to U \to V \to Y$ of maps in $\mathscr{B}_1$ such that $0 \to X \to U \to V \to Y \to 0$ is exact in $\mathscr{B}$.*

Proof. By duality, it will suffice to construct only half of the exact sequence. Suppose that $U \to V$ is the map induced in homology by the following translation of short sequences.

$$\begin{array}{ccccc} A_1 & \xrightarrow{f} & A_2 & \xrightarrow{g} & A_3 \\ \downarrow{\scriptstyle \alpha_1} & & \downarrow{\scriptstyle \alpha_2} & & \downarrow{\scriptstyle \alpha_3} \\ B_1 & \xrightarrow{h} & B_2 & \xrightarrow{k} & B_3 \end{array}$$

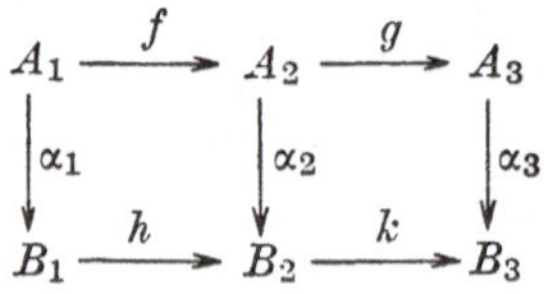

Then the desired $X \to U \to V$ is obtained by passing to homology in (6.2).

(6.2)
$$\begin{array}{ccccc}
A_1 \oplus B_1 & \xrightarrow{\begin{pmatrix} f & 0 \\ \alpha_1 & -1 \end{pmatrix}} & A_2 \oplus B_1 & \xrightarrow{\begin{pmatrix} \alpha_2 & -h \\ g & 0 \end{pmatrix}} & B_2 \oplus A_3 \\
\downarrow{\scriptstyle p_1} & & \downarrow{\scriptstyle p_1} & & \downarrow{\scriptstyle p_2} \\
A_1 & \xrightarrow{f} & A_2 & \xrightarrow{g} & A_3 \\
\downarrow{\scriptstyle \alpha_1} & & \downarrow{\scriptstyle \alpha_2} & & \downarrow{\scriptstyle \alpha_3} \\
B_1 & \xrightarrow{h} & B_2 & \xrightarrow{k} & B_3
\end{array}$$

Step 2. This step, in effect, adjoins to $\mathscr{B}_1$ the inverses of those maps of $\mathscr{B}_1$ which are isomorphisms in $\mathscr{B}$.

Lemma 6.3. *Let A, B be objects of $\mathscr{B}_1$, and let $x\colon A \to B$ be a map in $\mathscr{B}$. Then the following are equivalent.*

a) There exists a map $X \to A$ in $\mathscr{B}_1$ such that $0 \to X \to A \to 0$ is exact in $\mathscr{B}$, and the composition $X \to A \to B$ is in $\mathscr{B}_1$.

b) There exists $B \to Y$ in $\mathscr{B}_1$ such that $0 \to B \to Y \to 0$ is exact in $\mathscr{B}$, and $A \to B \to Y$ is in $\mathscr{B}_1$.

c) There exists $X \to A$ in $\mathscr{B}_1$ such that $X \to A \to 0$ is exact in $\mathscr{B}$, and $X \to A \to B$ is in $\mathscr{B}_1$.

d) There exists $B \to Y$ in $\mathscr{B}_1$ such that $0 \to B \to Y$ is exact in $\mathscr{B}$, and $A \to B \to Y$ is in $\mathscr{B}_1$.

e) There exist $X \to A$ and $B \to Y$ in $\mathscr{B}_1$ such that $X \to A \to 0$ and $0 \to B \to Y$ are exact in $\mathscr{B}$, and $X \to A \to B \to Y$ is in $\mathscr{B}_1$.

Proof. Clearly, a) $\Rightarrow$ c) $\Rightarrow$ e) and b) $\Rightarrow$ d) $\Rightarrow$ e). To show d) $\Rightarrow$ a), assume $y\colon B \to Y$ and the composition yx are in $\mathscr{B}_1$ and that $0 \to B \to Y$ is exact in $\mathscr{B}$. Let $k\colon X \to A \oplus B$ be a map in $\mathscr{B}_1$ such that

$$0 \to X \xrightarrow{k} A \oplus B \xrightarrow{(yx,-y)} Y$$

is exact in $\mathscr{B}$. Then, since y is a monomorphism,

$$0 \to X \xrightarrow{k} A \oplus B \xrightarrow{(x,-1)} B$$

is exact, and so

$$\begin{array}{ccc}
X & \xrightarrow{p_1 k} & A \\
{\scriptstyle p_2 k}\downarrow & & \downarrow{\scriptstyle x} \\
B & \xrightarrow{1_B} & B
\end{array}$$

is a pullback diagram in $\mathscr{B}$. $p_1 k$ is an isomorphism because 1_B is. Moreover, $x(p_1 k) = p_2 k$ belongs to $\mathscr{B}_1$, and a) is proved.

c) $\Rightarrow$ b) follows by a dual argument. To prove e) $\Rightarrow$ c), simply apply the preceding argument with x replaced by $X \to A \to B$.

Now adjoin to $\mathscr{B}_1$ all maps x in $\mathscr{B}$ that satisfy the above properties. The equivalences show that the enlarged set $\mathscr{C}$ is closed under composition, hence is a category.

Lemma 6.4. *$\mathscr{C}$ is an exact abelian subcategory of $\mathscr{B}$.*

Proof. We verify the axioms in [2]. $\mathscr{C}$ has kernels: Let $x: A \to B$ belong to $\mathscr{C}$. Then there is a $\mathscr{B}$-isomorphism $f: B \to C$ in $\mathscr{B}_1$ such that fx is in $\mathscr{B}_1$. Let $k: K \to A$ be a map in $\mathscr{B}_1$ such that $0 \to K \xrightarrow{k} A \xrightarrow{fx} C$ is exact in $\mathscr{B}$. Then $k = \ker(x)$. For, suppose $g: D \to A$ is in $\mathscr{C}$ and $xg = 0$. Then $fxg = 0$, and there is a unique map $y: D \to K$ in $\mathscr{B}$ such that $ky = g$. We must show that y belongs to $\mathscr{C}$. By Lemma 6.3, there exists a $\mathscr{B}$-monomorphism $z: A \to E$ in $\mathscr{B}_1$ such that zg is in $\mathscr{B}_1$. Now zk is a $\mathscr{B}$-monomorphism in $\mathscr{B}_1$ such that $(zk)y = zg$ lies in $\mathscr{B}_1$. Hence, by Lemma 6.3, y belongs to $\mathscr{C}$, and so $k = \ker(x)$.

Dually, $\mathscr{C}$ has cokernels. Note that the kernel (resp., cokernel) of x in $\mathscr{C}$ is its kernel (resp., cokernel) in $\mathscr{B}$.

Every $\mathscr{C}$-monomorphism is a kernel in $\mathscr{C}$: If $x: A \to B$ is a monomorphism in $\mathscr{C}$, then it is a monomorphism in $\mathscr{B}$, since its kernel in $\mathscr{C}$ is its kernel in $\mathscr{B}$. Let $y: B \to F$ be the $\mathscr{C}$-cokernel and (hence) the $\mathscr{B}$-cokernel of x. Then x is the $\mathscr{B}$-kernel of y, and so it is the $\mathscr{C}$-kernel of y.

Dually, every $\mathscr{C}$-epimorphism is a cokernel in $\mathscr{C}$. The argument for finite sums and products may easily be supplied by the reader. (It suffices to show the existence of row and column matrices in $\mathscr{C}$.) The lemma is proved.

Now each object of $\mathscr{C}$ is uniquely determined by a short sequence of the form

$$A_1 \xrightarrow{x} A_2 \xrightarrow{y} A_3 \tag{6.5}$$

in $\mathscr{B}_0$. Let $K \to A_2 = \ker(y)$ and $A_2 \to F = \operatorname{cok}(x)$. Then our $\mathscr{C}$-object is the image of the composition

$$K \to A_2 \to F. \tag{6.6}$$

Accordingly, we shall say that an object in $\mathscr{C}$ is "represented" by diagrams of the form (6.5) and (6.6), where it is understood that the latter diagram is obtained from the former as above. A morphism in $\mathscr{C}$ is induced by a translation of diagrams of the form (6.5), and hence by a translation of diagrams of the form (6.6).

Using the above, it is possible to show that each object of $\mathscr{C}$ is a subquotient of a direct sum of objects in the original diagram.

We shall now show that each $\mathscr{C}$-map between objects in the original diagram is a composition of matrices and inverses of matrices whose entries are maps in the diagram. In particular, this will be true of z^*, since z^* is in $\mathscr{C}$ by 5.

Let A, B be objects in z, and let $w : A \to B$ be a map in $\mathscr{C}$. By Lemma 6.3, w has the form gf^{-1}, where $f : X \to A$ and $g : X \to B$ are maps in $\mathscr{B}_1$, and f is an isomorphism in $\mathscr{B}$. Symbolically,

$$A \xrightarrow{w} B = A \xleftarrow{f} X \xrightarrow{g} B .$$

Let X be represented by the diagrams $X_1 \to X_2 \xrightarrow{y} X_3$ and $K \to X_2 \to F$. A, being an object in z, is represented by $0 \to A \to 0$ and $A \xrightarrow{1} A \xrightarrow{1} A$, and similarly for B. w is then determined by a commutative diagram of the form

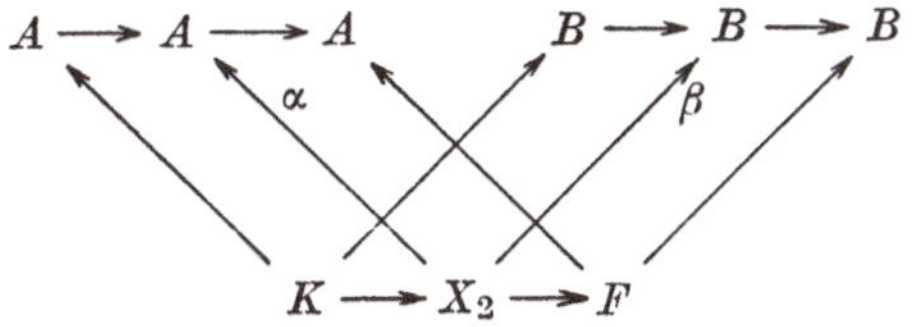

and is given by $\beta(1_{K\times K})\alpha^{-1}$, where $1_{K\times K}$ is the relation

$$u_1^{-1}\binom{1}{y} = X_2 \xrightarrow{\binom{1}{y}} X_2 \oplus X_3 \xleftarrow{u_1} X_2 ,$$

i.e., the "identity relation of X_2 restricted to $K \times K$". ($u_1 =$ the injection into the direct sum.) $\beta(1_{K\times K})\alpha^{-1}$ can be rewritten as the relation

$$u_3{}^{-1}\begin{pmatrix} 1 & -1 \\ y & 0 \\ \beta & 0 \end{pmatrix}(\alpha, 0)^{-1},$$

$$A \xleftarrow{(\alpha, 0)} X_2 \oplus X_2 \xrightarrow{\begin{pmatrix} 1 & -1 \\ y & 0 \\ \beta & 0 \end{pmatrix}} X_2 \oplus X_3 \oplus B \xleftarrow{u_3} B.$$

Since this last diagram consists of maps in $\mathscr{B}_0$, then X_2 and X_3 are direct sums of objects in z, and α, y, β, u_3, 1, -1 are matrices of maps in z. Hence w has the form $(\text{matrix})^{-1}$ (matrix) $(\text{matrix})^{-1}$. In particular, if $w = z^*$, we have shown that z^* is a DCS for $(z, \mathscr{C})$, and hence for $(z, \mathscr{B})$.

References

[0] FABER, R.: Adjoint Functors, Representations, and Fill-in Theorems, Thesis, Brandeis Univ., 1965.
[1] FREYD, P.: Functor Theory, Thesis, Princeton Univ., 1960.
[2] — Abelian Categories. New York: Harper & Row 1964.
[3] LUBKIN, S.: Imbedding of Abelian Categories. Trans. Amer. Math. Soc. **97**, 410—417 (1960).
[4] MITCHELL, B.: The Full Embedding Theorem. Amer. J. Math. **86**, 619—637 (1964).
[5] MACLANE, S.: Homology. New York: Academic Press 1963.
[6] PUPPE, D.: Korrespondenzen in Abelschen Kategorien. Math. Ann. **148**, 1—30 (1962).

Department of Mathematics
University of Pennsylvania
Philadelphia, Pennsylvania

Coherent Functors *

By

Maurice Auslander

Let $\mathscr{C}$ be an abelian category and F a (covariant) functor from $\mathscr{C}$ to abelian groups. We say that F is a coherent functor if there exists an exact sequence $(X, \_) \to (Y, \_) \to F \to 0$ where (X, A) denotes the maps from X to A. The main purpose of this paper is to initiate a study of the full subcategory $\check{\mathscr{C}}$ of coherent functors and give some applications to the theory of complexes in abelian categories as well as to some more specialized questions concerning modules over rings.

The first two sections of the paper are devoted to questions of notation and some of the more elementary questions concerning the category $\check{\mathscr{C}}$. For instance, it is shown that if $0 \to F_1 \to F_2 \to F_3 \to F_4 \to 0$ is an exact sequence of functors with F_2 and F_3 coherent, then F_1 and F_4 are also coherent. It is also shown that $\check{\mathscr{C}}$ is closed under extensions. Thus if X is a complex in C, then the cohomology functors $H^i(X, \_)$ are in $\check{\mathscr{C}}$. Since $\check{\mathscr{C}}$ has enough projectives, it makes sense to talk about the global dimension of $\check{\mathscr{C}}$. It is shown that the gl. dim $\check{\mathscr{C}} = 0$ or 2 and that gl. dim $\check{\mathscr{C}} = 0$ if and only if the gl. dim $\mathscr{C} = 0$.

We of course have the usual right exact functor $u: \mathscr{C} \to (\check{\mathscr{C}})^0$ given by $u(A) = (A, \_)$. If $\mathscr{D}$ is an abelian category, then we have the induced functor $u.: \mathscr{H}om(\mathscr{C}, \mathscr{D}) \to \mathscr{H}om((\check{\mathscr{C}})^0, \mathscr{D})$. Section two ends by showing that the functor $u.$ always has an adjoint $u^{\cdot}: \mathscr{H}om((\check{\mathscr{C}})^0, \mathscr{D}) \to$ $\to \mathscr{H}om(\mathscr{C}, \mathscr{D})$ having the following properties: a) if $F: \mathscr{C} \to \mathscr{D}$, then $(u^{\cdot} F)u = F$ and $u^{\cdot} F$ is left exact and $u^{\cdot} F$ is exact if F is right exact; b) if $\mathscr{C}$ has enough projectives and $F: \mathscr{C} \to \mathscr{D}$ is right exact, then $(L^i F)(A) = u^{\cdot} F(\operatorname{Ext}^i(A, \_))$, where $u^{\cdot} F$ is exact.

As seen above, the identity functor $I: \mathscr{C} \to \mathscr{C}$ can be factored through $(\check{\mathscr{C}})^0$ as $I = (u^{\cdot} I)u$ where $u^{\cdot} I: (\check{\mathscr{C}})^0 \to \mathscr{C}$ is exact. It is this functor $u^{\cdot} I$ which is studied in section three. Denoting by $\check{\mathscr{C}}_0$, the full subcategory of $\check{\mathscr{C}}$ such that $u^{\cdot} I$ sends the objects in $(\check{\mathscr{C}}_0)^0$ to zero, we have that $\check{\mathscr{C}}_0$ is a dense subcategory of $\check{\mathscr{C}}$ and that $\mathscr{C}$ is equivalent to $(\check{\mathscr{C}}/\check{\mathscr{C}}_0)^0$. Since

* Received September 13, 1965.

the objects of $\check{\mathscr{C}}_0$ can be described as those objects F in $\check{\mathscr{C}}$ such that $(F, G) = 0$ for all projective objects G in $\check{\mathscr{C}}$, it follows that up to equivalence of categories, the category $\check{\mathscr{C}}$ determines the category $\mathscr{C}$. An exact sequence of fundamental importance in this section and the rest of the paper is introduced in the course of the proofs of the above statements. Namely for F in $\check{\mathscr{C}}$, it is shown that the $R^0 F \approx (u^{\cdot} I(F), \_)$ and that there exists an exact sequence $0 \to F_0 \to F \to R^0 F \to F_1 \to 0$ with F_0 and F_1 in $\check{\mathscr{C}}_0$.

Section four is devoted to a discussion of half exact coherent functors. Under certain very mild conditions on $\mathscr{C}$ it is shown that if $\mathscr{C}$ has enough projectives, then a functor F is isomorphic to $\mathrm{Ext}^1(C, \_)$ for some C in $\mathscr{C}$ if and only if it is coherent, $(F, G) = 0$ for all representable functors G and F is half exact. From this it follows that if F is coherent and half exact, then we have for some A and B in $\mathscr{C}$ an exact sequence

$$0 \to \mathrm{Ext}^1(B, \_) \to F \to (A, \_) \to \mathrm{Ext}^2(B, \_).$$

Introducing a suitable equivalence relation in such exact sequences for a fixed A and B, it is shown that the elements of $\mathrm{Ext}^2(B, A)$ are in natural one to one correspondence with the equivalence classes of such exact sequences with half exact F. In particular, if the $\mathrm{gl.dim}\, \mathscr{C} \leqq 1$, we have that the half exact coherent functors are all of the form $\mathrm{Ext}^1(B, \_) + (A, \_)$ (direct sum).

Section five is devoted to applying some of these results to complexes in an abelian category. Among other things it is shown that if X is a complex and $F: \mathscr{C} \to \mathscr{Ab}$ is right exact, then

$$H_i(F(X)) \approx (H^i(X, \_), F)$$

for all i.

In section six and seven we apply the theory of coherent functors to rings. It is shown that $\otimes M$ is coherent if and only if M is a finitely presented R-module. And a new proof is given for our earlier result that if $0 \to A \to B \to C \to 0$ is an exact sequence of R-modules, with C finitely presented, which remains exact when tensored with arbitrary modules, then the exact sequence splits. We also show that if R is noetherian and F is a coherent functor, then $F \approx \mathrm{Tor}_1(\_, A)$ for some R-module A if and only if F is half exact, commutes with direct limits over directed sets, and $F(R) = 0$.

I would like to take this opportunity to thank my friends for many helpful discussions and suggestions, especially D. Buchsbaum, P. Freyd, and D. Kan. I would also like to thank the Conference for supplying the impetus to write this paper, as well as my friends in the Fulbright Commission and the Instituto de Matemática y Estadística in the University of Uruguay, Montevideo, for supplying such pleasent working conditions. And, of course, my indebtedness to the National Science Foundation for their generous support need hardly be mentioned.

1. Preliminaries

Let $\mathfrak{U}$ be a universe in the sense of GROTHENDIECK (see [2] for details). We shall say that a category $\mathscr{C}$ is a $\mathfrak{U}$-category if the objects of $\mathscr{C}$ and the maps of $\mathscr{C}$ are both sets which are elements of $\mathfrak{U}$. If $\mathscr{C}$ and $\mathscr{D}$ are $\mathfrak{U}$-categories then we shall denote by $\mathscr{H}om(\mathscr{C}, \mathscr{D})$ the category of covariant functors from $\mathscr{C}$ to $\mathscr{D}$ which is easily seen to be a $\mathfrak{U}$-category.

A particular interesting category associated with the universe $\mathfrak{U}$ is the category of $\mathfrak{U}$-sets $\mathscr{S}$. The objects of $\mathscr{S}$ are the elements of $\mathfrak{U}$ and for A and B in $\mathscr{S}$ the set $\mathscr{S}(A, B)$ of maps from A to B is defined to be ordinary set maps. Although $\mathscr{S}(A, B)$ is in $\mathfrak{U}$ for all A and B in $\mathscr{S}$, it can easily be seen that $\mathscr{S}$ is not a $\mathfrak{U}$-category. However if we let $\mathfrak{U}'$ be a universe containing $\mathfrak{U}$ as an element, then $\mathscr{S}$ is a $\mathfrak{U}'$-category. This suggests the following language which is sufficient for the purposes of this paper.

Let $\mathfrak{U}$ be a fixed universe and $\mathfrak{U}'$ the smallest universe containing $\mathfrak{U}$ as an element. By a category we shall mean a $\mathfrak{U}'$-category. Thus as observed above, if $\mathscr{C}$ and $\mathscr{D}$ are categories, then $\mathscr{H}om(\mathscr{C}, \mathscr{D})$, the category of covariant functors from $\mathscr{C}$ to $\mathscr{D}$ is, also a category. Also $\mathscr{S}$ as defined above is a category. Similarly $\mathscr{A}b$, the category of abelian groups whose underlying sets are elements of $\mathfrak{U}$ is a category in this sense.

Now let $\mathscr{I}$ and $\mathscr{C}$ be categories. Then we always have the covariant functor $i : \mathscr{C} \to \mathscr{H}om(\mathscr{I}, \mathscr{C})$ given by $i(C)(X) = C$ for all X in $\mathscr{I}$ and C in $\mathscr{C}$ and for $f : C \to D$ we have $i(f) : i(C)(X) \to i(D)(X)$ is the map f, for each X in $\mathscr{I}$. We shall say that $\mathscr{C}$ has $\mathscr{I}$-direct limits if the functor i has an adjoint, i.e. there exists a functor $\varinjlim : \mathscr{H}om(\mathscr{I}, \mathscr{C}) \to \mathscr{C}$ such that $\mathscr{C}(\varinjlim \alpha, C) \approx \mathscr{H}om(\mathscr{I}, \mathscr{C})\,(\alpha, i(C))$ for all α in $\mathscr{H}om(\mathscr{I}, \mathscr{C})$ and C in $\mathscr{C}$, the isomorphism being functorial in α and C. We shall say that $\mathscr{C}$ has direct limits if $\mathscr{C}$ has $\mathscr{I}$-direct limits for all $\mathfrak{U}$-categories $\mathscr{I}$ which are directed sets. If $\mathscr{C}$ is an abelian category, then $\mathscr{H}om(\mathscr{I}, \mathscr{C})$ is an abelian category. If $\mathscr{I}$-direct limits exist, then the functor $\varinjlim$ is right exact since it is an adjoint. We shall say that an abelian category $\mathscr{C}$ has exact direct limits if for each $\mathfrak{U}$-category $\mathscr{I}$ which is a directed set, $\mathscr{I}$-direct limits exist and the functor $\varinjlim : \mathscr{H}om(\mathscr{I}, \mathscr{C}) \to \mathscr{C}$ is exact. The category $\mathscr{A}b$ is an example of an abelian category with exact direct limits. Also if $\mathscr{C}$ is an abelian category with exact direct limits and $\mathscr{I}$ is an arbitrary category, then $\mathscr{H}om(\mathscr{I}, \mathscr{C})$ is an abelian category with exact direct limits.

Let $\mathscr{C}$ be a category. When we speak of a family of objects or maps in $\mathscr{C}$ we shall always mean a family which can be indexed by a set in $\mathfrak{U}$. If $(A_i)_{i\in I}$ is a family of objects in $\mathscr{C}$ and A is an object in $\mathscr{C}$, then a map of $(A_i)_{i\in I}$ to A is simply a family of maps $f_i : A_i \to A$. We shall say that $f_i : A_i \to A$ is an epimorphism if the induced maps $\mathscr{C}(A, X) \to \to \prod \mathscr{C}(A_i, X)$ are monomorphisms for all X in $\mathscr{C}$. An object A in $\mathscr{C}$ will be said to be finitely generated if given an epimorphism $f_i : A_i \to A$ with $i \in I$, there exists a finite subset J of I such that $f_j : A_j \to A$ (with $j \in J$) is an epimorphism.

Suppose now that $\mathscr{C}$ is an abelian category with exact direct limits. Let $(A_i)_{i\in I}$ be a family of objects in $\mathscr{C}$ and A an object in $\mathscr{C}$. Then a map from $(A_i)_{i\in I}$ to A is nothing more than a map of $\sum\limits_{i\in I} A_i$ to A, where $\sum\limits_{i\in I} A_i$ denotes the direct sum of the family $(A_i)_{i\in I}$. Also the condition that the map $(A_i)_{i\in I} \to A$ is an epimorphism is equivalent to the statement that the associated map $\sum\limits_{i\in I} A_i \to A$ is an epimorphism. Thus an object A in $\mathscr{C}$ is finitely generated if and only if given any family $(A_i)_{i\in I}$ in $\mathscr{C}$ and an epimorphism $f : \sum A_i \to A$, there exists a finite subset J of I such that f restricted to $\sum\limits_{j\in J} A_j$ is an epimorphism. The rest of this section is devoted to indicating how the usual terminology and fundamental properties of finitely generated objects in the category of modules over a ring can be extended to an abelian category $\mathscr{C}$ with exact direct limits.

Let $(A_i)_{i\in I}$ be a family of subobjects of an object A in $\mathscr{C}$. Then the $\operatorname{Im}(\sum A_i \to A)$ will be called the subobject of A generated by $(A_i)_{i\in I}$. We shall say that the family $(A_i)_{i\in I}$ generates A if the map $\sum A_i \to A$ is an epimorphism.

Thus A is finitely generated if and only if given any family of subobjects of A which generate A there is a finite subfamily which generates A.

Another useful formulation of when an object A is finitely generated is the following. We shall say that a family $(A_i)_{i\in I}$ of subobjects of A is directed if given i_0, i_1 in I there is an i_2 in I such that $A_{i_0} \subset A_{i_2}$ and $A_{i_1} \subset A_{i_2}$. Thus an object A is finitely generated if and only if given any directed family $(A_i)_{i\in I}$ of subobjects of A which generates A there is an $i \in I$ such that $A = A_i$. It should be observed that if $(A_i)_{i\in I}$ is a directed family of subobjects of A, then we can introduce the partial ordering $i_1 \leqq i_2$ if $A_{i_1} \subset A_{i_2}$ which makes I into a directed $\mathfrak{U}$-category $\mathscr{I}$. Then the subobject of A generated by $(A_i)_{i\in I}$ is $\varinjlim\limits_{i\in I} A_i$, where by $\varinjlim\limits_{i\in I} A_i$ we mean $\varinjlim F$ where F is the functor $\mathscr{I} \to \mathscr{C}$ which associates to i the object A_i.

These descriptions of finitely generated objects together with the following technical lemma, give most of the properties of finitely generated objects.

Lemma 1.1. *Let $\mathscr{C}$ be an abelian category with exact direct limits. Suppose $0 \to A' \xrightarrow{g} A \xrightarrow{f} A'' \to 0$ is an exact sequence in $\mathscr{C}$ and $(A_i)_{i\in I}$ is a directed set of subobjects of A. Then the families $(g^{-1}(A_i))_{i\in I}$ and $(f(A_i))_{i\in I}$ are directed sets of subobjects of A' and A'' and $(g^{-1}(A_i))_{i\in I}$ and $(f(A_i))_{i\in I}$ generate A' and A'' respectively if and only if $(A_i)_{i\in I}$ generates A.*

Proof. The fact that they are directed sets of subobjects is trivial to show. From the exact sequences

$$0 \to A'/g^{-1}(A_i) \to A/A_i \to A''/f(A_i) \to 0$$

we obtain passing to the limit, the exact sequence

$$0 \to A'/\varinjlim g^{-1}(A_i) \to A/\varinjlim A_i \to A''/\varinjlim f(A_i) \to 0 .$$

From this we see immediately that $A = \varinjlim A_i$ if and only if

$$A' = \varinjlim g^{-1}(A_i) \quad \text{and} \quad A'' = \varinjlim f(A_i),$$

which gives us our desired result.

As an immediate consequence we obtain

Proposition 1.2. *Let $\mathscr{C}$ be an abelian category with exact direct limits. Let $0 \to A' \to A \to A'' \to 0$ be an exact sequence in $\mathscr{C}$.*

a) If A is finitely generated, then A'' is finitely generated.

b) If A' and A'' are finitely generated, then A is finitely generated.

Corollary 1.3. *Let $\mathscr{C}$ be as above. Then any object which is generated by a finite number of finitely generated objects, is finitely generated.*

Proposition 1.4. *Let $\mathscr{C}$ be an abelian category with exact direct limits. Suppose $\mathscr{I}$ is a $\mathfrak{U}$-category which is a directed set, $F : \mathscr{I} \to \mathscr{C}$ a functor and $C = \varinjlim F$.*

If A is a finitely generated object in $\mathscr{C}$, then the map $\varinjlim \mathscr{C}(A, F(i)) \to \mathscr{C}(A, C)$ is a monomorphism and is an isomorphism if each of the maps $F(i) \to C$ is a monomorphism.

If A is a projective object in $\mathscr{C}$, then the following statements are equivalent:

a) A is finitely generated;

b) $\varinjlim \mathscr{C}(A, F(i)) = \varinjlim \mathscr{C}(A, C)$;

c) the functor $\mathscr{C}(A, \_)$ commutes with direct sums.

Proof. Suppose A is finitely generated. In order to show that the map $\varinjlim \mathscr{C}(A, F(i)) \to \mathscr{C}(A, C)$ is a monomorphism, it suffices to show that given any $f : A \to F(i)$ such that the composite $A \xrightarrow{f} F(i) \to C$ is zero, there is a $j \geqq i$ such that the composite $A \xrightarrow{f} F(i) \to F(j)$ is zero. Let J consist of all $j \in I$ such that $j \geqq i$. Then $\varinjlim F|J = C$. Thus

$$0 = \operatorname{Im}(A \to F(i) \to C) = \varinjlim_{J} \operatorname{Im}(A \to F(i) \to F(j))$$

$$= \varinjlim_{J} A/A_j \text{ where } A_j = \operatorname{Ker}(A \to F(i) \to F(j)).$$

Thus $(A_j)_{j \in J}$ is a directed set of subobjects of A which generate A. Therefore $A = A_{j_0}$ for some $j_0 \in J$, since A is finitely generated. Thus the map $A \to F(i) \to F(j_0)$ is the zero map, which gives us our desired result.

Suppose now that each $F(i) \to C$ is a monomorphism. Let $C_i = \operatorname{Im}(F(i) \to C)$. Then $(C_i)_{i \in I}$ is a directed set of subobjects of C which generate C. Let $f: A \to C$ and $A_i = f^{-1}(C_i)$. Then it is easily seen that $(A_i)_{i \in I}$ is a directed family of subobjects of A which generate A. Since A is finitely generated we know that $A = A_{i_0}$ for some $i_0 \in I$, i.e. $\operatorname{Im} f \subset A_{i_0}$. Therefore it follows that f is in the $\operatorname{Im}(\mathscr{C}(A, F(i_0)) \to \mathscr{C}(A, C))$. Thus we obtain that $\varinjlim \mathscr{C}(A, F(i)) \to \mathscr{C}(A, C)$ is an isomorphism, which finishes the proof of the first part of the proposition.

Suppose A is a projective object in $\mathscr{C}$. a) $\Rightarrow$ b). By the first part of the proposition we know that $\varinjlim \mathscr{C}(A, F(i)) \to \mathscr{C}(A, C)$ is a monomorphism. Let $C_i = \operatorname{Im}(F(i) \to C)$. Then $(C_i)_{i \in I}$ is a directed family of subobjects of C which generate C. Thus given $f: A \to C$ we know by the first part of the proposition that $\operatorname{Im} f \subset C_{i_0}$ for some $i_0 \in I$. Since A is projective we know there is a map $A \to F(i_0)$ such that $A \to F(i_0) \to C_{i_0} \to C$ is the map f. Therefore f is in the $\operatorname{Im}(\mathscr{C}(A, F(i_0)) \to \mathscr{C}(A, C))$ which shows that $\varinjlim \mathscr{C}(A, F(i)) = \mathscr{C}(A, C)$.

b) $\Rightarrow$ c). Trivial.

c) $\Rightarrow$ a). Suppose $(A_i)_{i \in I}$ is a family of subobjects of A which generate A. Then the map $\sum A_i \to A$ is an epimorphism. Since A is projective, we know there is a map $g: A \to \sum A_i$ such that the composite $A \xrightarrow{g} \sum A_i \to A$ is the identity. But $\mathscr{C}(A, \_)$ commutes with direct sums. Thus there is a finite subset J of I such that $\operatorname{Im} g \subset \sum_{i \in J} A_i$. From this it follows that $(A_i)_{i \in J}$ generate A and thus that A is finitely generated.

Let $\mathscr{C}$ be an arbitrary category. Then the set whose objects are of the form $\mathscr{C}(A, B)$ for all A and B in $\mathscr{C}$ is easily seen to be in $\mathfrak{U}'$. Similarly the union of all sets of the form set maps from $\mathscr{C}(A, B)$ to $\mathscr{C}(C, D)$ for all A, B, C, D in $\mathscr{C}$ is an element of $\mathfrak{U}'$. Thus $\mathscr{Set}(\mathscr{C})$, the "category" whose objects are $\mathscr{C}(A, B)$ for all A and B in $\mathscr{C}$ with ordinary set maps for maps is a category in the sense of this paper, i.e. is a $\mathfrak{U}'$-category.

Denoting the opposite category of $\mathscr{C}$ by $\mathscr{C}^0$, we have the usual functor $\mathscr{C}^0 \times \mathscr{C} \to \mathscr{S}(\mathscr{C})$ given by $(A, B) \mapsto \mathscr{C}(A, B)$.

We shall say that a full subcategory $\mathscr{I}$ of $\mathscr{C}$ is a generator of $\mathscr{C}$ if

a) The objects of $\mathscr{I}$ can be indexed by a set in $\mathfrak{U}$.

b) The functor $F: \mathscr{I}^0 \times \mathscr{C} \to \mathscr{S}ets(\mathscr{C})$ defined by $F(A, B) = \mathscr{C}(A, B)$ is isomorphic to a functor $G: \mathscr{I}^0 \times \mathscr{C} \to \mathscr{S}$ (the category of $\mathfrak{U}$-sets), i.e. for each pair A, B we are given isomorphisms $F(A, B) \approx G(A, B)$ which are functorial in A and B.

c) For all A and B in $\mathscr{C}$ the map

$$\mathscr{C}(A, B) \to \prod_{Y \in \mathscr{I}} \mathrm{Hom}(\mathscr{C}(Y, A), \mathscr{C}(Y, B))$$

is a monomorphism where Hom indicates set maps.

Suppose $\mathscr{I}$ is a generator for the category $\mathscr{C}$. Then it is not difficult to see that for each A in $\mathscr{C}$ the set of maps $f: Y \to A$ where Y ranges through $\mathscr{I}$ and f ranges through $\mathscr{C}(Y, A)$ is a family of maps which is an epimorphism. Also there exists a functor $\mathscr{C}^0 \times \mathscr{C} \to \mathscr{S}$ which is isomorphic to the functor $\mathscr{C}^0 \times \mathscr{C} \to \mathscr{S}ets(\mathscr{C})$ given by $(A, B) \mapsto \mathscr{C}(A, B)$. Thus if $\mathscr{C}'$ is a subcategory of $\mathscr{C}$ whose objects can be indexed by a set in $\mathfrak{U}$, then $\mathscr{C}'$ is isomorphic to a $\mathfrak{U}$-category.

If $\mathscr{C}$ is an abelian category, then for $\mathscr{S}ets(\mathscr{C})$ we substitute $\mathscr{A}b(\mathscr{C})$, the additive category of abelian groups whose objects are $\mathscr{C}(A, B)$ for all A and B in $\mathscr{C}$ and maps group homomorphisms. We shall say that a full subcategory $\mathscr{I}$ of $\mathscr{C}$ is a generator for $\mathscr{C}$ if it satisfies conditions a, b, c with $\mathscr{A}b(\mathscr{C})$ and $\mathscr{A}b$ substituted for $\mathscr{S}ets(\mathscr{C})$ and $\mathscr{S}$ respectively and group homomorphisms substituted for set maps. It is clear that if $\mathscr{I}$ satisfies these conditions, then the category $\mathscr{I}'$ whose objects are all finite direct sums of objects in $\mathscr{I}$ also satisfies the same conditions. For sake of convenience we shall assume in addition to the conditions a, b, c, that a generator for an abelian category has finite direct sums, i.e. is a full additive subcategory of $\mathscr{C}$ satisfying a, b, und c. If $\mathscr{C}$ is an abelian category with a generator, then $\mathscr{C}$ has, in addition to the properties sited above, the feature that the set of subobjects of an object in $\mathscr{C}$ can be indexed by a set in $\mathfrak{U}$ (see [*1*, pg. 336]).

Suppose $\mathscr{C}$ is an abelian category with exact direct limits and a generator $\mathscr{I}$ consisting of finitely generated objects in $\mathscr{C}$. Then an object A in $\mathscr{C}$ is finitely generated if and only if there exists an exact sequence $B \to A \to 0$ with B in $\mathscr{I}$. It follows from proposition 1.2 that if such an exact sequence exists, then A is finitely generated. On the other hand, since $\mathscr{I}$ is a generator, we know that there exists an exact sequence $\sum_{j \in J} B_j \to A \to 0$ with the B_j in $\mathscr{I}$. If A is finitely generated we know that there is a finite subset $J' \subset J$ such that $\sum_{j \in J} B_j \to A \to 0$ is exact, which gives us the other implication. Finally defining an object A in $\mathscr{C}$ to be finitely presented if it is finitely generated and given any exact sequence $0 \to X' \to X \to A \to 0$ with X finitely generated, we have that X' is finitely generated, we obtain the following usual result.

Proposition 1.5. *Let $\mathscr{C}$ be an abelian category. Suppose that $\mathscr{I}$ is a generator of $\mathscr{C}$ such that each object in $\mathscr{I}$ is a finitely generated projective object in $\mathscr{C}$. Also suppose $\mathscr{C}$ has exact direct limits. Then we have:*

a) An object X in $\mathscr{C}$ is finitely generated if and only if there exists an epimorphism $A \to X$ for some A in $\mathscr{I}$.

b) If X is an object in $\mathscr{C}$, then the following statements are equivalent:

i) X is finitely generated;

ii) there is an exact sequence $A_1 \to A_2 \to X \to 0$ with the A_i in $\mathscr{I}$.

iii) there is an exact sequence $A + A \to A \to X \to 0$ with A in $\mathscr{I}$.

Proof. a) Obvious.

b) i) $\Rightarrow$ ii) Obvious.

ii) $\Rightarrow$ iii) We imbed the exact sequence $A_1 \to A_2 \to X \to 0$ in the exact sequence $A_1 + A_1 \to A_1 + A_2 \to X \to 0$ by defining the map on the extra $A_1 \to A_1$ to be the identity. In turn we imbed this exact sequence in $A_1 + A_1 + A_2 + A_2 \to A_1 + A_2 \to X \to 0$ by sending $A_2 + A_2$ to zero. Thus letting $A = A_1 + A_2$ we obtain our desired exact sequence.

iii) $\Rightarrow$ ii) Obvious.

ii) $\Rightarrow$ i) Same as for modules and left as an excercise.

Corollary 1.6. *Let $\mathscr{C}$ and $\mathscr{I}$ be as in proposition 1.5 and let $\mathscr{C}'$ be the full subcategory of $\mathscr{C}$ whose objects are the finitely presented objects of $\mathscr{C}$. Then*

a) $\mathscr{C}'$ is equivalent to a $\mathfrak{U}$-category.

b) Suppose $\mathscr{I}$ is a $\mathfrak{U}$-category which is a directed set and $F: \mathscr{I} \to \mathscr{C}$, a functor. If A is in $\mathscr{C}'$, then $\varinjlim \mathscr{C}(A, F(i)) = \mathscr{C}(A, \varinjlim F)$.

Proof. a) Let $\mathscr{D}$ be the full subcategory of $\mathscr{C}$ consisting of representatives of the isomorphism classes of objects in $\mathscr{C}'$. For each object D in $\mathscr{D}$ choose a representation $A_1 \to A_2 \to D \to 0$ with A_1 and A_2 in $\mathscr{I}$. Then the objects of $\mathscr{D}$ are in one to one correspondence with a subset of $\bigcup \mathscr{C}(A_1, A_2)$ where A_1 and A_2 range through $\mathscr{I}$. Since the set of objects of $\mathscr{I}$ can be indexed by a set in $\mathfrak{U}$, so can the objects in $\mathscr{I}$. Since each $\mathscr{C}(A_1, A_2)$ is isomorphic to a $\mathfrak{U}$-set, we have that $\bigcup \mathscr{C}(A_1, A_2)$ is isomorphic to a $\mathfrak{U}$-set. Thus the objects in $\mathscr{D}$ can be indexed by a $\mathfrak{U}$-set. Therefore it follows from the remarks following the definition of a generator for a category, that the category $\mathscr{D}$ is isomorphic to a $\mathfrak{U}$-category. Since $\mathscr{D}$ is equivalent to $\mathscr{C}$, we are done.

b) Same proof as for modules.

2. Coherent Functors

Throughout the rest of this paper we will assume (unless stated to the contrary) that all categories and functors are additive. Let $\mathscr{C}$ denote a fixed abelian $\mathfrak{U}$-category. We shall denote by $\check{\mathscr{C}}$ the category $\mathscr{H}om(\mathscr{C}, \mathscr{A}b)$ of covariant (additive) functors from $\mathscr{C}$ to $\mathscr{A}b$, the category of abelian groups in $\mathfrak{U}$. We shall denote by $\hat{\mathscr{C}}$ the category $\mathscr{H}om(\mathscr{C}^0, \mathscr{A}b)$ of contravariant functors from $\mathscr{C}$ to $\mathscr{A}b$. It should be noted that since $\mathscr{A}b$ has exact direct limits, so do $\hat{\mathscr{C}}$ and $\hat{\mathscr{C}}^0$. The rest of this paper is devoted to studying the categories $\check{\mathscr{C}}$ and $\hat{\mathscr{C}}$, the full subcategories of finitely presented objects in $\check{\mathscr{C}}$ and $\hat{\mathscr{C}}$, respectively. Since $\hat{\mathscr{C}}^0 = \check{\mathscr{C}}$ and $\check{\mathscr{C}}^0 = \hat{\mathscr{C}}$ it suffices to state and prove theorems for one or the other of these categories. The analogous statements for the other category is left as an excercise for the reader, although such statements will be used freely in the paper when convenient.

Let $\mathscr{P}$ be the full subcategory of $\hat{\mathscr{C}}$ consisting of those functors of the form $\mathscr{C}(-, A)$ or more simply $(-, A)$ for each A in $\mathscr{C}$. Suppose G is in $\hat{\mathscr{C}}$ and $\alpha: (-, A) \to G$ in $\hat{\mathscr{C}}$. Then $\alpha_A: (A, A) \to G(A)$ and thus $\alpha_A(1_A)$ is an element in $G(A)$. It is well known that $\varphi: \hat{\mathscr{C}}((-, A), G) \to G(A)$ given by $\varphi(\alpha) = \alpha_A(1_A)$ is an isomorphism which is functorial in A and G. From this it readily follows that

a) $(-, A)$ is projective for each A in $\mathscr{C}$;

b) $\hat{\mathscr{C}}((-, A), -)$ commutes with direct limits and thus by proposition 1.4, each $(-, A)$ is finitely generated;

c) $(-, A) + (-, B) = (-, A + B)$ and thus $\mathscr{P}$ is closed under finite direct sums;

d) $\mathscr{P}$ is a generator for $\hat{\mathscr{C}}$.

Thus the generator $\mathscr{P}$ of $\hat{\mathscr{C}}$ satisfies the hypothesis of proposition 1.5. Therefore a functor $F \in \hat{\mathscr{C}}$ is finitely presented if and only if there exists an exact sequence $(-, A) \to (-, B) \to F \to 0$. Suppose $(-, A) \to (-, B) \to F \to 0$ is exact. Then the map $(-, A) \to (-, B)$ is induced by a map $A \to B$. If we let $A' = \operatorname{Ker}(A \to B)$, then we obtain the exact sequence

$$0 \to (-, A') \to (-, A) \to (-, B) \to F \to 0.$$

Thus the projective dimension of F (notation: pd F) is at most 2. Therefore we have shown that if $P_1 \to P_2$ is a map in $\mathscr{P}$, then the $\operatorname{Ker}(P_1 \to P_2)$ is in $\mathscr{P}$. Also that each $F \in \hat{\mathscr{C}}$ the category of finitely presented functors, has a projective resolution which is also in $\hat{\mathscr{C}}$ and that the pd $F \leqq 2$. Some additional properties of the category $\hat{\mathscr{C}}$ of finitely presented functors will become clear from the following general proposition.

Proposition 2.1. *Let $\mathscr{A}$ be an abelian category and $\mathscr{P}$ a full additive subcategory of $\mathscr{A}$ whose objects are projective objects in $\mathscr{A}$. Let $\mathscr{P}(\mathscr{A})$ be the full subcategory of $\mathscr{C}$ consisting of those objects A in $\mathscr{A}$ such that there exists an exact sequence $P_1 \to P_0 \to A \to 0$ in $\mathscr{A}$ with the P_i in $\mathscr{P}$. Then $\mathscr{P}(\mathscr{A})$ has the following properties:*

a) If $0 \to C_1 \to C_2 \to C_3 \to 0$ is exact in $\mathscr{A}$ with C_1 and C_3 in $\mathscr{P}(\mathscr{A})$, then C_2 is in $\mathscr{P}(\mathscr{A})$.

b) Suppose $0 \to C_1 \to C_2 \to C_3 \to C_4 \to 0$ is exact with C_2 and C_3 in $\mathscr{P}(\mathscr{A})$, then C_4 is in $\mathscr{P}(\mathscr{A})$. If $\mathscr{P}$ has the additional property that given $P_2 \to P_3$ in $\mathscr{P}$, there is an exact sequence $P_1 \to P_2 \to P_3$ with P_1 in $\mathscr{P}$, then C_1 is also in $\mathscr{P}(\mathscr{A})$.

c) The inclusion functor $u: \mathscr{P} \to \mathscr{P}(\mathscr{A})$ has the property that given any additive category $\mathscr{D}$ with cokernels, then the induced functor

$$u: \mathscr{H}om(\mathscr{P}(\mathscr{A}), \mathscr{D}) \to \mathscr{H}om(\mathscr{P}, \mathscr{D})$$

has a left adjoint $u^{\cdot}: \mathscr{H}om(\mathscr{P}, \mathscr{D}) \to \mathscr{H}om(\mathscr{P}(\mathscr{A}), \mathscr{D})$. This $u^{\cdot}$ and the functorial isomorphisms $(u^{\cdot}F, G) \approx (F, u_{.}G)$ for F in $\mathscr{H}om(\mathscr{P}, \mathscr{D})$ and G in $\mathscr{H}om(\mathscr{P}(\mathscr{A}), \mathscr{D})$ can be chosen so that:

i) $u^{\cdot}u.F = F$ for all F in $\mathscr{H}om(\mathscr{P}, \mathscr{D})$;

ii) $u^{\cdot}F$ is right exact for all F in $\mathscr{H}om(\mathscr{P}, \mathscr{D})$ (i.e. given an exact sequence $C_1 \to C_2 \to C_3 \to 0$ in $\mathscr{P}(\mathscr{A})$, then $F(C_1) \to F(C_2) \to F(C_3) \to 0$ is exact);

iii) $u.u^{\cdot}u.G = u.G$ for all $G \in \mathscr{H}om(\mathscr{P}(\mathscr{A}), \mathscr{D})$ and the usual map $u^{\cdot}u.G \to G$ is an isomorphism if and only if G is right exact, in which case $u^{\cdot}u.G = G$.

Proof. a) Follows easily from standard techniques of building resolutions for exact sequences.

b) Suppose $P_1 \to P_0 \to C_2 \to 0$ and $Q_1 \to Q_0 \to C_3 \to 0$ are exact with the P_i and Q_i in $\mathscr{P}$. Given the map $g: C_2 \to C_3$ we know that it can be lifted to a map of the complexes $g: P \to Q$ where P is the complex $\ldots 0 \to P_1 \to P_0 \to 0 \ldots$ and Q is the complex $\ldots 0 \to Q_1 \to Q_0 \to 0 \ldots$ Let $M(g)$ be the mapping cone of g, i.e. $M(g)_i = P_{i-1} + Q_i$ and $d: M(g)_i \to M(g)_{i-1}$ is $(-d_P, d_Q + g)$ where d_P and d_Q are the boundary maps in P and Q respectively. Then as is well known (see [*4*, ch. II, §4] for instance) we obtain an exact sequence

$(\star)$ $H_1(P) \to H_1(Q) \to H_1(M(g)) \to H_0(P) \to H_0(Q) \to H_0(M(g)) \to 0$

with $H_0(P) = C_2$ and $H_0(Q) = C_3$ and the map $H_0(P) \to H_0(Q)$ the map $g: C_2 \to C_3$. Thus we have that $H_0(M(g)) \approx \operatorname{Coker} g \approx C_4$. But we have the exact sequence $P_0 + Q_1 \to Q_0 \to H_0(M(g)) \to 0$ which shows that $H_0(M(g)) \in \mathscr{P}(\mathscr{A})$ since $\mathscr{P}$ is closed under finite direct sums. Thus the first part of b) is proven.

Suppose now that $\mathscr{P}$ has the additional property that given $Y \to Z$ in $\mathscr{P}$ there is an exact sequence $X \to Y \to Z$ with X in $\mathscr{P}$. This is the same thing as assuming that if $Y \to Z$ is in $\mathscr{P}$ then $\operatorname{Ker}(Y \to Z)$ is in $\mathscr{P}(\mathscr{A})$. Therefore it follows that $H_1(P)$ and $H_1(Q)$ are in $\mathscr{P}(\mathscr{A})$. Since we have shown that $\mathscr{P}(\mathscr{A})$ is closed under cokernels, we obtain from the exact sequence (*), the exact sequence

$$0 \to B \to H_1(M(g)) \to H_0(P) \to H_0(Q)$$

with B in $\mathscr{P}(\mathscr{A})$. Thus if we show that $H_1(M(g))$ is in $\mathscr{P}(\mathscr{A})$, then we will have again by part a), that $\operatorname{Ker}(H_0(P) \to H_0(Q)) \approx C_1$ is in $\mathscr{P}(\mathscr{A})$. But $Z_1(M(g)) = \operatorname{Ker}(M(g)_1 \to M(g)_0)$ which is in $\mathscr{P}(\mathscr{A})$ since $M(g)_1$ and $M(g)_0$ are in $\mathscr{P}$. Therefore $H_1(M(g)) = \operatorname{Coker}(M(g)_2 \to Z_1(M(g))$ is in $\mathscr{P}(\mathscr{A})$, which concludes the proof of b).

c) We first have to describe the functor

$$u^{\cdot}: \mathscr{H}om(\mathscr{P}, \mathscr{D}) \to \mathscr{H}om(\mathscr{P}(\mathscr{A}), \mathscr{D}).$$

For each C in $\mathscr{A}$ we pick a fixed exact sequence $P_1 \to P_0 \to C \to 0$ subject to the condition that if C is in $\mathscr{P}$, then we pick the sequence $0 \to C = C \to 0$. Suppose $F: \mathscr{P} \to \mathscr{D}$. Then define

$$u^{\cdot} F(C) = \operatorname{Coker}(F(P_1) \to F(P_0)).$$

If we have $C \xrightarrow{g} C'$, then we lift this to a map of $P \to P'$ which gives us a map $u^{\cdot}(F)(g): F(C) \to F(C')$ which is well known to be independent of the lifting used. Thus $u^{\cdot}$ is defined.

Suppose F is in $\mathscr{H}om(\mathscr{P}, \mathscr{D})$ and G is in $\mathscr{H}om(\mathscr{P}(\mathscr{C}), \mathscr{D})$. If P is in $\mathscr{P}$, then $u^{\cdot}(F(P) = F(P)$ and $u.\,G(P) = G(P)$. Thus given h in $(u^{\cdot} F, G)$, we have for each P in $\mathscr{P}$ a map $h_P: u^{\cdot} F(P) \to G(P)$ which is the same thing as a map $h_P: F(P) \to u.\,G(P)$. Thus we have a map $(u^{\cdot} F, G) \to (F, u.\,G)$ which is easily seen to be an isomorphism which is functorial in F and G. The properties i), ii), and iii) are easy to check and are left as excercises.

It should be observed that in the case $\mathscr{P}(\mathscr{A}) = \mathscr{A}$ and G is in $\mathscr{H}om(\mathscr{A}, \mathscr{D})$, then $u^{\cdot} u.\,G$ is usually denoted by $L_0 G$ and is called the 0-th left derived functor of G.

We know return to the categories $\widehat{\mathscr{C}}$ and $\hat{\mathscr{C}}$. Letting $\mathscr{P}$ be the full subcategory of $\hat{\mathscr{C}}$ whose objects are $(-, A)$ for all A in $\mathscr{C}$, we know that $\hat{\mathscr{C}} = \mathscr{P}(\hat{\mathscr{C}})$ and that if $P_2 \to P_3$ is in $\mathscr{P}$, then $\operatorname{Ker}(P_2 \to P_3) \in \mathscr{P}$. It then follows from a) and b) of proposition 2.1, that if $0 \to F_1 \to F_2 \to F_3 \to 0$ is exact with F_1 and F_2 finitely presented, then F_2 is finitely presented. Also if $0 \to F_1 \to F_2 \to F_3 \to F_4 \to 0$ is exact and F_2 and F_3 are finitely presented, then F_1 and F_4 are finitely presented. Thus $\hat{\mathscr{C}}$ is an abelian category and the inclusion functor $\hat{\mathscr{C}} \to \hat{\mathscr{C}}$ is fully faithful, exact and preserves projective objects.

Suppose $0 \to F' \to F \to F'' \to 0$ is exact with F finitely presented and F' finitely generated. Then we have an exact sequence $(-, A) \to F' \to 0$ and thus an exact sequence $(-, A) \to F \to F'' \to 0$. Since $(-, A)$ and F are finitely presented, we have that F'' is finitely presented and thus that F' is finitely presented. Thus every finitely generated subobject of a finitely presented functor is finitely presented. Such objects in a category are usually called coherent objects. Thus all finitely presented functors are coherent functors.

We now give some examples of coherent functors. Suppose X is a complex in $\mathscr{C}$. Then $(-, X)$ is a complex in $\hat{\mathscr{C}}$ and thus for each i, the functors $Z_i((-, X))$, $B_i((-, X))$ and $H_i((-,))$ are in $\hat{\mathscr{C}}$. Thus if $\mathscr{C}$ has sufficiently many injective objects, then $\operatorname{Ext}^i(-, A)$ is in $\hat{\mathscr{C}}$ for all i and all A in $\mathscr{C}$. On the other hand, suppose $\mathscr{C}$ has sufficiently many projective objects. If $P \to A \to 0$ is exact with P projective, then the functor $\underline{\Pi}(-, A) = \operatorname{Coker}((-, P) \to (-, A))$ introduced by Eckmann-Hilton is independent of the choice of P and is clearly in $\hat{\mathscr{C}}$. More generally, $\underline{\Pi}_n(-, A)$ is in $\hat{\mathscr{C}}$ for all n, where $\underline{\Pi}_{n+1}(-, A) = H_n((-, P))$ (for $n > 0$) and P is a projective resolution of A (see [2] for further details).

Similarly if X is a complex in $\mathscr{C}$, then the complex $(X, -)$ is in $\check{\mathscr{C}}$ and thus $Z_i((X, -))$, $B_i((X, -))$ and $H_i((X, -))$ are in $\check{\mathscr{C}}$. Thus if $\mathscr{C}$ has sufficiently many projective objects, then $\operatorname{Ext}^i(A, -)$ is in $\check{\mathscr{C}}$ for all i and A in $\mathscr{C}$. If $\mathscr{C}$ has sufficiently many injectives, then the injective homotopy functors $\overline{\Pi}_n(A, -)$ are also in $\check{\mathscr{C}}$ where for $n > 0$ the functor $\overline{\Pi}_{n+1}(A, -) = H_n((Q, -)$ where Q is an injective resolution of A. We shall give other examples later on.

We now return to applying part c) of proposition 2.1 to the categories of coherent functors. First we consider the category $\hat{\mathscr{C}}$. The canonical functor $\mathscr{C} \to \hat{\mathscr{C}}$ given by $A \to (-, A)$ gives us an isomorphism between the categories $\mathscr{C}$ and $\mathscr{P}$ which we will often consider an identification. Thus by proposition 2.1 c) we have that given any abelian category $\mathscr{D}$, the functor $u.\colon \mathscr{H}om(\hat{\mathscr{C}}, \mathscr{D}) \to \mathscr{H}om(\mathscr{C}, \mathscr{D})$ induced by $u\colon \mathscr{C} \to \hat{\mathscr{C}}$ has a left adjoint $u^{\cdot}\colon \mathscr{H}om(\mathscr{C}, \mathscr{D}) \to \mathscr{H}om(\hat{\mathscr{C}}, \mathscr{D})$ which has the following properties:

1. Suppose F is in $\mathscr{H}om(\mathscr{C}, \mathscr{D})$ and G is in $\mathscr{H}om(\hat{\mathscr{C}}, \mathscr{D})$. Then $u^{\cdot} F((-, A)) = F(A)$ and $u.\, G(A) = G((-, A))$ for all A in $\mathscr{C}$. Further, if $\alpha\colon u^{\cdot} F \to G$ is a map in $\mathscr{H}om(\hat{\mathscr{C}}, \mathscr{D})$, then for each A in $\mathscr{C}$, we get a map $F(A) \to u.\, G(A)$, namely $\alpha_A\colon u^{\cdot} F((-, A)) \to G((-, A))$. Thus we get an isomorphism $(u^{\cdot} F, G) \to (F, u.\, G)$ which is functorial in F and G.

2. $u. u^{\cdot} F = F$ for all F in $\mathcal{H}om(\mathcal{C}, \mathcal{D})$.

3. $u^{\cdot} F$ is right exact for all F in $\mathcal{H}om(\mathcal{C}, \mathcal{D})$.

4. The usual map $u^{\cdot} u. G \to G$ for all G in $\mathcal{H}om(\hat{\mathcal{C}}, \mathcal{D})$ has the properties that a) $u^{\cdot} u. G((-, A)) \to G((-, A))$ is the identity for A in $\mathcal{C}$ and b) $u^{\cdot} u. G \to G$ is an isomorphism if and only if G is right exact in which case $u^{\cdot} u. G = G$.

We have also one more property that is not simply a rephrasing of proposition 2.1.

5. Suppose F in $\mathcal{H}om(\mathcal{C}, \mathcal{D})$ is left exact. Then $u^{\cdot} F : \hat{\mathcal{C}} \to \mathcal{D}$ is exact. For let T be in $\hat{\mathcal{C}}$ and let $0 \to A_1 \to A_2 \to A_3$ be an exact sequence in $\mathcal{C}$ such that $0 \to (-, A_1) \to (-, A_2) \to (-, A_3) \to T \to 0$ is exact. Since $u^{\cdot} F((-, A_i)) = F(A_i)$ and F is left exact we have that

$$0 \to u^{\cdot} F((-, A_i)) \to u^{\cdot} F((-, A_2)) \to u^{\cdot} F((-, A_3))$$

is exact. Thus the derived functors $L^i u^{\cdot} F(T) = 0$ for $i > 0$. Since this is true for all T in $\hat{\mathcal{C}}$, it follows that $u^{\cdot} F$ is exact.

Suppose now that $F : \mathcal{C} \to \mathcal{D}$ is left exact and X is a complex in $\mathcal{C}$. Then $F(X) = u^{\cdot} F((-, X))$ and thus $H_i(F(X)) = u^{\cdot} F(H_i(-, X))$ for all i since $u^{\cdot} F$ is exact. In particular, suppose $\mathcal{C}$ has enough injectives and $F : \mathcal{C} \to \mathcal{D}$ is left exact. Then $R^i F(A) = u^{\cdot} F(\mathrm{Ext}^i(-, A))$ for all i and all A in $\mathcal{C}$.

We end this section by summarizing in the following theorems some of the more important facts concerning the categories of coherent functors established so far.

Theorem 2.2. *Let $\mathcal{C}$ be an abelian $\mathfrak{U}$-category and $\hat{\mathcal{C}}$ the category of contravariant finitely presented functors from $\mathcal{C}$ to $\mathcal{A}b$.*

a) A functor F in $\hat{\hat{\mathcal{C}}}$ is in $\hat{\mathcal{C}}$ if and only if there is an exact sequence $(-, A) \to (-, B) \to F \to 0$.

b) $\hat{\mathcal{C}}$ is an abelian category with enough projectives of global dimension at most 2. Also the inclusion functor from $\hat{\mathcal{C}}$ to $\hat{\hat{\mathcal{C}}}$ is a fully faithful, exact functor.

c) The functor $u : \mathcal{C} \to \hat{\mathcal{C}}$ given by $u(A) = (-, A)$ is a fully faithful, left exact functor.

d) If $\mathcal{D}$ is an abelian category, then the exact functor $u. : \mathcal{H}om(\hat{\mathcal{C}}, \mathcal{D}) \to \mathcal{H}om(\mathcal{C}, \mathcal{D})$ has a left adjoint $u^{\cdot}$ with the following properties:

1. If $F : \mathcal{C} \to \mathcal{D}$, then $(u^{\cdot} F)\, u = F$ and $u^{\cdot} F$ is right exact.

2. $F : \mathcal{C} \to \mathcal{D}$ is left exact if and only if $u^{\cdot} F$ is exact.

3. If $\mathcal{C}$ has enough injectives and $F : \mathcal{C} \to \mathcal{D}$ is left exact, then $(R^i F)(A) = u^{\cdot} F(\mathrm{Ext}^i(-, A))$ for all $i \geqq 0$ and all A in $\mathcal{C}$.

We state for the convenience of the reader the analogous theorem for $\check{\mathcal{C}}$.

Theorem 2.3. *Let $\mathcal{C}$ be an abelian $\mathfrak{U}$-category and $\check{\mathcal{C}}$ the category of finitely presented covariant functors from $\mathcal{C}$ to $\mathcal{A}b$.*

a) A functor F in $\check{\mathscr{C}}$ is in $\check{\mathscr{C}}$ if and only if there exists an exact sequence $(A, -) \to (B, -) \to F \to 0$.

b) $(\check{\mathscr{C}})^0$ *is an abelian category with enough injectives of global dimension at most 2. Further, the inclusion functor* $(\check{\mathscr{C}})^0 \to (\check{\mathscr{C}})^0$ *is a fully faithful, exact functor.*

c) The functor $u\colon \mathscr{C} \to (\check{\mathscr{C}})^0$ *given by* $u(A) = (A, -)$ *is a fully faithful, right exact functor.*

d) If $\mathscr{D}$ *is an abelian category, then the exact functor* $u_{\cdot}\colon \mathscr{H}om((\hat{\mathscr{C}})^0, \mathscr{D}) \to \mathscr{H}om(\mathscr{C}, \mathscr{D})$ *has a left adjoint* $u^{\cdot}\colon \mathscr{H}om(\mathscr{C}, \mathscr{D}) \to \mathscr{H}om((\check{\mathscr{C}})^0, \mathscr{D})$ *with the following properties:*

1. If $F\colon \mathscr{C} \to \mathscr{D}$*, then* $(u^{\cdot} F)\, u = F$ *and* $u^{\cdot} F$ *is a left exact.*

2. $F\colon \mathscr{C} \to \mathscr{D}$ *is right exact if and only if* $u^{\cdot} F$ *is exact.*

3. If $\mathscr{C}$ *has enough projectives and* F *is right exact, then* $L^i F(A) = u^{\cdot} F(\mathrm{Ext}^i(A, -))$ *for all* $i \geq 0$ *and all* A *in* $\mathscr{C}$.

3. The Categories $\hat{\mathscr{C}}_0$ and $\hat{\mathscr{C}}/\hat{\mathscr{C}}_0$

As in section § 2, we assume that $\mathscr{C}$ is a fixed abelian $\mathfrak{U}$-category and we continue our study of the categories $\hat{\mathscr{C}}$ and $\check{\mathscr{C}}$. Letting $u\colon \mathscr{C} \to \hat{\mathscr{C}}$ be the usual functor $C \mapsto (-, C)$ we know by theorem 2.2, that the functor $u_{\cdot}\colon \mathrm{Hom}(\hat{\mathscr{C}}, \mathscr{C}) \to \mathrm{Hom}(\mathscr{C}, \mathscr{C})$ has a left adjoint $u^{\cdot}$ which has certain properties. In particular, since the identity functor $I\colon \mathscr{C} \to \mathscr{C}$ is exact, we know that $u^{\cdot} I\colon \hat{\mathscr{C}} \to \mathscr{C}$ is exact and $(u^{\cdot} I)\, u = u$. We shall denote $u_{\cdot} I$ by v. It follows from the exactness of v, that $\hat{\mathscr{C}}_0$, the full subcategory of $\hat{\mathscr{C}}$ consisting of all F such that $v(F) = 0$, is a dense subcategory of $\hat{\mathscr{C}}$, i.e. if $0 \to F_1 \to F_2 \to F_3 \to 0$ is exact in $\hat{\mathscr{C}}$, then F_2 is in $\hat{\mathscr{C}}_0$ if and only if F_1 and F_2 are in $\hat{\mathscr{C}}_0$. Since the category $\hat{\mathscr{C}}_0$ plays an important role in our general theory, we will give various descriptions of the objects in $\hat{\mathscr{C}}_0$. However, before doing this we make the following trivial but useful observation.

Lemma 3.1. *Let F be in $\hat{\mathscr{C}}$ and G in $\hat{\mathscr{C}}$. Suppose* $0 \to A_0 \to A_1 \to A_2$ *is an exact sequence in $\mathscr{C}$ such that* $0 \to (-, A_0) \to (-, A_1) \to (-, A_2) \to F \to 0$ *is exact. Then* $\mathrm{Ext}^i(F, G) \approx H_i(G(A))$ *where $G(A)$ is the complex*

$$\ldots 0 \to G(A_2) \to G(A_1) \to G(A_0) \to 0 \ldots$$

Proof. Follows immediately from the fact that $((-, B), G) \approx G(B)$ for all B in $\mathscr{C}$.

Proposition 3.2. *Let F be in $\hat{\mathscr{C}}$. Then the following statements are equivalent:*

a) F is in $\hat{\mathscr{C}}_0$.

b) If $0 \to A_0 \to A_1 \to A_2$ *is exact in* $\mathscr{C}$ *such that* $0 \to (-, A_0) \to (-, A_1) \to (-, A_2) \to F \to 0$ is *exact, then* $0 \to A_0 \to A_1 \to A_2 \to 0$ *is exact.*

c) There exists an exact sequence $0 \to A_0 \to A_1 \to A_2 \to 0$ *in* $\mathscr{C}$ *such that* $0 \to (-, A_0) \to (-, A_1) \to (-, A_2) \to F \to 0$ *is exact.*

d) $\operatorname{Ext}^i(F, G) = 0$ *for* $i = 0{,}1$ *and any* G *in* $\widehat{\mathscr{C}}$ *which is left exact, i.e. if* $X \to Y \to Z \to 0$ *is exact in* $\mathscr{C}$, *then* $0 \to G(Z) \to G(Y) \to G(X)$ *is exact.*

e) $(F, G) = 0$ *for any* G *in* $\widehat{\mathscr{C}}$ *which is left exact.*

f) $(F, (-, A)) = 0$ *for all* A *in* $\mathscr{C}$.

Proof. a) $\Leftrightarrow$ b). Suppose $0 \to A_0 \to A_1 \to A_2$ is exact such that $0 \to (-, A_0) \to (-, A_1) \to (-, A_2) \to F \to 0$ is exact. Then applying v to this exact sequence, we obtain the exact sequence $0 \to A_0 \to A_1 \to A_2 \to v(F) \to 0$. Thus $v(F) = 0$ if and only if $0 \to A_0 \to A_1 \to A_2 \to 0$ is exact, which gives us our desired result.

b) $\Rightarrow$ c) Trivial.

c) $\Rightarrow$ d) Follows immediately from Lemma 3.1.

d) $\Rightarrow$ e) and e) $\Rightarrow$ f) are trivial.

f) $\Rightarrow$ b) Suppose $0 \to A_0 \to A_1 \to A_2$ is exact such that $0 \to (-, A_0) \to (-, A_1) \to (-, A_2) \to F \to 0$ is exact. Then it follows from Lemma 3.1, that $(F, (-, A)) = \operatorname{Ker}((A_2, A) \to (A_1, A))$. Thus $(F, (-, A)) = 0$ for all A if and only if $0 \to (A_2, A) \to (A_1, A)$ is exact for all A, i.e. if and only if $A_1 \to A_2 \to 0$ is exact. Thus f) $\Rightarrow$ b), which finishes the proof.

We now recall the definition of the $v\colon \widehat{\mathscr{C}} \to \mathscr{C}$. For each F in $\widehat{\mathscr{C}}$ we chose a fixed exact sequence $0 \to A_0 \to A_1 \to A_2$ and a fixed map $(-, A_i) \to F$ such that $0 \to (-, A_0) \to (-, A_1) \to (-, A_2) \to F \to 0$ is exact, subject only to the condition that if $F = (-, X)$, then we choose $A_0 = 0 = A_1$ and $A_2 = X$ with the map $(-, A_2) \to F$ the identity. Then $v(F) = A_3 = \operatorname{Coker}(A_1 \to A_2)$. Now if we let $B = \operatorname{Coker}(A_0 \to A_1) = \operatorname{Ker}(A_2 \to A_3)$, then from the exact sequence $0 \to A_0 \to A_1 \to B \to 0$ and $0 \to B \to A_2 \to A_3 \to 0$ we deduce the following commutative diagram with exact rows and columns

$$
\begin{array}{ccccccccccc}
 & & & & & & 0 & & & & \\
 & & & & & & \downarrow & & & & \\
0 & \to & (-, A_0) & \to & (-, A_1) & \to & (-, B) & \to & F_0 & \to & 0 \\
 & & \| & & \| & & \downarrow & & & & \\
0 & \to & (-, A_0) & \to & (-, A_1) & \to & (-, A_2) & \to & F & \to & 0 \\
 & & & & & & \downarrow & & & & \\
 & & & & & & (-, A_3) & & & & \\
 & & & & & & \downarrow & & & & \\
 & & & & & & F_1 & & & & \\
 & & & & & & \downarrow & & & & \\
 & & & & & & 0 & & & &
\end{array}
\tag{3.3}
$$

where F_0 and F_1 are defined by the exact sequences and are obviously in $\hat{\mathscr{C}}^0$. From this it follows that there is a unique sequence of maps

$$F_0 \to F \to (-, A_3) \to F_1$$

which makes the above diagram commutative. Further elementary diagram chasing shows that $0 \to F_0 \to F \to (-, A_3) \to F_1 \to 0$ is exact. Since $A_3 = v(F)$, this exact sequence can be rewritten as $0 \to F_0 \to F \to (-, v(F)) \to F_1 \to 0$. It is not difficult to see that: a) the map $F \to (-, v(F))$ is functorial in F; b) F_0 and F_1 are functorial in F; c) the exact sequence $0 \to F_0 \to F \to (-, v(F)) \to F_1 \to 0$ is functorial in F.

Let G be a left exact functor in $\hat{\mathscr{C}}$. Since F_i is in $\hat{\mathscr{C}}_0$ for $i = 0$ and 1, we know by proposition 3.2, that $\mathrm{Ext}^j(F_i, G) = 0$ for $j = 0$ and 1 and $i = 0$ and 1. From this it follows by easy direct computations that the map $((-, v(F)), G) \to (F, G)$ is an isomorphism. Thus given any map $F \to G$ there exists one and only one map $(-, v(F)) \to G$ which makes the diagram

$$\begin{array}{ccc} F & \to & (-, v(F)) \\ \| & & \downarrow \\ F & \to & G \end{array}$$

commutative.

As an application of the above observations, we obtain:

Proposition 3.4. *Let F be in $\hat{\mathscr{C}}$ and let $0 \to F' \to F \to G \to F'' \to 0$ be an exact sequence in $\hat{\mathscr{C}}$ with G left exact and F' and F'' in $\hat{\mathscr{C}}_0$. Then there exists one and only one map of exact sequences*

$$\begin{array}{ccccccccc} 0 & \to & F_0 & \to & F & \to & (-, v(F)) & \to & F_1 & \to 0 \\ & & \downarrow & & \| & & \downarrow & & \downarrow & \\ 0 & \to & F' & \to & F & \to & G & \to & F'' & \to 0 \end{array}$$

and this is an isomorphism.

Proof. The existence and uniqueness of the map has already been shown. A similar argument shows that since F' and F'' are in $\hat{\mathscr{C}}_0$, then the induced map $(G, G') \to (F, G')$ is an isomorphism whenever G' is left exact. Thus we obtain that there is one and only one map of exact sequences

$$\begin{array}{ccccccccc} 0 & \to & F' & \to & F & \to & G & \to & F'' & \to 0 \\ & & \downarrow & & \| & & \downarrow & & \downarrow & \\ 0 & \to & F_0 & \to & F & \to & (-, v(F)) & \to & F_1 & \to 0 \end{array}$$

and this map is easily shown to be the inverse af the previous map.

We now briefly recall the definition of the category $\hat{\mathscr{C}}/\hat{\mathscr{C}}_0$, the quotient category of $\hat{\mathscr{C}}$ by $\hat{\mathscr{C}}_0$ (see [*1*] for details). The objects of $\hat{\mathscr{C}}/\hat{\mathscr{C}}_0$ are the same as the objects in $\hat{\mathscr{C}}$. Given F and G in $\hat{\mathscr{C}}$ we define $\hat{\mathscr{C}}/\hat{\mathscr{C}}_0(F, G) =$

$\varinjlim_{F', G'} \hat{\mathscr{C}}(F', G/G')$ when F' and G' run through all subobjects of F and G respectively such that F/F' and G' are in $\hat{\mathscr{C}}_0$. Then it is well known that $\hat{\mathscr{C}}/\hat{\mathscr{C}}_0$ is an abelian category and that the canonical functor $\hat{\mathscr{C}} \to \hat{\mathscr{C}}/\hat{\mathscr{C}}_0$ given by $F \to F$ is exact. Since $v(\hat{\mathscr{C}}_0) = 0$, it follows from general facts that there is one and only one functor $v' : \hat{\mathscr{C}}/\hat{\mathscr{C}}_0 \to \mathscr{C}$ such that $\hat{\mathscr{C}} \to \hat{\mathscr{C}}/\hat{\mathscr{C}}_0 \to \hat{\mathscr{C}}$ is v and that v' is exact. From the definition of v' (see [*1*, page 369]) and the above discussion we see that $v' : \hat{\mathscr{C}}/\hat{\mathscr{C}}_0 (F, G) \to \mathscr{C}(v'(F), v'(G))$ is an isomorphism for all F and G in $\hat{\mathscr{C}}/\hat{\mathscr{C}}_0$. Thus the category $\hat{\mathscr{C}}/\hat{\mathscr{C}}_0$ is equivalent to $\mathscr{C}$.

It should be observed that the category $\hat{\mathscr{C}}_0$ can be described without reference to the category $\mathscr{C}$ or the functor $v : \hat{\mathscr{C}} \to \mathscr{C}$. For we have seen that F is in $\hat{\mathscr{C}}_0$ if and only if $(F, (-, A)) = 0$ for all A in $\mathscr{C}$. But since the projective objects in $\hat{\mathscr{C}}$ are precisely those isomorphic to $(-, A)$ for some A in $\mathscr{C}$ (i.e. they are the representable functors), we have that F is in $\hat{\mathscr{C}}_0$ if and only if $(F, G) = 0$ for all projective objects G in $\hat{\mathscr{C}}$. Thus if $\hat{\mathscr{C}}$ is equivalent to $\hat{\mathscr{D}}$, it then follows that $\hat{\mathscr{C}}_0$ is equivalent to $\hat{\mathscr{D}}_0$ under this equivalence and thus we get an induced equivalence of $\hat{\mathscr{C}}/\hat{\mathscr{C}}_0$ with $\hat{\mathscr{D}}/\hat{\mathscr{D}}_0$. Since $\mathscr{C}$ is equivalent to $\hat{\mathscr{C}}/\hat{\mathscr{C}}_0$ and $\mathscr{D}$ is equivalent to $\hat{\mathscr{D}}/\hat{\mathscr{D}}_0$, it follows that $\mathscr{C}$ is equivalent to $\mathscr{D}$. Thus, up to equivalence the category $\hat{\mathscr{C}}$ determines the category $\mathscr{C}$.

Finally we end this section by observing that the gl. dim $\hat{\mathscr{C}} = 0$ or 2 or in other words, the gl. dim $\hat{\mathscr{C}} \neq 1$. For suppose the gl. dim $\hat{\mathscr{C}} \leq 1$. Let $0 \to A' \to A \to A'' \to 0$ be an exact sequence in $\mathscr{C}$ and consider the exact sequence $0 \to (-, A') \to (-, A) \to (-, A'') \to F \to 0$. Now by Lemma 3.1 we know that $\mathrm{Ext}^2(F, (-, A')) = \mathrm{Coker}((A, A') \to (A', A'))$. Since we are assuming that the gl.dim $\hat{\mathscr{C}} \leq 1$, we know that

$$\mathrm{Ext}^2 (F, (-, A')) = 0 \quad \text{or that} \quad (A, A') \to (A', A') \to 0$$

is exact. Thus the sequence $0 \to A' \to A \to A'' \to 0$ splits. Since this is true for any exact sequence, it follows that if the gl. dim $\hat{\mathscr{C}} \leq 1$, then all exact sequences in $\mathscr{C}$ split, i.e. the gl. dim $\mathscr{C} = 0$. But from this it follows trivially that gl. dim $\hat{\mathscr{C}} = 0$ also, concluding our proof. Thus the gl. dim $\hat{\mathscr{C}} = 0$ if and only if gl. dim $\mathscr{C} = 0$. Further if the gl. dim $\mathscr{C} = 0$, then $\hat{\mathscr{C}}_0 = 0$ or $\hat{\mathscr{C}}$ is equivalent to $\mathscr{C}$.

Of course a similar discussion can be carried through for the category $\check{\mathscr{C}}$ and it is left as an excercise to the reader to actually do so.

4. Half-Exact Functors

As usual, we shall assume that $\mathscr{C}$ is an abelian $\mathfrak{U}$-category. In this section we shall be concerned with classifying the half exact coherent functors. Since most of the applications we have in mind are for covariant functors, we shall be concerned mainly with the category $\check{\mathscr{C}}$. For the sake of convenience we shall make the blanket assumption that $\mathscr{C}$ has enough projective objects, although there are some results which do not need this assumption.

We start by investigating the structure of the half exact functors in $\check{\mathscr{C}}_0$.

Lemma 4.1. *All functors which are isomorphic to* $\mathrm{Ext}^1(C, -)$ *for some* C *in* $\mathscr{C}$ *are in* $\check{\mathscr{C}}_0$. *Also if* G *is in* $\check{\mathscr{C}}_0$, *then there is a monomorphism* $G \to \mathrm{Ext}^1(C, -)$ *for some* C *in* $\mathscr{C}$.

Proof. Let $0 \to K \to P \to C \to 0$ be an exact sequence in $\mathscr{C}$ with P a projective object in $\mathscr{C}$. Then we have the exact sequence

$$0 \to (C, -) \to (P, C) \to (K, -) \to \mathrm{Ext}^1(C, -) \to 0,$$

which shows that $\mathrm{Ext}^1(C, -)$ is in $\check{\mathscr{C}}_0$ (see proposition 3.2).

Since G in $\check{C}_0$, we know there is an exact sequence

$$0 \to C_0 \to C_1 \to C_2 \to 0$$

in $\mathscr{C}$ such that

$$0 \to (C_2, -) \to (C_1, -) \to (C_0, -) \to G \to 0$$

is exact. But we also have the exact sequence

$$0 \to (C_2, -) \to (C_1, -) \to (C_0, -) \to \mathrm{Ext}^1(C_2, -) \to \cdots,$$

which shows that there is a monomorphism $G \to \mathrm{Ext}^1(C_2, -)$.

Lemma 4.2. *Let* F *be in* $\check{\mathscr{C}}$. *Then the following statements are equivalent:*

a) F is half exact;

b) $\mathrm{Ext}^1(G, F) = 0$ *for all* G *in* $\check{\mathscr{C}}_0$. *Thus if* F *is half exact, then the functor* $(-, F)\colon \check{\mathscr{C}}_0 \to \mathscr{Ab}$ *given by* $G \mapsto (G, F)$ *is exact.*

Proof. a) $\Leftrightarrow$ b). Let C be the exact sequence $0 \to C_0 \to C_1 \to C_2 \to 0$ and let $0 \to (C_2, -) \to (C_1, -) \to (C_0, -) \to G \to 0$ be exact. Then by lemma 3.1, we know that $\mathrm{Ext}^1(G, F) = H_1(F(C))$. From this it follows that $\mathrm{Ext}^1(G, F) = 0$ if and only if $F(C_0) \to F(C_1) \to F(C_2)$ is exact. Since as C runs through all possible short exact sequences in $\mathscr{C}$ we get all objects G in $\hat{\mathscr{C}}_0$ and the other way around, we have that a) $\Leftrightarrow$ b).

The last assertion follows trivially from the first.

Note. The fact that $\mathscr{C}$ has sufficiently many projectives was not used in this lemma.

As an immediate consequence of lemmas 4.1 and 4.2 we have

Proposition 4.3. *Let F be in $\check{\mathscr{C}}_0$. Then the following statements are equivalent:*

a) *F is half exact;*

b) $\operatorname{Ext}^1(G, F) = 0$ *for all G in $\check{\mathscr{C}}_0$;*

c) *F is injective in $\check{\mathscr{C}}_0$;*

d) *F is a direct summand of* $\operatorname{Ext}^1(C, \_)$ *for some C in $\mathscr{C}$.*

Thus we see that the problem of finding the half exact functors in $\check{\mathscr{C}}_0$ is the same as finding the direct summands of $\operatorname{Ext}^1(C, \_)$ for all C in $\mathscr{C}$. It seems reasonable to conjecture that if G is a direct summand of $\operatorname{Ext}^1(C, \_)$ for some C in $\mathscr{C}$, then $G \approx \operatorname{Ext}^1(C', \_)$ for some C' in $\mathscr{C}$.

To the best knowledge of the author, this is still an open question. We briefly summarize what is known at the present time.

It is a well known result of Hilton and Eckmann that the natural map

$$(C, D) \to (\operatorname{Ext}^1(D, \_), \operatorname{Ext}^1(C, \_))$$

induces an isomorphism

$$\underline{\Pi}(C, D) \approx (\operatorname{Ext}^1(D, \_), \operatorname{Ext}^1(C, \_)).$$

Thus we have that

$$\underline{\Pi}(C, C) \approx (\operatorname{Ext}^1(C, \_), \operatorname{Ext}^1(C, \_)).$$

If $(C, P) = 0$ for all projective objects P in $\mathscr{C}$, then $\underline{\Pi}(C, C) = (C, C)$ and thus the direct summands of $\operatorname{Ext}^1(C, \_)$ are given by $\operatorname{Ext}^1(D, \_)$ where D ranges over all direct summands of C and the conjecture holds in this case.

P. Freyd has shown that if $\mathscr{C}$ is closed under denumerable direct sums, then the conjecture is true for arbitrary C in $\mathscr{C}$. The reader is referred to his article in this publication for the proof.

We present now an independent proof in the case $\mathscr{C}$ has finite projective dimension, which does not require the existence of denumerable direct sums.

Lemma 4.4. *Let C be an arbitrary object in $\mathscr{C}$ and suppose* $\operatorname{Ext}^1(C, \_) \approx G_1 + G_2$ *(direct sum). If* $G_1 \approx \operatorname{Ext}^1(D_1, \_)$ *for some D_1 in $\mathscr{C}$, then* $G_2 \approx \operatorname{Ext}^1(D_2, \_)$ *for some D_2 in $\mathscr{C}$.*

Proof. This is essentially a result of Hilton and Rees [*3*, Theorem 2.4]. They show that since we have a monomorphism $0 \to \operatorname{Ext}^1(D, \_) \to \operatorname{Ext}^1(C, \_)$, then there is a projective object P in $\mathscr{C}$ and an epimorphism $C + P \xrightarrow{f} D_1 \to 0$ which splits such that the induced map $\operatorname{Ext}^1(D_1, \_) \to \operatorname{Ext}^1(C + P, \_) = \operatorname{Ext}^1(C, \_)$ is the original imbedding of G_1 in $\operatorname{Ext}^1(C, \_)$. If we let $D_2 = \operatorname{Ker} f$, then it follows that $G_2 \approx \operatorname{Ext}^1(D, \_)$.

Proposition 4.5. *Let* $0 \to P_1 \to P_0 \to A \to 0$ *be an exact sequence in* $\mathscr{C}$ *with* P_1 *a projective object in* $\mathscr{C}$. *Also suppose that* $(P_0, \_) \to (P_1, \_) \to G \to 0$ *is exact. Further suppose that we are given an exact sequence*

$$G \to \operatorname{Ext}^1(B, \_) \to G' \to 0.$$

Then there is a D *in* $\mathscr{C}$ *such that* $\operatorname{Ext}^1(D, \_) \approx G'$ *and* $\operatorname{Ext}^i(D, \_) \approx \approx \operatorname{Ext}^i(B, \_)$ *for* $i > 1$.

Proof. Let $0 \to Q_1 \to Q_0 \to B \to 0$ be exact with Q_0 projective. Then the map $G \to \operatorname{Ext}^1(B, \_)$ can be extended to a map of the exact sequences

$$\begin{array}{ccccccc} (P_0, \_) & \to & (P_1, \_) & \to & G & \to & 0 \\ \downarrow & & \downarrow & & \downarrow & & \\ (Q_0, \_) & \to & (Q_1, \_) & \to & \operatorname{Ext}^1(B, \_) & \to & 0 \end{array}$$

Applying the mapping cone we obtain an exact sequence

$$(Q_0, \_) + (P_1, \_) \to (Q_1, \_) \to G' \to 0.$$

Thus we have an exact sequence $Q_1 \to Q_0 + P_1 \to D \to 0$. Since G' is in $\hat{\mathscr{C}}_0$, it follows that $0 \to Q_1 \to Q_0 + P_1 \to D \to 0$ is exact (see proposition 3.2). Since Q_0 and P_1 are projective, we have that $G' \approx \operatorname{Ext}'(D, \_)$. Since $\operatorname{Ext}^{i+1}(B, \_) \approx \operatorname{Ext}^i(Q_1, \_) \approx \operatorname{Ext}^{i+1}(D, \_)$ for all $i \geqq 1$, we have the last assertion.

As an immediate consequence we have:

Corollary 4.6. *Let* A *and* B *be in* $\mathscr{C}$ *with the* pd $A \leqq 1$. *If* $\operatorname{Ext}^1(A, \_) \to \to \operatorname{Ext}^1(B, \_) \to G \to 0$ *is exact, then there is a* D *in* $\mathscr{C}$ *such that* $G \approx \approx \operatorname{Ext}^1(D, \_)$ *and* $\operatorname{Ext}^i(B, \_) \approx \operatorname{Ext}^i(D, \_)$ *for* $i > 1$.

We now combine the above remarks to obtain our desired result.

Proposition 4.7. *Let* C *be in* $\mathscr{C}$ *with the* pd $C < \infty$. *If* G *is a direct summand of* $\operatorname{Ext}^1(C, \_)$, *then there is a* D *in* $\mathscr{C}$ *such that* $G \approx \operatorname{Ext}'(D, \_)$.

Proof. By induction on $n = \operatorname{pd} C$. Suppose $n = 1$. Then there is an exact sequence $0 \to \operatorname{Ext}^1(C, \_) \to \operatorname{Ext}^1(C, \_) \to G \to 0$. Thus $G \approx \approx \operatorname{Ext}^1(D, \_)$ for some D in $\mathscr{C}$ by corollary 4.6.

Suppose true for $n = k \geqq 1$ and let $n = k + 1$. Since G is a direct summand of $\operatorname{Ext}^1(C, \_)$ we know that G is in $\hat{\mathscr{C}}_0$. Let $0 \to X \to Y \to Z \to 0$ be an exact sequence in $\mathscr{C}$ such that $0 \to (Z, \_) \to (Y, \_) \to (X, \_) \to G \to 0$ is exact and let $G \to \operatorname{Ext}^1(C, \_)$ be the imbedding of G as a direct summand. Thus we obtain the map $(X, \_) \to \operatorname{Ext}^1(C, \_)$. Let $0 \to X \to E \to C$ be the image of the identity $X \to X$ in $\operatorname{Ext}^1(C, X)$.

Then we obtain a commutative diagram with exact rows and columns

$$\begin{array}{ccccccccc} & & & & 0 & & & & \\ & & & & \downarrow & & & & \\ (Y,\_) & \to & (X,\_) & \to & G & \to & 0 & & \\ \downarrow & & \| & & \downarrow & & & & \\ (E,\_) & \to & (X,\_) & \to & \mathrm{Ext}^1(C,\_) & \to & \mathrm{Ext}^1(E,\_) & \to & \cdots \end{array}$$

from which it follows that

$$0 \to G \to \mathrm{Ext}^1(C,\_) \to \mathrm{Ext}^1(E,\_) \to \mathrm{Ext}^1(X,\_) \to \mathrm{Ext}^2(C,\_)$$

is exact. Since G is a direct summand of $\mathrm{Ext}^1(C,\_)$ it is half exact and thus injective in $\check{\mathscr{C}}_0$ (see proposition 4.3). Since each of the objects appearing in this exact sequence in $\check{\mathscr{C}}_0$ are half exact and thus injective in $\check{\mathscr{C}}_0$, it follows that the $\mathrm{Im}(\mathrm{Ext}^1(X,\_) \to \mathrm{Ext}^2(C,\_))$ is a direct summand of $\mathrm{Ext}^2(C,\_)$. But $\mathrm{Ext}^2(C,\_) \approx \mathrm{Ext}^1(C',\_)$ for some C' in $\mathscr{C}$ with $\mathrm{pd}\, C' = k$. Thus by the inductive hypothesis we have that

$$\mathrm{Ext}^1(E,\_) \to \mathrm{Ext}^1(X,\_) \to \mathrm{Ext}^1(X',\_) \to 0$$

is exact for some X' in $\mathscr{C}$. But then by lemma 4.4, we have that the $\mathrm{Ker}(\mathrm{Ext}^1(X,\_) \to \mathrm{Ext}^1(X',\_))$ is isomorphic to $\mathrm{Ext}^1(E',\_)$ for some E', since

$$\mathrm{Ext}^1(X,\_) \approx \mathrm{Ext}^1(X',\_) + \mathrm{Ker}(\mathrm{Ext}^1(X,\_) \to \mathrm{Ext}^1(X',\_)).$$

Proceeding in this way we conclude that $G \approx \mathrm{Ext}^1(D,\_)$ for some D in $\mathscr{C}$.

We summarize our remarks in

Theorem. 4.8. *Let $\mathscr{C}$ be an abelian $\mathfrak{U}$-category which either has denumerable sums or else each object in $\mathscr{C}$ has finite projective dimension. For a G in $\check{\mathscr{C}}$, the following statements are equivalent:*

a) $G \approx \mathrm{Ext}^1(D,\_)$ for some D in $\mathscr{C}$.

b) G is in $\check{\mathscr{C}}_0$ and is half exact.

c) G is in $\check{\mathscr{C}}_0$ and is injective in $\check{\mathscr{C}}_0$.

For the rest of this section we will assume that $\mathscr{C}$ not only has enough projectives, but also statisfies the conclusions of Theorem 4.8. In view of Theorem 4.8 we see that this is not a serious restriction.

We now turn our attention to arbitrary half exact functors in $\check{\mathscr{C}}$. If F is an arbitrary object in $\check{\mathscr{C}}$ not necessarily half exact, we have the exact sequence $0 \to F_0 \to F \to (w(F),\_) \to F_1 \to 0$ where $w: (\check{\mathscr{C}})^0 \to \mathscr{C}$ is the analogue of the functor $v: \hat{\mathscr{C}} \to \mathscr{C}$. From the results of § 3, it follows that $(w(F), \_) \approx R^0F$, the 0-th right derived functor of F and that $F \to (w(F), \_)$ is the usual functor from F to R^0F.

If we now assume that F is half exact, then it easily follows from the exact sequence $0 \to F_0 \to F \to (w(F), \_) \to F_1 \to 0$ that F_0 is also half exact and thus of the form $\operatorname{Ext}^1(B, \_)$ for some B in $\mathscr{C}$. Thus given a half exact functor F, associated with it is an exact sequence

$$0 \to \operatorname{Ext}^1(B, \_) \to F \to (w(F), \_) \to F_1 \to 0$$

with the F_1 in $\hat{\mathscr{C}}_0$. Therefore, if given two objects A and B in $\mathscr{C}$, we can classify the half exact functors F in $\hat{\mathscr{C}}$ such that there is an exact sequence $0 \to \operatorname{Ext}^1(B, \_) \to F \to (A, \_) \to F_1 \to 0$ with F_1 in $\check{\mathscr{C}}_0$, we will in effect have given a complete classification of all half exact functors F in $\check{\mathscr{C}}$.

Let $0 \to \operatorname{Ext}^1(B, \_) \to F \to (A, \_) \to F_1 \to 0$ and $0 \to \operatorname{Ext}^1(B, \_) \to \to G \to (A, \_) \to G_1 \to 0$ be exact with the F_1 and G_1 in $\check{\mathscr{C}}_0$. By a map from the first sequence to the second, we mean a map $F \to G$ which gives a commutative diagram

$$\begin{array}{ccccccccc} 0 \to & \operatorname{Ext}^1(B, \_) & \to & F & \to & (A, \_) & \to & F_1 & \to 0 \\ & \| & & \downarrow & & \| & & & \\ 0 \to & \operatorname{Ext}^1(B, \_) & \to & G & \to & (A, \_) & \to & G_1 & \to 0 . \end{array}$$

Given this commutative diagram it is trivial to verify: a) there is a unique map $F_1 \to G_1$ which makes the diagram commutative and this map is an epimorphism; b) $0 \to F \to G$ is exact; c) there is a map $G \to F_1$ which gives an exact sequence $0 \to F \to G \to F_1 \to G_1 \to 0$.

We shall say that the two sequences $0 \to \operatorname{Ext}^1(B, \_) \to F \to (A, \_) \to F_1 \to 0$ and $0 \to \operatorname{Ext}^1(B, \_) \to G \to (A, \_) \to G_1 \to 0$ are isomorphic if there is a map of sequences which is an isomorphism on F to G and thus on the whole sequence. Our primary interest will be in the isomorphism classes of these exact sequences.

Our first step in the classification of half exact functors is to characterize those exact sequences $0 \to \operatorname{Ext}^1(B, \_) \to F \to (A, \_) \to F_1 \to 0$ such that the F is a half exact functor. Before doing this we need the following technically important lemma.

Lemma 4.9. *Suppose we are given an exact sequence* $0 \to \operatorname{Ext}^1(B, \_) \to \to F \to (A, \_) \to F_1 \to 0$ *(with* F_1 *in* $\check{\mathscr{C}}_0$ *of course). Then*

a) There is an exact sequence $0 \to A \to X \to Y \to B \to 0$ *in* $\mathscr{C}$ *and exact sequences* $0 \to (B, \_) \to (Y, \_) \to (X, \_) \to F \to 0$ *and* $0 \to (B, \_) \to \to (Y, \_) \to (Z, \_) \to \operatorname{Ext}^1(B, \_) \to 0$, *where* $Z = \operatorname{Im}(X \to Y)$, *with the property that our given maps* $0 \to \operatorname{Ext}^1(B, \_) \to F \to (A, \_)$ *are the unique maps which complete the following commutative diagram*

$$\begin{array}{ccccccccc}
 & & & & 0 & & & & \\
 & & & & \downarrow & & & & \\
0 \to (B, _) & \to & (Y, _) & \to & (Z, _) & \to & \mathrm{Ext}^1(B, _) & \to & 0 \\
\| & & \| & & \downarrow & & & & \\
0 \to (B, _) & \to & (Y, _) & \to & (X, _) & \to & F & \to & 0 \\
 & & & & \downarrow & & & & \\
 & & & & (A, _) & = & (A, _) & & \\
 & & & & \downarrow & & & & \\
 & & & & 0 & & & &
\end{array}$$

b) If the Y in part a) is projective then F is half exact.

c) There exists a map of exact sequences

$$\begin{array}{ccccccccc}
0 \to \mathrm{Ext}^1(B, _) & \to & F & \to & (A, _) & \to & F_1 & \to & 0 \\
\| & & \downarrow & & \| & & \downarrow & & \\
0 \to \mathrm{Ext}^1(B, _) & \to & G & \to & (A, _) & \to & G_1 & \to & 0
\end{array}$$

with G half exact.

Proof. a) Since $w(F) \approx A$ we know we can find an exact sequence $0 \to A \to X \to V \to D \to 0$ such that $(V, \_) \to (X, \_) \to F \to 0$ is exact. Since the $\mathrm{Im}(\mathrm{Ext}^1(B, \_) \to F) = F_0$ an argument similar to that used in § 3 for deriving (3.3) shows that we have an exact sequence

$$0 \to (D, \_) \to (V, \_) \to (V', \_) \to \mathrm{Ext}^1(B, \_) \to 0$$

where $V' = \mathrm{Im}(X \to V)$, which gives us a commutative diagram

$$(*)\qquad \begin{array}{ccccccccc}
 & & & & 0 & & 0 & & \\
 & & & & \downarrow & & \downarrow & & \\
0 \to (D, _) & \to & (V, _) & \to & (V', _) & \to & \mathrm{Ext}^1(B, _) & \to & 0 \\
\| & & \| & & \downarrow & & \downarrow & & \\
0 \to (D, _) & \to & (V, _) & \to & (X, _) & \to & F & \to & 0 \\
 & & & & \downarrow & & \downarrow & & \\
 & & & & (A, _) & = & (A, _) . & &
\end{array}$$

Now we also have the commutative diagram

$$\begin{array}{ccccccccc}
 & & & & & & 0 & & \\
 & & & & & & \downarrow & & \\
0 \to (D, _) & \to & (U, _) & \to & (V', _) & \to & \mathrm{Ext}^1(B, _) & \to & 0 \\
\| & & \| & & \| & & \downarrow & & \\
0 \to (D, _) & \to & (U, _) & \to & (V', _) & \to & \mathrm{Ext}^1(D, _) & \to & \cdots .
\end{array}$$

From this it follows that there is a commutative diagram

$$\begin{array}{ccccccc}
0 \to V' & \to & V & \to & D & \to & 0 \\
\| & & \downarrow & & \downarrow & & \\
0 \to V' & \to & Y & \to & B & \to & 0
\end{array}$$

(with exact rows and columns), such that the induced map $\mathrm{Ext}^1(B,\_) \to \mathrm{Ext}^1(D,\_)$ agrees with our original one. Thus we see that we have a commutative diagram

$$\begin{array}{ccccccccc} 0 \to & (B,\_) & \to & (Y,\_) & \to & (V',\_) & \to & \mathrm{Ext}^1(B,\_) & \to 0 \\ & \downarrow & & \downarrow & & \| & & \| & \\ 0 \to & (D,\_) & \to & (V,\_) & \to & (V',\_) & \to & \mathrm{Ext}^1(B,\_) & \to 0 \end{array}$$

with exact rows and columns. It is now a straight forward matter to check from (*) that the exact sequence $0 \to A \to X \to Y \to B \to 0$ together with the maps $(V',\_) \to \mathrm{Ext}^1(B,\_)$ and $(X,\_) \to F$ given in (*) have our desired properties.

b) Follows trivially by diagram chasing from the exact sequence $(Y,\_) \to (X,\_) \to F \to 0$.

c) Let $0 \to A \to X \to Y \to B \to 0$ be an exact sequence having the properties described in a). Let $P \to Y \to 0$ be exact with P projective. Then it is well known that we have a commutative diagram

$$\begin{array}{ccccccccc} 0 \to & A & \to & S & \to & P & \to & B & \to 0 \\ & \| & & \downarrow & & \downarrow & & \| & \\ 0 \to & A & \to & X & \to & Y & \to & B & \to 0 \end{array}$$

with exact rows and columns. Thus we have the commutative diagrams

$$\begin{array}{ccccccccc} 0 \to & (B,\_) & \to & (Y,\_) & \to & (Y',\_) & \to & \mathrm{Ext}^1(B,\_) & \to 0 \\ & \| & & \downarrow & & \downarrow & & \| & \\ 0 \to & (B,\_) & \to & (P,\_) & \to & (P',\_) & \to & \mathrm{Ext}^1(B,\_) & \to 0 \end{array} \tag{1}$$

(where $P' = \mathrm{Im}(S \to P)$ and $Y' = \mathrm{Im}(X \to Y)$);

$$\begin{array}{ccccccccc} 0 \to & (B,\_) & \to & (Y,\_) & \to & (X,\_) & \to & F & \to 0 \\ & \| & & \downarrow & & \downarrow & & \downarrow & \\ 0 \to & (B,\_) & \to & (P,\_) & \to & (S,\_) & \to & G & \to 0. \end{array} \tag{2}$$

From this it easily follows that we have an exact sequence

$$0 \to \mathrm{Ext}^1(B,\_) \to G \to (A,\_) \to G_1 \to 0$$

with G_1 in $\check{\mathscr{C}}_0$ such that the map $F \to G$ described in (2) gives a map

$$\begin{array}{ccccccccc} 0 \to & \mathrm{Ext}^1(B,\_) & \to & F & \to & (A,\_) & \to & F_1 & \to 0 \\ & \| & & \downarrow & & \| & & \downarrow & \\ 0 \to & \mathrm{Ext}^1(B,\_) & \to & G & \to & (A,\_) & \to & G_1 & \to 0. \end{array}$$

Since P is projective we know by part b) that G is half exact, which finishes the proof of c).

Proposition 4.10. *Let* $0 \to \mathrm{Ext}^1(B,\_) \to F \to (A,\_) \to F_1 \to 0$ *exact (with* F_1 *in* $\check{\mathscr{C}}_0$*). Then the following statements are equivalent:*

a) F is half exact;
b) any map

$$\begin{array}{ccccccccc} 0 \to & \mathrm{Ext}^1(B, \_) & \to & F & \to & (A, \_) & \to & F_1 & \to 0 \\ & \| & & \downarrow & & \| & & \downarrow & \\ 0 \to & \mathrm{Ext}^1(B, \_) & \to & G & \to & (A, \_) & \to & G_1 & \to 0 \end{array}$$

is an isomorphism;

c) there exists an exact sequence $0 \to A \to X \to Y \to B \to 0$ *in* $\mathscr{C}$ *with* Y *projective which satisfies the properties of part a) of lemma 4.9.*

Proof. a) $\Rightarrow$ b). If we have such a map then we have an exact sequence $0 \to F \to G \to F_1 \to G_1 \to 0$ by our initial remarks concerning maps of sequences. Thus we have an exact sequence $0 \to F \to G \to H \to 0$ with H in $\check{\mathscr{C}}_0$. Since F is half exact, we know that $\mathrm{Ext}^1(H, F) = 0$ (see proposition 4.3). Thus the sequence $0 \to F \to G \to H \to 0$ splits, i.e. $G = F + H'$ with H' in $\check{\mathscr{C}}_0$. But then the

$$\mathrm{Ker}(G \to (A, \_)) \supset \mathrm{Im}(\mathrm{Ext}^1(B, \_) \to G) + H'.$$

Since the

$$\mathrm{Ker}(G \to (A, \_)) = \mathrm{Im}(\mathrm{Ext}^1(B, \_) \to G),$$

we have that $H' = 0$. Thus H is zero or $F \to G$ is an isomorphism.

b) $\Rightarrow$ c). The same construction used in proving part c) of lemma 4.9 shows that b) $\Rightarrow$ c).

c) $\Rightarrow$ a). Follows trivially from part b) of lemma 4.9.

Our classification of half exact functors will consist of showing that there is a natural one to one correspondence between the isomorphism classes of exact sequences $0 \to \mathrm{Ext}^1(B, \_) \to F \to (A, \_) \to F_1 \to 0$ (with F_1 in $\check{\mathscr{C}}_0$) and the elements of $\mathrm{Ext}^2(B, A)$. To achieve this, we need

Proposition 4.11. *Let* $0 \to G \to (A, \_) \to G_1 \to 0$ *be exact with* G_1 *in* $\check{\mathscr{C}}_0$. *Then we have an exact sequence*

$$0 \to \mathrm{Ext}^1(G, \mathrm{Ext}^i(B, \_)) \to \mathrm{Ext}^{i+1}(B, A) \quad \textit{for all} \quad i \geqq 0\,.$$

Proof. We first define the map $\mathrm{Ext}^1(G, \mathrm{Ext}^i(B, \_)) \to \mathrm{Ext}^{i+1}(B, A)$. We know that there is an exact sequence $C_0 \to C_1 \to 0$ in $\mathscr{C}$ such that $0 \to (C_1, \_) \to (C_0, \_) \to G \to 0$ is exact. Thus we have the exact sequence $0 \to (C_1, \_) \to (C_0, \_) \to (A, \_) \to G_1 \to 0$. Since G_1 is in $\check{\mathscr{C}}_0$, we know that the associated sequence $0 \to A \to C_0 \to C_1 \to 0$ is exact (see proposition 3.2). Thus we have the exact sequence

$$\mathrm{Ext}^i(B, C_0) \to \mathrm{Ext}^i(B, C_1) \to \mathrm{Ext}^{i+1}(B, A)\,.$$

But by lemma 3.1 we know that

$$\mathrm{Ext}^1(B, C_0) \to \mathrm{Ext}^1(B, C_1) \to \mathrm{Ext}^1(G, \mathrm{Ext}^i(B, \_)) \to 0$$

is exact. Thus we have our desired monomorphism

$$\mathrm{Ext}^1(G, \mathrm{Ext}^i(B, \_)) \to \mathrm{Ext}^{i+1}(B, A).$$

It is easily seen that this map does not depend on the particular resolution of G which was chosen.

Now suppose that given a map $F \to (A, \_)$ we denote by $\bar{F}$ the $\mathrm{Im}(F \to (A, \_))$. Now if $0 \to \mathrm{Ext}^1(B, \_) \to F \to (A, \_) \to F_1 \to 0$ is exact (with F_1 in $\check{\mathscr{C}}_0$), then $0 \to \mathrm{Ext}^1(B, \_) \to F \to \bar{F} \to 0$ is an element of $\mathrm{Ext}^1(\bar{F}, \mathrm{Ext}^1(B, \_))$ and this element depends only on the isomorphism class of $0 \to \mathrm{Ext}^1(B, \_) \to F \to (A, \_) \to F_1 \to 0$. Combining this map with the map of $\mathrm{Ext}^1(\bar{F}, \mathrm{Ext}^1(B, \_)) \to \mathrm{Ext}^2(B, A)$ given in proposition 4.11 we obtain a map of the isomorphism classes of exact sequences to the elements of $\mathrm{Ext}^2(B, A)$. Thus if we denote by $H(B, A)$ the set of isomorphism classes of exact sequences $0 \to \mathrm{Ext}^1(B, \_) \to$ $\to F \to (A, \_) \to F_1 \to 0$ with F half exact and F_1 in $\check{\mathscr{C}}_0$, we obtain a map $H(B, A) \to \mathrm{Ext}^2(B, A)$ which we wish to show is one to one and onto.

We now define a map $\mathrm{Ext}^2(B, A) \to H(B, A)$. It is well known and easily seen that any element in $\mathrm{Ext}^2(B, A)$ can be represented by an exact sequence $0 \to A \to X \to Y \to B \to 0$ with Y projective. From the commutative diagram

$$(4.12)\qquad \begin{array}{ccccccccc} & & & & 0 & & & & \\ & & & & \downarrow & & & & \\ 0 \to (B, \_) & \to & (Y, \_) & \to & (Y', \_) & \to & \mathrm{Ext}^1(B, \_) & & \\ \| & & \| & & \downarrow & & & & \\ 0 \to (B, \_) & \to & (Y, \_) & \to & (X, \_) & \to & G \to 0 & & \\ & & & & \downarrow & & & & \\ & & & & (A, \_) & & & & \\ & & & & \downarrow & & & & \\ & & & & \mathrm{Ext}^2(B, \_) & & & & \end{array}$$

with exact rows and columns where $Y' = \mathrm{Im}(X \to Y)$. From this, as usual, we deduce the unique exact sequence

$$0 \to \mathrm{Ext}^1(B, \_) \to G \to (A, \_) \to \mathrm{Ext}^2(B, \_)$$

which makes the above diagram commutative.

If we choose another representative $0 \to A \to X_1 \to Y_1 \to B \to 0$ of the element in $\mathrm{Ext}^2(B, A)$ represented by $0 \to A \to X \to Y \to B \to 0$, then it is not hard to see that we will have a map of the new sequence into the old which must be an isomorphism since all the functors are half exact (see proposition 4.10).

Thus we have obtained a map $\mathrm{Ext}^2(B, A) \to H(B, A)$. It is not difficult to see that the composite maps $H(B, A) \to \mathrm{Ext}^2(B, A) \to \to H(B, A)$ and $\mathrm{Ext}^2(B, A) \to H(B, A) \to \mathrm{Ext}^2(B, A)$ are identity maps which shows that the map $H(B, A) \to \mathrm{Ext}^2(B, A)$ is indeed a one to one correspondence.

The diagram 4.12 shows that if $0 \to \mathrm{Ext}^1(B, \_) \to F \to (A, \_) \to \to F_1 \to 0$ is exact with F half exact then the map $(A, \_) \to \mathrm{Ext}^2(B, \_)$ given by sending the identity of A to the element of $\mathrm{Ext}^2(B, A)$ determined by the original exact sequence gives us an exact sequence

$$0 \to \mathrm{Ext}^1(B, \_) \to F \to (A, \_) \to \mathrm{Ext}^2(B, A).$$

Thus we have established:

Theorem 4.13. *Let $\mathscr{C}$ be an abelian $\mathfrak{U}$-category with enough projectives and such that all half exact functors in $\check{\mathscr{C}}_0$ are of the form $\mathrm{Ext}^1(B, \_)$ for some B in $\mathscr{C}$. Let F be a half exact coherent functor. Then for some B and A in $\mathscr{C}$ we have an exact sequence*

$$0 \to \mathrm{Ext}^1(B, \_) \to F \to (A, \_) \to \mathrm{Ext}^2(B, \_).$$

Further the map which sends this exact sequence to the element of $\mathrm{Ext}^2(B, A)$ determined by the map $(A, \_) \to \mathrm{Ext}^2(B, \_)$ gives a one to one correspondence between $H(B, A)$ and $\mathrm{Ext}^2(B, A)$.

Corollary 4.14. *Suppose $\mathscr{C}$ is an abelian $\mathfrak{U}$-category with enough projectives and with $\mathrm{gl.dim}\,\mathscr{C} \leqq 1$. Then if F is a coherent functor and F_0 is half exact, then $F_0 \approx \mathrm{Ext}^1(B, \_)$ for some B in $\mathscr{C}$ and F_0 is a direct summand of F. Thus $F \approx F_0 + \bar{F}$ where $\bar{F} = \mathrm{Im}(F \to (w(F), \_)$. Further, F is half exact if and only if $F_0 \approx \mathrm{Ext}^1(B, \_)$ for some B and $\bar{F} = = (w(F), \_)$, in which case $F \approx \mathrm{Ext}^1(B, \_) + (w(F), \_)$ for some B in $\mathscr{C}$.*

Proof. We know we have an exact sequence

$$0 \to F_0 \to F \to (w(F), \_) \to F_1 \to 0$$

with F_1 in $\check{\mathscr{C}}_0$. Suppose $F_0 \approx \mathrm{Ext}^1(B, \_)$. Then we know that

$$\mathrm{Ext}^1(\bar{F}, \mathrm{Ext}^1(B, \_)) \subset \mathrm{Ext}^2(B, w(F))$$

by proposition 4.11. Since the $\mathrm{gl.dim}\,\mathscr{C} \leqq 1$, we know that $\mathrm{Ext}^2(B, \_) = 0$ and thus the extension $0 \to F_0 \to F \to \bar{F} \to 0$ splits. Thus the first part of the theorem is established.

The rest is a trivial consequence of theorem 4.13.

5. Complexes

In the last three sections of this paper we give some applications and illustrations of the general theory of coherent functors. In this section we will be concerned with complexes in an abelian $\mathfrak{U}$-category $\mathscr{C}$.

We recall that a covariant functor $F: \mathscr{C} \to \mathscr{Ab}$ is right exact if given any exact sequence $0 \to C' \to C \to C'' \to 0$, the sequence $F(C') \to \to F(C) \to F(C'') \to 0$ is exact. In section two we have shown that if $\mathscr{C}$ has enough projectives, then there is a functor $L^0: \check{\mathscr{C}} \to \check{\mathscr{C}}$ and a map $L^0 \to I$ (the identity on $\check{\mathscr{C}}$) such that $L^0 F$ is right exact for all F in $\check{\mathscr{C}}$ and $L^0 F(P) \to F(P)$ is an isomorphism for all projective objects P in $\mathscr{C}$ (see proposition 2.1 and the discussion following it). Also $L^0 F \to F$ is an isomorphism if and only if F is right exact.

Similarly we say that a contravariant functor $G: \mathscr{C} \to \mathscr{Ab}$ is right exact if given any exact sequence $0 \to C' \to C \to C'' \to \to 0$ the sequence $G(C'') \to G(C) \to G(C') \to 0$ is exact. By duality, if $\mathscr{C}$ has enough injectives then there is a functor $L^0: \hat{\mathscr{C}} \to \hat{\mathscr{C}}$ and a map $L^0 \to I$ (where I is the identity on $\hat{\mathscr{C}}$), such that $L^0 G$ is right exact and $L^0 G(Q) \to G(Q)$ is an isomorphism for all injective objetcs Q in $\mathscr{C}$. Also $L^0 G \to G$ is an isomorphism if and only if G is right exact.

As an immediate consequence of these definitions and lemma 3.1 we have

Lemma 5.1. *A covariant functor $F: \mathscr{C} \to \mathscr{Ab}$ is right exact if and only if the functor $\check{\mathscr{C}} \to \mathscr{Ab}$ given by $F' \mapsto (F', F)$ is exact.*

Similarly a contravariant functor $G: \mathscr{C} \to \mathscr{Ab}$ is right exact if and only if the functor $\hat{\mathscr{C}} \to \mathscr{Ab}$ given by $G' \mapsto (G', G)$ is exact.

We now state and prove the main result of the first part of this section. It can be viewed as being a sort of universal coefficient theorem.

Theorem 5.2. *Let X be a complex in $\mathscr{C}$ and $F: \mathscr{C} \to \mathscr{Ab}$ a covariant right exact functor. Then we have functorial isomorphisms $H_i(F(X)) \approx \approx \check{\mathscr{C}}(H_i((X, -)), F)$ for all i.*

Similarly if $G: \mathscr{C} \to \mathscr{Ab}$ is a contravariant right exact functor, then we have functorial isomorphisms $H_i(G(X)) \approx \hat{\mathscr{C}}(H_i((-, X)), G)$ for all i.

Proof. Since for all j we have that $F(X_j) = ((X, -), F)$, we have that $H_i(F(X)) \approx H_i(\hat{\mathscr{C}}((X, -), F)$. But since F is right exact, we know by lemma 5.1 that the functor $(-, F)$ is exact. Thus $H_i(\hat{\mathscr{C}}((X, -), F) \approx \approx \hat{\mathscr{C}}(H_i((X, -), F)$, which gives the desired result. The other half of the theorem is proved similarly.

Suppose now that $\mathscr{C}$ has enough projectives and that X is a projective complex in $\mathscr{C}$. Now let F be an arbitrary covariant functor from $\mathscr{C}$ to $\mathscr{Ab}$. Then we know that the map $L^0F \to F$ has the property that $L^0F(P) \approx F(P)$ for all projectives in $\mathscr{C}$. Thus we have that

$$H_i(F((X)) \approx H_i(L^0F(X)) \approx \check{\mathscr{C}}(H_i((X,-)), L^0F),$$

the last isomorphism being the one given in theorem 5.2. Thus we obtain

Corollary 5.3. *Suppose $\mathscr{C}$ has enough projectives and X is a projective complex. Let F be in $\check{\mathscr{C}}$. Then we have functorial isomorphisms $H_i(F(X)) \approx$ $\approx \check{\mathscr{C}}(H_i(X,-), L^0F)$ for all i.*

Similarly, suppose $\mathscr{C}$ has enough injectives and that X is an injective complex. Let G be in $\hat{\mathscr{C}}$. Then we have functorial isomorphisms $H_i(G(X)) \approx$ $\approx \check{\mathscr{C}}(H_i((-,X)), L^0G)$ for all i.

In particular we have the following well known result of Yoneda.

Corollary 5.4. *Suppose $\mathscr{C}$ has enough projectives and C is an object in $\mathscr{C}$. Let F be in $\check{\mathscr{C}}$. Then we have functorial isomorphisms*

$$(L^iF)(C) \approx \check{\mathscr{C}}(\mathrm{Ext}^i(C,-), L^0F)$$

for all i.

Similarly, suppose $\mathscr{C}$ has enough injectives and C is an object in $\mathscr{C}$. Let G be in $\hat{\mathscr{C}}$. Then we have functorial isomorphisms

$$(L^iF(C) \approx \hat{\mathscr{C}}(\mathrm{Ext}^i(-,C), L^0G)$$

for all i.

Another immediate consequence of corollaries 5.3 and 5.4 is

Corollary 5.5. *Suppose $\mathscr{C}$ has enough projectives and A and C are objects in $\mathscr{C}$. Then we have for all $i > 0$, the functorial isomorphisms*

$$\underline{\Pi}_{i+1}(C,A) \approx \check{\mathscr{C}}(\mathrm{Ext}^i(A,-), L^0(C,-))$$
$$\approx (L^i(C,-))(A).$$

Similarly, suppose $\mathscr{C}$ has enough injectives. Then for all $i > 0$, we have functorial isomorphisms

$$\overline{\Pi}_{i+1}(A,C) \approx \mathscr{C}(\mathrm{Ext}^i(-,A), L^0(-,C))$$
$$\approx (L^i(-,C))(A).$$

We now give some interpretation for these formulae in the case $\mathscr{C}$ is the category of modules over a ring.

We shall say that a ring R is a ring in $\mathfrak{U}$ if the underlying set of R is an element in $\mathfrak{U}$ and $\mathfrak{U}$ is not the smallest universe containing R as an element. If R is a ring in $\mathfrak{U}$, we shall denote by ${}_R\mathscr{M}$ and $\mathscr{M}_R$ the category of left and right R-modules respectively whose underlying sets are elements of the same universe in $\mathfrak{U}$ containing R. Thus ${}_R\mathcal{M}$ and $\mathcal{M}_R$ are $\mathfrak{U}$-categories if R is a ring in $\mathfrak{U}$ which we shall call the categories of left and right R-modules in $\mathfrak{U}$.

Suppose R is a ring in $\mathfrak{U}$ and ${}_R\mathcal{M}$ and $\mathcal{M}_R$ the categories of left and right R-modules in $\mathfrak{U}$. Suppose C is a finitely presented left R-module. We wish to find $L^0(C, \_)$. Let (C, R) be denoted by C^*. Then we have the usual map $C^* \otimes \_ \to (C, \_)$ where for each B in ${}_R\mathcal{M}$ the map $C^* \otimes B \to (C, B)$ is given by $(f \otimes b)(c) = f(c)b$ for all f in C^*, b in B and c in C. It is well known that $C^* \otimes P \to (C, P)$ is an isomorphism for all finitely generated free R-modules P. Since both $C^* \otimes\_$ and $(C, \_)$ commute with direct limits (remember C is finitely presented), we have that $C^* \otimes P \to (C, P)$ is an isomorphism for all free R-modules P and thus for all projective R-modules P. Since $C^* \otimes \_$ is also right exact, it easily follows that $L^0(C, \_)$ is isomorphic to $C^* \otimes \_$ in such way that we have a commutative diagram

$$\begin{array}{ccc} C^* \otimes \_ & \longrightarrow & (C, \_) \\ \wr\wr & & \| \\ L^0(C, \_) & \to & (C, \_) . \end{array}$$

Now if A is left R-module, we have by corollary 5.5, that

$$\underline{\Pi}_{i+1}(C, A) \approx L^i((C, \_))(A)$$

for all $i > 0$. Since $L^0(C, \_) \approx C^* \otimes \_$, we obtain that

$$\underline{\Pi}_{i+1}(C, A) \approx \operatorname{Tor}_i^R(C^*, A) .$$

Summarizing we have

Proposition 5.6. *Let R be a ring and C is a finitely presented R-module. Then $L^0(C, \_) \approx C^* \otimes \_$ and for each $i > 0$ the functors $\underline{\Pi}_{i+1}(C, \_)$ and $\operatorname{Tor}_i^R(C^*, \_)$ are functorially isomorphic.*

The last application in this direction that we give is the following:

Proposition 5.7. *Let R be a ring, X a complex of finitely generated projective left R-modules. If A is a left R-module, then for all i we have functorial isomorphisms*

$$H_i((X, A)) \approx \check{\mathcal{M}}_R(H_i(\_ \otimes X), \_ \otimes A) .$$

In particular, if B is a left R-module which has a projective resolution consisting of finitely generated R-modules, then for all left mogules A we have functorial isomorphisms

$$\operatorname{Ext}^i(B, A) \approx \check{\mathcal{M}}_R(\operatorname{Tor}_i(\_, B), \_ \otimes A)$$

for all i.

Proof. Since each of the components X_i in X is a finitely generated projective R-module, we know that we have maps $X_i^* \otimes \_ \to (X_i, \_)$ which are isomorphisms for all i. Thus the complexes $(X, \_)$ and $X^* \otimes \_$ are isomorphic. Thus for a particular A, we know that $(X, A) \approx X^* \otimes A$. Since $\_ \otimes A$ is a right exact functor we know by theorem 5.2, that

$$H_i(X^* \otimes A) \approx \check{\mathcal{M}}_R(H_i(X^*, \_), \_ \otimes A)$$

for all i. But the complex $(X^*, \_)$ and $\_ \otimes X$ are isomorphic. Thus we obtain our desired result that

$$H_i(X, A) \approx \check{\mathcal{M}}_R(H_i(\_ \otimes X), \_ \otimes A).$$

The second part of the proposition follows trivially from the first part.

We now return to the situation of an arbitrary abelian $\mathfrak{U}$-category $\mathscr{C}$ and give some results in different directions. These results are given in terms of covariant functors since this is the form we shall use them in the next section. Of course analogous results hold for contravariant functors.

Let X be a complex $\cdots \to X_{i+1} \to X_i \to X_{i-1} \to \cdots$ in $\mathscr{C}$. We shall denote the Coker $(X_{i+1} \to X_i)$ by $Z_i'(X)$ or, more simply, by Z_i'. Thus for each i we have an exact sequence

$$0 \to H_i(X) \to Z_i' \to X_{i-1} \to Z_{i-1}' \to 0.$$

From this we obtain the exact sequence

$$0 \to (Z_{i-1}', \_) \to (X_{i-1}, \_) \to (Z_i', \_) \to H^i(X, \_) \to 0$$

where $H^i(X, \_) = H_i((X, \_))$. Then by the results of section 4 following Theorem 4.8 we have that $w(H^i(X, \_)) = H_i(X)$, that $R^0(H^i(X, \_)) \approx$ $\approx (H_i(X), \_)$ and that the map $H^i(X, \_) \to (H_i(X), \_)$ has the usual properties of the map of a functor into its 0-th right derived functor.

If we assume further that $\mathscr{C}$ has enough projective objects and that X_{i-1} is a projective object, then it follows from the commutative diagram (3.3) and the definition of the functors $H^i(X, \_)_0$ and $H^i(X, \_)_1$, that $H^i(X, \_)_0 = \mathrm{Ext}^1(Z_{i-1}', \_)$ and that $H^i(X, \_)_1$ is a subfunctor of $\mathrm{Ext}^2(Z_{i-1}', \_)$. Further, if X_i is projective and $X_{i+1} = 0$, then we have that $Z_i' = X_i$ and thus that $H^i(X, \_)_1 = \mathrm{Ext}^2(Z_{i-1}', \_)$. Summarizing we have,

Proposition 5.8. *Let $\mathscr{C}$ be a $\mathfrak{U}$-category and $\check{\mathscr{C}}$ the category of coherent covariant functors.*

a) If F is in $\check{\mathscr{C}}$, then $R^0 F \approx (w(F), \_)$ and $F \to (w(F), \_)$ is the usual map $F \to R^0 F$.

b) If X is a complex in $\mathscr{C}$, then $w(H^i(X, \_ = H_i(X)$ and thus $R^0(H^i(X, \_)) = (H_i(X), \_)$.

c) If X is a projective complex, then for each i we have that

$$H^i(X,-)_0 = \mathrm{Ext}^1(Z'_{i-1},-) \text{ and that } H^i(X,-)_1 \subset \mathrm{Ext}^2(Z'_{i-1},-).$$

Thus we have exact sequence

$$0 \to \mathrm{Ext}^1(Z'_{i-1},-) \to H^i(X,-) \to (H_i(X),-) \to \mathrm{Ext}^2(Z'_{i-1},-)$$

which is functorial in X with X in the category of projective complexes. Further, if $X_{i+1} = 0$, then $H^i(X,-)_1 = \mathrm{Ext}^2(Z'_{i-1},-)$.

As a special case of proposition 5.7, we obtain the classical result,

Corollary 5.9. *If* gl.dim $\mathscr{C} \leqq 1$ *and X is a projective complex in $\mathscr{C}$, then we have exact sequences*

$$0 \to \mathrm{Ext}^1(H_{i-1}(X),-) \to H^i(X,-) \to (H_i(X),-) \to 0$$

which of course split.

Proof. Since we have the exact sequence $0 \to H_{i-1}(X) \to Z'_{i-1} \to X_{i-2}$ and the $\mathrm{Im}(Z'_{i-1} \to X_{i-2})$ is projective (remember that the gl.dim $\mathscr{C} \leqq 1$), we have that $Z'_{i-1} = H_{i-1}(X) + B_{i-2}$ where B_{i-2} is projective. Thus $\mathrm{Ext}^1(Z'_{i-1},-) = \mathrm{Ext}^1(H_{i-1}(X),-)$. Since the gl.dim $\mathscr{C} \leqq 1$, we know that $\mathrm{Ext}^2(Z'_{i-1},-) = 0$, which gives us our desired exact sequence. They split since $(H_i(X),-)$ is projective for each i.

It is a well known result of homological algebra that if X and Y are two projective resolutions of the same object, then for each i we have that $Z'_i(X) + P_i \approx Z'_i(Y) + Q_i$ where P_i and Q_i are projective objects in $\mathscr{C}$. We now give a generalization of this result.

Proposition 5.10. *Let X and Y be two projective complexes in $\mathscr{C}$ and $f\colon X \to Y$ a map and i an integer such that the induced map*

$$f^i\colon H^i(Y,-) \to H^i(X,-)$$

is an isomorphism. Then $f_i\colon H_i(X) \to H_i(Y)$ is an isomorphism and the induced map $g\colon Z'_{i-1}(X) \to Z'_{i-1}(Y)$ gives an isomorphism

$$\mathrm{Ext}^1(Z'_{i-1}(Y),-) \to \mathrm{Ext}^1(Z'_{i-1}(X),-).$$

Thus there exist projective objects P and Q such that

$$Z'_{i-1}(X) + P \approx Z'_{i-1}(Y) + Q.$$

Proof. The map $f\colon X \to Y$ induces a map of exact sequences

$$\begin{array}{ccccccccc} 0 \to & \mathrm{Ext}^1(Z'_{i-1}(Y),-) & \to & H^i(Y,-) & \to & (H_i(Y),-) & \to & \mathrm{Ext}^2(Z'_{i-1}(Y),-) \\ & \downarrow & & \downarrow & & \downarrow & & \downarrow \\ 0 \to & \mathrm{Ext}^1(Z'_{i-1}(X),-) & \to & H^i(X,-) & \to & (H_i(X),-) & \to & \mathrm{Ext}^2(Z'_{i-1}(X),-) \end{array}$$

since it induces maps of $Z'_{i-1}(X) \to Z'_{i-1}(Y)$ and $H_i(X) \to H_i(Y)$. Now since the left hand and right hand functors are in $\check{\mathscr{C}}_0$, we know by proposition 3.4 that this is the only map of exact sequences compatible with the given map $H^i(Y,-) \to H^i(X,-)$. Since the map $H^i(Y,-) \to H^i(X,-)$ is an isomorphism, it follows from proposition 3.4 that the rest of the maps are isomorphisms, which gives us our first results.

It is a well known result of ECKMANN and HILTON (see [2]), that $\mathrm{Ext}^1(D,-) \approx \mathrm{Ext}^1(D',-)$ for any objects D and D' if and only if there are projective objects P and Q such that $P + D \approx Q + D'$. Applying this to our special case gives the last result.

6. Tensor Products

We assume throughout this section that R is a ring in $\mathfrak{U}$ and that ${}_R\mathcal{M}$ and $\mathcal{M}_R$ are the categories of left and right R-modules in $\mathfrak{U}$, as defined in section 5. Since we will not change the ring R, we will suppress the R and denote the additive, subcategories of finitely presented left and right R-modules by $\ell(\mathcal{F})$ and $r(\mathcal{F})$ respectively. Also, we shall denote by $\mathcal{E}(\ell(\mathcal{F}))$ the additive full subcategory of ${}_R\check{\mathcal{M}}_0$ whose objects consist of all functors $G \approx \mathrm{Ext}^1(M,-)$ for some finitely presented left R-module M. A similar definition can be given for $\mathcal{E}(r(\mathcal{F}))$.

Now we have the usual contravariant functor $\ell(\mathcal{F}) \to \mathcal{E}(\ell(\mathcal{F}))$ given by $M \mapsto \mathrm{Ext}^1(M,-)$. It is well known that the maps

$$(M,N) \to (\mathrm{Ext}^1(N,-), \mathrm{Ext}^1(M,-))$$

are epimorphisms and that $M \to N$ gives the zero map

$$\mathrm{Ext}^1(N,-) \to \mathrm{Ext}^1(M,-)$$

if and only if it can be factored through a projective. Thus the maps

$$(M,N) \to (\mathrm{Ext}^1(N,-), \mathrm{Ext}^1(M,-))$$

induce maps

$$\Pi(M,N) \to (\mathrm{Ext}^1(N,-), \mathrm{Ext}^1(M,-))$$

which are isomorphisms. Therefore if we denote by $\Pi(\ell(\mathcal{F}))$, the additive category whose objects are the same as those of $\ell(\mathcal{F})$ and whose maps are $\Pi(M,N)$ for M and N in $\ell(\mathcal{F})$, we obtain a functor $\Pi(\ell(\mathcal{F})) \to$ $\to E(\ell(\mathcal{F}))$ which is an equivalence of categories. Having established these notational matters, we now turn our attention to tensor products.

Lemma 6.1. *Let A be a left R-module. Then the functor $-\otimes A : \mathcal{M}_R \to \mathcal{A}b$ is coherent if and only if A is a finitely presented module. Further, if $P_1 \to P_0 \to A \to 0$ is exact with P_i finitely presented projective R-modules, then $(P_1^*,-) \to (P_0^*,-) \to -\otimes A \to 0$ is exact, where $P_i^* = (P_i, R)$.*

Proof. Suppose A is a finitely presented R-module and $P_1 \to P_0 \to A \to 0$ is exact with the P_i finitely generated projective R-modules. Now we know that there exists a commutative diagram with exact rows and columns

$$\begin{array}{ccccc}
0 & & 0 & & \\
\downarrow & & \downarrow & & \\
-\otimes P_1 & \to & -\otimes P_0 & \to -\otimes A \to 0 & \\
\downarrow & & \downarrow & & \\
(P_1^*, -) & \to & (P_0^*, -) & & \\
\downarrow & & \downarrow & & \\
0 & & 0 & &
\end{array}$$

where $B \otimes P_i \to (P_i^*, B)$ is the usual map given by $(b \otimes p)(f) = f(p)(b)$ for all b in B, $p \in P_i$ and f in P_i^*. Thus we obtain the last part of the lemma.

Suppose $-\otimes A$ is coherent and let $F \to A \to 0$ be exact with a free R-module. Then $-\otimes F \to -\otimes A \to 0$ is exact. Since $-\otimes F$ is isomorphic to a direct sum of copies of $-\otimes R$ and $-\otimes A$ is coherent and thus finitely generated, we know that we can find a finitely generated free R-module G_0 such that $-\otimes G_0 \to -\otimes A \to 0$ is exact. Thus we obtain an epimorphism $G_0 \to A \to 0$ with G_0 a finitely generated free R-module. Suppose $G \to G_0 \to A \to 0$ is exact with G a free R-module. Then we know that $-\otimes G \to -\otimes G_0 \to -\otimes A \to 0$ is exact. Since $-\otimes A$ is coherent, we know that $H = \operatorname{Im}(-\otimes G \to -\otimes G_0)$ is finitely generated. Since $-\otimes G \to H \to 0$ is exact and $-\otimes G$ is isomorphic to a direct sum of copies of $-\otimes R$, we know there is a finitely generated free module G_1 such that $-\otimes G_1 \to H \to 0$ is exact, or what is the same thing,

$$-\otimes G_1 \to -\otimes G_0 \to -\otimes A \to 0$$

is exact. Thus we obtain an exact sequence $G_1 \to G_0 \to A \to 0$ which shows that A is finitely presented.

Thus for each M in $\ell(\mathscr{F})$, we know that the functor $-\otimes M$ is a coherent functor. Since $-\otimes M$ is certainly half exact we know by theorem 4.13, that $(-\otimes M)_0$ is in $\mathscr{E}(r(\mathscr{F}))$, i.e. for some N in $r(\mathscr{F})$ we have that $(-\otimes M)_0 \approx \operatorname{Ext}^1(N, -)$. Thus we obtain a functor $\ell(\mathscr{F}) \to \mathscr{E}(r(\mathscr{F}))$ by sending M to $(-\otimes M)_0$. It is our purpose to show next that this induces an equivalence of categories $\Pi\ell((\mathscr{F})) \to \mathscr{E}(r(\mathscr{F}))$. To this end, we first make some general observations.

Proposition 6.2. *Let $\mathscr{C}$ be an abelian $\mathfrak{U}$-category and let $\check{\mathscr{C}} \to \check{\mathscr{C}}_0$ be the functor $F \to F_0$. Then*

a) If $F \to G$ can be factored through a projective in $\check{\mathscr{C}}$, then the corresponding map $F_0 \to G_0$ is the zero map.

b) The functor $\check{\mathscr{C}} \to \check{\mathscr{C}}_0$ induces a functor $\Pi(\check{\mathscr{C}}) \to \check{\mathscr{C}}_0$.

c) If G is right exact, then the map $\Pi(\check{\mathscr{C}})(F, G) \to (F_0, G_0)$ is a monomorphism for all F in $\check{\mathscr{C}}$.

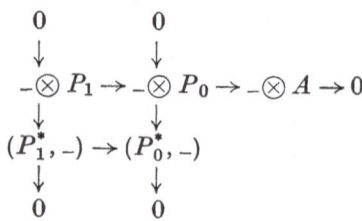

d) If G preserves epimorphisms, then the map $\Pi(\check{\mathscr{C}})(F, G) \to (F_0, G_0)$ is an epimorphism for all F in $\check{\mathscr{C}}$.

e) If G is right exact, then the map $\Pi(\check{\mathscr{C}})(F, G) \to (F_0, G_0)$ is an isomorphism for all F in $\check{\mathscr{C}}$.

Proof. The only projective objects in $\check{\mathscr{C}}$ are the representable functors. Since $(A, -)_0 = 0$ for all A in $\mathscr{C}$, we have established a).

b) Follows trivially from a).

c) Suppose G is half exact and suppose we have a map $F \to G$ such that the induced map $F_0 \to G_0$ is the zero map. Suppose

$$0 \to F_0 \to F \to (A, -) \to F_1 \to 0$$

is our usual exact sequence with the F_i in $\check{\mathscr{C}}_0$. Since $F_0 \to G_0$ is the zero map, we have a map $\bar{F} \to G$ such that $F \to \bar{F} \to G$ is the map $F \to G$ where $\bar{F} = \operatorname{Im}(F \to (A, -)$. From the exact sequence

$$0 \to \bar{F} \to (A, -) \to F_1 \to 0$$

we deduce the exact sequence $((A, -), G) \to (\bar{F}, G) \to \operatorname{Ext}^1(F_1, G)$. Since F_1 is in $\check{\mathscr{C}}_0$ and G is half exact, we know by proposition 3.2, that $\operatorname{Ext}^1(F_1, G) = 0$. Thus the map $\bar{F} \to G$ can be extended to $(A, -) \to G$ which shows that the map $F \to G$ can be factored through a projective. Thus we have that if G is half exact, then the map $\Pi(\check{\mathscr{C}}_0)(F, G) \to (F_0, G_0)$ is a monomorphism for all F in $\check{\mathscr{C}}$.

d) Suppose G preserves epimorphisms and let $0 \to F_0 \to F \to \bar{F} \to 0$ be as in c). Then we have the exact sequence $(F, G) \to (F_0, G) \to \operatorname{Ext}^1(\bar{F}, G)$. But it is easily seen that $\bar{F}$ has a resolution $0 \to (B, -) \to (A, -) \to \bar{F} \to 0$ where $A \to B \to 0$ is exact. From this and lemma 3.1, it follows that since G preserves epimorphisms, then $\operatorname{Ext}^1(\bar{F}, G) = 0$. Thus

$$(F, G) \to (F_0, G) \to 0$$

is exact. But the map $0 \to G_0 \to G$ induces an isomorphism

$$(F_0, G_0) \to (F_0, G).$$

From this it follows that the map $(F, G) \to (F_0, G_0)$ is an epimorphism, giving our desired result.

c) Trivial.

Returning to the category $\ell(\mathscr{F})$, we see that for all M and N in $\ell(\mathscr{F})$, the map $(-\otimes M, -\otimes N) \to ((-\otimes M)_0, (-\otimes N)_0)$ induces an isomorphism $\Pi(-\otimes M, -\otimes N) \to ((-\otimes M)_0, (-\otimes N)_0)$.

Now we also have the obvious maps $(M, N) \to (-\otimes M, -\otimes N)$ which can be easily seen to be an isomorphism for all M and N in $\ell(\mathscr{F})$ (actually in ${}_R\mathscr{M}$). Claim this isomorphism induces an isomorphism

$$\Pi(M, N) \to \Pi(-\otimes M, -\otimes N).$$

For if $M \to N$ can be factored through any projective it can be factored through any exact sequence $P \to N \to 0$ with P projective. Since N is finitely presented, we can choose the P to be a finitely generated projective R-module. Thus we obtain that the induced map $-\otimes M \to -\otimes N$ can be factored through $-\otimes P$. But $-\otimes P \approx (P^*, -)$ and is thus projective. Therefore we indeed have an epimorphism

$$\Pi(M, N) \to \Pi(-\otimes M, -\otimes N).$$

The fact that it is an isomorphism is essentially the same argument as above.

Combining the above observations we see that the functor

$$\ell(\mathscr{F}) \to \mathscr{E}(r(\mathscr{F})) \quad \text{given by} \quad M \mapsto (-\otimes M)_0$$

does indeed induce a functor $\Pi\,\ell((\mathscr{F})) \to \mathscr{E}(r(\mathscr{F})$ which is fully faithful. In order to establish that it is an equivalence, we have to show that given any finitely presented right R-module N, there is a left R-module M such that $\operatorname{Ext}^1(N, -) \approx (-\otimes M)_0$.

This is a trivial consequence of

Proposition 6.3. *Let $P_1 \to P_0 \to M \to 0$ be exact with the P_i finitely generated projective R-modules (left). Let $N = \operatorname{Coker}(P_0^* \to P_1^*)$. Then N is a finitely presented right R-module and $\operatorname{Ext}^1(N, -) \approx (-\otimes M)_0$. Further $R^0(-\otimes M) \approx (M^*, -)$ and we have an exact sequence*

$$0 \to \operatorname{Ext}^1(N, -) \to (-\otimes M) \to (M^*, -) \to \operatorname{Ext}^2(N, -) \to 0$$

where $(-\otimes M) \to (M^*, -)$ is the usual map.
Thus if $Q_1 \to Q_0 \to A$ is an exact sequence of finitely generated right R-modules with the Q_i projectives and $B = \operatorname{Coker}(Q_0^* \to Q_1^*)$, then $(-\otimes B)_0 \approx \operatorname{Ext}^1(A, -)$.

Proof. Since each of the P_i are finitely generated projective right R-modules, it follows that N is a finitely presented right R-module. From the exact sequence $0 \to M^* \to P_0^* \to P_1^* \to N \to 0$, we deduce the exact sequence

$$0 \to (N, -) \to (P_1^*, -) \to (P_0^{**}, -) \to -\otimes M \to 0$$

(see lemma 6.1). The first part of the proposition now follows easily from proposition 5.8.

The rest of the proposition is an immediate consequence of the first part, once one observes that finitely generated projective modules are reflexive, i.e. $Q_i \approx Q_i^{**}$ under the natural map.

Remark. It should be observed that this proposition gives a proof of the fact that $(\_\otimes M)_0$ is in $\mathscr{E}(r(\mathscr{F}))$ which is independent of the results of section 4.

Thus we have proven.

Theorem 6.4. *The functor* $\Pi\ell((\mathscr{F}))\to\mathscr{E}(r(\mathscr{F}))$ *given by* $M\mapsto(\_\otimes M)_0$ *is an equivalence of categories. Since* $\mathscr{E}(r(\mathscr{F}))$ *is contravariantly equivalent to* $\Pi(r(\mathscr{F}))$, *by means of the functor* $\Pi(r(\mathscr{F})\to\mathscr{E}(r(\mathscr{F}))$, *we obtain a contravariant equivalence between* $\Pi(\ell(\mathscr{F}))$ *and* $\Pi(r(\mathscr{F}))$.

As an easy consequence of this theorem we obtain,

Proposition 6.5. *Suppose* $0\to A\to B\to C\to 0$ *is an exact sequence of right R-modules with C finitely presented. Then the following are equivalent:*

a) The exact sequence $0\to A\to B\to C\to 0$ *splits.*

b) For every left R-module M, the sequence $0\to A\otimes M\to B\otimes M$ *is exact.*

c) For every finitely presented left R-module M, the sequence

$$0\to A\otimes M\to B\otimes M$$

is exact.

d) If M is a finitely presented left R-module such that

$$\mathrm{Ext}^1(C,\_)\approx(\_\otimes M)_0, \quad \textit{then} \quad 0\to A\otimes M\to B\otimes M$$

is exact.

Proof. Clearly the implications a) ⇒ b) ⇒ c) ⇒ d) are all trivial.

d) ⇒ a). That such M exist we know from theorem 6.4. From the exact sequence $0\to\mathrm{Ext}^1(C,\_)\to\_\otimes M\to(M^*,\_)$ (see proposition 6.3), we deduce the commutative diagram with exact rows and columns

$$\begin{array}{ccccc}
 & & 0 & & 0\\
 & & \downarrow & & \downarrow\\
 & & \mathrm{Ext}^1(C,A) & \to & \mathrm{Ext}^1(C,B)\\
 & & \downarrow & & \downarrow\\
 & & A\otimes M & \longrightarrow & B\otimes M\\
 & & \downarrow & & \downarrow\\
0 & \to & (M^*,A) & \longrightarrow & (M^*,B)
\end{array}$$

from which it easily follows that the

$$\mathrm{Ker}(\mathrm{Ext}^1(C,A)\to\mathrm{Ext}^1(C,B))=\mathrm{Ker}(A\otimes M\to B\otimes M).$$

Thus we have the exact sequence

$$\mathrm{Hom}(C,B)\to\mathrm{Hom}(C,C)\to A\otimes M\to B\otimes M.$$

Thus if $0\to A\otimes M\to B\otimes M$ is exact, then $\mathrm{Hom}(C,B)\to\mathrm{Hom}(C,C)\to 0$ is exact and thus $0\to A\to B\to C\to 0$ splits.

Remark. It should be observed that the argument given above actually yields the following more general result. Let $\mathscr{C}$ be a $\mathfrak{U}$-category satisfying the hypothesis that every half exact functor in $\check{\mathscr{C}}_0$ is isomorphic to $\mathrm{Ext}^1(C, \_)$ for some C in $\mathscr{C}$. Let F be a half exact coherent functor. Then F_0 is half exact and thus $F_0 \approx \mathrm{Ext}^1(C, \_)$ for some C in $\mathscr{C}$. Let $0 \to C_1 \to C_2 \to C_3 \to 0$ be an exact sequence in $\mathscr{C}$. Then we have an exact sequence $(C, C_2) \to (C, C_3) \to F(C_1) \to F(C_2)$. Thus if $C_3 \approx C$ and $0 \to F(C_1) \to F(C_2)$ is exact, then the sequence $0 \to C_1 \to C_2 \to C_3 \to 0$ splits.

As another application of the results of this section we prove

Proposition 6.6. *Let C be a right R-module such that there exists an exact sequence $P_2 \to P_1 \to P_0 \to C \to 0$ with the P_i finitely genented projective R-modules. Then the following statements are equivalent:*

a) $\mathrm{Ext}^1(C, R) = 0$.

b) There is a finitely presented left R-module M such that

$$\mathrm{Tor}_1(\_, M) \approx \mathrm{Ext}^1(C, \_).$$

c) If $0 \to A \to B \to C \to 0$ is exact and $0 \to \mathrm{Tor}_1(A, M) \to \mathrm{Tor}_1(B, M)$ is exact for all finitely presented left R-modules M, then

$$0 \to A \to B \to C \to 0$$

splits.

Proof. a) $\Rightarrow$ b). Since $\mathrm{Ext}^1(C, R) = 0$, we know that

$$0 \to C^* \to P_0^* \to P_1^* \to P_2^*$$

is exact. Let $M = \mathrm{Coker}(P_1^* \to P_2^*)$. Then M is a finitely presented R-module. Also we have that $H_1(\_ \otimes P^*) = \mathrm{Tor}_1(\_, M)$. But the complex $\_ \otimes P^*$ is isomorphic to the complex $(P, \_)$. Thus

$$\mathrm{Tor}_1(\_, M) \approx \mathrm{Ext}^1(C, \_).$$

b) $\Rightarrow$ c). Trivial.

c) $\Rightarrow$ a). Let $0 \to R \to E \to C \to 0$ be exact. Then certainly

$$0 \to \mathrm{Tor}_1(R, M) \to \mathrm{Tor}_1(E, M)$$

is exact for all finitely presented left R-modules M, since $\mathrm{Tor}_1(R, \_) = 0$. Thus $0 \to R \to E \to C \to 0$ splits or, what is the same thing,

$$\mathrm{Ext}^1(C, R) = 0.$$

7. Coherent Functors which are Tor

We make the same assumptions on the ring R as in section 6. We now describe another property possessed by the functor $\Pi\,\ell\,((\mathscr{F})) \to \mathscr{E}(r(\mathscr{F}))$, defined by $C \mapsto (\_ \otimes C)_0$.

Proposition 7.1. *Suppose C is a finitely presented left R-module and M is a finitely presented right R-module such that $(-\otimes C)_0 \approx \mathrm{Ext}^1(M, -)$. Let $C^* \otimes - \to (C, -)$ be the usual map. Then $L^0((C, -)) \approx C^* \otimes -$ and we have an exact sequence*

$$0 \to \mathrm{Tor}_2(M, -) \to C^* \otimes - \to (C, -) \to \mathrm{Tor}_1(M, -) \to 0 .$$

Proof. Since C is finitely presented, we know that $(C, -)$ commutes with direct sums. Thus $C^* \otimes P \to (C, P)$ is an isomorphism for all projective modules P. Combining this with the fact that $C^* \otimes -$ is right exact, we see immediately that $L^0((C, -)) \approx C^* \otimes -$.

Now suppose that $P_1 \to P_0 \to C \to 0$ is exact with the P_i finitely generated projective R-modules. Then from the exact sequence

$$0 \to C^* \to P_0^* \to P_1^* \to N \to 0$$

we get a commutative diagram

$$\begin{array}{ccccccc} 0 \to \mathrm{Tor}_2(N, -) \to & C^* \otimes - & \to & P_0^* \otimes - & \to & P_1^* \otimes - \\ & \downarrow & & \wr & & \wr \\ 0 \to & (C, -) & \to & (P_0, -) & \to & (P_1, -) \end{array}$$

with the obvious exactness properties. From this it easily follows that we have an exact sequence

$$0 \to \mathrm{Tor}_2(N, -) \to C^* \otimes - \to (C, -) \to \mathrm{Tor}_1(N, -) \to 0 .$$

But by it follows that $(- \otimes C)_0 \approx \mathrm{Ext}^1(N, -)$ and thus that

$$\mathrm{Ext}^1(N, -) \approx \mathrm{Ext}^1(M, -) .$$

Therefore $\mathrm{Tor}_i(N, -) \approx \mathrm{Tor}_i(M, -)$ for all i, which completes the proof.

It should be observed that since we know that if M is a finitely presented right module we can find a finitely presented left module C such that $(- \otimes C)_0 \approx \mathrm{Ext}^1(M, -)$ (see theorem 6.4), it follows that given any such right module M there is a finitely presented left module C such that

$$0 \to \mathrm{Tor}_2(M, -) \to C^* \otimes - \to (C, -) \to \mathrm{Tor}_1(M, -) \to 0$$

is exact.

In this section we are interested in generalizing proposition 7.1 to a larger class of functors in the case that R is a noetherian ring. For example, if $F: {}_R\mathscr{M} \to \mathscr{Ab}$ is a functor which commutes with direct sums, then there is a unique map $F(R) \otimes - \to F$ such that $R \otimes F(R) \to F(R)$ is the identity. From this it easily follows that $L^0 F \approx F(R) \otimes -$. It is our aim to describe the kernel and cokernel of such maps $F(R) \otimes - \to F$ where F is a coherent functor which commutes with direct limits and R is a noetherian ring. However, before restricting ourselves to the case of R noetherian, we make some useful observations.

In the category ${}_R\check{\mathscr{M}}$ let $\mathscr{D}_1$ be the full subcategory whose objects F have the property that for each module M we have that $\varinjlim F(M_i) = F(M)$, where (M_i) is the family of finitely generated submodules of M. It is easily checked that if $0 \to F_1 \to F_2 \to F_3 \to F_4 \to 0$ is exact with F_2 and F_3 in $\mathscr{D}_1$, then F_1 and F_4 are in $\mathscr{D}_1$.

If $0 \to F_1 \to F_2 \to F_3 \to 0$ is exact and F_1 and F_3 are in $\mathscr{D}_1$ so is F_2 in $\mathscr{D}_1$. Also $\mathscr{D}_1$ is closed under direct limits over directed sets. We now define $\varphi\colon {}_R\check{\mathscr{M}} \to \mathscr{D}_1$ as follows: $\varphi(F)(M) = \varinjlim F(M_i)$ where (M_i) is the family of finitely generated submodules of M. Since for each $M_i \subset M$ we have maps $F(M_i) \to F(M)$, this gives a map of $\varphi(F)(M) \to$ $\to F(M)$. Thus we have the map $\varphi(F) \to F$. It is easily seen that φ and $\varphi(F) \to F$ have the properties given below,

Lemma 7.2. *Let $\varphi\colon {}_R\check{\mathscr{M}} \to \mathscr{D}_1$ be the functor described above. Then*

a) φ is exact and commutes with direct limits;

b) For all G in $\mathscr{D}_1$ and F in ${}_R\check{\mathscr{M}}$, the map $(G, \varphi(F)) \to (G, F)$ induced by the map $\varphi(F) \to F$, is an isomorphism.

c) $\varphi(F)(M) \to F(M)$ is an isomorphism if M is finitely generated.

d) $\varphi(F) \to F$ is an isomorphism if and only if F is in $\mathscr{D}_1$.

We shall denote the full subcategory of ${}_R\check{\mathscr{M}}$ consisting of those functors which commute with direct limits taken over directed sets by $\mathscr{D}_0$. It is obvious that $\mathscr{D}_0 \subset \mathscr{D}_1$. Also it is easily seen that if

$$0 \to F_1 \to F_2 \to F_3 \to F_4 \to 0$$

is an exact sequence with F_2 and F_3 in $\mathscr{D}_0$, then F_1 and F_4 are in $\mathscr{D}_0$. And if $0 \to F_1 \to F_2 \to F_3 \to 0$ is exact with F_1 and F_3 in $\mathscr{D}_0$, then F_2 is in $\mathscr{D}_0$. And $\mathscr{D}_0$ is closed under direct limits over directed sets.

Proposition 7.3. *If R is a noetherian ring, then $\mathscr{D}_0 = \mathscr{D}_1$.*

Proof. Let F be in ${}_R\check{\mathscr{M}}$. Then we know that there exists an exact sequence

$$\sum_{j \in J} (A_j, -) \to \sum_{i \in I} (B_i, -) \to F \to 0.$$

Thus

$$\sum_{j \in J} \varphi((A_j, 6)) \to \sum_{i \in I} \varphi((B_i, -)) \to \varphi(F) \to 0$$

is exact since φ is exact and commutes with direct limits. Therefore if we show that $\varphi(B, -)$ is in $\mathscr{D}_0$ for all R-modules B, then we will have shown that $\varphi(F)$ is in $\mathscr{D}_0$ for all F in ${}_R\check{\mathscr{M}}$. Since $\varphi(F) \approx F$ for F in $\mathscr{D}_1$ this will show that $\mathscr{D}_0 = \mathscr{D}_1$.

Suppose $P_1 \to P_0 \to B \to 0$ is exact with the P_i projective, then we have the exact sequence $0 \to (B, -) \to (P_0, -) \to (P_1, -)$. From this it follows $0 \to \varphi((B, -)) \to \varphi((P_0, -) \to \varphi((P_1, -))$ is exact. Thus if we show that $\varphi((P_i, -)$ commutes with direct limits over directed sets, we will have that $\varphi((B, -))$ also does. Therefore in order to establish the proposition it suffices to show that $\varphi(P, -)$ is in $\mathcal{D}_0$ for projective modules P.

Suppose P is a projective module and $M = \varinjlim M_i$ (a directed direct limit). Now it is easily seen that $\varphi((P, -))(M)$ is nothing more than the maps from P to M whose images are contained in finitely generated submodules of M. Or, since R is noetherian, $\varphi((P, -))(M)$ is the set of maps from P to M whose images are finitely generated. Let $f: P \to M_i$ be a map such that the composite $P \to M_i \to M$ is zero. Since $f(P)$ is finitely generated and $f(P) \subset \mathrm{Ker}(M_i \to M)$, we know there is a $j \geqq i$ such that $f(P)$ is carried to zero under the map $M_i \to M_j$, i.e. the composite $P \to M_i \to M_j$ is zero. Thus we have shown that the map $\varinjlim \varphi(P, -)(M_i) \to \varphi((P, -))(M)$ is a monomorphism.

Suppose $f: P \to M$ and $f(P)$ is finitely generated. Then there is an M_i such that $f(P) \subset \mathrm{Im}(M_i \to M)$. Let N be a finitely generated submodule of M_i which goes onto $f(P)$. Then, since P is projective, there is a map $P \to N$ such that the composite $P \to N \to M$ is the map f. Thus we have shown that the map $\varinjlim \varphi((P, -))(M_i) \to \varphi((P, -))(M)$ is an epimorphism and thus an isomorphism. Therefore $\varphi((P, -))$ is in $\mathcal{D}_0$ which completes the proof of the proposition.

From now on we will assume that our ring R is a noetherian ring. It is a trivial matter to check

Lemma 7.4. *Let M be an R-module. Then,*

a) The map $\varphi((M, -)) \to (M, -)$ is a monomorphism.

b) $\varphi((M, -))$ is left exact and is exact if M is projective.

c) The map $M^* \otimes - \to (M, -)$ has a unique factorization $M^* \otimes - \to \to \varphi((M, -)) \to (M, -)$ and if M is projective, then $M^* \otimes - \to \varphi((M, -))$ is an isomorphism.

d) $L_0(\varphi((M, -)) \approx M^* \otimes -$.

Suppose F is a coherent half exact functor. Then by proposition 4.10 we know that there is an exact sequence $C \to D \to E \to 0$ with D a projective module such that $(D, -) \to (C, -) \to F \to 0$ is exact. Thus we have the exact sequence $\varphi((D, -)) \to \varphi((C, -)) \to \varphi(F) \to 0$. Therefore if we assume that F also commutes with direct limits over directed sets, then we have an exact sequence $\varphi((D, -)) \to \varphi((C, -)) \to F \to 0$ with the D a projective R-module. Since $\varphi((X, -))(R) = X^*$ we also obtain the exact sequence $D^* \to C^* \to F(R) \to 0$. From this we deduce the commutative diagram with exact rows and columns

$$\begin{array}{ccccccc}
 & & 0 & & & & \\
 & & \downarrow & & & & \\
 & & G_2 & = & G_2 & & \\
 & & \downarrow & & \downarrow & & \\
D^* \otimes - & \to & C^* \otimes - & \to & F(R) \otimes - & \to & 0 \\
\wr & & \downarrow & & \downarrow & & \\
\varphi((D,-)) & \longrightarrow & \varphi((C,-)) & \to & F & \to & 0 \\
 & & \downarrow & & \downarrow & & \\
 & & G_1 & = & G_1 & & \\
 & & & & \downarrow & & \\
 & & & & 0 & &
\end{array}$$

where G_1 and G_2 are the cokernel and kernel respectively of the map $C^* \otimes - \to \varphi((C,-))$. Thus if we obtain a description of G_1 and G_2 then we will have also described the cokernel and kernel of $F(R) \otimes - \to F$.

Let $P_1 \to P_0 \to C \to 0$ be exact with the P_i projective R-modules. Then we have the exact sequence $0 \to C^* \to P_0^* \to P_1^* \to N \to 0$. Since we know that P_i^* are flat (see lemma 7.4) we have a commutative diagram

$$\begin{array}{ccccccccc}
0 & \to & \mathrm{Tor}_2(N,-) & \to & C^* \otimes - & \to & P_0^* \otimes - & \to & P_1^* \otimes - \\
 & & & & \downarrow & & \wr & & \wr \\
 & & 0 & \to & \varphi((C,-)) & \to & \varphi((P_0,-)) & \to & \varphi((P_1,-)).
\end{array}$$

From this it follows as in the proof of proposition 7.1 that we have an exact sequence

$$0 \to \mathrm{Tor}_2(N,-) \to C^* \otimes - \to \varphi((C,-)) \to \mathrm{Tor}_1(N,-) \to 0.$$

Combining the above we obtain the promised generalization of proposition 7.1.

Theorem 7.5. *Let R be a noetherian ring and $P_1 \to P_0 \to C \to 0$ an exact sequence of R-modules with the P_i projective. If the right module N is defined by $P_0^* \to P_1^* \to N \to 0$, then we have the exact sequence*

$$0 \to \mathrm{Tor}_2(N,-) \to C^* \otimes - \to \varphi((C,-)) \to \mathrm{Tor}_1(N,-) \to 0.$$

More generally if F is a half exact coherent functor which commutes with directed direct limits then there is a right module N and an exact sequence

$$\mathrm{Tor}_2(N,-) \to F(R) \otimes - \to F \to \mathrm{Tor}_1(N,-) \to 0.$$

As an immediate consequence we have,

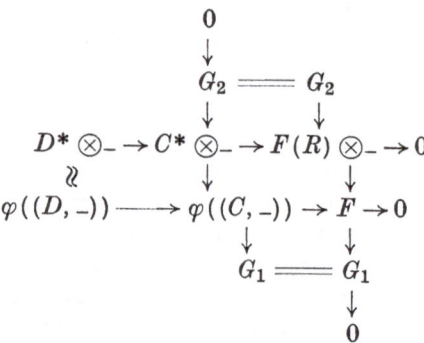

Corollary 7.6. *Let R be a noetherian ring and F a half exact functor*

which commutes with directed direct limits. Then $F \approx \mathrm{Tor}_1(N, -)$ *for some right R-module N if and only if* $F(R) = 0$.

References

[1] GABRIEL, P.: Des catégories abeliennes. Bull. Soc. Math. France 1962, pp. 323—448.
[2] HILTON, P. J.: Homotopy theory of modules and duality. Symposium internacional de Topologia Algebraica 1958, pp. 273—281.
[3] —, and D. REES: Natural maps of extension functors and a theorem of R. G. SWAN. Proc. Cambridge Philo. Soc. 1961, pp. 489—502.
[4] MACLANE, S.: Homology. New York: Academic Press 1963.

Department of Mathematics
Brandeis University
Waltham, Massachusetts

Epimorphisms and Dominions*, **

By

John R. Isbell

Introduction

According to Grothendieck's definition, which is by now standard in part of the literature, an *epimorphism* is a map $f: A \to B$ such that for any maps $g: B \to C$, $h: B \to C$, if $g \neq h$, then $gf \neq hf$. In groups, for example, epimorphisms are onto, but in rings, for example, they are not.

This paper grows out of an attack on the problem, given an algebra A (particularly a semigroup or ring), to describe all its epimorphic images. Freyd has shown [*1*, p. 93] that there is a bound on their size, hence on their number. It turns out that they cannot exceed A in power if the power of A is large enough ($\aleph_0$, unless there are more than $\aleph_0$ algebraic operations). The assumptions on the category of algebras are different; briefly, for Freyd's proof it must be elementarily definable, for mine it must be closed under certain constructions (colimits in a primitive category of algebras).

The problem splits in at least two places. First, one need only consider epimorphic embeddings, since $f: A \to B$ is epimorphic if and only if $f(A) \subset B$ is epimorphic, and the description of $A \to f(A)$ by means of congruence relations is well developed. Second, define an *equalizer* $E \subset B$ (= left equalizer = difference kernel) as a subalgebra of the form $\{x : g(x) = h(x)\}$, for any $g: B \to C$, $h: B \to C$. For any subalgebra A of B, the intersection of all equalizers containing A is an equalizer D_1, the *dominion* of A in B. A is epimorphically embedded if and only if $D_1 = B$. In general, domination is unstable; $A \subset D_1$ need not be epimorphic but may have a smaller dominion D_2. Transfinite descent leads to the *stable dominion* D_∞ in which A is epimorphically embedded. (D_∞ is the *coimage* of $A \subset B$ according to terminology suggested in a footnote of Grothendieck and used by Freyd [*1*], but which I think should be abandoned.)

In a suitable category of algebras, if the cardinal number of A is sufficiently large, no dominion of A can be larger. However, neither

* Supported by National Science Foundation Grant GP1791.
** Received June 12, 1965.

dominions nor epimorphic extensions behave like completions or compactifications. What happens is that every element d dominated by A is dominated by means of a finite system σ of relations. σ does not determine the algebraic nature of d, and two systems σ_1, σ_2 may not be simultaneously usable; but there is an exclusion principle, that two d's cannot occupy the same σ.

Having descended from epimorphisms to dominating systems of relations σ, we have an almost manageable problem for the cases of semigroups and rings. For rings B, the dominion of a subring A is the additive closure of the multiplicative dominion, i.e. of the dominion of A considered as a multiplicative semigroup. In semigroups, A dominates $d \in B - A$ if and only if for some $m > 0$ there are elements $a_0, a_1, \ldots, a_{2m}$ of A and $x_1, \ldots, x_m$, $y_1, \ldots, y_m$ of B such that $d = a_0 y_1 = x_m a_{2m}$, $a_0 = x_1 a_1$, $x_i a_{2i} = x_{i+1} a_{2i+1}$ for $i = 1, \ldots, m-1$, $a_{2m} = a_{2m-1} y_{2m}$, and $a_{2i} y_{i+1} = a_{2i-1} y_i$ for $i = 1, \ldots, m-1$.

The further results in this paper are examples, establishing the negative remarks above and some other negatives. For instance, the transfinite descent from D_1 to D_∞ may be arbitrarily long. A finite semigroup A may have an infinite epimorphic extension even if every element of A is idempotent.

Some affirmative results in semigroups will appear in a joint paper by J. M. Howie and me. It should be noted that the special form of dominating relations in commutative semigroups, which is not a corollary of the non-commutative result, was proved earlier by Howie. I am indebted to all the members of the Tulane algebra seminar, Howie, A. H. Clifford, John Dauns, and Karl Hofmann, for helpful remarks on this paper.

1. Bounds

An *algebra* here has the standard meaning of a set with a family of everywhere defined finitary operations Q_α, possibly infinite in number. The boundedness theorem for epimorphisms in primitive categories of algebras (described in the introduction) can be extended in at least two ways: one can permit infinitary operations, and one need not require quite a primitive category. The latter extension will be proved, as it does not seriously complicate the argument.

Some remarks about the argument are in order here. We want to associate in a one-to-many fashion elements d of an extension algebra which are dominated by the subalgebra A to certain finite constructs σ on A. Essentially this is just an application of the (extended) Gödel completeness theorem; domination of d is a restriction on the behavior of d throughout all models for a suitable theory θ of pairs of homomorphisms, and since it holds in all models it must follow from the axioms of θ by one or more proofs σ. The trouble is that it is awkward to have theorems about d unless θ includes a proper name for d, while if θ includes proper names for all elements of the arbitrarily large extension algebra, boundedness is lost. This seems to require changing the subject (replacing θ by a more artificial theory[1]); we choose instead to argue on the algebras, but we need some special *partial algebras*, which differ from algebras in that the n-ary operations may be undefined for some n-tuples.

A primitive class of algebras with given operations is a class closed under formation of direct products, subalgebras, and (onto) homomorphic images. By a classical theorem of G. Birkhoff, every primitive class is definable by axioms having the form of identically valid equations. Accordingly it has, and is closed under, a notion of coproduct (= free product), and coproducts are algebras of words in substantially the same way as in the familiar case of groups. We call a subclass of a primitive class *right closed* if it is closed under formation of coproducts and (onto) homomorphic images. *A right closed category of algebras* is the category formed by a right closed class in a primitive class of algebras, with the homomorphisms between them. It is a *primitive category of algebras* if the class is primitive.

Let $\mathscr{C}$ be a right closed category of algebras. We shall be considering subalgebras A of partial algebras P contained in algebras B belonging to $\mathscr{C}$. We define an element d of P to be *dominated* by A, with respect to $\mathscr{C}$, if every two homomorphisms $g: P \to C$, $h: P \to C$, C in $\mathscr{C}$, which coincide on A, must agree at d. Evidently, if A dominates d in P, then A dominates d also in every extension of P. There are some other obvious relationships, several of which are summed up in the statement that in algebras P, passage to the dominion (the set of all dominated elements) is a closure operation.

Lemma 1.1. *With respect to a right closed category of algebras including an algebra B, a necessary and sufficient condition for a subalgebra A to dominate an element d of B is as follows. Let $B * B$ be a coproduct of two copies of B, with coordinate injections i_1, i_2. There must exist a finite sequence $(w_0, w_1, \dots, w_n)$ of elements w_j of $B * B$, beginning $w_0 = i_1(d)$, ending $w_n = i_2(d)$, such that in $(B * B) \times (B * B)$ each element (w_j, w_{j+1}) lies in the subalgebra generated by all elements of the three forms (x, x) $(i_1(a), i_2(a))$, and $(i_2(a), i_1(a))$, $a \in A$.*

Proof. First, the conditions on $(w_0, \dots, w_n)$ say precisely that $i_1(d)$ and $i_2(d)$ are congruent in the least congruence relation ϱ which makes all pairs $i_1(a)$, $i_2(a)$ congruent. For we have an equivalence relation ϱ, obviously; and ϱ is a subalgebra because several sequences $(w_0^k, \dots, w_{n_k}^k)$

[1] Freyd [*1*] uses pairs of theories instead of pairs of homomorphisms, and gets a less sharp bound.

can be brought to the same length by repeating terms and operated on term by term. "Least" is obvious.

Since the category is right closed, it contains $B * B$ and the quotient by ϱ, $B *_A B$, with the quotient homomorphism $q: B * B \to B *_A B$. As qi_1 and qi_2 coincide on A, domination requires that they agree at d. Conversely, any two homomorphisms $g: B \to C$, $h: B \to C$ in this category determine $g * h: B * B \to C$ (by $(g * h) i_1 = g$, $(g * h) i_2 = h$), and if they coincide on A, $g * h$ is right divisible by q, so that $g(d) = h(d)$.

Now the apparatus of 1.1, the pairs (w_j, w_{j+1}) and their representations in terms of (x, x), $(i_1(a), i_2(a))$, $(i_2(a), i_1(a))$, amounts to a computational proof that A dominates d. Therefore this is a sufficient condition for domination in any partial algebra supporting the apparatus. We restate this in terms of the following notion. A *strictly finite extension* of an algebra U is a partial algebra P containing U, such that $P - U$ is finite, all but finitely many operations are defined only for n-tuples in U, and the remaining operations are defined for only finitely many other n-tuples besides those in U.

Theorem 1.2. *With respect to a right closed category of algebras including an algebra B, if the subalgebra A dominates the element d, then there is a strictly finite extension P of a finitely generated subalgebra U of A such that $d \in P$ and U dominates d in P.*

Corollary 1.3. *If A is an epimorphically embedded subalgebra of B, then B is a union of countable extensions of A in which A is epimorphically embedded.*

For this, each element d_1 of B is dominated by means of finitely many elements $d_2, \ldots, d_{n_1}$; d_2 is dominated by means of $d_{n_1+1}, \ldots, d_{n_2}$; and so on.

Corollary 1.4. *The dominion of a subalgebra A in B has cardinal power no greater than the sum of the powers of all inequivalent strictly finite extensions of A.*

Proof. We construct a one-to-one function from the dominion D_1 into a maximal disjoint union of inequivalent strictly finite extensions P_α as follows. For each d in D_1, select an extension $P \subset B$ as given by the theorem, and let d correspond to its image under an equivalence (isomorphism leaving A pointwise fixed) from $P \cup A$ to a P_α. If d_1 and d_2 have the same correspondent, we have an equivalence of P_1 to $P_2 \subset B$ taking d_1 to d_2. Since A dominates d_1 in P_1, $d_2 = d_1$ in B.

Corollary 1.5. *If the power of A is at least $\aleph_0$ and at least equal to the number of algebraic operations, then no dominion of A has larger power.*

Corollary 1.6. *In a right closed category of algebras, no algebra is the domain of a proper class of inequivalent epimorphisms.*

For the final corollaries, we remark that a right closed class of algebras is closed under uniting expanding families (incidentally, even under all passages to direct limits); for $\cup A_\alpha$ is a homomorphic image of $\Sigma^* A_\alpha$. One may note that until now only finite coproducts were needed.

Call an algebra A *saturated* if every epimorphic embedding $A \subset B$ is isomorphic. Call A *absolutely closed* if for every embedding $A \subset B$, the dominion of A is just A.

Corollary 1.7. *Every algebra can be epimorphically embedded in a saturated algebra.*

The proof is a routine Zorn's Lemma or transfinite induction argument.

Corollary 1.8. *Every algebra can be embedded in an absolutely closed algebra.*

Proof. Given $A = A_1$, a routine Zorn's Lemma argument yields an embedding $A_1 \subset A_2$, where the dominion of A_1 is D^1, such that in every extension of A_2 the dominion of A_1 is still just D^1. Construct a similar embedding $A_2 \subset A_3$, and so on for all n. Let $A_\omega = \cup A_n$. If A_ω were not absolutely closed, there would be an embedding $A_\omega \subset B$ in which A_ω dominates another element d. By 1.2, there must be a finitely generated algebra $G \subset B$ in which $G \cap A_\omega$ dominates d. But $G \cap A_\omega \subset A_n$ for some n; thus in the subalgebra generated by G and A_{n+1}, A_n dominates d, a contradiction.

The extensions constructed in 1.7 may be called *saturations*. The extensions of 1.8 should not be called absolute closures; one would like those to be absolutely closed extensions D dominated by A in some $B \supset D$, and I do not know if they exist.

From 1.5, if the power of A is sufficiently large, no saturation can be larger. Going back to the theorem, one sees that the extensions in 1.8 can also be constructed to have the same power.

2. Zigzags

Let us consider the category of all semigroups, and reexamine the domination lemma 1.1. The generation of (w_j, w_{j+1}) as described there yields a pair of expressions $u_1 u_2 \ldots u_k = w_j$, $v_1 v_2 \ldots v_k = w_{j+1}$, where for $i = 1, \ldots, k$, either $u_i = v_i$ or the unordered pair $\{u_i, v_i\}$ is a pair $\{i_1(a), i_2(a)\}$, a in A. Further, all u_i and v_i can be taken in $i_1(B) \cup i_2(B)$ (which generates $B * B$). One can also assume, by introducing intermediate steps, that each passage from $u_1 \ldots u_k$ to $v_1 \ldots v_k$ involves only one index i for which $u_i \neq v_i$. The various expressions $u_1 \ldots u_k$ and

$v_1 \dots v_k$ naturally fall into a roughly rectangular tableau which we can treat as made up of elements of B with attached marks $(+1, -1)$ indicating which factor of $B * B$ they belong in.

Lemma 2.1. *An element d of a semigroup B is in the dominion of a subsemigroup A if and only if there exist a tableau $(x_{ij}) = x : \mathscr{I} \to B$ of elements of B, on the indices $i = 1, 2, \dots, 2r$, and $j = 1, \dots, n_i$ $(r > 0, n_i > 0)$, and a function σ on the set $\mathscr{I}$ of index pairs to the set $\{1, -1\}$, satisfying the following conditions.*

(1) *$x(1,1)x(1,2)\dots x(1,n_1) = d$.*

(2) *$\sigma(1,j) = 1$ for all $j \leqq n_1$, and $\sigma(2r,j) = -1$ for all $j \leqq n_{2r}$.*

(3) *For each odd $i = 1, 3, \dots, 2r-1$, one has $n_{i+1} = n_i$; $x(i+1,j) = x(i,j)$ for all j; there is exactly one j such that $\sigma(i+1,j) \neq \sigma(i,j)$, and for that j, $x(i,j) \in A$.*

(4) *For each even $i = 2, 4, \dots, 2r-2$, the functions $\sigma(i,\ )$ and $\sigma(i+1,\ )$ have the same number k of intervals of constancy $I_1^i, \dots, I_k^i$, resp. $I_1^{i+1}, \dots, I_k^{i+1}$; $\sigma(i,1) = \sigma(i+1,1)$; and for $j = 1, \dots, k$, the product of the values of x over the ordered set I_j^i is the same as over I_j^{i+1}.*

The odd-numbered steps of (3) are the steps from $u_1 \dots u_k$ to $v_1 \dots v_k$ as above. The even-numbered steps (4) go from $w = v_1 \dots v_k$ to another expression $w = u_1' \dots u_m'$; the strong restrictions stated in (4) follow from the uniqueness of the decomposition of $w \in B * B$ into a product of factors taken alternately from the two factor semigroups [*3*, p. 412]. (Actually this property is used also to justify (2).)

The crucial step comes next, and can be described as follows. Suppose the letters x_{ij} written out in a distorted (topological) rectangle, x_{ij} being written in red if $\sigma(i,j) = 1$, in blue if $\sigma(i,j) = -1$. Then the whole top row is red, the whole bottom row is blue, and therefore the boundary between red and blue must contain a simple path joining the left and right sides.

It is a routine matter to write out a formal statement and proof for this step. I shall write just enough to establish some needed terminology and to indicate what constructs seem to enter. This appeal to a topological theorem, whose proof sends us back to combinatorics (with a subdivision theory), is certainly objectionable. It seems likely that the formalities could be reduced by using bookkeeping devices (such as are familiar in linear programming) instead of topology.

A colored tableau (x, σ) as in 2.1 will be called a *dominating tableau* for d over A in B. Given a dominating tableau (x, σ) on $\mathscr{I}$, its *combinatorial figure* is an "indexed cell complex" consisting of a set of elements b_{ij} indexed by $\mathscr{I}$ called *boxes*, a finite set of *edges* e_k, and a finite set of *vertices* v_m, all together called *cells*, together with a reflexive symmetric relation of *incidence* of cells. Informally, one writes out (x_{ij})

in $2r$ horizontal rows spaced to a constant length, row indices i increasing from top to bottom, second indices j from left to right. One encloses the array in a rectangle and separates rows by horizontal lines crossing the rectangle. Each row is divided into boxes b_{ij} containing x_{ij} by vertical segments. These segments are drawn two rows high, each $2n$-1st and $2n$-th row being vertically aligned. For $i = 2, \ldots, 2r - 2$, the intervals of constancy I^i_j, I^{i+1}_j are vertically aligned; the placement of vertical edges within I^{i+1}_j does not matter. (Let them be equally spaced, so that we define a unique combinatorial figure.) The resulting picture exhibits the cells and incidence, with additional structure, such as left-right and the coloring, all of which is determined by the formal data (in particular, the indexing of boxes). By using a topological counterpart figure, one concludes:

2.2. *The combinatorial figure of any dominating tableau contains a simple path of edges e_k crossing from the left to the right side, each of which edges is incident with two boxes b_{ij}, b_{mn}, such that $\sigma(i,j) = -\sigma(m,n)$.*

Next we examine, and collect in groups, consecutive edges in a simple path given by 2.2. There are only r horizontal edges in the figure across which σ changes, one under each odd-numbered row. Each of these edges e is the bottom of a box b_{ij} for which $x_{ij} = a \in A$. The i-th row presents a factorization of d finer than $d = xay$, where

$$x = x(i, 1) \ldots x(i, j-1) \quad \text{and} \quad y = x(i, j+1) \ldots x(i, n_i)\,;$$

there are important exceptions when j is 1 or n_i, and a trivial exception when $n_i = 1$, $d \in A$. Traversing the whole path from the left to the right side, e is joined to the next horizontal e' by a straight vertical path of edges.

Each vertical edge marks a factorization $d = pq$, where p is the product of the entries x_{ij} preceding the edge in its row and q the product of the remainder of the row. Vertical edges lying in a straight vertical path across which (i.e. across each edge of which) σ changes sign mark the same factorization; this follows by a trivial induction from 2.1 (3) and (4).

Returning to e and e', suppose both are traversed from left to right. Then $d = xay = x'a'y'$ and also $xa = x'$, $y = a'y'$. We deduce the factorization $d = x(aa')y'$, $aa' \in A$. A similar combination exists if e and e' are both traversed from right to left. Now consider the maximal portions of the path consisting of one or more horizontal edges traversed in the same direction and the verticals connecting these edges. There are an odd number $2m + 1$ of these, $m + 1$ being traversed from left to right. Considering finally the remaining $2m$ connecting straight verticals, we arrive at:

Theorem 2.3. *A subsemigroup A dominates an element d if and only if $d \in A$ or d has two factorizations $d = a_0 y_1 = x_m a_{2m}$ connected by $2m$ relations $a_0 = x_1 a_1$, $a_1 y_1 = a_2 y_2$, $x_1 a_2 = x_2 a_3, \ldots, a_{2m-1} y_m = a_{2m}$, for some $m \geqq 1$, with all a_i in A.*

Corollary 2.4. *An element b of a semigroup B which commutes with every element of a subsemigroup A commutes with every element of the dominion of A.*

Proof. $bd = ba_0 y_1 = a_0 b y_1 = x_1 a_1 b y_1 = x_1 b a_1 y_1 = \cdots = db$.

Corollary 2.5. *The center of a subsemigroup A is central in the dominion D_1 of A. If A is commutative, so is D_1.*

Proof. The first assertion is immediate from 2.4. For the second, consider the subsemigroup M generated by A and any element d of D_1. Since $A \subset M \subset D_1$, D_1 is the dominion of M. d is central in M; so also in D_1.

Corollary 2.6. *A (left) zero or unit of a subsemigroup is a (left) zero or unit of the dominion.*

Proof. $0d = 0a_0 y_1 = 0y_1 = 0a_1 y_1 = 0a_2 y_2 = 0y_2 = \cdots = 0$. $1d = 1a_0 y_1 = d$.

Another corollary, for use as a lemma below (in 3.2 and 3.4).

2.7. *Every element d of the dominion of a commutative subsemigroup A satisfies a relation $da = a'$ over A.*

Proof. Given the zigzag relations of 2.3, one may compute

$$d a_{2m-1} a_{2m-3} \ldots a_1 = a_0 a_2 \ldots a_{2m} .$$

Much the same results hold for rings, because of the following refinement lemma. In a coproduct $A * B$ of two rings, a *monomial* is a product of elements of A and elements of B.

2.8. *Let Σm_i, Σn_j, be equal sums of monomials in a coproduct of two rings. Then there exist monomials p_{ij} satisfying*

$$\Sigma_j p_{ij} = m_i , \quad \Sigma_i p_{ij} = n_j .$$

Proof. One needs a description of the coproduct ring $A * B$. The following is easily verified. Let a *reduced monomial* be a monomial expression free of zero factors and consecutive factors from the same factor ring. Call two reduced monomials *distributable* if they have the forms mcn, mdn, where m and n are monomial (expressions) and c and d are factors from the same factor ring. We may call a sum of reduced monomials *reduced* if it is free of distributable pairs. Two reduced sums may be equal; the test for this is simply that the complete elimination of distributable pairs from their difference yields 0.

Then the lemma is proved by induction on the total number of monomials occurring. The inductive step has two cases according as the distribution is across the equation $\Sigma m_i = \Sigma n_j$ or on one side, but each case is trivial.

Applying 2.8 to the expressions in 1.1, one finds that a dominated element d can be broken up into a sum of terms p each of which goes through as a monomial; that is:

Theorem 2.9. *In the category of rings, every element d of the dominion of a subring $A \subset B$ is a sum of elements each of which is dominated by the multiplicative semigroup of A relative to the category of semigroups.*

As the converse is obvious, we may say roughly that the problem of dominions in rings is reduced, as to the theorems, to the corresponding problem in semigroups. On the other hand, every example in semigroups yields an example in rings, by passage to semigroup rings; for this procedure evidently takes the dominion into the dominion, and it also takes equalizer subsemigroups to equalizer subrings.

Specifically, 2.4—2.6 for rings follow from 2.9.—2.7 holds too, but it may mean no more than $d0 = 0$.

We remark that 2.8, hence 2.9, carries over to any class of not necessarily associative rings defined by identities which do not involve addition, for example, to alternative or commutative associative rings.

3. Examples

All the examples are in semigroups. We begin with counterexamples to analogues of some theorems about compactness and completeness.

3.1. *Finite algebras need not be saturated.*

3.2. *The saturation of an algebra is not unique, and there is not a universal saturation which maps onto all others.*

3.3. *Saturated algebras need not be absolutely closed.*

3.4. *An epimorphism with saturated domain need not be onto*[2].

Proof of 3.1. Let X be a 3-element set $\{p, q, 0\}$, and let B be the semigroup of those 5 functions on X to X which take 0 and at least one other point to 0. If c and d are the non-idempotent elements of B,

[2] One wants "absolutely closed" here. The example given is absolutely closed; this follows from a theorem to appear in a joint paper by J. M. Howie and the present author.

$B - \{d\}$ is a subsemigroup dominating d by the zigzag

$$d = (dc)d = dcd = d(cd).$$

Proof of 3.2. Let Z^+ be a free semigroup on one generator a. Consider the extensions E_n of Z^+ obtained by choosing $a^n \in Z^+$ and adjoining b subject to $ba = ab$, $b^2a = b$, $ba^{n+1} = a^n$. Z^+ is epimorphically embedded in E_n, by the obvious zigzag around $b^{n+1}\, a^{2n+1}\, b^{n+1}$. In general, given $Z^+ \subset B$ and d dominated by Z^+, 2.7 yields a relation $da^n = a^m$; thence a homomorphism of the dominion into $Z (= E_1)$, d going to a^{m-n} (in group notation). Then (it is convenient but not necessary to use the assumption that $D_1 = B$) routine work shows that the dominion is Z^+ or some E_n. Thus the E_n are saturated, Z^+ has no other saturation, and 3.2 follows.

Proof of 3.3. Let A be a 3-element zero semigroup $\{x, y, 0\}$; every product is 0. We remark that every zero semigroup is saturated; the case we need here, at least, is very easy and is omitted. However, one can embed A in a 6-element commutative semigroup with additional elements t, t^2, t^2x, setting $tx = y$ and $t^3 = 0$. A dominates t^2x by a zigzag around txt.

Proof of 3.4. Let A be the semigroup of ordered pairs (n, f) where n is a positive integer and f is a function on the set of integers $\geqq n$ to Z_2; multiply elements by adding first coordinates and adding pointwise the appropriate restrictions of second coordinates. Obviously A projects epimorphically into, not onto, a cyclic group Z.

We sketch a proof that A is saturated. Suppose A is epimorphically embedded in B. One can assign a "first coordinate" $p(b)$ in $\{1, 2, \ldots, \infty\}$ to each b in B, the least upper bound of those n such that some (equivalently, all) (n, f) in A divide b. Evidently $p(xy) \geqq p(x) + p(y)$. The point of the restriction of functions f is that (because of it) $p((n, f)) = n$ on A. Then p is actually finite everywhere, because of 2.7. Choosing d in $B - A$ to minimize $p(d)$, we get $d = a_0 y_1$, $p(d) > p(y_1)$, which is absurd.

We have defined the dominion D_1 for any pair of algebras $A \subset B$. We remark that it is functorial; a homomorphism $B \to B'$ taking A into A' takes D_1 into D_1'. One can set it up as a pair-to-pair functor Δ taking (B, A) to (D_1, A). Then Δ^2 is meaningful. One can go on to transfinite powers $\Delta^\alpha(B, A) = (D_\alpha, A)$, since the D_α are decreasing in B, by defining D_α for a limit number α as the intersection of the preceding D_β. For each (B, A), the descent must stop at some $D_\lambda = D_{\lambda+1} = D_\infty$. We note that λ is not bounded.

3.5. *For every ordinal number α there is a pair (B, A) of commutative semigroups such that $\Delta^{\alpha+1}(B, A) \neq \Delta^\alpha(B, A)$.*

Proof sketch. First (for finite α) consider the semigroup consisting of 0 and symbols x^t, for t in the open interval (0,2), multiplied by $x^t x^u = x^{t+u}$ if $t + u < 2$, $x^t x^u = 0$ otherwise. Impose the conditions $x^{1-2\mu} \in A$, $x^{1-\mu} \in A$, $x^1 \in B - A$, for some irrational $\mu < 1/2$. One can add $x^\mu \in B - A$ and stop there, securing $x^1 \in D_1 - D_2$. Or one can continue, adjoining $x^{\mu-2\nu}$ and $x^{\mu-\nu}$ to A, x^ν to B, for suitable ν linearly independent of 1 and μ over the rationals; then $x^\mu \in D_1 - D_2$, $x^1 \in D_2 - D_3$. Evidently one can continue for any finite number of steps. If we take ω_0 steps like this, we get A epimorphically embedded in B, which is not what we want. Modify the setup, then, by introducing an unlimited number of symbols x^t_β (β ordinal, $0 < t < 2$). Select a rapidly decreasing sequence of numbers μ_n in (0,2) linearly independent over the rationals; for notational convenience we shall write μ for μ_n, ν for μ_{n+1}. For any suitable μ and β, for any $\gamma > \beta$, we consider the step St (γ, β, μ) of extending a semigroup B containing x^ν_γ, with a subsemigroup A not containing x^ν_γ, by adjoining $x^{\mu-2\nu}_\beta$ and $x^{\mu-\nu}_\beta$ to A and x^μ_β to $B - A$ and identifying every symbol $x^b_\beta\, x^c_\gamma$ such that $b + c \geqq \mu - \nu$ with x^{b+c}_β and with x^{b+c}_γ. Further identifications from (1) all symbols commute, and usually from (2) identifications previously specified in the definition of B. Now, having the α-th pair $(B^\alpha(\mu, \beta), A^\alpha(\mu, \beta))$ defined for every suitable μ and β, the $\alpha + 1$ st pair is constructed by applying St $(\beta + 1, \beta, \mu)$ to $(B^\alpha(\nu, \beta + 1), A^\alpha(\nu, \beta + 1))$. For a limit ordinal λ, for each β and μ, one selects a function γ on ordinals $< \lambda$ to ordinals $> \beta$ such that every two semigroups $B^\alpha(\nu, \gamma(\alpha))$, $\alpha < \lambda$, have no common element except 0. Then $(B^\lambda(\mu, \beta), A^\lambda(\mu, \beta))$ is constructed from the union of all $(B^\alpha(\nu, \gamma(\alpha)), A^\alpha(\nu, \gamma(\alpha)))$ $(\alpha < \lambda)$ by applying St $(\gamma(\alpha), \beta, \mu)$ to each of them; multiplications for which no rules have so far been given are done formally (with commuting symbols).

The full proof is considerably longer than this sketch, but because of the treelike form of the construction there is no difficulty.

Turning to special features of semigroups, several questions naturally suggest themselves. Perhaps first: what about groups? In the category of groups, epimorphisms are onto, by a theorem of SCHREIER; but beyond that, every group is an absolutely closed semigroup, by a theorem of HOWIE [2]. In a sequel to this paper, HOWIE and the present author will show that every inverse semigroup is absolutely closed. One definition of inverse semigroups is this: for every x there is a unique y such that $xyx = x$ and $yxy = y$. The next example (3.6) shows that the assumption cannot be radically weakened.

In the commutative case it suffices to assume that for every x there is at least one y such that $xyx = x$; this implies the inverse property. A natural second question is, how much are matters simplified in commutative semigroups? A crude answer: not enough. There seems to be no substantial simplification of the zigzag theorem 2.3 in the commutative case[3]. The preceding examples are commutative except for 3.1. (Finite commutative semigroups are saturated; this will be in the sequel.) Still it should be possible to go far toward a complete theory of epimorphism and dominion in commutative semigroups. This paper ends with two steps in that direction.

3.6. *There is an idempotent semigroup of 40 elements which has an epimorphically embedded proper subsemigroup.*

Proof. Let F be the free idempotent semigroup on four generators, x, a_1, a_2, a_3. Let G be the subsemigroup generated by a_1, a_2, a_3, xa_1, and a_3x. Let ϱ be the least congruence relation on F containing (a_2a_1, a_1), (a_1a_2, a_1), (a_2a_3, a_3), (a_3a_2, a_3), (x, xa_2x), (xa_3, xa_2), and (a_1x, a_2x). These seven elements of $F \times F$ will be called the *relators* of ϱ; the first five of them are the *relators* of a smaller congruence relation σ. The example B is the quotient of F by ϱ. We shall show that the image A of G dominates x (strictly speaking, the image of x) but does not include it.

We omit listing the 40 elements of B. Some readers might prefer to rest the proof on the list. The arguments are about equally long.

3.6.1. *If $(v, w) \in \varrho$ and v is a word beginning with x then w begins with x.*

For this, note that since F is free, there is no ambiguity about words beginning with x; every element is a product of generators in a unique way without repetition, and even if repetition is allowed, all representations or none begin with x. Then note that the smallest subalgebra of $F \times F$ containing the seven relators of ϱ, their converses, and the diagonal elements (u, u), has no element (v, w) in which just v begins with x. Hence, neither does its transitive completion ϱ.

3.6.2. *If $(v, w) \in \varrho$ and v ends with x then w ends with x.*

Just like 3.6.1.

Next we want the corresponding result for words beginning xa_1a_3. First pass to the quotient F' of F by σ. Every word of F', except a_2, is equal to a word which does not contain a_2 except possibly in initial a_2x and in final xa_2. For σ allows us to delete any a_2 which adjoins an a_i or is surrounded by x's. On the other hand, if H is the subsemigroup of F generated (freely) by x, a_1, and a_3, $\sigma \cap (H \times H)$ is the diagonal of $H \times H$; in fact, the subalgebra generated by the relators of σ, their converses, and the diagonal of $F \times F$ has no element (h, i), where $h \in H$ and $i \neq h$, except as i may differ from h by substitution of xa_2x, nor does the transitive completion σ. This means we know F' completely;

[3] We have not proved it in the commutative case. The proof will appear in the joint paper with J. M. Howie already cited.

its elements are exactly the *reduced words*, which are repetition-free non-empty words in x, a_1, and a_3, with the possible prefix or suffix a_2 adjoining an x, and with one exceptional reduced word a_2. Accordingly some details of arguments in F' will be omitted below.

Let ϱ' be the image of ϱ in $F' \times F'$, so that B is the quotient of F' by ϱ'.

3.6.3. *If* $(v, w) \in \varrho'$ *and* v *begins with* $x a_1 a_3$, *then* w *is equal to a reduced word beginning with* $x u a_1 a_3$, *where the word* u *may be empty.*

Proof. It suffices, as before, to prove that the subsemigroup R of $F' \times F'$ generated by the diagonal, (xa_3, xa_2), (xa_2, xa_3), $(a_1 x, a_2 x)$, and $(a_2 x, a_1 x)$, has no element (v, w) for which v begins $x u a_1 a_3$ and w does not. We have $v = f_1 \ldots f_n$, $w = g_1 \ldots g_n$, with each $g_i = f_i$ or (f_i, g_i) one of the four special elements of R (relators and converses). Where $g_i = f_i$ we may assume f_i is x, a_1, a_2, or a_3. From $v = x u a_1 a_3 z = f_1 \ldots f_n$ in F', considering the eight possible values for f_i, we conclude that for some j, $f_1 \ldots f_j = x u a_1 a_3$. Moreover, f_1 and (by 3.6.1) g_1 begin with x. f_j ends with a_3, and since $f_1 \ldots f_j = x u a_1 a_3$, f_j is not $x a_3$. Hence $f_j = g_j = a_3$. We may assume $f_{j-1} \neq a_3$. Then f_{j-1} ends with a_1, which means $f_{j-1} = a_1 = g_{j-1}$, and 3.6.3 is proved.

It follows now that no word w in G is ϱ-equivalent to x. For w must begin with x, and therefore with $x a_1$. Also w must end with x. Therefore a_3 occurs in w, and w has an initial factor $x a_1 r a_3$, where r is a word in a_1, a_2, and $x a_1$. Hence $(x a_1 r, x a_1) \in \varrho$. Thus x is equivalent to a word v beginning with $x a_1 a_3$, contradicting 3.6.3. Restating the conclusion, x is not in A.

However, the dominion of A in B contains $x = (x a_1) x = x (a_3 x)$, zigzagging by $(x a_1) = x a_1$, $x a_2 = x a_3$, $(a_3 x) = a_3 x$, $a_1 x = a_2 x$. Since A and x generate B, the proof is complete.

3.7. *There is an infinite semigroup which has an epimorphically embedded finite idempotent subsemigroup.*

Details of 3.7 will be omitted. One modifies 3.6 by requiring only the 36 elements of A to be idempotent. All the steps of the proof hold a fortiori; one need only verify that all powers of x are distinct.

3.7 shows in particular that identities valid in a subsemigroup need not (like commutativity, 2.5) carry over to the dominion. (That is, the proof of 3.7 shows that; I do not know whether an infinite semigroup dominated by a finite subsemigroup can be idempotent.) There is a simpler example for this point about identities, in 3.1; the subsemigroup satisfies $xyx^2yx = xyx$, but B does not. Next we note that 2.7 does not hold for all semi groups.

3.8. *The dominion of a subsemigroup can be a free extension of it.*

Proof. Let B be a free semigroup on three generators, x,y,z. Let A be the subsemigroup generated by xz,zy, and z. It dominates xzy, obviously. Thus the dominion of A contains the subsemigroup E consisting of all words which contain no x not followed by a z and no y not preceded by a z. Obviously E is freely generated by xz, zy, z, and xzy. Finally, to see that E is an equalizer, consider a semigroup C generated by x_1, x_2, y_1, y_2, and z subject to $x_1 z = x_2 z$, $z y_1 = z y_2$. Two obvious embeddings f_1, f_2 of B in C agree exactly on E.

Remarks. One easily checks that homomorphic images of the example in 3.8 yield examples in which the dominion of A is A with a unit or zero adjoined. This justifies the statement in the introduction that the dominating relations do not determine the algebraic nature of d. The associated statement that two systems of dominating relations may not be simultaneously usable can be illustrated by constructing a semigroup with two elements c, d, such that one can adjoin an inverse of c or an inverse of d but not both, e.g. because cd already has a left inverse.

Call a semigroup *(left) division-ordered* if for every c and d such that $c \neq d$ and whenever $ad = bd$, then $ac = bc$, and whenever $ad = d$, then $ac = c$, c is a multiple dx.[4]

3.9. *Every absolutely closed commutative semigroup is division-ordered.*

Proof. Given a commutative semigroup A with elements c,d, such that $ad = bd$ ($ad = d$) implies $ac = bc$ ($ac = c$), but c is not a multiple of d. Let A^1 be A with a unit adjoined. Form the direct product of A^1 and $(Z^+)^1$; its elements may be written as $x^n a$, $n \geqq 0$, $a \in A^1$. We define a *reduced word* in the product as a word $x^n a$ for which $n = 0$ or a is not a multiple of d. We define a congruence relation, beginning by defining the *first reduction* of a non-reduced word $x^n du$. Given $a = du$, u may not be uniquely determined, but cu is (that is, $dv = a = du$ implies $cv = cu$); the first reduction of $x^n du$ $(n > 0)$ is $x^{n-1} cu$. Iteration evidently must lead to a unique reduced word, the complete reduction of $x^n a$. Two words are equivalent if they have the same complete reduction. Since the complete reduction of a product $x^m a \cdot x^n b$ can be computed by first reducing $x^m a$ and $x^n b$, this is a congruence relation. The quotient is a commutative semigroup B whose elements may be identified with the reduced words; A^1 is isomorphic with the subsemigroup of words $x^0 a$. Now $p = x^1 c$ is reduced, $p \notin A$, but A dominates $p = x^1 d x^1$.

3.10. *Not all division-ordered commutative semigroups are absolutely closed.*

---

[4] K. H. Hofmann suggested the term. The clause about $ad = d$ is essential; consider a multiplicative semigroup $\{0,1\}$.

Proof. Generate a semigroup B by elements $g_1, \ldots, g_8$ as follows. Multiplication is commutative and all squares have the same value 0 — which is a zero in B. Next we note an element d which has five factorizations, $d = g_3 g_1 g_2 g_7 = g_5 g_1 g_2 g_7 = g_5 g_1 g_4 g_8 = g_6 g_3 g_4 g_8 = g_6 g_3 g_4 g_2$. Four other relations can be read off from these, by canceling respectively g_7, g_5, g_8, g_6. Note that these relations make a zigzag, so that d will be dominated by the subsemigroup A generated by g_1, g_2, g_3, g_4.

Another element e of B has three factorizations, $e = g_1 g_2 g_3 g_4 = (g_1 g_2 g_5) g_4 = g_1 (g_3 g_4 g_8)$. In all, B has 55 elements, which are 0, the 8 generators, the 16 values of the 20 products of three different generators occurring in the displayed factorizations of d and e, and finally d and e. Every product of generators is 0 except for the indicated 28 pairs, 20 triples, and 8 quadruples. Every product xy is computed by factoring x and y into generators. Non-unique factorization does not occur unless x or y has three factors and certainly does not matter if they have more than four factors between them (for every product of 5 factors is 0). There are 4 x's having nonunique three-factor factorizations; for each there is exactly one y such that $xy = d$, and for two of them there is z such that $xz = e$, but multiplication is always uniquely defined. From the form of the definition, it is commutative and associative. d is not in the 16-element "Grassmann" subsemigroup A, and A is obviously division-ordered.

Last remark. 3.10 is almost as bad as 3.8. The dominion of A is the 17-element semigroup $A \cup \{d\}$, and d is an annihilator.

References

[1] Freyd, P.: Abelian categories. New York: Harper and Row 1964.
[2] Howie, J. M.: Embedding theorems with amalgamation for semigroups. Proc. London Math. Soc. 3, 12, 511—534 (1962).
[3] Ljapin, E. S.: Semigroups. Amer. Math. Soc. Translations, Providence 1963.

Department of Mathematics
Tulane University of Louisiana
New Orleans, Louisiana

Categories of Mapping Filters*, **

By

Erwin Engeler

The subject of this paper is a topic in that branch of universal algebra called the theory of models. The main trait that distinguishes model theory from other approaches to algebra is the fact that in model theory the language in which theorems and definitions are to be coined is explicitly, indeed formally, specified. This gives, of course, a peculiar slant to the type of problems that are of immediate interest to model theorists. The generalities in which we are interested concern the exact extent of definability of mathematical notions and the characterizability of types of mathematical structures in various formal languages, the existence of structures with particular properties in formally characterizable classes of structures, formal descriptions of types of properties preserved under various mathematical constructions, and the like. In a nutshell, the difference between an algebraist and a model theorist is the following: To an algebraist two mathematical structures $\mathscr{A}$, $\mathscr{B}$ are "essentially the same" if they are isomorphic,

$$\mathscr{A} \cong \mathscr{B},$$

while to a model theorist they are essentially the same in case they are elementarily equivalent,

$$\mathscr{A} \equiv \mathscr{B},$$

(if $\mathscr{A}$ and $\mathscr{B}$ have the same first-order properties).

While the morphisms are well-known in the algebraic case (and are at the basis of a categorical treatment of algebra) a satisfactory notion of morphism in the model theoretic case has been missing.

1. Filtration of Categories

Recall that a filter on a set P is a family F of subsets of P such that, (a) if S, $T \in F$ then $S \cap T \in F$, (b) if $S \in F$ and $S \subseteq T \subseteq P$ then $T \in F$. A filter is *proper* if $\emptyset \notin F$. An *ultrafilter* is a proper filter which for every

* With partial support from NSF Grant GP 1612.

** Received September 16, 1965.

subset S of P contains either S or its complement S relative to P. If F_0 is a family of subsets of P in which no finite collection of elements has an empty intersection then the filter *generated* by F_0, denoted by $[F_0]$, consists of all sets $T \subseteq P$ such that T contains an intersection of finitely many elements of F_0. A filter F is called *principal* if there exists $p \in P$ such that $F = [\{p\}]$, (often abbreviated as $[p]$).

Let C be a category; we define the *filtration* C' of the *category* C to be the following family of filters: For every pair of objects $A, B \in C$ consider $C(A,B)$, the family of all maps (in C) from A to B. Let C' be the family of all proper filters F on $C(A,B)$ (for all A, B). First we define how to compose filters. Suppose F is a filter on $C(A,B)$, G a filter on $C(B,C)$, then the following set $G \cdot F$ will turn out to be a filter on $C(A,C)$:

$$S \in G \cdot F \text{ iff } \{q \in C(B,C) : \{p \in C(A,B) : qp \in S\} \in F\} \in G.$$

Indeed: (a) $\emptyset \notin G \cdot F$ if both G, F are proper; (b) If $S_1, S_2 \in G \cdot F$ then $S_1 \cap S_2 \in G \cdot F$. Namely: from $\{q : \{p : qp \in S_1\} \in F\} \in G$ and

$$\{q : \{p : qp \in S_2\} \in F\} \in G \quad \text{it follows that}$$

$$\{q : \{p : qp \in S_1\} \in F\} \cap \{q : \{p : qp \in S_2\} \in F\} \in G,$$

$$\{q : \{p : qp \in S_1\} \text{ and } \{p : qp \in S_2\} \in F\} \in G,$$

$$\{q : \{p : q\, p \in S_1\} \text{ and } qp \in S_2\} \in F\} \in G, \quad \text{and thus}$$

$$\{q : \{p : q\, p \in S_1 \cap S_2\} \in F\} \in G, \quad \text{i.e.} \quad S_1 \cap S_2 \in G \cdot F;$$

(c) If $S \in G \cdot F$ and $S \subseteq T \subseteq C(A,C)$ then $T \in G \cdot F$. Namely: $\{q : \{p : qp \in S\} \in F\} \subseteq \{p : qp \in T\} \in F\}$.

Since the former set belongs to G so does the latter, hence $T \in G \cdot F$. Thus, *if F, G are proper filters then so is $G \cdot F$*; moreover, as is easy to verify, if both F,G are ultrafilters then so is $G \cdot F$. This leads us to define the *ultrafiltration* C^u of a category C as the family of all ultrafilters on sets $C(A,B)$.

Observe next that $H \cdot (G \cdot F) = (H \cdot G) \cdot F$ for all $F \in C(A,B)$, $G \in C(B,C)$, $H \in C(C,D)$. Namely: $S \in H \cdot (G \cdot F)$ iff

$$\{r : \{s : rs \in S\} \in G \cdot F\} \in H, \quad \text{iff} \quad \{r : \{q : \{p : qp \in \{s : rs \in S\}\} \in F\} \in G\} \in H,$$

iff $$\{r : \{q : \{p : rqp \in S\} \in F\} \in G\} \in H,$$

similarly, $S \in (H \cdot G) \cdot F$ can be found equivalent to this condition.

Finally consider the family of principal filters in C', let it be denoted by C^*. C^* is a subcategory of C', in fact, if $F = [p_0]$, $G = [q_0]$ then $G \cdot F = [q_0 p_0]$. In particular, if e_A, e_B are the identities in C, on A, B respectively, then $E_A = [e_A]$, $E_B = [e_B]$ are the identities in C' on A, B. Indeed, if $F \in C'(A,B)$, then $E_B \cdot F = F \cdot E_A = F$ as can be easily verified. We have thus proved the following proposition: *For every category* $\boldsymbol{C}$ *its filtration* $\boldsymbol{C}'$ *and ultrafiltration* $\boldsymbol{C}^u$ *are categories with the same objects.*

Let $\boldsymbol{C}$ be a category, A, B, C objects in $\boldsymbol{C}$ and suppose that $\boldsymbol{C}(A,B)$, $\boldsymbol{C}(B,A)$, $\boldsymbol{C}(A,C)$, $\boldsymbol{C}(B,C)$ are all nonempty.

Triangle Lemma. *For any proper filters* U *on* $\boldsymbol{C}(A,B)$, V *on* $\boldsymbol{C}(B,A)$ *there exist ultrafilters* $F \in \boldsymbol{C}^u(A,C)$, $G \in \boldsymbol{C}^u(B,C)$ *such that* $F \supseteq G \cdot U$ *and* $G \supseteq F \cdot V$; moreover, given $F_0 \in \boldsymbol{C}'(A,C)$, $G_0 \in \boldsymbol{C}'(B,C)$ such that $F_0 \supseteq G_0 \cdot U, G_0 \supseteq F_0 \cdot V$ then F, G may be found as refinements of F_0, G_0.

To prove the first part of the lemma, let $\{R_\nu\}$, $(\nu < \alpha)$, be a (transfinite) enumeration of the set of all subsets of $\boldsymbol{C}(A,C)$ possibly with repetitions; similarly, let $\{S_\nu\}$, $(\nu < \alpha)$, be an enumeration of the set of all subsets of $\boldsymbol{C}(B,C)$. By transfinite recursion define F_ν, G_ν, $(\nu < \alpha)$, simultaneously as follows:

$F_0 = \{\boldsymbol{C}(A,C)\}$, $G_0 = \{\boldsymbol{C}(B,C)\}$. For $\nu > 0$ we let $F_\nu = \bigcup_{\mu<\nu} H_\mu$, $G_\nu = \bigcup_{\mu<\nu} K_\mu$, where

$$H_\mu = \begin{cases} [G_\mu \cdot U \cup \{R_\mu\}], & \text{if this filter is proper,} \\ G_\mu \cdot U & \text{otherwise;} \end{cases}$$

$$K_\mu = \begin{cases} [F_\mu \cdot V \cup \{S_\mu\}], & \text{if this filter is proper,} \\ F_\mu \cdot V & \text{otherwise.} \end{cases}$$

Induction on ν shows that each F_ν, G_ν is proper. Hence so are $F = \bigcup_{\nu<\alpha} F_\nu$, $G = \bigcup_{\nu<\alpha} G_\nu$. F and G are maximal filters by construction, hence F and G are ultrafilters. To prove $G \cdot U \subseteq F$ and $F \cdot V \subseteq G$ we need the following: whenever G_ν, $(\nu < \alpha)$, is an increasing chain of proper filters and U is a proper filter then $\left(\bigcup_{\nu<\alpha} G_\nu\right) \cdot U = \bigcup_{\nu<\alpha} (G_\nu \cdot U)$. Namely: $R \in \left(\bigcup_{\nu<\alpha} G_\nu\right) \cdot U$ iff $\{q : \{p : q p\} \in R\} \in U\} \in \bigcup_{\nu<\alpha} G_\nu$, iff $R \in G_\nu \cdot U$ for some $\nu < \alpha$, iff $R \in \bigcup_{\nu<\alpha} (G_\nu \cdot U)$.

Thus we may compute as follows:

$$G \cdot U = \left(\bigcup_{\nu<\alpha} G_\nu\right) \cdot U = \bigcup_{\nu<\alpha} (G_\nu \cdot U) = \bigcup_{\nu<\alpha} \left(\bigcup_{\mu<\nu} G_\mu \cdot U\right) \subseteq \bigcup_{\mu<\alpha} F_\nu = F,$$

and similarly for $F \cdot V \subseteq G$.

The proof of the second part of the lemma is an obvious modification of the above proof.

2. Relational Categories, their Filtration, and Model Theory

Subsets of the set A^n of all n-tuples of elements of a non-empty set A are called n-ary *relations* on A, $n = 1, 2, \ldots$. Every function p from a set B into a set A induces a *relational map* $\hat{p}$ from the set of all relations on A into the set of all relations on B as follows: for every $R \subseteq A^n$ and $\langle b_1, \ldots, b_n\rangle \in B^n$ let $\langle b_1, \ldots, b_n\rangle \in \hat{p} R$ iff $\langle p b_1, \ldots, p b_n\rangle \in R$. A family of sets together with a family of induced relational maps between them that form a category in the usual sense will be called a *relational category*.

The filtration $\boldsymbol{C}'$ of a relational category is again a category of relational maps. We have to indicate how a filter F on $\boldsymbol{C}(A, B)$ may be regarded as a relational map of A into B: Let R be an n-ary relation on A, then $F R$ is the n-ary relation on B such that $\langle b_1, \ldots, b_n\rangle \in F R$ iff $\{\hat{p} \in \boldsymbol{C}(A, B) : \langle b_1, \ldots, b_n\rangle \in \hat{p} R\} \in F$. This definition of mapping relations by means of filters agrees with our definition of composition of filters, indeed

$$G(F R) = (G \cdot F) R,$$

as can be easily verified.

Let A, B be any objects in a relational category $\boldsymbol{C}$, let $F \in \boldsymbol{C}'(A, B)$. An $n+1$-ary relation R on A is called a *function* on A if for any $a_1, \ldots, a_n \in A$ there exists a unique $a_{n+1} \in A$ such that $\langle a_1, \ldots, a_n, a_{n+1}\rangle \in R$. Observe that we do not necessarily have that $F(R)$ is a function if R is one. Observe also that the F-image of the identity relation on A is not necessarily the identity relation on B, if it is we shall call F a *normal* filter. Let Φ be a set of functions on A, if F is a normal filter and $F(f)$ is a function for all $f \in \Phi$ then we call F a *Φ-complete* filter. A filter which is Φ-complete for every Φ is called *functionally complete*, (see our paper [*1*] for characterizations and various properties of functionally complete filters). In the present paper we shall be interested in Φ-complete filters for particular Φ's, introduced in the theory of models by Skolem.

A *relational structure* $\mathscr{A} = \langle A, A_i\rangle_{i \in I}$ *of type* t in the sense of model theory consists of a set A and $t(i)$-ary relations A_i on A for all $i \in I$. The cardinality of $\mathscr{A}$ is that of A and the cardinality of the type t is that of I. If $K \supset I$ and s is an extension of t to K and S_k is an $s(k)$-ary relation on A such that $S_i = A_i$ for all $i \in I$ we shall denote by $(\mathscr{A}, S_k)_{k \in K-I}$ the relational structure $\langle A, S_k\rangle_{k \in K}$. $\mathscr{A}$ is said to be he restriction of $(\mathscr{A}, S_k)_{k \in K-I}$ to type t.

We use the applied first-order language associated with the type t to express properties of relational structures of type t, assuming familiarity with the basic notions of first-order logic, in particular with the concept of an assignment p (of elements of A to variables occurring free in φ) to *satisfy* a formula φ. In case $x_1, \ldots, x_n$ are all of the free variables of $\varphi(x_1, \ldots, x_n)$ we signal satisfaction by writing $\mathscr{A} \models \varphi(p(x_1), \ldots, p(x_n))$. Two relational structures $\mathscr{A}, \mathscr{B}$ of the same type are said to be *elementarily equivalent* (in symbols $\mathscr{A} \equiv \mathscr{B}$) if $\mathscr{A} \models \varphi$ exactly when $\mathscr{B} \models \varphi$ for all closed formulas of the language. If e is a function from A to B then e is said to be an *elementary embedding* of $\mathscr{A} = \langle A, A_i\rangle_{i \in I}$ into $\mathscr{B}$

$= \langle B, B_i\rangle_{i\in I}$ iff for every formula $\varphi(x_1, \ldots, x_n)$ (with all free variables displayed) and every $a_1, \ldots, a_n \in A$ we have $\mathscr{A} \models \varphi(a_1, \ldots, a_n)$ iff $\mathscr{B} \models \varphi(e(a_1), \ldots, e(a_n))$.

For every type t let $\boldsymbol{S}(t)$ be the category of all relational structures of type t with induced relational maps $\hat{p}$ between them defined as above. We shall be interested in the filtration $\boldsymbol{S}(t)'$ and ultrafiltration $\boldsymbol{S}(t)^u$ of $\boldsymbol{S}(t)$. Let $\langle a_1, \ldots, a_n\rangle \in R_\varphi^{\mathscr{A}}$ iff $\mathscr{A} \models \varphi(a_1, \ldots, a_n)$.

Theorem. *If $\mathscr{A}$ and $\mathscr{B}$ are elementarily equivalent then there exists a normal filtermap $F \in \boldsymbol{S}'(t)(\mathscr{A}, \mathscr{B})$ such that $F(R_\varphi^{\mathscr{A}}) = R_\varphi^{\mathscr{B}}$ for all formulas* φ (and vice versa).

Namely, consider the filter F on $\boldsymbol{S}(t)(\mathscr{A}, \mathscr{B})$ generated by the sets

$$Q_\varphi(b_1, \ldots, b_n) = \{\hat{p} : \langle b_1, \ldots, b_n\rangle \in \hat{p}(R_\varphi^{\mathscr{A}})\}$$

for all φ and $b_1, \ldots, b_n$ for which $\mathscr{B} \models \varphi(b_1, \ldots, b_n)$, and φ of the form $\varphi_1 \wedge \bigwedge_{i \neq j} x_i \neq x_j$. F has the desired properties as can be easily verified.

In general we call a filtermap *elementary* (of type t) if $F(R_\varphi^{\mathscr{A}}) = R_\varphi^{\mathscr{B}}$ for all formulas φ of type t. The above theorem thus indicates, in effect, that (elementary) filtermaps are the morphisms appropriate for model theory.

Let $\mathscr{A}$ be a relational structure of type t. For every formula of the form

$$(\exists x_{n+1})\, \varphi(x_1, \ldots, x_n, x_{n+1})$$

(with all free variables displayed), we introduce a corresponding n-ary function f_φ on A by selecting for any $a_1, \ldots, a_n \in A$ an element a_{n+1} such that

$$\mathscr{A} \models \varphi(a_1, \ldots, a_n, a_{n+1}) \quad \text{if} \quad \mathscr{A} \models (\exists x)\, \varphi(a_1, \ldots, a_n, x),$$

arbitrary otherwise. Let $\Phi_{\mathscr{A}}$ be the family of all such functions f_φ (called Skolem-functions).

Consider maps $F \in \boldsymbol{S}'(t)(\mathscr{A}, \mathscr{B})$ such that F is $\Phi_{\mathscr{A}}$-complete and $F(A_i) = B_i$ for all $i \in I$. To show that the *filtermaps F preserve all elementary properties*, we prove by induction on the structure of formulas φ: Whenever $R_\varphi^{\mathscr{A}}$, $R_\varphi^{\mathscr{B}}$ are the relations defined by the formula φ in $\mathscr{A}$, $\mathscr{B}$ respectively, then $F(R_\varphi^{\mathscr{A}}) = R_\varphi^{\mathscr{B}}$. For atomic formulas this is true by assumption. If φ is of the form $\varphi_1 \wedge \varphi_2$ we have $\langle b_1, \ldots, b_n\rangle \in F(R_\varphi^{\mathscr{A}})$ iff

$$\{\hat{p} : \langle b_1, \ldots, b_n\rangle \in \hat{p}(R_\varphi^{\mathscr{A}})\} \in F, \quad \text{iff} \quad \{\hat{p} : \langle p(b_1), \ldots, p(b_n)\rangle \in R_\varphi^{\mathscr{A}}\}$$
$$= \{\hat{p} : \mathscr{A} \models \varphi_1(p(b_1), \ldots, p(b_n)) \wedge \varphi_2(p(b_1), \ldots, p(b_n))\}$$
$$= \{\hat{p} : \mathscr{A} \models \varphi_1(p(b_1), \ldots, p(b_n)) \text{ and } \mathscr{A} \models \varphi_2(p(b_1), \ldots, p(b_n))\}$$
$$= \{\hat{p} : \mathscr{A} \models \varphi_1(p(b_1), \ldots, p(b_n))\} \cap \{\hat{p} : \mathscr{A} \models \varphi_2(p(b_1), \ldots, p(b_n))\} \in F,$$

iff both sets in the intersection belong to F, i.e. iff

$$\langle b_1,\dots,b_n\rangle \in F(R^{\mathscr{A}}_{\varphi_1}) = R^{\mathscr{B}}_{\varphi_1} \text{ and } \langle b_1,\dots,b_n\rangle \in F(R^{\mathscr{A}}_{\varphi_2}) = R^{\mathscr{B}}_{\varphi_2}$$

(induction assumption), iff $\langle b_1,\dots,b_n\rangle \in R^{\mathscr{B}}_{\varphi}$. If φ is of the form $\neg\varphi_1$ then $\langle b_1,\dots,b_n\rangle \in F(R^{\mathscr{A}}_{\varphi})$ iff $\{\hat{p}:\mathscr{A}\models\neg\varphi_1(p(b_1),\dots,p(b_n))\}\in F$, iff $\{\hat{p}:\mathscr{A}\models\varphi_1(p(b_1),\dots,p(b_n))\}\notin F$, iff $\langle b_1,\dots,b_n\rangle\notin F(R^{\mathscr{A}}_{\varphi_1}) = R^{\mathscr{B}}_{\varphi_1}$, iff $\langle b_1,\dots,b_n\rangle \in R^{\mathscr{B}}_{\varphi}$. Finally, if φ is of the form $(\exists x)\,\varphi_1(x,x_1,\dots,x_n)$ let $g = F(f_{\varphi_1})$ and let $Q = \{\hat{p}: pg(b_1,\dots,b_n) = f_{\varphi_1}(p(b_1),\dots,p(b_n))\}$. Since F is $\Phi_{\mathscr{A}}$-complete $Q\in F$ and we may argue as follows: $\langle b_1,\dots,b_n\rangle\in F(R^{\mathscr{A}}_{\varphi})$

iff $\{\hat{p}:\mathscr{A}\models(\exists x)\,\varphi_1(x,p(b_1),\dots,p(b_n))\}\in F$,

iff $\{\hat{p}:\mathscr{A}\models\varphi_1(f_{\varphi_1}(p(b_1),\dots,p(b_n)),p(b_1),\dots,p(b_n))\}\in F$,

iff $\{\hat{p}\in Q:\mathscr{A}\models\varphi_1(p(g(b_1,\dots,b_n)),p(b_1),\dots,p(b_n))\}\in F$,

iff $\langle g(b_1,\dots,b_n),b_1,\dots,b_n\rangle\in F(R^{\mathscr{A}}_{\varphi_1}) = R^{\mathscr{B}}_{\varphi_1}$, iff $\langle b_1,\dots,b_n\rangle\in R^{\mathscr{B}}_{\varphi}$.

Consider now two types t_1, t_2 with common restriction t_0 and some common extension t_3. Let $\mathscr{A}$ be an infinite relational structure of type t_1, $\mathscr{B}$ an infinite structure of type t_2, and let $\mathscr{A}_0, \mathscr{B}_0$ denote the restrictions of $\mathscr{A}, \mathscr{B}$ to type t_0. If $\mathscr{C}$ is a structure of type t_3 denote by $\mathscr{C}_1, \mathscr{C}_2$ its restrictions to t_1, t_2. The following theorem is due to ROBINSON [3]: *$\mathscr{A}_0 \equiv \mathscr{B}_0$ iff there exists a relational structure $\mathscr{C}$ of type t_3 such that $\mathscr{C}_1 \equiv \mathscr{A}$ and $\mathscr{C}_2 \equiv \mathscr{B}$*; by choosing appropriate additional relations in $\mathscr{A}, \mathscr{B}$ we may moreover obtain $\mathscr{C}_1, \mathscr{C}_2$ as elementary extensions of $\mathscr{A}, \mathscr{B}$.

Our proof of this result is immediate from the Triangle Lemma: Let S be a set of distinct symbols, let there be given an n-ary functional symbol f_i for each formula $(\exists x_{n+1})\,\varphi_i(x_1,\dots,x_{n+1})$ of the first-order language of type t_3, extended to contain the individual symbols from S and the functional symbols f_i. Let P be the set of all relational maps $\hat{p}$ for functions p from S to A. Extend P to T in the obvious fashion by associating to the functional symbol f_i the function f_{φ_i} on A in case φ_i is of type t_1, otherwise chose an arbitrary function on A (with the same number of variables). Let Q be found analogously (for $\mathscr{B}$ instead of $\mathscr{A}$). By our previous result there exist filtermaps $U\in \boldsymbol{S}(t_0)'(\mathscr{A}_0,\mathscr{B}_0)$ and $V\in\boldsymbol{S}(t_0)'(\mathscr{B}_0,\mathscr{A}_0)$ which preserve the relations $R^{\mathscr{A}}_{\varphi}$, $R^{\mathscr{B}}_{\varphi}$ for all of φ type t_0. U and V may be assumed ultrafilters. Now we apply the Triangle Lemma to find ultrafilters F_0, G_0 on P, Q respectively such that

$$F_0\cdot V\subseteq G_0,\quad G_0\cdot U\subseteq F_0.$$

F_0, G_0 fail to be $\Phi_{\mathscr{A}}, \Phi_{\mathscr{B}}$-complete in general since they may not be normal. Normality is achieved as follows: It is straightforward to verify that for any relations $R^{\mathscr{A}}, R^{\mathscr{B}}$ for which $U(R^{\mathscr{A}}) = R^{\mathscr{B}}$ and $V(R^{\mathscr{B}}) = R^{\mathscr{A}}$ we have $F_0(R^{\mathscr{A}}) = G_0(R^{\mathscr{B}})$. In particular, since U, V are normal, the images under F_0, G_0 of the identity relations on A, B respectively are the same, say $\approx$; $\approx$ is an equivalence relation. Let C be the set of all equivalence-classes of T modulo $\approx$. For every function p from T into A we now select a function p^* from C into A such that $p^*(c) = p(\tau)$ for some $\tau \in c$.

Let F consist of those sets M for which $\{\hat{p} : \hat{p^*} \in M\} \in F_0$. By construction F is a $\Phi_{\mathscr{A}}$-complete ultrafilter. Let G be defined analoguously. Observe that we still have $F \cdot V \subseteq G$, $G \cdot U \subseteq F$. Hence, if we define the relations on C according to type t_1 by the filtermap F and the relations according to type t_2 by G we know that these definitions agree on the relations of the common type t_0. Let the resulting relational structure of type t_3 be denoted by $\mathscr{C}$. Clearly $\mathscr{C}_1 \equiv \mathscr{A}$, $\mathscr{C}_2 \equiv \mathscr{B}$ since F, G are elementary filtermaps (of types t_1, t_2 respectively).

Notice that F, G do not necessarily belong to $\boldsymbol{S}(t_1)^u$, $\boldsymbol{S}(t_2)^u$; this may be remedied, if we wish, by using the second part of the triangle lemma.

In the above proof we used the artifice of Skolem-functions to assure elementary equivalence by the way of assuring $\Phi_{\mathscr{A}}$ and $\Phi_{\mathscr{B}}$-completeness of the mapping filters. The *direct power* construction (see [2]) is another such way, it has the advantage of leading to functional completeness, (to every function on a set A there corresponds a natural image-function in the direct power A^I of A).

References

[1] Engeler, E.: Structures defined by mapping filters. (To appear.)

[2] Frayne, T., A. C. Morel, and D. S. Scott: Reduced direct products. Fundamenta Mathematica **51**, 195—228 (1962); **53**, 117 (1963).

[3] Robinson, A.: A result on consistency and its applications to the theory of definition. Nederl. Akad. Wetensch. Proc. Ser. A, **59**, 47—58 (1956).

Department of Mathematics
University of Minnesota
Minneapolis, Minnesota

Correspondences and Exact Squares *

By

PETER HILTON

1. Introduction

The algebra of correspondences (= additive relations = homomorphic relations) in an abelian category was developed systematically[1] by MACLANE [*8*]. Later PUPPE [*11*] gave a set of necessary and sufficient conditions that a category $\tilde{\mathscr{A}}$ be the category of correspondences based on an abelian category $\mathscr{A}$. PUPPE also showed how certain standard notions of homological algebra (the connecting homomorphism, the differentials of a spectral sequence) could be obtained by considering relations instead of just morphisms in the category.

The basic point of view in the work of MACLANE and PUPPE and other authors [*2, 5, 6*] was that a relation Γ from the object A to the object B of the abelian category $\mathscr{A}$ was a subobject of $A \oplus B$. This generalized the classical notion of an additive relation in an abelian group; and the classical composition of relations could also be generalized to yield a composition of relations in $\mathscr{A}$ and hence a category $\tilde{\mathscr{A}}$. This composition was a little awkward to handle however.

Our point of view has been to provide a definition of the category $\tilde{\mathscr{A}}$ which would exploit to the full the standard notions of abelian category theory and which would render diagram-chasing in $\tilde{\mathscr{A}}$ as fully automatic as diagram-chasing in $\mathscr{A}$. Our starting point was the observation that a square

(1.1)
$$\begin{array}{ccc} R & \xrightarrow{\alpha} & A \\ {\scriptstyle\beta}\downarrow & & \downarrow{\scriptstyle\varphi} \\ B & \xrightarrow[\psi]{} & X \end{array}$$

which is commutative in $\mathscr{A}$ should be commutative in $\tilde{\mathscr{A}}$ (i.e., commutative when proceeding from *any* vertex to the diagonally opposite vertex) if and only if it is exact. This notion of an exact square is the

* Received May 10, 1965.

[1] It seems to have been first studied by LAMBEK; see [*3*]. [Added later: LAMBEK informs me that priority should be accorded to K. SHODA, Osaka Math. J. 1, 182—225 (1949).]

common generalization of cartesian square (= pull-back) and cocartesian square (= push-out); it is basic to any Mayer-Vietoris sequence and has appeared frequently in many contexts [9, 10]. Now if α, β in (1.1) are the components of a *monic* $R \rightarrowtail^{\{\alpha,\beta\}} A \oplus B$, then we may think of the pair α, β as constituting a relation in the classical sense. Moreover we may complete α, β to an exact square (1.1) and any two completions (φ_1, ψ_1), (φ_2, ψ_2) have the property that there are monics μ_1, μ_2 such that

$$\mu_1 \varphi_1 = \mu_2 \varphi_2, \quad \mu_1 \psi_1 = \mu_2 \psi_2. \tag{1.2}$$

Conversely, given any member (φ, ψ) of the equivalence class described by (1.2), we may construct a pull-back (1.1) and the pair (α, β) depends only on the equivalence class. We are thus led to *define* a correspondence from A to B as an equivalence class of morphism-pairs

$$A \xrightarrow{\varphi} X \xleftarrow{\psi} B \tag{1.3}$$

and to develop the algebra of correspondences from this standpoint. We remark that there is a strong analogy here with the classical procedure for passing from integers to fractions; the noncommutativity of composition leads to notions of left-fractions and right-fractions and we may regard the equivalence class of (1.3) as a right-fraction of morphisms of $\mathscr{A}$. Then we identify the right-fraction φ/ψ with the left-fraction $\alpha\backslash\beta$ precisely when (1.1) is exact. We then find that composition in $\tilde{\mathscr{A}}$ resembles multiplication of fractions and an addition may be defined in $M_{\tilde{\mathscr{A}}}(A, B)$ resembling the classical addition of fractions; moreover we make heavy use of the presence, within every equivalence class, of a minimal representative whose role is much like that of the expression of a fraction in its lowest terms[2].

The plan of the paper is as follows. After devoting a brief section to stating a few fundamental propositions of abelian categories, we enumerate the properties of exact squares in section 3. In section 4 we define the category $\tilde{\mathscr{A}}$ and obtain its main algebraic properties. We prove the theorem (see [8, 11]) that $\tilde{\mathscr{A}}$ is regular in the sense of von Neumann, that is, to each morphism Γ in $\tilde{\mathscr{A}}$ there is a morphism $\bar{\Gamma}$ such that $\Gamma\bar{\Gamma}\Gamma = \Gamma$. Indeed, $\Gamma \to \bar{\Gamma}$ is an anti-involution on the category $\tilde{\mathscr{A}}$ and is simply defined by $\overline{\varphi/\psi} = \psi/\varphi$. We characterize the monics and epics in $\tilde{\mathscr{A}}$ and show that every morphism Γ in $\tilde{\mathscr{A}}$ has an essentially unique decomposition as $\Gamma = ME$, with M monic, E epic in $\tilde{\mathscr{A}}$. We also show how — again, essentially uniquely — every morphism in $\tilde{\mathscr{A}}$ from A to B may be regarded as a morphism in $\mathscr{A}$ from a subobject of A to a quotient object of B; and, of course, conversely. In section 5, we define an ad-

[2] Indeed the process for obtaining it has already been described in the literature as that of finding the greatest common left divisor of φ and ψ.

dition in each $M_{\tilde{\mathscr{A}}}(A, B)$ extending the addition in $M_{\mathscr{A}}(A, B)$; we show that, with this addition, $M_{\tilde{\mathscr{A}}}(A, B)$ is an (additive) commutative semi-group with zero; and we examine the distributive laws. These turn out not to be universally valid: the relation $(\Gamma_1 + \Gamma_2)\Delta = \Gamma_1\Delta + \Gamma_2\Delta$ holds for all Γ_1, Γ_2 if Δ is a morphism in $\mathscr{A}$ or if Δ is epic in $\tilde{\mathscr{A}}$, but there are, in general, Δ for which it fails.

We do not give applications of the algebra in this paper; we hope to devote a subsequent paper to a further study of the properties of $\tilde{\mathscr{A}}$ and to give applications then. We adopt the terms monic, epic, for mono-morphism, epimorphism and the notations $\rightarrowtail$, $\twoheadrightarrow$ of [9]. We often refer to an isomorphism as a *unit* in view of our interest in the categorical algebra. We also follow [1] in writing $\{\alpha, \beta\}$ for the morphism into a product (= direct sum) with components α, β and $\langle\varphi, \psi\rangle$ for the mor-phism out of a coproduct (= direct sum) with components φ, ψ.

We would like to acknowledge several sources for the ideas described in this paper. First, we were much impressed by LAMBEK's very beau-tiful 2-square version of GOURSAT's Theorem [4] and the scope for "square-chasing" it subsequently revealed. Second, we wished to describe precisely what is meant by saying that the nth differential of the spectral sequence associated with the exact couple

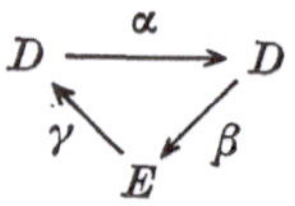

is induced by $\beta\alpha^{-n}\gamma$, and to provide a simple and familiar algorithm for proving it. Third, we were influenced — perhaps subconsciously — by the work of BEBERMAN's group at the University of Illinois in teaching fractions to children under the guise of "stretchers and shrinkers" [12].

2. Abelian Categories

Let $\mathscr{A}$ be an abelian category and let

(2.1) $$\cdot \xrightarrow{\beta} \cdot \xrightarrow{\alpha} \cdot$$

α, β be morphisms in $\mathscr{A}$ such that $\alpha\beta = 0$. We write

$\alpha|\beta$ if (2.1) is exact,
$\alpha[\beta$ if β is the kernel of α,
$\alpha]\beta$ if α is the cokernel of β,
$\alpha \,[\,]\, \beta$ if α, β are mutual annihilators.

Proposition 2.2. (i) *If μ is monic, then $\alpha|\beta \Leftrightarrow \mu\alpha|\beta$.*
(ii) *If $\alpha]\beta$ and $\xi\alpha|\beta$, then ξ is monic.*
(iii) *If $\alpha|\mu\beta'$ and μ is monic, then $\alpha\mu|\beta'$.*

Corollary 2.3. *Consider the diagram*

$$\begin{array}{ccccc} \cdot & \xrightarrow{\beta} & \cdot & \xrightarrow{\alpha} & \cdot \\ {\scriptstyle \varkappa}\uparrow & & \uparrow{\scriptstyle \mu} & & \uparrow{\scriptstyle \nu} \\ \cdot & \xrightarrow[\beta']{} & \cdot & \xrightarrow[\alpha']{} & \cdot \end{array}$$

where $\alpha \mid \beta$. *Then* (i) *if* ν *is monic*, $\alpha' \mid \beta'$, (ii) *if* $\alpha'] \beta'$, ν *is monic.*

Proof. By Proposition 2.2, $\alpha \mid \beta \Rightarrow \nu\alpha' \mid \beta'$.

3. Exact Squares

Let $\mathscr{A}$ be an abelian category and let

(3.1)
$$\begin{array}{ccc} R & \xrightarrow{\alpha} & A \\ {\scriptstyle \beta}\downarrow & & \downarrow{\scriptstyle \varphi} \\ B & \xrightarrow{\psi} & X \end{array}$$

$(\alpha, \beta; \varphi, \psi)$ be a commutative square in $\mathscr{A}$. We may associate with (3.1) the differential sequence

(3.2)
$$R \xrightarrow{\{\alpha, \beta\}} A \oplus B \xrightarrow{\langle \varphi, -\psi \rangle} X .$$

We say that

(3.1) is *exact*, and write $\operatorname{ex}(\alpha, \beta; \varphi, \psi)$, if (3.2) is exact;
(3.1) is *cartesian*, and write $\operatorname{car}(\alpha, \beta; \varphi, \psi)$, if $\langle \varphi, -\psi \rangle \,[\{\alpha, \beta\}$;
(3.1) is *cocartesian*, and write $\operatorname{coc}(\alpha, \beta; \varphi, \psi)$, if $\langle \varphi, -\psi \rangle]\, \{\alpha, \beta\}$;
(3.1) is *bicartesian*, and write $\operatorname{bic}(\alpha, \beta; \varphi, \psi)$, if $\langle \varphi, -\psi \rangle \,[\,]\, \{\alpha, \beta\}$.

Remarks. (i) The notions *cartesian, cocartesian* may evidently be characterized by universal mapping properties and are by no means confined to abelian categories.

(ii) The notions *exact, cartesian, cocartesian* were called *smooth, upper complete, lower complete* in [7], where most of the properties listed in this section also appeared. The notions *cartesian, cocartesian, bicartesian* were called *pull-back, push-out, push-me-pull-you* in [2].

(iii) Notice that if $\operatorname{ex}(\alpha, \beta; \varphi, \psi)$ and one of α, β is monic then $\operatorname{car}(\alpha, \beta; \varphi, \psi)$.

Given

$$\begin{array}{ccc} & & A \\ & & \downarrow{\scriptstyle \varphi} \\ B & \xrightarrow[\psi]{} & X \end{array}$$

we may construct α_0, β_0 such that $\operatorname{car}(\alpha_0, \beta_0; \varphi, \psi)$. Then plainly

$$\operatorname{ex}(\alpha, \beta; \varphi, \psi) \Leftrightarrow \exists \text{ epic } \varrho \text{ with } \alpha_0 \varrho = \alpha, \ \ \beta_0 \varrho = \beta .$$

We may then construct φ_0, ψ_0 such that $\mathrm{coc}(\alpha_0, \beta_0; \varphi_0, \psi_0)$, and there is a monic σ with $\sigma\varphi_0 = \varphi$, $\sigma\psi_0 = \psi$. Moreover

$$\mathrm{bic}(\alpha_0, \beta_0; \varphi_0, \psi_0).$$

Thus we have

Theorem 3.3. *Let* (3.1) *be a commutative square. Then*

$$\mathrm{ex}(\alpha, \beta; \varphi, \psi) \Leftrightarrow \exists \text{ epic } \varrho \text{ and monic } \sigma, \text{ and}$$
$$\alpha = \alpha_0 \varrho, \quad \beta = \beta_0 \varrho, \quad \varphi = \sigma\varphi_0, \quad \psi = \sigma\psi_0,$$

and $\mathrm{bic}(\alpha_0, \beta_0; \varphi_0, \psi_0)$.
Moreover $\alpha_0, \beta_0, \varphi_0, \psi_0$ *are determined (up to equivalence) by* $\alpha, \beta, \varphi, \psi$.

(3.4) 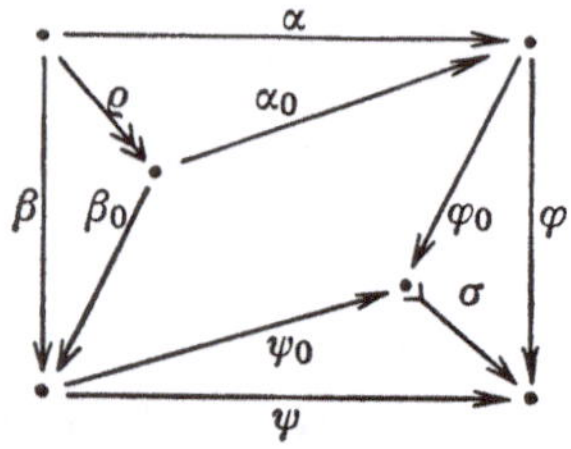

Theorem 3.5. *Suppose* (3.1) *is exact. Then the induced map* $\ker \beta \to \ker \varphi$ *is epic and the induced map* $\operatorname{coker} \beta \to \operatorname{coker} \varphi$ *is monic. If, moreover,* $\mathrm{car}(\alpha, \beta; \varphi, \psi)$, *then* $\ker \beta \rightarrowtail\!\!\twoheadrightarrow \ker \varphi$.

Theorem 3.5 *shows that, if* $\mathrm{ex}(\alpha, \beta; \varphi, \psi)$, *then*

$$\beta \text{ monic} \Rightarrow \varphi \text{ monic}$$
$$\varphi \text{ epic} \Rightarrow \beta \text{ epic},$$

and, if $\mathrm{car}(\alpha, \beta; \varphi, \psi)$, *then*

$$\beta \text{ monic} \Leftrightarrow \varphi \text{ monic}.$$

We may strengthen the last assertion as follows.

Theorem 3.6. (i) *Suppose* $\mathrm{car}(\alpha, \beta; \varphi, \psi)$ *with* φ *monic. Then* $\theta\psi [\beta$ *for any* θ *such that* $\theta [\varphi$.

(ii) *Conversely, given* φ, ψ *with* φ *monic, and* θ *such that* $\theta [\varphi$, *construct* β *such that* $\theta\psi [\beta$. *Then* $\exists | \alpha$ *such that* $\mathrm{car}(\alpha, \beta; \varphi, \psi)$.

We now prove two theorems about putting together exact squares and taking them apart.

Theorem 3.7. *Suppose given*

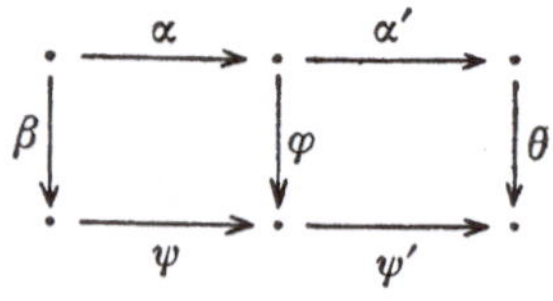

where each square is exact (cartesian, cocartesian, bicartesian). Then so is the composite square $(\alpha'\alpha, \beta; \theta, \psi'\psi)$.

Proof. If each square is cartesian, so is the composite — this is an elementary consequence of the universal mapping property. The case when the squares are cocartesian follows by duality and hence also the case when the squares are bicartesian.

Now suppose each square exact. We prove a lemma.

Lemma 3.8. *Consider the commutative diagram*

$$
\begin{array}{ccc}
\bullet & \xrightarrow{\ \alpha\ } & \bullet \\
\beta \downarrow & & \downarrow \varphi \\
\bullet & \xrightarrow{\ \psi\ } & \bullet
\end{array}
\quad
\begin{array}{c}
\searrow^{\varphi_1} \\
\bullet \\
\nearrow_{\varphi_0}
\end{array}
$$

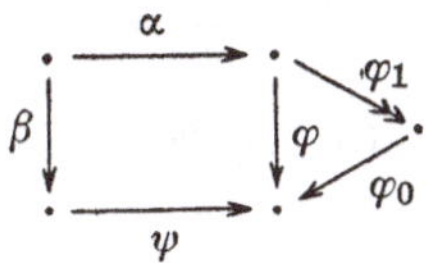

Then $\mathrm{ex}\,(\varphi_1\alpha,\,\beta;\,\varphi_0,\,\psi)$ *if* $\mathrm{ex}\,(\alpha,\,\beta;\,\varphi,\,\psi)$.

Proof. We have the commutative diagram

$$
\begin{array}{ccccc}
\bullet & \xrightarrow{\{\alpha,\,\beta\}} & \bullet & \xrightarrow{\langle\varphi,\,-\psi\rangle} & \bullet \\
\Vert & & \downarrow{\varphi_1\oplus 1} & & \Vert \\
\bullet & \xrightarrow{\{\varphi_1\alpha,\,\beta\}} & \bullet & \xrightarrow{\langle\varphi_0,\,-\psi\rangle} & \bullet
\end{array}
$$

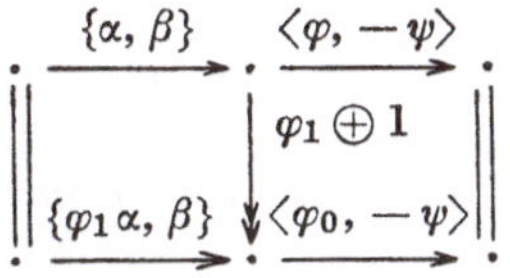

where the top row is exact. We apply Corollary 2.3 (i) d.

Returning to the theorem we construct the diagram

(3.9)

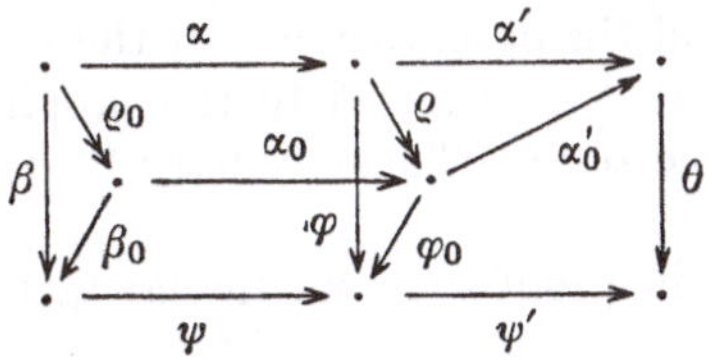

as follows. First we construct α_0', φ_0, ϱ so that $\mathrm{car}\,(\alpha_0',\,\varphi_0;\,\theta,\,\psi')$ (see the observations preceding Theorem 3.3). We then construct α_0, β_0, ϱ_0 so that $\mathrm{car}\,(\alpha_0,\,\beta_0;\,\varphi_0,\,\psi)$. We note that ϱ_0 is epic since $\mathrm{ex}\,(\varrho\alpha,\,\beta;\,\varphi_0,\,\psi)$ by Lemma 3.8.

Then $\mathrm{car}\,(\alpha_0,\,\beta_0;\,\varphi_0,\,\psi)$ and $\mathrm{car}(\alpha_0',\,\varphi_0;\,\theta,\,\psi')$ so that $\mathrm{car}\,(\alpha_0'\alpha_0,\,\beta_0;\,\theta,\,\psi'\psi)$, whence $\mathrm{ex}\,(\alpha_0'\alpha_0\varrho_0,\,\beta_0\varrho_0;\,\theta,\,\psi'\psi)$ or $\mathrm{ex}\,(\alpha'\alpha,\,\beta;\,\theta,\,\psi'\psi)$.

Theorem 3.10. *Consider again the diagram of Theorem 3.7. Then if* α', ψ' *are monic and* $(\alpha'\alpha,\,\beta;\,\theta,\,\psi'\psi)$ *is exact (cartesian) so is* $(\alpha,\,\beta;\,\varphi,\,\psi)$.

Proof. We have the commutative diagram

$$
\begin{array}{ccccc}
\bullet & \xrightarrow{\{\alpha'\alpha,\,\beta\}} & \bullet & \xrightarrow{\langle\theta,\,-\psi^1\,\psi\rangle} & \bullet \\
\Vert & & \uparrow{\alpha'\otimes 1} & & \downarrow{\psi'} \\
\bullet & \xrightarrow{\{\alpha,\,\beta\}} & \bullet & \xrightarrow{\langle\varphi,\,-\psi\rangle} & \bullet
\end{array}
$$

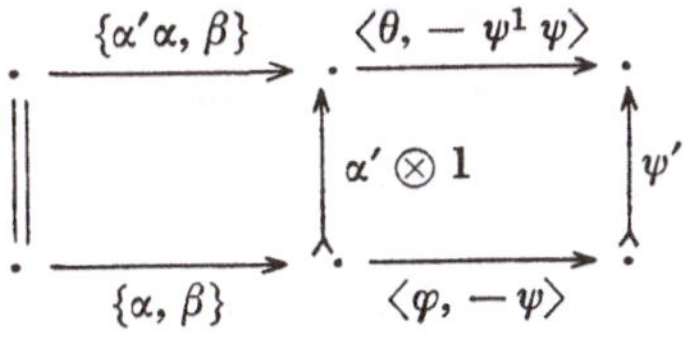

where the top row is exact. Thus, by Corollary 2.3 (i), the bottom row is exact. Moreover it is plain that $\{\alpha, \beta\}$ is monic if $\{\alpha'\alpha, \beta\}$ is monic.

Corollary 3.11. *Let*

$$\begin{array}{ccc} \cdot & \xrightarrow{\alpha} & \cdot \\ {\scriptstyle\beta}\downarrow & & \downarrow{\scriptstyle\varphi} \\ \cdot & \xrightarrow[\psi]{} & \cdot \end{array} \tag{3.12}$$

be a commutative square and let β, φ be split to obtain

$$\begin{array}{ccc} \cdot & \xrightarrow{\alpha} & \cdot \\ {\scriptstyle\varepsilon_\beta}\downarrow & & \downarrow{\scriptstyle\varepsilon_\varphi} \\ \cdot & \xrightarrow{\gamma} & \cdot \\ {\scriptstyle\mu_\beta}\downarrow & & \downarrow{\scriptstyle\mu_\varphi} \\ \cdot & \xrightarrow[\psi]{} & \cdot \end{array} \tag{3.13}$$

Then if (3.12) *is exact (cartesian, cocartesian, bicartesian) so is each square in* (3.13).

Proof. The conclusion if (3.12) is exact follows from Theorem 3.10 and the duality principle. The conclusion if (3.12) is cartesian follows from Theorem 3.10 and the observation that the bottom square in (3.13) is certainly cartesian if it is exact. A further application of the duality principle completes the proof. (This corollary leads to a simple proof of Theorem 3.5 above.)

We now list without proof some particular facts about commutative squares.

Proposition 3.14. *The square*

$$\begin{array}{ccc} & \xrightarrow{\varphi} & \\ {\scriptstyle 1}\downarrow & & \downarrow{\scriptstyle 1} \\ & \xrightarrow[\varphi]{} & \end{array}$$

is bicartesian for any φ.

Proposition 3.15. *Let the square*

$$\begin{array}{ccc} \cdot & \xrightarrow{\alpha} & \cdot \\ {\scriptstyle\alpha}\downarrow & & \downarrow{\scriptstyle\varphi} \\ \cdot & \xrightarrow[\psi]{} & \cdot \end{array}$$

be cocartesian. Then $\exists \pi$ such that $\pi\varphi = \pi\psi = 1$.

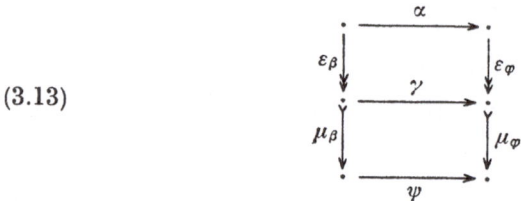

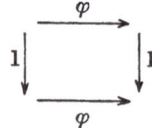

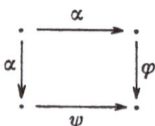

Proposition 3.16. *Let the square*

$$\begin{array}{ccc} \cdot & \xrightarrow{\alpha} & \cdot \\ {\scriptstyle\alpha}\downarrow & & \downarrow{\scriptstyle\varphi} \\ \cdot & \xrightarrow[\varphi]{} & \cdot \end{array}$$

be exact. Then α *is epic,* φ *is monic.*

Proposition 3.17. *Consider the commutative squares*

$$\begin{array}{ccc} R & \xrightarrow{\alpha} & A \\ {\scriptstyle\beta}\downarrow & & \downarrow{\scriptstyle\varphi} \\ B & \xrightarrow[\psi]{} & X \end{array} \qquad \begin{array}{ccc} R\oplus R & \xrightarrow{\langle 1, 1\rangle} & R \\ \downarrow{\scriptstyle\alpha\oplus\beta} & & \downarrow{\scriptstyle\gamma} \\ A\oplus B & \xrightarrow[\langle\varphi,\psi\rangle]{} & X \end{array}$$

where $\gamma = \varphi\alpha = \psi\beta$. *Then one square is exact (cocartesian) if the other is.*

4. The Category $\tilde{\mathscr{A}}$

Let us fix A, B in $\mathscr{A}$ and consider the collection of morphism-pairs (φ, ψ) where

$$A \overset{\varphi}{\to} X \overset{\psi}{\leftarrow} B,$$

X being an object of $\mathscr{A}$. We generate an equivalence relation in this collection by declaring $(\varphi, \psi) \sim (\mu\varphi, \mu\psi)$, where μ is monic. In fact, this equivalence relation is given explicitly by

$$(\varphi_1, \psi_1) \sim (\varphi_2, \psi_2) \text{ if and only if } \exists \text{ monic } \mu_1, \mu_2 \text{ with}$$
$$(\mu_1\varphi_1, \mu_1\psi_1) = (\mu_2\varphi_2, \mu_2\psi_2).$$

We write φ/ψ for the equivalence class containing (φ, ψ), and $\varphi/\psi\colon A \tilde{\to} B$.

Remark. It follows from the observations preceding Theorem 3.3 that every equivalence class has a *minimal representative* which is unique up to a unit in $\mathscr{A}$. That is, given φ/ψ, we construct $\operatorname{car}(\alpha, \beta; \varphi, \psi)$ and then $\operatorname{coc}(\alpha, \beta; \varphi_0, \psi_0)$. Then $\varphi/\psi = \varphi_0/\psi_0$ and $\varphi = \sigma\varphi_0$, $\psi = \sigma\psi_0$ with σ monic. We may liken φ_0/ψ_0 to a fraction in its "lowest terms".

Now let $\varphi/\psi\colon A \tilde{\to} B$, $\varkappa/\lambda\colon B \tilde{\to} C$. We construct the diagram

$$\begin{array}{ccccc} A & \xrightarrow{\varphi} & X & \xrightarrow{\varkappa'} & Z \\ & & {\scriptstyle\psi}\uparrow & & \uparrow{\scriptstyle\psi'} \\ & & B & \xrightarrow[\varkappa]{} & Y \\ & & & & \uparrow{\scriptstyle\lambda} \\ & & & & C \end{array}$$

where $\operatorname{ex}(\varkappa, \psi; \psi', \varkappa')$.

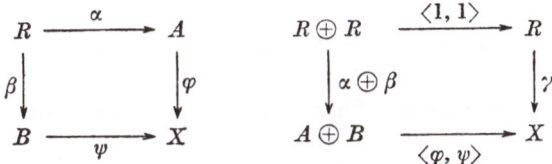

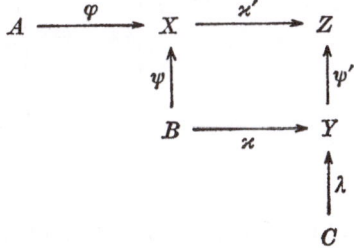

Proposition 4.1. *$\varkappa' \varphi/\psi' \lambda$ depends only on φ/ψ and $\varkappa/\lambda$.*

Proof. It is plainly permissible to suppose $\mathrm{coc}(\varkappa, \psi; \psi', \varkappa')$. Let $\mu: X \rightarrowtail X_0$, $\nu: Y \rightarrowtail Y_0$, and suppose $\mathrm{ex}(\nu\varkappa, \mu\psi; \psi_0', \varkappa_0')$. Consider the diagram

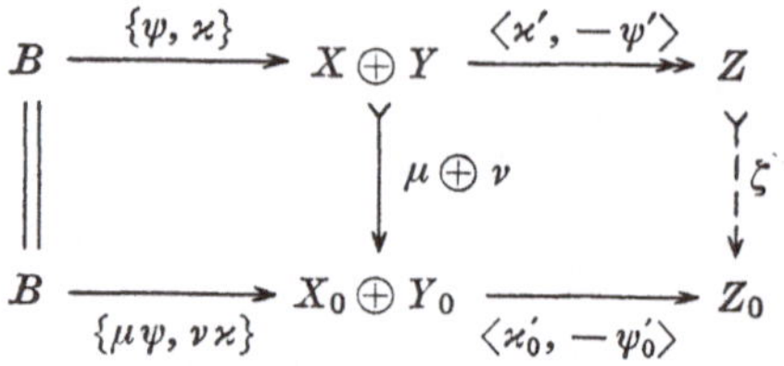

where the rows are exact. Then $\exists$ unique ζ making the right square commutative,

$$\zeta \varkappa' = \varkappa_0' \mu, \quad \zeta \psi' = \psi_0' \nu,$$

and ζ is monic by Corollary 2.3 (ii). Thus

$$\varkappa' \varphi/\psi' \lambda = \zeta \varkappa' \varphi/\zeta \psi' \lambda = \varkappa_0'(\mu \varphi)/\psi_0'(\nu \lambda)$$

and the proposition is proved. We write (temporarily)

(4.2) $$\varkappa' \varphi/\psi' \lambda = \varkappa/\lambda \,\tilde{\circ}\, \varphi/\psi .$$

Theorem 4.3. *There is a category $\tilde{\mathscr{A}}$ such that* $\mathrm{Ob}\,\tilde{\mathscr{A}} = \mathrm{Ob}\,\mathscr{A}$; $M_{\tilde{\mathscr{A}}}(A, B)$ *is the set of equivalence classes φ/ψ; and the law of composition is given by* (4.2).

Proof. First the identity morphisms of $\tilde{\mathscr{A}}$ are just the classes $1_A/1_A$; this follows immediately from Proposition 3.14.

Second we must prove the associativity of composition. We are given φ_1/ψ_1, φ_2/ψ_2, φ_3/ψ_3 and we construct the diagram

(4.4)

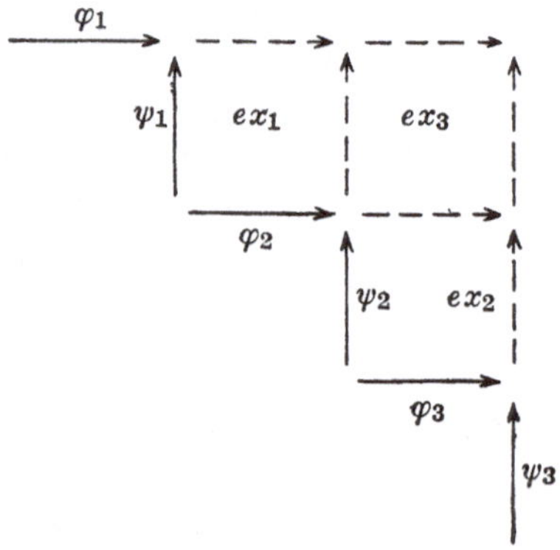

where we first construct the exact squares ex_1 and ex_2 and then construct ex_3. It follows from Theorem 3.7 that the pair consisting of the horizontal morphism and the vertical morphism in (4.4) represents both

$$\varphi_3/\psi_3 \mathbin{\tilde{\circ}} (\varphi_2/\psi_2 \mathbin{\tilde{\circ}} \varphi_1/\psi_1)$$

and

$$(\varphi_3/\psi_3 \mathbin{\tilde{\circ}} \varphi_2/\psi_2) \mathbin{\tilde{\circ}} \varphi_1/\psi_1 .$$

We will henceforth permit ourselves to represent composition in $\tilde{\mathscr{A}}$ by juxtaposition.

Theorem 4.5. (i) *We may (and will) embed $\mathscr{A}$ in $\tilde{\mathscr{A}}$ by the functor $\varphi \to \varphi/1$.*

(ii) *There is an anti-involution on $\tilde{\mathscr{A}}$ given by*[3] $\overline{\varphi/\psi} = \psi/\varphi$.

(iii) $\varphi/\psi = \bar{\psi}\,\varphi$.

We omit the proof; but remark that $(\varphi, 1)$ is a minimal representative for φ in $\tilde{\mathscr{A}}$. Indeed any (φ, ψ) with φ or ψ epic is a minimal representative of its class.

As we have set up the category $\tilde{\mathscr{A}}$ there is an apparent asymmetry between the roles played by monics and epics in $\mathscr{A}$; we wish now to dissipate this impression. To this end we construct a category $\tilde{\tilde{\mathscr{A}}}$ by a process dual to that used for constructing $\tilde{\mathscr{A}}$. That is, we consider pairs (α, β),

$$A \overset{\alpha}{\leftarrow} R \overset{\beta}{\rightarrow} B$$

and generate an equivalence relation in such pairs by declaring $(\alpha, \beta) \sim (\alpha\varepsilon, \beta\varepsilon)$, where ε is epic. We need not detail the development, but it is evident that we do obtain a category $\tilde{\tilde{\mathscr{A}}}$ in this way; we write the morphism of $\tilde{\tilde{\mathscr{A}}}$ containing (α, β) as $\alpha\backslash\beta$, and we embed $\mathscr{A}$ in $\tilde{\tilde{\mathscr{A}}}$ by $\varphi \to 1\backslash\varphi$.

Theorem 4.6. *There is an isomorphism $\Omega\colon \tilde{\mathscr{A}} \to \tilde{\tilde{\mathscr{A}}}$, respecting the embeddings of $\mathscr{A}$, given by*

$$\Omega(\varphi/\psi) = \alpha\backslash\beta ,$$

where $\operatorname{ex}(\alpha, \beta; \varphi, \psi)$. *Moreover, Ω^{-1} is given by*

$$\Omega^{-1}(\alpha\backslash\beta) = \beta\bar{\alpha} .$$

This theorem enables us to identify $\tilde{\mathscr{A}}$ with $\tilde{\tilde{\mathscr{A}}}$ and thus restore the parity (i.e., duality) in the roles played by epics and monics in $\mathscr{A}$. We notice that the essential observation is that the square

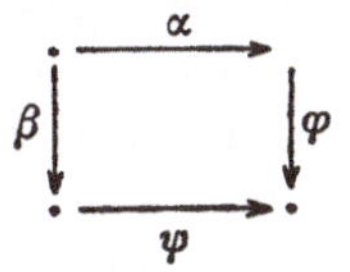

[3] We call $\bar{\Gamma}$ the *reverse* of Γ.

is fully commutative in $\tilde{\mathscr{A}}$ (i.e., commutative from any corner to the diagonally opposite corner) if and only if it is exact.

We now proceed to study the structure of the category $\tilde{\mathscr{A}}$; we will largely concentrate attention on monics and epics in $\tilde{\mathscr{A}}$.

Theorem 4.7. *The following statements about the morphism μ in $\mathscr{A}$ are equivalent;*

(i) *μ is monic in $\mathscr{A}$;*
(ii) *μ is monic in $\tilde{\mathscr{A}}$;*
(iii) *$\bar{\mu}$ is epic in $\tilde{\mathscr{A}}$;*
(iv) $\bar{\mu}\,\mu = 1$.

Proof. (i) ⇒ (iv). For, if μ is monic in $\mathscr{A}$, $(1,1) \sim (\mu, \mu)$.
(iv) ⇒ (ii). (iii). Trivial.
(ii) ⇒ (i). Trivial.

That (iii) ⇒ (iv) will follow immediately from the corollary to the next theorem, or by invoking the anti-involution on $\tilde{\mathscr{A}}$.

Theorem[4] **4.8.** *For any morphism Γ in $\tilde{\mathscr{A}}$,*

$$\Gamma \bar{\Gamma} \Gamma = \Gamma .$$

Proof. Take a minimal representative φ/ψ for Γ. Then $\exists\, \alpha, \beta$ with $\mathrm{bic}(\alpha, \beta; \varphi, \psi)$. Consider the diagram

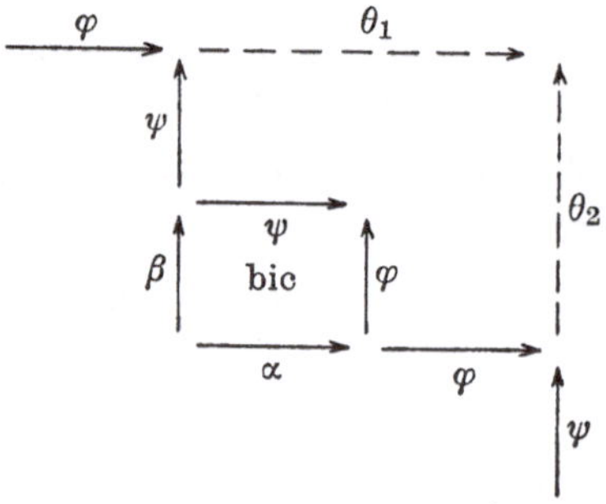

We observe that we may obtain a representative of $\Gamma \bar{\Gamma} \Gamma$ by taking θ_1, θ_2 so that

$$\mathrm{coc}(\varphi\alpha,\ \psi\beta,\ \theta_2, \theta_1);$$

then $\Gamma \bar{\Gamma} \Gamma = \theta_1 \varphi / \theta_2 \psi$ and we must show that

$$\theta_1 \varphi / \theta_2 \psi = \varphi / \psi . \qquad (4.9)$$

Now $\langle \varphi, -\psi \rangle \square \{\alpha, \beta\}$, and $\theta_1 \varphi \alpha - \theta_2 \psi \beta = 0$ since $\varphi\alpha = \psi\beta$ and $\theta_1 \psi \beta = \theta_2 \varphi \alpha$. Thus it remains to show that if $\langle \theta_1 \varphi, -\theta_2 \psi \rangle \{\gamma, \delta\} = 0$ then also $\langle \varphi, -\psi \rangle \{\gamma, \delta\} = 0$; that is, we must infer $\varphi\gamma = \psi\delta$ from $\theta_1 \varphi \gamma = \theta_2 \psi \delta$. This we do immediately by invoking Proposition 3.15.

[4] See [*8, 11*].

Corollary 4.9. (i) Γ is monic if and only if $\overline{\Gamma}\Gamma = 1$.

(ii) Γ is epic if and only if $\Gamma\overline{\Gamma} = 1$.

Of course, Corollary 4.9 (ii) completes the proof of Theorem 4.7. With that theorem (and its dual) we see that among the monics of $\widetilde{\mathscr{A}}$ are the monics of $\mathscr{A}$ and the reverses of the epics of $\mathscr{A}$. We now show that these generate all the monics of $\widetilde{\mathscr{A}}$.

Theorem 4.10. *The following statements about the morphism Γ in $\widetilde{\mathscr{A}}$ are equivalent:*

(i) *Γ is monic;*

(ii) *$\overline{\Gamma}\Gamma = 1$;*

(iii) *$\Gamma = \overline{\psi}\,\varphi$ with φ monic, ψ epic;*

(iv) *$\Gamma = \beta\,\bar{\alpha}$ with α epic, β monic.*

Proof. We already know (i) $\Leftrightarrow$ (ii). That (iii) $\Leftrightarrow$ (iv) is a ready consequence of Theorem 4.6 and Theorem 3.5. That (iii) $\Rightarrow$ (i) has already been observed. Thus it remains to show that (ii) $\Rightarrow$ (iii). We again take a minimal representative (φ, ψ) for Γ (we observe that for no other representative could it be true that ψ is epic). Then we have the diagram

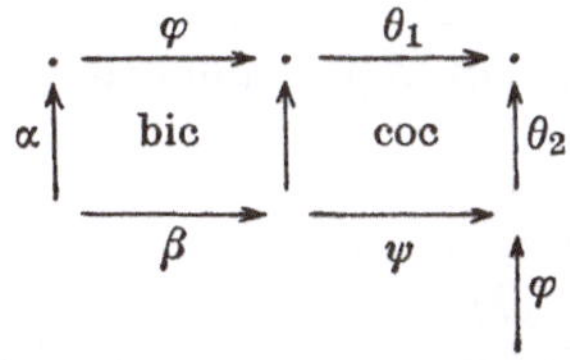

and $\theta_1\varphi/\theta_2\varphi = 1$. This last relation means that $\theta_1\varphi = \theta_2\varphi$ and is monic, so we immediately conclude that φ is monic. But also $\theta_1\psi = \theta_2\psi$ and so, since bic $(\alpha, \beta; \varphi, \psi)$, we must have $\theta_1 = \theta_2$. Proposition 3.16 then asserts that, since ex $(\psi, \psi; \theta_2, \theta_1)$, it follows that ψ is epic.

Theorem 4.10 and its dual give us a complete description of the monics and epics in $\widetilde{\mathscr{A}}$. We wish also to infer that every morphism in $\widetilde{\mathscr{A}}$ is expressible, essentially uniquely, as an epic followed by a monic, but we do not proceed directly to this goal. Instead, we establish the precise connection between a morphism of $\widetilde{\mathscr{A}}$ and a relation in $\mathscr{A}$, in the sense of a morphism in $\mathscr{A}$ from a subobject of A to a quotient object of B.

Now plainly given $\mu\colon A_0 \rightarrowtail A$, $\varepsilon\colon B \twoheadrightarrow B_0$, $\theta\colon A_0 \to B_0$, there is an associated morphism $\bar{\varepsilon}\,\theta\,\bar{\mu}\colon A \overset{\cdot}{\to} B$. Conversely, consider $\varphi/\psi\colon A \overset{\cdot}{\to} B$. We split ψ as $\mu_\psi \varepsilon_\psi$ and obtain the diagram

$$
\begin{array}{ccc}
A_0 & \overset{\mu}{\rightarrowtail} & A \\
{\scriptstyle\theta}\downarrow & \text{car} & \downarrow{\scriptstyle\varphi} \\
B \underset{\varepsilon}{\twoheadrightarrow} B_0 & \overset{\mu_\psi}{\rightarrowtail} & X
\end{array} , \quad \varepsilon = \varepsilon_\psi
$$

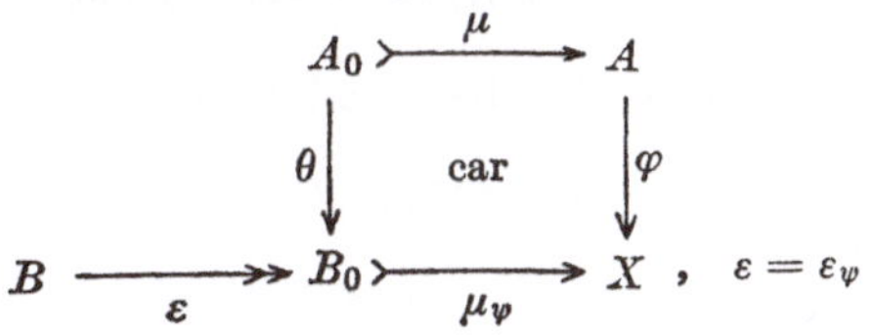

Then $\varphi/\psi = \overline{\psi}\varphi = \overline{\mu_\psi \varepsilon}\varphi = \overline{\varepsilon}\,\overline{\mu}_\psi\varphi = \overline{\varepsilon}\theta\overline{\mu}$, so that every morphism of $\widetilde{\mathscr{A}}$ may be written in this form. We now prove the uniqueness of the representation. Precisely, we show

Theorem 4.11. *If* $\overline{\varepsilon}_1\theta_1\overline{\mu}_1 = \overline{\varepsilon}_2\theta_2\overline{\mu}_2$, *then there are units* ω, ω' *in* $\mathscr{A}$ *such that*

$$\varepsilon_1 = \omega\,\varepsilon_2, \quad \mu_1 = \mu_2\,\omega', \quad \theta_1 = \omega\,\theta_2\,\omega'.$$

Proof. (i) First suppose that $\mu_1 = \mu_2 = 1$. Then $\overline{\varepsilon}_1\theta_1 = \overline{\varepsilon}_2\theta_2$. But then $(\theta_1, \varepsilon_1)$ and $(\theta_2, \varepsilon_2)$ are both minimal representatives so that $\exists$ unit ω with $\varepsilon_1 = \omega\,\varepsilon_2$, $\theta_1 = \omega\,\theta_2$.

(ii) Return to the general case and construct $\operatorname{car}(\nu_1, \nu_2; \mu_1, \mu_2)$ with $\mu = \mu_1\nu_1$. Then $\overline{\mu}_1\mu = \overline{\mu}_1\mu_1\nu_1 = \nu_1$, $\overline{\mu}_2\mu = \nu_2$ so that, composing on the right with μ,

$$\overline{\varepsilon}_1\,\theta_1\,\nu_1 = \overline{\varepsilon}_2\,\theta_2\,\nu_2\,.$$

By (i) $\exists\ \omega$ with $\varepsilon_1 = \omega\,\varepsilon_2$. Substituting back in the original relation and cancelling $\overline{\varepsilon}_1$ (which is monic in $\widetilde{\mathscr{A}}$) we have

$$\theta_1\,\overline{\mu}_1 = \omega\,\theta_2\,\overline{\mu}_2\,.$$

We now dualize the argument in (i) to infer that $\exists$ unit ω' with $\mu_1 = \mu_2\,\omega'$, $\theta_1 = \omega\,\theta_2\,\omega'$.

Theorem 4.12. (i) *Every morphism* Γ *in* $\widetilde{\mathscr{A}}$ *is expressible as* $\Gamma = ME$ *with* E *epic,* M *monic in* $\widetilde{\mathscr{A}}$;

(ii) *if* $M_1E_1 = M_2E_2$ *then* $\exists$ *unit* χ *in* $\mathscr{A}$ *with*

$$M_1 = M_2\,\chi$$
$$\chi\,E_1 = E_2\,.$$

Proof. (i) We may express Γ as $\overline{\varepsilon}\theta\,\overline{\mu}$. Then θ splits as $\mu_\theta\,\varepsilon_\theta$, so that

$$\Gamma = (\overline{\varepsilon}\,\mu_\theta)\,(\varepsilon_\theta\,\overline{\mu})\,,$$

and this is a splitting of Γ of the desired kind.

(ii) By Theorem 4.10 and its dual we may write

(4.13) $$M_i = \overline{\varepsilon}_i\,\xi_i\,, \quad E_i = \eta_i\,\overline{\mu}_i\,, \quad i = 1, 2,$$

with ξ_i, μ_i monic, η_i, ε_i epic. Then

$$\overline{\varepsilon}_1(\xi_1\,\eta_1)\,\overline{\mu}_1 = \overline{\varepsilon}_2(\xi_2\,\eta_2)\,\overline{\mu}_2\,.$$

Thus by Theorem 4.11, $\exists$ units ω, ω' with

(4.14) $$\varepsilon_1 = \omega\,\varepsilon_2\,, \quad \mu_1 = \mu_2\,\omega'$$

(4.15) $$\xi_1\,\eta_1 = \omega\,\xi_2\,\eta_2\,\omega'.$$

But it follows from (4.15) and the uniqueness of the decomposition in $\mathscr{A}$ that $\exists\,\chi$, unit in $\mathscr{A}$, such that

(4.16) $$\xi_1 = \omega\,\xi_2\,\chi\,, \quad \chi\,\eta_1 = \eta_2\,\omega'\,.$$

Then

$$M_1 = \bar{\varepsilon}_1\,\xi_1 = \bar{\varepsilon}_2\,\omega^{-1}\,\omega\,\xi_2\,\chi = \bar{\varepsilon}_2\,\xi_2\,\chi = M_2\,\chi\,,$$

and similarly $\chi E_1 = E_2$.

Corollary 4.17. *The units of $\tilde{\mathscr{A}}$ are just the units of $\mathscr{A}$.*

Proof. Plainly a unit of $\mathscr{A}$ is a unit of $\tilde{\mathscr{A}}$. Conversely if Γ is a unit of $\tilde{\mathscr{A}}$ then $\Gamma = \Gamma 1 = 1 \Gamma$ are two decompositions so that there exists a unit χ in $\mathscr{A}$ with $\Gamma = \chi$.

Thus the procedure of passing from $\mathscr{A}$ to $\tilde{\mathscr{A}}$ does not render invertible any morphisms not already invertible and does not enlarge the stock of units.

There are many further properties of the category $\tilde{\mathscr{A}}$ that should be investigated. It is, of course, a crucial fact about $\tilde{\mathscr{A}}$ that it is not a pointed category. For, if it were, then the points of $\tilde{\mathscr{A}}$ would have to be the points of $\mathscr{A}$; but there are many morphisms in $\tilde{\mathscr{A}}$ from A to N, where N is a point of $\mathscr{A}$ (they are in one-one correspondence with the quotients of A). Indeed one may show that the only morphisms of $\tilde{\mathscr{A}}$ for which the zeros of $\mathscr{A}$ constitute a system of two-sided zeros are the morphisms of $\mathscr{A}$. We defer consideration of these and other questions to a later paper and devote the final section here to a short discussion of addition in $M_{\tilde{\mathscr{A}}}(A, B)$.

5. Addition in $\tilde{\mathscr{A}}$

Our object here is to extend the addition in $\tilde{\mathscr{A}}$ to an addition in $\tilde{\mathscr{A}}$ in a way which corresponds to a natural procedure for adding relations[5]. We give, in fact, three definitions and show that they coincide.

Suppose given φ_1/ψ_1, φ_2/ψ_2: $A \xrightarrow{\sim} B$. We construct the diagram

(5.1)

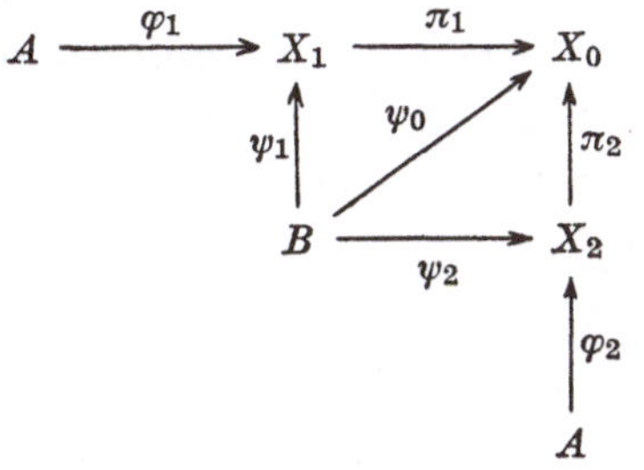

[5] It also corresponds to the procedure for adding fractions.

with $\mathrm{ex}(\psi_2, \psi_1; \pi_2, \pi_1)$, and define

(5.2) $$\varphi_1/\psi_1 + \varphi_2/\psi_2 = \varphi_0/\psi_0, \quad \text{where} \quad \varphi_0 = \pi_1 \varphi_1 + \pi_2 \varphi_2, \quad \psi_0 = \pi_1 \psi_1 = \pi_2 \psi_2 .$$

Dually, we may suppose given $\alpha_1 \backslash \beta_1, \alpha_2 \backslash \beta_2 \colon A \overset{\approx}{\rightarrow} B$ and construct the diagram

(5.1d)

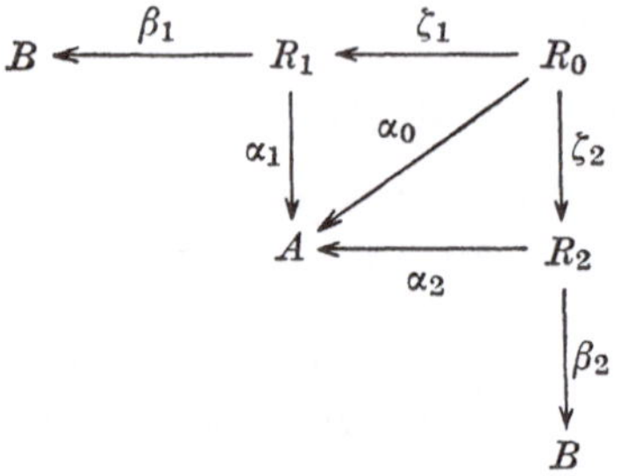

with $\mathrm{ex}(\zeta_1, \zeta_2; \alpha_1, \alpha_2)$, and define

(5.2d) $$\alpha_1 \backslash \beta_1 + \alpha_2 \backslash \beta_2 = \alpha_0 \backslash \beta_0, \quad \text{where} \quad \alpha_0 = \alpha_1 \zeta_1 = \alpha_2 \zeta_2, \quad \beta_0 = \beta_1 \zeta_1 + \beta_2 \zeta_2 .$$

We will show that these two definitions[6] coincide under the canonical identification of $\widetilde{\mathscr{A}}$ with $\overset{\approx}{\mathscr{A}}$ (we remark that it is easily seen that these definitions extend the law of addition in $\mathscr{A}$).

To this end we make a third definition. We consider morphisms $\bar{\varepsilon}_i \theta_i \bar{\mu}_i \colon A \tilde{\rightarrow} B$, $i = 1, 2$, and give the rule for adding them. Namely, we construct the diagram

(5.3)

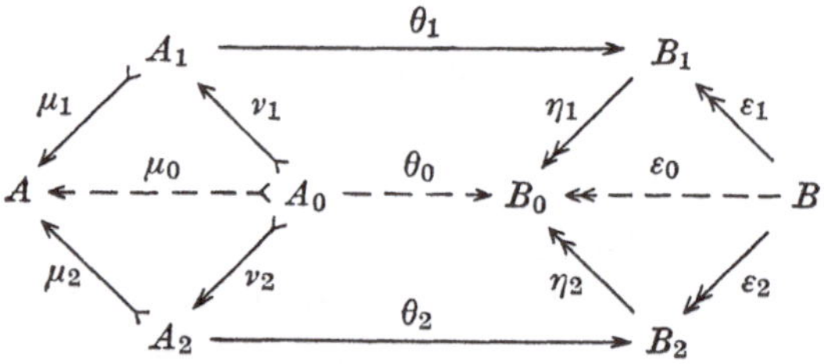

where $\mathrm{car}(\nu_1, \nu_2; \mu_1, \mu_2)$, $\mathrm{coc}(\varepsilon_1, \varepsilon_2; \eta_1, \eta_2)$, and define

(5.4) $$\bar{\varepsilon}_1 \theta_1 \bar{\mu}_1 + \bar{\varepsilon}_2 \theta_2 \bar{\mu}_2 = \bar{\varepsilon}_0 \theta_0 \bar{\mu}_0, \quad \text{where} \quad \varepsilon_0 = \eta_1 \varepsilon_1 = \eta_2 \varepsilon_2, \quad \mu_0 = \mu_1 \nu_1 = \mu_2 \nu_2, \quad \text{and} \quad \theta_0 = \eta_1 \theta_1 \nu_1 + \eta_2 \theta_2 \nu_2 .$$

Theorem 5.5. *The three definitions given coincide.*

Proof. It is plainly sufficient to show that the definitions (5.2) and (5.4) coincide; since addition is clearly well-defined by (5.4), this will also establish that it is well-defined by (5.2) (and (5.2d)).

[6] We have not even shown that addition is well-defined by (5.2) or (5.2d).

We recall how φ/ψ is identified with $\bar{\varepsilon}\,\theta\,\bar{\mu}$. We construct, in fact,

$$\begin{array}{ccccc} & & \cdot & \overset{\mu}{\rightarrowtail} & \cdot \\ & & {\scriptstyle\theta}\Big\downarrow & \text{car} & \Big\downarrow{\scriptstyle\varphi} \\ & \underset{\varepsilon}{\twoheadrightarrow} & \cdot & \underset{\lambda}{\rightarrowtail} & \cdot \end{array}$$

with $\lambda\,\varepsilon = \psi$. Thus we wish to show that if

$$\begin{array}{ccccc} & & A_i & \overset{\mu_i}{\rightarrowtail} & A \\ & & {\scriptstyle\theta_i}\Big\downarrow & \text{car} & \Big\downarrow{\scriptstyle\varphi_i} \\ B & \underset{\varepsilon_i}{\twoheadrightarrow} & B_i & \underset{\lambda_i}{\rightarrowtail} & X_i \end{array} \qquad \psi_i = \lambda_i\,\varepsilon_i\,, \tag{5.6}$$

for $i = 1,2$, then (5.6) also subsists for $i = 0$, with a suitably defined λ_0. We now define λ_0; consider the diagram

$$\begin{array}{ccccc} B & \xrightarrow{\{\varepsilon_1,\,-\varepsilon_2\}} & B_1 \oplus B_2 & \overset{\langle\eta_1,\,\eta_2\rangle}{\twoheadrightarrow} & B_0 \\ \Big\| & & \Big\downarrow{\scriptstyle\lambda_1 \oplus \lambda_2} & & \Big\downarrow{\scriptstyle\lambda_0} \\ B & \xrightarrow[\{\psi_1,\,-\psi_2\}]{} & X_1 \oplus X_2 & \xrightarrow[\langle\pi_1,\,\pi_2\rangle]{} & X_0 \end{array}$$

where the rows are exact. Then there is a unique λ_0 making the right hand square commutative, and moreover λ_0 is monic by Corollary 2.3 (ii). Further it follows by a slight extension of Theorem 3.6 (ii) d that the right hand square is exact and hence cartesian, since $\lambda_1 \oplus \lambda_2$ is monic.

We now prove (5.6) holds with $i = 0$. First,

$$\psi_0 = \pi_1\,\psi_1 = \pi_1\,\lambda_1\,\varepsilon_i = \lambda_0\,\eta_1\,\varepsilon_1 = \lambda_0\,\varepsilon_0\,.$$

Thus it remains to show $\text{car}(\mu_0, \theta_0; \varphi_0, \lambda_0)$. We consider the diagram

$$\begin{array}{ccccccc} A_0 & \overset{\{\nu_1,\,\nu_2\}}{\rightarrowtail} & A_1 \oplus A_2 & \xrightarrow{\theta_1 \oplus \theta_2} & B_1 \oplus B_2 & \xrightarrow{\langle\eta_1,\,\eta_2\rangle} & B_0 \\ \Big\downarrow{\scriptstyle\mu_0} & & \Big\downarrow{\scriptstyle\mu_1 \oplus \mu_2} & & \Big\downarrow{\scriptstyle\lambda_1 \oplus \lambda_2} & & \Big\downarrow{\scriptstyle\lambda_0} \\ A & \overset{\{1,\,1\}}{\rightarrowtail} & A \oplus A & \xrightarrow{\varphi_1 \oplus \varphi_2} & X_1 \oplus X_2 & \xrightarrow{\langle\pi_1,\,\pi_2\rangle} & X_0 \end{array}$$

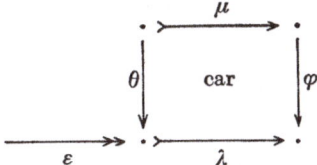

Plainly the composite of the morphisms in the top row is $\eta_1\theta_1\nu_1 + \eta_2\theta_2\nu_2 = \theta_0$; and the composite of the morphisms in the bottom row is $\pi_1\varphi_1 + \pi_2\varphi_2 = \varphi_0$. Moreover, the left square is cartesian by Proposition 3.17 d, the centre square is cartesian as the direct sum of cartesian squares, and we have just observed that the right square is cartesian. By Theorem 3.7 the composite square is cartesian and so the theorem is proved.

Theorem 5.7. *The addition defined in* $M_{\tilde{\mathscr{A}}}(A, B)$ *is associative and commutative and the zero of* $M_{\mathscr{A}}(A, B)$ *acts as a zero.*

We omit the proof, contenting ourselves with the remark that associativity follows immediately from the associativity of the "push-out".

Theorem 5.8. *The left distributive law*

$$(\Gamma_1 + \Gamma_2)\,\Delta = \Gamma_1\Delta + \Gamma_2\Delta$$

holds if (i) $\Delta \in \mathscr{A}$ *or* (ii) Δ *is epic.*

Proof. (i) Let $\Delta = \tau\colon T \to A$. It is then clear from Definition (5.2) that

$$\varphi_1\tau/\psi_1 + \varphi_2\tau/\psi_2 = (\pi_1\varphi_1\tau + \pi_2\varphi_2\tau)/\psi_0 = \varphi_0\tau/\psi_0 .$$

This proves (i). To prove (ii) it is sufficient to suppose $\Delta = \bar{\sigma}$, with σ monic. It is then clear from Definition (5.4) that

$$\bar{\varepsilon}_1\theta_1\bar{\mu}_1\bar{\sigma} + \bar{\varepsilon}_2\theta_2\bar{\mu}_2\bar{\sigma} = \bar{\varepsilon}_1\theta_1\overline{\sigma\mu_1} + \bar{\varepsilon}_2\theta_2\overline{\sigma\mu_2} = \bar{\varepsilon}_0\theta_0\overline{\sigma\mu_0} ,$$

since $\operatorname{car}(\nu_1, \nu_2; \sigma\mu_1, \sigma\mu_2)$ if $\operatorname{car}(\nu_1, \nu_2; \mu_1, \mu_2)$. But

$$\bar{\varepsilon}_0\theta_0\overline{\sigma\mu_0} = \bar{\varepsilon}_0\theta_0\bar{\mu}_0\bar{\sigma} = (\bar{\varepsilon}_1\theta_1\bar{\mu}_1 + \bar{\varepsilon}_2\theta_2\bar{\mu}_2)\,\bar{\sigma} ,$$

and (ii) is proved.

We point out finally that the distributive law does not hold in general. As an example, let $\varphi\colon A \to B$ and let N be a point in $\mathscr{A}$. Then one may show that

$$\varphi/0 + (-\varphi)/0 = 0/0\colon A \tilde{\to} N$$

if and only if $\varphi = 0$. Thus there are further properties of the addition whose elucidation must await a later paper.

References

[1] Eckmann, B., and P. J. Hilton: Group-like structures in general categories I. Multiplication and Comultiplication. Math. Annal. **145**, 227–255 (1962).

[2] Freyd, P. J.: Abelian Categories. New York: Harper and Row 1964.

[3] Lambek, J.: Goursat's theorem and the Zassenhaus lemma. Can. J. Math. **10**, 45–56 (1958).

[4] — Goursat's theorem and homological algebra. Can. Math. Bull. **7**, 597–608 (1964).

[5] Leicht, J. B.: Remarks to the axiomatic theory of additive relations. Can. J. Math. (to appear).

[6] — Outlines of a theory of additive relations, Mimeographed notes. University of Toronto.

[7] Leicht, J. B.: On commutative squares. Can. J. Math. **15**, 59—79 (1963).
[8] MacLane, S.: An algebra of additive relations. Proc. Nat. Acad. Sci., USA **47**, 1043—1051 (1961).
[9] — Homology. Berlin-Göttingen-Heidelberg: Springer 1963.
[10] Olum, P.: On non-abelian cohomology and van Kampen's Theorem. Ann. Math. **68**, 658—668 (1958).
[11] Puppe, D.: Korrespondenzen in abelsche Kategorien. Math. Annal. **148**, 1—30 (1962).
[12] University of Illinois Committee on School Mathematics, Mathematics Project (1965).

Department of Mathematics
Cornell University
Ithaca, New York

Canonical Categories *

By

JOHANN SONNER

It is the purpose of this expository paper to review the present situation with respect to such notions as sub- and quotient structures, images and coimages, and to look into the possibility of rectangular decompositions of morphisms which will replace the canonical decompositions of homomorphisms or continuous functions in the classical case.

0. In deriving new mathematical structures from a given family of structures and morphisms, the following procedures are well established:

a) construction of projective and inductive limits of functors;

b) construction of images and coimages of morphisms.

The discussion of the latter principle splits into two parts. In the first one, we imitate the classical situation of a concrete category over the category of sets and functions, and define (extrinsically) *substructures* and *quotient structures*. Unfortunately, a canonical decomposition of a function in the lower category does not always lift to a rectangular decomposition in the upper category. In the second part, we decompose in a rectangular fashion morphisms in a category without the help of a base category, and define (intrinsically) *subgadgets* and *quotient gadgets*. These form the natural test morphisms in the formulation of injectives and projectives. Rectangular decompositions of morphisms are guaranteed in so-called *canonical categories*. On the one hand, they are poorer in axioms than exact categories; on the other hand, they can be endowed with Isbell bicategory structures.

Part I. Substructures and Quotient Structures

1. Let $\mathscr{A}$ and $\mathscr{B}$ be categories, F a functor from $\mathscr{A}$ into $\mathscr{B}$, and a and b units of $\mathscr{A}$. In what follows we designate be $\mathscr{A}(a, b)$ (resp. $\mathscr{A}(a, \cdot)$, $\mathscr{A}(\cdot, b)$, $\mathscr{A}_0$, $\mathscr{A}^*$) the set of the $\mathscr{A}$-morphisms with source a and target b (resp. with source a, with target b, which are units, which are invertible). Frequently, the source and the target of an $\mathscr{A}$-morphism f are abbreviated by $S(f)$ and $T(f)$ respectively. In order to avoid confusion we denote, for f, g in $\mathscr{A}$, by $\widetilde{\mathscr{A}}(f, g)$ the mapping $u \mapsto fug$ from $\mathscr{A}(T(f)$,

* Received August 27, 1965.

$S(g))$ into $\mathscr{A}(S(f), T(g))$; thus, for a, b in $\mathscr{A}_0$, $\tilde{\mathscr{A}}(a, b)$ is the identity mapping on $\mathscr{A}(a, b)$. A similar meaning can be attached to the symbols $\tilde{\mathscr{A}}(f, \cdot)$ and $\tilde{\mathscr{A}}(\cdot, g)$. Since $u \in \mathscr{A}(a, b)$ (resp. $u \in \mathscr{A}(a, \cdot)$, $u \in \mathscr{A}_0$, $u \in \mathscr{A}^*$) implies $F(u) \in \mathscr{B}(F(a), F(b))$ (resp. $F(u) \in \mathscr{B}(F(a) \cdot)$, $F(u) \in \mathscr{B}_0$, $F(u) \in \mathscr{B}^*$), F induces, by passing to subsets, a mapping from $\mathscr{A}(a, b)$ into $\mathscr{B}(F(a), F(b))$ (resp. from $\mathscr{A}(a, \cdot)$ into $\mathscr{B}(F(a), \cdot)$, from $\mathscr{A}_0$ into $\mathscr{B}_0$, from $\mathscr{A}^*$ into $\mathscr{B}^*$) which we shall denote by F_{ab} (resp. $F_{a.}$, F_0, F^*). Note that the family (F_{ab}) indexed by the units of $\mathscr{A}^0 \times \mathscr{A}$ is a natural morphism from the functor $(f, g) \mapsto \tilde{\mathscr{A}}(f, g)$ into the functor $(f, g) \mapsto \tilde{\mathscr{B}}(F(f), F(g))$, both defined in $\mathscr{A}^0 \times \mathscr{A}$ with values in the category of the sets and functions of type $\mathfrak{U}$ (where $\mathfrak{U}$ is a sufficiently large universe). Likewise the family $(F_{a.})$ indexed by the units of $\mathscr{A}$ is a natural morphism from the functor $f \mapsto \tilde{\mathscr{A}}(f, \cdot)$ into the functor $f \mapsto \tilde{\mathscr{B}} F((f), \cdot)$ both defined in $\mathscr{A}^0$. Note further that F_0 is a functor from the discrete category $\mathscr{A}_0$ into the discrete category $\mathscr{B}_0$, while F^* is a functor from the groupoid $\mathscr{A}^*$ into the groupoid $\mathscr{B}^*$.

Definition 1. The relation «for all $a \in \mathscr{A}_0$, for all $b \in \mathscr{A}_0$, F_{ab} is injective (resp. surjective, bijective)» is denoted by «*F is almost injective* (resp. *surjective, bijective*)». The relation «for all $a \in \mathscr{A}_0$, $F_{a.}$ is injective (resp. surjective, bijective)» is denoted by «*F is locally injective* (resp. *surjective, bijective*)». The relation $(\exists u)\ (u \in \mathscr{A}(a, b) \wedge F(u) = v)$ is denoted by «*v is an F-morphism from a into b*».

Remarks. 1. In case $\mathscr{A}$ is a subcategory of $\mathscr{B}$ and F the canonical injection from $\mathscr{A}$ into $\mathscr{B}$, the relations «F is almost surjective» and «$\mathscr{A}$ is full» are equivalent. Also the relations «F^* is locally surjective» and «$\mathscr{A}$ is saturated» are equivalent [*4*, def. 7, p. 5].

2. A faithful (resp. fully faithful) functor is a functor F which is almost injective (resp. bijective). A transportable functor is a functor F such that F^* is locally bijective [*6*, p. I.4).

3. An injective functor is almost and locally injective. Not every surjective functor is almost or locally surjective.

4. Every locally injective functor is almost injective; the converse is not true. There exist functors which are locally but not almost surjective, and there exist functors which are almost but not locally surjective.

5. While a functor F from $\mathscr{A}$ into $\mathscr{B}$ preserves, but not necessarily reflects sections and retractions, an almost injective functor F reflects, but not necessarily preserves monomorphisms and epimorphisms. Indeed, the relations $uv = a \in \mathscr{A}_0$ imply $F(u)\, F(v) = F(a) \in \mathscr{B}_0$; whence the first assertion. For the proof of the second assertion assume for example

$fu = gu$ in $\mathscr{A}$ with $F(u)$ monic in $\mathscr{B}$. Then f and g have same source and same target, and $F(f) = F(g)$. We conclude $f = g$.

Example. ANDREOTTI proved in [*1*, th., p. 1—11]: Let F be a functor from the category $\mathscr{A}$ into the category $\mathscr{B}$. The following properties are equivalent:

a) There exists a functor G from $\mathscr{B}$ into $\mathscr{A}$ such that $G \circ F$ and $F \circ G$ are naturally isomorphic to the respective identity mappings.

b) F is almost bijective, and each unit b of $\mathscr{B}$ is isomorphic to an element of the form $F(a)$ where a is a unit of $\mathscr{A}$.

Proposition 1. *Let F be a functor from the category $\mathscr{A}$ into the category $\mathscr{B}$, F' a functor from $\mathscr{B}$ into the category $\mathscr{C}$, and F'' a functor from $\mathscr{A}$ into $\mathscr{C}$ such that $F'' = F' \circ F$. Then:*

a) *If F and F' are almost injective, then F'' is almost injective.*

b) *If F and F' are almost surjective, then F'' is almost surjective.*

c) *If F'' is almost injective, then F is almost injective.*

d) *If F'' is almost surjective and F_0 surjective, then F' is almost surjective.*

e) *If F'' is almost surjective and F' almost injective, then F is almost surjective.*

f) *If F'' is almost injective, F almost surjective and F_0 surjective, then F' is almost injective.*

The proposition remains true if «almost» is everywhere replaced by «locally».

The truth of the above statements follows immediately from the definitions.

It is convenient to term a functor F from $\mathscr{A}$ into $\mathscr{B}$ *forgetful* if F is almost injective and F^* locally bijective.

Example. The mapping $(G, U, X, U', X') \mapsto (G, X, X')$ from the category of the groups and homomorphisms of type $\mathfrak{U}$ into the category of the sets and functions of type $\mathfrak{U}$ (where $\mathfrak{U}$ is a universe) is a forgetful functor. Likewise for most classical categories. Cf. [*11*, p. 172].

2. In what follows let F be a functor from $\mathscr{A}$ into $\mathscr{B}$.

The elements (a, u) of $\mathscr{A}_0 \times \mathscr{B}$ such that $F(a) = S(u)$ form a subset $\mathscr{L}$ of $\mathscr{A}_0 \times \mathscr{B}$. Each element (a, u) of $\mathscr{L}$ is said to be *a left structure on* $T(u)$. The category $\mathscr{A}$ operates on $\mathscr{L}$ from the left by means of the external composition law

$$(s, (a, u)) \to (S(s), F(s)\, u) \qquad (T(s) = a);$$

the category $\mathscr{B}$ operates on $\mathscr{L}$ from the right by means of the external composition law

$$(v, (a, u)) \to (a, uv) \qquad (T(u) = S(v)).$$

Thus $\mathscr{L}$ becomes an $(\mathscr{A}; \mathscr{B})$-biset [*12*, def. 3, p. 203]. Write $s \perp (a, u)$ instead of $(S(s), F(s)u)$ for convenience.

The elements (a, v) of $\mathscr{A}_0 \times \mathscr{B}$ such that $F(a) = T(v)$ form a subset $\mathscr{R}$ of $\mathscr{A}_0 \times \mathscr{B}$. Each element (a, v) of $\mathscr{R}$ is said to be *a right structure on* $S(v)$. The category $\mathscr{A}$ operates on $\mathscr{R}$ from the right by means of the external composition law

$$(t, (a, v)) \mapsto (T(t), vF(t)) \qquad (S(t) = a);$$

the category $\mathscr{B}$ operates on $\mathscr{R}$ from the left by means of the external composition law

$$(u, (a, v)) \mapsto (a, uv) \qquad (T(u) = S(v)).$$

Thus $\mathscr{R}$ becomes *a* $(\mathscr{B}; \mathscr{A})$ — biset. Write $(a, v) \perp t$ instead of $(T(f), vF(t))$ for convenience.

Definition 2. Let $b \in \mathscr{B}_0$. We say «*the families* $(x_\iota, u_\iota)_{\iota \in I}$ *of left structures on* b *and* $(y_\varkappa, v_\varkappa)_{\varkappa \in K}$ *of right structures on* b *are compatible*» instead of «for all $\iota \in I$, for all $\varkappa \in K$, $u_\iota v_\varkappa$ is an F-morphism from x_ι into $y_\varkappa$ (def. 1)».

The above definition can in particular be applied to the case where I or K is reduced to one element. The left structures (x, u) on b which are compatible with a given family $(y_\varkappa, v_\varkappa)_{\varkappa \in K}$ of right structures on b form *a* stable subset $\mathscr{L}_1$ of the left $\mathscr{A}$-set $\mathscr{L}$. Following a suggestion by J. W. Duskin, we are interested in the left universal solutions of b in $\mathscr{L}_1$. More precisely:

Definition 3a. We denote by «(x, u) *is an initial structure on* b *with respect to the family* $(y_\varkappa, v_\varkappa)$ *of right structures on* b» the conjunction of the following relations:

(i) (x, u) is *a* left structure on b which is compatible with the family $(y_\varkappa, v_\varkappa)$.

(ii) Whenever (x', u') is a left structure on b which is compatible with the family $(y_\varkappa, v_\varkappa)$, then there exists one and only one morphism s of $\mathscr{A}$ such that $s \perp (x, u)$ is defined and equal to (x', u').

The right structures (y, v) on b which are compatible with a given family $(x_\iota, u_\iota)_{\iota \in I}$ of left structures on b form a stable subset $\mathscr{R}_1$ of the right $\mathscr{A}$-set $\mathscr{R}$. By considering right universal solutions of b in $\mathscr{R}_1$ one is led to:

Definition 3b. We denote by «(y, v) *is a final structure on* b *with respect to the family* (x_ι, u_ι) *of left structures on* b» the conjunction of the following relations:

(i) (y, v) is a right structure on b which is compatible with the family (x_ι, u_ι).

(ii) Whenever (y', v') is a right structure on b which is compatible with the family (x_ι, u_ι), then there exists one and only one morphism t of $\mathscr{A}$ such that $(y, v) \perp t$ is defined and equal to (y', v').

Remarks. 1. As is the case with all universal solutions, initial (resp. final) structures on b are determined uniquely up to isomorphism of $\mathscr{A}$ operating from the left (resp. right) [*12*, prop. 3, p. 204].

2. Initial and final structures may also be described in terms of representations of or universal points for a suitable set functor [*9*, th. 7.1, pp. 53].

Proposition 2. *Let F be a forgetful functor from $\mathscr{A}$ into $\mathscr{B}$. Furthermore, let $b \in \mathscr{B}_0$ and let $(y_\varkappa, v_\varkappa)_{\varkappa \in K}$ be a family of right structures on b. In order that b admit an initial structure with respect to the family $(y_\varkappa, v_\varkappa)$ of the form (x, b) it is necessary and sufficient that b admit an initial structure (x', u) with respect to the family $(y_\varkappa, v_\varkappa)$ where $u \in \mathscr{B}^*$.*

Since $\mathscr{B}_0 \subset \mathscr{B}^*$ the necessity of the condition is clear. Let (x', u) be an initial structure on b with respect to the family $(y_\varkappa, v_\varkappa)$ where $u \in \mathscr{B}^*$. By hypothesis F^* is locally bijective. Hence there exists one (and only one) $\bar{u} \in \mathscr{A}^*$ such that $S(\bar{u}) = x'$ and $F(\bar{u}) = u$. Denote $T(\bar{u})$ by x. Because $\bar{u} \perp (x, b) = (x', u)$ and since $\bar{u} \in \mathscr{A}^*$, (x, b) is also an initial structure on b with respect to the family $(v_\varkappa, y_\varkappa)$.

Remark. A simple analysis shows that the initial structures of the form (x, b) are precisely the initial structures in the sense of N. Bourbaki [*3*, p. 32] in categorical language. Consequently, b admits an initial structure in the sense of N. Bourbaki if and only if it admits an initial structure (x', u) in the sense of J. W. Duskin where $u \in \mathscr{B}^*$ (everything with respect to the family $(y_\varkappa, v_\varkappa)$). Analogous considerations hold for final structures. For another approach compare [*5*].

We say the family $(y_\varkappa, v_\varkappa)$ of right structures on b admits a *strict initial* structure on b if it admits an initial structure of the form (x, b). We say the family (x_ι, u_ι) of left structure on b admits a *strict final* structure on b if it admits a final structure of the form (y, b).

3. We study now the case where both I and K are reduced to one element, and where the functor F from $\mathscr{A}$ into $\mathscr{B}$ is almost injective.

Proposition 3. *Let (y, v) be a right structure on $b \in \mathscr{B}_0$, and let (x, u) be an initial structure on b with respect to (y, v). If v is monic, then the unique $\bar{v} \in \mathscr{A}\,(x, y)$ such that $F(\bar{v}) = uv$ is monic.*

Let s and t be elements of $\mathscr{A}$ such that $s\bar{v} = t\bar{v}$. Then s and t are morphisms from some $x' \in \mathscr{A}_0$ into x. Moreover, the equations

$$F(s)\,uv = F(s)\,F(\bar{v}) = F(s\bar{v}) = F(t\bar{v}) = F(t)\,F(\bar{v}) = F(t)\,uv$$

hold. Since v is monic by hypothesis, $F(s)u = F(t)u$. The last equation reads $s \perp (x, u) = t \perp (x, u)$ and implies $s = t$ because (x, u) is an initial structure.

In the situation of prop. 3, $\bar{v}$ (resp. x) is said to be a *substructure* (resp. *Substructure*) *of* y *over* v. In the dual situation one speaks of a *quotient structure* (resp. *Quotient structure*) *of* x *over* u. As mentioned earlier the substructures of y (resp. quotient structures of x) on b over v (resp. u) are determined uniquely up to isomorphisms of $\mathscr{A}$. One also speaks of *strict sub-*, *Sub-*, *quotient* and *Quotient structures* if the initial resp. final structures involved are strict.

Proposition 4. *Let f be an $\mathscr{A}$-morphism from x into y, and let $F(f) = g_1 v$ be a decomposition of $F(f)$ in $\mathscr{B}$ where v is monic. Suppose that the right structure (y, v) on $S(v)$ admits an initial structure on $S(v)$ of the form (y', v'); let $\bar{v}$ be the unique $\mathscr{A}$-morphism from y' into y such that $F(\bar{v}) = v'v$. Then $\bar{v}$ is monic, and there exists one and only one $f_1 \in \mathscr{A}(x, y')$ such that $f = f_1 \bar{v}$.*

For the proof, consider the diagrams

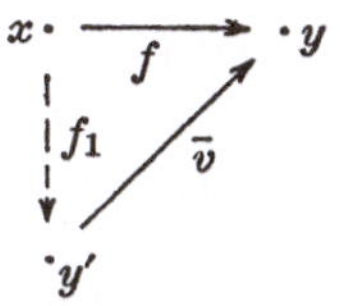

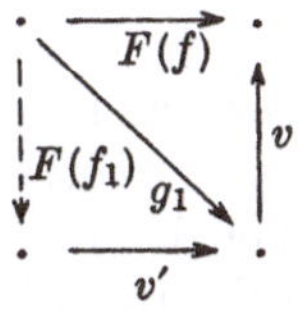

where (x, g_1) is a left structure on $S(v)$ which is compatible with the right structure (y, v) due to the fact that $F(f) = g_1 v$. Hence there exists $f_1 \in \mathscr{A}(x, y')$ such that $F(f_1)v' = g_1$. We wish to show that in fact $f_1 \bar{v} = f$. Note that both $f_1 \bar{v}$ and f are $\mathscr{A}$-morphisms from x into y. The assertion follows now from the equations

$$F(f_1 \bar{v}) = F(f_1) F(\bar{v}) = F(f_1) v' v = g_1 v = F(f)$$

and the almost injectiveness of F. The remaining statements are consequences of the hypothesis and prop. 3.

Theorem 1. *Let f be an $\mathscr{A}$-morphism from x into y, and let $F(f) = ugv$ be a decomposition of $F(f)$ in $\mathscr{B}$ where u is epic and v monic. Suppose that y admits a substructure over v, say $\bar{v}$, and that x admits a quotient structure over u, say $\bar{u}$, one of which is strict. Then there exists one and only one $\mathscr{A}$-morphism $\bar{f}$ such that $f = \bar{u}\bar{f}\bar{v}$.*

To fix the ideas assume that $\bar{v}$ is a strict substructure of y over v; in other words there is an initial structure (y', v') on $S(v)$ with respect to (y, v) such that $F(\bar{v}) = v'v$ and $v' \in \mathscr{B}^*$. Without loss of generality we may suppose $v' = S(v) \in \mathscr{B}_0$ (prop. 2). With the notation of the proof of prop. 4 one has a decomposition $f = f_1 \bar{v}$ where $F(f_1) = ug$. Thus ug is an F-morphism from x into y', and (y', g) becomes a right structure on $T(u)$ which is compatible with the left structure (x, u). However, $\bar{u}$ is a quotient structure of x over u; in other words there is a final structure (x', u') on $T(u)$ with respect to (x, u) such that $F(\bar{u}) = uu'$. We infer the existence of an $\bar{f} \in \mathscr{A}(x', y')$ such that $u'F(\bar{f}) = g$. The last equation of the theorem results from the computation

$$F(\bar{u}\bar{f}\bar{v}) = F(\bar{u})F(\bar{f})F(\bar{v}) = u\,u'F(\bar{f})v'v = u\,g\,v = F(f)$$

and the almost injectiveness of F.

Remark. $\bar{f}$ need not be a bimorphism of $\mathscr{A}$ even if g is a bimorphism of $\mathscr{B}$.

Examples. 1. Let $\mathscr{A}$ be the category of the topological spaces and continuous functions, $\mathscr{B}$ the category of the sets and functions, both of type $\mathfrak{U}$ ($\mathfrak{U}$ a universe) and F the forgetful functor. Sub- and quotient structures always exist; every continuous function f has a rectangular decomposition in the sense of th. 1 over the canonical decomposition of $F(f)$; in fact, all sub- and quotient structures are strict.

2. Let $\mathscr{A}$ be the category of the categories and functors, $\mathscr{B}$ the category of the sets and functions, both of type $\mathfrak{U}$, and F the forgetful functor. Sub- and quotient structures do not always exist. Let Y be a category of type $\mathfrak{U}$, and v the canonical injection from a subset B of $F(Y)$ into $F(Y)$. In order that Y admits a substructure over v it is necessary and sufficient that B contains the composite $\varphi\psi$ of each pair (φ, ψ) of elements of B composable in Y whose sources and targets also belong to B; in order that Y admits a strict substructure over v it is necessary and sufficient that B contains source and target of each element φ of B, and that B contains the composite $\varphi\psi$ of each pair (φ, ψ) of elements of B composable in Y. Let X be a category of type $\mathfrak{U}$, and u the canonical surjection from $F(X)$ into a quotient set B of $F(X)$. In order that X admits a quotient structure over u it suffices that B is the quotient of $F(X)$ modulo a congruence relation R [*6*, th. 1.2, p. 567]; in order that X admits a strict quotient structure over u it suffices that B is the quotient of $F(x)$ modulo a congruence relation R which, in addition, verifies the following replacement condition: If α is a unit of X and φ a morphism of X such that α is equivalent mod. R to the source of φ, then there exists a morphism φ' of X equivalent mod. R to φ whose source is α [*4*, prop. 13, p. 13 bis].

Part II. Subgadgets and Quotient Gadgets

4. Let $\mathscr{A}$ be a category. We are going to introduce preorder relations in the sets $\mathscr{A}^m$ of the monomorphisms and $\mathscr{A}^e$ of the epimorphisms of $\mathscr{A}$. For subsets X and Y of $\mathscr{A}$, we denote by XY the set of the morphisms of the form fg where (f, g) is a composable pair with $f \in X$, $g \in Y$; we write fX in the place of $\{f\}X$, etc.

Definition 4. By abuse of the language, we say «*u is contained in v*» and write $u \subset v$, instead of «$u \in \mathscr{A}^m \wedge v \in \mathscr{A}^m \wedge u \in \mathscr{A} v$»; we say «*$u$ is coarser than v*» and write $u \sqsubset v$, instead of «$u \in \mathscr{A}^e \wedge v \in \mathscr{A}^e \wedge u \in v\mathscr{A}$».

Remarks. 1. If $u \subset v$, then there exists one and only one $\mathscr{A}$-morphism f such that $u = fv$; moreover, f is monic, and one has $T(u) = T(v)$. Trivially $u \subset T(u)$ for each monomorphism u.

2. The relations $u \subset u \Leftrightarrow u \in \mathscr{A}^m$, and $u \subset v \wedge v \subset w \Rightarrow v \subset w$ are theorems. In other words, the relation $u \subset v$ is a preorder relation in the set $\mathscr{A}^m$. As is well-known, it leads to an equivalence relation in $\mathscr{A}^m$, namely «$u \subset v \wedge v \subset u$», denoted by «*$u$ and v are mono-equivalent*» and to an order relation in the associated quotient set. For our purposes, however, it is more convenient to form an ordered set by picking representatives. Note that u and v are mono-equivalent if and only if they are monic and verify $u \in \mathscr{A}^* v$. In the dual case we say «*u and v are epi-equivalent*».

In the spirit of the last remark, the term $\tau_x(x \subset u \wedge u \subset x)$ will be denoted by (u) or by «*the canonical monomorphism of $\mathscr{A}$ associated with u*». We say that $\mathfrak{u}$ *is a canonical monomorphism of $\mathscr{A}$* if $\mathfrak{u}$ is equal to the canonical monomorphism of $\mathscr{A}$ associated with some monomorphism of $\mathscr{A}$. Dually one speaks of *canonical epimorphisms*, and uses the notation $[u]$.

Proposition 5. 1° *Let u be an $\mathscr{A}$-monomorphism. Then u and (u) are mono-equivalent: inparticular, (u) is monic and has the same target as u.*

2° *In order that the $\mathscr{A}$-monomorphisms u and v be mono-equivalent it is necessary and sufficient that their associated canonical monomorphisms (u) and (v) be equal.*

3° *If $\mathfrak{u}$ is a canonical monomorphism then $\mathfrak{u}$ and $(\mathfrak{u})$ are equal.*

4° *Mono-equivalent canonical monomorphisms are equal.*

Denote by R or $x \equiv x'$ the equivalence relation «$x \subset x' \wedge x' \subset x$». Denote further by T the letter u, by T' the letter v. Apply criteria C4 of the appendix and note that $\vartheta\{T\}$[1] is (u), $\vartheta\{T'\}$ is (v), to obtain prop. 5.

Corollary. *The set $\mathscr{A}^{mc}$ of the canonical monomorphisms of $\mathscr{A}$ is a subset of $\mathscr{A}^m$, and, endowed with the order induced by $u \subset v$, an ordered set.*

Follows from 1° and 4°.

Proposition 6. *Let $u \in \mathscr{A}^m(a, b)$. Then $\mathfrak{v} \in \mathscr{A}^{mc}(\cdot, a)$ implies $(\mathfrak{v}u) \in \in \mathscr{A}^{mc}(\cdot, b)$ and $(\mathfrak{v}u) \subset u$. The mapping $\mathfrak{v} \mapsto (\mathfrak{v}u)$ from $\mathscr{A}^{mc}(\cdot, a)$ into the set of the $\mathfrak{w} \in \mathscr{A}^{mc}(\cdot, b)$ such that $\mathfrak{w} \subset u$ is bijective.*

---

[1] For typographical reasons braces rather than wavy lines are used to emphasize a letter in an assemblage.

Let $\mathfrak{v} \in \mathscr{A}^{mc}(\cdot, a)$. Then the pair $(\mathfrak{v}, u)$ is composable. Because $(\mathfrak{v}u)$ and $\mathfrak{v}u$ are mono-equivalent one obtains $(\mathfrak{v}u) \subset \mathfrak{v}u \subset u$ and $T((\mathfrak{v}u)) = T(\mathfrak{v}u) = T(u) = b$. We now have a mapping as indicated in the statement. If $(\mathfrak{v}u) = (\mathfrak{v}'u)$ then there exists an isomorphism f such that $\mathfrak{v}'u = f\mathfrak{v}u$. But u is monic; hence $v' = f\mathfrak{v}$ which signifies the mono-equivalence of $\mathfrak{v}$ and $\mathfrak{v}'$. Since $\mathfrak{v}$ and $\mathfrak{v}'$ are canonical we infer $\mathfrak{v} = \mathfrak{v}'$. Thus our function is injective. Now, let $\mathfrak{w} \in \mathscr{A}^{mc}(\cdot, b)$ be given subject to the condition $\mathfrak{w} \subset u$. By definition, there exists a morphism g such that $\mathfrak{w} = gu$, and g is monic. Write $\mathfrak{v}$ in place of (g). Then $T(\mathfrak{v}) = T(g) = S(u) = a$, and $\mathfrak{v} \in \mathscr{A}^{mc}(\cdot, a)$. We assert that the equation $(\mathfrak{v}u) = \mathfrak{w}$ is true. Since $\mathfrak{w}$ is canonical, it suffices to show that $\mathfrak{v}u$ and $\mathfrak{w}$ are mono-equivalent. By construction there is an isomorphism k such that $\mathfrak{v} = kg$; hence $\mathfrak{v}u = kgu = k\mathfrak{w}$. We infer that $\mathfrak{v}u$ and $\mathfrak{w}$ are mono-equivalent. Thus the function in question is surjective.

5. Let $\mathscr{A}$ be a category. Recall that $\mathscr{A}^m$ (resp. $\mathscr{A}^e$) denotes the set of the monomorphisms (resp. epimorphisms) of $\mathscr{A}$, and that $u \subset v$ (resp. $u \sqsubset v$) abbreviates the relation «$u \in \mathscr{A}^m \wedge v \in \mathscr{A}^m \wedge u \in \mathscr{A}\, v$» (resp. «$u \in \mathscr{A}^e \wedge v \in \mathscr{A}^e \wedge u \in v\,\mathscr{A}$»).

Definition 5a. Let $f \in \mathscr{A}$. We say «*v is an image of f in $\mathscr{A}$*» in place of the conjunction of the following relations:

(i) $v \in \mathscr{A}^m \wedge f \in \mathscr{A}^e v$.

(ii) $(\forall v')\,(v' \in \mathscr{A}^m \wedge f \in \mathscr{A}^e v' \Rightarrow v' \subset v)$.

Remarks. 1. Let f and v' be $\mathscr{A}$-morphisms, and let v be an image of f. In order that v' be an image of f it is necessary and sufficient that v and v' be mono-equivalent.

2. Let $f = uv = u'v'$ be decompositions of f where u, u' are epic, v, v' are monic. If v is an image of f, then there exists a unique bimorphism t of $\mathscr{A}$ such that $u = u't$ and $v' = tv$, and one has $u \sqsubset u'$, $v' \subset v$.

The term τ_x (x is an image of f in $\mathscr{A}$) is denoted by $\operatorname{im}(f)$, the term $S(\operatorname{im}(f))$ by $\operatorname{Im}(f)$.

Proposition 7. *Let $f \in \mathscr{A}$; suppose that f admits an image. Then:*

1° $\operatorname{im}(f)$ *is an image of f.*

2° $\operatorname{im}(f)$ *is the largest canonical monomorphism of $\mathscr{A}$ with the property $f \in \mathscr{A}^e \operatorname{im}(f)$* (relative to the order relation $\mathfrak{u} \subset \mathfrak{v}$).

Denote by P the relation «x is an image of f in $\mathscr{A}$», and, as previously, by R or $x \equiv x'$ the equivalence relation «$x \subset x' \wedge x' \subset x$». By hypothesis the relation $(\exists x)\, P$, which is $P\{\operatorname{im}(f)\}$ or «$\operatorname{im}(f)$ is an image of f in $\mathscr{A}$», is a theorem; whence 1°. In particular, one has according to def. 5a:

(i) $\operatorname{im}(f) \in \mathscr{A}^m \wedge f \in \mathscr{A}^e \operatorname{im}(f)$.

(ii) $(\forall v')\,(v' \in \mathscr{A}^m \wedge f \in \mathscr{A}^e v' \Rightarrow v' \subset \operatorname{im}(f))$.

To finish the proof of 2° it suffices therefore to show that im (f) is actually canonical. By remark 1, the relations

$$P\{x\} \wedge R\{x, x'\} \Rightarrow P\{x'\} \quad \text{and} \quad P\{x\} \wedge P\{x'\} \Rightarrow R\{x, x'\}$$

are theorems; as already mentioned $(\exists x)\, P$ is also a theorem. In short, P is *compatible* with R in x, and is *R-functional* in x. Application of crit. C5 results in the theorem

$$\tau_x(P) = \vartheta\{\mathfrak{a}\} \wedge \mathfrak{a} \in \mathscr{A}^m$$

where $\mathfrak{a}$ is an auxiliary constant. However, $\tau_x(P)$ is $\mathrm{im}(f)$.

For completeness' sake we give:

Definition 5b. Let $f \in \mathscr{A}$. We say «*u is a coimage of f in $\mathscr{A}$*» in place of the conjunction of the following relations:

(i) $u \in \mathscr{A}^e \wedge f \in u\mathscr{A}^m$.

(ii) $(\forall u')\,(u' \in \mathscr{A}^e \wedge f \in u'\mathscr{A}^m \Rightarrow u' \sqsubset u)$.

One also introduces the abbreviations $\mathrm{coim}(f)$ and $\mathrm{Coim}(f)$.

Theorem 2. *Let f be a morphism of the category $\mathscr{A}$. Assume that f admits an image and a coimage. Then there exists one and only one $\mathscr{A}$-morphism g such that $f = \mathrm{coim}(f) \cdot g \cdot \mathrm{im}(f)$; besides g is a bimorphism.*

By prop. 7 $\mathrm{im}(f)$ is monic, $\mathrm{coim}(f)$ epic; hence there exists at most one g with the desired property. Consider the diagram

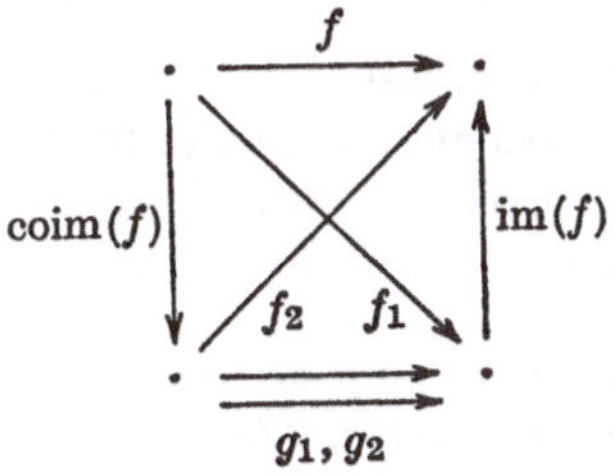

in which $f = f_1\,\mathrm{im}(f)$ and $f = \mathrm{coim}(f)\, f_2$ are the decompositions of f associated with $\mathrm{im}(f)$ and $\mathrm{coim}(f)$ respectively; in particular, f_1 is epic, f_2 monic. By definition of $\mathrm{im}(f)$, $f_2 \subset \mathrm{im}(f)$; so there exists a monomorphism g_2 such that $f_2 = g_2 \cdot \mathrm{im}(f)$ (remark 1 after def. 4). By definition of coim (f), $f_1 \sqsubset \mathrm{coim}(f)$; so there exists an epimorphism g_1 such that $f_1 = \mathrm{coim}(f)\, g_1$. We obtain the equations $f = f_1\,\mathrm{im}(f) = \mathrm{coim}(f)\, g_1\, \mathrm{im}(f)$ and $f = \mathrm{coim}(f)\, f_2 = \mathrm{coim}(f)\, g_2\, \mathrm{im}(f)$. By the preceding remark, $g_1 = g_2$, which shows that g_1 is not only epic but also monic.

In connexion with theorem 2, we call a category *canonical* if every morphism admits an image and a coimage. Each morphism f of a canonical category has a unique decomposition $f = \mathrm{coim}(f)\, \bar{f}\, \mathrm{im}(f)$, called *the canonical decomposition of f*, in which $\bar{f}$ is a bimorphism.

In canonical categories, one can characterize monomorphisms and epimorphisms in terms of coimages and images.

Proposition 8. *Let f be a morphism of a canonical category $\mathcal{A}$ with source a and target b.*

a) *f is monic if and only if a is a coimage of f.*

b) *f is epic if and only if b is an image of f.*

Indeed, let $f = \mathrm{coim}(f)\bar{f}\,\mathrm{im}(f)$ be the canonical decomposition of f. If f is monic, then $f = af$ is a decomposition with $a \in \mathcal{A}^e$, $f \in \mathcal{A}^m$, so that $a \sqsubset \mathrm{coim}(f) \sqsubset a$. In other words a and $\mathrm{coim}(f)$ are epi-equivalent. Conversely, if a is a coimage of f then $\mathrm{coim}(f)$ and a are epi-equivalent which renders $\mathrm{coim}(f)$ an isomorphism. By theorem 2, $\bar{f}$ is a bimorphism; thus f is the composite of monomorphisms.

6. In canonical categories, one can define direct images and reciprocal coimages under morphisms.

Definition 6. Let $f \in \mathcal{A}(a, b)$. For each $u \in \mathcal{A}^m(\cdot, a)$, the term $\mathrm{im}(uf)$ is denoted by $f(u)$ and called *the direct image of u under f*. For each $u \in \mathcal{A}^e(b, \cdot)$ the term $\mathrm{coim}(fu)$ is denoted by $f^{-1}[u]$ and called *the reciprocal coimage of u under f*.

Remark. For $f \in \mathcal{A}(a, b)$ one has immediately the formulae

$$\mathrm{im}(f) = f(a), \quad \mathrm{coim}(f) = f^{-1}[b].$$

One obtains more information about images and coimages if one requires, as is the case in abelian categories, that every bimorphism is an isomorphism. We term a category with this property a category *of compact type*. (For a different description see next section.)

Theorem 3. *Let $\mathcal{A}$ be a category of compact type in which every morphism admits an image or a coimage. Then $\mathcal{A}$ is canonical. More precisely, if $f = uv$ is a decomposition of f where v is an image or u a coimage of f, then v is an image and u a coimage of f.*

To fix the ideas let $f = uv$ be the decomposition of f associated with an image v. We claim that u is a coimage of f. Clearly, u is epic. Let $f = u'v'$ be a decomposition with u' epic, v' monic. According to remark 2 after def. 5a, there is a bimorphism t such that $u = u't$ and $v' = tv$. By hypothesis, t belongs to $\mathcal{A}^*$ which yields $u' \sqsubset u$. Therefore the conditions (i) and (ii) of def. 5b are fulfilled.

Remark. In a canonical category $\mathcal{A}$ of compact type, the «rectangular» canonical decomposition $f = \mathrm{coim}(f) \cdot \bar{f} \cdot \mathrm{im}(f)$ of $f \in \mathcal{A}$ can be replaced by a «triangular» decomposition $f = uv$ where u is a coimage, v an image of f, which however, is not canonical anymore.

Proposition 9. *Let $\mathscr{A}$ be a canonical category. $\mathscr{A}$ is of compact type if and only if, for each $\mathscr{A}$-morphism f, the bimorphism $\bar{f}$ in the canonical decomposition $f = \operatorname{coim}(f)\bar{f}\operatorname{im}(f)$ is invertible.*

The necessity is clear; the sufficiency follows from prop. 8.

Proposition 10. *Let $\mathscr{A}$ be a canonical category of compact type; furthermore let (f, g) be a composable pair of morphisms of $\mathscr{A}$. Then*

a) $\operatorname{im}(fg) \subset \operatorname{im}(g)$, *and equality holds if f is epic.*

b) $\operatorname{coim}(fg) \sqsubset \operatorname{coim}(f)$, *and equality holds if g is monic.*

Consider the commutative diagram

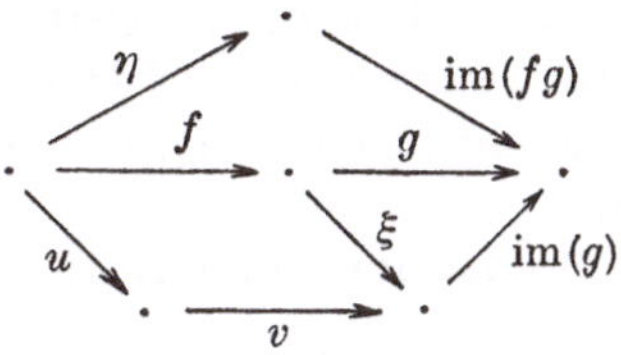

in which ξ is uniquely determined by $\operatorname{im}(g)$, η by $\operatorname{im}(fg)$, in which v is an image of $f\xi$, and in which u is uniquely determined by v. In view of theorem 3, ξ, η, u are coimages of $g, fg, f\xi$ respectively. Because $fg = u\, v\operatorname{im}(g)$ is a decomposition of fg where u is epic, $v\operatorname{im}(g)$ monic, one has $u \sqsubset \eta$ and $\operatorname{im}(fg) \subset v\operatorname{im}(g) \subset \operatorname{im}(g)$ (remark 2 after def. 5a). If, in addition, f is epic then $fg = f\xi\operatorname{im}(g)$ is a decomposition of fg where $f\xi$ is epic, $\operatorname{im}(g)$ monic, and the relation $\operatorname{im}(g) \subset \operatorname{im}(fg)$ results.

Corollary. *Let f be an $\mathscr{A}$-morphism from a into b. Then the mappings $i \mapsto f(i)$ from $\mathscr{A}^m(\cdot, a)$ into $\mathscr{A}^m(\cdot, b)$ and $q \mapsto f^{-1}[q]$ from $\mathscr{A}^e(b, \cdot)$ into $\mathscr{A}^e(a, \cdot)$ are increasing* (relative to $u \subset v$ and $u \sqsubset v$ respectively).

Suppose for example $i' \subset i$. Then $i' = ti$ for some monomorphism t. Hence $f(i') = \operatorname{im}(i' f) = \operatorname{im}(t i f) \subset \operatorname{im}(i f) = f(i)$.

In order to be able to talk about reciprocal images and direct coimages under morphisms one has to require that the category $\mathscr{A}$ *admits fibred products of pairs* (pull-backs) *and amalgamated sums of pairs* (push-outs). For the remainder of this section we assume those requirements to be satisfied, and refer to $\mathscr{A}$ as a *category with pull-backs and push-outs.* We shall need:

Lemma. *Let $(x_\iota)_{\iota\in I}$ be a family of $\mathscr{A}$-morphisms $x_\iota\colon a_\iota \to a_0$; furthermore, let $(P, (u_\iota)_{\iota\in I\cup\{0\}})$ be a fibred product of the family (x_ι), where u_0 is such that $u_\iota x_\iota = u_0$ for all $\iota \in I$. Let $\varkappa \in I$. If, for all $\iota \in I$ such that $\iota \neq \varkappa$, the morphisms x_ι are monic, then $u_\varkappa$ is monic.*

Let the $\mathscr{A}$-morphisms s, t be such that $s u_\varkappa = t u_\varkappa$. By the uniqueness condition incorporated in the definition of limits it sufficies to show $s u_\iota = t u_\iota$ for all $\iota \in I$ in order to obtain $s = t$. Note that, for $\iota \in I$, one has $su_\iota x_\iota = su_0 = su_\varkappa x_\varkappa = tu_\varkappa x_\varkappa = tu_0 = tu_\iota x_\iota$. However, for $\iota \neq \varkappa$, x_ι is monic; thus $su_\iota = tu_\iota$ in this case, and trivially for $\iota = \varkappa$.

Definition 7. Let $f \in \mathscr{A}(a, b)$. For each $u \in \mathscr{A}^m(\cdot, b)$, the term $\tau_x((\exists f')((x, f')$ is a fibred product of (f, u) in $\mathscr{A}))$ is denoted by $f^{-1}(u)$ and called the *reciprocal image of u under f*. For each $u \in \mathscr{A}^e(a, \cdot)$, the term $\tau_x((\exists f')((x, f')$ is an amalgamated sum of (f, u) in $\mathscr{A}))$ is denoted by $f[u]$ and called *the direct coimage of u under f*.

Remarks. 1. Under our hypothesis on the category $\mathscr{A}$, for $f \in \mathscr{A}(a, b)$, $j \in \mathscr{A}^m(\cdot, b)$, $p \in \mathscr{A}^e(a, \cdot)$, there exist $\mathscr{A}$-morphism f' and f'' such that $(f^{-1}(j), f')$ is a fibred product of (f, j), and that $(f[p], f'')$ is an amalgamated sum of (f, p). In particular, one has $f^{-1}(j) \cdot f = f'j$, $f \cdot f[p] = pf''$; $f^{-1}(j)$ is a canonical monomorphism, $f[p]$ a canonical epimorphism (Crit. C5 of the appendix).

2. For $f \in \mathscr{A}(a, b)$, one has immediately the formulae $(a) = f^{-1}(b)$, $[b] = f[a]$.

Proposition 11. *Let $\mathscr{A}$ be a category with pull-backs and push-outs, and let $f \in \mathscr{A}(a, b)$. Then the mappings $j \mapsto f^{-1}(j)$ from $\mathscr{A}^m(\cdot, b)$ into $\mathscr{A}^m(\cdot, a)$ and $p \mapsto f[p]$ from $\mathscr{A}^e(a, \cdot)$ into $\mathscr{A}^e(b, \cdot)$ are increasing* (relative to $u \subset v$ and $u \sqsubset v$ respectively).

Suppose for example $j' \subset j$, so that $j' = tj$ for some $t \in \mathscr{A}$. Let $(f^{-1}(j), g)$ be a fibred product of (f, j), $(f^{-1}(j'), g')$ a fibred product of (f, j'). Then $f^{-1}(j') f = g'j' = g'tj$. Hence there exists $s \in \mathscr{A}$ such that $sf^{-1}(j) = f^{-1}(j')$ and $sg = g't$. In particular, one has $f^{-1}(j') \subset f^{-1}(j)$.

Proposition 12. *Let $\mathscr{A}$ be a canonical category with pull-backs and push-outs, of compact type. In the situation*

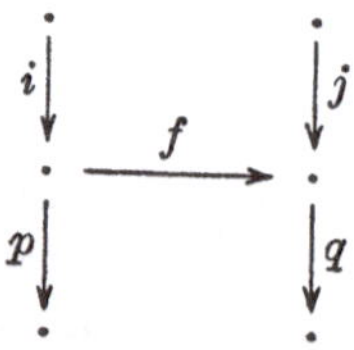

where i and j are monic, p and q are epic, one has:

a) *$i \subset f^{-1}(f(i))$, and equality holds if f is monic, i canonical.*

b) *$q \sqsubset f[f^{-1}[q]]$, and equality holds if f is epic, q canonical.*

c) *$f(f^{-1}(j)) \subset j$, and equality holds if f is a retraction, j canonical.*

d) *$f^{-1}[f[p]] \sqsubset p$, and equality holds if f is a section, p canonical.*

a) Write k for $f(i)$. Since $if = f'k$ for some epimorphism f', there is $t \in \mathscr{A}$ such that $tf^{-1}(k) = i$ and $tf'' = f'$ where $(f^{-1}(k), f'')$ is a fibred product of (f, k). In particular one has $i \subset f^{-1}(k)$. Assume now f monic, i canonical. In this case, f' is an isomorphism (prop. 8, th. 3); write t for the inverse of f'. If (u, v) is a pair of $\mathscr{A}$-morphisms such that $uf = vk$, then necessarily $vtf' = v$; besides one has $vt\, if = vtf'k = vk = uf$. But f is monic; so $vti = u$. We have shown, that in this case (i, f') is a fibred product of (f, k). We conclude that $i = f^{-1}(k)$.

c) Write k for $f^{-1}(j)$. For some $\mathscr{A}$-morphism f', the pair (k, f') is a fibred product of (f, j). f' has a decomposition $f' = f_1 f_2$ where f_1 is epic, f_2 monic, and $kf = f_1 f_2 j$ is a decomposition of kf with f_1 epic, $f_2 j$ monic. There is another decomposition of kf, namely $kf = f'' f(k)$ where f'' is a coimage of kf, $f(k)$ monic (theorem 3). We infer that $f_1 \sqsubset f''$ and $f(k) \subset f_2 j \subset j$ (remark 2 after def. 5a.). Assume now that f is a retract, j canonical. We claim that f' is also a retract. Indeed, let g be a section associated with f. One has $jgf = j = S(j)j$. Since (k, f') is a fibred product of (f, j), there is an $\mathscr{A}$-morphism g' such that $g'k = jg$ and $g'f' = S(j)$. In particular, $kf = f'j$ is a decomposition of kf where f' is epic, j monic. It follows that $j \subset f(k)$.

7. In slight modification of Isbell's suggestion [*8*, no 1.2., p. 6], we pose, for a category $\mathscr{A}$:

Definition 8a. The relation «$v \in \mathscr{A}^m \wedge (\forall h)\, (h \in \mathscr{A}^e \wedge v \in h\mathscr{A}^m \Rightarrow h \in \mathscr{A}^*)$» is denoted by «*$v$ is an extremal monomorphism of $\mathscr{A}$*».

Definition 8b. The relation «$u \in \mathscr{A}^e \wedge (\forall h)\, (h \in \mathscr{A}^m \wedge u \in \mathscr{A}^e h \Rightarrow h \in \mathscr{A}^*)$» is denoted by «*$u$ is an extremal epimorphism of $\mathscr{A}$*».

We denote by $\mathscr{A}^{me}$ the set of the extremal monomorphisms, by $\mathscr{A}^{ee}$ the set of the extremal epimorphisms of $\mathscr{A}$. In the situation $v \in \mathscr{A}^{me}(a,b)$, we say that v (resp. a) is *a subgadget* (resp. *Subgadget*) *of* b; in the situation $u \in \mathscr{A}^{ee}(a, b)$ we say that u (resp. b) is *a quotient gadget* (resp. *Quotient gadget*) *of* a.

Remarks. 1. In a *decomposable* category, i.e. a category $\mathscr{A}$ such that $\mathscr{A} = \mathscr{A}^e \mathscr{A}^m$, and especially in a canonical category, v is an extremal monomorphism if and only if it verifies the relation «$v \in \mathscr{A}^m \wedge (\forall h)$ $(h \in \mathscr{A}^e \wedge v \in h\mathscr{A} \Rightarrow h \in \mathscr{A}^*)$». In this case, our and Isbell's notions coincide, and the relations $uv \in \mathscr{A}^{me} \Rightarrow u \in \mathscr{A}^{me}$ and $uv \in \mathscr{A}^{ee} \Rightarrow v \Rightarrow \mathscr{A}^{ee}$ are true.

2. Sections are extremal monomorphisms; retractions are extremal epimorphisms. Indeed, let $fg = a \in \mathscr{A}_0$, and let $f = hk$ where $h \in \mathscr{A}^e$, $k \in \mathscr{A}^m$. Then $hkg = a$ so that h becomes an epic section, and hence an isomorphism. One obtains now immediately the equations

$$\mathscr{A}^e \cap \mathscr{A}^{me} = \mathscr{A}^{ee} \cap \mathscr{A}^m = \mathscr{A}^*.$$

Proposition 13. *Let $\mathscr{A}$ be a category. The following properties are equivalent:*

a) *Every monomorphism of $\mathscr{A}$ is an extremal monomorphism.*

b) *Every epimorphism of $\mathscr{A}$ is an extremal epimorphism.*

c) *$\mathscr{A}$ is of compact type.*

Using remark 2 above, one has $\mathscr{A}^e \cap \mathscr{A}^m = \mathscr{A}^e \cap \mathscr{A}^{me} = \mathscr{A}^*$, proving the implication a) $\Rightarrow$ c). In order to prove the implication $c) \Rightarrow a)$, let v be a monomorphism, and $v = hk$ a decomposition of v with h epic, k monic. Then h is also monic, and hence invertible.

Remark. In a category of compact type, the notions extremal monomorphism, pure monomorphism, copure monomorphism and monomorphism coincide. For def. see [*8*, pp. 6 and 7].

Theorem 4. *Let $f \in \mathscr{A}$, and assume that f admits an image. If $v \in \mathscr{A}^{me}$ and $f \in \mathscr{A}^e v$, then v is an image of f.*

Let v' be an image of f, and $f = u'v'$ the corresponding decomposition of f with u' epic. Then $v \subset v'$ so that $v = tv'$ for some $t \in \mathscr{A}$. One knows that t is a bimorphism (remark 2 after def. 5a). But v is extremal; thus t is an isomorphism, which makes v an image of f (remark 1 after def. 5a).

Proposition 14. *For an $\mathscr{A}$-morphism v the following properties are equivalent:*

a) *v is an extremal monomorphism.*

b) *v is its own image.*

c) *v is an image of some $f \in \mathscr{A}$.*

a) $\Rightarrow$ b) Obviously $v = S(v) \cdot v$ where $S(v)$ is epic, v monic. Now let $v = u'v'$ be a decomposition where u' is epic, v' monic. Then $u' \in \mathscr{A}^*$, and hence $v' \subset v$.

b) $\Rightarrow$ c) is immediate.

c) $\Rightarrow$ a) Let v be an image of the $\mathscr{A}$-morphism f, and let $f = uv$ be the corresponding decomposition with u epic. Further, let $v = \varphi\psi$ be a decomposition of v where φ is epic, ψ monic. We wish to show that φ is an isomorphism. However, $f = u\varphi\psi$ is a decomposition of f where $u\varphi$ is epic, ψ monic. So there exists $t \in \mathscr{A}$ such that $\psi = tv$, and one gets the equations $\varphi t v = \varphi\psi = v = S(v) \cdot v$. Since v is monic, this reduces to $\varphi t = S(v)$. In other words, φ is an epic section, and hence $\varphi \in \mathscr{A}^*$.

Remark. In view of prop. 14 one could dispose of the phrases «v is an extremal monomorphism», and «v is a subgadget of some $b \in \mathscr{A}_0$», using instead the terminology «v is an image». For example, one could restate remark 2 before prop. 13 as: Every section is an image. Dually: Every retraction is a coimage. Furthermore: Every retract of $a \in \mathscr{A}_0$ is a Subgadget and a Quotientgadget of a.

Examples. 1. Let $\mathscr{A}$ be the category of the groups and homomorphisms of type $\mathfrak{U}$ ($\mathfrak{U}$ a universe). Monomorphisms are injective, epimorphisms are surjective. Every bimorphism is invertible, so that $\mathscr{A}$ is of compact type.

Every homomorphism f from X into Y has an image, namely the canonical injection from $f\langle X\rangle$ into Y, and a coimage, namely the mapping from X into $F\langle X\rangle$ induced by f; hence $\mathscr{A}$ is canonical.

2. Let $\mathscr{A}$ be the category of the categories and functors of type $\mathfrak{U}$. The questions posed in [7, p. 575] with respect to images and coimages in $\mathscr{A}$ have an easy solution: Monomorphisms are injective, epimorphisms are mappings $f\colon X \to Y$ such that Y is the subcategory of Y generated by $f\langle X\rangle$. Every bimorphism is invertible [7, p. 574], so that $\mathscr{A}$ is of compact type. Every functor f from X into Y has an image, namely the canonical injection from the subcategory Y' of Y generated by $f\langle X\rangle$, into Y, and a coimage, namely the mapping from X into Y' induced by f; hence $\mathscr{A}$ is canonical.

3. Let $\mathscr{A}$ be the category of the Hausdorff spaces and continuous functions of type $\mathfrak{U}$. Monomorphisms are injective, epimorphisms are mappings $f\colon X \to Y$ such that $f\langle X\rangle$ is dense in Y. Not every bimorphism is invertible; hence $\mathscr{A}$ is not of compact type. Every continuous mapping from X into Y has an image, namely the canonical injection from $\overline{f\langle X\rangle}$ into Y, and a coimage, namely the canonical surjection from X into X/f. The canonical injections from closed subspaces A of Y into Y are examples of images in the category $\mathscr{A}$, all other images being mono-equivalent to them. The canonical surjections from X into quotient spaces Q of X are examples of coimages in the category $\mathscr{A}$, all other coimages being epi-equivalent to them.

8. In order to arrive at a useful definition of injectives and projectives in categories not necessarily of compact type one has to restrict the monomorphisms and epimorphisms employed in the abelian case. A good choice is offered by the extremel monomorphisms and extremal epimorphisms.

Let $\mathscr{A}$ be a category.

Definition 9a. The relation «$x \in \mathscr{A}_0 \wedge (\forall u)\,(u \in \mathscr{A}^{me} \Rightarrow \tilde{\mathscr{A}}(u, x)$ is a surjection)» is denoted by «*x is an injective of $\mathscr{A}$*». (Cf. notation introduced before def. 1.)

In other words, x is an injective if every $\mathscr{A}$-morphism f with target x can be prolonged along images.

Proposition 15. *Every retract of an injective of $\mathscr{A}$ is an injective.*

Indeed, let $u\colon a \to b$ be an extremal monomorphism, and let $q\colon x \to y$ be a retraction of the injective x. Then $\tilde{\mathscr{A}}(b, q)\,\tilde{\mathscr{A}}\,(u, y) = \tilde{\mathscr{A}}\,(u, q) = \tilde{\mathscr{A}}(u, x)\,\tilde{\mathscr{A}}(a, q)$, where $\tilde{\mathscr{A}}(u, x)$ is a surjection by hypothesis, while $\tilde{\mathscr{A}}(a, q)$ is a retraction. Thus the right side of the above equation is a surjection, which implies the surjectiveness of $\tilde{\mathscr{A}}\,(u, y)$.

Corollary. *Let x and y be isomorphic units of $\mathscr{A}$. Then x is an injective if and only if y is an injective.*

Proposition 16. *Let x be a family of injectives of $\mathscr{A}$ indexed by I, and suppose that x admits a product, say (P, π). Then P is an injective of $\mathscr{A}$.*

Let $u: a \to b$ be an extremal monomorphism, and let (P, λ) be the representation of the functor $f \mapsto \prod_{\iota \in I} \tilde{\mathscr{A}}(f, x_\iota)$ from $\mathscr{A}^\circ$ into $\mathscr{F}$ (where $\mathscr{F}$ is the category of the sets and functions of type $\mathfrak{U}$ for a sufficiently large universe $\mathfrak{U}$) corresponding to the universal point (P, π) for the same functor according to [*9*, th. 7.1, p. 53]. Each function $\tilde{\mathscr{A}}(u, x_\iota)$ is surjective by hypothesis, hence a retraction. Thus $\prod_\iota \tilde{\mathscr{A}}(u, x_\iota)$ is a retraction, and the same is true of $\tilde{\mathscr{A}}(u, P)$ by virtue of the natural isomorphisms $\lambda(a)$ and $\lambda(b)$. We infer that $\tilde{\mathscr{A}}(u, P)$ is a surjection.

Definition 9b. The relation «$x \in \mathscr{A}_0 \wedge (\forall u)\, (u \in \mathscr{A}^{ee} \Rightarrow \tilde{\mathscr{A}}(x, u)$ is a surjection)» is denoted by «*x is a projective of $\mathscr{A}$*».

There are obvious dualizations of prop. 15 and 16.

Definition 10a. We denote by «*x is a strict injective of $\mathscr{A}$*» the conjunction of the following relations:

(i) x is an injective of $\mathscr{A}$.

(ii) $(\forall u)\, (u \in \mathscr{A}^e \wedge \tilde{\mathscr{A}}(u, x)$ is a surjection $\Rightarrow u \in \mathscr{A}^*)$.

Furthermore, we shall adhere to the following terminology relative to a universe $\mathfrak{U}$. A categroy $\mathscr{A}$ is called $\mathfrak{U}$-small, or a *$\mathfrak{U}$-category* if $\mathscr{A}(a, b)$ belongs to $\mathfrak{U}$ for all pairs of units a, b of $\mathscr{A}$. Note that there always exists the least universe $\mathfrak{U}$ such that $\mathscr{A}$ is $\mathfrak{U}$-small. We say that $\mathscr{A}$ is *left $\mathfrak{U}$-complete* (resp. *right $\mathfrak{U}$-complete, $\mathfrak{U}$-complete*) if every $\mathscr{A}$-valued functor F from a category Φ whose underlying set belongs to $\mathfrak{U}$ admits limits (resp. colimits, limits and colimits). Remark that completeness may also be defined by requiring that the set underlying Φ is equipotent to an element of $\mathfrak{U}$. Remark further, that left $\mathfrak{U}$-completeness implies the existence of terminal objects, right $\mathfrak{U}$-completeness the existence of coterminal objects (for $\mathfrak{U} \neq \emptyset$). It is wellknown that $\mathscr{A}$ is left $\mathfrak{U}$-complete if and only if $\mathscr{A}$ admits $\mathfrak{U}$-products (i.e. products of families indexed by an element of $\mathfrak{U}$) and equalizers of pairs.

Closely related is the concept of a *left self-complete* (resp. *right self-complete, self-complete*) category by which we mean a category $\mathscr{A}$ such that every $\mathscr{A}$-valued functor F from a category Φ whose underlying set is a subset of some set of the form $\mathscr{A}(a, b)$ admits limits (resp. colimits, limits and colimits). It can readily be seen that a left $\mathfrak{U}$-complete $\mathfrak{U}$-category is left self-complete.

Definition 11. Let $a \in \mathscr{A}_0$. We say «*x is an a-direct object of $\mathscr{A}$*» instead of «$(\exists I)\, (\exists u)\, (x, u)$ is a product of the (constant) family $I \times \{a\}$».

We say «*x is an a-free object of $\mathscr{A}$*» instead of «$(\exists I)\,(\exists u)\,((x, u)$ is a coproduct of the (constant) family $I \times \{a\})$».

In more familiar terms, an a-direct unit x is isomorphic to a Product aI of copies of a, an a-free unit x is isomorphic to a Coproduct $a(I)$ of copies of a.

Theorem 5 (Semadeni [*10*, th., p. 64]). *Let $\mathscr{A}$ be a category which is left self-complete and canonical; let a be a strict injective of $\mathscr{A}$. Then, for all $x \in \mathscr{A}_0$, one has:*

a) *If x is a-direct, then x is an injective.*

b) *x is a Subgadget of some a-direct object.*

c) *x is an injective if and only if x is a retract of some a-direct object.*

d) *x is an injective if and only if x is an absolute subretract.*

a) follows from prop. 16.

b) Denote by I the set $\mathscr{A}(x, a)$. By hypothesis the (constant) family $I \times \{a\}$ admits a product, say (P, π). Because Δ_I is a family of morphisms $u: x \to a$, indexed by I, there exists a unique $\mathscr{A}$-morphism f such that, for all $u \in I$, $f\pi_u$ is defined and equal to u. Let $f = \operatorname{coim}(f)\bar{f}\operatorname{im}(f)$ be the canonical decomposition of f. Then $\operatorname{coim}(f)\bar{f}$ is epic, and $\operatorname{im}(f)$ a subgadget of P. Hence $u \in I$ implies $\operatorname{coim}(f)\bar{f}\operatorname{im}(f)\,\pi_u = f\pi_u = u$, and the function $\mathscr{A}(\operatorname{coim}(f)\bar{f}, a)$ is surjective. Because a is a strict injective, $\operatorname{coim}(f)\bar{f}$ is an isomorphism. We infer that f is also a subgadget of P, and its source is x.

c) If x is a retract of an a-direct object then x is an injective by the already proved part a) and by prop. 15. Assume now that x is an injective. In virtue of b), there is an extremal monomorphism $f: x \to P$ where P is a-direct. Hence x can be prolonged along f, resulting in an $\mathscr{A}$-morphism g such that $fg = x$. Thus x is a retract of P.

d) If x is an injective and the source of a subgadget f of some y then x can be prolonged along f resulting in an $\mathscr{A}$-morphism g such that $fg = x$. Hence x is an absolute subretract. Conversely assume that x is an absolute subretract. According to the already proved part b), there is an extremal monomorphism $f: x \to P$ where P is a-direct, and consequently x is a retract of P. In view of c), x is an injective.

Corollary. *Let $\mathscr{A}$ be a category which is left self-complete and canonical. If $\mathscr{A}$ possesses a strict injective, then $\mathscr{A}$ possesses enough injectives.*

Remark. It becomes clear from the proof, that th. 5 already holds whenever $\mathscr{A}$ admits products of the (constant) families $I \times \{a\}$ where I is of the form $\mathscr{A}(b. c)$. It was for convenience that we required left self-completeness.

9. By a *system of null-morphisms* for the category $\mathscr{A}$ we mean a family ω indexed by $\mathscr{A}_0 \times \mathscr{A}_0$ verifying the following properties:

(i) $(a, b) \in \mathscr{A}_0 \times \mathscr{A}_0$ implies $\omega_{ab} \in \mathscr{A}(a, b)$.

(ii) The relations $f \in \mathscr{A}(a', a)$ and $g \in \mathscr{A}(b, b')$ imply the relation $f\omega_{ab}g = \omega_{a'b'}$.

A category possesses at most one system of null-morphisms. Let ω be such a system for $\mathscr{A}$. Frequently we omit the subscripts in ω_{ab}. One can define kernels and cokernels in the usual way. Let $f \in \mathscr{A}$. We say «*k is a kernel of f in* $\mathscr{A}$» instead of «k is an equalizer of the pair $(f, \omega_{S(f)T(f)})$ in $\mathscr{A}$»; we say «l *is a cokernel of f in* $\mathscr{A}$» instead of «l is a coequalizer of the pair $(f, \omega_{S(f)T(f)})$ in $\mathscr{A}$». We agree to write $\ker(f)$ for the term τ_x (x is a kernel of f in $\mathscr{A}$), coker (f) for the term τ_x (x is a cokernel of f in $\mathscr{A}$).

Remarks. 1. If f admits a kernel, then $\ker(f)$ is a canonical monomorphism and a kernel of f.

2. The relations: a) u is a null-morphism (i.e. of the form ω_{ab}); b) $S(u)$ is a kernel of u; c) $T(u)$ is a cokernel of u; are equivalent.

3. If u is a monomorphism of $\mathscr{A}$, then $fu = \omega$ implies $f = \omega$.

Proposition 17. *Let f be a morphism of the category $\mathscr{A}$ with null-morphisms. Assume that f admits a kernel and a cokernel; assume further, that* $\ker(f)$ *admits a cokernel, and that* coker(f) *admits a kernel. Then there exists one and only one $\mathscr{A}$-morphism g such that* $f = \text{coker}(\ker(f)) \cdot g \cdot \ker(\text{coker}(f))$.

Since f coker$(f) = \omega$, there is $f_1 \in \mathscr{A}$ such that $f = f_1 \ker(\text{coker}(f))$. Then one has $\ker(f) f_1 \ker(\text{coker}(f)) = \ker(f) f = \omega$, so that $\ker(f) f_1 = \omega$; for ker (coker(f)) is monic (remarks 1 and 3). Consequently there exists g such that $f_1 = \text{coker}(\ker(f))\, g$. The computation coker (ker(f)) $\times\, g \cdot \ker(\text{coker}(f)) = f_1\, ker\, (\text{coker}(f)) = f$ ends the existence proof. The uniqueness of g is clear since coker (ker(f)) is epic, ker(coker(f)) monic (remark 1).

Proposition 18. *We preserve the notation and hypotheses of prop. 17. If the unique morphism $\bar{f}$ in the decomposition*

$$f = \text{coker}(\ker(f)) \cdot \bar{f} \cdot \ker(\text{coker}(f))$$

of f is epic, then ker(coker(f)) *is an image of f.*

Assume $\bar{f}$ is epic. Then $f = \text{coker}(\ker(f)) \cdot \bar{f} \cdot \ker(\text{coker}(f))$ is a decomposition of f with coker (ker(f)) $\bar{f}$ epic, ker (coker(f)) monic. Take now a decomposition $f = uv$ of f with u epic, v monic. Then uv coker (f) $= f$ coker$(f) = \omega$; so v coker $(f) = \omega$ since u is epic. Hence there exists $t \in \mathscr{A}$ such that $v = t \ker(\text{coker}(f))$. In other words, one has $v \subset \ker(\text{coker}(f))$.

By an *exact category* we mean a category $\mathscr{A}$ with null-morphisms such that every morphism f of $\mathscr{A}$ admits a kernel and a cokernel, and that the morphism $\bar{f}$ determined uniquely by the equation

$$f = \operatorname{coker}(\ker(f)) \cdot \bar{f} \cdot \ker(\operatorname{coker}(f))$$

is a bimorphism.

Example. Every abelian category is exact.

Theorem 6. *Let $\mathscr{A}$ be an exact category. Then $\mathscr{A}$ is a canonical category. More precisely, for $f \in \mathscr{A}$ one has:*
$\operatorname{im}(f) = \ker(\operatorname{coker}(f))$;
$\operatorname{coim}(f) = \operatorname{coker}(\ker(f))$.

The theorem is a corollary to prop. 17 and 18, using the fact that the above equations are built up of canonical monomorphisms and canonical epimorphisms respectively.

By an *Isbell bicategory* we mean a category $\mathscr{A}$ which, in addition to the graph of the internal composition law, is endowed with a pair $(\mathscr{S}, \mathscr{I})$ verifying the following axioms:
(i) $\mathscr{S}$ is a stable subset of $\mathscr{A}^e$; $\mathscr{I}$ is a stable subset of $\mathscr{A}^m$.
(ii) $\mathscr{S} \cap \mathscr{I} = \mathscr{A}^*$.
(iii) $\mathscr{S}\mathscr{I} = \mathscr{A}$.
(iv) $(uv = u'v' \wedge u \in \mathscr{S} \wedge u' \in \mathscr{S} \wedge v \in \mathscr{I} \wedge v' \in \mathscr{I})$
$\Rightarrow (\exists t)(t \in \mathscr{A}^* \wedge u = u't \wedge v' = tv)$.

Theorem 7. *Let $\mathscr{A}$ be a canonical category. If $\mathscr{A}^{ee}$ or $\mathscr{A}^{me}$ is a stable subset of $\mathscr{A}$, then $\mathscr{A}$ can be made into an Isbell bicategory.*

To fix the ideas assume that $\mathscr{A}^{me}$ is a stable subset of $\mathscr{A}$. Take $\mathscr{A}^e$ for $\mathscr{S}$, $\mathscr{A}^{me}$ for $\mathscr{I}$. Condition (i) is evident, (ii) follows from remark 2 after def. 8b, (iii) signifies the existence of images, (iv) results from th. 4.

Corollary. *Every canonical category of compact type can be made into an Isbell bicategory. In an Isbell bicategory of compact type, the pair $(\mathscr{S}, \mathscr{I})$ is uniquely determined by the composition law; more precisely one has $\mathscr{S} = \mathscr{A}^e$ and $\mathscr{I} = \mathscr{A}^m$.*

For the stability of $\mathscr{A}^{me}$ and $\mathscr{A}^{ee}$ see prop. 13. As far as the uniqueness is concerned, it suffices to outline the proof of the proposition $\mathscr{A}^m \subset \mathscr{I}$. If $f = uv$ is one of the guaranteed decompositions of the monomorphism f with $u \in \mathscr{S}$, $v \in \mathscr{I}$, then u is a bimorphism and hence an isomorphism. We conclude that $f = uv \in \mathscr{I}$ using the properties (i) and (ii).

Appendix. Classes of Equivalent Objects.

10. We consider ourselves working in a mathematical theory $\mathscr{T}$ stronger than the strong set theory in the sense of [*11*, no. 3, p. 167]. It is our aim to generalize the deductive criteria C45 and C46 of [*2*, p. 47] and to apply these generalizations to classes of equivalent objects as defined in [2, p. 123]. In what follows we denote by x, x', y, z distinct letters which are not constants of $\mathscr{T}$, by R, $R\{x, x'\}$ or $x \equiv x'$ an equivalence relation of $\mathscr{T}$ in the letters x, x', and by P or $P\{x\}$ a relation of $\mathscr{T}$. We assume that the letters y, z do not appear in R, and that the letters x', y, z do not appear in P. We assume furthermore that the relation P is compatible (relative to x) with the equivalence relation R. The last requirement means that $(\forall y)\,(\forall z)\,(P\{y\} \wedge R\{y, z\} \Rightarrow P\{z\})$ is a theorem of $\mathscr{T}$.

We say that P is *R-univocal* in x in $\mathscr{T}$ if the relation

$$(\forall y)\,(\forall z)\,(P\{y\} \wedge P\{z\} \Rightarrow R\{y, z\})$$

is a theorem of $\mathscr{T}$.

C 1. *If P is R-univocal in x in $\mathscr{T}$, then $P \Rightarrow (x \equiv \tau_x(P))$ is a theorem of $\mathscr{T}$. Conversely, if, for a term T of $\mathscr{T}$ not containing x, $P \Rightarrow (x \equiv T)$ is a theorem of $\mathscr{T}$, then P is R-univocal in x in $\mathscr{T}$.*

Suppose that P is R-univocal in x in $\mathscr{T}$, and prove that

$$P \Rightarrow (x \equiv \tau_x(P))$$

is a theorem of $\mathscr{T}$. Adjoin the hypothesis P. Then $(\tau_x(P) | x) P$ is true; hence «$P\{x\} \wedge P\{\tau_x(P)\}$ is true. Now, since P is R-univocal in x, $x \equiv \tau_x(P)$ is a theorem of $\mathscr{T}$.

Conversely suppose that $P \Rightarrow (x \equiv T)$ be a theorem of $\mathscr{T}$. Since x is not a constant of $\mathscr{T}$ and does not appear in T, the relations

$$P\{y\} \Rightarrow (y \equiv T) \quad \text{and} \quad P\{z\} \Rightarrow (z \equiv T)$$

are theorems of $\mathscr{T}$. Adjoin the hypotheses $P\{y\}$ and $P\{x\}$. Then $y \equiv T$ and $z \equiv T$ are true, hence $y \equiv z$ is true.

We say that P is *solvable* in x in $\mathscr{T}$, if the relation $(\exists x)\, P$ is a theorem of $\mathscr{T}$.

C 2. *If P is solvable in x in $\mathscr{T}$, then $(x \equiv \tau_x(P)) \Rightarrow P$ is a theorem of $\mathscr{T}$. Conversely, if, for a term T of $\mathscr{T}$ not containing x, and equivalent to itself mod. R, $(x \equiv T) \Rightarrow P$ is a theorem of $\mathscr{T}$, then P is solvable in x in $\mathscr{T}$.*

Suppose that P is solvable in x in $\mathscr{T}$, and prove that $(x \equiv \tau_x(P)) \Rightarrow P$ is a theorem of $\mathscr{T}$. According to compatibility, the relation

$$«P\{\tau_x(P)\} \wedge \tau_x(P) \equiv x \Rightarrow P\{x\}»$$

is a theorem of $\mathscr{T}$. If we adjoin the hypothesis $x \equiv \tau_x(P)$, one sees that $P\{x\}$ is true. Hence $(x \equiv \tau_x(P)) \Rightarrow P$ is a theorem of $\mathscr{T}$.

Conversely, suppose that $(x \equiv T) \Rightarrow P$ is a theorem of $\mathscr{T}$. Then $(T \equiv T) \Rightarrow P\{T\}$ is a theorem of $\mathscr{T}$. Hence $P\{T\}$, and consequently $(\exists x)\, P$, are theorems of $\mathscr{T}$.

We say that P is *R-functional* in x in $\mathscr{T}$ if P is both R-univocal and solvable in x. By combining the deductive criteria C 1 and C 2 we obtain:

C 3. *If P is R-functional in x in $\mathcal{T}$, then $P \Leftrightarrow (x \equiv \tau_x(P))$ is a theorem of $\mathcal{T}$. Conversely, if, for a term T of $\mathcal{T}$ not containing x and equivalent to itself mod. R, $P \Leftrightarrow (x \equiv T)$ is a theorem of $\mathcal{T}$, then P is R-functional in x in $\mathcal{T}$.*

The case in which the relation P is of the form $R\{x, T\}$ for some term T not containing x deserves special attention. For such T the term $\tau_x(R\{x, T\})$ is denoted by $\vartheta\{T\}$ and called *the class of objects equivalent to* T (with regard to the equivalence relation R).

C 4. *Let T and T' be terms of $\mathcal{T}$ not containing x and equivalent to themselves mod. R respectively. The following relations are theorems of $\mathcal{T}$:*

α) $T \equiv \vartheta\{T\}$.

β) $T \equiv T' \Leftrightarrow \vartheta\{T\} = \vartheta\{T'\}$.

γ) $\vartheta\{T\} = \vartheta\{\vartheta\{T\}\}$.

δ) $\vartheta\{T\} \equiv \vartheta\{T'\} \Rightarrow \vartheta\{T\} = \vartheta\{T'\}$.

α) Adjoin the hypotheses $R\{y, T\}$ and $R\{y, z\}$. Then $R\{z, T\}$ is a theorem; in other words, the relation $R\{x, T\}$ is compatible (relative to x) with the relation R. The deductive criterion C3 applied to the relation $R\{x, T\}$ instead of P yields $R\{x, T\} \Leftrightarrow x \equiv \vartheta\{T\}$ as a theorem of $\mathcal{T}$. If one replaces x by T and recalls that T is equivalent to itself mod. R, one obtains the theorem $T \equiv \vartheta\{T\}$.

β) Observe that, according to α), the relations $T \equiv \vartheta\{T\}$, $T' \equiv \vartheta\{T'\}$ and, due to scheme S6 of [2, depliant],

$$(\vartheta\{T\} = \vartheta\{T'\}) \Rightarrow (T \equiv \vartheta\{T\} \Leftrightarrow T \equiv \vartheta\{T'\})$$

are theorems of $\mathcal{T}$. If we adjoin the hypothesis $\vartheta\{T\} = \vartheta\{T'\}$, one sees that $T \equiv T'$ is a theorem of $\mathcal{T}$. Conversely, note that the relation $R\{T, T'\}$ implies $R\{x, T\} \Leftrightarrow R\{x, T'\}$, hence also the relation $(\forall x)(R\{x, T\} \Leftrightarrow R\{x, T'\})$. According to scheme S7 [loc. cit], one sees that the relation $T \equiv T'$ implies $\vartheta\{T\} = \vartheta\{T'\}$.

γ) Follows now by letting T' be $\vartheta\{T\}$.

δ) Follows now by letting T be $\vartheta\{T\}$ and T' be $\vartheta\{T'\}$.

It is useful to know that the privileged solution of a relation P which is R-functional in x, is in fact a class of equivalent objects. More precisely:

C 5. *If P is R-functional in x in $\mathcal{T}$, then the relation*

$$(\exists y)(R\{y, y\} \wedge \tau_x(P) = \vartheta\{y\})$$

is a theorem of $\mathcal{T}$.

Let T be the term $\tau_x(P)$. The relation $P \Leftrightarrow x \equiv T$ is a theorem by virtue of C3. Since x is not a constant, $(\forall x)(P \Leftrightarrow x \equiv T)$ is also a theorem of $\mathcal{T}$. Application of scheme S7 [loc. cit.] results in the theorem $\tau_x(P) = \vartheta\{T\}$. In view of the hypothesis $(\exists x)\, P$ and of C3, T is equivalent to itself. We have shown that «$R\{T, T\} \wedge \tau_x(P) = \vartheta\{T\}$» and hence $(\exists y)(R\{y, y\} \wedge \tau_x(P) = \vartheta\{y\}$ are theorems of $\mathcal{T}$.

References

[1] Andreotti, A.: Généralités sur les catégories abéliennes. Séminaire A. Grothendieck 1957.
[2] Bourbaki, N.: Théorie des ensembles, chap. 1 et 2, 2. ed. Paris: Hermann 1960.
[3] — Théorie des ensembles, chap. 4. Paris: Hermann 1957.
[4] Ehresmann, C.: Catégories différentiables et géométrie différentielle. Lecture notes, Université de Montréal 1961.
[5] — Structures quotient. Comment. Math. Helv. **38**, 219—283. (1964).
[6] Eilenberg, S.: Foundations of fiber bundles. Lecture notes. University of Chicago, 1957.
[7] Isbell, J. R.: Some remarks concerning categories and subspaces. Canad. J. Math. **9**, 563—577 (1957).
[8] — Subobjects, adequacy, completeness, and categories of algebras. Rozprawy Mat. **36**, 3—32 (1964).
[9] MacLane, S.: Categorical algebra. Bull. Amer. Math. Soc. **71**, 40—106 (1965).
[10] Semandeni, Z.: Free and direct objects. Bull. Amer. Math. Soc. **69**, 63—65 (1963).
[11] Sonner, J.: On the formal definition of categories. Math. Z. **80**, 163—176 (1962).
[12] — Universal and special problems. Math. Z. **82**, 200—211 (1963).

Department of Mathematics
University of South Carolina
Columbia, South Carolina

Operational Categories *

By

OSWALD WYLER

1. Introduction

The present paper is an attempt to formulate at least part of the BOURBAKI theory of "espèces de structures" (see [1]) in categorical terms. While our theory is far from including all "espèces de structures" found in Mathematics — for instance, categories of manifolds or of fiber bundles are excluded — it does include all or almost all "espèces de structures" found in Algebra, and many "espèces de structures" found in Topology.

An operational category is, roughly speaking, a concrete category in which objects are determined by their underlying sets and the operations on these sets, maps are essentially mappings compatible with the operations, and subobjects are formed from subsets closed under the operations. To illustrate this, let us consider groups. A group A is determined by its underlying set $\bar{A}$ and its group operation $m_A\colon \bar{A} \times \bar{A} \to \bar{A}$. It allows also an operation $\iota_A\colon \bar{A} \to \bar{A}$ of forming inverses. A group homomorphism $f\colon A \to B$ is essentially a mapping $\bar{f}\colon \bar{A} \to \bar{B}$ of the underlying sets for which the diagrams

$$\begin{array}{ccc} \bar{A}\times\bar{A} & \underset{m_A}{\to} & \bar{A} \\ \downarrow \bar{f}\times\bar{f} & & \downarrow \bar{f} \\ \bar{B}\times\bar{B} & \underset{m_B}{\to} & \bar{B} \end{array} \qquad\qquad \begin{array}{ccc} \bar{A} & \underset{\iota_A}{\to} & \bar{A} \\ \downarrow \bar{f} & & \downarrow \bar{f} \\ \bar{B} & \underset{\iota_B}{\to} & \bar{B} \end{array}$$

are commutative. A subgroup C of a group A is given by its underlying set $\bar{C}$, a non-empty subset of $\bar{A}$ and closed under the operations m_A and ι_A, and by its operations m_C and ι_C, the restrictions to $\bar{C}$ of m_A and ι_A.

The two diagrams of the preceding paragraph show that the operations m_A and ι_A on groups A define natural transformations $m\colon U \times U \to U$ and $\iota\colon U \to U$, where U denotes the "forgetful" functor which assigns to every group A its underlying set $\bar{A}$. These and many similar examples lead to a definition of an operation as a natural transformation. In the present paper, we have modified this definition to some extent, primarily

* Received September 22, 1965.

in order to include operations which are not "everywhere defined" or not single-valued.

Our operational categories include the equational categories considered by Linton and others (see e.g. [*10*]), the categories of Ω-algebras considered by P. M. Cohn (see [*3*]), and the "algebraic theories" of Lawvere (see [*9*]). Since operational categories are more general than these other classes of categories, their theory is less rich. It is still possible, however, to discuss e.g. lattices of subobjects and of quotients, and direct and inverse limits, for operational categories in general. This we have done in the present paper. It is also possible to generalize the Jordan-Hölder theorem, in the form given to it by Goldie (see [*6*]), to general operational categories. This has been done in [*11*]; we do not treat this topic in the present paper.

An "espèce de structures" *à* la Bourbaki is characterized by its operations and its axioms, whereas we use only operations in our definition of operational categories. The axioms are not, however, left out entirely in our theory. An algebraic "espèce de structures" has in general two kinds of axioms: formal laws and existence axioms. As Eckmann and Hilton (see [*4*]) have shown, formal laws can be expressed as commutative diagrams, with operations as arrows. This is outside the scope of the present paper. An existence axiom, on the other hand, usually leads to an additional operation (example: inverses in group theory), and this is within the scope of our paper. See § 5 for examples.

After preliminaires (§ 2) and a section on inclusion functors (§ 3), we define operations and operational categories (§ 4), and we give examples (§ 5). In §§ 6—10, we discuss the general theory of operational categories: inverse limits, lattices of subobjects, direct and inverse images, injections and projections, direct limits, and related questions.

Apart from the omitted general Jordan-Hölder theorem, this is the present state of our theory. However, some natural questions about operational categories remain unanswered. For example, when is an operational category weakly exact (in the sense of [*12*]) ? More generally, when and how are kernels of maps in an operational category $\mathscr{C}$ defined (these kernels may be in a category larger than $\mathscr{C}$) ?

We shall write composition of maps in a category "from left to right", so that fg means: "first f, then g". Similarly, the result of applying a functor T to a map f is denoted by fT. If A, B are objects of a category $\mathscr{C}$, then $\mathscr{C}(A, B)$ denotes the set of all maps $f\colon A \to B$ of $\mathscr{C}$. We shall refer by $(m\,.\,n)$ to the nth item of section m.

2. Preliminaries

2.1. We denote by $\mathscr{S}$ the category of sets and mappings, with sets as objects and mappings as maps. $\mathscr{R}$ denotes the larger category of sets and relations, with sets as objects and relations as maps. A relation f from a set A to a set B, or $f\colon A \to B$, may be considered as a triple $f = (A, \Gamma, B)$, where Γ is a subset of $A \times B$, called the *graph* of f. Composition of relations is defined by

$$(A, \Gamma, B)\,(B, \Delta, C) = (A, \Gamma \circ \Delta, C),$$

where $\Gamma \circ \Delta$ is defined as the set of all pairs $(x, z) \in A \times C$ such that $(x, y) \in \Gamma$ and $(y, z) \in \Delta$ for some $y \in B$.

2.2. An important subcategory of $\mathscr{R}$ is the category of sets and functional relations which we denote by $\mathscr{Q}$. Its objects are sets, its maps are functional relations, where $f = (A, \Gamma, B)$ is called *functional* if for any $x \in A$ there is at most one $y \in B$ such that $(x, y) \in \Gamma$. Clearly $\mathscr{S}$ may be considered as a subcategory of $\mathscr{Q}$.

2.3. If $f = (A, \Gamma, B)$ is a relation, we often write $x f y$ for $(x, y) \in \Gamma$, and also $y = xf$ if f is functional.

If $f = (A, \Gamma, B)$ and $g = (A, \Delta, B)$ are relations with the same source and the same target, we put $f \leqq g$ if $\Gamma \subset \Delta$. This defines a (partial) ordering of relations.

We define the *inverse relation* $f^{\#} = (B, \Gamma^{\#}, A)$ of a relation $f = (A, \Gamma, B)$ by putting $(y, x) \in \Gamma^{\#} \Leftrightarrow (x, y) \in \Gamma$, for $(x, y) \in A \times B$. This obviously defines a contravariant functor on $\mathscr{R}$.

2.4. We denote by $\boldsymbol{P}$ the *direct image functor* and by $^{\#}\boldsymbol{P}$ the *inverse image functor*. For a set A, we denote by $A\boldsymbol{P}$ the set of all subsets of A. For a relation $f\colon A \to B$ and a subset X of A, we put

$$X(f\boldsymbol{P}) = \{y \in B \mid x f y \quad \text{for some} \quad x \in X\}.$$

This defines $\boldsymbol{P}$. We define $^{\#}\boldsymbol{P}$ by putting $f(^{\#}\boldsymbol{P}) = (f^{\#})\,\boldsymbol{P}$ for a relation f (so that parentheses become superfluous), and $A\,(^{\#}\boldsymbol{P}) = A\,\boldsymbol{P}$ for a set A. For $f\colon A \to B$ and subsets X of A and Y of B, the sets $X(f\boldsymbol{P})$ and $Y(f^{\#}\boldsymbol{P})$ are often written as Xf and $Yf^{\#}$ (or Yf^{-1}) respectively.

$\boldsymbol{P}$ is a covariant functor, and $^{\#}\boldsymbol{P}$ a contravariant functor, from $\mathscr{R}$ to $\mathscr{S}$. We shall denote by the same letters $\boldsymbol{P}$ and $^{\#}\boldsymbol{P}$ the functors form $\mathscr{S}$ to $\mathscr{S}$ obtained by restricting the domain of $\boldsymbol{P}$ and of $^{\#}\boldsymbol{P}$.

2.5. If $f = (A, \Gamma, B)$ is a relation, and if $X \subset A$ and $Y \subset B$, with inclusion mappings $j\colon X \to A$ and $k\colon Y \to B$, then the relation $jfk^{\#}\colon X \to Y$ is called the *restriction* of f to X and Y. Its graph is $\Gamma \cap (X \times Y)$.

The following three statements are logically equivalent.
(i) $X(f\boldsymbol{P}) \subset Y$. (ii) $X \subset Y(f^{\#}\boldsymbol{P})$. (iii) There is a relation $f_1\colon X \to Y$ such that $jf = f_1 k$. If these statements are true, then $f_1 = jfk^{\#}$, and f_1 is a mapping if f is a mapping.

2.6. We extend the product notation from sets and mappings (or relations) to functors with values in $\mathscr{S}$ (or in $\mathscr{R}$) as follows. Let $(T_i)_{i \in I}$ be a family of functors from a category $\mathscr{C}$ to the category $\mathscr{S}$ (or to $\mathscr{R}$).

We define a *product functor* ΠT_i by putting.

$$A(\Pi T_i) = \Pi(A T_i), \quad f(\Pi T_i) = \Pi(f T_i),$$

for an object A and a map f of $\mathscr{C}$. More general inverse and direct limits of functors can of course also be defined, but we shall not need them. We put $\Pi T_i = T^I$ if $T_i = T$ for all $i \in I$.

3. Inclusion Functors

Definition 3.1. *An inclusion functor (on $\mathscr{S}$) is a covariant functor $T: \mathscr{S} \to \mathscr{S}$ with the following property. Whenever X is a subset of A, with inclusion mapping $j: X \to A$, then XT is a subset of AT, and $jT: XT \to AT$ is the inclusion mapping.*

See (3.6) and (3.7) for examples of inclusion functors.

Proposition 3.2. *Let T be an inclusion functor, let $f: A \to B$ be a mapping, and let $X \subset A$ and $Y \subset B$. If $X(f\mathbf{P}) \subset Y$ (or equivalently $X \subset Y(f^{\#}\mathbf{P})$), then $(XT)(fT\mathbf{P}) \subset YT$ (and $XT \subset (YT)(fT^{\#}\mathbf{P})$).*

Proof. $X(f\mathbf{P}) \subset Y$ if and only if there is a commutative diagram

$$\begin{array}{ccc} X & \xrightarrow{j} & A \\ \downarrow f_1 & & \downarrow f \\ Y & \xrightarrow{k} & B \end{array}$$

with inclusion mappings j and k, and with a mapping f_1 at left. T preserves such a diagram, hence the Proposition.

We have in particular

$$(XT)(fT\mathbf{P}) \subset (X(f\mathbf{P}))T$$

for $X \subset A$, and

$$(Y(f^{\#}\mathbf{P}))T \subset (YT)(fT^{\#}\mathbf{P})$$

for $Y \subset B$. We say that T *preserves direct images* if always

$$(XT)(fT\mathbf{P}) = (X(f\mathbf{P}))T$$

for $f: A \to B$ and $X \subset A$, and we say that T *preserves inverse images* if always

$$(Y(f^{\#}\mathbf{P}))T = (YT)(fT^{\#}\mathbf{P})$$

for $f: A \to B$ and $Y \subset B$.

Proposition 3.3. *If T is an inclusion functor, then $(\bigcap X_i)T \subset \bigcap(X_i T)$ for any non-empty family $(X_i)_{i \in I}$ of sets.*

Proof. $\bigcap X_i \subset X_i$ for each $i \in I$, and hence $(\bigcap X_i)T \subset X_i T$ for each $i \in I$. The Proposition follows.

We say that T *preserves intersections* if $(\bigcap X_i)\, T = \bigcap (X_i T)$ for any non-empty family $(X_i)_{i \in I}$ of sets.

3.4. If $f, g: A \to B$ are mappings with the same source and the same target, we put

$$E(f, g) = \{x \in A \mid xf = xg\},$$

and we call $E(f, g)$ the *equalizer* of f and g.

Proposition. *If T is an inclusion functor, then $(E(f, g))\, T \subset E(fT, gT)$ for any two mappings $f, g: A \to B$.*

Proof. If $X \subset A$, with inclusion mapping $j: X \to A$, then $X \subset E(f, g)$ if and only if $jf = jg$. This equality is preserved by T, and thus $X \subset E(f, g)$ always implies $XT \subset E(fT, gT)$. For $X = E(f, g)$, the Proposition follows.

We say that T *preserves equalizers* if always $(E(f, g))\, T = E(fT, gT)$ for mappings $f, g: A \to B$ with the same source and the same target.

3.5. We say that a functor T on $\mathscr{S}$ *preserves products* if for any product A of a family $(A_i)_{i \in I}$ of sets, with projections $p_i: A \to A_i$, the set AT is a product of the family $(A_i T)_{i \in I}$ of sets, with projections $p_i T$.

More generally, we say that a functor T on $\mathscr{S}$ *preserves inverse limits* if for any inverse limit $\varprojlim V$ of a functor V from a small category to $\mathscr{S}$, $(\varprojlim V)\, T$ is an inverse limit of the functor VT.

Since all inverse limits in $\mathscr{S}$ can be constructed from products, equalizers, and intersections, an inclusion functor T preserves inverse limits if (and only if) T preserves products, equalizers, and intersections.

3.6. (i) The identity functor $\mathbf{1}$ on $\mathscr{S}$ is an inclusion functor which of course preserves everything.

(ii) If S and T are inclusion functors, then ST is an inclusion functor. If S and T both preserve direct images (inverse images, intersections, equalizers, products), then ST also preserves direct images (inverse images, intersections, equalizers, products).

(iii) If $(T_i)_{i \in I}$ is a family of inclusion functors, then the product functor ΠT_i is an inclusion functor. If each functor T_i preserves direct images (inverse images, intersections, equalizers, products), then ΠT_i also preserves direct images (inverse images, intersections, equalizers, products).

Examples 3.7. All inclusion functors listed below preserve direct images, and intersections. Except for constant functors, they do not preserve equalizers or products.

(i) For any set K, we define a *constant functor* C_K on $\mathscr{S}$ by putting $AC_K = K$ for any set A, and $fC_K = 1_K$ for any mapping f. This is an inclusion functor which preserves equalizers. C_K preserves products only if K is a singleton or the empty set.

(ii) The direct image functor $\boldsymbol{P}$ is an inclusion functor.

(iii) We define the *filter basis functor* $\boldsymbol{B}$ as follows. For a set A, let $A\,\boldsymbol{B}$ be the set of all filter bases on A. For a mapping $f: A \to B$ and a filter basis Φ on A, let $\Phi(f\boldsymbol{B})$ be the filter basis $\{X(f\boldsymbol{P}) \mid X \in \Phi\}$ on B. This defines $f\boldsymbol{B}: A\,\boldsymbol{B} \to B\,\boldsymbol{B}$. The functor $\boldsymbol{B}$ thus defined is an inclusion functor. A filter functor $\boldsymbol{F}$ can also be defined, but this is not an inclusion functor.

(iv) We define the *ultrafilter basis functor* $\boldsymbol{UB}$ as follows. For a set A, let $A\,\boldsymbol{UB}$ be the set of all bases of ultrafilters on A. For a mapping $f: A \to B$, let $f\,\boldsymbol{UB}$ be the restriction of $f\,\boldsymbol{B}$ to $A\,\boldsymbol{UB}$ and $B\,\boldsymbol{UB}$. This is a mapping. The functor $\boldsymbol{UB}$ thus defined on $\mathcal{S}$ is an inclusion functor.

4. Operational Categories

4.1. We define a *concrete category* as a pair $(\mathcal{C}, U)$, where $\mathcal{C}$ is a category and U a faithful functor from $\mathcal{C}$ to $\mathcal{S}$, i.e. if $f, g: A \to B$ are maps of $\mathcal{C}$ with the same source and the same target, and if $fU = gU$, then always $f = g$.

By a standard "abus de langage", we usually write $\mathcal{C}$ for $(\mathcal{C}, U)$. If $(\mathcal{C}, U)$ is a concrete category, we call U the *forgetful functor* from $\mathcal{C}$ to $\mathcal{S}$. If A is an object and f a map of $\mathcal{C}$, we call $A\,U$ the *underlying set* of A and fU the *underlying mapping* of f, and we usually write $\bar{A}$ for $A\,U$ and $\bar{f}$ for fU.

Definition 4.2. *An operation ω on a concrete category $\mathcal{C}$ is given by two inclusion functors D_ω and R_ω, called domain functor and range functor of ω, and by a correspondence which assigns to every object A of $\mathcal{C}$ a relation $\omega_A: \bar{A}\,D_\omega \to \bar{A}\,R_\omega$, with the following property. For any map $f: A \to B$ of $\mathcal{C}$, we have*

$$\omega_A(\bar{f}R_\omega) \leqq (\bar{f}D_\omega)\,\omega_B\,. \tag{4.3}$$

In other words, we require that $u\,\omega_A v$ always implies $u'\,\omega_B v'$ for

$$u' = u(\bar{f}D_\omega) \quad \text{and} \quad v' = v(\bar{f}R_\omega)\,.$$

For examples of operations, see § 5. In all our examples, the range functor R_ω is either $\mathbf{1}$ or a constant functor, but the general theory does not change if more general range functors are allowed.

4.4. We call an operation ω on a concrete category $\mathcal{C}$ *single-valued* if ω_A is a functional relation for every object A of $\mathcal{C}$, and we call ω a *strict* operation if all relations ω_A are, in fact, mappings. We say that ω is a *proper* operation if we have

$$\omega_A(\bar{f}R_\omega) = (\bar{f}D_\omega)\,\omega_B\,, \tag{4.5}$$

instead of (4.3), for every map f of $\mathcal{C}$. A strict operation is always proper, and a proper operation may be regarded as a natural transformation.

Occasionally, e.g. when considering places in Commutative Algebra, one considers maps and "operations" for which (4.3) is replaced by the reverse inequality (see [2], § 2, no. 1).

4.6. Let $\mathscr{C}$ be a concrete category, Ω a class of operations on $\mathscr{C}$, and A and B objects of $\mathscr{C}$. A map $f: A \to B$ of $\mathscr{C}$ is called *Ω-proper* if f satisfies (4.5), instead of (4.3), for every $\omega \in \Omega$. A mapping $\bar{f}: \bar{A} \to \bar{B}$ is called *Ω_{AB}-compatible*, or *properly Ω_{AB}-compatible*, if $\bar{f}$ satisfies (4.3), or (4.5) respectively, for every $\omega \in \Omega$. A subset X of $\bar{A}$ is called *Ω_A-closed* if

$$(X D_\omega)(\omega_A \mathbf{P}) \subset X R_\omega ,$$

for every $\omega \in \Omega$. An object C of $\mathscr{C}$ is called an Ω-subobject of A if $\bar{C}$ is a subset of $\bar{A}$, and there is an Ω-proper map $j: C \to A$ of $\mathscr{C}$ such that $j: \bar{C} \to \bar{A}$ is the inclusion mapping. We call this map j the *inclusion map* from C to A.

For any map $f: A \to B$ of $\mathscr{C}$, the underlying mapping $\bar{f}$ is Ω_{AB}-compatible, and $\bar{f}$ is properly Ω_{AB}-compatible if, and only if, f is Ω-proper. If C is an Ω-subobject of A, then $\bar{C}$ is Ω_A-closed, and it follows from (2.5) that ω_C is the restriction of ω_A to $\bar{C} D_\omega$ and $\bar{C} R_\omega$, for every $\omega \in \Omega$.

Definition 4.7. An *operational category* is a pair $(\mathscr{C}, \Omega)$ where $\mathscr{C}$ is a concrete category and Ω a class of operations on $\mathscr{C}$ which satisfies the following three conditions.

(OC 1) If A and B are objects of $\mathscr{C}$ such that $AU = BU$, and if $\omega_A = \omega_B$ for every $\omega \in \Omega$, then $A = B$.

(OC 2) If A and B are objects of $\mathscr{C}$, then for every Ω_{AB}-compatible mapping $\bar{f}: AU \to BU$, there is a map $f: A \to B$ of $\mathscr{C}$ such that

$$\bar{f} = fU .$$

(OC 3) If A is an object of $\mathscr{C}$ and X an Ω_A-closed subset of AU, then there is an Ω-subobject C of A such that $X = CU$.

We also say that $\mathscr{C}$ is an *Ω-category* if $(\mathscr{C}, \Omega)$ is an operational category. See § 5 for examples of operational categories.

4.8. By (OC 1), an object A of an Ω-category $\mathscr{C}$ is determined by its underlying set $\bar{A}$ and its operations ω_A, $\omega \in \Omega$. Let now A and B be objects of an Ω-category $\mathscr{C}$. By (OC 2) and (4.6), there is a natural bijective correspondence, induced by the forgetful functor U, between maps $f: A \to B$ of $\mathscr{C}$ and Ω_{AB}-compatible mappings $\bar{f}: \bar{A} \to \bar{B}$. By (OC 3), (OC 1), and (4.6), there is a natural bijective correspondence, also induced by U, between Ω-subobjects of A and Ω_A-closed subsets of $\bar{A}$.

4.9. A concrete category $\mathscr{C}$ may be an Ω-category for more than one class Ω of operations; see (5.1). We identify operational categories $(\mathscr{C}, \Omega)$

and $(\mathscr{C}, \Omega')$, with the same underlying category $\mathscr{C}$, if Ω-proper maps of $\mathscr{C}$ are the same as Ω'-proper maps. If this is the case, then Ω-subobjects of an object A of $\mathscr{C}$ are the same as Ω'-subobjects of A.

If $\mathscr{C}$ is an Ω-category, there is a largest class $\overline{\Omega}$ of operations on $\mathscr{C}$ such that Ω-proper maps are the same as $\overline{\Omega}$-proper maps. $\overline{\Omega}$ consists of all operations ω on $\mathscr{C}$ (including all $\omega \in \Omega$) for which every Ω-proper map f of $\mathscr{C}$ satisfies (4.5) rather than (4.3).

5. Examples of Operational Categories

We indicate the range functor R_ω of an operation ω only if R_ω is not the identity functor $\mathbf{1}$ on $\mathscr{S}$.

5.1. Groups and quasigroups. For quasigroups, we let $\Omega = \{\gamma, \lambda, \varrho\}$, with $D_\gamma = D_\lambda = D_\varrho = \mathbf{1} \times \mathbf{1}$. γ is the "composition law", with $(a, b)\,\gamma_A = ab$ for a multiplicative quasigroup A and a, b in $\bar{A}$. λ is left division and ϱ is right division, with $x = (a, b)\,\lambda_A \Leftrightarrow (x, b)\,\gamma_A = a$, and $y = (a, b)\,\varrho_A \Leftrightarrow (b, y)\,\gamma_A = a$, for a quasigroup A and a, b in $\bar{A}$. Maps are quasigroup homomorphisms, subobjects are subquasigroups. For every quasigroup A, the empty set $\emptyset$ is an Ω_A-closed subset of $\bar{A}$, and thus we must admit an empty quasigroup so that (OC 3) can be satisfied.

For groups, we introduce two additional operations ι and ε, with $D_\iota = \mathbf{1}$ and $D_\varepsilon = C_1$, where $1 = \{\emptyset\}$. For a group A and $a \in \bar{A}$, $a\,\iota_A$ is the inverse of a in A, and $\emptyset\,\varepsilon_A$ is the neutral element of A. Groups form an Ω-category for $\Omega = \{\gamma, \iota, \varepsilon\}$, as well as for $\Omega = \{\lambda, \varepsilon\}$, or $\Omega = \{\varrho, \varepsilon\}$. If we omit ε from the list of operations, we must admit an empty group in order to satisfy (OC 3). If we consider only γ and ε, omitting ι, then Ω_A-closed subsets of $\bar{A}$, for a group A, correspond to subsemigroups of A (with the neutral element of A as neutral element), and not to subgroups of A.

5.2. Rings and fields. For rings, we consider four operations $\gamma, \mu, \nu, \varepsilon$, with $D_\gamma = D_\mu = \mathbf{1} \times \mathbf{1}$, $D_\nu = \mathbf{1}$, and $D_\varepsilon = C_1$ (where $1 = \{\emptyset\}$). γ and μ are addition and multiplication, ν is the operation of forming additive inverses, and $\emptyset\,\varepsilon_A$, for a ring A, is the zero element of A. Maps are ring homomorphisms, and subobjects are subrings.

For rings with unit element, we introduce an additional operation η with $D\eta = C_1$, where $\emptyset\,\eta_A$ is the unit element of a ring A with unit element. Now maps $f: A \to B$ are ring homomorphismus which map the unit element of A into the unit element of B, and subobjects of a ring A are subrings of A which contain the unit element of A.

For fields, we introduce a sixth operation ι, with $a\,\iota_A = a^{-1}$ for a field A and $a \neq 0$ in $\bar{A}$. Now maps correspond to field extensions, and subobjects are subfields.

5.3. Vector spaces, modules, inner product spaces. We use the operations γ, ν, ε of (5.2), and a fourth operation μ with $D_\mu = C_K \times \mathbf{1}$, where K is a given field or ring. Objects are vector spaces or modules over K, subobjects are subspaces or submodules, and maps are linear transformations or mappings.

For inner product spaces, we let K be one of the three fields $\boldsymbol{R}$ (real numbers), $\boldsymbol{C}$ (complex numbers), or $\boldsymbol{H}$ (quaternions), and we introduce an additional operation σ, with $D_\sigma = \mathbf{1} \times \mathbf{1}$ and $R_\sigma = C_K$. For a space A and a, b in $\bar{A}$, $(a, b)\sigma_A$ is the inner product of a and b. Subobjects are again subspaces. Maps are unitary (or orthogonal) linear transformations.

5.4. Lattices, complete lattices, σ-complete lattices. We use two operations, inf and sup, with the same domain functor. For lattices, this functor is $\mathbf{1} \times \mathbf{1}$, for complete lattices, it is $\boldsymbol{P}$, and for σ-complete lattices it is $\mathbf{1}^{\boldsymbol{N}}$, where $\boldsymbol{N}$ is the set of natural numbers.

5.5. Small categories. We put $\Omega = \{\alpha, \beta, \gamma\}$, with $D_\gamma = \mathbf{1} \times \mathbf{1}$ and $D_\alpha = D_\beta = \mathbf{1}$. For a small category A, we let $A\,U = \bar{A}$ be the set of all maps of A, and γ_A is composition of maps in A. For $a \in \bar{A}$, $a\alpha_A$ and $a\beta_A$ are the right identity and the left identity of a. Subobjects are subcategories, and maps are functors.

5.6. Metric spaces. We use a single operation d, with $D_d = \mathbf{1} \times \mathbf{1}$ and $R_d = C_{\boldsymbol{R}}$, where $\boldsymbol{R}$ is the field of real numbers. For a metric space A, d_A is the metric of A. Subobjects are subspaces, and maps are isometric mappings.

5.7. Topological spaces and convergence spaces. We consider a single operation q, with $D_q = \boldsymbol{B}$ (see (3.7), (iii)). For a space A, a point x of A, and a filter basis Φ on A, we put $\Phi q_A x$ if and only if Φ converges to x in A. This operation is in general not single-valued; it becomes single-valued if we restrict ourselves to Hausdorff spaces. Maps are continuous mappings, and subobjects are closed subspaces.

If we consider only compact Hausdorff spaces, we may replace q by its restriction q' to ultrafilter bases, with $D_{q'} = \boldsymbol{UB}$. This has the advantage that q' is a strict operation.

6. Inverse Limits

Let $\mathscr{C}$ be an Ω-category. We show in this section that the forgetful functor U preserves those inverse limits which are preserved by the range functors R_ω, $\omega \in \Omega$. It follows that inverse limits in $\mathscr{C}$ are obtained from inverse limits in $\mathscr{S}$ by providing the latter with the appropriate operations.

Proposition 6.1. *Assume that all range functors R_ω, $\omega \in \Omega$, preserve intersections. If A is an object of $\mathscr{C}$ and $(X_i)_{i \in I}$ a family of Ω_A-closed subsets of $\bar{A}$, then the set intersection $\bigcap X_i$ is also Ω_A-closed.*

Proof. Let $u\,\omega_A\,v$, with $u \in (\bigcap X_i)\,D_\omega$. By (3.3), we have $u \in X_i D_\omega$ for all $i \in I$, and it follows that $v \in \bigcap (X_i R_\omega) = (\bigcap X_i)\,R_\omega$. Thus $\bigcap X_i$ is Ω_A-closed.

Proposition 6.2. *Assume that all operations $\omega \in \Omega$ are single-valued, and that all range functors R_ω, $\omega \in \Omega$, preserve equalizers. If $f, g : A \to B$ are two maps of $\mathscr{C}$ with the same source and the same target, then $E(fU, gU)$ is an Ω_A-closed subset of AU.*

Proof. Let $X = E(\bar{f}, \bar{g})$, and let $u\,\omega_A\,v$, with $u \in X$. Let $u' = u(\bar{f} D_\omega)$, $u'' = u(\bar{g} D_\omega)$, $v' = v(\bar{f} R_\omega)$, $v'' = v(\bar{g} R_\omega)$. Then $u'\,\omega_B\,v'$ and $u''\,\omega_B\,v''$. By (3.4), $u' = u''$, and hence $v' = v''$ since ω is single-valued. But then $v \in E(\bar{f} R_\omega, \bar{g} R_\omega) = X R_\omega$, and X is Ω_A-closed.

6.3. We call $\mathscr{C}$ an *Ω-category with products* if $\mathscr{C}$ has the following properties.

(i) All range functors R_ω, $\omega \in \Omega$, preserve products.

(ii) For any family $(A_i)_{i \in I}$ of objects of $\mathscr{C}$, there is an object P of $\mathscr{C}$ with Ω-proper maps $p_i : P \to A_i$, one for each $i \in I$, such that $\bar{P}$ is a product in $\mathscr{S}$ of the sets $\bar{A}_i$, $i \in I$, with projections $\bar{p}_i : \bar{P} \to \bar{A}_i$.

With these notations, we have

$$(\bar{p}_i D_\omega)\,\omega_{A_i} = \omega_P\,(\bar{p}_i R_\omega)$$

for $i \in I$, $\omega \in \Omega$. Since the mappings $\bar{p}_i R_\omega$ are projections of products, these equations determine the relations ω_P uniquely, for all $\omega \in \Omega$.

Proposition 6.4. *Let $\mathscr{C}$ be an Ω-category with products. With the notations of (6.3), P is a product in $\mathscr{C}$ of the family $(A_i)_{i \in I}$ of objects of $\mathscr{C}$, with projections p_i.*

If follows that the forgetful functor U preserves products.

Proof. Let $f_i : B \to A_i$ be maps of $\mathscr{C}$, one for each $i \in I$. If $f_i = f p_i$ in $\mathscr{C}$, with $f : B \to P$ and for all $i \in I$, then also $\bar{f}_i = \bar{f} \bar{p}_i$ for all $i \in I$, and this determines $\bar{f}$, and hence f, uniquely. On the other hand, there is a mapping $\bar{f} : \bar{B} \to \bar{P}$ such that $\bar{f}_i = \bar{f} \bar{p}_i$ for all $i \in I$, and $f_i = f p_i$ for all $i \in I$ and a map f if and only if the mapping $\bar{f}$ is Ω_{BP}-compatible. As the mappings $p_i : P \to A_i$ are Ω-proper, we have

$$\begin{aligned}(\bar{f} D_\omega)\,\omega_P\,(\bar{p}_i R_\omega) &= (\bar{f} D_\omega)\,(\bar{p}_i D_\omega)\,\omega_{A_i} = (\bar{f}_i D_\omega)\,\omega_{A_i} \\ &\geqq \omega_B\,(\bar{f}_i R_\omega) = \omega_B\,(\bar{f} R_\omega)\,(\bar{p}_i R_\omega)\,,\end{aligned}$$

for all $i \in I$. As the mappings $\bar{p}_i R_\omega$ are the projections of a product, it follows that $(\bar{f} D_\omega)\,\omega_P \geqq \omega_B\,(\bar{f} R_\omega)$, and $\bar{f}$ is indeed Ω_{BP}-compatible.

Theorem 6.5. *Let $\mathscr{C}$ be an Ω-category with products. If all range functors R_ω, $\omega \in \Omega$, preserve intersections and equalizers as well as products, and if all operations $\omega \in \Omega$ are single-valued, then any functor V from a small category to $\mathscr{C}$ has an inverse limit $\varprojlim V$ in $\mathscr{C}$, with the property that $(\varprojlim V)\,U$ is an inverse limit in $\mathscr{S}$ of the functor VU.*

In other words, inverse limits can be lifted from $\mathscr{S}$ to $\mathscr{C}$ if $\mathscr{C}$ satisfies the conditions of the Theorem.

Proof. Let $V: I \to \mathscr{C}$, let I_0 be the set of objects of I, and let $i\,V = V_i$ for $i \in I_0$. Let P be a product of the objects V_i of $\mathscr{C}$, with projections $p_i: P \to V_i$. For a map $u: i \to j$ of I, let $X_u = E(\bar{p}_i(uVU), \bar{p}_j)$, and let $X = \bigcap X_u$, for all maps u of I. By (6.1) and (6.2), X is Ω_P-closed, and thus $X = \bar{M}$ for an Ω-subobject M of P. Let $m: M \to P$ be the inclusion map, and let $m_i = m\,p_i$ for $i \in I_0$. The maps m_i are Ω-proper, $\bar{M}$ is an inverse limit in $\mathscr{S}$ of the functor VU, with projections $\bar{m}_i$, and $\bar{M}R_\omega$ is an inverse limit in $\mathscr{S}$ of the functor VUR_ω, for $\omega \in \Omega$, with projections $\bar{m}_i R_\omega$.

Now let maps $f_i: B \to V_i$ be given, one for each $i \in I_0$, such that $f_i(uV) = f_j$ for any map $u: i \to j$ of I. There is a unique mapping $\bar{f}: \bar{B} \to \bar{M}$ such that $\bar{f}\bar{m}_i = \bar{f}_i$ for every $i \in I_0$. Thus there is a map $f: B \to M$ such that $f m_i = f_i$, for every $i \in I_0$, if and only if the mapping $\bar{f}$ is Ω_{BM}-compatible, and then f is uniquely determined. The Ω_{BM}-compatibility of $\bar{f}$ follows as in the proof of (6.4), with P and the maps p_i replaced by M and the maps m_i.

7. Properties of Subobjects

We assume in this section that $\mathscr{C}$ is an Ω-category, and that all range functors R_ω, $\omega \in \Omega$, preserve intersections and inverse images. We show that Ω-subobjects of an object A of $\mathscr{C}$ form a complete lattice, and we study direct and inverse images of Ω-subobjects under maps of $\mathscr{C}$.

7.1. We denote by $A\,P_\Omega$ the set of all Ω-subobjects of an object A of $\mathscr{C}$, and we put $A_2 \leqq A_1$, for A_1, A_2 in $A\,P_\Omega$, if A_2 is an Ω-subobject of A_1. This defines an order relation in $A\,P_\Omega$.

The order relation just defined has the following alternative characterizations (we omit the trivial proofs).

(i) $A_2 \leqq A_1$ if and only if $\bar{A}_2 \subset \bar{A}_1$.

(ii) If $j_i: A_i \to A$ $(i = 1, 2)$ are the inclusion maps, then $A_2 \leqq A_1$ if and only if $j_2 = j\,j_1$ for a map $j: A_2 \to A_1$ of $\mathscr{C}$. The map j thus determined is the inclusion map.

Proposition 7.2. *With the order relation $\leqq$ of* (7.1), *the set $A\,P_\Omega$ of Ω-subobjects of an object A of $\mathscr{C}$ is a complete lattice, with* $(\inf A_i)\,U = \bigcap (A_i U)$ *for any non-empty family $(A_i)_{i \in I}$ of Ω-subobjects of A.*

Proof. $A' \leqq A_i$ for all $i \in I$ if and only if $\bar{A}' \subset \bar{A}_i$ for all i, by (7.1). By (6.1) and (OC 3), $\bigcap \bar{A}_i = \bar{C}$ for an Ω-subobject C of A, and then $A' \leqq A_i$ for all $i \in I$ if and only if $A' \leqq C$. Thus $\inf A_i = C$, with $\bar{C} = \bigcap \bar{A}_i$. Since $\inf A_i$ is defined for any non-empty family $(A_i)_{i \in I}$ of elements of $A P_\Omega$, and $\sup(A P_\Omega) = A$ is defined, $A P_\Omega$ is a complete lattice.

Proposition 7.3. *Let $f: A \to B$ be a map of $\mathscr{C}$. If Y is an Ω_B-closed subset of $\bar{B}$, then $Y(\bar{f}^{\#}\boldsymbol{P})$ is an Ω_A-closed subset of $\bar{A}$.*

Proof. Let $Y(\bar{f}^{\#}P) = X$, let $u \,\omega_A v$, with $u \in X D_\omega$, and let

$$u' = u(\bar{f} D_\omega), \quad v' = v(\bar{f} R_\omega).$$

Then $u' \omega_B v'$ since f is a map, and $u' \in Y D_\omega$ by (3.2). Thus $v' \in Y R_\omega$, and $v \in (Y R_\omega)(\bar{f} R_\omega{}^{\#}\boldsymbol{P}) = X R_\omega$, which shows that X is Ω_A-closed.

7.4. Let A be an object of $\mathscr{C}$. If $X \subset \bar{A}$, we denote by $X\,\mathrm{sp}_A$ the infimum, in the lattice structure of $A P_\Omega$, of all Ω-subobjects A_1 of A such that $X \subset \bar{A}_1$, and we call $X\,\mathrm{sp}_A$ the Ω-subobject of A *spanned* by X. This defines a mapping $\mathrm{sp}_A : A U \boldsymbol{P} \to A P_\Omega$. We put $(X\,\mathrm{sp}_A) U = X^*$ for $X \subset \bar{A}$ whenever this notation is convenient.

By (7.2), X^* is the intersection of all Ω_A-closed subsets of $\bar{A}$ which contain X, for $X \subset \bar{A}$. Thus $X \subset X^*$, and $X \subset \bar{A}_1$, for $A_1 \in A P_\Omega$, if and only if $X^* \subset \bar{A}_1$. In particular, $\bar{A}_1\,\mathrm{sp}_A = A_1$ for any $A_1 \in A P_\Omega$.

Proposition 7.5. *Let A be an object of $\mathscr{C}$. If $(X_i)_{i \in I}$ is a family of subsets of $\bar{A}$, then*

$$\sup(X_i\,\mathrm{sp}_A) = (\bigcup X_i)\,\mathrm{sp}_A$$

in the lattice structure of $A P_\Omega$. In particular,

$$\sup A_i = (\bigcup \bar{A}_i)\,\mathrm{sp}_A$$

for a family $(A_i)_{i \in I}$ of Ω-subobjects of A.

Proof. $\sup(X_i\,\mathrm{sp}_A) \leqq A_1$, for $A_1 \in A P_\Omega$, if and only if $X_i \subset \bar{A}_1$ for all $i \in I$, and hence if and only if $(\bigcup X_i)\,\mathrm{sp}_A \leqq A_1$, and the result follows. For $X_i = \bar{A}_i$, with $X_i\,\mathrm{sp}_A = A_i$, we obtain the second part of the Proposition.

7.6. For a map $f: A \to B$ of $\mathscr{C}$, we define mappings

$$f P_\Omega : A P_\Omega \to B P_\Omega, \quad f^{\#} P_\Omega : B P_\Omega \to A P_\Omega,$$

by putting

$$A_1(f P_\Omega) = (\bar{A}_1(\bar{f}\boldsymbol{P}))\,\mathrm{sp}_B$$

for an Ω-subobject A_1 of A, and

$$B_1(f^{\#} P_\Omega) = (\bar{B}_1(\bar{f}^{\#}\boldsymbol{P}))\,\mathrm{sp}_A$$

for an Ω-subobject B_1 of B.

It follows immediately that

$$(A_1(f P_\Omega))\, U = (\bar{A}_1(\bar{f} \boldsymbol{P}))^*$$

for $A_1 \in A P_\Omega$. For $B_1 \in B P_\Omega$, we have

$$(B_1(f^{\#} P_\Omega))\, U = \bar{B}_1(\bar{f}^{\#} \boldsymbol{P}),$$

since $\bar{B}_1(\bar{f}^{\#} \boldsymbol{P})$ is Ω_A-closed by (7.3).

Proposition 7.7. *If $f: A \to B$ is a map of $\mathscr{C}$, then*

$$X(\bar{f}\boldsymbol{P}) \subset X^*(\bar{f}\boldsymbol{P}) \subset (X(\bar{f}\boldsymbol{P}))^*,$$

and

$$(X\, \mathrm{sp}_A)(f P_\Omega) = (X(\bar{f}\boldsymbol{P}))\, \mathrm{sp}_B,$$

for any subset X of $\bar{A}$.

Proof. $X(\bar{f}\boldsymbol{P}) \subset X^*(\bar{f}\boldsymbol{P})$ follows from $X \subset X^*$. Let now $(X(\bar{f}\boldsymbol{P}))^* = Y$. Then $X \subset Y(\bar{f}^{\#}\boldsymbol{P})$, and $Y(\bar{f}^{\#}\boldsymbol{P})$ is Ω_A-closed by (7.3), so that also $X^* \subset \subset Y(\bar{f}^{\#}\boldsymbol{P})$ and $X^*(\bar{f}\boldsymbol{P}) \subset Y$. It follows that

$$X(\bar{f}\boldsymbol{P})\, \mathrm{sp}_B = X^*(\bar{f}\boldsymbol{P})\, \mathrm{sp}_B = (X\, \mathrm{sp}_A)(f P_\Omega).$$

Proposition 7.8. *If $f: A \to B$ and $g: B \to C$ are maps of $\mathscr{C}$, then $(fg) P_\Omega = (f P_\Omega)(g P_\Omega)$, and $(fg)^{\#} P_\Omega = (g^{\#} P_\Omega)(f^{\#} P_\Omega)$.*

In other words, P_Ω is a covariant functor, and ${}^{\#}P_\Omega$ a contravariant functor, from $\mathscr{C}$ to $\mathscr{S}$.

Proof. If $A_1 \in A P_\Omega$ and $Y = \bar{A}_1(\bar{f}\boldsymbol{P})$, then

$$\begin{aligned} A_1((fg) P_\Omega) &= \bar{A}_1((\bar{f}\bar{g})\boldsymbol{P})\, \mathrm{sp}_C = Y(\bar{g}\boldsymbol{P})\, \mathrm{sp}_C = Y\, \mathrm{sp}_B (g P_\Omega) \\ &= A_1(f P_\Omega)(g P_\Omega), \end{aligned}$$

by the definitions and (7.7). If $C_1 \in C P_\Omega$ and $B_1 = C_1(g^{\#} P_\Omega)$, then

$$\begin{aligned} B_1(f^{\#} P_\Omega) &= \bar{B}_1(\bar{f}^{\#}\boldsymbol{P})\, \mathrm{sp}_A = \bar{C}_1(\bar{g}^{\#}\boldsymbol{P})(\bar{f}^{\#}\boldsymbol{P})\, \mathrm{sp}_A \\ &= \bar{C}_1((\bar{f}\bar{g})^{\#}\boldsymbol{P})\, \mathrm{sp}_A = C_1((fg)^{\#} P_\Omega), \end{aligned}$$

by the definitions and the last part of (7.6).

Proposition 7.9. *If $f: A \to B$ is a map of $\mathscr{C}$, then*

$$(\sup A_i)(f P_\Omega) = \sup(A_i(f P_\Omega)),$$

for any family $(A_i)_{i \in I}$ of Ω-subobjects of A.

Proof. $(\sup A_i)(f P_\Omega) = (\bigcup \bar{A}_i)\, \mathrm{sp}_A (f P_\Omega)$
$= (\bigcup \bar{A}_i)(\bar{f}\boldsymbol{P})\, \mathrm{sp}_B = (\bigcup \bar{A}_i(\bar{f}\boldsymbol{P}))\, \mathrm{sp}_B = \sup(\bar{A}_i(\bar{f}\boldsymbol{P})\, \mathrm{sp}_B) = \sup(A_i(f P_\Omega))$, by (7.5), (7.7), and the definitions.

Proposition 7.10. *If all domain functors D_ω, $\omega \in \Omega$, preserve direct images, and if $f: A \to B$ is an Ω-proper map of $\mathscr{C}$, then $(A_1(f P_\Omega))\, U = (A_1 U)(f U \boldsymbol{P})$ for any Ω-subobject A_1 of A.*

Proof. If $Y = \bar{A}_1(\bar{f}\boldsymbol{P})$, we must only show that Y is Ω_B-closed. Let $u'\omega_B v'$, with $u' \in Y D_\omega$. As D_ω preserves direct images, $u' = u(\bar{f}D_\omega)$ for some $u \in \bar{A}_1 D_\omega$. Since f is Ω-proper, we have $u\omega_A v$ and $v' = v(\bar{f}R_\omega)$ for some $v \in \bar{A}_1 R_\omega$. By (3.2), $v' \in Y R_\omega$.

8. Injections and Projections

We assume again that $\mathscr{C}$ is an Ω-category, and that all range functors R_ω, $\omega \in \Omega$, preserve intersections and inverse images. We show in this section that $\mathscr{C}$ has almost all properties of a bicategory, in the sense of [*8*].

Definition 8.1. *Let $f\colon A \to B$ be a map of $\mathscr{C}$. We say that f is Ω-closed if $X(\bar{f}\boldsymbol{P})$ is Ω_B-closed for every Ω_A-closed subset X of $\bar{A}$. We call f an Ω-injection of $\mathscr{C}$ if f is Ω-closed and Ω-proper, and $\bar{f}$ injective. We call f an Ω-projection of $\mathscr{C}$ if $A(fP_\Omega) = B$.*

Under the assumptions of (7.10), any Ω-proper map of $\mathscr{C}$ is Ω-closed, and our definition of an Ω-injection becomes somewhat redundant. The author has at present no example of an operational category which does not satisfy these assumptions.

On the other hand, an Ω-closed map f need not be Ω-proper, even if $\bar{f}$ is bijective. For example, let $\mathscr{C}$ be the category of limit spaces (see [*5*]), with the operation q of (5.7). Let A be a limit space which is not topological, and let A_t be the set $\bar{A}$, with the limit structure (Limitierung in [*5*]) induced by the topology defined by the open sets of A. Then the embedding map $f\colon A \to A_t$ is not Ω-proper, although $\bar{f} = 1_A$ is bijective. Since Ω_A-closed subsets of $\bar{A} = \bar{A}_t$ are the same as Ω_{A_t}-closed subsets, f is Ω-closed.

If f is an Ω-projection, $\bar{f}$ need not be surjective. For example, if $\mathscr{C}$ is the category of topological spaces, or of Hausdorff spaces, then $f\colon A \to B$ is a projection if and only if $\bar{A}(\bar{f}\boldsymbol{P})$ is dense in B, and this may well be the case even if $\bar{f}$ is not surjective.

Lemma 8.2. (i) *A map $f\colon A \to B$ of $\mathscr{C}$ is isomorphic if and only if f is Ω-proper and $\bar{f}$ bijective.* (ii) *Isomorphisms and inclusion maps are Ω-injections.*

Proof. If f is isomorphic with inverse g, then $\bar{f}$ is bijective with inverse $\bar{g}$, and we have

$$(1) \qquad \omega_B(\bar{g}R_\omega) \leqq (\bar{g}D_\omega)\,\omega_A\,.$$

If we multiply this with $\bar{f}D_\omega$ on the left and with $\bar{f}R_\omega$ on the right, we obtain

$$(2) \qquad (\bar{f}D_\omega)\,\omega_B \leqq \omega_A(\bar{f}R_\omega)\,,$$

so that f is Ω-proper. Conversely, we have (2) if f is Ω-proper. If $\bar{f}$ is bijective, with inverse $\bar{g}$, we obtain (1) by multiplying (2) with $\bar{g} D_\omega$ on the left and with $\bar{g} R_\omega$ on the right. But then $\bar{g} = g U$ for a map $g: B \to A$ of $\mathscr{C}$, and it is easily seen that f is isomorphic, with inverse g.

If $f: A \to B$ is isomorphic with inverse g, and if $X \subset \bar{A}$ is Ω_A-closed, then $X(\bar{f} \boldsymbol{P}) = X(\bar{g}^{\#} \boldsymbol{P})$ is Ω_B-closed by (7.3). Thus f is Ω-closed, and hence an Ω-injection. If $j: A' \to A$ is an inclusion map, then an $\Omega_{A'}$-closed set X is Ω_A-closed, and $X(j \boldsymbol{P}) = X$. Since j is Ω-proper and $\bar{j}$ injective, j is an Ω-injection.

Proposition 8.3. *Any map f of $\mathscr{C}$ has a unique factorization $f = pj$ such that p is an Ω-projection and j an inclusion map. If f is an Ω-injection, then p is an isomorphism.*

Proof. Let $f: A \to B$. If $f = pj$ for an Ω-projection $p: A \to C$ and an inclusion map $j: C \to B$, then $A(p P_\Omega) = C = C(j P_\Omega)$, so that $C = A(f P_\Omega)$ by (7.8). This determines j, and then also p since j is monomorphic.

Conversely, let $C = A(f P_\Omega)$, and let $j: C \to B$ be the inclusion map. Since $\bar{A}(\bar{f} \boldsymbol{P}) \subset \bar{C}$, there is a factorization $\bar{f} = \bar{p} \bar{j}$ for a (unique) mapping $\bar{p}: \bar{A} \to \bar{C}$. But then

$$(\bar{p} D_\omega)\, \omega_C (\bar{j} R_\omega) = (\bar{p} D_\omega)(\bar{j} D_\omega)\, \omega_B = (\bar{f} D_\omega)\, \omega_B$$
$$\geqq \omega_A (\bar{f} R_\omega) = \omega_A (\bar{p} R_\omega)(\bar{j} R_\omega),$$

and $\omega_A(\bar{p} R_\omega) \leqq (\bar{p} D_\omega)\, \omega_C$ follows. Thus $\bar{p}$ underlies a map $p: A \to C$ of $\mathscr{C}$, with $f = pj$. Now

$$A(p P_\Omega) = A(p P_\Omega)(j P_\Omega) = A(f P_\Omega) = C,$$

so that p is an Ω-projection.

If f is an Ω-injection, and hence Ω-proper, then the inequalities of the preceding paragraph become equalities, so that p is Ω-proper. Also, $\bar{A}(\bar{p} \boldsymbol{P}) = \bar{A}(\bar{f} \boldsymbol{P})$ is Ω_B-closed and spans C. Thus $\bar{A}(\bar{p} \boldsymbol{P}) = \bar{C}$, and $\bar{p}$ is surjective. Since $\bar{f} = \bar{p} \bar{j}$ is injective, $\bar{p}$ is also injective. Now p is an isomorphism of $\mathscr{C}$ by (8.2).

8.4. A *bicategory* in the sense of [*8*] (or a *strict bicategory* in the sense of [*12*]) is a category $\mathscr{C}$ with two fixed classes of maps, called the class of *injections* of $\mathscr{C}$ and the class of *projections* of $\mathscr{C}$, which satisfy the following three conditions.

(BC 1) Injections and projections form subcategories of $\mathscr{C}$. Every isomorphism of $\mathscr{C}$ is both an injection and a projection.

(BC 2) Every injection is monomorphic, every projection epimorphic, in $\mathscr{C}$.

(BC 3) Any map f of $\mathscr{C}$ has a factorization $f = pj$, where p is a projection and j an injection. If $f = p'j'$ and $f = p''j''$ are two such factorizations, then $j'' = vj'$, and $p'' = p'v^{-1}$, for an isomorphism v of $\mathscr{C}$.

We call a factorization $f = pj$ a *decomposition* of f if p is a projection, and j an injection, of $\mathscr{C}$.

Theorem 8.5. *Let $\mathscr{S}$ be an Ω-category such that all range functors R_ω, $\omega \in \Omega$, preserve intersections and inverse images. With Ω-injections as injections, and Ω-projections as projections, $\mathscr{C}$ satisfies conditions* (BC 1) *and* (BC 3) *of* (8.4), *and injections are monomorphic.*

We cannot assert, however, that all projections are epimorphic (but see (8.6)).

Proof. The first part of (BC 1) follows immediately from the definitions, with (7.8). An isomorphism $f: A \to B$ is an Ω-injection by (8.2). Since $\bar{A}(\bar{f}\boldsymbol{P}) = \bar{B}$ for the bijective mapping $\bar{f}$, we have $A(fP_\Omega) = B$, and f is an Ω-projection. An Ω-injection j is monomorphic since $\bar{j} = jU$ is monomorphic and U a faithful functor.

Now we must verify (BC 3). By (8.3), with (8.2), any map f of $\mathscr{C}$ has a decomposition. If $f = p'j'$ and $f = p''j''$ are two decompositions, let $j' = p_1 j_1$ and $j'' = p_2 j_2$, where j_1 and j_2 are inclusion maps, p_1 and p_2 projections. By (8.3), p_1 and p_2 are isomorphisms. Now $p'p_1 j_1 = f = p'' p_2 j_2$. By (8.3) and (BC 1), $j_1 = j_2$ and $p' p_1 = p'' p_2$. Thus $j'' = vj'$, and $p'' = p'v^{-1}$, for the isomorphism $v = p_2 p_1^{-1}$ of $\mathscr{C}$.

8.6. If $\mathscr{C}$ is the category of topological spaces, then a projection $p: A \to B$ is a continuous mapping which maps A onto a dense subset of B. If B is not a Hausdorff space, such a mapping need not be epimorphic.

This troublesome situation cannot occur if the conditions of the following proposition are satisfied.

Proposition. *If all operations $\omega \in \Omega$ are single-valued, and all range functors R_ω, $\omega \in \Omega$, preserve equalizers, then all Ω-projections of $\mathscr{C}$ are epimorphic in $\mathscr{C}$.*

Proof. Let $p: A \to B$ be an Ω-projection. If f, g are maps such that $pf = pg$, then $\bar{A}(\bar{p}\boldsymbol{P}) \subset E(\bar{f}, \bar{g})$. On the other hand, $\bar{A}(\bar{p}\boldsymbol{P})$ spans B, and $E(\bar{f}, \bar{g})$ is Ω_B-closed by (6.2). It follows that $E(\bar{f}, \bar{g}) = \bar{B}$. But then $\bar{f} = \bar{g}$, and hence $f = g$.

9. Direct Limits

As the examples of § 5 indicate, the construction of direct limits in operational categories is much more complicated than the construction of inverse limits. There is, however, one general result which we shall discuss in this section. This result is valid for any category with the properties listed in (9.3).

9.1. We need some facts about projections of a (strict) bicategory $\mathscr{C}$.

If p, p' are projections of $\mathscr{C}$, with the same source A, we put $p' \leqq p$ if $p' = pv$ for a map v of $\mathscr{C}$ (which must be a projection). This is a pre-order relation. We put $p \cong p'$ if $p' \leqq p$ and $p \leqq p'$. This is an equivalence relation, and $p \cong p'$ if and only if $p' = pv$ for an isomorphism v of $\mathscr{C}$. In particular, $p \cong 1_A$ if and only if p is an isomorphism of $\mathscr{C}$.

We shall use the following axiom.

(Q 1) For every object A of $\mathscr{C}$, there is a set, denoted by AQ, of projections with source A, with the following properties. (i) $1_A \in AQ$. (ii) $A\,Q$ contains exactly one representative from each equivalence class, for the relation $\cong$, of projections with source A.

We call a projection $p \in A\,Q$ a *quotient* of A. If $f\colon B \to A$ is a map of $\mathscr{C}$ and $p \in A\,Q$, then fp has a unique decomposition $fp = qj$ with $q \in B\,Q$ (and j an injection). We put $q = p(f\,Q)$. This defines a mapping $f\,Q\colon A\,Q \to B\,Q$, and it is easily seen that we have, in fact, defined a contravariant functor Q from $\mathscr{C}$ to $\mathscr{S}$.

The pre-order relation $\leqq$ for projections with source A induces in $A\,Q$ an order relation which we also denote by $\leqq$. We shall require the following two axioms.

(Q 2) $A\,Q$, with the order relation $\leqq$, is a complete lattice.

(Q 3) For any map $f\colon B \to A$ and any family $(p_i)_{i\in I}$ of quotients of A, we have $\sup(p_i(f\,Q)) = (\sup p_i)\,(f\,Q)$.

We extend our notations by putting $p(f\,Q) = p'(f\,Q)$ for $f\colon B \to A$ and $p \cong p'$, $p' \in A\,Q$, and $\sup p_i = \sup p_i'$ if $p_i \cong p_i'$, $p_i' \in A\,Q$, for every i. It is easily seen that (Q 3) remains valid with these extended notations.

9.2. For an operational category $\mathscr{C}$, quotients are in a sense dual to subobjects, and (Q 2) and (Q 3) are dual to (7.2) and (7.9). This duality could be improved by using inclusion maps in lieu of subobjects, but this seems hardly worthwhile.

Axiom (Q 1) is purely formal; its only purpose is to make our work fit into any underlying axiomatic set theory. Axioms (Q 2) and (Q 3) are valid for any (strict) bicategory with products which satisfies (Q 1). This is proved e.g. in [7]. We also refer to [7] for properties of quotients which follow from the definitions and axioms of (9.1).

9.3. We shall require the following conditions.

(i) $\mathscr{C}$ is a concrete category, and also a (strict) bicategory.

(ii) For any injection j of $\mathscr{C}$, the mapping $j\,U$ is injective. If $j\,U$ is bijective, then j is an isomorphism of $\mathscr{C}$.

(iii) $\mathscr{C}$ satisfies axioms (Q 1), (Q 2), (Q 3) of (9.1).

(iv) The forgetful functor $U\colon \mathscr{C} \to \mathscr{S}$ has a left adjoint functor $F\colon \mathscr{S} \to \mathscr{C}$.

An Ω-category $\mathscr{C}$ which satisfies the conditions of (8.5) and (8.6) also satisfies (i) and (ii) above. If $\mathscr{C}$ is an Ω-category with products, then $\mathscr{C}$

also satisfies (iii). If $\mathscr{C}$ satisfies (iv), then $X F$, for a set X, is called the *free object* of $\mathscr{C}$ generated by X.

9.4. Adjointness of F to U means that for any set X and any object B of $\mathscr{C}$, there is a bijective correspondence

$$\varphi_{X,B}\colon \mathscr{S}(X, B U) \to \mathscr{C}(X F, B),$$

natural in X and in B. It follows from this that for any object A of $\mathscr{C}$, there is a map $\beta_A = 1_{AU}\,\varphi_{AU,A}\colon A U F \to A$ such that

$$(f U)\,\varphi_{AU,B} = \beta_A f = (f U F)\,\beta_B$$

for any map $f\colon A \to B$ of $\mathscr{C}$ with source A. Thus the maps β_A define a natural transformation, from the functor $U F$ to the identity functor of $\mathscr{C}$.

Lemma. *β_A is a projection.*

This is the only place where we use condition (ii) of (9.3).

Proof. Let $\beta_A = p j$ be a decomposition. From the naturality of φ^{-1}, we obtain

$$1_{AU} = \beta_A \varphi^{-1} = (p\,\varphi^{-1})\,(j\,U).$$

Thus $j U$ is surjective, and then j is an isomorphism of $\mathscr{C}$ by (9.3), (ii).

Theorem 9.5. *If $\mathscr{C}$ is a category which satisfies the four conditions of* (9.3), *then any functor $V\colon I \to \mathscr{C}$ from a small category to $\mathscr{C}$ has a direct limit in $\mathscr{C}$.*

Proof. We use the notations of (9.4), except that we omit all subscripts for φ, we denote by I_0 the set of all objects of I, and we put $i V = V_i$ and $\beta_{iV} = \beta_i$ for $i \in I_0$.

Let $M = \varinjlim (V U)$ in $\mathscr{S}$, with injections $m_i\colon \bar{V}_i \to M$ for $i \in I_0$. For a map $u\colon i \to j$ of I, we have:

$$m_i = (u\,V\,U)\,m_j, \quad \beta_i(u\,V) = (u\,V\,U\,F)\,\beta_j.$$

We consider the family of all quotients p_α of $M F$ for which $p_\alpha((m_i F)Q) \leqq \beta_i$ for all $i \in I_0$, and we let $\sup p_\alpha = r\colon M F \to R$ for this family. By (Q 3), $r((m_i F)\,Q) \leqq \beta_i$ for all $i \in I_0$, so that there is a factorization

$$(m_i F)\,r = \beta_i r_i$$

for every $i \in I_0$. We shall see that R is the desired direct limit, with injections $r_i\colon V_i \to R$.

First we have

$$\beta_i(u\,V)\,r_j = (u\,V\,U\,F)\,\beta_j r_j = (u\,V\,U\,F)\,(m_j F)\,r = (m_i F)\,r = \beta_i r_i,$$

and hence $(u\,V)\,r_j = r_i$, for any map $u\colon i \to j$ of I.

Now let maps $f_i\colon V_i \to A$ of $\mathscr{C}$ be given so that $(u\,V)\,f_j = f_i$ for any map $u\colon i \to j$ of I. There is, by definition of $\varinjlim (V U)$, a (unique) mapping

$g: M \to \bar{A}$ such that $\bar{f}_i = m_i g$ for all $i \in I_0$. Now

$$\beta_i f_i = \bar{f}_i \varphi = (m_i F)(g\varphi)$$

in $\mathscr{C}$. If $g\varphi = pj$ is a decomposition, it follows that $p((m_i F)Q) \leqq \beta_i$ for all $i \in I_0$. Thus $p \leqq r$, and there is a factorization $g\varphi = rf$ in $\mathscr{C}$. Then

$$\beta_i f_i = (m_i F) r f = \beta_i r_i f,$$

and hence $f_i = r_i f$, for all $i \in I_0$.

Finally, if $f_i = r_i f$ for all $i \in I_0$ and $rf = g\varphi$, then f is determined by g. But then

$$\beta_i r_i f = (m_i F)(g\varphi) = (m_i g)\varphi$$

for all $i \in I_0$. This determines the mappings $m_i g$, $i \in I_0$, and hence g. Thus the equations $f_i = r_i f$ determine f.

10. Complements

10.1. We examine in this section the question whether the injections and projections of an operational category $\mathscr{C}$ are determined by the categorical structure of $\mathscr{C}$. Only partial answers to this question are known, and several open questions remain.

We shall need the following definition. We call a monomorphism m of an arbitrary category $\mathscr{C}$ *strict* (in [*12*], *regular* is used) if $m = m'e'$, with e' epimorphic (and of course also monomorphic) and m' monomorphic, always implies that e' is an isomorphism of $\mathscr{C}$.

Proposition 10.2. *Let $\mathscr{C}$ be an Ω-category which satisfies the following conditions.* (i) *All range functors R_ω, $\omega \in \Omega$, preserve intersections, inverse images, and equalizers.* (ii) *All domain functors D_ω, $\omega \in \Omega$, preserve direct images.* (iii) *All operations $\omega \in \Omega$ are single-valued.* (iv) *Any epimorphism e of $\mathscr{C}$ is a projection, i.e $A(eP_\Omega) = B$ for $e: A \to B$. Then the injections of $\mathscr{C}$ are the strict monomorphisms of $\mathscr{C}$.*

Proof. If m is a strict monomorphism, with the decomposition $m = pj$ of (8.3), then p is an isomorphism, and thus m an injection. Conversely, let $j: A \to B$ be an injection, and let $j = em$ for an epimorphism $e: A \to C$ and a monomorphism $m: C \to B$. If

$$u' \omega_C v' \quad \text{for} \quad u' = u(eD_\omega) \quad \text{and} \quad v' = v(eR_\omega),$$

and if $\quad u'(mD_\omega) = u(jD_\omega) = u'' \quad$ and $\quad v'(mR_\omega) = v(jR_\omega) = v''$,

then $u'' \omega_B v''$, hence also $u \omega_A v$ as j is Ω-proper. Thus e is an Ω-proper map. Now $\bar{e}\bar{m} = \bar{j}$ is injective, so that $\bar{e}$ is injective. By condition (iv) and (7.9), $\bar{C} = (A(eP_\Omega))U = \bar{A}(\bar{e}\boldsymbol{P})$, and $\bar{e}$ is surjective. Thus e is an isomorphism by (8.2).

10.3. Conditions (i), (ii), (iii) of (10.2) are satisfied by all examples of § 5, except that condition (iii) is satisfied in (5.7) only if we restrict ourselves to Hausdorff spaces. Condition (iv) is not as easily verified, and the problem of characterizing operational categories which satisfy it is unsolved. There is a standard method for proving condition (iv); we reproduce it below.

We need some definitions. Let $\mathscr{C}$ be a concrete category with direct limits, and let $V: I \to \mathscr{C}$ be a functor from a small category I to $\mathscr{C}$. With the notations of (9.4) and (9.5), let $M = \varinjlim (VU)$ in $\mathscr{S}$, with injections $m_i: \bar{V}_i \to M$, and let $R = \varinjlim V$ in $\mathscr{C}$, with injections $r_i: V_i \to R$. By definition of a direct limit, there is a unique mapping $\mu: M \to \bar{R}$ such that $\bar{r}_i = m_i \mu$ for all objects i of I. We say that $\varinjlim V$ is *free* if μ is injective.

Let now $\mathscr{C}$ be an Ω-category, and let A and B be objects of $\mathscr{C}$ with a common subobject C, with inclusion maps $j: C \to A$ and $j': C \to B$. If R is defined by a pushout diagram

$$\begin{array}{ccc} C & \xrightarrow{j} & A \\ \downarrow j' & & \downarrow r \\ B & \xrightarrow[r']{} & R \end{array}$$

then R is called the *amalgamated coproduct* of A and B *over* C. This is a direct limit.

Lemma 10.4. *If the amalgamated coproduct of* (10.3) *is free, then the pushout diagram defining it is also a pullback diagram. In particular, if* $A = B$ *and* $j = j'$, *then* j *is the equalizer of* r *and* r' *in* $\mathscr{C}$.

Proof. If M is the amalgamated coproduct of $\bar{A}$ and $\bar{B}$ over $\bar{C}$ in $\mathscr{S}$, with injections m and m', then $fm = f'm'$ for mappings f and f' if and only if $f = g\bar{j}$, $f' = g\bar{j}'$ for a mapping g, so that the pushout diagram defining M is also a pullback diagram. Now let $\bar{r} = m\mu$, $\bar{r}' = m'\mu$, with μ injective. If $fr = f'r'$ in $\mathscr{C}$, then $\bar{f}m\mu = \bar{f}'m'\mu$, and hence $\bar{f}m = \bar{f}'m'$. Thus $\bar{f} = \bar{g}\bar{j}$, $\bar{f}' = \bar{g}\bar{j}'$ for a (unique) mapping $\bar{g}$. We conclude as in the proof of (8.3) that $\bar{g}$ underlies a map g, with $gj = f$ and $gj' = f'$. This shows that the diagram is also a pullback diagram. The second part of the Lemma follows immediately from the first part, by specializing to the case $f' = f$.

Proposition 10.5. *Let* $\mathscr{C}$ *be an* Ω*-category which satisfies condition* (i) *of* (10.2), *and assume further that all amalgamated coproducts are defined in* $\mathscr{C}$, *and free. Then every epimorphism of* $\mathscr{C}$ *is a projection.*

Proof. Let $f: A' \to A$, and suppose that $A'(fP_\Omega) = C \neq A$, with inclusion map $j: C \to A$. If R is the amalgamated coproduct of A and A

over C with injections r and r', then j is the equalizer of r and r', by (10.4). Since j is not an isomorphism, $r \neq r'$. But f has a factorization $f = pj$ by (8.5), and hence $fr = fr'$. Thus f is not epimorphic.

We note that a free object functor F, left adjoint to the forgetful functor U, is defined in almost all examples of § 5. Fields, inner product spaces, and metric spaces are exceptions. In all these examples, with the exception already noted for (5.7), the conditions of (9.3) are satisfied, and it follows in particular that amalgamated coproducts are defined. Moreover, in every case examined by the author, amalgamated coproducts are free, so that (10.5), and hence also (10.2), applies. However, there seems to be, at least at present, no satisfactory characterization of operational categories in which amalgamated coproducts are free.

10.6. The results of §§ 8 and 9 do not apply directly to categories of topological spaces and of convergence spaces, unless we restrict ourselves to Hausdorff spaces, since the operation q of (5.7) is not single-valued without this restriction. However, these results can be salvaged as follows.

We call C a *subspace* of A if $\bar{C}$ is a subset of $\bar{A}$ and q_C the restriction of q_A to $\bar{C}\,\boldsymbol{B}$ and $\bar{C}$ (where $\boldsymbol{B}$ is the filter basis functor). We denote by $A\,P_*$ the set of all subspaces of A. For topological spaces, this is equivalent to the usual definition. For $f: A \to B$ and $A' \in A\,P_*$, we denote by $A'(f\,P_*)$ the subspace of B with underlying set $(A'\,U)\,(\bar{f}\boldsymbol{P})$. This defines a mapping $f\,P_*: A\,P_* \to B\,P_*$, and we have a functor P_*. Any map $f: A \to B$ has a unique factorization $A \xrightarrow{p} A\,(f\,P_*) \xrightarrow{j} B$, where j is the inclusion map. If we call f an *injection* whenever p is an isomorphism (homeomorphism), and a *projection* whenever $\bar{f}$ is surjective, we obtain a bicategory.

The categories of topological spaces and of convergence spaces have free objects: $X\,F$, for a set X, is the set X with the discrete topology or convergence structure. Thus the conditions of (9.3) are satisfied, and we have direct limits. It is easily verified that all direct limits are free (in fact, the mappings μ of (10.3) are bijective). Thus the results of this section also apply, with the obvious minor changes.

References

[1] Bourbaki, N.: Eléments de mathématique. Livre I, Théorie des ensembles, chap. 4. Paris 1957.

[2] — Eléments de mathématique. Algèbre commutative, chap. 6. Paris 1964.

[3] Cohn, P. M.: Universal Algebra. New York 1965.

[4] Eckmann, B., and P. J. Hilton: Group-like structures in general categories I. Math. Ann. **145**, 227—255 (1962).

[5] Fischer, H. R.: Limesräume. Math. Ann. **137**, 269—303 (1959).

[6] Goldie, A. W.: The Jordan-Hölder Theorem for general abstract algebras. Proc. London Math. Soc. (2) **52**, 107—131 (1951).

[7] Hofmann, F.: Über eine die Kategorie der Gruppen umfassende Kategorie. S.ber. Bayer. Akad. Wiss., Math.-Naturw. Klasse, 1960, 163—204.
[8] Isbell, J. R.: Some remarks concerning categories and subspaces. Canad. J. Math. **9**, 563—577 (1957).
[9] Lawvere, F. W.: Functorial Semantics of Algebraic Theories. Thesis, Columbia University, 1963.
[10] Linton, F. E. J.: The Functorial Foundations of Measure Theory. Thesis, Columbia University, 1963.
[11] Wyler, O.: Categories of Structures. Univ. of New Mexico Technical Report No. 32, April 1963.
[12] — Weakly exact categories. To appear in Archiv der Mathematik.

Department of Mathematics
Carnegie Institute of Technology
Pittsburgh 13, Penn.

Transparent Categories and Categories of Transition Systems *, **

By

YEHOSHAFAT GIVE'ON

1. Introduction

Several recent results in automata theory (in particular, HARTMANIS and STEARNS 1964, ZEIGER 1964) give evidence of the importance of homomorphisms in the study of transition systems and automata. It is natural therefore to inquire how much information can be retrieved from the algebra of homomorphism compositions with respect to transition systems. The natural mathematical framework for the discussion of this problem is categorical algebra.

We define a category $\mathscr{A}_W$ of the transition systems with input W, where W is any arbitrary fixed monoid, and with arbitrary sets of states. A preliminary study of $\mathscr{A}_W$ (GIVE'ON 1964) shows that one can reconstruct the internal structure of any transition system from the way homomorphisms (i.e., the morphisms of $\mathscr{A}_W$) behave around it.

In this paper we show that $\mathscr{A}_W$ has a generator, M_W (which is W operating on itself as a transition system) and that there exists a functor $\mathrm{Mor}: \mathscr{A}_W \to \mathscr{A}_W$ naturally equivalent to the identity functor of $\mathscr{A}_W$ which factors through $\mathrm{Hom}_{\mathscr{A}_W}(M_W, -)$.

A general exposition of the nature of properties which are retrievable from the "morphism-behavior" in an arbitrary category is presented so that it provides a rigorous general basis for studying "retrievable" properties and categories in which every structural property of objects and morphisms is "retrievable".

Finally, we prove that for a very broad class of input monoids, which includes all the types of input-monoids encountered in automata theory, the categories $\mathscr{A}_W$ are *transparent*. That is, anything which can be said about the structure of transition systems with input W, can be said by referring to their homomorphisms only. In particular, all the auto-

---

* This research was supported by the Office of Naval Research under Contract No. Nonr 1224 (21).

** Received June 23, 1965.

morphisms of $\mathscr{A}_W$, for this type of W, are naturally equivalent to the identity functor of $\mathscr{A}_W$.

Some elementary acquaintance with categorical algebra is needed. In particular, we shall make use of the following notions:

(i) *Category*, it *objects* and its *morphisms*.

(ii) *Epic*, *monic*, and *invertible* morphisms versus *surjective*, *injective*, and *bijective* functions.

(iii) *Initial* and *terminal* objects.

(iv) *Functors*, *natural transformations* and *natural equivalences* of functors.

(v) *Embedding* functors, *automorphism* functors, and *adjoint* functors.

The reader who is not familiar with these notions is referred to the literature (KAN 1958, FREYD 1964, and MACLANE 1965). Additional issues of categorical algebra with reference to automata theory are discussed in (GIVE'ON 1965).

Finally, I wish to thank P. FREYD for his encouragement and his interest in my research. And in particular for the invaluable discussions I had with him during the Conference on Categorical Algebra, La Jolla, California, June 1965.

2. Categories of Transition Systems

2.1. Let W be a fixed monoid. We denote by $\mathscr{A}_W$ the category specified as follows.

The objects of $\mathscr{A}_W$ are *transition-systems* with input W. That is, systems of the form

$$A = (S(A) \times W \xrightarrow{\lambda_A} S(A))$$

where:

(i) $S(A)$ is any set, the *set of states* of A;

(ii) $\lambda_A : S(A) \times W \to S(A)$ is a function, the *transition function* of A, with the following properties (we write $s \cdot \omega$ for $\lambda_A(s, \omega)$):

(iii) $s \cdot 1_W = s$ for all $s \in S(A)$, where 1_W is the identity element of W;

(iv) $s \cdot (\omega_1 \omega_2) = (s \cdot \omega_1) \cdot \omega_2$ for all $s \in S(A)$ and all $\omega_1, \omega_2 \in W$.

The morphisms of $\mathscr{A}_W$ are of the form

$$A \xrightarrow{f} B$$

where $f : S(A) \to S(B)$ is a function satisfying $f(s \cdot \omega) = f(s) \cdot \omega$ for all $s \in S(A)$ and all $\omega \in W$. (Note that $s \cdot \omega$ on the left hand of this equation refers to the transition function of A, while $f(s) \cdot \omega$ refers to the transition function of B.)

The composition of the morphisms of $\mathscr{A}_W$ is determined in an obvious manner by the composition of the functions which underly the morphisms.

That is, $(C \xrightarrow{g} D)\,(A \xrightarrow{f} B)$ is defined *only when* $B = C$ and then it is equal to $A \xrightarrow{gf} D$.

2.2. As in many other "natural" categories of mathematical systems, we have a *forgetful* functor $S: \mathscr{A}_W \to \mathscr{S}$ from $\mathscr{A}_W$ to $\mathscr{S}$ the category of sets where $S(A)$ is the set of states of A and

$$S(A \xrightarrow{f} B) = (f: S(A) \to S(B)).$$

Note that $\mathscr{A}_W$ contains, among its objects, an *empty* object to be denoted by $\emptyset_A$. Here we adopt the useful convention that for any set T there exists a unique function which is injective (i.e., one-one into) from $\emptyset$, the empty set, into T. Thus, the transition function of $\emptyset_W$ is this "empty" function: $\emptyset \times W \to \emptyset$ $(\emptyset \times W = \emptyset)$ and for any object A of $\mathscr{A}_W$ there exists a unique morphism $\emptyset_W \xrightarrow{\sigma_A} A$ which is determined by the "empty" $\emptyset = S(\emptyset_W) \to S(A)$.

The forgetful functor $S: \mathscr{A}_W \to \mathscr{S}$ has an adjoint (cf. Kan 1958, MacLane 1965), the functor $\mathrm{Fr}: \mathscr{S} \to \mathscr{A}_W$, which assigns to each set T, an object $\mathrm{Fr}(T)$ of $\mathscr{A}_W$ which is free on $T \subseteqq S(\mathrm{Fr}(T))$.

The functor $\mathrm{Fr}: \mathscr{S} \to \mathscr{A}_W$ can be specified as follows. For any set T, $\mathrm{Fr}(T)$ is the transition system defined by:

$$S(\mathrm{Fr}(T)) = T \times W,$$
$$(t, \omega_1) \cdot \omega_2 = (t, \omega_1 \omega_2).$$

For any function $f: T_1 \to T_2$ there exists a unique morphism

$$\mathrm{Fr}(T_1) \xrightarrow{\mathrm{Fr}(f)} \mathrm{Fr}(T_2)$$

such that for any $t \in T_1: [\mathrm{Fr}(f)]\,(t, 1_W) = (f(t), 1_W)$. Hence $\mathrm{Fr}(T)$ is "free on $T \times \{1_W\}$". We identify the elements of $T \times \{1_W\}$ with the elements of $T: t \equiv (t, 1_W)$.

If T_1 and T_2 are sets which have the same cardinality, then $\mathrm{Fr}(T_1)$ and $\mathrm{Fr}(T_2)$ are isomorphic (i.e., there exists an invertible morphism $\mathrm{Fr}(T_1) \to \mathrm{Fr}(T_2)$ of $\mathscr{A}_W$). In particular, if T is a single-element set then we denote $\mathrm{Fr}(T)$ by M_W.

M_W serves a very important role in $\mathscr{A}_W$ as we shall see later. Note that M_W may be defined as W operating on itself. That is,

$$S(M_W) = W,$$
$$\omega_1 \cdot \omega_2 = \omega_1 \omega_2.$$

2.3. $\mathscr{A}_W$ shares with the abelian "natural" categories, e.g., of groups or of modules, (cf. Freyd 1964) the property that the monic (respectively, the epic, and the invertible) morphisms are precisely those morphisms of $\mathscr{A}_W$ whose underlying functions are injective (respectively, surjective and bijective). The arguments that establish these facts are similar to the arguments employed in the category of groups for the same end.

The existence of the forgetful functor $S: \mathscr{A}_W \to \mathscr{S}$ implies that a morphism $A \xrightarrow{f} B$ is invertible in $\mathscr{A}_W$ iff f is bijective. Since S is an embedding functor, every morphism of $\mathscr{A}_W$ whose underlying function is injective (respectively, surjective) must be monic (respectively, epic).

In order to prove the converse (for monic and epic morphisms of $\mathscr{A}_W$) we need some additional observations about $\mathscr{A}_W$. These observations will be incorporated in the proofs of the following lemmata.

Lemma 2.3.1. *If $A \xrightarrow{e} B$ is an epic morphism of $\mathscr{A}_W$ then $e: S(A) \to S(B)$ is surjective.*

Proof. The image of $e: S(A) \to S(B)$ is a subset $e(S(A))$ of $S(B)$ such that for any $\omega \in W$ and any $s \in e(S(A))$, $s \cdot \omega \in e(S(A))$. Hence $e(S(A))$ is a transition system $e(A)$ which is a sub-system of B.

We define a new object $B/e(A)$ of $\mathscr{A}_W$ by:

$$S(B/e(A)) = (S(B) - e(S(A))) \cup \{s_*\} \quad \text{where} \quad s_* \notin S(B);$$

the transition function of $B/e(A)$ is the same as of B except for the cases where $s \cdot \omega \in e(S(A))$; in these cases we set $s \cdot \omega = s_*$, and for all $\omega \in W$ we set $s_* \cdot \omega = s_*$.

Obviously, $B/e(A)$ is formed from B by contracting $e(S(A))$ to a single state s_*. This contraction takes the form of a canonical morphism $B \xrightarrow{q_e} B/e(A)$, where $q_e: S(B) \to S(B/e(A))$ is identical on $S(B) - e(S(A))$ and it maps all of $e(S(A))$ onto s_*.

In addition to $B \xrightarrow{q_e} B/e(A)$, we have another morphism $B \xrightarrow{z} B/e(A)$ which maps all of $e(S(A))$ onto s_*. Clearly e is surjective iff $q_e = z$.

Obviously $ze = q_e e$, since both map all of $S(A)$ onto s_*. But e is epic and therefore $ze = q_e e$ implies $q_e = z$.

Lemma 2.3.2. *If $A \xrightarrow{j} B$ is a monic morphism of $\mathscr{A}_W$ then $j: S(A) \to S(B)$ is injective.*

Proof. Assume that for $s_1, s_2 \in S(A)$ we have $j(s_1) = j(s_2)$. We define two morphisms $M_W \xrightarrow{f_1} A$ and $M_W \xrightarrow{f_2} A$ by $f_1(1_W) = s_1$, and $f_2(1_W) = s_2$. Obviously, $jf_1 = jf_2$, and since j is monic, it follows that $f_1 = f_2$; i.e., $s_1 = s_2$.

2.4. For any object A of $\mathscr{A}_W$ any any subset $T \subseteqq S(A)$, we define $A(T)$, *the subsystem of A generated by T*, as follows:

$$S(A(T)) = T \cdot W = \{t \cdot \omega : t \in T \text{ and } \omega \in W\},$$

$$(t \cdot \omega_1) \cdot \omega_2 = t \cdot (\omega_1 \omega_2).$$

A subset T of $S(A)$ is said to *generate* A iff $A(T) = A$; i.e., iff $T \cdot W = S(A)$. In particular A is said to be *monogenic* iff A is generated by a single-element subset of $S(A)$.

For example, M_W is monogenic since $\{1_W\}$ generates M_W (obviously for any $\omega \in W: 1_W \cdot \omega = \omega$). More generally, M_W is generated by $\{u\}$ iff there exists $v \in W$ such that $uv = 1_W$.

Note that an object A of $\mathscr{A}_W$ is monogenic iff for any $T \subseteqq S(A)$ which generates A there exists $t \in T$ such that $\{t\}$ generates A.

Lemma 2.4.1. *An object A of $\mathscr{A}_W$ is monogenic iff for any family $\{A_j\}$ of subsystems of A indexed by a set J, $U\{S(A_j) : j \varepsilon J\} = S(A)$ implies $S(A_j) = S(A)$ for some $j \in J$.*

Proof. Assume that A is monogenic and generated by $\{s_0\}$. If $U S(A_j) = S(A)$ then $s_0 \in S(A_j)$ for some $j \in J$, and $S(A_j) = S(A)$.

Assume that for any family $\{A_j\}$ indexed by a set J, $U S(A_j) = S(A)$ implies $S(A_j) = S(A)$ for some $j \in J$. Define the family $\{A_s\}$ for all $s \in S(A)$, where A_s is the subsystem of A generated by $\{s\}$. Obviously $U S(A_s) = S(A)$ and therefore there exists $s_0 \in S(A)$ for which $S(A_{s_0}) = S(A)$. Hence A is generated by $\{s_0\}$.

Corollary 2.4.2. *For any monogenic object A of $\mathscr{A}_W$ and any automorphism $F: \mathscr{A}_W \to \mathscr{A}_W$ of $\mathscr{A}_W$, $F(A)$ is also a monogenic object of $\mathscr{A}_W$.*

Proof. We recall that an automorphism F of $\mathscr{A}_W$ is a functor $F: \mathscr{A}_W \to \mathscr{A}_W$ for which there exists a functor $G: \mathscr{A}_W \to \mathscr{A}_W$ such that both $F \circ G$ and $G \circ F$ are equal to the identity functor of $\mathscr{A}_W$.

The families of subsystems of A are represented faithfully by the families of monic morphisms of $\mathscr{A}_W$ with range A. Given a set J of monic morphisms $A_j \xrightarrow{j} A$ we define a category $\mathscr{J}$ (which is a subcategory of $\mathscr{A}_W$) whose objects are all the monic morphisms $B \xrightarrow{b} A$ such that for any j $\in J$ there exists a monic $A_j \xrightarrow{b_J} B$ with $b b_j = j$. The morphisms of $\mathscr{J}$ are of the form

$$(B_1 \xrightarrow{b_1} A) \xrightarrow{f} (B_2 \xrightarrow{b_2} A)$$

where $B_1 \xrightarrow{f} B_2$ is a morphism of $\mathscr{A}_W$ with $b_2 f = b_1$.

For any set J of monic morphisms of A_W with range A, the category $\mathscr{J}$ has an initial object $U(J)$, which is unique up to an isomorphism of $\mathscr{J}$ (which is an equivalence of monic morphisms in $\mathscr{A}_W$ (cf. Freyd 1964, MacLane 1965)). $U(J)$ is a monic morphism of A_W with range A and whose image is precisely the union of the images of the morphism in J.

We can rephrase now Lemma 2.4.1: An object A of $\mathscr{A}_W$ is monogenic iff for any set J of monic morphisms of $\mathscr{A}_W$ with range A, if $U(J)$ is an invertible morphism of $\mathscr{A}_W$ (i.e., an isomorphism) then there is a $j \in J$ which is invertible.

Since this characterization of the monogenic objects in $\mathscr{A}_W$ is preserved under the automorphisms of $\mathscr{A}_W$ the proof follows.

2.5. In the proof of Cor. 2.4.2 we have shown that the property of being a monogenic object of $\mathscr{A}_W$, which was defined originally by "looking inside A", is in fact definable by means of general properties of morphisms in categories. Knowing the way morphisms behave around an object A is sufficient in order to determine whether A contains a state from which all the rest of the states of A are accessible. In other words the property of being a monogenic object in $\mathscr{A}_W$ is *categorical*. In Chapter 4 we shall present a rigorous explication of this notion. The properties of M_W, that we shall derive in the next chapter, will yield the result that all properties of objects of $\mathscr{A}_W$ (which are invariant under isomorphisms in $\mathscr{A}_W$) are categorical (provided that W belong to a very broad class of monoids). That is, if W satisfies some weak conditions, then all the properties of the transition systems with input W can be derived from the categorical-algebra study of $\mathscr{A}_W$.

3. A Study of M_W

Lemma 3.1. *An object A of $\mathscr{A}_W$ is monogenic iff there exists an epic morphism $M_W \xrightarrow{e} A$.*

Proof. Assume that A is monogenic and generated by $\{s_0\}$. Define a morphism $M_W \xrightarrow{f_{s_0}} A$ by $f_{s_0}(1_W) = s_0$ (recall that M_W is free on 1_W). Obviously, $f_{s_0}: W \to S(A)$ is surjective and therefore $M_W \xrightarrow{f_{s_0}} A$ is epic.

On the other hand, if $M_W \xrightarrow{e} A$ is epic then A is generated by $e(1_W)$ since $e(\omega) = e(1_W) \cdot \omega$.

3.2. We define a functor $H_{M_W}: \mathscr{A}_W \to \mathscr{S}$ by:

$$H_{M_W}(A) = \mathrm{Hom}_{\mathscr{A}_W}(M_W, A),$$

$$H_{M_W}(A \xrightarrow{f} B) = (l_f: \mathrm{Hom}_{\mathscr{A}_W}(M_W, A) \to \mathrm{Hom}_{\mathscr{A}_W}(M_W, B))$$

where

$$l_f(M_W \xrightarrow{g} A) = (M_W \xrightarrow{fg} B).$$

We define a transformation of functors $\varrho: S \to H_{M_W}$ as follows: For any object A of $\mathscr{A}_W$, $\varrho(A): S(A) \to H_{M_W}(A)$ is given by

$$[\varrho(A)](s): W \to S(A): \omega \to s \cdot \omega.$$

In other words, $[\varrho(A)](s)$ is the morphism $M_W \xrightarrow{f_s} A$ which is determined by $f_s(1_W) = s$.

The function $\varrho(A): S(A) \to H_{M_W}(A)$ is bijective. It is injective since $f_{s_1} = f_{s_2}$ implies $f_{s_1}(1_W) = f_{s_2}(1_W)$. It is surjective since for any morphism $M_W \xrightarrow{g} A$ we have $f_{g(1_W)} = g$.

Furthermore, for any morphism $A \xrightarrow{g} B$ of $\mathscr{A}_W$, and for any $s \in S(A)$ we have

$$g\, f_s = f_{g(s)},$$

where $f_{g(s)} = [\varrho(B)](g(s))$.

For we clearly have

$$(g\, f_s)(\omega) = g(s \cdot \omega) = g(s) \cdot \omega = f_{g(s)}(\omega).$$

From this follows directly, that for any morphism $A \xrightarrow{g} B$ of $\mathscr{A}_W$, the following diagram is commutative.

$$\begin{array}{ccc} S(A) & \xrightarrow{g} & S(B) \\ \varrho(A)\downarrow & & \downarrow\varrho(B) \\ H_{M_W}(A) & \xrightarrow[H_{M_W}(g)]{} & H_{M_W}(B) \end{array}$$

Thus we have proved:

Proposition 3.2.1. *The transformation $\varrho: S \to H_{M_W}$ is a natural equivalence of functors.*

3.2.2. The pair (M_W, ϱ^{-1}) is a *representation* of the forgetful functor $S: \mathscr{A}_W \to \mathscr{S}$ (cf. MacLane 1965).

3.2.3. Since $S: \mathscr{A}_W \to \mathscr{S}$ is an embedding functor (i.e., one-one on the morphisms) it follows that H_{M_W} is also an embedding and therefore M_W is a *generator* of $\mathscr{A}_W$ (cf. Freyd 1964).

Corollary 3.2.4. *M_W is a projective object of $\mathscr{A}_W$. An object P of a category $\mathscr{C}$ is projective iff for any morphism $P \xrightarrow{g} B$ and any epic morphism $A \xrightarrow{e} B$ of $\mathscr{C}$ there exists a morphism $P \xrightarrow{f} A$ of $\mathscr{C}$ for which the following diagram is commutative:*

$$\begin{array}{ccc} & & P \\ & f\swarrow & \downarrow g \\ A & \xrightarrow[e]{} & B \end{array}$$

Proof. It is sufficient (and necessary) to show that if $A \xrightarrow{e} B$ is an epic morphism of $\mathscr{A}_W$ then $H_{M_W}(e): H_{M_W}(A) \to H_{M_W}(B)$ is surjective. From the commutative diagram for $\varrho: S \to H_{M_W}$ we derive $H_{M_W}(e) = \varrho(B)\, e(\varrho(A))^{-1}$. Hence $H_{M_W}(e)$ is surjective.

Proposition 3.3.3. *The bijection* $\varrho(M_W): W \to H_{M_W}(M_W)$ *determines an isomorphism of monoids*

$$W \xrightarrow{R} \mathrm{End}_{\mathscr{A}_W}(M_W)$$

where $\mathrm{End}_{\mathscr{A}_W}(M_W)$ *is the monoid of the morphisms* $M_W \to M_W$ *of* $\mathscr{A}_W$ *with respect to the composition of morphisms in* $\mathscr{A}_W$.

Proof. Since $f_\omega(\omega') = \omega\omega'$, it follows that $f_{\omega_1} f_{\omega_2} = f_{\omega_1 \omega_2}$.

3.4. From Prop. 3.3.3 it follows that for any object A of $\mathscr{A}_W$, the set $H_{M_W}(A)$ enjoys a structure of a transition system with input W by combining $H_{M_W}(A)$ with $W \xrightarrow{R} \mathrm{End}_{\mathscr{A}_W}(M_W)$.

Formally, we define a functor Mor: $\mathscr{A}_W \to \mathscr{A}_W$, where for any object A of $\mathscr{A}_W$ we define Mor(A) by:

$$S(\mathrm{Mor}(A)) = H_{M_W}(A),$$
$$f_s \cdot \omega = f_s f_\omega = f_{s\cdot\omega} \quad \text{for any } M_W \xrightarrow{f_s} A \text{ and } \omega \in W.$$

For any morphism $A \xrightarrow{g} B$ we define $\mathrm{Mor}(A) \xrightarrow{\mathrm{Mor}(g)} \mathrm{Mor}(B)$ by $\mathrm{Mor}(g) = H_{M_W}(g)$.

An immediate verification shows that Mor(A) is an object of $\mathscr{A}_W$, and that $H_{M_W}(g)$ determines in fact a morphism of $\mathscr{A}_W$. Furthermore, it follows directly from the fact that H_{M_W} is a functor that Mor: $\mathscr{A}_W \to \mathscr{A}_W$ is also a functor. Likewise, the transformation $\varrho: S \to H_{M_W}$ determines directly a transformation $\varrho_{\mathscr{A}_W}: I \to \mathrm{Mor}$ from the identity functor of $\mathscr{A}_W$ to Mor, and we have:

Theorem 3.4.1. *The transformation* $\varrho_{\mathscr{A}_W}: I \to \mathrm{Mor}$ *is a natural equivalence of functors.*

3.4.2. Intuitively speaking, the functor Mor constructs the "internal structure" of any object A of $\mathscr{A}_W$ from a part of the category $\mathscr{A}_W$ which lies around M_W and between M_W and A. Hence it is intuitively clear, that if M_W can be recognized in $\mathscr{A}_W$ (up to an isomorphism) by means of some categorical predicate, then the "internal structure" of any object can be reconstructed "categorically", and therefore any property of the transition systems with input W can be determined "categorically" as well.

Lemma 3.5. *If W is a unit-commutative monoid (i.e., if $uv = 1_W$ in W then $vu = 1_W$) then every epic morphism $A \xrightarrow{e} M_W$ of $\mathscr{A}_W$, where A is monogenic, is an isomorphism.*

Proof. If W is a unit-commutative monoid and A is generated by $\{s_0\}$, then $\{e(s_0)\}$ must generate M_W, that is, $e(s_0)\,v = 1_W$ for some $v \in W$, and therefore $v\,e(s_0) = 1_W$.

Assume that $e(s_0 \cdot \omega_1) = e(s_0 \cdot \omega_2)$ for some $\omega_1, \omega_2 \in W$, then we have $\omega_1 = ve(s_0)\,\omega_1 = ve(s_0 \cdot \omega_1) = ve(s_0 \cdot \omega_2) = ve(s_0)\,\omega_2 = \omega_2$, and therefore $s \cdot \omega_1 = s \cdot \omega_2$, which shows that e is also injective.

Corollary 3.5.1. *If W is a unit-commutative monoid, then for any automorphism F of $\mathscr{A}_W$, $F(M_W)$ is isomorphic to M_W.*

Proof. From lemma 3.5 it follows that an object M of $\mathscr{A}_W$ is isomorphic to M_W iff

(i) M is monogenic, and

(ii) for any monogenic object A of $\mathscr{A}_W$ there exists an epic morphism $M \xrightarrow{e} A$ of $\mathscr{A}_W$.

Since these properties of morphisms and objects of $\mathscr{A}_W$ are preserved under the automorphisms of $\mathscr{A}_W$, the corollary follows.

3.5.2. Note that the class of unit-commutative monoids is broad enough to cover all the classes of monoids which are employed in automata theory. For example, the left-cancellative and the right-cancellative monoids are all unit-commutative. Hence the free monoids and the groups are unit-commutative. Note also that the cartesian products of unit-commutative monoids are unit-commutative, and therefore we can apply our results to multi-input transition systems as well. Finally, note that every finite monoid is unit-commutative.

4. Categorical Predicates and Transparent Categories

4.1. A subcategory $\mathscr{D}$ of $\mathscr{C}$ is said to be *very full* iff for any morphism h of $\mathscr{D}$ and for any morphisms f and g of $\mathscr{C}$ such that $fg = h$, $fh = g$ or $hf = g$ holds in $\mathscr{C}$ it follows that f and g belong to $\mathscr{D}$.

A functor $T\colon \mathscr{D} \to \mathscr{C}$ is said to be a *very full embedding* iff T is injective and the image of T is a very full subcategory of $\mathscr{C}$.

Let $\mathscr{D}$ be a category and D a class of morphisms of $\mathscr{D}$, we denote by $(\mathscr{D}, D, \mathscr{C})$ the class of all the images of the morphisms in D under any very full embedding $T\colon \mathscr{D} \to \mathscr{C}$. That is, $f \in (\mathscr{D}, D, \mathscr{C})$ iff there are a morphism $d \in D$ and a very full embedding $T\colon \mathscr{D} \to \mathscr{C}$ such that $f = T(d)$.

A class K of morphisms of $\mathscr{C}$ is said to be *categorical (in $\mathscr{C}$)* iff there is a category $\mathscr{D}$ such that $K = (\mathscr{D}, D, \mathscr{C})$ for some class D of morphisms of $\mathscr{D}$.

Proposition 4.2. *A class K is categorical in $\mathscr{C}$ iff it is closed under all the automorphisms of $\mathscr{C}$.*

Proof. Since for any very full embedding functor $T\colon \mathscr{D} \to \mathscr{C}$ and for any automorphism F of $\mathscr{C}$, $F \circ T\colon \mathscr{D} \to \mathscr{C}$ is also a very full embedding, it follows that every categorical class in $\mathscr{C}$ is closed under all the automorphisms of $\mathscr{C}$.

On the other hand, let K be a class of morphisms of $\mathscr{C}$ which is closed under all automorphisms of $\mathscr{C}$. Denote by $\mathscr{D}(K)$ the minimal very full subcategory of $\mathscr{C}$ which includes K, then $K = (\mathscr{D}(K), K, \mathscr{C})$.

In order to see this, let $T: \mathscr{D}(K) \to \mathscr{C}$ be any very full embedding and define $F_T: \mathscr{C} \to \mathscr{C}$ by

$$F_T(f) = \begin{cases} T(f) & \text{if } f \in \mathscr{D}(K), \\ f & \text{otherwise.} \end{cases}$$

Since $\mathscr{D}(K)$ is a very full subcategory of $\mathscr{C}$, F_T maps $\mathscr{D}(K)$ into itself, and because T is a very full embedding, F_T maps $\mathscr{D}(K)$ onto itself in an injective manner. Furthermore, F_T must be a functor and it has an inverse, hence it is an automorphism of $\mathscr{C}$.

Now, since K is closed under automorphisms it follows that $(\mathscr{D}(K), K, \mathscr{C}) \subseteqq K$, and since clearly $K \subseteqq (\mathscr{D}(K), K, \mathscr{C})$ we have the desired equality.

Corollary 4.2.1. *A class K is categorical in $\mathscr{C}$ iff $K = (\mathscr{D}, D, \mathscr{C})$ for some very full subcategory $\mathscr{D}$ of $\mathscr{C}$.*

Corollary 4.2.2. *Let $\mathscr{D}$ be any category and D a class of some morphisms of $\mathscr{D}$, then the class of all values of the morphisms in D under all embedding functors $\mathscr{D} \to \mathscr{C}$ is categorical in $\mathscr{C}$.*

Corollary 4.2.3. *The class of all values of the morphisms in D under all functors $\mathscr{D} \to \mathscr{C}$ is categorical in $\mathscr{C}$.*

4.2.4. Note that we cannot dispense with the requirement of employing very full embeddings in the definition of the categorical classes in any arbitrary category. For example in the category $\mathscr{N}$ of natural numbers where the morphisms represent the natural partial order of natural numbers, every set of morphisms is categorical. However, the categorical classes achieved by means of 4.2.2 or 4.2.3 are always infinite or empty.

4.3.1. A class of morphisms of $\mathscr{C}$ is said to be *natural* iff it is closed under all those automorphisms of $\mathscr{C}$ which are naturally equivalent to $I_{\mathscr{C}}$, the identity functor of $\mathscr{C}$.

Obviously, by Prop. 4.2 we have that every categorical class is natural. Note that a class of identity morphisms of $\mathscr{C}$ which is closed under the isomorphisms within $\mathscr{C}$ is always natural.

4.3.2. A category $\mathscr{C}$ is said to be *transparent* if all the natural classes in $\mathscr{C}$ are categorical.

4.4. Obviously, if all the automorphisms of $\mathscr{C}$ are naturally equivalent (i. e., to $I_{\mathscr{C}}$) then $\mathscr{C}$ is transparent.

Let us call a category $\mathscr{C}$ *autotrivial* iff all the automorphisms of $\mathscr{C}$ are naturally equivalent.

It is not known whether all transparent categories are autotrivial. All "natural" categories of mathematical systems that are known to be transparent are in fact autotrivial as well.

The equivalence between the notion of transparent categories and that of autotrivial categories, in a special case, takes the form of the following problem in group theory:

Do all groups whose automorphisms are all (conjugate) class preserving have only inner automorphisms?

Any example of a group all of whose automorphisms are class preserving and which has an outer automorphism, yields a transparent category (with a single object and all its morphisms are invertible and in one-one correspondence with the elements of the group) which is not autotrivial).

5. Categorical Constructions and the Transparence of $\mathscr{A}_W$

5.1. We proceed in formalizing the ideas expressed intuitively in 3.4.2, and give an explication of the notion "categorical construction". For our immediate needs we restrict this notion by the following definition:

A functor $T: \mathscr{C}_1 \to \mathscr{C}_2$ is said to be *categorical* iff for any automorphism $F: \mathscr{C}_1 \to \mathscr{C}_1$ of $\mathscr{C}_1$, $T: \mathscr{C}_1 \to \mathscr{C}_2$, and $T \circ F: \mathscr{C}_1 \to \mathscr{C}_2$ are naturally equivalent.

5.1.1. The justification for our definition must be evident.

Generally speaking, a functor $T: \mathscr{C}_1 \to C_2$ can be viewed as a class K_T of morphisms in the product category $\mathscr{C}_1 \times \mathscr{C}_2$. If K_T is a categorical class in $\mathscr{C}_1 \times \mathscr{C}_2$ then obviously T is categorical according to our definition in 5.1. However, except for very trivial categories $\mathscr{C}_2$, a categorical class in $\mathscr{C}_1 \times \mathscr{C}_2$, is never functional. Furthermore, categorical classes in $\mathscr{C}_1 \times \mathscr{C}_2$ depend equally on the structure of $\mathscr{C}_1$ and on the structure of $\mathscr{C}_2$.

It should be clear that there exist constructions which are "evidently categorical" without being functorial. For example, the construction which assigns to each object of a category $\mathscr{C}$, the group of the automorphisms of this object within $\mathscr{C}$, is "evidently categorical". At the same time it is not necessarily functorial.

5.2. It follows immediately that a category $\mathscr{C}$ is autotrivial iff $I_{\mathscr{C}}$ is categorical. Furthermore, any functor which is naturally equivalent to a categorical functor must be categorical as well. Hence $\mathscr{C}$ is autotrivial iff it has some functor $T: \mathscr{C} \to \mathscr{C}$ which is naturally equivalent to $I_{\mathscr{C}}$ and categorical.

5.3. A functor $R: \mathscr{C} \to \mathscr{D}$ is said to *reflect natural equivalences* iff any pair of functors $T_1, T_2: \mathscr{C} \to \mathscr{C}$ are naturally equivalent whenever $R \circ T_1$ and $R \circ T_2$ are.

For example, as one may easily verify, the forgetful functor of $\mathscr{A}_W$, and therefore $H_{M_W}: \mathscr{A}_W \to \mathscr{S}$ reflect natural equivalences.

Proposition 5.3.1. *Let $R: \mathscr{C} \to \mathscr{D}$ be a functor which reflects natural equivalences; then $\mathscr{C}$ is autotrivial iff R is categorical.*

Proof. If $\mathscr{C}$ is autotrivial then any functor from $\mathscr{C}$ must be categorical. If $R: \mathscr{C} \to \mathscr{D}$ reflects natural equivalences and it is categorical, then for any automorphism F of $\mathscr{C}$ we have that $R \circ F$ is naturally equivalent to $R \circ I_{\mathscr{C}}$, and therefore F is naturally equivalent to $I_{\mathscr{C}}$.

5.4. In many "natural" categories we have a similar situation to the one we found to exist in $\mathscr{A}_W$. This situation can be described in the following general form:

Given a category $\mathscr{C}$, which is provided with a forgetful functor $S: \mathscr{C} \to \mathscr{S}$ which reflects natural equivalences, one may find in $\mathscr{C}$ an object M with the following properties. There exists a functor $T_M: \mathscr{C} \to \mathscr{C}$ which is naturally equivalent to $I_{\mathscr{C}}$ and $S \circ T_M = \mathrm{Hom}_{\mathscr{C}}(M, -)$.

Since S reflects natural equivalences, and T_M is naturally equivalent to $I_{\mathscr{C}}$, it follows that $\mathrm{Hom}_{\mathscr{C}}(M, -)$ is naturally equivalent to S and therefore it also reflects natural equivalences. By 5.2 and 5.3.1 we know now that $\mathscr{C}$ is autotrivial iff either one of the three functors S, T_M, and $\mathrm{Hom}_{\mathscr{C}}(M, -)$ is categorical.

5.4.1. An object M of $\mathscr{C}$ is said to be a *natural reflector* iff $H_M: \mathscr{C} \to \mathscr{S}$ (where $H_M = \mathrm{Hom}_{\mathscr{C}}(M, -)$) reflects natural equivalences.

M is said to be *distinguishable in* $\mathscr{C}$ iff for any automorphism F of $\mathscr{C}$, M and $F(M)$ are isomorphic within $\mathscr{C}$. That is, iff the isomorphism type of M in $\mathscr{C}$ is categorical (cf. 4.3.1).

Theorem 5.5. *A category $\mathscr{C}$ with a natural reflector is autotrivial iff $\mathscr{C}$ is transparent. In particular, if M is a natural reflector of $\mathscr{C}$ then $\mathscr{C}$ is autotrivial iff M is distinguishable in $\mathscr{C}$.*

Proof. If $\mathscr{C}$ is autotrivial then it is transparent and therefore M, like any other object of $\mathscr{C}$, is distinguishable in $\mathscr{C}$ (cf. 4.3.1).

On the other hand, let $G: \mathscr{C} \to \mathscr{C}$ be the inverse functor of an automorphism F of $\mathscr{C}$. Then F and G are adjoint and in particular $H_M \circ F$ is naturally equivalent to $H_{G(M)}$. If M is distinguishable in $\mathscr{C}$, then M and $G(M)$ are isomorphic within $\mathscr{C}$, and therefore H_M and $H_{G(M)}$ are naturally equivalent. Hence H_M must be categorical and by 5.2.1 the theorem follows.

5.6. Now, to turn back to $\mathscr{A}_W$, we know that M_W is a natural reflector of $\mathscr{A}_W$. Hence $\mathscr{A}_W$ is autotrivial iff it is transparent, and in particular, $\mathscr{A}_W$ is autotrivial iff M_W is distinguishable in $\mathscr{A}_W$. By 3.5.1 we know that if W is finite or unit-commutative then M_W is distinguishable in $\mathscr{A}_W$. Thus we have:

Theorem 5.6.1. *If W is a unit-commutative monoid then $\mathscr{A}_W$ is auto-trivial (and therefore transparent).*

6. Discussion

6.1. Our result as expressed by Theorem 5.6.1 implies that, for a very broad class of monoids, the categorical study of a domain of all transition systems with input monoid of this class, is equivalent *in principle* to the "complete" study (or the "inside" study) of these systems. However, only experience may show us that in fact there is a psychological advantage to the categorical approach in the study of these systems.

6.2. If G is a group then $\mathscr{A}_G$ is the category of all *representations of G as operating on sets*. Since every group is in particular a unit-commutative monoid, we have that the categorical study of the representations of a fixed arbitrary group G is sufficient *in principle* for producing all the algebraic properties of the representations of G.

6.3. Our results so far, give rise to some general problems that deserve attention. For example, is it true that for any monoid W, M_W is distinguishable in $\mathscr{A}_W$?

More generally, what additional properties of natural reflectors of categories, if any at all, are necessary in order to insure that they are distinguishable? In particular, is it true that every natural reflector which is also a projective generator is distinguishable?

6.3.1. Note that we have defined reflection of natural equivalences (cf. 5.3) in a restricted manner. More generally, a functor R is said to reflect natural equivalences iff for *any* two functors T_1, and T_2 such that $R \circ T_1$ and $R \circ T_2$ are well defined and are naturally equivalent, it follows that T_1 and T_2 are naturally equivalent.

6.4. Important and much more interesting categories of transition systems are those of *finite-state* transition systems (i.e., transition systems whose sets of states are finite). If the input monoid W is finite then our results remain valid since M_W is also finite. If however W is infinite, as it is the case in the ordinary theory of finite automata (where the input monoids are finitely generated free monoids) then M_W is no longer applicable.

6.5. Another interesting restriction of $\mathscr{A}_W$ is to *abelian* transition systems. A transition system A is said to be *abelian* iff $s \cdot \omega_1 \omega_2 = s \cdot \omega_2 \omega_1$ holds for all $s \in S(A)$ and $\omega_1, \omega_2 \in W$.

For any arbitrary monoid W there exists a homomorphism of monoids $W \xrightarrow{ab} W^{ab}$ where W^{ab} is an abelian monoid with the following universality property: any homomorphism of W into an abelian monoid factors uniquely through ab. The direct construction of W^{ab} must be evident (W^{ab} is "W made abelian").

Denote by M_W^{ab} the following object of $\mathscr{A}_W$:

$$S(M_W^{ab}) = W^{ab},$$
$$ab(\omega_1) \cdot \omega_2 = ab(\omega_1 \omega_2).$$

Obviously, M_W^{ab} is abelian. If we denote by $\mathscr{A}_W^{ab}$ the full subcategory of $\mathscr{A}_W$ of abelian objects, one can easily follow the example of M_W in $\mathscr{A}_W$ and show that M_W^{ab} is a natural reflector of $\mathscr{A}_W^{ab}$. Furthermore, for any arbitrary monoid W, M_W^{ab} is distinguishable by the same properties which distinguish M_W in $\mathscr{A}_W$ in the case of unit-commutative input monoid W (cf. 3.5.1), hence for any arbitrary monoid W, $\mathscr{A}_W^{ab}$ is transparent and autotrivial.

An equivalent proof of these properties of $\mathscr{A}_W^{ab}$ follows directly from the fact that $\mathscr{A}_W^{ab}$ is a category isomorphic to $\mathscr{A}_{W ab}$.

References

FREYD, P.: Abelian categories: An introduction to the theory of functors. Harper's Series in Modern Mathematics. New York 1964.

GIVE'ON, Y.: Toward a homological algebra of automata I: The representation and completeness theorem for categories of abstract automata. Technical Report, Department of Communication Sciences, ORA. The University of Michigan 1964.

— Toward a homological algebra of automata. Technical Report, Department of Communication Sciences, ORA, The University of Michigan 1965.

HARTMANIS, J., and R. E. STEARNS: Pair algebra and automata theory. Inform. Control **7**, 485—507 (1964).

KAN, D. M.: Adjoint functors. Trans. Amer. Math. Soc. **87**, 295—329 (1958).

MACLANE, S.: Categorical algebra. Bull. Amer. Math. Soc. **71**, 40—106 (1965).

ZEIGER, H. P.: Loop-free synthesis of finite state machines. Thesis, MIT (1964).

Logic of Computers Group
Department of Communication Sciences
University of Michigan
Ann Arbor, Michigan
Currently: at the
Aiken Computation Laboratory
Harvard University
Cambridge, Massachusetts

A Homology Theory for Small Categories *, **

By

Charles E. Watts

The notion of derived functors is well-established as being a powerful and profound tool in diverse areas of mathematics. The theme of Cartan-Eilenberg [1] is that almost all algebraic homology theories are examples thereof; and Grothendieck [5] displays Cech cohomology of topological spaces as a further deep application.

We shall sketch here methods by which derived functors yield still other classical homology theories, chief among them being the singular homology and cohomology theories of topology. Other applications are to general simplicial and semi-simplicial complexes, and a special case is the homology theory of partially ordered sets, as developed in Deheuvels [2]. Details of this work will appear in Watts [7]. The author would like to express his gratitude to Professor Saunders MacLane for his help and encouragement.

Let $\mathcal{M}$ be any small category. We employ the following notational devices: The set of objects of $\mathcal{M}$ is denoted by $0(\mathcal{M})$, and each element of $0(\mathcal{M})$ is thought of as an identity morphism of $\mathcal{M}$. If $u, v \in 0(\mathcal{M})$, then $M(u, v)$ denotes the set of morphisms from u to v. Thus, for example, $u \in M(u, u)$ for any $u \in 0(\mathcal{M})$.

A covariant functor from a small category $\mathcal{M}$ into the category of abelian groups will be called a *left $\mathcal{M}$-module*; this is in agreement with the fact that if $\mathcal{M}$ is a monoid (i.e., if $0(\mathcal{M})$ has only one element) then such a functor is precisely a left module over the integral monoid algebra. For a similar reason, a contravariant functor from $\mathcal{M}$ to abelian groups will be called a *right $\mathcal{M}$-module*; a right $\mathcal{M}$-module is a left $\mathcal{M}^*$-module where $\mathcal{M}^*$ is the category opposite to $\mathcal{M}$. If A is a left $\mathcal{M}$-module, and $u \in 0(\mathcal{M})$, we write Au for the abelian group associated with u, and if $x \in M(u, v)$, then

$$Ax: Au \to Av$$

is the associated group homomorphism; similarly for right $\mathcal{M}$-modules.

From now on, unless otherwise indicated, $\mathcal{M}$ is an arbitrary small category and "module" means "left module".

---

* This work was supported in part by NSF grant GP 4040.

** Received August 30, 1965.

Let A and B be $\mathcal{M}$-modules; an *$\mathcal{M}$-map* $f\colon A \to B$ is an additive natural transformation of functors. Thus for each $u \in 0(\mathcal{M})$ we have a homomorphism $fu\colon Au \to Bu$ such that if $x \in M(u, v)$, then

$$fv \circ Ax = Bx \circ fu.$$

The set of all $\mathcal{M}$-maps from A to B is an abelian group which we denote by $\mathrm{Hom}_{\mathcal{M}}(A, B)$ and the class of all $\mathcal{M}$-modules and $\mathcal{M}$-maps is a category which we call $\mathrm{Mod}(\mathcal{M})$.

If G is any abelian group, we identify it with an $\mathcal{M}$-module, also denoted by G, such that $Gu = G$ for all $u \in 0(\mathcal{M})$ and $Gx =$ identity on G for all morphisms x of $\mathcal{M}$. In particular, for instance, we have the *zero* $\mathcal{M}$-module 0.

If $f\colon A \to B$ is an $\mathcal{M}$-map, we define $\mathcal{M}$-modules $\mathrm{Ker}\, f$, $\mathrm{Im}\, f$ in the obvious way and we then have the notion of an *exact sequence* of $\mathcal{M}$-modules. Fortunately, a sequence $A \to B \to C$ of $\mathcal{M}$-modules is exact if and only if the sequence $Au \to Bu \to Cu$ of abelian groups is exact for each $u \in 0(\mathcal{M})$.

It is easy to check that the category $\mathrm{Mod}(\mathcal{M})$ satisfies Buchsbaum's Axioms I through V ([*1*], Appendix). Less obvious is the fact that $\mathrm{Mod}(\mathcal{M})$ has enough projectives and injectives, i.e., satisfies Axioms VI and VI* of the above. The proofs are not difficult but are somewhat tedious; they will appear in [*7*]. Here is an outline of the proof that there are enough projectives: Let A be any $\mathcal{M}$-module. For each $u \in 0(\mathcal{M})$, choose an exact sequence

$$F(u) \to Au \to 0$$

where $F(u)$ is a free abelian group. Then define

$$Pu = \sum \{F(Dy) \mid y \text{ is a morphism of } \mathcal{M} \text{ with } Ry = u\},$$

where Dy, Ry are the domain and range of y, respectively. If $x \in M(u, v)$, define $Px\colon Pu \to Pv$ by letting the restriction of Px to $F(Dy)$ be the map identifying $F(Dy)$ with the summand $F(Dxy)$ of Pv. Finally, one verifies that

1. P is a projective $\mathcal{M}$-module;

2. There is an exact sequence $P \to A \to 0$.

That there are enough injectives can be proved similarly.

These existence theorems out of the way, we are now in a position to consider derived functors — provided that we can first construct appropriate functors to derive. Two of these are already at hand: $\mathrm{Hom}_{\mathcal{M}}$ is contravariant in the first variable and covariant in the second. As in classical homological algebra, one can show that for $\mathcal{M}$-modules A and B, the groups $H^p(\mathrm{Hom}_{\mathcal{M}}(X, B))$ and $H^p(\mathrm{Hom}_{\mathcal{M}}(A, Y))$ are isomorphic and independent of the projective resolution X of A and the injective resolution Y of B. We denote these groups by $\mathrm{Ext}^p_{\mathcal{M}}(A, B)$.

We also need a tensor product of $\mathcal{M}$-modules. Thus let A be a *right* $\mathcal{M}$-module, B a *left* $\mathcal{M}$-module. We first define an abelian group T to be the direct sum of the groups $Au \otimes_{\mathbb{Z}} Bu$, u ranging over $0(\mathcal{M})$. If $y \in M(u, v)$, then $Ay: Av \to Au$ and $By: Bu \to Bv$. Now if $a \in Av$ and $b \in Bu$, we set

$$d(a, y, b) = Ay(a) \otimes b - a \otimes By(b);$$

this is an element of T. We let N be the subgroup of T generated by all elements of the form $d(a, y, b)$ and define

$$A \otimes_{\mathcal{M}} B = T/N.$$

That this is the correct definition is evidenced by the fact that it is right-exact and covariant in each variable and solves the appropriate universal lifting problem. Moreover, it is useful; for example, suppose $\mathcal{M}$ is a *partially ordered set*, i.e., a small category with $M(u, v)$ either empty or a singleton for all $u, v \in 0(\mathcal{M})$. Let A be a right $\mathcal{M}$-module; A can be thought of as either a direct or inverse system of groups over $\mathcal{M}$. (Notice that we do not require that $\mathcal{M}$ be directed.) Then $A \otimes_{\mathcal{M}} Z$ is precisely the direct limit of the system of groups, where Z denotes the group of integers considered as a left $\mathcal{M}$-module. It is also true that if B is a left $\mathcal{M}$-module, then $\mathrm{Hom}_{\mathcal{M}}(Z, B)$ is the inverse limit of the system of groups B. Thus our theory includes the study of derived functors of limits (cf. Nöbeling [6]). Incidentally, the exactness of the direct limit functor in case $\mathcal{M}$ is directed is equivalent to Z being $\mathcal{M}$-flat in that case; it is amusing to prove that Z is indeed a direct limit of projective $\mathcal{M}$-modules indexed by $\mathcal{M}$ itself.

Now let $\mathcal{S}$ be the small category whose objects are the natural numbers $0, 1, 2, \ldots$ and where an element of $S(p, q)$ is a monotone function on the set $\{0, \ldots, p\}$ into $\{0, \ldots, q\}$; composition is ordinary composition of functions. A left $\mathcal{S}$-module is called a *semisimplicial* (or *FD*) *cochain complex*, and a right $\mathcal{S}$-module is called a *semi-simplicial chain complex*. With each semi-simplicial cochain (chain) complex A there is associated an ordinary cochain (chain) complex of abelian groups (see [4]) whose (co)homology groups we denote by $H^p(A)$ $(H_p(A))$.

Theorem 1. *For any semi-simplicial cochain (chain) complex A, there are natural isomorphisms*

$$H^p(A) \approx \mathrm{Ext}^p_S(\mathbb{Z}, A)$$
$$(H_p(A) \approx \mathrm{Tor}^S_p(A, \mathbb{Z}))$$

for each $p = 0, 1, 2, \ldots$

(Tor^S_p denotes, of course, the p-th derived functor of $\otimes_{\mathcal{S}}$.)

The proofs of this theorem and the following one will appear in [7].

Now the singular homology and cohomology groups of a topological space are in fact obtained from semi-simplicial complexes. Therefore, Theorem 1 shows that these groups are computable by resolutions in an appropriate category, Mod$(\mathscr{S})$. There is a second way to obtain the singular groups and others, as follows.

Let K be a *simplicial* complex, not necessarily finite. By $T = T(K)$ we denote the set of simplexes of K, partially ordered by the face relationship and thus considered as a small category.

Theorem 2. *For any coefficient group G, there are natural isomorphisms*

$$H^p(K; G) \approx \mathrm{Ext}_p^T(\mathbb{Z}, G),$$

$$H_p(K; G) \approx \mathrm{Tor}_p^T(G, \mathbb{Z}),$$

valid for all $p = 0, 1, 2, \ldots$

Applying Theorem 2 to the singular complex of a space, we get another method of obtaining singular theory by resolutions, this time in a category which depends on the space. Considering the cohomology isomorphism of Theorem 2 in the case $G = \mathbb{Z}$, we are led to a conceptually easy definition of products in simplicial cohomology: just take Yoneda composition products. And in fact it is easy to show that these yield precisely the classical cup-products. (The theory can be modified to give products over any commutative ring of coefficients.) Theorem 2 also yields a nice conceptual proof that the ordered and alternating simplicial homologies coincide.

The statement of Theorem 2 includes the assumption that G is a "trivial" T-module, i.e., is essentially just an abelian group. But if G is non-trivial (i.e., is an arbitrary system of groups indexed by the singular simplexes of the space) then the formula suggests a viable way of defining singular homology and cohomology groups with coefficients in what might be called "singular sheaves" and "singular cosheaves". We are investigating these notions and other topological byproducts of the theory, and hope to have more information later.

We feel, however, that the natural habitat of this homology theory is the abstract small category and that topological and algebraic applications are just that: applications.

References

[1] CARTAN, H., and S. EILENBERG: Homological algebra. Princeton: Princeton University Press 1956.

[2] DEHEUVELS, R.: Homologie des ensembles ordonnés et des espaces topologiques. Bull. Soc. Math. de France **90**, 261–321 (1962).

[3] EILENBERG, S., and S. MACLANE: On the groups $H(\pi, n)$ I. Ann. Math. **58**, 55–106 (1953).

[4] Godement, R.: Théorie des Faisceaux. Paris: Hermann et Cie. 1958.
[5] Grothendieck, A.: Sur quelques points d'algèbre homologique. Tohoku Math. J. **9**, 119—221 (1957).
[6] Nöbeling, C.: Über die Derivierten des Inversen und des Direkten Limes einer Modulfamilie. Topology **1**, 47—61 (1962).
[7] Watts, C.: Homological algebra of categories I (to appear).

Department of Mathematics
The University of Rochester
Rochester, N.Y.

Acyclic Models and Triples *, **

By

Michael Barr and Jon Beck

1. Introduction

We shall prove two theorems on the "triple" cohomology of algebras [*1*] using a method of acyclic models suggested by H. Appelgate. Specifically, we show that the triple cohomology coincides with slight modifications of the usual theories (the same modifications as used in [*8*] in the cases of groups and associative algebras). We also prove a direct sum theorem for the cohomology of a coproduct of algebras, subject to a certain condition.

These theorems are proved by setting up cochain equivalences between standard cochain complexes for the theories involved. Unfortunately, algebra cohomology cannot in general be viewed as a derived functor, or equivalently, the standard complexes from which it comes need not be resolutions. Thus the usual techniques for comparing resolutions, using acyclicity, are inapplicable. However, it is usually possible to prove that the standard complexes become acyclic when applied to free algebras. This suggests using an acyclic models approach, with free algebras as models. Note that since the free functor from the underlying category to the category of algebras has an adjoint, the main theorem on extension of maps and homotopies is simpler than normally appears in topology [*2*].

2. Triple Cohomology

In this section we present a short discussion of tripleable categories and cohomology. For detailed accounts, see [*1*], [*4*]. Let $\mathscr{A}$ be a category. $\boldsymbol{T} = (T, \eta, \mu)$ is a *triple* in $\mathscr{A}$ if $T: A \to A$ is a functor, $\eta: 1_A \to T$ and $\mu: TT \to T$ are morphisms of functors, called the *unit* and *multiplication* of $\boldsymbol{T}$ respectively, such that

$$T\eta \cdot \mu = \eta T \cdot \mu = 1_T : T \to T, \quad T\mu \cdot \mu = \mu T \cdot \mu : TTT \to T.$$

* During the preparation of this paper, both authors where partially supported by NSF Contract GP 730.

** Received August 26, 1965.

(X, ξ) is a ***T**-algebra* if X is an object in $\mathscr{A}$ and $\xi : XT \to X$ is a morphism in $\mathscr{A}$ such that

$$X\eta \cdot \xi = 1_X : X \to X, \quad \xi T \cdot \xi = X\mu \cdot \xi : XTT \to X.$$

ξ is called the ***T**-structure* of the algebra. $f : (X, \xi) \to (Y, \theta)$ is a *morphism* of $\boldsymbol{T}$-algebras if $fT \cdot \theta = \xi f$. The category of $\boldsymbol{T}$-algebras will be denoted by $\mathscr{A}^{\boldsymbol{T}}$. Categories of the form $\mathscr{A}^{\boldsymbol{T}}$ are called *tripleable.*

There is an adjoint pair of functors $F \dashv U$ where $F : \mathscr{A} \to \mathscr{A}^{\boldsymbol{T}}$ is the *free* $\boldsymbol{T}$-algebra functor given by $XF = (XT, X\mu)$ and $U : \mathscr{A}^{\boldsymbol{T}} \to \mathscr{A}$ is the *forgetful* functor $(X, \xi)U = X$. Clearly $T = FU$, and η and μ are derivable from the adjointness morphisms $1 \to FU$, $UF \to 1$. Let $G = UF : \mathscr{A}^{\boldsymbol{T}} \to \mathscr{A}^{\boldsymbol{T}}$. By adjointness, G is a *cotriple* in $\mathscr{A}^{\boldsymbol{T}}$. Explicitly, $(X, \xi)G = (XT, X\mu)$, the counit

$$\varepsilon : G \to 1_{\mathscr{A}^{\boldsymbol{T}}} \quad \text{is} \quad (X, \xi)\varepsilon = \xi : (XT, X\mu) \to (X, \xi),$$

and the comultiplication

$$\delta : G \to GG \quad \text{is} \quad (X, \xi)\delta = X\eta T : (XT, X\mu) \to (XTT, XT\mu).$$

The cotriple $\boldsymbol{G} = (G, \varepsilon, \delta)$ now gives rise to a cohomology theory in $\mathscr{A}^{\boldsymbol{T}}$. The theory will have coefficients in an X-module, for a $\boldsymbol{T}$-algebra X (omitting the $\boldsymbol{T}$-structure from the notation, for brevity). To define X-modules, let $(\mathscr{A}^{\boldsymbol{T}}, X)$ be the category whose objects are $\boldsymbol{T}$-algebra morphisms $W \to X$ and whose morphisms are commutative triangles $X \leftarrow W \to W' \to X$. $(\mathscr{A}^{\boldsymbol{T}}, X)$ is called the category of $\boldsymbol{T}$-algebras and morphisms *over* X. $Y \to X$ is an *X-module* if it is an abelian group object in the category $(\mathscr{A}^{\boldsymbol{T}}, X)$, that is, if the functor $\mathrm{Hom}_X(-, Y)$ which *a priori* has values in the category of sets, factors through the category of abelian groups (Hom_X stands for Hom in the category $(\mathscr{A}^{\boldsymbol{T}}, X)$). If $Y \to X$ is an X-module, there is a $\boldsymbol{T}$-algebra morphism $X \to Y$, the *zero section,* such that $X \to Y \to X$ is the identity. Y will therefore be a "split extension" of X, with some abelian structure "fiberwise". In effect, we confuse modules with split extensions.

Now we define cohomology groups $H^n(W, Y)_X$ where $W \to X$ is a $\boldsymbol{T}$-algebra over X and $Y \to X$ is an X-module. (The cohomology of X itself is retrievable as $H^n(X, Y)_X$, regarding $X \to X$ as an algebra over X by the identity morphism.) The cotriple $\boldsymbol{G}$ acts in $(\mathscr{A}^{\boldsymbol{T}}, X)$ according to the rule $(W \to X)G = WG \to X$ where if the first morphism is p, the second is $pG \cdot X\varepsilon$. Let G^{n+1} be the $(n+1)$-st iterate of G, and let $\varepsilon_i = G^i \varepsilon G^{n-i} : G^{n+1} \to G^n$, $0 \leqq i \leqq n$. Then we get a simplicial object

$$\cdots \overrightarrow{\overrightarrow{\overrightarrow{\rightarrow}}} WG^3 \overrightarrow{\overrightarrow{\rightarrow}} WG^2 \underset{\varepsilon_1}{\overset{\varepsilon_0}{\rightrightarrows}} WG \xrightarrow{\varepsilon} W$$

in the category $(\mathscr{A}^{\boldsymbol{T}}, X)$. That is, each WG^{n+1} is an algebra over X, by the above, and each face operator ε_i is a morphism over X. (Note that WG^{n+1} is in dimension n, and that degeneracy operators could also be defined using $\delta : G \to G^2$.) If $Y \to X$ is an X-module, we can form the cochain complex of abelian groups

$$0 \to \mathrm{Hom}_X(WG, Y) \to \cdots \to \mathrm{Hom}_X(WG^{n+1}, Y) \to \cdots$$

with coboundary operators $d = \sum(-1)^i \mathrm{Hom}_X(W\varepsilon_i, Y)$. $H(W, Y)_X$ is the cohomology of this complex.

3. Acyclic Models

Let $\mathscr{C}$ be a category and let K, L be standard cochain complexes $\mathscr{C}^* \to \mathscr{A}b$. This means $K = \{K^n\}_{n \geq -1}$, there are morphisms $d : K^n \to \to K^{n+1}$, $dd = 0$, and each K^n is an ordinary functor $\mathscr{C}^* \to \mathscr{A}b$; the same for L. $\mathscr{A}b$ is the category of abelian groups. In this paper we only use contravariant standard complexes, but the dual is obvious.

Let $G : \mathscr{C} \to \mathscr{C}$ be a functor with a counit $\varepsilon : G \to 1_{\mathscr{C}}$. The standard cochain complex K is *G-acyclic* if there is a functorial contracting homotopy s (of degree -1) in the "composite" complex GK. L is *G-representable* if there are morphisms $\theta^n : GL^n \to L^n$ such that $\varepsilon L^n \cdot \theta^n = 1$, for $n \geqq 0$. (The θ's do not have to commute with coboundaries.)

Theorem 3.1. *Let K be G-acyclic and L G-representable. Then any morphism of functors $f^{-1} : K^{-1} \to L^{-1}$ can be extended to a natural cochain transformation $f : K \to L$. If $f, g : K \to L$ and $f^{-1} = g^{-1}$, then there exists a natural cochain homotopy $\Phi : f \simeq g$.*

Proof. f^n is constructed inductively as the composition

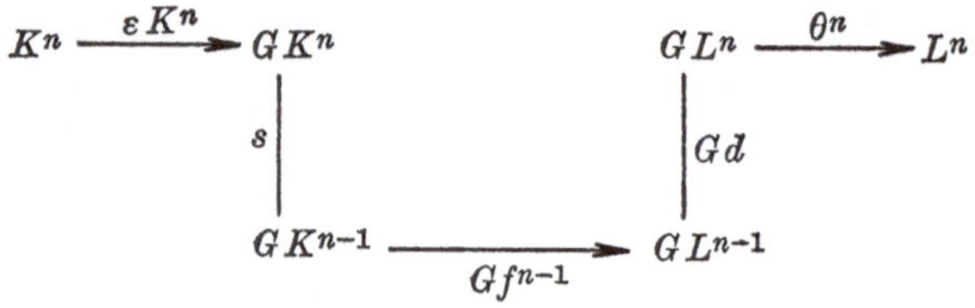

$\Phi^{-1} = \Phi^0 = 0$ and Φ^n is the upper minus the lower composition in the diagram

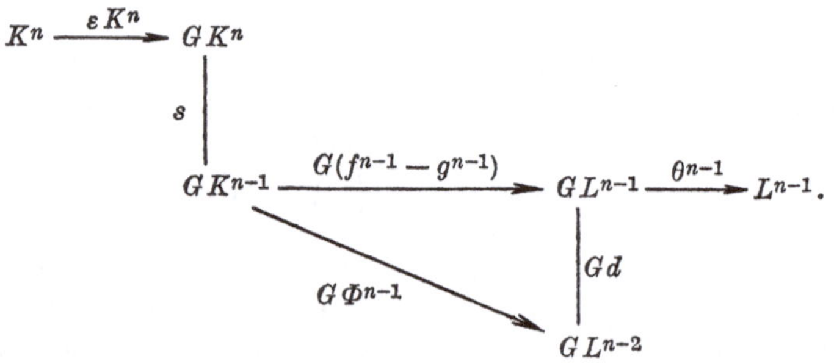

Corollary 3.2. *If K, L are both G-acyclic and G-representable, and $K^{-1} \cong L^{-1}$, then $K \cong L$.*

Corollary 3.3. *If K, L are both G-acyclic and G-representable, $f: K \to L$, and $f^{-1}: K^{-1} \cong L^{-1}$, then f is a cochain equivalence.*

Remark. In our applications, $\mathscr{C}$ is always a tripleable category, G is the "free" cotriple on $\mathscr{C}$, and θ comes from the unit η of the triple. It should therefore be possible to sharpen (3.1) in the cases we are interested in.

4. Cohomology of Groups and Algebras

Let $\mathscr{A}$ be the category of sets and $\mathscr{G}$ the category of groups. There is an adjoint pair of functors $F \dashv U$ where $F: \mathscr{A} \to \mathscr{G}$ is the free group functor and $U: \mathscr{G} \to \mathscr{A}$ is the forgetful functor. Let $T = FU: \mathscr{A} \to \mathscr{A}$. By adjointness there are morphisms

$$\eta: 1_{\mathscr{A}} \to T = FU \quad \text{and} \quad \mu: TT = F(UF)U \to FU = T$$

which make $\boldsymbol{T} = (T, \eta, \mu)$ into a triple. If X is a set, there is a $1-1$ correspondence between group laws on X and $\boldsymbol{T}$-structures $XT \to X$. Hence the category of $\boldsymbol{T}$-algebras $\mathscr{A}^{\boldsymbol{T}}$ is isomorphic to the category of groups $\mathscr{G}$.

If π is a group, let $Y \to \pi$ be a π-module as defined in § 2. Using the zero section $\pi \to Y$ one shows that $Y \to \pi$ is isomorphic to the split extension of π by the ordinary left π-module $M = \ker Y \to \pi$. Thus $Y \cong M \times \pi$ as a set and the multiplication in Y is

$$(m_1, x_1)(m_2, x_2) = (m_1 + x_1 m_2, x_1 x_2),$$

in terms of the left π-operators on M. It follows that if $W \to \pi$ is a group over π, there is a natural isomorphism $\mathrm{Hom}_\pi(W, Y) \cong \mathrm{Der}(W, M)$, where a *derivation* $f: W \to M$ satisfies $(w_1 w_2)f = w_1 f + w_1(w_2 f)$. Here W acts on M via the given morphism $W \to \pi$.

Theorem 4.1. *There is a natural isomorphism*

$$H^n(W, Y)_\pi \to \begin{cases} \mathrm{Der}(W, M), & n = 0, \\ H^{n+1}(W, M), & n > 0, \end{cases}$$

where $H^{n+1}(W, M)$ is the Eilenberg-MacLane cohomology group.

Proof. Let $K: (\mathscr{G}, \pi)^* \to \mathscr{Ab}$ be the cochain complex

$$WK^n = \mathrm{Hom}_\pi(WG^{n+1}, Y), \quad n \geqq -1,$$

as in § 2. Here $G: \mathscr{G} \to \mathscr{G}$ is the cotriple $WG = WUF$, the free group generated by the underlying set of W, with the natural group epimorphism $\varepsilon: WG \to W$ as its counit, and if W is a group over π, so is

WG. (Note that G^{n+1} is the *iterate* of the functor G, not the $(n+1)$-fold cartesian power.)

The form of the bar construction which we use is $L: (\mathcal{G}, \pi)^* \to \mathcal{A}b$ where WL^n is the abelian group of functions from the cartesian power $W^{n+1} \to M$, $n \geqq 0$, and $WL^{-1} = \operatorname{Der}(W, M)$. The coboundary $d: WL^n \to WL^{n+1}$ is

$$\begin{aligned}(w_0, \ldots, w_{n+1})(fd) = w_0 p \cdot (w_1, \ldots, w_{n+1}) f \\ + \sum_{i=1}^{n+1} (-1)^i (w_0, \ldots, w_{i-1} w_i, \ldots, w_{n+1}) f \\ + (-1)^{n+2} (w_0, \ldots, w_n) f\end{aligned}$$

where $n \geqq 0$, p is the morphism $W \to \pi$, and $d: WL^{-1} \to WL^0$ is the obvious inclusion. Then $H(L)$ is the cohomology on the right in (4.1). In effect, we have just cut off the bottom of the cochain complex given by EILENBERG and MACLANE [3].

It follows from what was said leading up to (4.1) that $K^{-1} \cong L^{-1}$. We shall apply (3.2) to prove that there is a natural cochain equivalence $K \simeq L$. The functor with counit used to compare K and L will be the free group cotriple G itself, acting in $(\mathcal{G}, \pi)$.

G-acyclicity. A contracting homotopy $s: GK^n \to GK^{n-1}$ is induced by $W\delta G^n: WG^{n+1} \to WG^{n+2}$.

As to the homotopy in GL, note that there is a natural cochain equivalence $L \simeq N$, where WN^n consists of those functions $W^{n+1} \to M$ vanishing when any argument equals 1 ("normalized" cochains, [7]). A natural contracting homotopy in GN has essentially been constructed in [6]. Therefore there exists a natural contracting homotopy s in GL. An explicit homotopy (probably different) is as follows. Let $f \in WGL^n$. Then $(g_0, \ldots, g_{n-1})(fs)$ is defined by induction on the length of the word $g_0 \in WG$. Spell the words in WG in letters (w) where $w \in W$. Then:

1. If $g_0 = (w) g$ where $g \in WG$, let
$$\begin{aligned}(g_0, \ldots, g_{n-1})(fs) = w p \cdot (g, \ldots, g_{n-1})(fs) \\ - ((w), g, \ldots, g_{n-1}) f.\end{aligned}$$
2. If $g_0 = (w)^{-1} g$ where $g \in WG$, let
$$\begin{aligned}(g_0, \ldots, g_{n-1})(fs) = w^{-1} p \cdot (g, \ldots, g_{n-1})(fs) \\ + w^{-1} p \cdot ((w), g_0, \ldots, g_{n-1}) f.\end{aligned}$$
3. If $g_0 = 1$, let $(1, g_1, \ldots, g_{n-1})(fs) = (1, 1, g_1, \ldots, g_{n-1}) f$.

Note that this homotopy s is natural with respect to morphisms $W \to W'$ in $(\mathcal{G}, \pi)$.

G-representability. $\theta^n: GK^n \to K^n$ is also induced by

$$W \delta G^n \to WG^{n+1} \to WG^{n+2}, \quad n \geqq 0.$$

θ^n: $WL^n \to L^n$ is given by

$$(w_0, \ldots, w_n)(f\theta) = ((w_0), \ldots, (w_n))f.$$

This completes the proof of (4.1).

As another application, the category of associative K-algebras with unit is tripleable over the category $\mathscr{A}$ of K-modules, using the tensor algebra as the triple. A similar argument shows:

Theorem 4.2. *There is a natural isomorphism*

$$H^n(\Gamma, Y)_\Lambda \to \begin{cases} \operatorname{Der}(\Gamma, M), & n = 0, \\ H^{n+1}(\Gamma, M), & n > 0, \end{cases}$$

where $\Gamma \to \Lambda$ is an algebra over Λ, $Y \to \Lambda$ is a Λ-module, i.e., split K-algebra extension by a kernel M with $M^2 = 0$, Λ operates on both sides of M, hence also Γ, and $H^{n+1}(\Gamma, M)$ is the Hochschild (K-relative) cohomology group.

5. Cohomology of a Coproduct

We return to the general situation outlined in §2, and assume:

(5.1) $\mathscr{A}$ has pullbacks (which implies that $\mathscr{A}^T$ has pullbacks).

(5.2) $\mathscr{A}^T$ has coproducts (denoted $X_1 * X_2$).

(5.3) A natural morphism u: $X_1 * X_2 \to X_1G * X_2G$ exists such that $u(X_1\varepsilon * X_2\varepsilon) = 1$.

The validity of these assumptions will be discussed below.

Let $Y \to X_1 * X_2$ be an $X_1 * X_2$-module, and let i_j: $X_j \to X_1 * X_2$ be the natural morphisms into the coproduct, $j = 1,2$. There are pullback diagrams

$$\begin{array}{ccc} Y_j & \dashrightarrow & Y \\ \downarrow & & \downarrow \\ X_j & \xrightarrow[i_j]{} & X_1 * X_2 \end{array}$$

One easily sees that Y_j is an X_j-module. Hence if $W_j \to X_j$ is in $(\mathscr{A}^T, X_j)$, there is a cohomology morphism still denoted i_j:

$$i_j: H(W_1 * W_2, Y)_{X_1 * X_2} \to H(W_j, Y_j)_{X_j}$$

Theorem 5.4.

$$(i_1, i_2): H(W_1 * W_2, Y)_{X_1 * X_2} \to H(W_1, Y_1)_{X_1} + H(W_2, Y_2)_{X_2}$$

is an isomorphism.

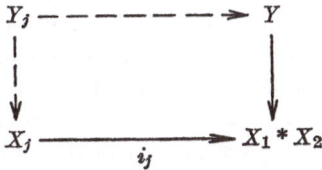

Proof. Define complexes $L, S\colon (\mathscr{A}^{\boldsymbol{T}}, X_1)^* \times (\mathscr{A}^{\boldsymbol{T}}, X_2)^* \to \mathscr{Ab}$ by
$(W_1, W_2)\,L^n = \mathrm{Hom}_{X_1 * X_2}((W_1 * W_2)\,G^{n+1}, Y)$,
$(W_1, W_2)\,S^n = \mathrm{Hom}_{X_1}(W_1 G^{n+1}, Y_1) + \mathrm{Hom}_{X_2}(W_2\,G^{n+1}, Y_2)$,
with coboundary operators as in the triple complex in §2. The inclusions into the coproduct induce a cochain morphism $(i_1, i_2)\colon L \to S$, hence in cohomology induce the morphism referred to in (5.4). It remains to show that $(i_1, i_2)\colon L \to S$ is a cochain equivalence. First, $(i_1, i_2)^{-1}$ is an isomorphism, since

$$\begin{aligned}\mathrm{Hom}_{X_1 * X_2}(W_1 * W_2, Y) &\cong \mathrm{Hom}_{X_1 * X_2}(W_1, Y) + \mathrm{Hom}_{X_1 * X_2}(W_2, Y)\\ &\cong \mathrm{Hom}_{X_1}(W_1, Y_1) + \mathrm{Hom}_{X_2}(W_2, Y_2).\end{aligned}$$

To extend $(i_1, i_2)^{-1}$ inverse to a homotopy inverse for (i_1, i_2), we apply (3.3), using the cotriple $\boldsymbol{G} \times \boldsymbol{G}$ which acts in the category $(\mathscr{A}^{\boldsymbol{T}}, X_1) \times \times (\mathscr{A}^{\boldsymbol{T}}, X_2)$ in the obvious way. L and S are acyclic on models by taking coproducts and sums of the contracting homotopy used for the triple complex in §4. S is also representable, in a similar manner. The representing morphism for L is the composition

$$\begin{aligned}(W_1 G, W_2 G)\,L^n &= \mathrm{Hom}((W_1 G * W_2 G)\,G^{n+1}, Y)\\ \longrightarrow\quad & \mathrm{Hom}((W_1 * W_2)\,G^{n+1}, Y)\\ &= (W_1, W_2)\,L^n\end{aligned}$$

where we use (5.3) and all the Hom's are in the category $(A^{\boldsymbol{T}}, X_1 * X_2)$. This completes the proof of (5.4).

As to the assumptions used in proving (5.4), (5.1) and (5.2) are routine and hold in all the usual algebraic categories. Assumption (5.3) is more delicate. Here are some cases in which it is known to hold: 1. $\mathscr{A}$ = sets, $\boldsymbol{T}$ = the free group triple, $\mathscr{A}^{\boldsymbol{T}}$ = the category of groups (§4). Thus the coproduct theorem holds for group cohomology (see also [8] for a proof of this fact). 2. $\mathscr{A}$ = K-modules, $\boldsymbol{T}$ = the tensor algebra triple, $\mathscr{A}^{\boldsymbol{T}}$ = the category of associative K-algebras with unit (§4), and the coproduct theorem holds for Hochschild cohomology. 3. $\mathscr{A}$ = K-modules, $\boldsymbol{T}$ = the symmetric algebra triple, $\mathscr{A}^{\boldsymbol{T}}$ = the category of commutative K-algebras (associative, with unit). The cohomology is probably Harrison's ([5], when K is a field). Assumption (5.3) fails in the case: $\mathscr{A}$ = sets, $\boldsymbol{T}$ = the triple given by XT = the commutative polynomial K-algebra generated by the elements of the set X. $\mathscr{A}^{\boldsymbol{T}}$ is again the category of commutative K-algebras, but the cohomology is different. For a counterexample take $K = \boldsymbol{Z}$, $W_1 = W_2 = X_1 = X_2 = \boldsymbol{Z}_2$ (integers mod 2), $Y = \boldsymbol{Z}_2 + \boldsymbol{Z}_2 \to X_1 * X_2 = \boldsymbol{Z}_2 \otimes \boldsymbol{Z}_2 \cong \boldsymbol{Z}_2$ by the projection, $\boldsymbol{Z}_2$ acting on itself as kernel in the usual way. Then in dimension 1 all three groups in (5.4) have the value $\boldsymbol{Z}_2$, which is a contradiction. It would be interesting to know conditions on $\mathscr{A}$ and $\boldsymbol{T}$ guaranteeing the validity of (5.3).

References

[1] Beck, J.: Triples, algebras and cohomology. To appear.

[2] Eilenberg, S., and S. MacLane: Acyclic models. Amer. J. Math. **75**, 189–199 (1953).

[3] — Cohomology theory in abstract groups, I. Ann. Math. **48**, 51–78 (1947).

[4] —, and J. C. Moore: Adjoint functors and triples. Ill. J. Math. **9**, 381–398 (1965).

[5] Harrison, D. K.: Commutative algebras and cohomology. Trans. AMS. **104**, 191–204 (1962).

[6] Lyndon, R. C.: New proof for a theorem of Eilenberg and MacLane. Ann. Math. **50**, 731–735 (1949).

[7] MacLane, S.: Homology. Berlin: Springer 1963.

[8] Barr, M., and G. Rinehart: Cohomology as the derived functor of derivations. To appear in Trans. AMS.

Department of Mathematics
University of Illinois
Urbana, Illinois

Department of Mathematics
Cornell University
Ithaca, New York

Cohomology in Tensored Categories *, **

By

Michael Barr

1. Introduction

Both MacLane [3] and Beck [2] have recently defined cohomology theories for algebras in abstract categorical settings. MacLane's theory is an abstract formalization of the (normalized) bar construction (see [4, p. 144] for example). Since, however, his proof of the normalization theorem [4, p. 236] remains perfectly valid, the normalized bar construction can be replaced by the un-normalized one. It is the purpose of this paper to show that under reasonable conditions Beck's cohomology is naturally equivalent to a slight modification of MacLane's. All notation not explicitly defined is taken from [3].

The proof given here is mainly just a formalization of that of [1]. However §§ 2, 3, 4 are devoted to showing that the techniques used there can all be applied to the general categorical situation.

2. Algebras

If $\mathscr{D}$ is a tensored category $\mathrm{Alg}(\mathscr{D})$ denotes the category of $\mathscr{D}$-algebras as described in [3, p. 79]. The purpose of this section is to show:

Theorem 2.1. *Suppose $\mathscr{D}$ has countable coproducts which commute with the $\otimes$. Then there is a triple $\boldsymbol{T} = (T, \eta, \mu)$ on $\mathscr{D}$ such that* $\mathrm{Alg}(\mathscr{D}) \cong \mathscr{D}^{\boldsymbol{T}}$.

Proof. We will show first that the underlying functor $U\colon \mathrm{Alg}(\mathscr{D}) \to \mathscr{D}$ has a left adjoint F. Then if $\eta\colon 1 \to UF$ and $\varepsilon\colon FU \to 1$ are the adjointness morphisms, it is known that $\boldsymbol{T} = (UF, \eta, U\varepsilon F)$ is a triple [1], [2]. Then we complete the proof by exhibiting a natural equivalence $\mathrm{Alg}(\mathscr{D}) \cong \mathscr{D}^{\boldsymbol{T}}$.

Definition 2.2. *Let $A_1, \ldots, A_n$ be objects of $\mathscr{D}$. We define $A_1 \otimes \cdots \otimes A_n$ inductively to be K if $n = 0$ and $(A_1 \otimes \cdots \otimes A_{n-1}) \otimes A_n$ for $n > 0$. If each $A_i = A$ we will also denote this by $A^{(n)}$. It follows from the definition*

---

* Research partially supported by NSF Contract GP 730.

** Received August 26, 1965.

of a tensored category that there is a unique isomorphism which we denote by $\sigma(m, n)\colon A^{(m)} \otimes A^{(n)} \to A^{(m+n)}$.

We now let $F(A) = \sum_{n \geq 0} A^{(n)}$. To describe an algebra structure on $F(A)$, let $\alpha_n\colon A^{(n)} \to F(A)$ be the inclusion. Since $\otimes$ commutes with this coproduct, a map $\pi_A\colon\ F(A) \otimes F(A) \cong \sum_{n\geq 0, m\geq 0} A^{(n)} \otimes A^{(m)} \to F(A)$ is determined by requiring that $\pi(\alpha_n \otimes \alpha_m) = \alpha_{n+m}\sigma(m, n)$. Then I claim that $F(A)$, π and α_0 (as unit) form an object of $\mathrm{Alg}(\mathscr{D})$. We must show three identities [*3*, p. 79]. The first is that if $e_{F(A)}\colon K \otimes F(A) \to F(A)$ is the MacLane isomorphism, then $\pi(\alpha_0 \otimes 1) = e_{F(A)}$. But $\sigma(0, m) = e_A(m)$, and by the naturality of e we have

$$\pi(\alpha_0 \otimes 1)(1 \otimes \alpha_m) = \pi(\alpha_0 \otimes \alpha_m) = \alpha_m\sigma(0, m) = e_{F(A)}(1 \otimes \alpha_m).$$

Since the $1 \otimes \alpha_m$ are inclusions to a direct sum, this gives the result. The second is derived similarly. The third is that

$$\pi(1 \otimes \pi) = \pi(\pi \otimes 1)\colon F(A) \otimes F(A) \otimes F(A) \to F(A).$$

But $F(A) \otimes F(A) \otimes F(A)$ is a direct sum

$$\sum_{m\geq 0,\, n\geq 0,\, p\geq 0} A^{(m)} \otimes A^{(n)} \otimes A^{(p)}.$$

$$\pi(1 \otimes \pi)(\alpha_m \otimes \alpha_n \otimes \alpha_p) = \pi(\alpha_m \otimes \pi(\alpha_n \otimes \alpha_p)) = \pi(\alpha_m \otimes \alpha_{n+p}\sigma(n, p))$$
$$= \pi(\alpha_m \otimes \alpha_{n+p})(1 \otimes \sigma(n, p)) = \alpha_{m+n+p}\sigma(m, n+p)(1 \otimes \sigma(n, p)).$$

But $\sigma(m, n+p)(1 \otimes \sigma(n, p))\colon A^{(m)} \otimes A^{(n)} \otimes A^{(p)} \to A^{(m+n+p)}$ is a MacLane isomorphism which by uniqueness must equal

$$\sigma(m+n, p)(\sigma(m, n) \otimes 1).$$

Then we get

$$\alpha_{m+n+p}\sigma(m+n, p)(\sigma(m, n) \otimes 1) = \pi(\alpha_{m+n} \otimes \alpha_p)(\sigma(m, n) \otimes 1)$$
$$= \pi(\alpha_{m+n}\sigma(m, n) \otimes \alpha_p) = \pi(\pi(\alpha_m \otimes \alpha_n) \otimes \alpha_p)$$
$$= \pi(\pi \otimes 1)(\alpha_m \otimes \alpha_n \otimes \alpha_p).$$

Since the $\alpha_m \otimes \alpha_n \otimes \alpha_p$ are a direct family of inclusions the result follows.

Now suppose $(\Lambda, p_\Lambda, u_\Lambda)$ is in $\mathrm{Alg}(\mathscr{D})$ and $f\colon A \to \Lambda$ is in $\mathscr{D}$. We prove adjointness by showing that there is a unique map $f^*\colon F(A) \to \Lambda$ in $\mathrm{Alg}(\mathscr{D})$ such that $f^*\alpha_1 = f$. (At least this proves adjointness with $\alpha_1\colon A \to UF(A)$ as one of the adjointness morphisms [cf. *3*, p. 58].) We define f^* by requiring that $f^*\alpha_0 = u_\Lambda$, $f^*\alpha_1 = f$ and $f^*\alpha_i = p_\Lambda(f^*\alpha_{i-1} \otimes f)$, for $i > 1$. To show this in $\mathrm{Alg}(\mathscr{D})$ we must show that it preserves the unit which follows from the definition of f^*a_0 and that $f^*\pi = p_\Lambda(f^* \otimes f^*)$. By induction on $n + m$,

$$f^*\pi(\alpha_n \otimes \alpha_m) = f^*\alpha_{n+m}\sigma(m, n) = p_\Lambda(f^*\alpha_{n+m-1} \otimes f)\sigma(m, n)$$
$$= p_\Lambda(f^*\alpha_{n+m-1} \otimes f)(\sigma(m, n-1) \otimes 1) = p_\Lambda(f^*\alpha_{n+m-1}\sigma(m, n-1) \otimes f)$$
$$= p_\Lambda(f^*\pi(\alpha_n \otimes \alpha_{m-1}) \otimes f) = p_\Lambda(p_\Lambda(f^* \otimes f^*)(\alpha_n \otimes \alpha_{m-1}) \otimes f)$$
$$= p_\Lambda(p_\Lambda(f^*\alpha_n \otimes f^*\alpha_{m-1}) \otimes f)$$

$$= p_\wedge (p_\wedge \otimes 1)(f^* \alpha_n \otimes f^* \alpha_{m-1} \otimes f)$$
$$= p_\wedge (1 \otimes p_\wedge)(f^* \alpha_n \otimes f^* \alpha_{m-1} \otimes f)$$
$$= p_\wedge (f^* \alpha_n \otimes p_\wedge (f^* \alpha_{m-1} \otimes f))$$
$$= p_\wedge (f^* \alpha_n \otimes f^* \alpha_m) = p_\wedge (f^* \otimes f^*)(\alpha_n \otimes \alpha_m).$$

Again, this implies that $f^* \pi = p_\wedge (f^* \otimes f^*)$. Suppose g is another map with $g\alpha_1 = f$. Then $g\alpha_0 = u_\wedge = f^* \alpha_0$ by the unitary condition, $g\alpha_1 = f = f^* \alpha_1$ by assumption and if we assume $g\alpha_{i-1} = f^* \alpha_{i-1}$ then

$$f^* \alpha_i = p_\wedge (f^* \alpha_{i-1} \otimes f) = p_\wedge (g\alpha_{i-1} \otimes g\alpha_1) = p_\wedge (g \otimes g)(\alpha_{i-1} \otimes \alpha_1)$$
$$= g\pi(\alpha_{i-1} \otimes \alpha_1) = g\sigma(i-1,1)\alpha_i = g\alpha_i$$

(since $\sigma(i-1,1)$ is the identity isomorphism). Hence $f^* \alpha_i = g\alpha_i$ for each i, so $f^* = g$.

This shows adjointness, and therefore gives rise to a triple as noted above. Moreover we have from [2] a natural functor Ψ: $\mathrm{Alg}(\mathscr{D}) \to \mathscr{D}^{\boldsymbol{T}}$ which takes $(\Lambda, p_\wedge, u_\wedge)$ to (Λ, λ) where λ: $F(\Lambda) \to \Lambda$ is the unique algebra map extending the identity map of $\Lambda \to \Lambda$. Specifically if β_i: $\Lambda^{(i)} \to F(\Lambda)$ are the direct system, then $\lambda\beta_0 = u_\wedge$, $\lambda\beta_1 = 1_\wedge$ and $\lambda\beta_i = p_\wedge(\lambda\beta_{i-1} \otimes 1_\wedge)$. Notice that $\lambda\beta_2 = p_\wedge$ which tells us how to construct an inverse functor. Specifically if (X, ξ) is a $\boldsymbol{T}$-algebra we let $u_X = \xi\gamma_0$ and $p_X = \xi\gamma_2$ where γ_i: $X^{(i)} \to F(X)$ are the inclusions. If these give X the structure of a $\mathscr{D}$-Algebra then they clearly define a functor Φ which is a right inverse to Ψ. Let φ_i: $F(X)^{(i)} \to F(F(X))$ be the inclusion. Then we know that $\xi\gamma_1 = 1_X$ by the definition of a $\boldsymbol{T}$-algebra. The other condition says that $\xi \cdot F\xi = \xi \cdot U\varepsilon F(X)$. But $U\varepsilon F(A)$: $F(F(A)) \to F(A)$ is just the map coming out of the algebra structure on $F(A)$. Hence $U\varepsilon F(X) \cdot \varphi_2 = \pi_X$: $F(X) \otimes F(X) \to F(X)$. On the other hand $F\xi \cdot \varphi_2$: $F(X) \otimes F(X) \to F(X)$ is just $\gamma_2(\xi \otimes \xi)$. Hence $\xi\gamma_2(\xi \otimes \xi) = \xi\pi_X$: $F(X) \otimes F(X) \to X$. Then

$$\xi\gamma_2(1 \otimes \xi\gamma_2) = \xi\gamma_2(\xi\gamma_1 \otimes \xi\gamma_2) = \xi\gamma_2(\xi \otimes \xi)(\gamma_1 \otimes \gamma_2)$$
$$= \xi\pi_X(\gamma_1 \otimes \gamma_2) = \xi\gamma_3\sigma(1,2) = \xi\gamma_3\sigma(2,1)\sigma(1,2)$$
$$= \xi\pi_X(\gamma_2 \otimes \gamma_1)\sigma(1,2) = \xi\gamma_2(\xi \otimes \xi)(\gamma_2 \otimes \gamma_1)\sigma(1,2)$$
$$= \xi\gamma_2(\xi\gamma_2 \otimes \xi\gamma_1)\sigma(1,2) = \xi\gamma_2(\xi\gamma_2 \otimes 1)\sigma(1,2)$$

which, strictly speaking, is the precise statement that $\xi\gamma_2$ is an associative law of composition. The unitary laws are left to the reader. Now we must show that Φ is a left inverse to Ψ as well. To do this we start with (X, ξ) in $\mathscr{D}^{\boldsymbol{T}}$ and make it into an algebra using $\xi\gamma_2$: $X \otimes X \to F(X) \to X$ as the rule of composition and $\xi\gamma_0$ as the unit. We now define a new map ζ: $F(X) \to X$ by setting $\zeta\gamma_0 = \xi\gamma_0$, $\zeta\gamma_1 = 1 = \xi\gamma_1$ and $\zeta\gamma_i = \xi\gamma_2(\zeta\gamma_{i-1} \otimes 1)$ for $i > 1$. Assuming, however, that $\zeta\gamma_{i-1} = \xi\gamma_{i-1}$, we have

$$\zeta\gamma_i = \xi\gamma_2(\zeta\gamma_{i-1} \otimes 1) = \xi\gamma_2(\xi\gamma_{i-1} \otimes 1) = \xi\gamma_2(\xi\gamma_{i-1} \otimes \xi\gamma_1)$$
$$= \xi\gamma_2(\xi \otimes \xi)(\gamma_{i-1} \otimes \gamma_1) = \xi\pi_X(\gamma_{i-1} \otimes \gamma_1) = \xi\gamma_i\sigma(i-1,1) = \xi\gamma_i.$$

Hence $\zeta = \xi$. This completes the proof of (2.1).

3. Derivations

Let Λ be in $\mathrm{Alg}(\mathscr{D})$ and M be a $\Lambda - \Lambda^0$ bimodule. This means that there are morphisms $p_M\colon \Lambda \otimes M \to M$ and $q_M\colon M \otimes \Lambda \to M$ satisfying the rules for right and left modules, and that $p_M(1 \otimes q_M) = q_M(p_M \otimes 1)$.

Definition 3.1. *A derivation of* Λ *to* M *is a morphism* $d\colon \Lambda \to M$ *such that* $d\, p_\wedge = p_M(1 \otimes d) + q_M(d \otimes 1)\colon \Lambda \otimes \Lambda \to M$. *It is clear that the set of derivations of* Λ *to* M *is a subgroup of* $\mathrm{Hom}_{\mathscr{D}}(\Lambda, M)$. *We denote this abelian group by* $\mathrm{Der}(\Lambda, M)$.

Proposition 3.2. *Let* $C\Lambda = \{C_n \Lambda\}$, $n \geqq 0$, *be the complex in* $\mathscr{D}$ *such that* $C_n \Lambda = \Lambda^{(n+2)}$, *and boundary operator* $\partial = \sum(-1)^i (1 \otimes 1 \otimes \cdots \otimes \otimes p_\wedge \otimes 1 \otimes \cdots \otimes 1)$. *Then* C *is an acyclic complex over* Λ, *i.e.*

$$\cdots \to C_n \to \cdots \to C_0 \overset{p_\wedge}{\to} \Lambda \to 0$$

is exact.

Proof. For let $s_n\colon C_n \Lambda \to C_{n+1}\Lambda$ by $s_n = ((u \otimes 1)\, e_\wedge^{-1}) \otimes 1 \otimes \cdots \otimes 1$, for $n \geqq 0$ and $s_{-1}\colon \Lambda \to C_0$ by $(u \otimes 1)\, e_\wedge^{-1}$. Then in the usual way it is easily shown that $p_\wedge s_{-1} = 1$ and $s_{n-1}\partial_n + \partial_{n+1}s_n = 1$ for $n \geqq 0$.

Definition 3.3. $J\Lambda = \ker p_\wedge \approx \operatorname{coker} \partial_2$. *Since by* [3, p. 81] *the category of* Λ*-bimodules is abelian,* $J\Lambda$ *is a* Λ*-bimodule also. If* $\psi\colon \Lambda^{(3)} \to J\Lambda$ *is the cokernel of* ∂_2, *then* $\mathrm{Hom}(\psi, -)\colon \mathrm{Hom}_{\wedge^e}(J\Lambda, -) \to \mathrm{Hom}_{\wedge^e}(\Lambda^{(3)}, -)$ is a monomorphism onto the subgroup of the latter, consisting of those maps whose composition with ∂_2 is 0. We can map

$$\varphi\colon \mathrm{Der}(\Lambda, -) \to \mathrm{Hom}_{\wedge^e}(\Lambda^{(3)}, -)$$

by restriction of the isomorphism $\mathrm{Hom}_{\mathscr{D}}(\Lambda, -) \cong \mathrm{Hom}_{\wedge^e}(\Lambda^{(3)}, -)$, whence φ is a monomorphism also.

Proposition 3.4. $\mathrm{Im}\,\varphi = \mathrm{Im}\,\mathrm{Hom}(\varepsilon, -)$ *and establishes a natural equivalence between* $\mathrm{Der}(\Lambda, -)$ *and* $\mathrm{Hom}_{\wedge^e}(J\Lambda, -)$.

Proof. After sorting through all the identifications made, it is easily seen that what I claim amounts to showing that $d\colon \Lambda \to M$ is a derivation if and only if $p_M(1 \otimes q_M)(1 \otimes d \otimes 1)\,\partial_2 = 0$. If we recall the defining identities of bimodules, it is a straightforward computation that

$$p_M(1 \otimes q_M)(1 \otimes d \otimes 1)\,\partial_2 = \\ = p_M(1 \otimes q_M)(1 \otimes [p_M(1 \otimes d) - d\,p_\wedge + q_M(d \otimes 1)] \otimes 1).$$

This gives one implication immediately. On the other hand, for any $f\colon \Lambda \to M$, $p_M(1 \otimes q_M)(1 \otimes f \otimes 1)(u_\wedge \otimes 1 \otimes u_\wedge)$
$= p_M(1 \otimes q_M)(u_\wedge \otimes f \otimes u_\wedge) = p_M(u_\wedge \otimes q_M(f \otimes u_\wedge))$
$= p_M(u_\wedge \otimes q_M(1 \otimes u_\wedge)(f \otimes 1)) = p_M(u_\wedge \otimes e_M(f \otimes 1))$
$= p_M(u_\wedge \otimes f e_\wedge) = p_M(u_\wedge \otimes 1)(1 \otimes f e_\wedge) = e_M(1 \otimes f e_\wedge) = f e_\wedge\, e_{\wedge \otimes K}$.
Since these e's are isomorphisms, the other implication follows.

Definition 3.5. *Let M be a Λ-bimodule as before. We let M^+ denote the algebra whose underlying $\mathscr{D}$ object is $\Lambda + M$ and whose multiplication is that map from*

$(\Lambda + M) \otimes (\Lambda + M) \cong \Lambda \otimes \Lambda + \Lambda \otimes M + M \otimes \Lambda + M \otimes M \to \Lambda + M$ *whose matrix is*

$$\begin{Vmatrix} p_\Lambda & 0 & 0 & 0 \\ 0 & p_M & q_M & 0 \end{Vmatrix}.$$

Let $\pi_1 \colon M^+ \to \Lambda$ and $\pi_2 \colon M^+ \to M$ be the coordinate projections. It is easily seen that π_1 is an algebra epimorphism.

Proposition 3.6. *The mapping $f \to \pi_2 f$ establishes an equivalence between $\{f \in \mathrm{Hom}_{\mathrm{Alg}(\mathscr{D})}(\Lambda, M^+) \mid \pi_1 f = 1\}$ and $\mathrm{Der}(\Lambda, M)$.*

Proof. Let f be such a mapping. Then f has a matrix

$$\begin{Vmatrix} f_1 \\ f_2 \end{Vmatrix}$$

where $f_1 = \pi_1 f = 1$ and $f_2 = \pi_2 f$. Hence there is a $1 - 1$ correspondence between such f and maps $f_2 \colon \Lambda \to M$ such that $\begin{Vmatrix} 1 \\ f_2 \end{Vmatrix}$ is an algebra homomorphism. This condition becomes

$$\begin{Vmatrix} p_\Lambda & 0 & 0 & 0 \\ 0 & p_M & q_M & 0 \end{Vmatrix} \begin{Vmatrix} 1 \otimes 1 \\ 1 \otimes f_2 \\ f_2 \otimes 1 \\ f_2 \otimes f_2 \end{Vmatrix} = \begin{Vmatrix} 1 \\ f_2 \end{Vmatrix} p_\Lambda,$$

which reduces to $p_\Lambda(1 \otimes 1) = p_\Lambda$, which is clear; and $p_M(1 \otimes f_2) + q_M(f_2 \otimes 1) = f_2 p_\Lambda$, which is the defining equation of a derivation.

Definition 3.7. *Let* $(\mathrm{Alg}(\mathscr{D}), \Lambda)$ *denote the category whose objects are* $\mathrm{Alg}(\mathscr{D})$ *morphisms $\gamma \colon \Gamma \to \Lambda$ and whose morphisms are commutative triangles $\Lambda \overset{\gamma}{\leftarrow} \Gamma \overset{\alpha}{\to} \Gamma' \overset{\gamma'}{\to} \Lambda$ where α is a morphism in* $\mathrm{Alg}(\mathscr{D})$. *We may always consider $\pi_1 \colon M^+ \to \Lambda$ as such an object. In that case (3.6) becomes*

Proposition 3.8. *If $\pi_1 \colon M^+ \to \Lambda$ is as above then*

$$\mathrm{Hom}_{(\mathrm{Alg}(\mathscr{D}), \Lambda)}(1_\Lambda, \pi_1) \cong \mathrm{Der}(\Lambda, M).$$

4. Modules

Suppose that $\gamma \colon \Gamma \to \Lambda$ is in $\mathrm{Alg}(\mathscr{D})$. We define functors

$$P = P_\gamma \colon \Lambda - \mathrm{Mod}(\mathscr{D}) \to \Gamma - \mathrm{Mod}(\mathscr{D}) \quad \text{and}$$
$$Q = Q_\gamma \colon \Gamma - \mathrm{Mod}(\mathscr{D}) \to \Lambda - \mathrm{Mod}(\mathscr{D}) \quad \text{by}$$

setting, for a Λ-module (M, p_M) and a Γ-module (N, p_N),
$P(M, p_M) = (M, p_M(\gamma \otimes 1))$, $Q(N, p_N) = (P^0(\Lambda) \otimes_\Gamma N,\ P^0(p_\Lambda) \otimes_\Gamma p_N)$,

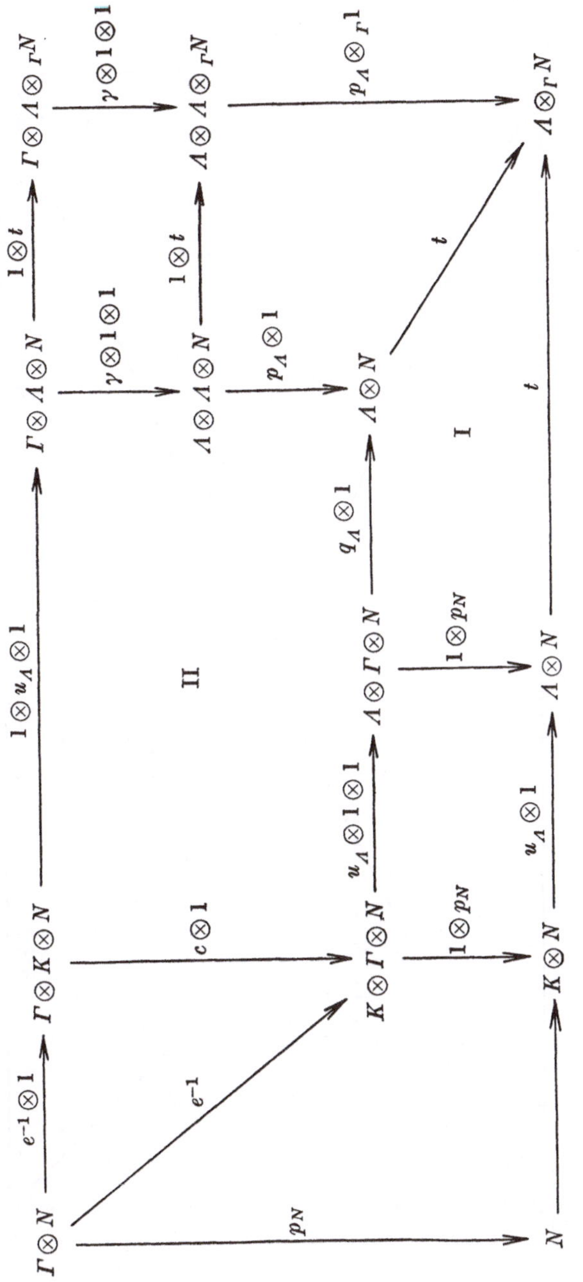
$\Gamma\otimes N$
$e^{-1}\otimes 1$
$\Gamma\otimes K\otimes N$
$1\otimes u_\Lambda\otimes 1$
$\Gamma\otimes\Lambda\otimes N$
$1\otimes t$
$\Gamma\otimes\Lambda\otimes_\Gamma N$
e^{-1}
$c\otimes 1$
$\gamma\otimes 1\otimes 1$
$\Lambda\otimes\Lambda\otimes N$
$1\otimes t$
$\Lambda\otimes\Lambda\otimes_\Gamma N$
II
p_N
$K\otimes\Gamma\otimes N$
$u_\Lambda\otimes 1\otimes 1$
$\Lambda\otimes\Gamma\otimes N$
$q_\Lambda\otimes 1$
$\Lambda\otimes N$
$p_\Lambda\otimes 1$
$p_\Lambda\otimes_\Gamma 1$
$1\otimes p_N$
I
t
N
$K\otimes N$
$u_\Lambda\otimes 1$
$\Lambda\otimes N$
$\Lambda\otimes_\Gamma N$

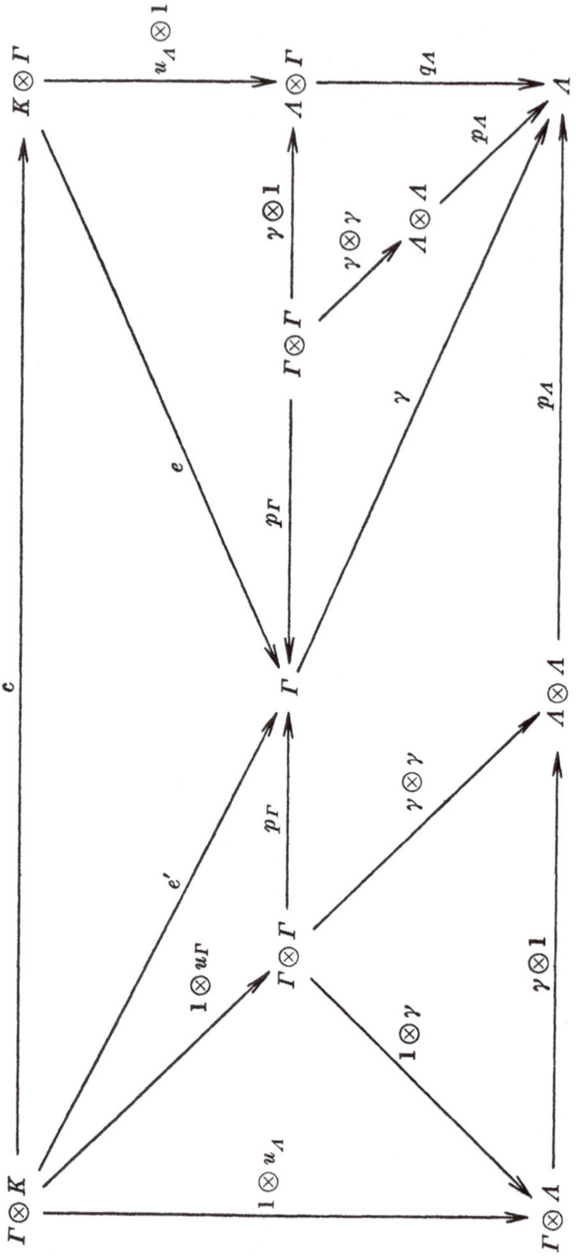
K ⊗ Γ
u_Λ ⊗ 1
Λ ⊗ Γ
q_Λ
Λ
γ ⊗ 1
γ ⊗ γ
Λ ⊗ Λ
p_Λ
Γ ⊗ Γ
γ
e
p_Γ
p_Λ
c
Γ
Λ ⊗ Λ
γ ⊗ γ
p_Γ
e′
Γ ⊗ Γ
1 ⊗ u_Γ
γ ⊗ 1
1 ⊗ γ
1 ⊗ u_Λ
Γ ⊗ K
Γ ⊗ Λ

where P^0 is the functor analogous to P for right Λ-modules.

Theorem 4.1. $Q \dashv P$.

Proof. First we define a morphism $\eta\colon 1 \to PQ$. By abuse of notation $\eta_N\colon N \to \Lambda \otimes_\Gamma N$ is defined by the composition $t(u_\Lambda \otimes 1)\, e_N^{-1}$ where $t\colon \Lambda \otimes N \to \Lambda \otimes_\Gamma N$ as defined in [3, p. 82]. We must show that

$$\eta_N p_N = p_{PQ(N)}(1 \otimes \eta_N)\colon \Gamma \otimes N \to \Lambda \otimes_\Gamma N .$$

To do this we refer to figure 1. In figure 1, every subdiagram except those labeled I and II commutes by either a coherence or a naturality condition. $q_\Lambda = p_\Lambda(1 \otimes \gamma)\colon \Lambda \otimes \Gamma \to \Lambda$ is just the operation of Γ on Λ and I commutes by definition of $\otimes_\Gamma$. As for II, we refer to figure 2. In this every subdiagram is commutative either because of coherence or naturality or because γ is a morphism of algebras. Tensoring the outer square with N gives the desired result. To complete the proof we must show that given a Γ-linear map $\alpha\colon N \to PM$ we can find a unique Λ-linear map $\beta\colon QN \to M$ such that $P\beta \cdot \eta_N = \alpha$. $\Lambda \otimes_\Gamma N$ is the cokernel of $q_\Lambda \otimes 1 - 1 \otimes p_N\colon \Lambda \otimes \Gamma \otimes N \to \Lambda \otimes N$.

$p_M(1 \otimes \alpha)(q_\Lambda \otimes 1) = p_M(p_\Lambda \otimes 1)(1 \otimes \gamma \otimes 1)(1 \otimes 1 \otimes \alpha)$
$= p_M(1 \otimes p_M)(1 \otimes \gamma \otimes 1)(1 \otimes 1 \otimes \alpha) = p_M(1 \otimes p_M(\gamma \otimes 1)(1 \otimes \alpha))$
$= p_M(1 \otimes \alpha p_N) = p_M(1 \otimes \alpha)(1 \otimes p_N)$.

Hence $p_M(1 \otimes \alpha)$ induces a map $\beta\colon \Lambda \otimes_\Gamma N \to M$. To show this is Λ-linear we must show that $p_M(1 \otimes \beta) = \beta(p_\Lambda \otimes_\Gamma 1)$. But since these are cokernels it reduces to showing that

$p_M[1 \otimes p_M(1 \otimes \alpha)] = p_M(1 \otimes p_M)(1 \otimes 1 \otimes \alpha)$
$= p_M(p_\Lambda \otimes 1)(1 \otimes 1 \otimes \alpha) = p_M(1 \otimes \alpha)(p_\Lambda \otimes 1)$.

The fact that $P\beta \cdot \eta_N = \alpha$ follows from the commutativity of

$$\begin{array}{ccccccc}
N & \longrightarrow & K \otimes N & \longrightarrow & \Lambda \otimes N & \xrightarrow{\ t\ } & \Lambda \otimes_\Gamma N \\
\downarrow{\scriptstyle\alpha} & & \downarrow{\scriptstyle 1 \otimes \alpha} & & \downarrow{\scriptstyle 1 \otimes \alpha} & & \downarrow{\scriptstyle\beta} \\
M & \longrightarrow & K \otimes M & \longrightarrow & \Lambda \otimes M & \xrightarrow{\ p_\Lambda\ } & M
\end{array}$$

together with the fact that the bottom row is just the identity. Now suppose $\beta'\colon \Lambda \otimes_\Gamma N \to M$ is another Λ-linear map with $P\beta' \cdot \eta_N = \alpha$. Then $\beta' t\colon \Lambda \otimes N \to M$ is a Λ-linear map such that $\alpha = \beta' t(u_\Lambda \otimes 1)e_N^{-1}$. Since $\beta' t$ is Λ-linear, $\beta' t(p_\Lambda \otimes 1) = p_M(1 \otimes \beta' t)$. From the former, we get

$$\begin{aligned}
1 \otimes \alpha &= (1 \otimes \beta' t)(1 \otimes u_\Lambda \otimes 1)(1 \otimes e_N^{-1}), \\
p_M(1 \otimes \alpha) &= p_M(1 \otimes \beta' t)(1 \otimes u_\Lambda \otimes 1)(1 \otimes e_N^{-1}) \\
&= \beta' t(p_\Lambda \otimes 1)(1 \otimes u_\Lambda \otimes e_N^{-1}) \\
&= \beta' t(e'_\Lambda \otimes e_N^{-1}) .
\end{aligned}$$

But by coherence, $e'_\wedge \otimes e_N^{-1} : \Lambda \otimes N \to \Lambda \otimes K \otimes N \to \Lambda \otimes N$ must be the identity, so $\beta' t = p_M(1 \otimes \alpha) = \beta t$. Since t is an epimorphism, the result follows.

Henceforth, we will not distinguish between M and $P(M)$. We have already noted that if M is a Λ-module then

$$\mathrm{Hom}_{(\mathrm{Alg}(\mathscr{D}),\wedge)}(1_\wedge, \pi_1) \cong \mathrm{Der}(\Lambda, M).$$

Now, if $\gamma : \Gamma \to \Lambda$ is in $(\mathrm{Alg}(\mathscr{D}), \Lambda)$, the fact that

$$\mathrm{Der}(\Gamma, M) \cong \mathrm{Hom}_{(\mathrm{Alg}(\mathscr{D}),\wedge)}(\gamma, \pi_1)$$

can be proved in exactly the same way. Thus I have shown half of the following:

Theorem 4.2. *$\pi_1 : M^+ \to \Lambda$ is an abelian object of $(\mathrm{Alg}(\mathscr{D}), \Lambda)$. If $\gamma : \Gamma \to \Lambda$ is an abelian object of that category and $N = \ker \gamma$ then $N = P(M)$ for suitable M and $\gamma \cong \pi_1$.*

Proof. If $\gamma : \Gamma \to \Lambda$ is an abelian object, then $\mathrm{Hom}(1, \gamma)$ is also an abelian group. If i denotes its identity, then $i : \Lambda \to \Gamma$ is a right inverse of γ so that γ is an epimorphism. It is easily seen that $\ker \gamma = N$ is a Γ-bimodule (for $\gamma : \Gamma \to \Lambda$ is also a morphism of Γ-bimodules) and so $M = P_i(N)$ is a Λ-module. Essentially i induces a Λ-structure on N. Now considered as objects of D, γ has a right inverse so that $\Gamma \cong \Lambda + M$ and p_Γ has a representation as a matrix

$$\begin{Vmatrix} \alpha_{11} & \alpha_{12} & \alpha_{13} & \alpha_{14} \\ \alpha_{21} & \alpha_{22} & \alpha_{23} & \alpha_{24} \end{Vmatrix}$$

with respect to the basis indicated in the expansion

$$(\Lambda + M) \otimes (\Lambda + M) = \Lambda \otimes \Lambda + \Lambda \otimes M + M \otimes \Lambda + M \otimes M.$$

The fact that with respect to these bases i has matrix $\begin{Vmatrix} 1 \\ 0 \end{Vmatrix}$ and is a morphism in $(\mathrm{Alg}(\mathscr{D}), \Lambda)$ shows that $\alpha_{21} = 0$. Since $\gamma : \Lambda + M \to \Lambda$ has matrix $\|1 \; 0\|$ and is in $(\mathrm{Alg}(\mathscr{D}), \Lambda)$, a similar computation shows that $\alpha_{12} = \alpha_{13} = \alpha_{14} = 0$. Also if p_M and q_M are the Λ-operations on M induced by i, then $p_M = \alpha_{22}$, $q_M = \alpha_{23}$. We will be finished if we show that $\alpha = \alpha_{24} = 0$. Since $\gamma : \Gamma \to \Lambda$ is an abelian object there is a mapping of $\gamma \times \gamma \to \gamma$ where the product is in $(\mathrm{Alg}(\mathscr{D}), \Lambda)$. It is easily checked that $\gamma \times \gamma$ has the underlying object $\Sigma = \Lambda + M + M'$ where $M' = M$ is given a different name for purpose of keeping track of an ordered basis. This algebra has multiplication map

$$\begin{aligned} p_\Sigma : \Sigma \otimes \Sigma \cong \Lambda \otimes \Lambda + \Lambda \otimes M + \Lambda \otimes M' + M \otimes \Lambda + M \otimes M \\ + M \otimes M' + M' \otimes \Lambda + M' \otimes M + M' \otimes M' \to \Sigma \end{aligned}$$

whose matrix is

$$\left\| \begin{matrix} p_\Lambda & 0 & 0 & 0 & 0 & 0 & 0 & 0 & 0 \\ 0 & p_M & 0 & q_M & \alpha & 0 & 0 & 0 & 0 \\ 0 & 0 & p_M & 0 & 0 & 0 & q_M & 0 & \alpha \end{matrix} \right\| .$$

If $i_1 : \Gamma \to \Sigma$ and $i_2 : \Gamma \to \Sigma$ are the injections (with matrices

$$\left\| \begin{matrix} 1 & 0 \\ 0 & 1 \\ 0 & 0 \end{matrix} \right\| \quad \text{and} \quad \left\| \begin{matrix} 1 & 0 \\ 0 & 0 \\ 0 & 1 \end{matrix} \right\|$$

respectively), then a group law on Γ would be a map $\theta : \Sigma \to \Gamma$ in $(\mathrm{Alg}(\mathscr{D}), \Lambda)$ satisfying (among other conditions) $\theta i_1 = \theta i_2 = 1$. Writing out θ in matrix form these conditions imply that $\alpha = 0$.

Remark 4.3. The meaning of this theorem is that the categories of Λ-modules for the $\mathscr{D}$-algebra Λ [3, p. 81] and Λ-modules for the $\boldsymbol{T}$-algebra Λ [2], [1] are isomorphic.

5. Cohomology

Theorem 5.1 *If* $\gamma : \Gamma \to \Lambda$ *is in* $(\mathrm{Alg}(\mathscr{D}), \Lambda), \varphi = \gamma \otimes \gamma^0 : \Gamma^e \to \Lambda^e$ *and* M^+ *is an abelian object in that category, then there is an isomorphism*

$$H^n(\Gamma, M^+)_\Lambda \to \begin{cases} \mathrm{Der}(\Gamma, P_\varphi(M)), & n = 0 \\ H^{n+1}(\Gamma, P_\varphi(M)), & n > 0 \end{cases}$$

where the left hand side refers to the triple cohomology and the right hand side is the cohomology of MacLane's bar construction.

Proof. We use the method of acyclic models as explained in detail in [1]. We define standard cochain complexes L and S: $(\mathrm{Alg}(\mathscr{D}), \Lambda) \to \mathscr{Ab}$, the category of abelian groups, by $L^n(\Gamma, M) = \mathrm{Hom}_{\Gamma^e}(\Gamma^{(n+3)}, M)$ where again, we are letting M stand for some $P(M)$ and

$$S^n(\Gamma, M) = \mathrm{Hom}_{(\mathrm{Alg}(\mathscr{D}), \Lambda)}(G\Gamma^{(n+1)}, M^+) .$$

The boundary operator in L takes f to

$$p_M(1 \otimes f) + \sum(-1)^i (1 \otimes \cdots \otimes p_\Gamma \otimes \cdots \otimes 1) + (-1)^{n+1} q_M(f \otimes 1)$$

where p_Γ is its multiplication map and p_M and q_M are the bimodule operations on M. The coboundary in S takes f to $\sum(-1)^i \mathrm{Hom}(f \cdot G^i \varepsilon G^{n-i}, M)$. Then it is clear that L and S are standard complexes for the modified MacLane cohomology appearing on the right of (4.1) and the triple cohomology, respectively. Using η we may show, exactly as in [1], that S is G-representable and G-acyclic on models. If we can show the same for L and that $L^{-1} \cong S^{-1}$, we will be through. If $\zeta = \eta\Gamma : \Gamma \to G\Gamma$ is the adjointness morphism (in $\mathscr{D}$), it induces a map $\zeta^{(n+1)}: \Gamma^{(n+1)} \to$ $\to G\Gamma^{(n+1)}$. We have,

$$\begin{aligned} L^n(\Gamma, M) &= \mathrm{Hom}_{\Gamma^e}(\Gamma^{(n+3)}, M) \cong \mathrm{Hom}(\Gamma^{(n+1)}, M) \to \\ &\xrightarrow{\mathrm{Hom}(\zeta^{(n+1)}, M)} \mathrm{Hom}(G\Gamma^{(n+1)}, M) \cong \mathrm{Hom}_{G\Gamma^e}(G\Gamma^{(n+3)}, M) \\ &= L^n(G\Gamma, M), \end{aligned}$$

a G-representation of L. To show that L is G-acyclic, we choose $f \in$ $\in L^n(G\Gamma, M)$ and define sf in $L^{n-1}(G\Gamma, M)$ as follows. Let

$$\alpha(i_0, \ldots, i_{n+1}): \Gamma^{(i_0)} \otimes \cdots \otimes \Gamma^{(i_{n+1})} \to G\Gamma^{(n+2)}$$

denote the inclusion and define sf by,

$$\begin{aligned} & sf \cdot \alpha(i_0, 0, i_n, \ldots, i_{n+1}) \\ & \qquad = f \cdot \alpha(i_0, 0, 0, i_n, \ldots, i_{n+1}) \times (1 \otimes e_K^{-1} 1 \otimes \cdots \otimes 1) \end{aligned}$$

for $i_1 = 0$ and inductively for $i_1 > 0$ by

$$\begin{aligned} & sf \cdot \alpha(i_0, i_1, \ldots, i_{n+1}) \\ & = sf \cdot \alpha(i_0 + 1, i_1 - 1, i_2, \ldots, i_n)[(\sigma(i_0, 1) \otimes 1)(1 \otimes \sigma(1, i_1 - 1)^{-1}) \otimes \\ & \qquad \otimes 1 \otimes \cdots \otimes 1] \\ & - f \cdot \alpha(i_0, 1, i_1 - 1, i_2, \ldots, i_n)(1 \otimes \sigma(1, i_1 - 1)^{-1} \otimes 1 \otimes \cdots \otimes 1). \end{aligned}$$

Then, just as in [*1*], s may be shown to be a contraction in $L(G\Gamma, M)$.

$$L^{-1}(\Gamma, M) \cong \mathrm{Hom}_{\Gamma^e}(J\Gamma, M) \cong \mathrm{Der}(\Gamma, M) \cong \mathrm{Hom}_{(\mathrm{Alg}(\mathscr{D}), \wedge)}(\Gamma, M^+)$$

by the results of § 3 and it is shown in [*2*] that the latter is

$$H^0(S(\Gamma, M)) \cong S^{-1}(\Gamma, M).$$

This completes the proof.

References

[*1*] Barr, M., and J. Beck: Acyclic models and triples, these proceedings, pp. 336—343.
[*2*] Beck, J.: Triples, algebras and cohomology. Dissertation, Columbia University,
[*3*] MacLane, S.: Categorical algebra. Bull. Amer. Math. Soc. **71**, 40—106, (1965).
[*4*] — Homology. Berlin: Springer 1963.

Department of Mathematics
University of Illinois
Urbana, Illinois

Extraordinary Homology and Chain Complexes*, **

By

Alex Heller

Introduction

We shall be concerned here with homology theories, i.e. what used to be called "extraordinary" homology theories, on the category of finite CW-complexes. Our approach is categorical; it is thus important to define homology theories as functors on the appropriate category, in this case the "stable-homotopy" category constructed from finite CW-complexes. The exactness axiom must then be expressed in terms of an additional structure on the category, namely that of the cofibration triangles, introduced first by Puppe [*6*].

Our principal result asserts that, under a certain restriction, a homology theory may be computed by means of chain complexes. To be more explicit, there is a functor with values in the category of chain complexes and chain-homotopy-classes of maps such that its composition with the ordinary homology functor of chain complexes is the original homology theory.

The restriction has to do with the universal coefficient theorem of Dold [*2*] for "extraordinary" homology. It is indeed just the assertion that the universal-coefficient sequence splits. However this needs to be investigated only in one special case; we get accordingly an amplification of Dold's theorem as a by-product (Theorem 4.6).

Homology theories satisfying this restriction, then, may be represented by chain-complexes. We prove, indeed, rather more. If the chain-complex-valued functors are suitably restricted too, once more by a condition related to the universal coefficient theorem, then the categories of chain-complex-valued functors and of homology theories are equivalent (Theorem 5.5).

We conclude with a list of unsolved problems which arise in this context.

The following notation will be used. If $\mathscr{A}$ is a graded category $\mathscr{A}_0$

---

* This research was in part supported by a National Science Foundation contract.

** Received Sept. 13, 1965.

is the subcategory containing the morphisms of degree 0. If $\mathscr{A}, \mathscr{B}$ are graded additive categories $\mathscr{F}(\mathscr{A}, \mathscr{B})$ is the graded additive category of graded additive functors. If $G: \mathscr{A}' \to \mathscr{A}$, $H: \mathscr{B} \to \mathscr{B}'$ we shall abbreviate the „composition" functors $\mathscr{F}(G, 1): \mathscr{F}(\mathscr{A}, \mathscr{B}) \to \mathscr{F}(\mathscr{A}', \mathscr{B})$ and $\mathscr{F}(1, H)$: $\mathscr{F}(\mathscr{A}, \mathscr{B}) \to \mathscr{F}(\mathscr{A}, \mathscr{B}')$ by $G^{\bullet}, H_{\bullet}$. Finally, if $F: \mathscr{A}_1 \times \ldots \times \mathscr{A}_n \to \mathscr{B}$ and $1 \leq i_1 < \cdots < i_k \leqq n$, $1 \leqq j_1 < \cdots < j_{n-k} \leqq n$ are complementary sets we write

$$F_{i_1, \ldots, i_k}: \mathscr{A}_{i_1} \times \cdots \times \mathscr{A}_{i_k} \to \mathscr{F}(\mathscr{A}_{j_1} \times \cdots \times \mathscr{A}_{j_{n-k}} \mathscr{B})$$

for the functor

$$(F_{i_1, \ldots, i_k}(A_{i_1}, \ldots, A_{i_k}))(A_{j_1}, \ldots, A_{j_{n-k}}) = F(A_1, \ldots, A_n).$$

1. The Stable Homotopy Category

We denote by $\mathscr{T}$ the category of finite CW-complexes with basepoint, $\mathscr{T}(X, Y)$ being the set of continuous basepoint-preserving maps. The category $\mathscr{T}^b$ has the same objects but is provided with homotopy-classes of maps in $\mathscr{T}$ as morphisms. The smash-product in $\mathscr{T}$ and $\mathscr{T}^b$ will be written $\otimes$, so that in T, $X \otimes Y = X \times Y/X \vee Y$. The cone-functor c and the suspension functor S are thus, respectively, $I \otimes -$ and $S^1 \otimes -$.

By a *cofibration in* $\mathscr{T}$ we mean a sequence $A \to X \to X/A$ where A is a subcomplex. To such a sequence we associate a unique morphism $X/A \to SA$ in $\mathscr{T}^b$ by commutativity in

$$\begin{array}{ccc} X \cup cA & \to & (X \cup cA)/X = SA \\ \downarrow & \nearrow & \\ (X \cup cA)/cA = X/A & & \end{array}$$

observing that the vertical map is an isomorphism in $\mathscr{T}^b$. By a *cofibration in* $\mathscr{T}^b$ we mean a sequence $A \to X \to X/A \to SA$ obtained in this way, or any sequence in $\mathscr{T}^b$ isomorphic to one of these.

It is well known that the smash product $\otimes$ is coherently associative and commutative with identity S^0, and that for any Y the functor $Y \otimes -$ preserves cofibrations.

The stable-homotopy category $\mathscr{S}$ is the graded additive category with objects $ob\,\mathscr{T} \times \mathbf{Z}$ and

$$\mathscr{S}((X, n), (Y, m)) = \varinjlim_k \mathscr{T}^b(S^{k+g-n} X, S^{k-m} Y),$$

the composition resulting from that of $\mathscr{T}^b$. The smash-product $\otimes$ in $\mathscr{S}$ is given by $(X, m) \otimes (Y, n) = (X \otimes Y, m + n)$ and adds degrees of morphisms. The suspension $S(X, n) = (X, n + 1)$ is of course isomorphic by a morphism of degree 0 to $(S^1, 0) \otimes -$ and the identity gives an isomorphism $s: 1_S \to S$ of degree 1.

There is an obvious functor $\mathscr{T}^b \to \mathscr{S}$ taking X into $(X, 0)$, which we shall henceforth also denote by X. This functor associates to a cofibration $X' \to X \to X'' \to SX$ in $\mathscr{T}^b$ a triangle $X' \xrightarrow{0} X \xrightarrow{0} X'' \xrightarrow{-1} X'$ in $\mathscr{S}$ with maps of the indicated degrees. The triangle so obtained as well as those isomorphic to them, constitute the *cofibrations in* $\mathscr{S}$.

Puppe [6] has listed some properties of the class $\boldsymbol{T}$ of cofibrations in $\mathscr{S}$ which we repeat here with inessential changes.

($P0$) $\boldsymbol{T}$ consists of triangular diagrams $X' \xrightarrow{x''} X \xrightarrow{x'} X'' \xrightarrow{x} X'$ with $\deg x'' = \deg x' = 0$, $\deg x = -1$; any such triangle isomorphic to one in $\boldsymbol{T}$ is itself in $\boldsymbol{T}$.

($P1$) For any $X \in \mathscr{S}$, $0 \to X \xrightarrow{1} X \to 0 \to 0$ is in $\boldsymbol{T}$.

($P2$) For any $f\colon A \to B$ in $\mathscr{S}$ there is a $C \to A \xrightarrow{f} B \to C$ in $\boldsymbol{T}$.

($P3$) If $X' \xrightarrow{x''} X \xrightarrow{x'} X'' \xrightarrow{x} X'$ is in $\boldsymbol{T}$ then so is $X \xrightarrow{x'} X'' \xrightarrow{sx} SX' \xrightarrow{x''s^{-1}} X$.

($P4$) If the rows of the commutative diagram

$$\begin{array}{ccccccc} X' & \to & X & \to & X'' & \to & X' \\ & & \downarrow & & \downarrow & & \\ Y' & \to & Y & \to & Y'' & \to & Y' \end{array}$$

are in $\boldsymbol{T}$ then there is a morphism $X' \to Y'$ such that the diagram augmented by this morphism still commutes.

Following Puppe we undertake a similar development for chain complexes[1]. Let Λ be a subring (containing 1) of the rational numbers. We write $\mathscr{C}_\Lambda$ for the category of finitely generated Λ-modules, $\mathscr{C}_\Lambda^\infty$ for the category of graded modules finitely generated in each degree. The (graded additive) category $\mathscr{K}_\Lambda$ consists of chain complexes over $\mathscr{C}_\Lambda$ which are (forgetting their derivations ∂) projective in $\mathscr{C}_\Lambda^\infty$, with chain-maps as morphisms. $\mathscr{K}_\Lambda^b$ has the same objects and is provided with chain-homotopy-classes of maps as morphisms. The tensor product $\otimes_\Lambda$ has in both cases its usual sense. If W is the complex with $W_1 = \Lambda$, $W_q = 0$, $q \neq 1$ then the suspension S is $W \otimes_\Lambda -$.

By a *cofibration in* K_Λ we mean a sequence $0 \to X' \to X \to X'' \to 0$ with maps of degree 0 which is split-exact in $\mathscr{C}_\Lambda^\infty$; to such a sequence we associate a morphism $X'' \to X'$ of degree -1 in $\mathscr{K}_\Lambda^b$, namely that containing $\varphi\,\partial\varphi^*\colon X'' \to X'$, where $\varphi\colon X \to X'$, $\varphi^*\colon X'' \to X$ constitute a splitting of the sequence in $\mathscr{C}_\Lambda^\infty$. (Alternatively, the same morphism could have been constructed exactly as it was in $\mathscr{T}^b$.) A *cofibration in* $\mathscr{K}_\Lambda^b$ is a triangular diagram isomorphic to one produced in the way just described. Puppe has shown that the class $\boldsymbol{T}$ of cofibrations in $\mathscr{K}_\Lambda^b$ satisfies conditions $P0$ —4 above.

[1] We do this in a very special case only. The same constructions could be applied over any abelian category, or even over a DG-category.

2. Triangulated Categories and Homological Functors

A *triangulation* of a graded additive category $\mathscr{A}$ is a collection $\boldsymbol{T}$ of triangular diagram s in $\mathscr{A}$ satisfying condition $P0$ above. The pair $(\mathscr{A}, \boldsymbol{T})$ is a *triangulated category* which will as usual be denoted simply by $\mathscr{A}$ when there is only one triangulation in question.

If $\mathscr{A}$ has the property that for any $A \in \mathscr{A}$, $n \in \boldsymbol{Z}$ there is an isomorphism $A \to B$ of degree n then we may define, uniquely up to isomorphism of functors, a "suspension" functor $S: \mathscr{A} \to \mathscr{A}$ and an isomorphism $s: 1_{\mathscr{A}} \to S$ of degree 1. If a triangulation $\boldsymbol{T}$ satisfies $P1-4$ we shall call it a *Puppe triangulation*. It is obvious that this notion is independent of the choice of S, s.

Not all graded additive categories admit Puppe triangulations (cf. Freyd [*3*]); a category which does may be called *Puppe-triangulable*, or perhaps just *triangulable*.

Among the properties of a Puppe triangulation $\boldsymbol{T}$ the following are of especial importance (Puppe, [*6*]). (2.1) If $A' \to A \to A'' \to A'$ is in $\boldsymbol{T}$ then it is *preexact*, i.e. for any $B \in \mathscr{A}$ the triangles.

$$\mathscr{A}(B, A') \to \mathscr{A}(B, A) \to \mathscr{A}(B, A'') \to \mathscr{A}(B, A')$$

$$\mathscr{A}(A', B) \to \mathscr{A}(A, B) \to \mathscr{A}(A'', B) \to \mathscr{A}(A', B)$$

of graded abelian groups are exact.

(2.2) If the rows of the commutative diagram

$$\begin{array}{ccccccc} A' & \to & A & \to & A'' & \to & A' \\ \downarrow & & \downarrow & & \downarrow & & \downarrow \\ B' & \to & B & \to & B'' & \to & B' \end{array}$$

are in $\boldsymbol{T}$ and any two of the vertical maps are isomorphisms then so also is the third.

It should be remarked that a triangulable category has in general more than one Puppe triangulation. In view of 2.2 this is equivalent to the fact that not every preexact triangle belongs to a given triangulation. For example, there corresponds to the exact sequence $0 \to \boldsymbol{Z} \to \boldsymbol{Z} \to \boldsymbol{Z}/n\boldsymbol{Z} \to 0$ the cofibration triangle $X \xrightarrow{n} X \xrightarrow{\alpha} Y \xrightarrow{\beta} X$ of projective resolutions in $\mathscr{K}^b_{\boldsymbol{Z}}$. Clearly $X \xrightarrow{-n} X \xrightarrow{\alpha} Y \xrightarrow{\beta} X$ is also preexact, but not a cofibration for $n > 2$.

If $(\mathscr{A}, \boldsymbol{T})$ is a triangulated category a *homological* functor is a graded additive functor $F: \mathscr{A} \to \mathscr{C}^{\infty}_{\wedge}$ such that for any $A' \to A \to A'' \to A'$ in $\boldsymbol{T}$ the triangle $FA' \to FA \to FA'' \to FA'$ in $\mathscr{C}^{\infty}_{\wedge}$ is exact[2]. The corresponding full subcategory of the functor category is denoted by $\mathfrak{H}((\mathscr{A}, \boldsymbol{T}), \mathscr{C}_{\wedge})$

[2] Homological functors and homology theories may obviously be defined more generally, with any graded abelian category for range.

or $\mathfrak{H}(\mathscr{A}, \mathscr{C}_\wedge)$ if $\boldsymbol{T}$ need not be indicated. A *homology theory* is simply a homological functor on $\mathscr{S}$, where $\mathscr{S}$ is the stable-homotopy category discussed above, supplied with the cofibrations as a triangulation.

3. The Moore Subcategories

The Moore subcategory $\mathscr{M}\mathscr{S}$ of $\mathscr{S}$ contains those objects for which the ordinary integral homology H is concentrated in degree 0, and all morphisms between them of degrees $\leqq 0$. These objects, or rather the spaces in $\mathscr{T}$ representing them, were first studied by MOORE in [5]. The notation $L(\pi, n)$ for a space whose (reduced) homology consists only of π in dimension n is familiar. But it is important to notice that $L(\pi,n)$ is not functorial in π and we shall accordingly avoid the notation.

$\mathscr{M}\mathscr{S}$ is to be thought of as triangulated by those cofibrations of $\mathscr{S}$ which lie in it. Notice that if $A' \to A \to A'' \to A'$ is a cofibration in $\mathscr{M}\mathscr{S}$ then $0 \to H_0 A' \to H_0 A \to H_0 A'' \to 0$ is exact.

We shall call a $W \in \mathscr{M}\mathscr{S}$ *free* if it is a coproduct of copies of S^0, i.e. if it is isomorphic in $\mathscr{M}\mathscr{S}$ to a discrete space. By a *free resolution* of an $A \in \mathscr{M}\mathscr{S}$ we mean a cofibration $W' \to W \to A \to W'$ in $\mathscr{M}\mathscr{S}$ with W, W' free. Clearly every $A \in \mathscr{M}\mathscr{S}$ has a free resolution.

The graded ring $\mathscr{S}(S^0, S^0)$ is the stable homotopy ring of the sphere; since in $\mathscr{M}\mathscr{S}$ we consider only morphisms of nonpositive degree, $\mathscr{M}\mathscr{S}(S^0, S^0)$ is concentrated in degree 0, where it is $\boldsymbol{Z}$. If W, V are free then by additivity the only nonzero term of $\mathscr{M}\mathscr{S}(W, V)$ is

$$\mathscr{M}\mathscr{S}(W, V)_0 = \operatorname{Hom}(H_0 W, H_0 V).$$

We have, however,

$$\mathscr{S}(W, V)_1 = (\boldsymbol{Z}/2\boldsymbol{Z}) \otimes_{\boldsymbol{Z}} \operatorname{Hom}(H_0 W, H_0 V), \text{ since } \mathscr{S}(S^0, S^0)_1 = \boldsymbol{Z}/2\boldsymbol{Z}.$$

If $V' \to V \to B \to V'$ is a free resolution we have, for W free,

$$\begin{aligned}(\boldsymbol{Z}/2\boldsymbol{Z}) \otimes_{\boldsymbol{Z}} \operatorname{Hom}(H_0 W, H_0 V') &\to (\boldsymbol{Z}/2\boldsymbol{Z}) \otimes_{\boldsymbol{Z}} \operatorname{Hom}(H_0 W, H_0 V)\\ &\to \mathscr{S}(W, B)_1 \to \operatorname{Hom}(H_0 W, H_0 V') \to \operatorname{Hom}(H_0 W, H_0 V)\\ &\to \mathscr{M}\mathscr{S}(W, B)_0 \to 0\end{aligned}$$

exact by 2.7 and thus canonical isomorphisms

$$\begin{aligned}\mathscr{S}(W, B)_1 &\approx (\boldsymbol{Z}/2\boldsymbol{Z}) \otimes_{\boldsymbol{Z}} \operatorname{Hom}(H_0 W, H_0 B),\\ \mathscr{M}\mathscr{S}(W, B)_0 &\approx \operatorname{Hom}(H_0 W, H_0 B).\end{aligned}$$

If also $W' \to W \to A \to W'$ is a free resolution we have accordingly

$$\begin{aligned}(\boldsymbol{Z}/2\boldsymbol{Z}) \otimes_{\boldsymbol{Z}} \operatorname{Hom}(H_0 W, H_0 B) &\to (\boldsymbol{Z}/2\boldsymbol{Z}) \otimes_{\boldsymbol{Z}} \operatorname{Hom}(H_0 W', H_0 B)\\ &\to \mathscr{M}\mathscr{S}(A, B)_0 \to \operatorname{Hom}(H_0 W', H_0 B)\\ &\to \operatorname{Hom}(H_0 W', H_0 B) \to \mathscr{M}\mathscr{S}(A, B)_{-1} \to 0\end{aligned}$$

exact. We thus have a canonical isomorphism

$$\alpha : \mathcal{M}\mathcal{S}(A,B)_{-1} \approx \mathrm{Ext}^1_{\mathbf{Z}}(H_0 A, H_0 B) \tag{3.1}$$

and a canonical exact sequence

$$0 \to (\mathbf{Z}/2\mathbf{Z}) \otimes_{\mathbf{Z}} \mathrm{Ext}^1_{\mathbf{Z}}(H_0 A, H_0 B) \to \mathcal{M}\mathcal{S}(A,B)_0 \to \mathrm{Hom}(H_0 A, H_0 B) \to 0\,. \tag{3.2}$$

The extension in 3.2 is determined, in view of additivity, by the condition that it is never trivial unless an end term is 0 (BARRATT, [1]). $\mathcal{M}\mathcal{S}(A,B)_q$ is of course 0 for $q < -1$.

The Moore subcategory $\mathcal{M}\mathcal{K}^b_{\mathbf{Z}}$ of $\mathcal{K}^b_{\mathbf{Z}}$ is the full subcategory containing the complexes whose homology is concentrated in degree 0. These are of course just projective resolutions of the modules in $\mathcal{C}_\wedge$. We have accordingly

$$\mathcal{M}\mathcal{K}^b_\wedge(A,B)_q = \mathrm{Ext}^{-q}_\wedge(H_0 A, H_0 B) \tag{3.3}$$

which is 0 for $q \neq 0, -1$.

$\mathcal{M}\mathcal{K}^b_{\mathbf{Z}}$ is also to be triangulated by the cofibrations lying in it. We may easily identify these as follows:

Proposition 3.4. *A triangle $A' \to A \to A'' \xrightarrow{a} A'$ in $\mathcal{M}\mathcal{K}^b_{\mathbf{Z}}$ is a cofibration if and only if $0 \to H_0 A' \to H_0 A \to H_0 A'' \to 0$ is exact with extension class $a \in \mathrm{Ext}^1_{\mathbf{Z}}(H_0 A'', H_0 A')$.*

We must next compare $\mathcal{M}\mathcal{S}$ with $\mathcal{M}\mathcal{K}^b_{\mathbf{Z}}$.

Proposition 3.5. *There is a functor $M : \mathcal{M}\mathcal{S} \to \mathcal{M}\mathcal{K}^b_{\mathbf{Z}}$, unique up to isomorphism, such that for $A \in \mathcal{M}\mathcal{S}$ we have $H_0 M A = H_0 A$, for $f \in \mathcal{M}\mathcal{S}(A, B_0)$ we have $H_0 M f = H_0 f$ and for $f \in \mathcal{M}\mathcal{S}(A,B)_{-1}$ we have $M f = \alpha f \in \mathrm{Ext}^1_{\mathbf{Z}}(H_0 A, H_0 B)$.*

We may choose for each $A \in \mathcal{M}\mathcal{S}$ a free resolution $W' \to W \to A \to W'$ and define MA as the chain complex $H_0 W' \to H_0 W$. If $f \in \mathcal{M}\mathcal{S}(A,B)_0$ and $V' \to V \to B \to V'$ is the resolution of B then since $\mathcal{S}(W, V')_{-1} = 0$ there is a commutative diagram

$$\begin{array}{cccccccc} W' & \to & W & \to & A & \to & W' \\ \downarrow g' & & \downarrow g & & \downarrow f & & \downarrow g' \\ V' & \to & V & \to & B & \to & V' \end{array}$$

and Hg, Hg' are the components of a chain map $MA \to MB$. A different choice of g, g' would lead to a homotopic chain map. We define Mf to be the homotopy class. If $f : A \to B$ is of degree -1 then since

$$S(W,B)_{-1} = 0 \quad \text{and} \quad S(W',V')_{-1} = 0$$

there is a map $k: W' \to V$ such that the composition $A \to W' \xrightarrow{k} V \to B$ is f. Then Hk gives a map $MA \to MB$ of degree -1. Its homotopy class, which is of course independent of the choice of k, is Mf.

Proposition 3.6. *Any object in $\mathcal{M}\mathcal{K}^b_{\mathbf{Z}}$ is isomorphic to one in the image of M. Further, M is full and its kernel, which is generated by $2_P = 2\,1_P$, where $H_0 P = \mathbf{Z}/2\mathbf{Z}$, is of square 0. Moreover M preserves and reflects cofibrations.*

The first two observations are obvious. For the kernel, because of the additivity, it is enough to show that if $A, B \in \mathcal{M}\mathcal{S}$, $H_0 A$ and $H_0 B$ are cyclic and $H_0 A$ 2-primary, $f: A \to B$ and $Mf = 0$ then f factors through 2_P. We may suppose that $\mathrm{Hom}(H_0 A, H_0 B)$ is cyclic of order 2^n so that $\mathcal{M}\mathcal{S}(A, B_0)$ is cyclic of order 2^{n+1}. But then $f = 2^n f'$ for some $f': A \to B$. Let $A' \xrightarrow{a''} A \to P \xrightarrow{p} A'$ be a cofibration in $\mathcal{M}\mathcal{S}$. Clearly $fa'' = 0$ so that $f = gp$ for $g: P \to B$. But $H_0 A \to H_0 P$ is surjective, so that $Hg = 0$ and $g = g'^0\, 2_P$.

The fact that M preserves cofibrations follows from the observation that cofibrations in $\mathcal{K}^b_{\mathbf{Z}}$ may be defined by analogy to the definition in $\mathcal{S}$; that it reflects them is a consequence of the fact that the kernel of M is radical, so that M reflects isomorphisms.

Corollary 3.7. *A graded additive functor F on $\mathcal{M}\mathcal{S}$ factors through M if and only if $F(2_P) = 0$.*

If we define $\mathfrak{H}_2(\mathcal{M}\mathcal{S}, \mathcal{C}_\wedge) \subset \mathfrak{H}(\mathcal{M}\mathcal{S}, C_\wedge)$ to be the full subcategory containing those F for which $F(2_P) = 0$ then 3.6, 3.7 have the following consequence.

Proposition 3.8. $\mathbf{M}^\bullet$: *$\mathfrak{H}(\mathcal{M}\mathcal{K}^b_{\mathbf{Z}}, \mathcal{C}_{\mathbf{Z}}) \to \mathfrak{H}_2(\mathcal{M}\mathcal{S}, \mathcal{C}_{\mathbf{Z}})$ is an equivalence of categories.*

4. Homological Functors on $\mathcal{M}\mathcal{S}$ and $\mathcal{M}\mathcal{K}^b_{\mathbf{Z}}$

We begin with a universal coefficient theorem due, essentially, to Dold [2].

Proposition 4.1. *For $F \in \mathfrak{H}(\mathcal{M}\mathcal{S}, \mathcal{C}_\wedge)$, $A \in \mathcal{M}\mathcal{S}$ there is an exact universal coefficient sequence*

$$(U) \qquad 0 \to F S^0 \otimes_{\mathbf{Z}} H_0 A \xrightarrow{\lambda} F A \xrightarrow{\mu} \mathrm{Tor}_1^{\mathbf{Z}}(F S^0, H_0 A) \to 0$$

with deg $\lambda = 0$, deg $\mu = -1$. This sequence is functorial on

$$\mathfrak{H}(\mathcal{M}\mathcal{S}, \mathcal{C}_\wedge) \times (\mathcal{M}\mathcal{S})_0.$$

Observe first that if $W \in \mathcal{M}\mathcal{S}$ is free then, by additivity,

$$F W \approx F S^0 \otimes_{\mathbf{Z}} H_0 W.$$

Now for any $A \in \mathcal{M}\mathcal{S}$ let $W' \to W \to A \to W'$ be a free resolution. Then there is a diagram

$$\begin{array}{ccccccccc}
0 \to \mathrm{Tor}_1^{\mathbf{Z}}(F S^0, H_0 A) & \to & F S^0 \otimes_{\mathbf{Z}} H_0 W' & \to & F S^0 \otimes_{\mathbf{Z}} H_0 W & \to & F S^0 \otimes_{\mathbf{Z}} H_0 A \to 0 \\
& & \approx \updownarrow & & \approx \updownarrow & & \\
& & F W' & \longrightarrow & F W & & \\
& & \mu' \nwarrow & & \swarrow \lambda' & & \\
& & & F A & & &
\end{array}$$

in which the rectangle commutes and the top row and the triangle are exact. The maps λ and μ are the canonical maps of the image λ' into FA and of FA onto the image of μ'. The familiar arguments with resolutions show the functoriality and thus the independence of the resolutions.

Corollary 4.3. *On* $\mathfrak{H}(\mathscr{S}, \mathscr{C}_{\mathbf{Z}}) \times \mathscr{S} \times (\mathscr{M}\mathscr{S})_0$ *there is a functorial exact sequence.*

$$(U') \qquad 0 \to h X \otimes_{\mathbf{Z}} H_0 A \to h(X \otimes A) \to \mathrm{Tor}_1^{\mathbf{Z}}(h X, H_0 A) \to 0\,.$$

For $A \to h(X \otimes A)$ is of course a homological functor on $\mathscr{M}\mathscr{S}$.

Dold described $h(X \otimes A)$ as homology with coefficients $H_0 A$. Since it is not functorial in $H_0 A$ it seems preferable to say that coefficient for homology lie in $\mathscr{M}\mathscr{S}$.

The sequence U is of course analogous to the corresponding one for homological functors on $\mathscr{M}\mathscr{K}^b_{\mathbf{Z}}$, *viz.*

$$(U'') \qquad 0 \to F\mathbf{Z} \otimes_{\mathbf{Z}} H_0 A \to F A \to \mathrm{Tor}_1^{\mathbf{Z}}(F\mathbf{Z}, H_0 A) \to 0$$

for $F \in \mathfrak{H}(\mathscr{M}\mathscr{K}_{\mathbf{Z}}, \mathscr{C}_\wedge)$. But a difference is indicated by the following observation.

Lemma 4.4. *The sequence* U'' *splits as a sequence of functors on* $A \in (\mathscr{M}\mathscr{K}^b_{\mathbf{Z}})_0$.

For $\wedge = \mathbf{Z}$ *this is proved in* [4]; *though the general case is not discussed there the same argument works and we refrain from repeating it.*

The sequence U *on the contrary does not always split. For example in* U' *take* $h = \mathscr{S}(P, -)$, $X = S^0$, $A = P$. *Then* U' *becomes*

$$0 \to \mathbf{Z}/2\mathbf{Z} \to \mathbf{Z}/4\mathbf{Z} \to \mathbf{Z}/2\mathbf{Z} \to 0\,.$$

But this is in some sense the universal example, as the following result shows.

Proposition 4.5. *If* $F \in \mathfrak{H}(\mathscr{M}\mathscr{S}, \mathscr{C}_\wedge)$ *and* $F P$ *is of exponent* 2 *then the sequence* U *splits as a sequence of functors on* $(\mathscr{M}\mathscr{S})_0$.

It follows from 3.7 that $F \approx F' M$ with $F' \in \mathfrak{H}(\mathscr{M}\mathscr{K}^b_{\mathbf{Z}}, \mathscr{C}_\wedge)$. Making this substitution we see that U becomes

$$0 \to F'\mathbf{Z} \otimes_{\mathbf{Z}} H_0 M A \to F' M A \to \mathrm{Tor}_1^{\mathbf{Z}}(F'\mathbf{Z}, H_0 M A) \to 0\,.$$

But this is just $M^\bullet$ applied to a split exact sequence (i.e. U'' with F' instead of F) and accordingly is itself split.

Theorem 4.6. *If* $h \in \mathfrak{H}(\mathcal{S}, \mathcal{C}_\Lambda)$, $X \in \mathcal{S}$ and $h(X \otimes P)$ *is of exponent* 2 *then, for any* $A \in \mathcal{M}\mathcal{S}$, *the sequence* U' *splits, so that*

$$h(X \otimes A) \approx hX \otimes H_0 A \otimes \mathrm{Tor}_1^{\mathbf{Z}}(hX, H_0 A).$$

Now suppose $F : \mathcal{M}\mathcal{S} \to \mathcal{C}_{\mathbf{Z}}$ is homological. Then, for any torsion free Λ, so is $\Lambda \otimes_{\mathbf{Z}} F$. But if $\mathbf{Z}_p$ is the localization at an odd prime p then $\mathbf{Z}_p \otimes_{\mathbf{Z}} FP = 0$. Thus for any odd prime p the sequence

$$0 \to \mathbf{Z}_p \otimes_{\mathbf{Z}} F S^0 \otimes_{\mathbf{Z}} H_0 A \to \mathbf{Z}_p \otimes_{\mathbf{Z}} FA \to \mathbf{Z}_p \otimes_{\mathbf{Z}} \mathrm{Tor}_1^{\mathbf{Z}}(FS^0, H_0 A) \to 0$$

splits, and we have the following addendum to 4.6.

Proposition 4.7. *For any* $h \in \mathfrak{H}(\mathcal{S}, \mathcal{C}_{\mathbf{Z}})$, $X \in \mathcal{S}$, $A \in \mathcal{M}\mathcal{S}$ *the odd torsion in* $h(X \otimes A)$ *is the direct sum of the odd torsion in* $hX \otimes_{\mathbf{Z}} H_0 A$ *and* $\mathit{Tor}_1^{\mathbf{Z}}(hX, H_0 A)$.

We conclude by indicating a connection between homological functors on $\mathcal{M}\mathcal{S}$ and chain complexes. We may define a functor

$$\Phi : \mathcal{K}^b_\Lambda \to \mathfrak{H}(\mathcal{M}\mathcal{K}^b_{\mathbf{Z}}, \mathcal{C}_\Lambda) \quad \text{by} \quad (\Phi X) A = H(X \otimes_{\mathbf{Z}} A).$$

Proposition 4.8. Φ *is an equivalence of categories; hence so also is* $M^\bullet \Phi : \mathcal{K}^b_\Lambda \to \mathfrak{H}_2(\mathcal{M}\mathcal{S}, \mathcal{C}_\Lambda)$.

For $\Lambda = \mathbf{Z}$ the first assertion is proved in [*4*]; the same argument proves the general case.

We shall have occasion to make use of a functor $\mathfrak{H}_2(\mathcal{M}\mathcal{S}, \mathcal{C}_\Lambda) \to \mathcal{K}^b_\Lambda$ which is inverse, up to isomorphism of functors, to $M^\bullet \Phi$. We chose one (recalling that it is unique up to isomorphism) and denote it by Θ.

5. Prehomological Functors and the Representation Theorem

If $\mathcal{A}$ is a triangulated category a functor $C : \mathcal{A} \to \mathcal{K}^b_\Lambda$ is *prehomological* if HC is homological. The full subcategory of $\mathcal{F}(\mathcal{A}, \mathcal{K}^b_\Lambda)$ containing the prehomological functors is $\mathfrak{P}(\mathcal{A}, \mathcal{C}_\Lambda)$. Then

$$H_\bullet : \mathfrak{P}(\mathcal{A}, \mathcal{C}_\Lambda) \to \mathfrak{H}(\mathcal{A}, \mathcal{C}_\Lambda).$$

Our principal interest is the case $\mathcal{A} = \mathcal{S}$. To say that a homology theory is representable by a chain-complex-valued functor means precisely that it lies, up to isomorphism, in the image of $H_\bullet$.

We denote by $\psi : \mathfrak{H}(\mathcal{S}, \mathcal{C}_\Lambda) \times \mathcal{S} \times \mathcal{M}\mathcal{S} \to \mathcal{C}_\Lambda$ the functor $\psi(h, X, A) = h(X \otimes A)$. Thus $\psi_{12} : \mathfrak{H}(\mathcal{S}, \mathcal{C}_\Lambda) \times \mathcal{S} \to \mathfrak{H}(\mathcal{M}\mathcal{S}, \mathcal{C}_\Lambda)$. Let $\mathcal{R} \subset \mathfrak{H}(\mathcal{S}, \mathcal{C}_\Lambda) \times \mathcal{S}$ be the full subcategory of (h, X) such that $h(1_X \otimes 2_P) = 0$. Then $\psi_{12} | \mathcal{R} : \mathcal{R} \to \mathfrak{H}_2(\mathcal{M}\mathcal{S}, \mathcal{C}_\Lambda)$. We set

$$\Gamma = \Theta(\psi_{12} | \mathcal{R}) : \mathcal{R} \to \mathcal{K}^b_\Lambda.$$

Interpreting this construction, we get the following statement.

Lemma 5.1. *The functors* $(h, X, A) \to h(X \otimes A)$ *and* $(h, X, A) \to \to H(\Gamma(h, X) \otimes_Z MA)$ *on* $\mathscr{R} \times \mathscr{M}\mathscr{S}$ *are isomorphic.*

We might also observe that by the Künneth theorem

$$H(\Gamma(h, X) \otimes_Z MA) \approx H(\Gamma(h, X) \otimes_Z H_0 A).$$

If for $h \in \mathfrak{H}(\mathscr{S}, \mathscr{C}_\wedge)$ we define $\mathscr{S}_h$ to be the full subcategory of $\mathscr{S}$ consisting of X such that $(h, X) \in \mathscr{R}$ then $\Gamma(h, -) : \mathscr{S}_h \to K_\wedge$. If in 5.1 we put $A = S^0$ we have

Lemma 5.2. $H\Gamma(h, -) = h | \mathscr{S}_h$.

Thus we have, at least on a subcategory of $\mathscr{S}$, represented h by a chain-complex-valued functor. Of course if $\mathscr{S}_h = \mathscr{S}$ then the problem can be solved as originally posed. We shall call homology theories with this property *regular* and denote by $\mathfrak{H}_r(\mathscr{S}, \mathscr{C}_\wedge)$ the full subcategory of $\mathfrak{H}(\mathscr{S}, \mathscr{C}_\wedge)$ containing them. We remark that if $\frac{1}{2} \in \wedge$ then every $h \in \mathfrak{H}(\mathscr{S}, \mathscr{C}_\wedge)$ is regular.

By restriction we have $\Gamma : \mathfrak{H}_r(\mathscr{S}, \mathscr{C}_\wedge) \times \mathscr{S} \to \mathscr{K}^b_\wedge$. We define $\mathfrak{C} = \Gamma_1$ so that $\mathfrak{C} : \mathfrak{H}_r(\mathscr{S}, \mathscr{C}_\wedge) \to \mathfrak{P}(\mathscr{S}, \mathscr{C}_\wedge)$.

Lemma 5.3. $H_0 \mathscr{C} \approx 1$ *on* $\mathfrak{H}_r(\mathscr{S}, \mathscr{C}_\wedge)$.

This follows immediately from 5.2.

The prehomological functors $C = \mathfrak{C}h$ all satisfy, obviously, the following condition:

(5.4) *the functors* $(X, A) \to HC(X \otimes A)$, $(X, A) \to H(CX \otimes_Z MA)$ *on* $\mathscr{S} \times \mathscr{M}\mathscr{S}$ *to* $\mathscr{C}_\wedge$ *are isomorphic.*

We shall call such prehomological functors *regular* and denote the corresponding subcategory of $\mathfrak{P}(\mathscr{S}, \mathscr{C}_\wedge)$ by $\mathfrak{P}_r(\mathscr{S}, \mathscr{C}_\wedge)$.

Theorem 5.5. $H_\bullet : \mathfrak{P}_r(\mathscr{S}, \mathscr{C}_\wedge) \to \mathfrak{H}_r(\mathscr{S}, \mathscr{C}_\wedge)$ *is an equivalence of categories with* $\mathfrak{C}$ *as its inverse.*

The latter assertion follows of course from the former and 5.3.

Observe first that if $C \in \mathfrak{P}_r(\mathscr{S}, \mathscr{C}_\wedge)$ then HC is regular; indeed for any $X \in \mathscr{S}$ we have $HC(X \otimes P) \approx H(CX \otimes_Z MP)$ which by the Künneth theorem is of exponent 2.

We must show next that every $C \in \mathfrak{P}_r(\mathscr{S}, \mathscr{C}_\wedge)$ is isomorphic to one of the form $\mathfrak{C}h$ with $h \in \mathfrak{H}_r(\mathscr{S}, \mathscr{C}_\wedge)$. But if C satisfies 5.4 we see that $C \approx \mathfrak{C}(HC)$. For $\mathfrak{C}(HC)X$ is defined, up to a unique isomorphism, as the chain complex such that the functors $A \to H(\mathfrak{C}(HC)X \otimes_Z MA)$, $A \to HC(X \otimes A)$ are isomorphic, a condition satisfied by CX itself.

We complete the proof by showing that $H_\bullet$ is faithful on $\mathfrak{P}_r(\mathscr{S}, \mathscr{C}_\wedge)$. Suppose $\varphi : C \to C'$. Then for each $X \in \mathscr{S}$, $A \in \mathscr{M}\mathscr{S}$ we have a commutative diagram

$$\begin{array}{ccc} HC(X \otimes A) & \approx & H(CX \otimes_Z MA) \\ \downarrow H\varphi_{X \otimes A} & & \downarrow H(\varphi \otimes 1_{MA}) \\ HC'(X \otimes A) & \approx & H(C'X \otimes_Z MA). \end{array}$$

If $H\varphi = 0$ then $H(\varphi_X \otimes_{\mathbf{Z}} 1_{MA}) = 0$ for all X, A. That is, $\Phi\varphi = 0$. By 4.8, then, $\varphi = 0$.

6. Concluding Remarks

The foregoing investigation raises, and leaves unanswered, several questions. We list some of these here.

First, if h is a homology theory its deviation from regularity is measured by $h(1_X \otimes 2_P)$. How may the extension in the universal coefficient sequence U' for $h(X \otimes A)$ be computed from this information?

It seems implausible that for every prehomological functor C the homological functor HC should be regular. Moreover even if HC is regular, it seems implausible that C is necessarily regular. However we lack examples for either of these phenomena. Can there be found, and can the relevant invariants be elucidated?

Since both $\mathscr{S}$ and $\mathscr{K}^b_\wedge$ are triangulated by cofibrations it is reasonable to consider functors $C: \mathscr{S} \to \mathscr{K}^b_\wedge$ which preserve cofibrations. These are obviously prehomological. For such a C moreover HC is always regular, since for any $X \in \mathscr{S}$, $CX \xrightarrow{2} CX \to C(X \otimes P) \to CX$ is a cofibration. Need C be regular?

If h is a regular homology theory is $\mathfrak{C}h$ necessarily cofibration-preserving? If $\mathfrak{C}h$ does preserve cofibrations and X is a CW-complex with skeletons X^q then

$$(\mathfrak{C}h)\, X^{q-1} \to (\mathfrak{C}h)\, X^q \to (\mathfrak{C}h)\,(X^q/X^{q-1}) \to (\mathfrak{C}h)\, X^{q-1}$$

are cofibrations. But X^q/X^{q-1} is a wedge of q-spheres. Thus $(\mathfrak{C}h)\,(X^q/X^{q-1})$ may be taken to be $H(X^q/X^{q-1}) \otimes_{\mathbf{Z}} (\mathfrak{C}h)\, S^0$. It follows that $(\mathfrak{C}h)\, X$ may be taken to be filtered with the associated graded complex just indicated. This would seem to suggest that computation with such homology theories might be especially easy.

References

[1] Barratt, M. G.: Track groups (II). Proc. London Math. Soc. 5, 285—329 (1955).
[2] Dold, A.: Relations between ordinary and extraordinary homology. Colloquium on Algebraic Topology, Aarhus 1962, 2—9.
[3] Freyd, P. J.: (Proceedings of this conference.)
[4] Heller, A.: Homological functors. Math. Zeitschr. 87, 283—298 (1965).
[5] Moore, J.: On homotopy groups of spaces with a single non-vanishing homology group. Ann. Math. 59, 549—557 (1954).
[6] Puppe, D.: On the formal structure of stable homotopy theory. Colloquium on Algebraic Topology, Aarhus, 1962, 65—71.

Department of Mathematics
University of New York
New York, N. Y.

Direct Decompositions of Radicals*

By

SPENCER DICKSON

1. Radicals and Torsion Subfunctors

In the past I have been concerned with the possibility of an extension of the primary decomposition of torsion abelian groups to some standing in arbitrary abelian categories satisfying at least the A.B.5 axiom of GROTHENDIECK [7] and having complete subobject- and factor object lattices which are sets. Such a primary decomposition theorem has been proved in such a general setting only under separate hypotheses and conditions.

In [4] I have called an object A of an abelian category $\mathscr{C}$ *torsion-free*, if it has no simple subobjects (recall an object S is simple in an abelian category if its subobject lattice has exactly two elements) and *torsion*, if each morphism to a torsion-free object is zero. If now $\mathscr{S}$ is a complete set (or class, as the case may be) of non-isomorphic representatives of the simple objects of $\mathscr{C}$, then for each $S \in \mathscr{S}$ we call an object A *S-primary*, if each non-zero factor object of A contains a subobject which is isomorphic to S. It follows that each object A has a unique maximum torsion subobject A_T and a unique maximum S-primary subobject A_S for each $S \in \mathscr{S}$. Moreover, the join subobject of all the A_S is isomorphic to the direct sum (coproduct) $\sum A_S$ of the A_S ($S \in \mathscr{S}$). Moreover, A_T is an essential extension of $\sum A_S$ (i.e., for any subobject H of A_T with $H \cap \sum A_S = 0$ we have $H = 0$). The question of the equality of A_T and $\sum A_S$ for categories of modules has been the main object of the investigations [5] and [6].

Equality does not hold in general even if $\mathscr{C}$ is the category of right modules over a principal (non-commutative) right ideal domain. However, if R is a right Noetherian ring such that for each two non-isomorphic simple right R-modules S_1 and S_2, $\mathrm{Ext}^1_R(S_1, S_2) = 0$ holds, then the equality holds [5]. The paper [6] was completed subsequent to this conference, and gives necessary and sufficient ideal-theoretic conditions for the primary decomposition in the above sense for modules over a commutative ring, thus answering a question raised in my talk. The con-

* Received August 30, 1965.

dition is satisfied for Noetherian rings, rings of analytic functions, and some other standard examples of commutative rings.

In this paper we shall work in the category of left modules over an unspecified ring R with unit, although to use the language of abelian categories one need only specify enough conditions, namely, those in the first sentence of this paper, and the proofs will go through like those in [*4*]. We shall not restrict ourselves to the above definition of "torsion module", for a glance at the literature reveals several other possible definitions of torsion module, (cf. [*8*], [*9*], [*11*] and [*4*]) all of which reduce to the standard abelian-group torsion for Dedekind domains. What these have in common is that the class $\mathcal{I}$ of "torsion" modules in each case is a *radical class* of modules in the sense of Amitsur and Kurosch (cf. [*1*], [*10*]), i.e., $\mathcal{I}$ is closed under homomorphic images, and for each module A there is a unique maximum "torsion" submodule $T(A) \in \mathcal{I}$ such that $A/T(A)$ has the zero submodule as its maximum "torsion" submodule.

It follows that T is a subfunctor of the identity such that i) $T^2 = T$ and ii) $T(A/T(A)) = 0$. If T is an arbitrary subfunctor of the identity we shall call T *idempotent* if T satisfies i) and a *radical* if T satisfies ii), in accordance with the terminology of Maranda [*12*]. These conditions are independent (cf. [*4*]). If T satisfies both conditions, T will be called an *idempotent radical*. If moreover, T is left exact, we shall call T a *torsion radical*, or *hereditary radical*, for T being left exact is equivalent to the condition that for each $A \subseteqq B$ with $T(B) = B$ it follows that $T(A) = A$.

In this paper we shall be concerned with the decomposition of a torsion radical T into a direct sum of torsion radicals $\{T_\alpha\}_{\alpha \in I}$, where I is some index set (or class). This phenomenon is linked with the vanishing of the right derived functors of the T_α, and this connection is discussed below. We compute the right derived functors of a given torsion functor, and also certain relative derived functors which are more closely linked with the decomposition.

Kurosch has shown (in [*10*] the setting was the category of rings or algebras over a fixed operator domain) that a radical class $\mathcal{I}$ is characterized by the following two properties:

I. $\mathcal{I}$ is closed under homomorphic images.

II. Given a module A such that each non-zero homomorphic image of A contains non-zero submodules from the class $\mathcal{I}$, it follows that $A \in \mathcal{I}$.

Kurosch gives a transfinite procedure whereby any class $\mathcal{I}_0$ can be enlarged to a smallest radical class containing the class $\mathcal{I}_0$, which is called the "lower radical". Due to the fact that the relation "ideal of" is not transitive in rings and algebras, examples show that the transfinite procedure of Kurosch will not always terminate after two steps, although for associative rings, this procedure involves at most countably many steps [2]. As the relation "submodule of" is transitive, the Kurosch procedure terminates after two steps, and these are as follows (we give the modernized version of this process to be found in [2]): Given an arbitrary class $\mathscr{I}_0$ of modules, the first step is to form the class $\mathscr{I}_1$ of all homomorphic images of modules in $\mathscr{I}_0$. The next step is to form $\mathscr{I}_2$ by throwing in all modules A such that each non-zero homomorphic image of A has non-zero submodules in one of the previous classes. This second step is repeated at each succeeding ordinal, and an ascending sequence of classes is obtained whose union is the lower radical. For modules one verifies (using transitivity of submodules) that the classes are not enlarged once $\mathscr{I}_2$ is obtained. Thus in the example above, the class $\mathscr{S}$ of representatives of the simple left R-modules is homomorphically closed, so that a module is torsion in the sense of [4] if and only if each nonzero homomorphic image has non-zero socle.

AMITSUR has given an alternate approach to the lower radical [1]. Given a class $\mathscr{I}_1$ which is homomorphically closed, the radical is defined locally as follows: For a module A define $N_1(A)$ to be the join of all the submodules of A which are members of $\mathscr{I}_1$. For an ordinal α define $N_\alpha(A)$ by the formula $N_\alpha(A)/N_{\alpha-1}(A) = N_1(A/N_{\alpha-1}(A))$ if $\alpha - 1$ exists and otherwise by the formula $N_\alpha(A) = \bigcup_{\alpha>\beta} N_\beta(A)$. Then the $N_\alpha(A)$ form a well-ordered increasing sequence and hence for some ordinal τ we must have $N_\tau(A) = N_{\tau+1}(A)$, and this particular submodule is called the "upper radical" of A, but we shall not use this terminology. In the category of rings, or in the category of algebras over a fixed operator domain, this ideal $N_\tau(A)$ turns out to be the maximal $\mathscr{I}$-ideal of A (where $\mathscr{I}$ is the lower radical containing $\mathscr{I}_1$) only under the additional hypothesis that the class $\mathscr{S}\mathscr{I}_1$ of rings (algebras) having only the zero ideal in $\mathscr{I}_1$ has the hereditary property for ideals. Again transitivity of submodules comes to our aid and for radicals of modules one obtains that the stationary $N_\tau(A)$ is always the maximal $\mathscr{I}$-submodule of A, where $\mathscr{I}$ is the lower radical containing the homomorphically closed class $\mathscr{I}_1$ of modules.

Proposition 1. *If $\mathscr{I}_1$ is a homomorphically closed class of left R-modules, where R is any ring with unit, then for each ordinal α, N_α is a subfunctor of the identity.*

Proof. By transfinite induction one can prove [1] that if $f: A \to B$ is a homomorphism, then $f(N_\alpha(A)) \subseteqq N_\alpha(f(A))$. So we will be through if we show that N_α preserves inclusions, for then we would have

$$f(N_\alpha(A)) \subseteqq N_\alpha(f(A)) \subseteqq N_\alpha(B).$$

We induct on the ordinal α. For $\alpha = 1$, $A \subseteqq B$ clearly implies $N_1(A) \subseteqq$

$\subseteqq N_1(B)$ (transitivity is used). Assume the assertion for all β less than α. Then $N_\alpha(A) \subseteqq N_\alpha(B)$ is immediate if α is a limit ordinal. If $\alpha - 1$ exists, there is a natural mapping $\varphi: A/N_{\alpha-1}(A) \to B/N_{\alpha-1}(B)$, from which it follows that

$$(N_\alpha(A) + N_{\alpha-1}(B))/N_{\alpha-1}(B) = \varphi(N_1(A/N_{\alpha-1}(A))) \subseteqq \\ \subseteqq N_1[(A + N_{\alpha-1}(B))/N_{\alpha-1}(B)] \subseteqq N_1(B/N_{\alpha-1}(B))$$

from which the inclusion $N_\alpha(A) \subseteqq N_\alpha(B)$ is immediate.

If R is the ring of integers, then for example, if $\mathscr{I}_1$ consists of the single abelian group cyclic of prime order p, then the N_α constructed over $\mathscr{I}_1$ become stationary at the first infinite ordinal.

Freyd has raised the question at this conference whether there is an ordinal ascending chain condition on subfunctors of the identity in the category of abelian groups. This appears unlikely in view of the large number of possible homomorphically closed classes of abelian groups, although we have no definite answer to this question. The above proposition may be helpful in construction of a counter-example.

Corollary 1. *If $\mathscr{I}_1$ is a homomorphically closed class of left R-modules, then the subfunctor N_1 constructed as above over $\mathscr{I}_1$ is contained in a smallest idempotent radical N, where $N(A)$ is given by the first $N_\tau(A)$ with $N_\tau(A) = N_{\tau+1}(A)$.*

Corollary. 2. *If $\{T_\alpha\}_{\alpha\in I}$ is a family of idempotent radicals then there is a unique smallest idempotent radical T containing them. Moreover, if $T_\alpha(A) = 0$ for each $\alpha \in I$, then $T(A) = 0$.*

Proof. If $\mathscr{I}_1$ denotes the class of modules A such that $T_\alpha(A) = A$ for some $\alpha \in I$, then $\mathscr{I}_1$ is homomorphically closed, and hence there is a smallest idempotent radical $T = N$ constructed over $\mathscr{I}_1$ as above. The second statement follows by an easy transfinite induction.

Proposition 2. *If $\mathscr{I}_1$ is a homomorphically closed class which is closed under submodules, then the lower radical $\mathscr{I}$ is hereditary (i.e., the constructed idempotent radical N is a torsion radical).*

Proof. Here we use a criterion developed in [4]. It suffices to show that the resulting torsion-free class is closed under essential extensions. By a transfinite induction, one can show that $N(A) = 0$ if and only if $N_1(A) = 0$. Let F be a module such that $N(F) = 0$, and let $E(F)$ be an essential extension of F. If A is a submodule of $E(F)$ with $A \in \mathscr{I}_1$ then $A \cap F \in \mathscr{I}_1 \subseteqq \mathscr{I}$ by hypothesis, and also, $N(A \cap F) = 0$ since $N_1(A \cap F) \subseteqq N_1(F) = 0$. Hence $A \cap F = 0$ and it follows that $A = 0$. This concludes the proof.

Corollary. *If $\{T_\alpha\}_{\alpha\in I}$ are torsion subfunctors of the identity then the smallest idempotent radical containing them is hereditary, i.e., is also a torsion subfunctor.*

2. Hom-orthogonal Subfunctors of the Identity

A family $\{T_\alpha\}_{\alpha\in I}$ of subfunctors of the identity is called *hom-orthogonal*, if for any module A we have $\mathrm{Hom}_R(T_\alpha(A),\ T_\beta(A)) = 0$ for $\alpha \neq \beta \in I$.

Lemma 1. *Let $\{T_\alpha\}_{\alpha\in I}$ be a hom-orthogonal family of torsion subfunctors of the identity. Then*

i) $T_\beta(\sum_{\alpha\in I} T_\alpha(A)) = T_\beta(A)$ for each $\beta\in I$;

ii) the family $\{T_\alpha(A)\}_{\alpha\in I}$ of submodules generates the direct sum $\sum_{\alpha\in I} T_\alpha(A)$ as a submodule of A.

Proof. For i), we clearly have $T_\beta(A) \subseteqq T_\beta(\sum_{\alpha\in I} T_\alpha(A))$, and the reverse inclusion will follow from the natural isomorphism

$$\sum_{\alpha\neq\beta} T_\alpha(A) \approx (\sum_{\alpha\in I} T_\alpha(A))/T_\beta(A)$$

if we show $T_\beta(\sum_{\alpha\neq\beta} T_\alpha(A)) = 0$. To see this, note that if $B \subseteqq \sum_{\alpha\neq\beta} T_\alpha(A)$ with $B = T_\beta(B)$ then the monomorphism $\sum_{\alpha\neq\beta} T_\alpha(A) \to \prod_{\alpha\neq\beta} T_\alpha(A)$ yields the exact sequence

$$0 \to \mathrm{Hom}_R(B, \sum_{\alpha\neq\beta} T_\alpha(A)) \to \mathrm{Hom}_R(B, \prod_{\alpha\neq\beta} T_\alpha(A)) \approx$$
$$\approx \prod_{\alpha\neq\beta} \mathrm{Hom}_R(B, T_\alpha(A))$$

but the right hand group is zero, so that $B = 0$. We leave the proof of ii) to the reader.

The following result was the motivation for the following sections, and its proof may be obtained with the aid of the above lemma, following the proofs of theorems 4.10 and 5.3 in [*4*].

Theorem 1. *Let $\{T_\alpha\}_{\alpha\in I}$ be a hom-orthogonal set of torsion subfunctors of the identity. Then if T is the smallest idempotent radical containing them, T is a torsion subfunctor and for each module $A, T(A)$ is an essential extension of the direct sum $\sum_{\alpha\in I} T_\alpha(A)$ and equality holds if and only if for each $\alpha\in I$ the functor T_α is exact on the full subcategory $\mathcal{I}$ of modules A such that $T(A) = A$.*

3. Absolute Derived Functors of Torsion Subfunctors

We shall use the following notation:

$$E^0(A) = A, \quad E^{n+1}(A) = E(E^n(A))/E^n(A) \quad \text{for} \quad n \geqq 1,$$

where $E(B)$ denotes the injective envelope of the module B. Then we resolve a module A as follows:

$$0 \to A \to E(E^0(A)) \xrightarrow{g^0} E(E^1(A)) \xrightarrow{g^1} E(E^2(A)) \xrightarrow{g^2} \cdots,$$

where the map $g_k : E(E^k(A)) \to E(E^{k+1}(A))$ for $k \geqq 0$ is the composition (of) $f_k : E(E^k(A)) \to E(E^k(A))/E^k(A) = E^{k+1}(A)$ followed by the injection $j_k : E^{k+1}(A) \to E(E^{k+1}(A))$. To obtain the absolute derived functors we first apply the torsion subfunctor T to obtain a complex (with restriction maps) where we denote $T(E(A))$ by $E_T(A)$:

$$0 \to A_T \to E_T(E^0(A)) \xrightarrow{g_T^0} E_T(E^1(A)) \xrightarrow{g_T^1} \cdots.$$

The homology for $n \geqq 1$ is then $R_T^n(A) = \operatorname{Ker} g_T^n/\operatorname{Im} g_T^{n-1}$. Now from the diagram (by inspection)

$$E(E^n(A)) \xrightarrow{g^n} E(E^{n+1}(A))$$
$$\searrow_{f^n} \quad _{j^n}\nearrow$$
$$E(E^n(A))/E^n(A)$$

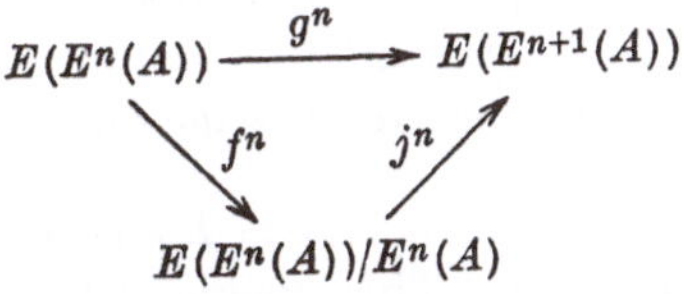

it follows that $\operatorname{Ker} g_T^n = E_T(E^n(A)) \cap E^n(A)$. From examination of the diagram for $n-1$ we see that

$$\operatorname{Im} g_T^{n-1} = [E_T(E^{n-1}(A)) + E^{n-1}(A)]/E^{n-1}(A),$$

so that the general formula for an idempotent radical T is

(1) $$R_T^n(A) = \{E_T(E^n(A)) \cap E^n(A)\}/\{[E_T(E^{n-1}(A)) + E^{n-1}(A)]/E^{n-1}(A)\}.$$

Since T was also to be left exact, the numerator reduces to the expression $[E^n(A)]_T$, and if this is $N/E^{n-1}(A)$ we have

(2) $$R_T^n(A) = N/(E_T(E^{n-1}(A)) + E^{n-1}(A)).$$

In order to obtain a further simplification of this formula we prove the following lemma.

Lemma 2. If A is T-torsion (i.e., $T(A) = A$) for the torsion functor T, then $R_T^1(A) = 0$.

Proof. We have $R_T(A) = (E(A)/A)_T/((E_T(A) + A)/A)$, which by [12, lemma 1, p. 110] is

$$(E_T(A)/A)/((E_T(A) + A)/A) = (E_T(A)/A)/(E_T(A)/A) = 0.$$

In the general case the exact sequence

$$0 \to A \to E(A) \to E^1(A) \to 0$$

yields the exact sequence for $n \geqq 1$:

$$0 = R^n_T(E(A)) \to R^n_T(E^1(A)) \to R^{n+1}_T(A) \to R^{n+1}(E(A)) = 0$$

so that $R^{n+1}_T(A) \approx R^n_T(E^1(A)) = R^n_T(E(A)/A)$. Now for $n = 1$ the formula (2) reduces to

(3) $$R^1_T(A) = [E(A)/A]_T \text{ if } A \text{ is } T\text{-torsion-free.}$$

In general the recursion formula just obtained yields

Theorem 2. *If T is a torsion subfunctor of the identity and A is any module, then*

(4) $$R^n_T(A) = R^1_T\,(E^{n-1}(A)) \quad \textit{for all} \quad n \geqq 1\,.$$

Corollary. *If T is the usual torsion subfunctor for modules over a commutative integral domain R, then $R^n_T(A) = 0$ for any module A and any $n \geqq 2$. $R^1_T(A) = Q/R \otimes_{\mathbb{Z}} A$, where Q is the quotient field of R.*

Proof. For $n \geqq 2$ we have $R^n_T(A) = R^1_T\,(E^{n-1}(A))$, but $E^{n-1}(A)$ is T-torsion. hence $R^n_T(A) = 0$. This yields $R^1_T(A) = R^1_T(A/A_T)$. Now $E\,(A/A_T) = Q \otimes A/A_T$, so that $R^1_T(A) = (Q \otimes A/A_T)/(R \otimes A/A_T) \approx Q/R \otimes A/A_T \approx Q/R \otimes A$, since A/A_T is flat. The reader may then check that from an exact sequence $0 \to A \to B \to C \to 0$, the (familiar) connected sequence of functors results (see [3], p. 130):

$$0 \to A_T \to B_T \to C_T \to Q/R \otimes A \to Q/R \otimes B \to Q/R \otimes C \to 0\,.$$

4. Relative Derived Functors of Torsion Subfunctors

In this section T will denote the smallest torsion subfunctor containing the set $\{T_\alpha\}_{\alpha \in I}$ of torsion functors. The class $\mathcal{I}$ of modules held fixed by T is a torsion class (radical class) and the modules of the form Q_T where Q is injective form relative T-injectives, in the sense that they are injective for diagrams in which each module is T-torsion. In taking a resolution by injectives to calculate the right derived functors of T_α for $\alpha \in I$, we shall use T-injectives. In this way we shall arrive at a correct first derived functor whose vanishing implies the exactness of the functor T_α when restricted to the full subcategory $\mathcal{I}$.

For $A \in \mathcal{I}$ we take a resolution by T-torsion injectives as follows:

$$0 \to A \to E_T(A) \to E_T(E_T(A)/A) \to \cdots.$$

We introduce the notation $E^n_T(A)$ inductively: $E^0_T(A) = A$, and for $n \geqq 1$, $E^n_T(A) = (E_T(E^n_T(A))/E^n_T(A))$. The resolution becomes

$$0 \to A \to E_T(E^0_T(A)) \to E_T(E^1_T(A)) \to E_T(E^2_T(A)) \to \cdots,$$

and yields the complex

$$0 \to A_{T_\alpha} \to E_{T_\alpha}(E^0_T(A)) \to E_{T_\alpha}(E^1_T(A)) \to \cdots,$$

and we find similar to the previous calculation that the first relative derived functor $\mathrm{Rel}^1_{T_\alpha}(A)$ is given by the formula

(5) $$\mathrm{Rel}^1_{T_\alpha}(A) = [E_T(A)/A]_{T_\alpha}/[(E_{T_\alpha}(A) + A)/A];$$

so that if A is T_α-torsion-free then $\mathrm{Rel}^1_{T_\alpha}(A) = (E^1_T(A))_{T_\alpha}$. The exact sequence $0 \to A \to E_T(A) \to E^1_T(A) \to 0$ yields (as above)

(6) $\mathrm{Rel}^n_{T_\alpha}(A) = \mathrm{Rel}^{n-1}_{T_\alpha}(E^1_T(A))$ for $n \geqq 1$ so we have proved

Theorem 3. *Let T be the smallest idempotent radical (hence torsion subfunctor) containing the family $\{T_\alpha\}_{\alpha\in I}$ of torsion subfunctors. Then the n-th right derived functor (relative to $\mathscr{I}$) of T_α is given by*

(7) $$\mathrm{Rel}^n_{T_\alpha}(A) = \mathrm{Rel}^1_{T_\alpha}(E^{n-1}_T(A)) \quad \text{for} \quad n \geqq 1.$$

As an application we prove a generalization of a result in [5].

Theorem 4. *Let R be a left Noetherian ring and let T be the smallest torsion subfunctor containing the hom-orthogonal family $\{T_\alpha\}_{\alpha\in I}$ of torsion subfunctors. If for each left R-module A we have the equality $E_T(T_\alpha(A)) = E_{T_\alpha}(T_\alpha(A))$, then for any module A, $T(A) = \sum_{\alpha\in I} T_\alpha(A)$.*

Proof. As direct sums of injectives are injective when R is left Noetherian, it follows by a standard argument that $\mathrm{Rel}^1_{T_\alpha}$ commutes with direct sums. We show for each $\alpha \in I$ that $\mathrm{Rel}^1_{T_\alpha}(A) = 0$ for each A with $T(A) = A$, and we will be through by theorem 1. The exact sequence

$$0 \to \sum_{\alpha\in I} T_\alpha(A) \to A \to A/\sum_{\alpha\in I} T(A) \to 0$$

yields the exact sequence for each $\beta \in I$

$$\mathrm{Rel}^1_{T_\beta}(\sum_{\alpha\in I} T_\alpha(A)) \to \mathrm{Rel}^1_{T_\beta}(A) \to \mathrm{Rel}^1_{T_\beta}(A/\sum_{\alpha\in I} T_\alpha(A)).$$

We have

$$\mathrm{Rel}^1_{T_\beta}(\sum_{\alpha\in I} T_\alpha(A)) = \sum_{\alpha\neq\beta} \mathrm{Rel}^1_{T_\beta}(T_\alpha(A)),$$

and for $\alpha \neq \beta$,

$$\mathrm{Rel}^1_{T_\beta}(T_\alpha(A)) = [E_T(T_\alpha(A))/T_\alpha(A)]_{T_\beta} = [E_{T_\alpha}(T_\alpha(A))/T_\alpha(A)]_{T_\beta} = 0.$$

Also, $\mathrm{Rel}^1_{T_\beta}(A/\sum_{\alpha\in I} T_\alpha(A)) = \mathrm{Rel}^1_{T_\beta}(A/\sum_{\alpha\neq\beta} T_\alpha(A))$ from the exact sequence

$$0 \to T_\beta(A) \to A/\sum_{\alpha\neq\beta} T_\alpha(A) \to A/\sum_{\alpha\in I} T_\alpha(A) \to 0,$$

for we have the exact sequence

$$0 = \mathrm{Rel}^1_{T_\beta}(T_\beta(A)) \to \mathrm{Rel}^1_{T_\beta}(A/\sum_{\alpha \neq \beta} T_\alpha(A)) \to \mathrm{Rel}^1_{T_\beta}(A/\sum_{\alpha \in I} T_\alpha(A)) \to \mathrm{Rel}^2_{T_\beta}(A),$$

and the last member vanishes, considering the sequence

$$0 \to T_\beta(A) \to E_T(T_\beta(A)) = E_{T_\beta} T_\beta(A)) \to E_{T_\beta}(T_\beta(A))/T_\beta(A) \to 0.$$

But $A/\sum_{\alpha \neq \beta} T_\alpha(A)$ is contained in $E_T(T_\beta(A)) = E_{T_\beta}(T_\beta(A))$, hence the right hand term is also zero, and we are through.

References

[1] Amitsur, S. A.: General theory of radicals II: Radicals in rings and bicategories. Amer. J. Math. **76**, 100—125 (1954).

[2] Anderson, T., N. Divinski, and A. Sulinski: Lower radical properties for associative and alternative rings (to appear).

[3] Cartan, H., and S. Eilenberg: Homological algebra. Princeton: Princeton University Press 1956.

[4] Dickson, S.: A torsion theory for abelian categories (to appear: Trans. Amer. Math. Soc.).

[5] — Decomposition of modules I: Classical rings (to appear: Math. Z.).

[6] — Decomposition of modules II: Rings without chain conditions (submitted for publication).

[7] Grothendieck, A.: Sur quelques points d'algèbre homologique. Tôhoku Math. J. (2) **9**, 119—221 (1957).

[8] Hattori, A.: A foundation of torsion theory for modules over general rings. Nagoya Math. J. **17**, 147—158 (1960).

[9] Jans, J. P.: Some aspects of torsion (to appear).

[10] Kurosch, A. G.: Radicals in rings and algebras. Mat. Sb. (N.S.) **33**, 13—26 (1953) (Russian).

[11] Levy, L.: Torsion-free and divisible modules over Noetherian integral domains. Can. J. Math. **15**, 132—151 (1963).

[12] Maranda, J.-M.: Injective structures. Trans. Amer. Math. Soc. **110**, 88—135 (1964).

[13] MacLane, S.: Homology. Berlin-Göttingen-Heidelberg: Springer 1963.

Department of Mathematics
University of Nebraska
Lincoln, Nebraska

Abelian Extensions and a Cohomology Theory of Harrison *, **

By

STEPHEN U. CHASE

Let k be a field, K be a separable closure of k, and Π be the group of all automorphisms of K leaving k pointwise fixed. Π is a compact, totally disconnected group in the topology for which a neighborhood base at the identity is the set of subgroups of Π which correspond, via the fundamental theorem of Galois theory, to the finite separable extensions of k.

New information concerning the group Π has recently been obtained in an interesting and provocative paper of D. K. HARRISON [*13*]. Given a commutative ring R and an abelian group J, HARRISON defines a cochain complex $\mathscr{H}(R, J) = \{U(R(J^p)), \delta^p\}$ $(p = 0, 1, \ldots)$. Here $J^0 = \{1\}$, $J^{p+1} = J \times J^p$, $R(\ )$ is the group algebra of the group ( ) with coefficients in R, and $U(\ )$ is the multiplicative group of units of the ring ( ). $\delta^p : U(R(J^p)) \to U(R(J^{p+1}))$ is defined by the formula

$$\delta^p(u) = \lambda(u)\left\{\prod_{i=1}^{p} \Delta_i(u)^{(-1)^i}\right\} \varrho(u)^{(-1)^{p+1}} \qquad (u \text{ in } U(R(J^p))) \tag{0.1}$$

the homomorphisms $\lambda, \varrho, \Delta_i : R(J^p) \to R(J^{p+1})$ being defined by the conditions

$$\begin{aligned} \lambda\{(\sigma_1, \ldots, \sigma_p)\} &= (1, \sigma_1, \ldots, \sigma_p) \\ \varrho\{(\sigma_1, \ldots, \sigma_p)\} &= (\sigma_1, \ldots, \sigma_p, 1) \end{aligned} \qquad (\sigma_1, \ldots, \sigma_p \text{ in } J)$$

$$\Delta_i\{(\sigma_1, \ldots, \sigma_p)\} = (\sigma_1, \ldots, \sigma_{i-1}, \sigma_i, \sigma_i, \sigma_{i+1}, \ldots, \sigma_p) \qquad (i = 1, \ldots, p).$$

HARRISON proves the following theorem: If k is an arbitrary field and J is a torsion abelian group of which every finite subgroup is cyclic, then there exists an isomorphism

$$\operatorname{Hom}(\Pi, J) \approx H^2(k, J) \tag{0.2}$$

where $H^p(R, J)$ is the p'th cohomology group of the complex defined above, and the left side of (0.2) denotes the group of continuous homo-

* Written with the support of National Science Foundation Grant GP 3665.
** Received September 13, 1965.

morphisms of Π into the discrete group J. For the case in which J is a finite cyclic group of order n, k has characteristic relatively prime to n, and contains all n'th roots of 1, (0.2) yields, via computations implicit in [*18*, Erster Teil, § 20], the following isomorphism of classical Kummer theory

(0.3) $$\mathrm{Hom}(\Pi, J) \approx U(k)/U(k)^n$$

the superscript n denoting the subgroup of all n'th powers [*17*, p. 163].

In this paper we push further the ideas of [*13*], [*14*], [*18*] and [*10*] to obtain additional information concerning the cohomology theory of Harrison and its relation to abelian extensions of commutative rings. In Section 1 we expand some remarks of [*13*] to construct a cohomology theory of groups in a category, of which the Harrison cohomology theory and the ordinary cohomology theory of groups are special cases. For the case in which k is a field of characteristic relatively prime to the exponent of the finite abelian group J, our point of view leads easily to a description of the groups $H^p(k, J)$ in terms of Γ-equivariant cohomology for a suitably chosen finite group Γ. For the definition of equivariant cohomology we refer the reader to [*10*], where the same result is proved by direct computations.

In Sections 2 and 3 we use techniques based on those of [*13*] and [*10*] to describe the abelian extensions of a connected commutative ring R in terms of the categories of projective modules over certain group algebras with coefficients in R.[1] Our results specialize to (0.2) for the case in which R is a field and J is a finite cyclic group. The material of these sections is closely related to a recent unpublished theorem of M. Orzech, which provides a cohomological generalization of (0.2) to connected commutative rings.

Throughout this paper R will denote a commutative ring with unit,[2] and unadorned $\otimes$ will mean $\otimes_R$. The symbol $\mathscr{A}^0$ will represent the dual of the category $\mathscr{A}$. If A and B are objects of $\mathscr{A}$, the set of $\mathscr{A}$-morphisms from A to B will usually (but not always) be denoted by $\mathscr{A}(A, B)$. The category of abelian groups will be represented by the symbol Ab. As regards category theory, we shall for the most part adhere to the notation and terminology of [*11*], and unexplained deviations from this rule will be apparent from the context. For the benefit of the reader we have omitted many easy proofs.

The author is indebted to Alex Rosenberg for many stimulating conversations regarding the material of this paper.

---

[1] A commutative ring is called *connected* if it has no idempotents other than 0 and 1.

[2] Throughout this paper we shall assume, unless explicitly stated otherwise, that every ring has a multiplicative identity which is preserved by every homomorphism, and every module is unital.

1. Cohomology of Groups in a Category

Throughout this section $\mathscr{A}$ will be a category with finite direct products and a terminal object $*$.

Let G be a group in $\mathscr{A}$. A *left G-object* is a pair (X, α), where X is an object of $\mathscr{A}$, $\alpha : G \times X \to X$, and the following diagrams commute

(1.1a)
$$\begin{array}{ccc} G \times G \times X & \xrightarrow{1 \times \alpha} & G \times X \\ \Big\downarrow{\scriptstyle \mu \times 1} & & \Big\downarrow{\scriptstyle \alpha} \\ G \times X & \xrightarrow{\alpha} & X \end{array}$$

(1.1b)
$$\begin{array}{ccc} * \times X & \xrightarrow{\varepsilon \times 1} & G \times X \\ & {\scriptstyle p}\searrow \quad \swarrow{\scriptstyle \alpha} & \\ & X & \end{array}$$

where $\mu : G \times G \to G$ is the multiplication map, $\varepsilon : * \to G$ is the identity of the group G, and $p : * \times X \to X$ is the projection. The left G-objects constitute a category $\mathscr{A}^G$, a morphism $f : (X, \alpha) \to (Y, \beta)$ in $\mathscr{A}^G$ being an $\mathscr{A}$-morphism $f : X \to Y$ which converts α to β in the obvious sense. Note that $(*, t)$, with $t : G \times * \to *$, is a left G-object, and is a terminal object of $\mathscr{A}^G$; it will be called the *trivial G-object*. Furthermore, (G, μ) is a left G-object, for in this case (1.1) reduces to the axioms for a group in $\mathscr{A}$ specifying associativity of multiplication and existence of a left identity. We shall often denote a given left G-object (X, α) simply by X if the map α is clearly indicated by the context.

The categories $\mathscr{A}$ and $\mathscr{A}^G$ are related by functors $S : \mathscr{A}^G \to \mathscr{A}$ and $T : \mathscr{A} \to \mathscr{A}^G$. S is the forgetful functor, whereas T is defined as follows: $T(X) = (G \times X, \mu \times 1)$ for X an object of $\mathscr{A}$, and $T(f) = 1 \times f : G \times \times X \to G \times Y$ for $f : X \to Y$ an $\mathscr{A}$-morphism. S and T are, in turn, related by natural transformations $\zeta : I_{\mathscr{A}} \to ST$ and $\eta : TS \to I_{\mathscr{A}^G}$. If X is an object of $\mathscr{A}$, then $\zeta_X : X \to G \times X$ is the composite

$$X \xrightarrow{(x,1)} * \times X \xrightarrow{\varepsilon \times 1} G \times X$$

where $x : X \to *$; whereas, if (X, α) is a left G-object, then

$$\eta_{(X,\alpha)} = \alpha : G \times X \to X.$$

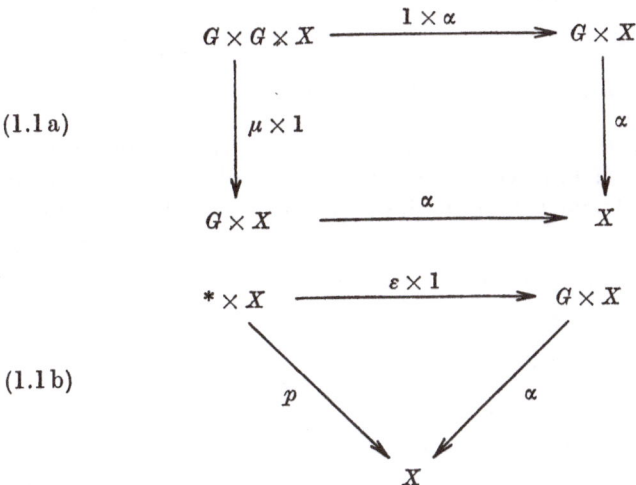

Proposition 1.2. *The following diagrams of functors and natural transformations commute.*

(a) 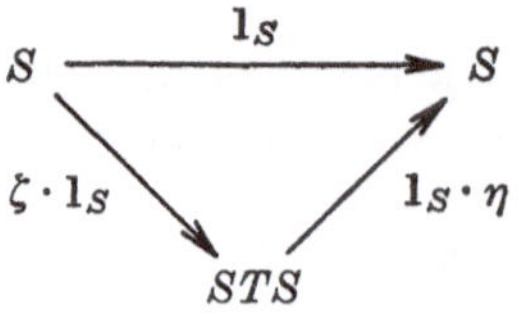

(b) 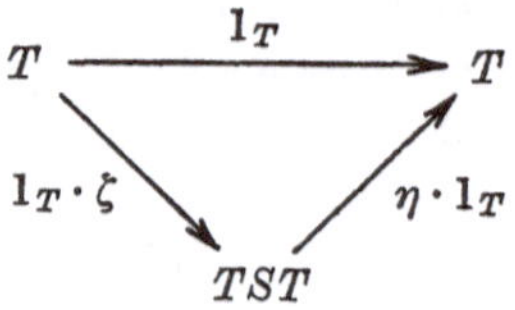

Hence, by [15, Thm. 4.1], T is a left adjoint of S.

Proof. The argument is routine. We remark only that commutativity of a) follows from (1.1b), whereas commutativity of b) rests upon the fact that ε is a right identity of G; i.e., the diagram below commutes

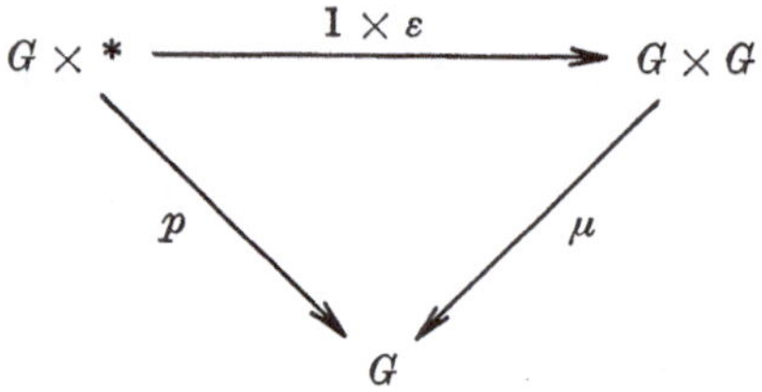

where p is the projection.

Corollary 1.3. *a) The triple (X_p, d_p^i, s_p^i) (for $p \geqq 0$; $0 \leqq i \leqq p$) is a semi-simplicial object in $\mathscr{A}^G$, where X_p, $d_p^i : X_p \to X_{p-1}$, and $s_p^i : X_p \to$ $\to X_{p+1}$ are defined as follows:*

$$X_p = (TS)^{p+1}(*) = \underbrace{G \times \cdots \times G}_{p+1} \times *$$

$$d_p^i = (TS)^i\,(\eta_{(TS)^{p-1}(*)}) = \begin{cases} \underbrace{1 \times \cdots \times 1}_{i} \times \mu \times \underbrace{1 \times \cdots \times 1}_{p-i-1} \times 1_* & \text{if } 0 \leqq i < p \\ \underbrace{1 \times \cdots \times 1}_{p} \times t & \text{if } i = p \end{cases}$$

where $t : G \times * \to *$.

$$s_p^i = (TS)^i[T(\zeta_{S(TS)^{p-i-1}(*)})]$$

$$= \begin{cases} \underbrace{1 \times \cdots \times 1}_{i+1} \times \zeta_G \times \underbrace{1 \times \cdots \times 1}_{p-i-1} \times 1_* & \text{if } \quad 0 \leqq i < p \\ \underbrace{1 \times \cdots \times 1}_{p+1} \times \zeta_* & \text{if } \quad i = p \end{cases}$$

b) Let $F: \mathscr{A}^G \to \mathrm{Ab}$ *be a contravariant functor. Then we obtain a cochain complex* $C(G, F)$ *of abelian groups*

$$F(X_0) \xrightarrow{\delta^0} F(X_1) \xrightarrow{\delta^1} F(X_2) \xrightarrow{\delta^2} \cdots$$

where

$$\delta^p = \sum_{i=0}^{p+1} (-1)^i F(d^i_{p+1}) : F(X_p) \to F(X_{p+1}).$$

Proof. (a) follows easily from the theory of standard constructions; see e.g. [*15*], in particular Theorem 4.2 and the discussion of Section 2. (b) is a consequence of [*16*, (5.13), p. 235].

Remark 1.4. In order to obtain the complex $C(G, F)$, one could just as well use, rather than the category $\mathscr{A}$, any full subcategory $\mathscr{A}'$ of $\mathscr{A}$ satisfying the following conditions:

(a) $\mathscr{A}'$ is closed under finite direct products;

(b) G and $*$ are objects of $\mathscr{A}'$.

Example 1.5. Let $\mathscr{A}$ be the category of sets. A group G in $\mathscr{A}$ is an ordinary group, and a left G-object is a left G-set in the usual sense. Now let M be a left G-module. Viewing M in the obvious way as an object of $\mathscr{A}^G$, we obtain a contravariant functor $F: \mathscr{A}^G \to \mathrm{Ab}$ defined by $F(\cdot) = \mathscr{A}^G(\cdot, M)$. ($F(X)$ is, of course, an abelian group under pointwise addition of functions.) A routine computation establishes that $C(G, F)$ is the Eilenberg-Mac Lane non-homogeneous complex for the group G with coefficients in the G-module M, and the cohomology groups of this complex are the Eilenberg-Mac Lane groups $H^p(G, M)$ [*7*, Chapter XII].

Example 1.6. Let Γ be an (ordinary) group, and let $\mathscr{A}$ be the category of left Γ-sets. A group G in $\mathscr{A}$ is a "Γ-group"; i.e., an ordinary group with Γ as group of operators. A left G-object is a left Γ-set X which is at the same time a left G-set such that the following condition holds for all x in X, σ in G, γ in Γ:

$$\gamma(\sigma x) = \gamma(\sigma)\gamma(x). \tag{1.7}$$

Let M be a left Γ-module which is at the same time a left G-module such that (1.7) holds for all x in M. Then M may be viewed as an object of $\mathscr{A}^G$, and we obtain a contravariant functor $F: \mathscr{A}^G \to \mathrm{Ab}$ defined by $F(\cdot) = \mathscr{A}^G(\cdot, M)$. It is easily verified that $C(G, F)$ is the analogue of the Eilenberg-Mac Lane non-homogeneous complex in the "Γ-equivariant" cohomology theory of the group G with coefficients in the G-module M [*10*]; the resulting cohomology groups will be denoted by $H^p(G, M)$.

Example 1.8. Let R be a commutative ring, $\mathrm{Alg}(R)$ be the category of commutative R-algebras, and $\mathscr{A} = \mathrm{Alg}(R)^0$. Since $\otimes$ is the direct sum in $\mathrm{Alg}(R)$, finite direct products exist in $\mathscr{A}$. Since R is an initial object of $\mathrm{Alg}(R)$, R^0 is a terminal object of $\mathscr{A}$.

Now let J be a finite abelian group, and $R(J)$ be the group algebra of J with coefficients in R. Let $\varepsilon : R(J) \to R$ and $\Delta : R(J) \to R(J) \otimes \otimes R(J)$ be the augmentation and diagonal maps, respectively; i.e., $\varepsilon(\sigma) = 1$ and $\Delta(\sigma) = \sigma \otimes \sigma$ for σ in J. It is then easily verified that $G = R(J)^0$ is a group in $\mathscr{A}$, with multiplication given by $\Delta^0 : G \times G = (R(J) \otimes R(J))^0 \to R(J)^0 = G$ and identity defined by $\varepsilon^0 : * = R^0 \to \to R(J)^0 = G$. Note that any left G-object has the form (A^0, α^0), where A is a commutative R-algebra and $\alpha : A \to R(J) \otimes A$ is an R-algebra homomorphism.

We now define a contravariant functor $F : \mathscr{A}^G \to \mathrm{Ab}$ as follows. Given an object $X = (A^0, \alpha^0)$ of $\mathscr{A}^G$, set $F(X) = \{u \text{ in } U(A)/\alpha(u) = 1 \otimes u\}$. If $f : X \to Y$ is a morphism of left G-objects, then $Y = (B^0, \beta^0)$ and $f = \varphi^0$ with $\beta : B \to R(J) \otimes B$ and $\varphi : B \to A$ in $\mathrm{Alg}(R)$. We define $F(f) : F(Y) \to F(X)$ by $F(f)(v) = \varphi(v)$ for v in $F(Y) \subsetneqq U(B)$. $\varphi(v)$ lies in $F(X)$ because $\alpha\varphi = (1 \otimes \varphi)\beta$, f being a morphism in $\mathscr{A}^G$.

It turns out that $C(G, F)$ is simply $\mathscr{H}(R, J)$, the Harrison complex of R as defined in the introduction. This is an easy consequence of routine calculations, together with the following observation.

Lemma 1.9. *Let $\mathscr{A}$ and G be as in Example 1.7, and define the functor $T : \mathscr{A} \to \mathscr{A}^G$ as in Proposition 1.2 and preceding discussion. Then $FT \approx U^0$ as functors from $\mathscr{A}$ to* Ab. *(Here U^0 is the contravariant functor induced in the obvious way by $U : Alg(R) \to$* Ab.*)*

Proof. If A is a commutative R-algebra, then $T(A^0) = ((R(J) \otimes A)^0, (\Delta \otimes 1)^0)$, and hence $FT(A^0) = \{u \text{ in } U(R(J) \otimes A)/(\Delta \otimes 1)(u) = 1 \otimes u\}$. We claim that $FT(A^0) = U(1 \otimes A)$ as subsets of $U(R(J) \otimes A)$. Clearly $U(1 \otimes A) \subsetneqq FT(A^0)$. Now, any u in $R(J) \otimes A$ can be written uniquely in the form $u = \sum_{\sigma \text{ in } J} \sigma \otimes u_\sigma$ with u_σ in A, in which case $(\Delta \otimes 1)(u) = \sum_{\sigma \text{ in } J} \sigma \otimes \sigma \otimes u_\sigma$. If u is in $FT(A^0)$, then $\sum_{\sigma \text{ in } J} \sigma \otimes \sigma \otimes u_\sigma = \sum_{\sigma \text{ in } J} \sigma\, 1 \otimes \sigma \otimes u_\sigma$, and comparison of coefficients then yields $u_\sigma = 0$ for $\sigma \neq 1$. Hence $u = 1 \otimes u_1$ is in $U(1 \otimes A)$, and so $FT(A^0) = U(1 \otimes A)$. Since $1 \otimes A \approx A$, the lemma follows readily.

In the remainder of this section we shall analyze the cohomology theory of Example 1.8 for a special case which we now describe. Let k be a field, n be a positive integer relatively prime to the characteristic of k, and $K = k(\zeta)$ with ζ a primitive n'th root of 1. Set $U_n = \{x \text{ in } K/x^n = 1\}$, a cyclic subgroup of $U(K)$ of order n. Note that K is a normal separable extension of k with Galois group Γ naturally isomorphic to a subgroup of the multiplicative group of residue classes modulo n which are prime to n. Specifically, if γ is in Γ, then there exists an integer a_γ prime to n, uniquely determined modulo n, such that

$$\gamma(x) = x^{a_\gamma} \quad \text{for all } x \text{ in } U_n. \tag{1.10}$$

We note finally that $Z|nZ$ is a Γ-group, in the sense of Example 1.6, via the rule

$$\gamma(\nu + nZ) = a_\gamma \nu + nZ \tag{1.11}$$

with γ in Γ, ν in Z, and a_γ as in (1.10).

Theorem 1.12. *Let $H^*(k, U_n)$ denote the Harrison cohomology groups of the field k and the abelian group U_n, and $H^*_\Gamma(Z/nZ, U(K))$ denote the Γ-equivariant cohomology groups of Example 1.6, with the Γ-group $Z|nZ$ operating trivially on the Γ-group $U(K)$. Then, with the above hypotheses, there exist isomorphisms $H^p(k, U_n) \approx H^p_\Gamma(Z|nZ, U(K))$.*

Before proving the theorem we exhibit a corollary.

Corollary 1.13. *Let k be a field containing all n'th roots of 1, with n relatively prime to the characteristic of k. Then*

$$H^p(k, U_n) \approx \begin{cases} U(k)/U(k)^n & p \text{ even and } > 0 \\ U_n & p \text{ odd} \end{cases}$$

where $U_n = \{x \text{ in } k/x^n = 1\}$, a cyclic group of order n.

Proof. In this case $K = k$ and $\Gamma = \{1\}$, and so Theorem 1.12 reduces to isomorphisms $H^p(k, U_n) \approx H^p(Z|nZ, U(k))$, the Eilenberg-Mac Lane cohomology groups of the group $Z|nZ$ with coefficients in the trivial $Z|nZ$-module $U(k)$. The corollary then follows easily from the material of [7, Chapter VII]; for details we refer the reader to [*10*].

We turn now to the proof of Theorem 1.12. Let $\mathscr{A}_1$ be the full subcategory of $\mathscr{A} = \mathrm{Alg}(k)^0$, the objects of which are all objects of $\mathscr{A}$ of the form A^0, with A isomorphic as a k-algebra to a direct product of finitely many subfields of K. $\mathscr{A}_1$ is closed under finite direct products, and our hypotheses on n, k, and K ensure that $k(U_n)^0$ is an object of $\mathscr{A}_1$, as is $* = k^0$. Hence Remark 1.4 applies, and we may compute the Harrison cohomology groups $H^p(k, U_n)$ via the method described earlier applied to the category $\mathscr{A}_1$ and the group $G = (k(U_n)^0, \Delta^0)$ in $\mathscr{A}_1$. We then have

Lemma 1.14. *Let $\mathscr{A}_2$ be the category of finite left Γ-sets. Then there exists a category isomorphism $\theta : \mathscr{A}_1 \approx \mathscr{A}_2$ such that $\theta(G) \approx Z|nZ$ as a group in $\mathscr{A}_2$. Furthermore, let $F_1 : \mathscr{A}_1 \to \mathrm{Ab}$ be defined as in Example 1.8, and $F_2(\cdot) = \mathscr{A}_2^{Z|nZ}(\cdot, U(K)) : \mathscr{A}_2^{Z|nZ} \to \mathrm{Ab}$ (where, as above, $U(K)$ is viewed as an object of $\mathscr{A}_2^{Z|nZ}$ via trivial action). Then there exists a natural equivalence $\varepsilon : F_1 \sim F_2\bar{\theta}$, where $\bar{\theta} : \mathscr{A}_1^G \to \mathscr{A}_2^{Z|nZ}$ is the category isomorphism induced by θ.*

Proof. Given an object A^0 of $\mathscr{A}_1$, let $\theta(A^0)$ be the set of all k-algebra homomorphisms from A to K, with Γ-action induced by the action of Γ on K; our hypotheses on A ensure that $\theta(A^0)$ is an object of $\mathscr{A}_2$. If $f^0: A^0 \to B^0$ in $\mathscr{A}_1$ (with $f: B \to A$ a homomorphism of k-algebras) and $\xi: A \to K$ is in $\theta(A^0)$, then we set $\theta(f^0)(\xi) = \xi f: B \to K$, an element of $\theta(B^0)$. Thus we obtain a map $\theta(f^0): \theta(A^0) \to \theta(B^0)$ which is easily seen to commute with the action of Γ. This defines the functor θ.

We define the inverse functor $\varphi: \mathscr{A}_2 \to \mathscr{A}_1$ as follows. Given an object X of $\mathscr{A}_2$, set $\varphi(X) = A^0$, where A is the set of all Γ-maps from X to K, viewed as a k-algebra with point-wise operations. If $f: X \to Y$ in $\mathscr{A}_2$ and $\varphi(Y) = B^0$, then we define $\bar{f}: B \to A$ by the condition $\bar{f}(\beta)(x) = \beta(f(x))$ (β in B, x in X) and set $\varphi(f) = \bar{f}^0: \varphi(X) \to \varphi(Y)$. It is then an easy exercise in the Galois theory of fields to verify that φ is a well-defined functor from $\mathscr{A}_2$ to $\mathscr{A}_1$, and that $\varphi\theta$ and $\theta\varphi$ are naturally equivalent to the identity functors on $\mathscr{A}_1$ and $\mathscr{A}_2$, respectively [*1*, p. 5].

$\theta(k(U_n)^0)$ is simply the set of all group homomorphisms from U_n to $U(K)$. Since U_n is the unique subgroup of $U(K)$ of order n, the image of such a homomorphism must lie in U_n, and so we obtain a set isomorphism $\theta(k(U_n)^0) \approx \mathrm{Ab}(U_n, U_n) \approx Z|nZ$. The mapping $Z|nZ \to \theta(k(U_n)^0)$ is obtained as follows: Given a coset $\nu + nZ$ in $Z|nZ$, its image in $\theta(k(U_n)^0)$ is the k-algebra homomorphism $f_\nu: k(U_n) \to K$ defined by the condition that $f_\nu(x) = x^\nu$ for all x in U_n.

Given γ in Γ, select a_γ as in (1.11). If $\nu + nZ$ and f_ν are as above and x is in U_n, we have from (1.10) that $\gamma(f_\nu)(x) = \gamma(f_\nu(x)) = \gamma(x^\nu) = x^{a_\gamma \nu} = f_{a_\gamma \nu}(x)$, and hence $\gamma(f_\nu) = f_{a_\gamma \nu}$, the image in $\theta(k(U_n)^0)$ of $a_\gamma \nu + nZ = \gamma(\nu + nZ)$. This establishes that $\theta(k(U_n)^0) \approx Z|nZ$ as objects of $\mathscr{A}_2$.

θ of course preserves products, and thus

$$\theta((k(U_n) \otimes k(U_n))^0) \approx Z|nZ \times Z|nZ$$

as objects of $\mathscr{A}_2$. In fact, the image in $\theta((k(U_n) \otimes k(U_n))^0)$ of $(\mu + nZ, \nu + nZ)$ in $Z|nZ \times Z|nZ$ is the k-algebra homomorphism

$$f_{\mu,\nu}: k(U_n) \otimes k(U_n) \to K$$

which satisfies the condition $f_{\mu,\nu}(x \otimes y) = f_\mu(x)\, f_\nu(y)$ for x, y in $k(U_n)$. Thus, if x is in U_n, we have that

$$(\theta(\Delta^0)(f_{\mu,\nu}))(x) = f_{\mu,\nu}(\Delta(x)) = f_{\mu,\nu}(x \otimes x) = f_\mu(x)\, f_\nu(x) = x^{\mu+\nu} = f_{\mu+\nu}(x),$$

and so $\theta(\Delta^0)(f_{\mu,\nu}) = f_{\mu+\nu}$, the image in $\theta(k(U_n)^0)$ of $\mu + \nu + nZ$. This establishes that $\theta(G) \approx Z|nZ$ as groups in $\mathscr{A}_2$.

Now let X be an object of $\mathscr{A}_2^{Z|nZ}$; then $\bar{\theta}^{-1}(X) = (A^0, \alpha^0)$, where A is the k-algebra of all Γ-maps from X to K, and $\alpha: A \to k(U_n) \otimes A$ is defined as follows. Note first that our preceding discussion allows us to identify $k(U_n) \otimes A$ with the k-algebra of all Γ-maps from $Z|nZ \times X$ to K in such a way that —

(1.15) $$(\zeta \otimes a)(\nu + nZ, x) = \zeta^{\nu} a(x)$$

for ζ in U_n, a in A, $\nu + nZ$ in $Z|nZ$, and x in X. Then —

(1.16) $$\alpha(a)(\nu + nZ, x) = a((\nu + nZ)x)$$

Recall that $F_1\bar{\theta}^{-1}(X) = F_1(A^0, \alpha^0) = \{u \text{ in } U(A) \,|\, \alpha(u) = 1 \otimes u\}$. If u is in A, then u is in $U(A)$ if and only if $\mathrm{Im}(u) \subseteqq U(K)$. Furthermore, it follows immediately from (1.15) and (1.16) that u is in $F_1\bar{\theta}^{-1}(X)$ if and only if $u((\nu + nZ)x) = u(x)$ for all x in X and $\nu + nZ$ in $Z|nZ$, and so $F_1\bar{\theta}^{-1}(X) = \mathscr{A}_2^{Z|nZ}(X, U(K)) = F_2(X)$, with $Z|nZ$ operating trivially on $U(K)$. The definition and desired properties of $\varepsilon\colon F_1 \approx F_2\bar{\theta}$ are then apparent, and the proof of the lemma is complete.

Returning now to the proof of Theorem 1.12, we obtain easily from Lemma 1.14 an isomorphism of complexes $C(G, F_1) \approx C(Z|nZ, F_2)$, which gives rise to the desired cohomology isomorphisms.

Remark 1.17. The cohomology isomorphisms of Theorem 1.12 are given explicitly in [*10*], with a direct proof. A proof of Corollary 1.13 is, in essence, contained in [*18*, Erster Teil, § 20].

2. Galois Extensions, Algebras and Coalgebras

In this section we focus our attention on the second cohomology group $H^2(R,J)$ of the cohomology theory of Example 1.8, and exploit some ideas of [*13*] to exhibit the relation between this group and the theory of Galois algebras initiated by Hasse and his school [*18*]. Our treatment parallels that of [*10*].

We begin with some remarks concerning the following situation. Let R be a commutative ring, J be a finite group, and A be simultaneously an R-algebra and a left $R(J)$-module such that each element of J acts upon A as a ring automorphism. Let $\varkappa\colon A \otimes A \to A$ be the contraction map, defined by $\varkappa(a \otimes b) = ab$. We shall denote by $E(A)$ the set of all functions $f\colon J^2 \to A$ such that $f(\sigma\tau, \sigma\varrho) = \sigma(f(\tau, \varrho))$. That is, $E(A) = \mathrm{Hom}_{R(J)}(R(J^2), A)$, where $R(J^2)$ is viewed as a left $R(J)$-module via the diagonal map $\Delta\colon R(J) \to R(J^2)$. $E(A)$ is a left $R(J^2)$-module via the formula $((\sigma,\tau)f)(\alpha,\beta) = f(\alpha\sigma, \beta\tau)$ for $\alpha, \beta, \sigma, \tau$ in J. Finally, we have the mappings $h\colon A \otimes A \to E(A)$ and $\nabla_A\colon E(A) \to A$ defined by —

$$(h(x \otimes y))(\sigma, \tau) = \sigma(x)\,\tau(y)$$

$$\nabla_A(f) = f(1).$$

We omit the easy verification of the following lemma.

Lemma 2.1. *h is an $R(J^2)$-homomorphism and the following diagram commutes*

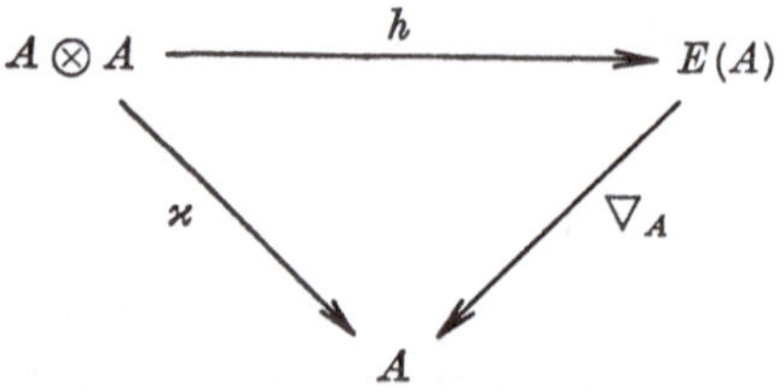

If $A \otimes A$ and $E(A)$ are viewed as left $R(J)$-modules via the diagonal map $\Delta: R(J) \to R(J^2)$, then $\varkappa$ and Δ are $R(J)$-homomorphisms.

We now shift our point of view in the following way. Denote by * the functor Hom$(\cdot, R)$. If M is a left $R(G)$-module with G a group, then M^* is likewise with the following action

$$(\sigma f)(m) = f(\sigma^{-1} m) \quad (\sigma \text{ in } G, f \text{ in } M^*, m \text{ in } M).$$

We set $C = A^*$. If A is finitely generated and projective as an R-module, we may identify $(A \otimes A)^*$ with $C \otimes C$ via the usual isomorphism to obtain the mappings $\varkappa^*: C \to C \otimes C$ and $h^*: E(A)^* \to C \otimes C$. C is a coalgebra with comultiplication $\varkappa^*$, in the sense of [*10*, Section 4].

We proceed to scrutinize $E(A)^*$ more closely. An important role will be played by the functor $\Delta(\cdot) = R(J^2) \otimes_{R(J)} (\cdot)$ from the category of left $R(J)$-modules to the category of left $R(J^2)$-modules, where $R(J^2)$ is viewed as an $(R(J^2), R(J))$-bimodule via left multiplication and the diagonal map $\Delta: R(J) \to R(J^2)$. If M is a left $R(J)$-module we define $\Delta^M: M \to \Delta(M)$ by $\Delta^M(m) = 1 \otimes m$; note that $\Delta^M(\sigma m) = (\sigma, \sigma)\Delta^M(m)$ for σ in J.

Lemma 2.2. *Let J, A, etc. be as above, and assume that A is a finitely generated projective R-module. Define $i_A: \Delta(C) \to E(A)^*$ and $\alpha: \Delta(C) \to C \otimes C$ by the formulae*

$$i_A\{(\sigma, \tau) \otimes f\}(g) = f(g(\sigma^{-1}, \tau^{-1}))$$
$$\alpha = h^* i_A$$

for σ, τ in J; f in C; g in $E(A)$. Then

(a) i_A is an $R(J^2)$-homomorphism, and is an isomorphism if A is a projective $R(J)$-module.

(b) The following diagrams commute

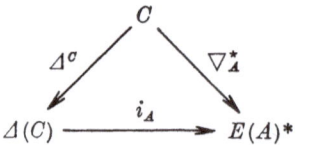

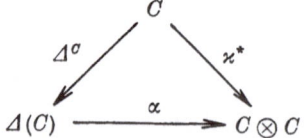

Proof. The first part of (a) is routine; in fact, i_A is a well-defined $R(J^2)$-homomorphism for any left $R(J)$-module A. We show first that i_A is an isomorphism for the special case in which A is the module $R(J)$.

Define a mapping $g_1 : R(J^2) \to R(J)$ by $g_1(\sigma, \tau) = \tau\delta_{\sigma\tau}$, where σ, τ are in J and $\delta_{\sigma\tau}$ denotes the Kronecker delta. It is easily verified that g_1 is in $E(A)$; furthermore, if g is in $E(A)$ and $g(\sigma, 1) = \sum_{\sigma \text{ in } J} a_{\sigma\tau}\tau$ with $a_{\sigma\tau}$ in R, then $g = \sum_{\sigma, \tau \text{ in } J} a_{\tau\sigma^{-1},\tau}(\sigma,\tau)g_1$. Hence g_1 generates $E(A)$ as an $R(J^2)$-module. Now define θ_1 in $E(A)^*$ by $\theta_1(g) = f_1(g(1,1))$ for g in $E(A)$, where f_1 in C is defined by $f_1(\sigma) = \delta_{1,\sigma}$ for σ in J. Observe now that, if $\theta = \sum_{\sigma, \tau \text{ in } J} c_{\sigma\tau}(\sigma,\tau)\ \theta_1$ in $E(A)^*$ with $c_{\sigma\tau}$ in R, then $\theta\{(\sigma, \tau)\ g_1\} = c_{\sigma\tau}$ for all σ, τ in J. Furthermore, if θ' is in $E(A)^*$ and

$$\theta'' = \sum_{\sigma, \tau \text{ in } J} \theta'\{(\sigma,\tau)\,g_1\}\,(\sigma,\tau)\,\theta_1, \quad \text{then} \quad \theta''((\sigma,\tau)\,g_1) = \theta'((\sigma,\tau)\,g_1)$$

for all σ, τ in J. Since the set $\{(\sigma,\tau)\,g_1/\sigma,\tau \text{ in } J\}$ generates $E(A)$ as an R-module, we may conclude that $E(A)^*$ is a free left $R(J^2)$-module with generator θ_1. Since $\Delta(C)$ is a free left $R(J^2)$-module with generator $(1,1) \otimes f_1$ and $i_A\{(1,1) \otimes f_1\} = \theta_1$, it follows that i_A is an isomorphism for the special case in which A is the module $R(J)$. A standard direct sum argument then establishes that i_A is an isomorphism for any finitely generated projective $R(J)$-module A. This completes the proof of (a).

As for (b), the commutativity of the left-most diagram is an easy computation, and commutativity of the other diagram follows from that of the first via Lemma 2.1.

Next we introduce a concept which will play a crucial role in this section.

Definition 2.3. *A is called a Galois extension of R with Galois group J if it is finitely generated, faithul and projective as an $R(J)$-module and the mapping $h: A \otimes A \to E(A)$ of Lemma 2.1 is an isomorphism.*

Remark 2.4. A is a Galois extension of R with Galois group J if and only if C is a finitely generated faithful projective $R(J)$-module and the mapping $\alpha: \Delta(C) \to C \otimes C$ of Lemma 2.2 is an isomorphism.

Proof. This is an easy consequence of Lemma 2.2 and well-known properties of the functor *.

Lemma 2.5. *If A is a Galois extension of R with Galois group J, then $A^J = R$ (here A^J denotes the set of elements of A left fixed by J, and we identify R with its image in A). Hence, if A is commutative, then A satisfies the conditions of* [8, *Theorem 1.3*].

Proof. If x is in A^J, then

$$h(x \otimes 1)(\sigma, \tau) = \sigma(x) = x = \tau(x) = h(1 \otimes x)(\sigma, \tau)$$

for all σ, τ in J, and so $h(x \otimes 1) = h(1 \otimes x)$ in $E(A)$. Our hypotheses then imply that $x \otimes 1 = 1 \otimes x$ in $A \otimes A$. Definition 5.3 guarantees that A is a faithfully flat R-algebra, and so we may apply [*9*, Lemma 3.8] to conclude that x is in R. The remainder of the Lemma follows from the easily verified observation that, if A is commutative, our requirement that h be an isomorphism is equivalent to condition (e) of [*8*, Theorem 1.3].

Lemma 2.5 shows that a commutative Galois extension, as defined here, is likewise a Galois extension in the sense of [*8*]. Conversely, if A is a Galois extension a la [*8*], then [*8*, Theorem 4.2] shows that A is finitely generated, faithful and projective as an $R(J)$-module; one then obtains from an easy argument using [*8*, Theorem 1.3 (e)] that A is a Galois extension as defined here. Thus the two definitions coincide for the special case in which A is commutative.

We now introduce the category $\mathscr{E}(J)$, the analysis of which is the main task of this section. The objects of $\mathscr{E}(J)$ are the Galois extensions of R with Galois group J. If A and B are such, then a morphism $\varphi: A \to B$ in $\mathscr{E}(J)$ is an R-algebra *isomorphism* which is, at the same time, an $R(J)$-module isomorphism. Our first result essentially describes $\mathscr{E}(J)$ in terms of the categories $\mathscr{P}(J^p)$ of finitely generated faithful projective left $R(J^p)$-modules and *isomorphisms* of such ($p = 1, 2, \ldots$) and the functors $\Delta_i: \mathscr{P}(J^p) \to \mathscr{P}(J^{p+1})$ ($i = 1, \ldots, p$). Here $\Delta_i(\cdot) = R(J^{p+1}) \otimes_{R(J^p)}(\cdot)$, where $R(J^{p+1})$ is viewed as a right $R(J^p)$-module via the „diagonal" map $\Delta_i: R(J^p) \to R(J^{p+1})$ of the introduction. For the case $p = 1$, we shall often write $\Delta_1 = \Delta$, as we have done earlier.

Note that the equality $\Delta_{i+1}\Delta_i = \Delta_i\Delta_i: R(J^p) \to R(J^{p+1})$ gives rise to a natural equivalence of functors $\Delta_{i+1}\Delta_i \sim \Delta_i\Delta_i: \mathscr{P}(J^p) \to \mathscr{P}(J^{p+2})$. We shall identify these functors via this equivalence. We also have, for each object X of $\mathscr{P}(J^p)$, an R-homomorphism $\Delta_i^X: X \to \Delta_i(X)$ defined by $\Delta_i^X(x) = 1 \otimes x$ for x in X. Note that $\Delta_i^X(\sigma x) = \Delta_i(\sigma)\Delta_i^X(x)$ for σ in J^p. Observe finally that, if X, Y are objects of $\mathscr{P}(J)$, then $X \otimes Y$ may be viewed in the obvious way as an object of $\mathscr{P}(J^2)$, and there exist isomorphisms $\Delta_1(X \otimes Y) \approx \Delta(X) \otimes Y$, $\Delta_2(X \otimes Y) \approx X \otimes \Delta(X)$ which are natural in X and Y. We shall treat these isomorphisms as identifications.

Theorem 2.6. *There exists an isomorphism* $T: \mathscr{E}(J) \to H\mathscr{P}(J)$ *of categories, where*

(*a*) *The objects of* $H\mathscr{P}(J)$ *consist of all pairs* (X, α), *with* X *an object of* $\mathscr{P}(J)$ *and* $\alpha: \Delta(X) \to X \otimes X$ *a morphism in* $\mathscr{P}(J^2)$ *such that the diagram below commutes*

$$(2.7)\qquad \begin{array}{ccc}
\Delta_2(X\otimes X) = X\otimes \Delta(X) & \xrightarrow{1\otimes\alpha} & X\otimes X\otimes X \\
\uparrow{\scriptstyle \Delta_2(\alpha)} & & \uparrow{\scriptstyle \alpha\otimes 1} \\
 & & \Delta(X)\otimes X \\
\Delta_2\Delta(X) & & \| \\
\diagdown\!\!\diagdown & & \\
\Delta_1\Delta(X) & \xrightarrow{\Delta_1(\alpha)} & \Delta_1(X\otimes X)
\end{array}$$

A morphism $\psi:(X,\alpha)\to(Y,\beta)$ in $H\mathscr{P}(J)$ is a morphism $\psi\colon Y\to X$ in $\mathscr{P}(J)$ such that the diagram below commutes

$$\begin{array}{ccc}
\Delta(Y) & \xrightarrow{\beta} & Y\otimes Y \\
\downarrow{\scriptstyle \Delta(\psi)} & & \downarrow{\scriptstyle \psi\otimes\psi} \\
\Delta(X) & \xrightarrow{\alpha} & X\otimes X
\end{array}$$

(*b*) *If A is an object of $\mathscr{E}(J)$, then $T(A) = (C,\alpha)$, where (C,α) are as in Lemma 2.2. and the preceding discussion. If $\varphi\colon A_1\to A_2$ is a map in $\mathscr{E}(J)$, then $T(\varphi)\colon (C_1,\alpha_1)\to(C_2,\alpha_2)$ is defined by $T(\varphi)=\varphi^*\colon C_2\to C_1$.*

Proof. We omit the trivial verification that $H\mathscr{P}(J)$ is a category. If A is an object of $\mathscr{E}(J)$, then we have from Remark 2.4 that $C = A^*$ is finitely generated and projective as an $R(J)$-module, and $\alpha\colon \Delta(C)\to C\otimes C$ is an isomorphism. Furthermore, setting $\gamma=\alpha\Delta^C$, we have from Lemma 2.2(b) that $\gamma=\varkappa^*\colon C\to C\otimes C$, and so the fact that A is an associative algebra then guarantees that the diagram below commutes

$$(2.8)\qquad \begin{array}{ccc}
C & \xrightarrow{\gamma} & C\otimes C \\
\downarrow{\scriptstyle \gamma} & & \downarrow{\scriptstyle 1\otimes\gamma} \\
C\otimes C & \xrightarrow{\gamma\otimes 1} & C\otimes C\otimes C
\end{array}$$

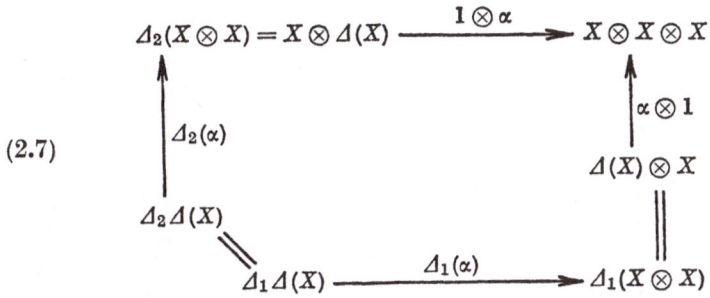

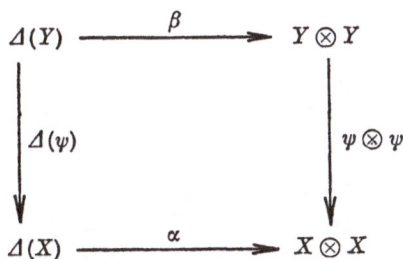

This gives rise to the diagram

(2.9)

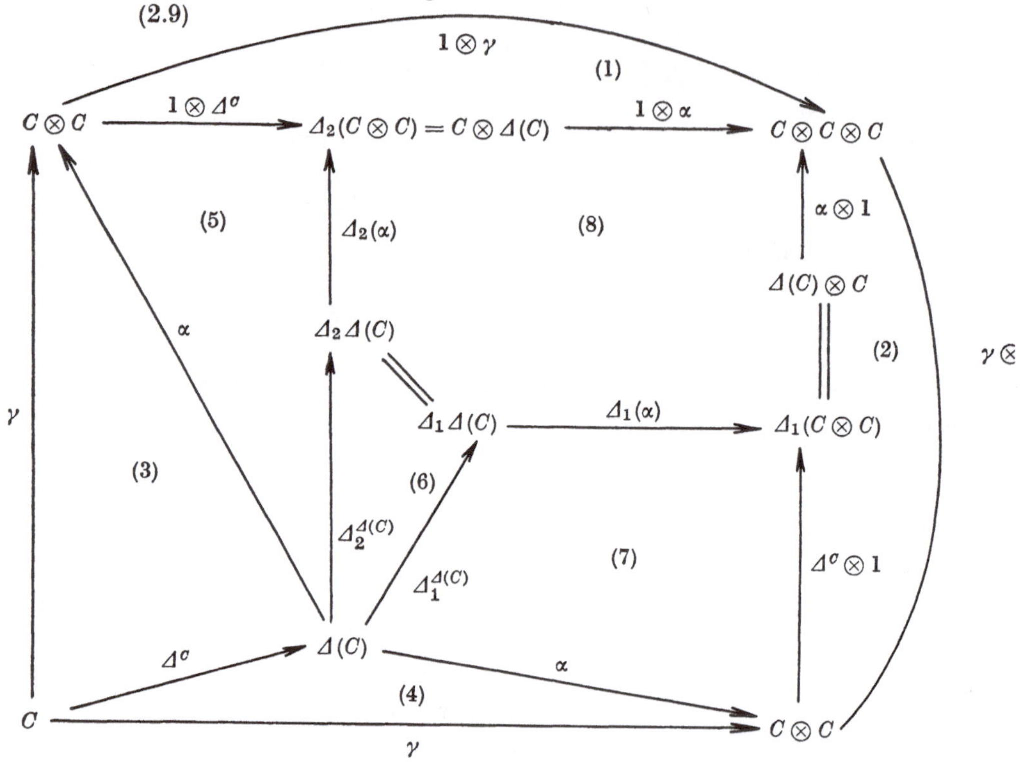

Now, parts (1) through (4) of (2.9) commute by definition of α, and (5) through (7) commute by easy computations. Hence the $\mathscr{P}(J^2)$-morphisms $(1 \otimes \alpha)\, \Delta_2(\alpha)$ and $(\alpha \otimes 1)\, \Delta_1(\alpha)$ coincide on the image in

$$\Delta_1 \Delta(C) = \Delta_2 \Delta(C) \quad \text{of} \quad \Delta_1^{\Delta(C)} \Delta^C = \Delta_2^{\Delta(C)} \Delta^C.$$

Since $\Delta_1 \Delta(C) = \Delta_2 \Delta(C)$ is generated as an $R(J^2)$-module by this subset, it follows that part (8) of (2.9) commutes. We may then conclude that $T(A) = (C, \alpha)$ is an object of $H\mathscr{P}(J)$.

If $\varphi\colon A_1 \to A_2$ in $\mathscr{E}(J)$, then our hypotheses on φ guarantee that the diagram below commutes

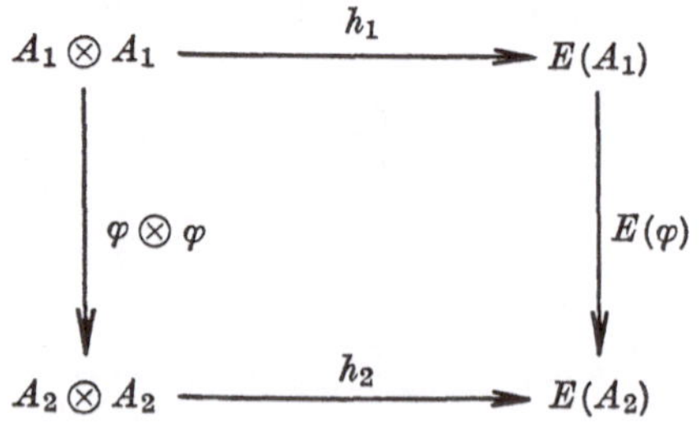

where $(E(\varphi)(f))(\sigma,\tau) = \varphi(f(\sigma,\tau))$. This, together with another easy computation, establishes the commutativity of the diagram

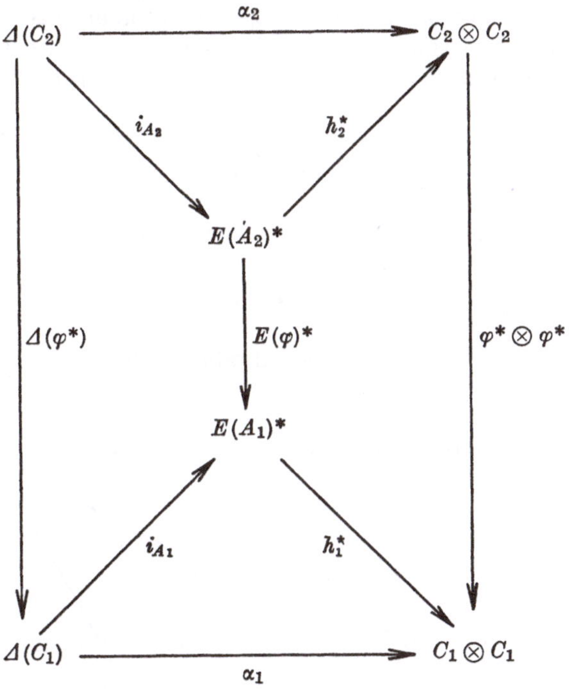

Hence $T(\varphi)$ is a morphism in $H\mathscr{P}(J)$. A routine argument then shows that T is a well-defined functor from $\mathscr{E}(J)$ to $H\mathscr{P}(J)$.

Now we define a functor $T'\colon H\mathscr{P}(J) \to \mathscr{E}(J)$ as follows. Given an object (X, α) of $H\mathscr{P}(J)$, we set $A = X^*$ and define $\varkappa = \gamma^*\colon A \otimes A \to A$, where $\gamma = \alpha\Delta^X\colon X \to X \otimes X$ and we identify $(X \otimes X)^*$ with $A \otimes A$ via the usual isomorphism, the existence of which follows from the fact that A is a finitely generated projective R-module. The commutativity of (2.7) then implies that of (2.9) (with X replacing C). Dualizing then gives the commutative diagram

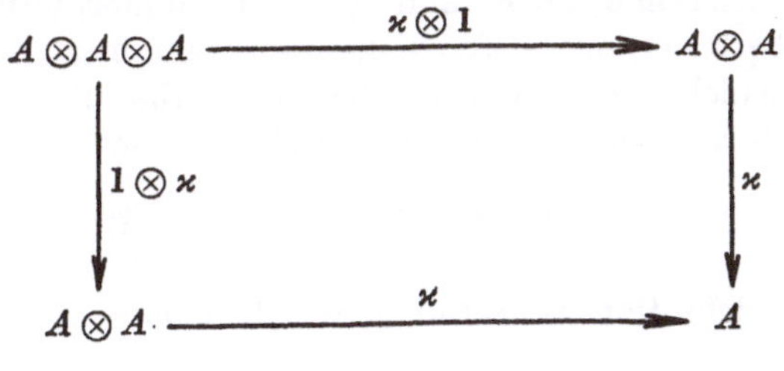

Hence, if we write $\varkappa(a \otimes b) = ab$ for a, b in A, A becomes an associative R-algebra (possibly without unit) with contraction map $\varkappa$. It follows from the definition of γ that $\gamma(\sigma x) = (\sigma \otimes \sigma)(\gamma(x))$ for σ in J, x in X; thus each σ in J acts on A as an R-algebra automorphism.

We claim that A is an object of $\mathscr{E}(J)$. Our first task is to show that A possesses a unit. To this end we introduce the R-module homomorphisms $\theta_X, \theta'_X : \Delta(X) \to X$ defined by

$$\theta_X((\sigma,\tau) \otimes x) = \tau x, \quad \theta'_X((\sigma,\tau) \otimes x) = \sigma x$$

with σ, τ in J and x in X. It is easily verified that

$$\theta_X \Delta^X = 1_X = \theta'_X \Delta^X. \tag{2.10}$$

Setting $\eta = \theta_X \alpha^{-1}$, $\eta' = \theta'_X \alpha^{-1}$, we obtain the diagram

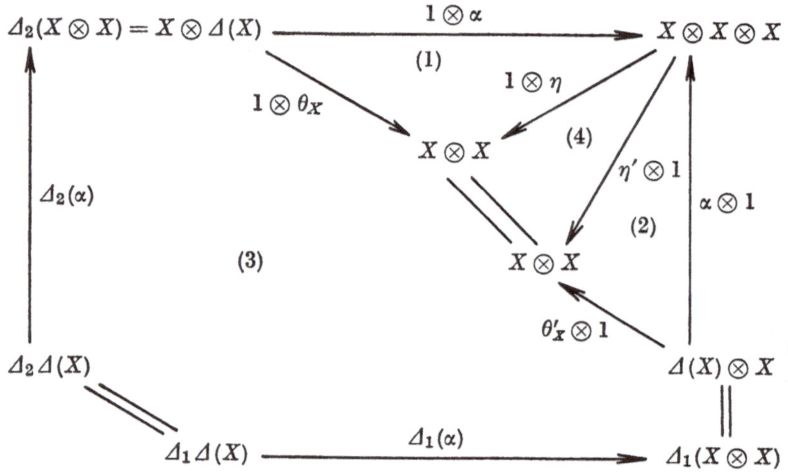

Parts (1) and (2) of this diagram are commutative by definition of η, η', and a routine computation establishes that part (3) is likewise. Since $\Delta_i(\alpha)$ $(i = 1,2)$ are isomorphisms and θ_X, θ'_X are epimorphisms, we obtain from (2.7) that part (4) of the above diagram commutes. We then apply Lemma 2.11, which is exhibited at the end of this proof, to conclude that there exists an R-homomorphism $\varepsilon: X \to R$ such that

$$\eta = \varepsilon \otimes 1: X \otimes X \to R \otimes X = X.$$

Recalling the definition of η and γ, we then have the commutative

diagram

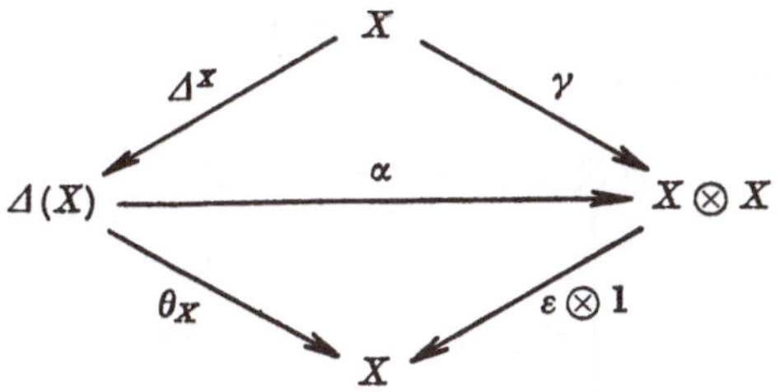

which gives rise, via 2.10 and dualization, to the commutative diagram

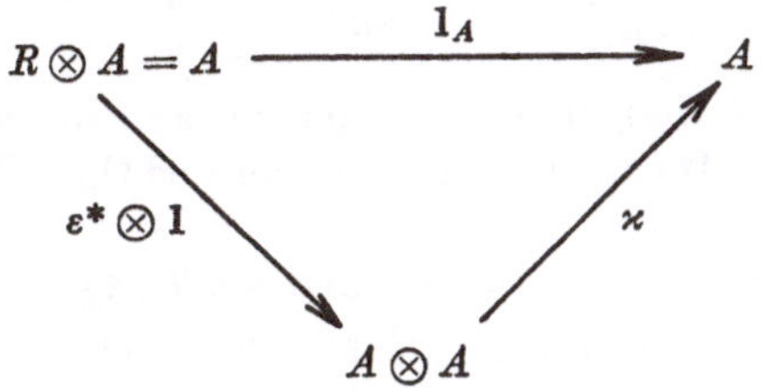

In other words, if $e = \varepsilon^*(1)$ in A, then e is a left identity for the R-algebra A. By symmetry A possesses a right identity e', and then $e' = ee' = e$. That is, A is an R-algebra with unit.

Now define $h: A \otimes A \to E(A)$ and $i_A: \Delta(X) \to E(A)^*$ as in Lemmas 2.1 and 2.2, respectively, and set $\alpha' = h^* i_A: \Delta(X) \to X \otimes X$. Then we have the diagrams

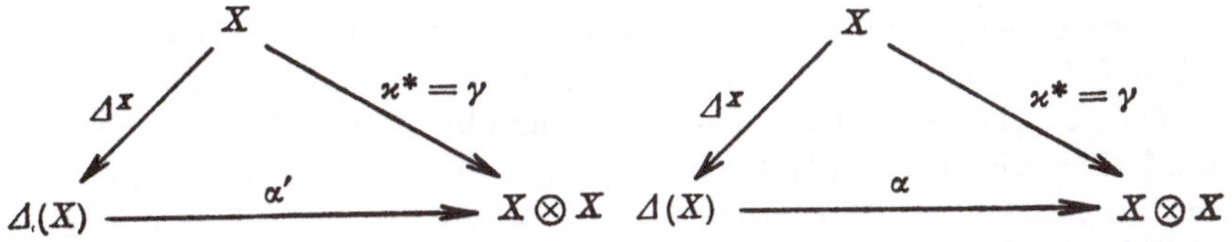

the commutativity of which are guaranteed by Lemma 2.2(b) and the definition of γ, respectively. Since $\Delta(X)$ is generated as an $R(J^2)$-module by $\mathrm{Im}(\Delta^X)$, it follows that $\alpha' = \alpha$, and is hence an isomorphism. Therefore, by Remark 2.4, $A = T'(X, \alpha)$ is an object of $\mathscr{E}(J)$.

Let $\psi: (X, \alpha) \to (Y, \beta)$ be a map in $H\mathscr{P}(J)$. We then have the commutative diagram

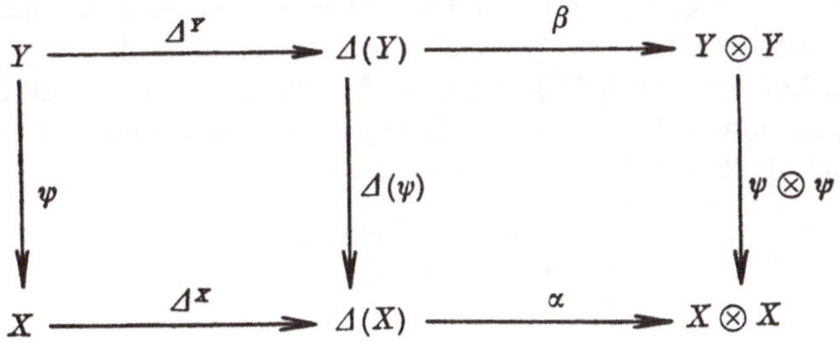

which gives rise, via dualization and the definition of the R-algebras A and B, to the commutative diagram

$$\begin{array}{ccc} A \otimes A & \xrightarrow{\varkappa_A} & A \\ \Big\downarrow{\scriptstyle \psi^* \otimes \psi^*} & & \Big\downarrow{\scriptstyle \psi^*} \\ B \otimes B & \xrightarrow{\varkappa_B} & B \end{array}$$

Therefore ψ^* is an R-algebra isomorphism, and hence a map in $\mathscr{E}(J)$. We set $T'(\psi) = \psi^*$. It can then be easily verified that $T' : H\mathscr{P}(J) \to \mathscr{E}(J)$ is a functor.

Now let A be an object of $\mathscr{E}(A)$, and set $T(A) = (C,\alpha)$. Then

$$T'\,T(A) = T'(C,\alpha) = C^* = A^{**};$$

here C^* is an algebra via the contraction map $\gamma^* \colon C^* \otimes C^* \to C^*$, where $\gamma = \alpha\Delta^C \colon C \to C \otimes C$. The natural identification of an object of $\mathscr{P}(J)$ with its second dual then gives rise to a $\mathscr{P}(J)$-morphism $\xi_A \colon A \to C^* = T'\,T(A)$. Let $\varkappa \colon A \otimes A \to A$ be the contraction map; then, by Lemma 2.5(b), $\varkappa^* = \alpha\Delta^C = \gamma$, from which it follows that ξ_A is an isomorphism in $\mathscr{E}(J)$. We then obtain easily a natural equivalence $\xi \colon I_{\mathscr{E}(J)} \sim T'\,T$.

Finally, let (X, α) be an object of $H\mathscr{P}(J)$, and set $A = T'(X,\alpha) = X^*$. Then $T\,T'(X,\alpha) = T(A) = (C,\alpha')$, with $C = A^* = X^{**}$ and $\alpha' \colon \Delta(C) \to$ $\to C \otimes C$. The natural identification of an object of $\mathscr{P}(J)$ with its second dual then yields a $\mathscr{P}(J)$-morphism $\zeta_{(X,\alpha)} \colon X \to C$. Let $\varkappa \colon A \otimes A \to A$ be the contraction map; then $\varkappa = \gamma^*$, where $\gamma = \alpha\Delta^X : X \to X \otimes X$. But, by Lemma 2.5(b), $\varkappa^* = \alpha'\Delta^C \colon C \to C \otimes C$. Since $\Delta(X)$ is generated as an $R(J^2)$-module by $\operatorname{Im}(\Delta^X)$, etc., we conclude as before that $\alpha' = \alpha$; i.e., $\zeta_{(X,\alpha)}$ is a morphism in $H\mathscr{P}(J)$. From this we obtain easily a natural equivalence $\zeta \colon I_{H\mathscr{P}(J)} \sim T\,T'$, completing the proof of the theorem.

Lemma 2.11. Let P be a finitely generated faithful projective R-module, and $\eta,\eta' \colon P \otimes P \to P$ be R-homomorphisms such that $\eta' \otimes 1 = 1 \otimes \eta \colon P \otimes P \otimes P \to P \otimes P$. Then there exists an R-homomorphism $\varepsilon \colon P \to R$ such that $\eta = \varepsilon \otimes 1 \colon P \otimes P \to R \otimes P = P$.

Proof. Let $S = \operatorname{End}_R(P)$ and $Q = P^*$. Then P and Q may be viewed as an (R, S)-bimodule and an (S, R)-bimodule, respectively. Define maps $\alpha \colon P \otimes_S Q \to R$ and $\beta \colon Q \otimes_R P \to S$ by the conditions

$$\alpha(x \otimes y) = \text{value of } y \text{ on } x$$
$$(x,z \text{ in } P;\ y \text{ in } Q)$$
$$z\beta(y \otimes x) = \alpha(z \otimes y)\,x$$

Then it is easily verified that $\alpha\,(\beta)$ is a homomorphism of (R, R)-bimodules ((S,S)-bimodules) and the following diagrams commute

(2.12)
$$\begin{array}{ccc} P\otimes_S Q\otimes_R P & \xrightarrow{\alpha\otimes 1} & R\otimes_R P \\ \downarrow{\scriptstyle 1\otimes\beta} & & \downarrow \\ P\otimes_S S & \longrightarrow & P \end{array}$$

$$\begin{array}{ccc} Q\otimes_R P\otimes_S Q & \xrightarrow{\beta\otimes 1} & S\otimes_S Q \\ \downarrow{\scriptstyle 1\otimes\alpha} & & \downarrow \\ Q\otimes_R R & \longrightarrow & Q \end{array}$$

(Hence the rings R, S and the bimodules P, Q form a Morita context in the sense of [*5*].) We may then apply [*3*, Proposition A.3 and Theorem A.2] to conclude that α and β are isomorphisms.

Clearly $\eta'\otimes 1 = 1\otimes\eta\colon P\otimes_R P\otimes_R P \to P\otimes_R P$ is a homomorphism of (R,S)-bimodules, the S-module structures arising from the right-most factors. Now, the preceding discussion ensures, via [*5*] or an easy direct argument, that there exists an R-epimorphism $f\colon \bar{P}\to R$, where $\bar{P}$ is a direct sum of finitely many copies of P. Also, the properties of η just discussed guarantee that $1\otimes\eta\colon \bar{P}\otimes_R P\otimes_R P \to \bar{P}\otimes_R P$ is a homomorphism of (R,S)-bimodules, where the S-module structures again arise from the right-most factors. Hence, if u and x are in $\bar{P}$ and $P\otimes P$, respectively, then $u\otimes\eta(xs) = u\otimes\eta(x)\,s$ for s in S. If u is chosen so that $f(u)=1$, then the map $R\to\bar{P}$ defined by $r\to ru$ is monic, from which it follows via the R-projectivity of P that the map $P\to\bar{P}\otimes_R P$ defined by $v\to u\otimes v$ is likewise monic. Therefore $\eta(xs)=\eta(x)s$ for x in P, s in S; i.e., η is an (R,S)-bimodule homomorphism.

Since α is an isomorphism, there exists a unique R-homomorphism $\varepsilon\colon P\to R$ rendering the diagram below commutative

$$\begin{array}{ccccc} P\otimes_R P\otimes_S Q & & \xrightarrow{\eta\otimes 1} & & P\otimes_S Q \\ \downarrow{\scriptstyle 1\otimes\alpha} & & & & \downarrow{\scriptstyle\alpha} \\ P\otimes_R R & \longrightarrow & P & \xrightarrow{\varepsilon} & R \end{array}$$

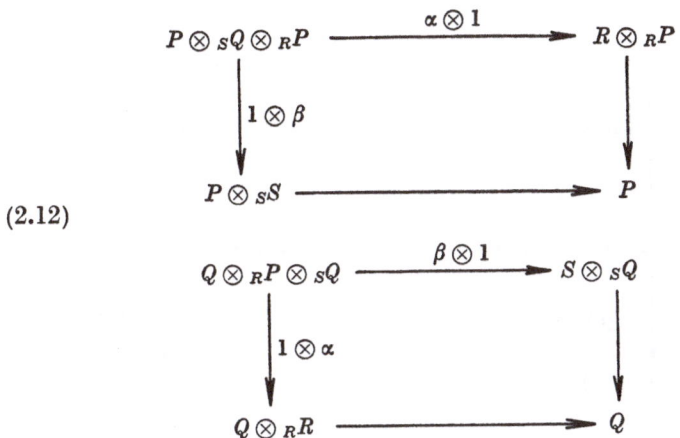

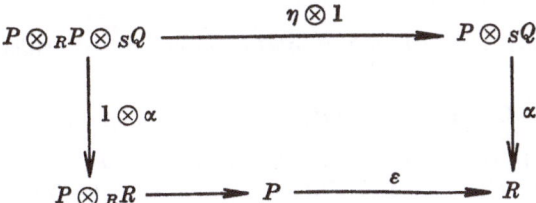

This in turn gives rise via (2.12) to the commutative diagram

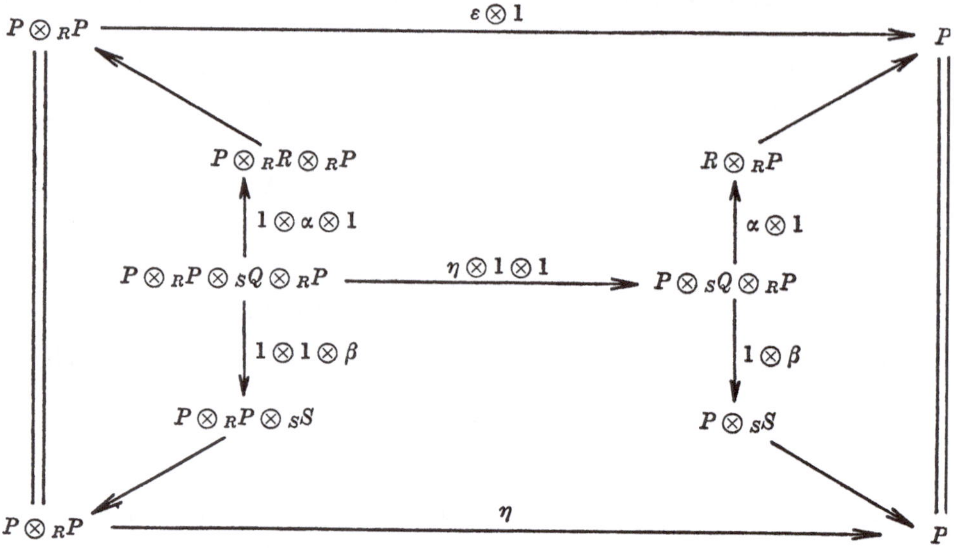

i.e., $\eta = \varepsilon \otimes 1 : P \otimes_R P \to P$, completing the proof.

In order to relate Theorem 2.6 to the cohomology theory of HARRISON, we first introduce a useful construction. Let $f: A \to C$ be a homomorphism of abelian groups. We define a category $\mathscr{C}(f)$ as follows. The objects of $\mathscr{C}(f)$ are the elements of C. If c_1, c_2 are such, then a morphism $a: c_1 \to c_2$ in $\mathscr{C}(f)$ is an element a of A such that $c_2 = c_1 + f(a)$. If also $a': c_2 \to c_3$ is a morphism in $\mathscr{C}(f)$, then the composite morphism is defined to be $a + a': c_1 \to c_3$. It is easily verified that $\mathscr{C}(f)$ is a category in which every morphism is an isomorphism.

If $\mathscr{A}$ is a category, we shall denote by $\overline{\mathscr{A}}$ the collection of isomorphism classes of objects of $\mathscr{A}$. If X is an object of $\mathscr{A}$, we shall represent the isomorphism class of X by $cl(X)$.

Remark 2.13. If $f: A \to C$ and $\mathscr{C}(f)$ are as above, then

(a) $\overline{\mathscr{C}(f)} = \mathrm{Coker}(f)$ as sets.

(b) $\mathrm{Aut}(0) = \mathrm{Ker}(f)$ as groups, 0 denoting the zero element of C viewed as an object of $\mathscr{C}(f)$.

Theorem 2.14. *Let J be a finite abelian group, and consider the Harrison complex*

$$U(R(J)) \xrightarrow{\delta^1} U(R(J^2)) \xrightarrow{\delta^2} U(R(J^3)) \longrightarrow \cdots$$

Let $\overline{\delta}: U(R(J)) \to Ker(\delta^2)$ be the homomorphism induced by δ^1. Then there exists a functor $F: \mathscr{C}(\overline{\delta}) \to H\mathscr{P}(J)$ satisfying the following conditions.

(a) $F(u) = (R(J), \alpha_u)$ *for u an element of* $Ker(\bar{\delta})$, *where*

$$\alpha_u : \Delta(R(J)) \to R(J) \otimes R(J)$$

is the composite

$$\Delta(R(J)) = R(J^2) \xrightarrow{m_u} R(J^2) = R(J) \otimes R(J).$$

(Here m_u *denotes multiplication by u, and the equality signs denote the obvious natural isomorphisms.) If* $v: u_1 \to u_2$ *is a morphism in* $\mathscr{C}(\bar{\delta})$, *then* $F(v): F(u_1) \to F(u_2)$ *is the* $R(J)$*-homomorphism* $R(J) \to R(J)$ *obtained from multiplication by v.*

(b) If u_1, u_2 *are objects of* $\mathscr{C}(\bar{\delta})$ *and* $\varphi: F(u_1) \to F(u_2)$ *is a morphism in* $H\mathscr{P}(J)$, *then there exists a unique morphism* $v: u_1 \to u_2$ *in* $\mathscr{C}(\bar{\delta})$ *such that* $F(v) = \varphi$.

(c) If (X, α) *is an object of* $H\mathscr{P}(J)$, *then* $(X, \alpha) \approx F(u)$ *for some object u of* $\mathscr{C}(\bar{\delta})$ *if and only if* $X \approx R(J)$ *as an* $R(J)$*-module.*

Before proving the theorem, we exhibit a corollary which is implicit in [*13*] (see also [*10*, Proposition 4.1 and Theorem 4.3]). First a definition.

Definition 2.15. *Let J be a finite group, and A be a Galois extension of R with Galois group J. A is called a Galois* (R, J)*-algebra if* $A \approx R(J)$ *as a left* $R(J)$*-module (i.e., if A possesses a normal base in the sense of classical Galois theory).*

Note that our definition of a Galois algebra is different from that of [*18*] or [*10*], in that these authors do not require that a Galois algebra be a Galois extension.

Corollary 2.16. *Let J be a finite abelian group. Then there exists a monomorphism of sets* $j: H^2(R, J) \to \overline{\mathscr{E}(J)}$, *where* $\overline{\mathscr{E}(J)}$ *is the set of isomorphism classes of Galois extensions of R with Galois group J (recall that an isomorphism of Galois extensions is an R-algebra isomorphism which is at the same time an* $R(J)$*-module isomorphism). j satisfies the following conditions.*

(a) If u is a 2-cocycle in $U(R(J^2))$ *and* $\bar{u}$ *is its cohomology class, then* $j(\bar{u}) = cl(T^{-1}F(u))$, *where the functors*

$$\mathscr{C}(\bar{\delta}) \xrightarrow{F} H\mathscr{P}(J) \xleftarrow{T} \mathscr{E}(J)$$

are as in Theorems 2.14 and 2.6, respectively.

(b) $\mathrm{Im}(j)$ *is the subset of* $\overline{\mathscr{E}(J)}$ *consisting of all isomorphism classes of Galois* (R, J)*-algebras. Hence j is an isomorphism if and only if every Galois extension of R with Galois group J is a Galois* (R, J)*-algebra.*

Proof. By Remark 2.13 (a), $\overline{\mathscr{C}(f)} = \mathrm{Coker}(\bar{\delta}) = H^2(R, J)$ as a set; hence the mapping $j: H^2(R, J) \to \overline{\mathscr{E}(J)}$ as given in (a) is well-defined.

It is one-to-one in virtue of Theorem 2.14 (b). (b) above then follows immediately from Theorem 2.14 (c) and the definition of T.

We turn now to the proof of Theorem 2.14. Let u be in $\operatorname{Ker}(\delta^2)$. We claim that the diagram (2.7) commutes with $F(u) = (R(J), \alpha_u)$ playing the role of (X, α). A routine computation reduces the proof of this assertion to a verification of the equation $\lambda(u)\Delta_2(u) = \varrho(u)\Delta_1(u)$, which holds in virtue of the fact that J is abelian and

$$\delta^2(u) = \lambda(u)\Delta_1(u^{-1})\Delta_2(u)\varrho(u) = 1$$

(the formula for $\delta^p : U(R(J^p)) \to U(R(J^{p+1}))$ is given in the introduction). Therefore $F(u)$ is an object of $H\mathscr{P}(J)$.

Let now $v: u_1 \to u_2$ be a map in $\mathscr{C}(\bar{\delta})$. That is, u_1 and u_2 are in $\operatorname{Ker}(\delta^2)$, v is in $U(J)$, and $u_2 = u_1\delta^1(v) = u_1\lambda(v)\Delta(v^{-1})\varrho(v)$. This leads, via an easy computation and the fact that J is abelian, to a proof that the second diagram of Theorem 2.6 (a) commutes, with $F(u_1)$, $F(u_2)$, $F(v)$ playing the roles of (X, α), (Y, β), and ψ, respectively. Thus $F(v): F(u_1) \to F(u_2)$ is a morphism in $H\mathscr{P}(J)$, from which it follows easily that F is a functor.

Now suppose given a morphism $\psi: F(u_1) \to F(u_2)$ in $H\mathscr{P}(J)$, with u_1, u_2 objects of $\mathscr{C}(\bar{\delta})$. Since $\psi: R(J) \to R(J)$ is an isomorphism of $R(J)$-modules, there exists a unique v in $U(J)$ such that $\psi = m_{v^{-1}}$. The fact that ψ is an $H\mathscr{P}(J)$-morphism then leads, via a computation similar to one above, to the equation $u_2 = u_1\delta(v)$; i.e., $v: u_1 \to u_2$ in $\mathscr{C}(\bar{\delta})$ and $\psi = F(v)$. If also $\psi = F(\bar{v})$ with $\bar{v}$ in $U(J)$, then it is clear that $\psi = m_{v^{-1}} = m_{\bar{v}^{-1}}$, and so $\bar{v} = v$. This establishes (b).

Turning finally to (c), let (X, α) be an object of $H\mathscr{P}(J)$. If $(X, \alpha) \approx F(u)$ for some object u of $\mathscr{C}(\bar{\delta})$, then there exists an $R(J)$-isomorphism $\psi: R(J) \to X$. Conversely, assume there is such a ψ. Then there exists u in $U(J^2)$ such that the diagram below commutes

$$\text{(2.17)} \qquad \begin{array}{ccc} R(J^2) & \xrightarrow{\;m_u\;} & R(J^2) \\ \| & & \| \\ \Delta(R(J)) & \xrightarrow{\;\alpha_u\;} & R(J)\otimes R(J) \\ \downarrow{\scriptstyle \Delta(\psi)} & & \downarrow{\scriptstyle \psi\otimes\psi} \\ \Delta(X) & \xrightarrow{\;\alpha\;} & X\otimes X \end{array}$$

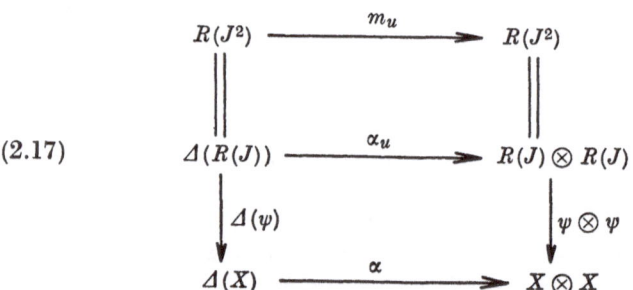

If we can show that u is an object of $\mathscr{C}(\bar{\delta})$, then (2.19) will imply that $\psi: (X, \alpha) \to F(u)$ is an isomorphism in $H\mathscr{P}(J)$. Since (X, α) is an object of $H\mathscr{P}(J)$, the commutativity of the lower square of (2.19) leads, via a final easy computation, to the fact that $(R(J), \alpha_u)$ is an object of $H\mathscr{P}(J)$. But then an argument similar to one appearing earlier in this proof shows that $\delta^2(u) = 1$, and so u is an object of $\mathscr{C}(\bar{\delta})$. This completes the proof of (c) and the theorem.

Remark 2.20. If 1 is the unit of $U(R(J^2))$, viewed as an object of $\mathscr{C}(\bar{\delta})$, then $F(1) = (R(J), \alpha_1)$ in $H\mathscr{P}(J)$, where $\alpha_1 : \Delta(R(J)) \to R(J) \otimes \otimes R(J)$ is the usual isomorphism (explicitly, $\alpha_1\{(\sigma, \tau) \otimes \varrho\} = \sigma\varrho \otimes \tau\varrho$ for σ, τ, ϱ in J). It is easy to see that $(R(J), \alpha_1)$ is the image under the functor $T : \mathscr{E}(J) \to H\mathscr{P}(J)$ of the Galois (R, J)-algebra E, which is defined to be the set of all functions $f : J \to R$, with algebra structure given by the pointwise operations, and the action of J given by the equation $(\sigma f)(\tau) = f(\tau\sigma)$ for σ, τ in J and f in E. Hence, if the mapping $j : H^2(R, J) \to \overline{\mathscr{E}(J)}$ is as in Corollary 2.16, then j carries the trivial cohomology class into the isomorphism class of E. E is called the *trivial* Galois algebra with group J.

We shall discuss later the sense in which j preserves the group structure of $H^2(R, J)$.

3. Harrison Cohomology and the Fundamental Group

We begin with a brief description of the fundamental group of a commutative ring as defined by Grothendieck [*12*, § 8]. Let $\mathscr{A}$ be the category of which the objects are all commutative separable R-algebras [*2*] which are finitely generated faithful projective R-modules and the morphisms are all R-algebra homomorphisms. Given a homomorphism $p : R \to K$ of R into an algebraically closed field K such that $\mathrm{Ker}(p)$ is a maximal ideal of R, we define a contravariant functor $V_p : \mathscr{A} \to \mathscr{S}$ as follows, where $\mathscr{S}$ denotes the category of *finite* sets. If A is an object of $\mathscr{A}$, then $V_p(A)$ is the set of all ring homomorphisms $\alpha : A \to K$ such that the restriction of α to R is p; that $V_p(A)$ is finite is guaranteed by the fact that A is a finitely generated R-algebra. If $f : A \to B$ is a morphism in $\mathscr{A}$ and β is in $V_p(B)$, we set $V_p(f)(\beta) = \beta f$. One checks easily that V_p is a well-defined functor.

Now we scrutinize the group $\Pi(p) = \mathrm{Aut}(V_p)$. Given an object A of $\mathscr{A}$, let $H(A) = \{\pi \text{ in } \Pi(p) \mid \pi_A = 1_A\}$. $H(A)$ is a normal subgroup of $\Pi(p)$ of finite index, and it is easily verified that $\Pi(p)$ becomes a topological group if we take $\{H(A) \mid A \text{ an object of } \mathscr{A}\}$ to be a fundamental system of open neighborhoods of 1. $\Pi(p)$ is called the *fundamental group of R relative to the "base point" p.*

Let $\mathscr{S}^{\Pi(p)}$ denote the category of finite left $\Pi(p)$-sets on which $\Pi(p)$ acts continuously, each such set being given the discrete topology. We obtain a (covariant) functor $V_p' : \mathscr{A}^0 \to \mathscr{S}^{\Pi(p)}$ as follows. If A is an object of $\mathscr{A}$, then we set $V_p'(A^0) = V_p(A)$, with the action of $\Pi(p)$ given by the formula $\pi(x) = \pi_A(x)$ for π in $\Pi(p)$, x in $V(A)$. If $f: A \to B$ is a morphism in $\mathscr{A}$, then $V_p'(f^0) = V(f): V_p'(B^0) \to V_p'(A^0)$. That this is a morphism in $\mathscr{S}^{\Pi(p)}$ follows immediately from the definition of $\Pi(p)$. One checks easily that the diagram below commutes up to natural equivalence

$$\begin{array}{ccc} \mathscr{A}^0 & \xrightarrow{\quad V_p' \quad} & \mathscr{S}^{\Pi(p)} \\ & {\scriptstyle V_p^0} \searrow \quad \swarrow & \\ & \mathscr{S} & \end{array} \tag{3.1}$$

the unlabeled arrow denoting the forgetful functor.

The main properties of the fundamental group are summarized in the following theorem which we state without proof (see [*12*, § 8]).

Theorem 3.2. *a) $\Pi(p)$ is a compact, totally disconnected topological group.*

b) If R is connected, then the functor $V_p': \mathscr{A}^0 \to \mathscr{S}^{\Pi(p)}$ is an isomorphism of categories.

c) If R is connected and $p_i: R \to K_i$ $(i = 1, 2)$ are base points of R, then $V_{p_1} \sim V_{p_2}$ as functors from $\mathscr{A}$ to $\mathscr{S}$. Hence $\Pi(p_1) \approx \Pi(p_2)$ as topological groups.

d) If R is a field, then $\Pi(p)$ is isomorphic as a topological group to the Galois group of the separable closure of R.

In the remainder of this paper we shall assume that R is connected. Then part c) of Theorem 3.2 allows us to denote $\Pi\ (p)$ simply by Π. We shall also write V for the functor $V_p': \mathscr{A}^0 \to \mathscr{S}$ of (3.1).

Proposition 3.3. *Let A be an object of $\mathscr{A}$, and J be a finite group acting as R-algebra automorphisms of A. Then*

a) $V(A^0)$, in addition to being a left Π-set, is also a right J-set via the formula $x\sigma = V(\sigma^0)x$ (x in $V(A^0)$, σ in J).

b) $V(A^0)$ is a (Π, J)-set, in the sense that $(\pi x)\sigma = \pi(x\sigma)$ for all π in Π, x in $V(A^0)$, σ in J.

c) A is a Galois extension of R with Galois group J if and only if $V(A^0) \approx J$ as a right J-set[3].

Proof. a) is trivial, and b) follows immediately from the fact that, if σ is in J, then $V(\sigma^0)$ is a morphism of Π-sets. We turn now to c).

---

[3] We note that a Galois extension of R which is a commutative R-algebra is an object of $\mathscr{A}$ [*8*, Theorem 1.3].

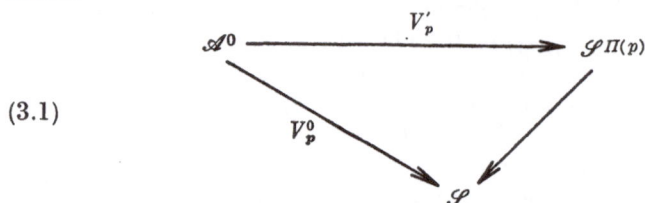

Suppose that A is a Galois extension of R with Galois group J. Viewing K as an R-algebra via the homomorphism $p: R \to K$, we have from [*8*, Lemma 1.7] that $K \otimes A$ is a Galois extension of K with Galois group J. Since K is algebraically closed, it follows that $K \otimes A$ is isomorphic to the direct product of n copies of K, where $n = \dim_K(K \otimes A) = [J:1]$, by [*8*, Lemma 4.1]. Since the factors in this direct product representation of $K \otimes A$ are in one-to-one correspondence with the elements of $V(A^0)$, it follows that $V(A^0)$ possesses $[J:1]$ elements. Thus, in order to establish that $V(A) \approx J$ as J-sets, we need only show that, if $\sigma \neq 1$ is in J and α is in $V(A^0)$, then $\alpha\sigma \neq \alpha$. But this is an immediate consequence of the uniqueness part of [*8*, Theorem 3.1].

Conversely, suppose that $V(A^0) \approx J$ as J-sets. We claim first that, if a is in A and $\sigma(a) = a$ for all σ in J, then a is in R. For, if a is not in R, it is easy to find α, β in $V(A^0)$ such that $\alpha(a) \neq \beta(a)$. Our hypotheses on $V(A^0)$ then guarantee that $\alpha\sigma = \beta$ for some σ in J, from which it follows that $\sigma(a) \neq a$.

We claim finally that, if $e \neq 0$ is an idempotent in A and $\sigma \neq \tau$ in J, then there exists a in A such that $\sigma(a)e \neq \tau(a)e$. For, since $e \neq 0$, it is easy to find α in $V(A^0)$ such that $\alpha(e) = 1$. The J-set isomorphism $V(A^0) \approx J$ then ensures that $\alpha\sigma \neq \alpha\tau$, and so there exists a in A such that $\alpha(\sigma(a)e) = \alpha(\sigma(a)) \neq \alpha(\tau(a)) = \alpha(\tau(a)e)$, implying that $\sigma(a)e \neq \tau(a)e$. We may then conclude that the elements of J give strongly distinct automorphisms of A in the sense of [*8*, Theorem 1.3(a)], and hence, by that theorem, A is a Galois extension of R with Galois group J. This completes the proof.

Now let J be a finite abelian group. As in the introduction, we shall denote by $\mathrm{Hom}(\Pi, J)$ the group of continuous homomorphisms from Π to the discrete group J. If χ is in $\mathrm{Hom}(\Pi, J)$, we define a (Π, J)-set $J(\chi)$ by the conditions below

(3.4a) $$J(\chi) = J \quad \text{as a right } J\text{-set},$$

(3.4b) $$\pi \cdot \sigma = \chi(\pi)\sigma \quad \text{for } \pi \text{ in } \Pi,\ \sigma \text{ in } J$$

the dot denoting the action of π on $J(\chi)$.

The next result was proved by D. K. Harrison, who used a different definition of the fundamental group [*14*, Theorem 8].

Theorem 3.5. *Let R be connected, J be a finite abelian group, and $\mathscr{E}_c(J)$ denote the full subcategory of $\mathscr{E}(J)$ of which the objects are all objects of $\mathscr{E}(J)$ which are commutative as R-algebras. Then there exists an isomorphism $i: \mathrm{Hom}(\Pi, J) \approx \overline{\mathscr{E}_c(J)}$ of sets. If χ is in $\mathrm{Hom}(\Pi, J)$, then $i(\chi) = cl(A)$, where A is an object of $\mathscr{E}_c(J)$ such that $V(A^0) \approx J(\chi)$ as (Π, J)-sets.*

Proof. The existence of a set mapping $i : \mathrm{Hom}(\Pi, J) \to \overline{\mathscr{E}_c(J)}$ satisfying the above conditions is an immediate consequence of Theorem 3.2 and Proposition 3.3. We need only show that i is one-to-one and onto.

Suppose that $i(\chi) = i(\chi')$, with χ, χ' in $\mathrm{Hom}(\Pi, J)$. Then by Theorem 3.2 (b), there exists an isomorphism $t : J(\chi) \approx J(\chi')$ of (Π, J)-sets. Let $\tau = t(1)$ in $J(\chi') = J$; then for π in Π we have that

$$\tau\chi(\pi) = t(\chi(\pi)) = t(\pi \cdot 1) = \pi \cdot \tau = \chi'(\pi)\tau$$

in J. Since J is abelian, we may conclude that $\chi = \chi'$. Thus i is one-to-one.

Let finally A be an object of $\mathscr{E}_c(J)$; then $V(A^0) \approx J$ as a right J-set by Proposition 3.3, and we identify the two sets via this isomorphism. Define a function $\chi : \Pi \to J$ by the condition $\chi(\pi) = \pi \cdot 1$. An easy computation then establishes that χ is in $\mathrm{Hom}(\Pi, J)$ and $V(A^0) = J(\chi)$, from which it follows that $i(\chi) = cl(A)$. Therefore i is an isomorphism of sets, and the proof is complete.

Corollary 3.6. *Let R be connected and J be a finite cyclic group. Then the mapping $\bar{T}i : \mathrm{Hom}(\Pi, J) \to \overline{H\mathscr{P}(J)}$ is an isomorphism of sets, where $i : \mathrm{Hom}(\Pi, J) \to \mathscr{E}_c(J)$ is as in Theorem 3.5 and $\bar{T} : \overline{\mathscr{E}(J)} \to \overline{H\mathscr{P}(J)}$ is induced by the functor $T : \mathscr{E}(J) \to H\mathscr{P}(J)$ of Theorem 2.6.*

Proof. This is an immediate consequence of Theorems 2.6 and 3.5 and Proposition 3.7 below, a proof of which can be found in [*13*] for the special case in which R is a field; for the general case see [*10*, Theorem 5.3][4].

Proposition 3.7. *If J is a finite cyclic group, then $\mathscr{E}_c(J) = \mathscr{E}(J)$. That is, every Galois extension of R with Galois group J is a commutative R-algebra.*

Finally, we exhibit a formulation of Harrison's theorem (0.2) (see also [*14*, Comment, p. 12]).

Theorem 3.8. *Let R be connected, and assume that every Galois extension of R is a Galois algebra (this will be the case if, for example, R is a semi-local ring; see [8, Theorem 4.2]). Then, if J is a finite cyclic group, there exists an isomorphism $\theta : \mathrm{Hom}(\Pi, J) \approx H^2(R, J)$ of groups. If χ is in $\mathrm{Hom}(\Pi, J)$ then $j(\theta(\chi)) = i(\chi)$, where*

$$\mathrm{Hom}(\Pi, J) \xrightarrow{i} \overline{\mathscr{E}(J)} \xleftarrow{i} H^2(R, J)$$

are as in Theorem 3.5 and Corollary 2.16, respectively.

Proof. The existence of a set isomorphism θ satisfying the above conditions is an immediate consequence of Corollary 2.16, Theorem 3.5, and

[4] Added in Proof: See also De Meyer, F.: Some notes on the general Galois theory of rings. Osaka J. Math. **2**, 117 – 127 (1965).

Proposition 3.7. We thus need only show that θ is a homomorphism of groups, for which task some preliminary remarks are necessary.

Define the homomorphism $\nabla : J^2 \to J$ by $\nabla(\sigma, \tau) = \sigma\tau$, and set $H = \operatorname{Ker}(\nabla)$. If X and Y are (Π, J)-sets which, as left Π-sets, are objects of $\mathcal{S}^{\Pi}$, we define $X \circ Y = X \times Y/\underset{H}{\sim}$, where $\underset{H}{\sim}$ is the equivalence relation defined by the condition that $(x_1, y_1) \underset{H}{\sim} (x_2, y_2)$ if and only if $(x_2, y_2) = (x_1\sigma, y_1\tau)$ for some (σ, τ) in H. We shall denote by $x \circ y$ the equivalence class in $X \circ Y$ of (x, y) in $X \times Y$. One checks easily that $X \circ Y$ is a (Π, J)-set which, as a Π-set, is an object of $\mathcal{S}^{\Pi}$, the (Π, J)-structure being defined by the condition $\pi(x \circ y)\sigma = (\pi x\tau) \circ (\pi x\varrho)$ (x in X; y in Y; π in Π; σ, τ, ϱ in J; $\nabla(\tau, \varrho) = \sigma$). We then have

Lemma 3.9. *If χ_1, χ_2 are in $\operatorname{Hom}(\Pi, J)$, then*

$$J(\chi_1\chi_2) \approx J(\chi_1) \circ J(\chi_2)$$

as (Π, J)-sets.

Proof. Using the fact that $J(\chi_1) = J(\chi_2) = J(\chi_1\chi_2) = J$ as J-sets, we define the mapping $\nu : J(\chi_1) \circ J(\chi_2) \to J(\chi_1\chi_2)$ by $\nu(\sigma \circ \tau) = \sigma\tau$ (σ, τ in J). An easy computation establishes that ν is an isomorphism of (Π, J)-sets.

Lemma 3.10. *Let R be connected, and A be a Galois extension of R with Galois group G. Let H be a subgroup of G, and $B = A^H = \{a \text{ in } A \mid \sigma(a) = a$ for all σ in $H\}$. Then B is an object of $\mathcal{A}$, and $V(B^0) \approx VA^0)/\underset{H}{\sim}$, where $\underset{H}{\sim}$ is the equivalence relation defined by $\alpha \sim \alpha'$ if and only if $\alpha' = \alpha\sigma$ for some σ in H.*

Proof. That B is an object of $\mathcal{A}$ follows from [8, Theorem 2.2]. Let $\mu : B \to A$ be the inclusion map; then it is clear that the pair (B, μ) is characterized in the category $\mathcal{A}$ by the universal property embodied in the following statements

a) $\sigma\mu = \mu$ for all σ in H.

b) Given a morphism $\mu' : B' \to A$ in $\mathcal{A}$ such that $\sigma\mu' = \mu'$ for all σ in H, then there exists a unique morphism $f : B' \to B$ in $\mathcal{A}$ such that $\mu f = \mu'$.

It is also easy to see that the pair (X, ε), with $X = V(A^0)/\underset{H}{\sim}$ and $\varepsilon : V(A^0) \to X$ the canonical morphism, is characterized in the category $\mathcal{S}^{\Pi}$ by the universal property dual to that above. The lemma then follows from Theorem 3.2(b).

Returning now to the proof of Theorem 3.8, let χ_k be in $\operatorname{Hom}(\Pi, J)$ and $\theta(\chi_k) = \bar{u}_k$, where $k = 1, 2$ and u_k is a cocycle in $U(J^2)$, — meaning "cohomology class of". Let $j(\theta(\chi_k)) = i(\chi_k) = cl(A_k)$, with A_k an object of $\mathcal{E}(J)$, and set $A_1 \circ A_2 = (A_1 \otimes A_2)^H$, where $H = \operatorname{Ker}(\nabla : J^2 \to J)$.

By [4, Theorem A.8] $A_1 \otimes A_2$ is a Galois extension of R with Galois group J^2, the J^2-action being defined in the obvious way; then, by [8, Theorem 2.2], $A_1 \circ A_2$ is a Galois extension of R with Galois group J, where $\sigma(a) = (\tau, \varrho)(a)$ for a in $A_1 \circ A_2$; σ, τ, ϱ in J; and $\nabla(\tau, \varrho) = \sigma$. We may then apply Theorems 3.2, 3.5 and Lemmas 3.9, 3.10 to conclude that $j(\theta(\chi_1 \chi_2)) = cl(A_1 \circ A_2)$ (an easy computation is needed to verify that the J-action on $A_1 \circ A_2$ is the correct one). Since j is an isomorphism, we then need only show that $j(\bar{u}_3) = cl(A_1 \circ A_2)$ in $\overline{\mathscr{E}(J)}$ where $u_3 = u_1 u_2$. That is, if $j(\bar{u}_3) = cl(A_3)$ in $\overline{\mathscr{E}(J)}$, then $A_3 \approx A_1 \circ A_2$ in the category $\mathscr{E}(J)$.

Now, we see from a brief review of Theorem 2.14 and Corollary 2.16 that $T(A_k) \approx F(u_k) = (R(J), \alpha_{u_k})$ in $H\mathscr{P}(J)$, where $k = 1, 2, 3$ and

$$\mathscr{E}(J) \xrightarrow{T} H\mathscr{P}(J) \xleftarrow{F} \mathscr{C}(\bar{\delta})$$

are as in Theorems 2.6 and 2.14. Then we may assume that $A_k = R(J)^*$, with contraction map $\varkappa_k = \overset{*}{\gamma_k}: A_k \otimes A_k \to A_k$, where $\gamma_k : R(J) \to \to R(J) \otimes R(J)$ is defined by the equation $\gamma_k = \Delta^{R(J)} \alpha_{u_k}$. Explicitly, $\gamma_k(\sigma) = (\sigma \otimes \sigma) u_k$ for σ in J, where we make the usual identification $R(J) \otimes R(J) = R(J^2)$.

We next define an R-epimorphism $\varphi : R(J) \otimes R(J) \to R(J)$ by $\varphi(\sigma \otimes \tau) = \sigma\tau$ for σ, τ in J; this gives rise to an R-monomorphism

$$\varphi^* : A_3 = R(J)^* \to (R(J) \otimes R(J))^* = A_1 \otimes A_2 .$$

A routine computation establishes that φ^* is a homomorphism of R-algebras and $\operatorname{Im}(\varphi^*) \subseteqq (A_1 \otimes A_2)^H = A_1 \circ A_2$. Now, since φ is an R-module epimorphism, $\operatorname{Im}(\varphi^*)$ is an R-direct summand of $A_1 \otimes A_2$; comparison of ranks then yields that $\operatorname{Im}(\varphi^*) = A_1 \circ A_2$. Thus we have constructed an R-algebra isomorphism $A_3 \approx A_1 \circ A_2$, which a final easy computation shows is also an isomorphism of $R(J)$-modules. Therefore $A_3 \approx A_1 \circ A_2$ in $\mathscr{E}(J)$, and the proof of Theorem 3.8 is complete.

Remark 3.11. Implicit in the proof of Theorem 3.8 is the fact that, if J is a finite abelian group, then $\mathscr{E}(J)$ and $H\mathscr{P}(J)$ are categories with composition in the sense of [6] relative to suitably defined "composition" functors $\circ : \mathscr{E}(J) \times \mathscr{E}(J) \to \mathscr{E}(J)$, etc. It is not difficult to show that these laws of composition render $\overline{\mathscr{E}(J)}$ and $\overline{H\mathscr{P}(J)}$ into abelian groups and, using the notation of [6], there exist natural isomorphisms

$$K_0(\mathscr{E}(J)) \approx \overline{\mathscr{E}(J)}, \quad K_0(H\mathscr{P}(J)) \approx \overline{H\mathscr{P}(J)};$$

furthermore, if R is connected then $K_1(\mathscr{E}(J)) \approx K_1(H\mathscr{P}(J)) \approx J$. The functor $T : \mathscr{E}(J) \to H\mathscr{P}(J)$ is then an isomorphism of categories with composition, and the induced mapping $\bar{T} : \overline{\mathscr{E}(J)} \approx \overline{H\mathscr{P}(J)}$ is an isomorphism of groups, as are the mappings $i : \operatorname{Hom}(\Pi, J) \approx \overline{\mathscr{E}_c(J)}$ and

$\bar{T}i : \mathrm{Hom}(\Pi, J) \approx \overline{H\mathscr{P}(J)}$ of Theorem 3.5 and Corollary 3.6, respectively.

The group $\overline{\mathscr{E}(J)}$ is denoted by $T(J, R)$ in [*14*], and the exact sequence of [*14*, Theorem 6] is an easy consequence of the five-term K-theory exact sequence of [*6*].

We note finally that Theorem 3.8, together with Corollary 1.13, gives the isomorphism (0.3) of classical Kummer theory.

References

[*1*] Artin, M.: Grothendieck topologies, mimeographed notes. Harvard University, Cambridge, Mass.

[*2*] Auslander, M., and D. Buchsbaum: On ramification theory in noetherian rings. Amer. J. Math. **81**, 749—765 (1959).

[*3*] —, and O. Goldman: Maximal orders. Trans. Amer. Math. Soc. **97**, 1—24 (1960).

[*4*] — — The Brauer group of a commutative ring. Trans. Amer. Math. Soc. **97**, 367—409 (1960).

[*5*] Bass, H.: The Morita theorems, mimeographed notes. Eugene, Ore.: University of Oregon 1962.

[*6*] — K-theory and quadratic forms. Proceedings of the Conference on Categorical Algebra, 1965.

[*7*] Cartan, H., and S. Eilenberg: Homological algebra. Princeton: Princeton University Press 1956.

[*8*] Chase, S. U., D. K. Harrison, and Alex Rosenberg: Galois theory and Galois cohomology of commutative rings. Memoirs Amer. Math. Soc. **52**, (1965).

[*9*] —, and A. Rosenberg: Amitsur cohomology and the Brauer group. Memoirs Amer. Math. Soc. **52**, 34—79 (1965).

[*10*] — — Galois algebras and a theorem of Harrison. To appear in Nagoya Math. J.

[*11*] Freyd, P.: Abelian categories. New York: Harper and Roe 1964.

[*12*] Grothendieck, A.: Geometrie formelle et geometrie algebrique. Seminaire Bourbaki **11**, (1958/59), Expose 182.

[*13*] Harrison, D. K.: Abelian extension of arbitrary fields. Trans. Amer. Math. Soc. **97**, 230—235 (1960).

[*14*] — Abelian extensions of commutative rings. Memoirs Amer. Math. Soc. **52**, 1—14 (1965).

[*15*] Huber, P. J.: Homotopy theory in general categories. Math. Annalen **144**, 361—385 (1961).

[*16*] Mac Lane, S.: Homology. New York: Academic Press 1963.

[*17*] Serre, J. P.: Corps locaux. Paris: Hermann, 1962 (Act. Scient. et Ind. 1296).

[*18*] Wolf, P.: Algebraische Theorie der Galoisschen Algebren, Math. Forschungsberichte III. Berlin: Deutscher Verlag der Wissenschaften 1956.

Department of Mathematics
Cornell University
Ithaca, New York

Quotient Categories of Modules *, **

By

Carol L. Walker and Elbert A. Walker

Gabriel and Popesco [4] have proved the following:

Theorem (*). *Let $\mathscr{C}$ be an Abelian category with exact direct limits. Let $U \in \mathscr{C}$, $E = \mathrm{Hom}_{\mathscr{C}}(U, U)$, $\mathscr{M}_E$ be the category of right E modules, and $S : \mathscr{C} \to \mathscr{M}_E : M \to \mathrm{Hom}_{\mathscr{C}}(U, M)$. Let $T : \mathscr{M}_E \to \mathscr{C}$ be an adjoint of S (one exists) and Φ the associated natural transformation from $T \circ S$ to the identity functor on $\mathscr{C}$. The following are equivalent.*

i. *U is a generator of $\mathscr{C}$.*

ii. *S is completely faithful.*

iii. *Φ is an isomorphism and T is exact.*

iv. *T is exact and induces an equivalence of $\mathscr{M}_E/\mathrm{Ker}\, T$ and $\mathscr{C}$.*

In [3] it was shown that if R is a ring and $\mathscr{S}$ is a localizing subcategory of $\mathscr{M}_R$, then $\mathscr{M}_R/\mathscr{S}$ is Abelian with a generator and exact direct limits. The point of *Theorem* (*) is that the converse holds. Namely, if $\mathscr{C}$ is Abelian with exact direct limits and a generator, then $\mathscr{C}$ is equivalent to the category of all modules over a ring, modulo a localizing subcategory. *Gabriel and Popesco's theorem* is non-trivial even for the case $\mathscr{C} = \mathscr{M}_R$ for some ring R.

Our purpose here is to examine this situation, or more generally, the situation when $\mathscr{C}$ is a full exact Abelian subcategory of $\mathscr{M}_R$ having a generator and exact direct limits.

Let $\mathscr{C}$ be such a category. In the notation of *Theorem* (*), $T = (\cdot \otimes_E U)$, and Section 1 is concerned with identifying $\mathrm{Ker}\, T$. Now $\mathrm{Ker}\, T$ is a localizing subcategory of $\mathscr{M}_E$ [3] (i.e., $\mathrm{Ker}\, T$ is closed under submodules homomorphic images, extensions, and arbitrary direct sums), and there is a natural one-one correspondence between localizing subcategories of $\mathscr{M}_E$ and non-empty sets $\mathscr{J}$ of right ideals of E satisfying

a) $I \in \mathscr{J}$, J a right ideal of E, $J \supset I$ imply $J \in \mathscr{J}$.

---

* This work was partially supported by the National Science Foundation Grant No. GP-3581. A major portion of the work on this paper was done while the first author held an NSF postdoctoral fellowship and the second author an NSF senior postdoctoral fellowship.

** Received September 13, 1965.

b) $I, J \in \mathscr{J}$ imply $I \cap J \in \mathscr{J}$.
c) $I \in \mathscr{J}$, $x \in E$, imply $I : x = \{e \in E \mid xe \in I\} \in \mathscr{J}$.
d) J a right ideal of E, $I \in \mathscr{J}$, $J : i \in \mathscr{J}$ for all $i \in I$ imply $J \in \mathscr{J}$.

(See [3] or [7].)

Such an $\mathscr{J}$ is called an idempotent topologizing filter.

If $\mathscr{S}$ is a localizing subcategory of $\mathscr{M}_E$, the associated $\mathscr{J}$ is just $\mathscr{J}(\mathscr{S}) = \{I \mid E/I \in \mathscr{S}\}$. The point is that localizing subcategories are determined by the cyclic modules they contain. a)—d) above specify which sets of cyclics can be the cyclics of a localizing subcategory. The sets $\mathscr{J}$ satisfying a)—d) can be quite complicated, but we show in Theorem 1.7 that if $\mathscr{C} = \mathscr{M}_R$ then $\mathscr{J}(\mathrm{Ker}\, T)$ is just those right ideals of E containing the trace ideal t of U in E. (t is the image of the trace map $\tau : U \otimes_R \mathrm{Hom}(U, E) \to E : u \otimes f \to f(u)$.)

It follows that $\mathrm{Ker}\, T = \{M \in \mathscr{M}_R \mid Mt = 0\}$. Several other properties of T are derived in Section 1. One amusing result of these is the following. If D is a division ring and V is a right vector space over D, then $\mathscr{M}_D$ is equivalent to $\mathscr{M}_E$ ($E = \mathrm{Hom}_D(V, V)$) modulo the localizing subcategory of those M in $\mathscr{M}_E$ annihilated by the minimum two-sided ideal $\{f \in E \mid \dim f(V) < \aleph_0\}$.

Section 2 is concerned with the extent to which the ring $E = \mathrm{Hom}_R(U, U)$ determines $\mathscr{C}$, U a generator in $\mathscr{C}$. Suppose U has the property that an inclusion map $A \to U$ induces an isomorphism $\mathrm{Hom}_{\mathscr{C}}(U, U) \to \mathrm{Hom}_{\mathscr{C}}(A, U)$ if and only if $A = U$. Call such a generator a *proper* generator.

Every Abelian category $\mathscr{C}$ with exact direct limits and a generator has a proper generator. In fact, if U is a generator of $\mathscr{C}$, then the injective envelope Q of the direct sum of all quotients of U is a co-generator, and so $U \oplus Q$ is a proper generator of $\mathscr{C}$. One of the principal results in Section 2 is the following. If V is a proper generator in an Abelian category $\mathscr{C}'$ with exact direct limits, and if $E \approx E' = \mathrm{Hom}_{\mathscr{C}'}(V, V)$, then there exists an equivalence $F : \mathscr{C} \to \mathscr{C}'$ such that $F(U) = V$. Let $\mathscr{E}$ be the class of rings that are endomorphism rings of proper generators in Abelian categories with exact direct limits. The result above partitions $\mathscr{E}$: two rings in $\mathscr{E}$ are in the same member of the partition if and only if they are endomorphism rings of proper generators in equivalent Abelian categories with exact direct limits. This should have some ring theoretical significance. It is shown in Corollary 2.3 that a ring R is in $\mathscr{E}$ if and only if the set of right ideals I such that the natural map $\mathrm{Hom}_R(R, R) \to \mathrm{Hom}_R(I, R)$ is an isomorphism satisfies properties a)—d) above for idempotent topologizing filters. The class $\mathscr{E}$ contains all commutative rings (Theorem 2.5) and all self-injective rings. Several other properties of $\mathscr{E}$ are exhibited in Section 2.

Using the notation in *Theorem* (*) above it is shown in Section 3 that E is right self-injective if and only if U is injective. Also, the injective envelope of U is $T(\bar{E})$, where $\bar{E}$ is the injective envelope of E in $\mathcal{M}_E$. In particular, if U is a generator in the category $\mathcal{M}_R$ of right R-modules, the injective envelope of U is $\bar{E} \otimes_E U$.

The category of Abelian p-groups is Abelian with a generator and exact direct limits. This category is the topic of Section 4. An Abelian p-group G such that G modulo its maximum divisible subgroup is unbounded turns out to be flat as a module over its endomorphism ring (Theorem 4.1). This fact is of some group theoretical interest [*6*]. It seems to be a bit difficult to determine which groups are proper generators in the category of Abelian p-groups.

1. The Kernel of T

Throughout, $\mathcal{C}$ will be an Abelian category with a generator U and exact direct limits, and E will be the ring $\operatorname{Hom}_{\mathcal{C}}(U, U)$. The category $\mathcal{C}$ has infinite sums and injective envelopes, and the functor

$$S = \operatorname{Hom}_{\mathcal{C}}(U, \cdot) : \mathcal{C} \to \mathcal{M}_E$$

has an adjoint [*3*] which will be denoted by T.

Definition. *Let I be a right ideal in E. Then IU is the image of the map*

$$\sum_{i \in I} U \xrightarrow{\Sigma i} U.$$

As pointed out in the introduction, to identify $\operatorname{Ker} T$ is the same as identifying $\mathcal{J}$, the set of right ideals I of E such that $T(E/I) = 0$. The following proposition does this.

Proposition 1.1. *A right ideal I of E is in $\mathcal{J}$ (i.e., $T(E/I) = 0$) if and only if $IU = U$.*

Proof. Since T is right exact, $T(E/I) = 0$ if and only if the inclusion $I \to E$ induces an epimorphism $T(I) \to T(E)$. But the image of this map is just IU. To see this, let $i \in I$. Then $S(i) : E \to E : e \to ie$. Since $S(i)(E) \subset I$, $TS(i) : T(E) \to T(E)$ can be factored through $T(I)$. In fact $\operatorname{Im}\left(\sum_{i \in I} T(E) \xrightarrow{\Sigma TS(i)} T(E)\right) = T(I)$, since T is exact and commutes with direct sums. In the commutative diagram

$$\begin{array}{ccc} \sum_{i \in I} T(E) & \xrightarrow{\Sigma TS(i)} & T(E) \\ \downarrow \Sigma \Phi_u & & \downarrow \Phi_u \\ \sum_{i \in I} U & \xrightarrow{\Sigma i} & U \end{array}$$

the vertical arrows are isomorphisms, by *Theorem* (*) *of Gabriel and Popesco*. It follows that the image of the composition $T(I) \to T(E) \xrightarrow{\Phi_u} U$ is IU.

Proposition 1.2. *If* $I \in \mathscr{J}$, *then* $\mathrm{Hom}_E(E/I, E) = 0$ *and* $\mathrm{Ext}^1_E(E/I, E) = 0$ *in* $\mathscr{M}_E$.

Proof. $\mathrm{Hom}_E(E/I, E) = \mathrm{Hom}_E(E/I, S(U)) \approx \mathrm{Hom}_{\mathscr{C}}(T(E/I), U) = 0$ since $T(E/I) = 0$. In the commutative diagram with exact row

$$\begin{array}{ccccc} \mathrm{Hom}_E(E, E) & \to & \mathrm{Hom}_E(I, E) & \to & \mathrm{Ext}^1_E(E/I, E) \to 0 \\ \downarrow & & \downarrow & & \\ \mathrm{Hom}_{\mathscr{C}}(T(E), U) & \to & \mathrm{Hom}_{\mathscr{C}}(T(I), U) & & \end{array}$$

the vertical maps and the lower horizontal map are isomorphisms. It follows that $\mathrm{Ext}^1_E(E/I, E) = 0$.

Corollary 1.3. *If* $I \in \mathscr{J}$, *then the left annihilator of* I *is zero and* I *is essential in* E *as a right ideal.*

Proof. The left annihilator of I is isomorphic to $\mathrm{Hom}_E(E/I, E)$, which is zero by Proposition 1.2 above. Let J be a right ideal of E and suppose $I \in \mathscr{J}$ such that $I \cap J = 0$. Then $I \oplus J \in \mathscr{J}$, since $\mathscr{J}$ is a filter, and by Proposition 1.2 above, the sequence

$$0 = \mathrm{Hom}_E(E/I, E) \to \mathrm{Hom}_E((I \oplus J)/I, E) \to \mathrm{Ext}^1_E(E/(I \oplus J), E) = 0$$

is exact, implying $\mathrm{Hom}_E((I \oplus J)/I, E) = 0$. It follows immediately that $J = 0$.

If U is a generator in $\mathscr{M}_R$ and A is a right R-module then

$$\mathrm{Ext}^1_R(U, A) \otimes_E U = 0\,.$$

A more general statement is the following.

Proposition 1.4. *If* U *is a generator in* $\mathscr{C}$, $A \in \mathscr{C}$ *then* $\mathrm{Ext}^1_{\mathscr{C}}(U, A) \in \mathrm{Ker}\, T$.

Proof. Let $0 \to A \to Q \to C \to 0$ be exact in $\mathscr{C}$ with Q injective. This yields an exact sequence

$$0 \to \mathrm{Hom}_{\mathscr{C}}(U, A) \to \mathrm{Hom}_{\mathscr{C}}(U, Q) \to \mathrm{Hom}_{\mathscr{C}}(U, C) \to \mathrm{Ext}^1_{\mathscr{C}}(U, A) \to 0$$

of right E-modules. Since $T: \mathscr{M}_E \to \mathscr{C}$ is exact, and $T \circ S \sim 1_{\mathscr{C}}$, this leads to an exact commutative diagram

$$\begin{array}{ccccccccc} 0 \to & T \circ S(A) & \to & T \circ S(Q) & \to & T \circ S(C) & \to & T(\mathrm{Ext}^1_{\mathscr{C}}(U, A)) & \to 0 \\ & \downarrow & & \downarrow & & \downarrow & & \downarrow & \\ 0 \to & A & \to & Q & \to & C & \to & 0 & \end{array}$$

with the vertical arrows isomorphisms. In particular, $T(\mathrm{Ext}^1_{\mathscr{C}}(U, A)) = 0$.

Definition. *Let R be a ring, $U \in \mathcal{M}_R$. The trace ideal of U in the ring $E = \mathrm{Hom}_R(U, U)$ is the image of the trace map*

$$U \otimes_R \mathrm{Hom}_E(U, E) \to E : u \otimes f \to f(u).$$

The trace map is a two-sided E map, so the trace ideal is a two-sided ideal of E. Note that the trace ideal of U in E is also the image of the map

$$\sum_{f \in \mathrm{Hom}_E(U,E)} U \xrightarrow{\Sigma f} E.$$

If $\mathscr{C}$ is a full exact subcategory of $\mathcal{M}_R$, then the trace ideal of the generator U of $\mathscr{C}$ is defined. In this case, the trace ideal of U in E will be denoted by t.

Proposition 1.5. *Let $\mathscr{C}$ be a full, exact subcategory of $\mathcal{M}_R$ for some ring R. Then $t \subseteqq \bigcap_{I \in \mathscr{J}} I$.*

Proof. Let $f \in \mathrm{Hom}_E(U, E)$ and $I \in \mathscr{J}$. Then

$$f(U) = f(IU) = If(U) \subseteqq I.$$

Let $\Gamma = \bigcap_{I \in \mathscr{J}} I$. In the case $\mathscr{C} = \mathcal{M}_R$, it is shown below that $t \in \mathscr{J}$, so that $t = \Gamma$. In Section 4 it is shown that $t = \Gamma$ in the case $\mathscr{C}$ is the category of all Abelian p-groups for some prime p. It is not known whether it is always the case that $t = \Gamma$. If this should be true, it would permit a generalization of the trace ideal to the case of a generator in an Abelian category with exact direct limits. The following proposition gives a faint hint that these two ideals might be the same.

Proposition 1.6. *The ideal $\Gamma = \bigcap_{I \in \mathscr{J}} I$ is a two-sided ideal of E.*

Proof. Let $x \in \Gamma$, $e \in E$, $I \in \mathscr{J}$. Then $I : e \in \mathscr{J}$ implies $x \in I : e$, so that $ex \in I$.

Theorem 1.7. *If U is a generator in $\mathcal{M}_R$, then the trace ideal t of U in E has the following properties.*

i. $tU = U$. *(Thus $E/I \in \mathrm{Ker}\, T$ if and only if $I \supset t$.)*
ii. $t^2 = t$.
iii. *The left annihilator of t is* 0.
iv. *t is finitely generated as a two-sided ideal.*
v. *t is essential as a right ideal.*

Proof. By the Dual Basis Lemma, there exist maps

$$f_1, \ldots, f_n \in \mathrm{Hom}_E(U, E)$$

and elements $u_1, \ldots, u_n \in U$ such that for any $u \in U$,

$$u = \sum_{i=1}^{n} f_i(u)(u_i).$$

But Im $f_i \subseteqq t$ for each i, so $u \in tU$ for any $u \in U$. Thus $tU = U$. The remainder of i. follows from Propositions 1.1 and 1.5. Now iii. and v. follow from Corollary 1.3, since $t \in \mathscr{J}$. Parts ii. and iv. appear in [*1*]. However, $t^2 = t$ follows immediately from the fact that $t^2 \in \mathscr{J}$ and $t \subseteqq \bigcap_{I \in \mathscr{J}} I$.

To restate some of the results above, if U is a generator in $\mathscr{M}_R$ and t is the trace ideal of U in E, then for $M \in \mathscr{M}_E$, $M \otimes_E U = 0$ if and only if $Mt = 0$, and $\mathscr{M}_R$ is isomorphic to the category $\mathscr{M}_E$ modulo the class of modules M such that $Mt = 0$.

It is interesting to note that U is a generator in $\mathscr{M}_E$ if and only if T is an equivalence $\mathscr{M}_E \to \mathscr{M}_R$.

Some of the members of $\mathscr{J}$ can be described in terms of families of sub-objects of U, and in the special case $\mathscr{C} = \mathscr{M}_R$ and U is projective the trace ideal can be described in terms of the set of finitely generated sub-modules of U. This is shown in the next two propositions.

Proposition 1.8. *Let γ be a set of subobjects of U which generate U and let I_γ be the right ideal of E generated by $\{f \in E \mid f(U) \subset S$ for some $S \in \gamma\}$. Then $I_\gamma \in \mathscr{J}$.*

Proof. Since U is a generator,

$$\sum_{f(U) \subset S} U \xrightarrow{\Sigma f} S$$

is an epimorphism for each S, and by the hypothesis, the inclusion maps $S \to U$ induce an epimorphism

$$\sum_{S \in \gamma} S \to U.$$

The composition of these maps

$$\sum_{S \in \gamma} \Big(\sum_{f(U) \subset S} U \Big) \xrightarrow{\Sigma f} \sum_{S \in \gamma} S \to U$$

is an epimorphism with image $I_\gamma U$. Thus $I_\gamma \in \mathscr{J}$.

This proposition, when U is an R-module, says the set of all endomorphisms which carry U into cyclic submodules of U generate a member of $\mathscr{J}$. Also the set of endomorphisms which carry U into finitely generated submodules of U generate a member of $\mathscr{J}$.

Proposition 1.9. *Suppose U is a projective generator in $\mathscr{M}_R$. Let Λ be* *the right ideal of E generated by $\{f \in E \mid f(U) \subset$ cyclic submodule of $U\}$ and $\Omega = \{f \in E \mid f(U) \subset$ finitely generated submodule of $U\}$. Then $t = \Lambda = \Omega$.*

Proof. From previous propositions, $t \subset \Lambda$. Clearly $\Lambda \subset \Omega$. Suppose $f \in E$ with $f(U) \subset uR$ for some $u \in U$. Since U is projective and $R \to uR: r \to ur$ is an epimorphism, there exists a homomorphism $g: U \to R$ such that $ug(x) = f(x)$ for all $x \in U$. Define $\sigma: U \to E$ by $\sigma(y)(x) = yg(x)$ for $x, y \in U$. Then for $e \in E$, $\sigma(e(y))(x) = e(y)\,g(x) = e(yg(x)) = e(\sigma(y)(x))$. It follows easily that σ is a left E homomorphism. Then $f = \sigma(u)$, implying $f \in t$. Thus $\Lambda \subset t$, so $\Lambda = t$.

Let $f \in \Omega$, $f(U) \subset u_1 R + \cdots + u_n R$. Since U is projective, f can be lifted to a map $g: U \to u_1 R \oplus \cdots \oplus u_n R$. Let e_i be the composition of g with the i-th projection. Then $f = e_1 + \cdots + e_n \in \Lambda$, since $e_i \in \Lambda$ for each i.

Corollary 1.10. *Let R be a division ring, $U \in \mathcal{M}_R$, $E = \mathrm{Hom}_R(U, U)$. Let I be the minimum two-sided ideal $\{e \in E \mid \dim e(U) < \aleph_0\}$, and let $\mathcal{S}$ be the subcategory $\{M \in \mathcal{M}_E \mid MI = 0\}$. Then there exists an equivalence*

$$\mathcal{M}_R \sim \mathcal{M}_E / \mathcal{S}.$$

Proof. By Proposition 1.9 above, $I = t$, and an application of Theorem 1.7 concludes the proof.

2. Endomorphism rings of proper generators

A ring can be the endomorphism ring of generators U und U' of Abelian categories $\mathcal{C}$ and $\mathcal{C}'$ with exact direct limits without $\mathcal{C}$ and $\mathcal{C}'$ being equivalent. For example, if R is a division ring and U is an infinite dimensional vector space in $\mathcal{M}_R$, then its endomorphism ring E is also the endomorphism ring of the generator $E \in \mathcal{M}_E$. However, $\mathcal{M}_R$ and $\mathcal{M}_E$ are not equivalent. The point is that E is not a proper generator in $\mathcal{M}_E$. The main objective here is to show that if U und U' are proper generators of Abelian categories $\mathcal{C}$ and $\mathcal{C}'$ with exact direct limits, and if U and U' have isomorphic endomorphism rings then $\mathcal{C}$ and $\mathcal{C}'$ are equivalent.

Lemma 2.1. *Let R be a ring and let $\mathcal{I}$ be the set of right ideals I of R such that the inclusion $I \to R$ induces an isomorphism $Hom_R(R, R) \to Hom_R(I, R)$. If $\mathcal{I}$ is an idempotent topologizing filter then R is a proper generator in the quotient category $\mathcal{D} = \mathcal{M}_R / \mathcal{S}(\mathcal{I})$.*

Proof. Let $\bar{f}: M \to R$ be an inclusion map in $\mathcal{D}$ which induces an isomorphism $\mathrm{Hom}_{\mathcal{D}}(R, R) \to \mathrm{Hom}_{\mathcal{D}}(M, R)$. Let $f: M' \to R/S$ be a representative of this map in $\mathcal{M}_R$, where $M/M' \in \mathcal{S}(\mathcal{I})$ and $S \in \mathcal{S}(\mathcal{I})$. Since $\mathrm{Hom}_R(R/I, R) = 0$ for all $I \in \mathcal{I}$, R has no non-zero submodules in $\mathcal{S}(\mathcal{I})$, so $S = 0$. Let $J = f(M')$. Then $M \approx R$ in $\mathcal{D}$ if and only if

$J \in \mathscr{I}$, for since f is a monomorphism, Ker $f \in \mathscr{S}$ and $M \approx J$ in $\mathscr{D}$. Consider the exact commutative diagram

$$\begin{array}{ccccccccc} 0 \to \mathrm{Hom}_R(R/J, R) & \to & \mathrm{Hom}_R(R, R) & \to & \mathrm{Hom}_R(J, R) & \to & \mathrm{Ext}^1_R(R/J, R) \to 0 \\ & & \downarrow & & \downarrow & & \\ 0 \quad \to & & \mathrm{Hom}_{\mathscr{D}}(R, R) & \to & \mathrm{Hom}_{\mathscr{D}}(J, R) \to 0 . & & \end{array}$$

The map $\mathrm{Hom}_R(R,R) \to \mathrm{Hom}_{\mathscr{D}}(R,R)$ is an isomorphism, from the definition of $\mathscr{I}$, so that $\mathrm{Hom}_R(R/J,R) = 0$. Since the diagram commutes, the other vertical arrow is an isomorphism so $\mathrm{Ext}^1_R(R/J,R) = 0$ and $J \in \mathscr{I}$.

Proposition 2.2. *Let $\mathscr{I}(E)$ be the set of right ideals of E such that $\mathrm{Hom}_E(E/I,E) = 0$ and $\mathrm{Ext}^1_E(E/I,E) = 0$. Then U is a proper generator in $\mathscr{C}$ if and only if $\mathscr{I}(E) = \mathscr{J}$.*

Proof. If $\mathscr{I}(E) = \mathscr{J}$, Lemma 2.1 above shows that U is a proper generator in $\mathscr{C}$ (using the equivalence $\mathscr{M}_E/\mathscr{S}(\mathscr{J}) \sim \mathscr{C}$ of Gabriel-Popesco's Theorem (*)).

Suppose U is a proper generator in $\mathscr{C}$. If I is a right ideal of E, the sequences

$$0 \to T(I) \to T(E) \to T(E/I) \to 0$$

and

$$0 \to IU \to U \to U/IU \to 0$$

are equivalent, by Proposition 1.1. Thus the adjoint maps lead to a commutative diagram

$$\begin{array}{ccccccc} 0 \to \mathrm{Hom}_{\mathscr{C}}(U/IU, U) & \to & \mathrm{Hom}_{\mathscr{C}}(U, U) & \to & \mathrm{Hom}_{\mathscr{C}}(IU, U) & & \\ \downarrow & & \downarrow & & \downarrow & & \\ 0 \to \mathrm{Hom}_E(E/I, E) & \to & \mathrm{Hom}_E(E, E) & \to & \mathrm{Hom}_E(I, E) & \to & \mathrm{Ext}^1_E(E/I, E) \to 0 \end{array}$$

with the vertical maps isomorphisms and the rows exact. Hence $I \in \mathscr{I}(E)$ if and only if $\mathrm{Hom}_{\mathscr{C}}(U,U) \to \mathrm{Hom}_{\mathscr{C}}(IU,U)$ is an isomorphism, and this map is an isomorphism if and only if $I \in \mathscr{J}$ (since U is proper).

From the two previous propositions one has immediately

Corollary 2.3. *A ring R is the endomorphism ring of a proper generator of an Abelian category with exact direct limits if and only if the set of right ideals I of R for which the map $\mathrm{Hom}_R(R,R) \to \mathrm{Hom}_R(I,R)$ is an isomorphism is an idempotent topologizing filter.*

If R is self-injective it is easy to see that $\mathscr{I}(R)$ is an idempotent topologizing filter. In this case, $\mathscr{I}(R)$ is the set of right ideals I such that $\mathrm{Hom}_R(R/I,R) = 0$. The next proposition allows us to describe other classes of rings which are included in $\mathscr{E}$, the class of all rings which are endomorphism rings of proper generators.

Proposition 2.4. *Let R be a ring. If $I \in \mathscr{I}(R)$ implies $I:x \in \mathscr{I}(R)$ for all $x \in R$, then $\mathscr{I}(R)$ is an idempotent topologizing filter.*

Proof. If $I \in \mathscr{I} = \mathscr{I}(R)$ and $J \supset I$, J a right ideal of R, the exact sequence $R/I \to R/J \to 0$ induces an exact sequence $0 \to \mathrm{Hom}_R(R/J, R) \to \mathrm{Hom}_R(R/I, R) = 0$. Thus $\mathrm{Hom}_R(R/J, R) = 0$.

Let $I \in \mathscr{I}$ and suppose J is a right ideal of R such that $J:x \in \mathscr{I}$ for all $x \in I$. Since $R/J:x \approx (xR + J)/J$, $\mathrm{Hom}_R((xR + J)/J, R) = 0$ for all $x \in I$. Thus $\mathrm{Hom}_R((I + J)/J, R) = 0$. The exact sequence $0 \to (I + J)/J \to R/J \to R/(I + J) \to 0$ induces an exact sequence

$$\mathrm{Hom}_R(R/(I + J), R) \to \mathrm{Hom}_R(R/J, R) \to \mathrm{Hom}_R((I + J)/J, R) = 0$$

But the first term is also zero by the argument of the preceding paragraph, so $\mathrm{Hom}_R(R/J, R) = 0$.

Let $I \in \mathscr{I}$ and $J \supset I$. The exact sequence $0 \to J/I \to R/I \to R/J \to 0$ induces the exact sequence

$$0 = \mathrm{Hom}_R(R/I, R) \to \mathrm{Hom}_R(J/I, R) \to \mathrm{Ext}^1_R(R/J, R) \to 0,$$

and since for $x \in J$, $I:x \in \mathscr{I}$ and

$$R/(I:x) \approx (xR + I)/I, \quad \mathrm{Hom}_R((xR + I)/I, R) = 0$$

for all $x \in J$. It follows that $0 = \mathrm{Hom}_R(J/I, R)$ and hence $0 = \mathrm{Ext}^1_R(R/J, R)$.

Let $I, J \in \mathscr{I}$. The exact sequence

$$0 \to (I + J)/J \to R/(I \cap J) \to R/I \to 0$$

yields an exact sequence

$$0 = \mathrm{Hom}_R(R/I, R) \to \mathrm{Hom}_R(R/(I \cap J), R) \to$$
$$\to \mathrm{Hom}_R((I + J)/J, R) \to \mathrm{Ext}^1_R(R/I, R) = 0$$

and the exact sequence

$$0 \to (I + J)/J \to R/J \to R/(I + J) \to 0$$

yields an exact sequence

$$0 = \mathrm{Hom}_R(R/J, R) \to \mathrm{Hom}_R((I + J)/J, R) \to \mathrm{Ext}^1_R(R/(I + J), R)$$
$$\to \mathrm{Ext}^1_R(R/J, R) = 0.$$

By the previous paragraph, $\mathrm{Ext}^1_R(R/(I + J), R) = 0$, so that

$$\mathrm{Hom}_R(R/(I \cap J), R) = 0.$$

The exact sequence

$$0 \to (I + J)/J \to R/(I \cap J) \to R/I \to 0$$

yields an exact sequence

$$0 = \mathrm{Ext}^1_R(R/I, R) \to \mathrm{Ext}^1_R(R/(I \cap J), R) \to \mathrm{Ext}^1_R((I + J)/J, R).$$

Now for

$x \in R,\ J:x \in \mathscr{I}$ so $\operatorname{Ext}^1_R(R/(J:x), R) = 0$, so $\operatorname{Ext}^1_R((xR + J)/J, R) = 0$.

Now consider the exact sequence

$$0 \to K \to \sum_{x \in I} (xR + J)/J \to (I + J)/J \to 0$$

where the maps $(xR + J)/J \to (I + J)/J$ are the inclusion maps. This yields an exact sequence

$$0 = \operatorname{Hom}_R(K, R) \to \operatorname{Ext}^1_R((I + J)/J, R) \to \operatorname{Ext}^1_R\Big(\sum_{x \in I} (xR + J)/J, R\Big) \approx$$

$$\approx \prod_{x \in I} \operatorname{Ext}^1_R((xR + J)/J, R) = 0 .$$

Thus $\operatorname{Ext}^1_R(R/(I \cap J), R) = 0$. Let $I \in \mathscr{I}$ and suppose $J:x \in \mathscr{I}$ for all $x \in I$, and some right ideal J of R. As in the above paragraph,

$$\operatorname{Ext}^1_R((I + J)/J, R) = 0 .$$

The exact sequence

$$0 \to ((I + J)/J \to R/J \to R/(I + J) \to 0$$

leads to an exact sequence

$$\operatorname{Ext}^1_R(R/(I + J), R) \to \operatorname{Ext}^1_R(R/J, R) \to 0 = \operatorname{Ext}^1_R(R/(I + J), R) .$$

But the first term is also zero, by the third paragraph of this proof. Thus $\operatorname{Ext}^1_R(R/J, R) = 0$.

Remark. Suppose $I \in \mathscr{I}$, $J \supset I$ imply $\operatorname{Ext}^1_R(R/J, R) = 0$. Let $I \in \mathscr{I}$, $x \in R$. The exact sequence

$$0 \to R/(I:x) \to R/I \to R/(xR + I) \to 0$$

yields an exact sequence

$$0 \to \operatorname{Hom}_R(R/I:x, R) \to \operatorname{Ext}^1_R(R/(xR + I), R) = 0 .$$

It follows that the hypothesis

"$I \in \mathscr{I}$, $J \supset I$, $x \in R$ imply $\operatorname{Ext}^1_R(R/J, R) = 0$ and $\operatorname{Ext}^1_R(R/I:x, R) = 0$"

would also imply that $\mathscr{I}(R)$ is an idempotent topologizing filter.

It is interesting to note that if $\mathscr{I}(R)$ is a filter then every $I \in \mathscr{I}(R)$ is an essential right ideal of R. To see this, let $I \in \mathscr{I}(R)$ and suppose J is a right ideal of R with $I \cap J = 0$. The exact sequence $0 \to J \to R/I \to R/(I \oplus J) \to 0$ yields an exact sequence

$$0 = \operatorname{Hom}_R(R/I, R) \to \operatorname{Hom}_R(J, R) \to \operatorname{Ext}^1_R(R/(I \oplus J), R) .$$

If $\mathscr{I}(R)$ is a filter then $I \oplus J \in \mathscr{I}(R)$, implying $\operatorname{Ext}^1_R(R/(I \oplus J), R) = 0$.

It follows that $\mathrm{Hom}_R(J, R) = 0$, which can happen only if $J = 0$.

Theorem 2.5. *If R is a commutative ring then R is the endomorphism ring of a proper generator.*

Proof. Let $I \in \mathscr{I}(R)$, $x \in R$. Since $I{:}x \supset I$, $\mathrm{Hom}(R/I{:}x, R) = 0$. For $y \in I{:}x/I$, $(yR + I)/I \approx R/(I{:}y)$, so $\mathrm{Hom}_R((yR + I)/I, R) = 0$. It follows that $\mathrm{Hom}_R((I{:}x)/I, R) = 0$. Now the exact sequence

$$0 \to ((I{:}x)/I) \to R/I \to (R/(I{:}x)) \to 0$$

yields the exact sequence

$$0 = \mathrm{Hom}_R((I{:}x)/I, R) \to \mathrm{Ext}^1_R(R/I{:}x, R) \to \mathrm{Ext}^1_R(R/I, R) = 0\,.$$

Thus $\mathrm{Ext}^1_R(R/(I{:}x), R) = 0$.

Proposition 2.6. If R is a principal ideal domain then R is a proper generator in $\mathscr{M}_R$.

Proof. Let I be an ideal of R and suppose $0 \neq I \neq R$. Let Q be the quotient field of R. Since R is a principal ideal domain and $I \neq R$, $I = xR$ with $x^{-1} \notin R$. Now $f: I \to R : xr \to r$ is an R-homomorphism which cannot be extended to an endomorphism of R, so $\mathrm{Hom}_R(R, R) \to \mathrm{Hom}_R(I, R)$ is not an isomorphism. It follows that $\mathscr{I}(R) = \{R\}$, and R is a proper generator in $\mathscr{M}_R$.

Example. Let F be a field, $R = F[x,y]$ and let I be the ideal in R generated by x and y. Any homomorphism $I \to R$ is a multiplication by an element p/q of the quotient field, where $p,q \in R$. If $(p/q)\,x = r$ and $(p/q)y = s$ with $r,s \in R$, then $px = rq$ and $py = sq$. It may be assumed that x does not divide both p and q. Suppose x does not divide p. Then x does not divide py, so x does not divide q. Thus, in any case, x does not divide q. But x does divide rq, hence x divides r, and $p/q \in R$. Thus any homomorphism $I \to R$ is a multiplication by an element of R. It follows that $\mathrm{Ext}^1_R(R/I, R) = 0$. Also $\mathrm{Hom}_R(R/I, R) = 0$ since R is an integral domain, so $I \in \mathscr{I}(R)$, $I \neq R$. Thus *not every integral domain R is a proper generator in $\mathscr{M}_R$.*

Proposition 2.7. If R is a hereditary ring then R is the endomorphism ring of a proper generator if and only if $I \in \mathscr{I}(R)$, $J \supseteq I$ imply $J \in \mathscr{I}(R)$.

Proof. Let $I \in \mathscr{I}(R)$, $x \in R$. The exact sequence

$$0 \to R/(I : x) \to R/I \to R/(xR + I) \to 0$$

yields an exact sequence

$$\mathrm{Hom}_R(R/I, R) = 0 \to \mathrm{Hom}_R(R/(I : x), R) \to \mathrm{Ext}^1_R(R/(xR + I), R) \to$$

$$\mathrm{Ext}^1_R(R/I, R) = 0 \to \mathrm{Ext}^1_R(R/(I:x), R) \to \mathrm{Ext}^2_R(R/(xR + I), R) = 0\,.$$

Since $xR + I \supset I$, if one assumes $\mathscr{I}(R)$ includes every right ideal of R which contains some member of $\mathscr{I}(R)$, $\mathrm{Ext}^1_R(R/(xR + I), R) = 0$. Thus every term of the above sequence is zero and $I : x \in \mathscr{I}(R)$. By Proposition 2.4, $\mathscr{I}(R)$ is an idempotent topologizing filter. Apply Corollary 2.3. The converse also follows from Corollary 2.3.

Theorem 2.8. *Let $\mathscr{C}$ and $\mathscr{C}'$ be Abelian categories with exact direct limits, and let U and U' be proper generators of $\mathscr{C}$ and $\mathscr{C}'$, respectively. If the rings* $\mathrm{Hom}_{\mathscr{C}}(U, U)$ *and* $\mathrm{Hom}_{\mathscr{C}'}(U', U')$ *are isomorphic, then there exists a categorical equivalence $F : \mathscr{C} \to \mathscr{C}'$ such that $F(U) = U'$.*

Proof. By Proposition 2.2, $\mathscr{I} = \mathscr{I}(E)$, so by *Gabriel and Popesco's theorem* (*), T induces an equivalence $\mathscr{M}_E/\mathscr{S}(\mathscr{I}(E)) \to \mathscr{C}$. Let T^{-1} denote the inverse of this equivalence. Then $T^{-1}(U) = E$. Similarly there is an equivalence $T' : \mathscr{M}_{E'}/\mathscr{S}(\mathscr{I}(E')) \to \mathscr{C}'$ with $T'(E') = U'$. The isomorphism $E \approx E'$ induces an equivalence $G : \mathscr{M}_E \sim \mathscr{M}_{E'}$ with $G(E) = E'$. Now $I \in \mathscr{I}(E)$ if and only if $G(I) \in \mathscr{I}(E')$, for since G is exact, $E'/G(I) \approx G(E/I)$, so that $\mathrm{Hom}_E(E/I, E) \approx \mathrm{Hom}_{E'}(E'/G(I), E')$ and

$$\mathrm{Ext}^1_E(E/I, E) \approx \mathrm{Ext}^1_{E'}(E'/G(I), E')\,.$$

Thus G induces an equivalence $G' : \mathscr{M}_E/\mathscr{S}(\mathscr{I}(E)) \sim \mathscr{M}_{E'}/\mathscr{S}(\mathscr{I}(E'))$ with $G'(E) = E'$. Now the composition $T' \circ G' \circ T^{-1}$ yields the desired equivalence.

3. Injectives and Injective Envelopes

Here the behavior of injectives under the action of the functors $S : \mathscr{C} \to \mathscr{M}_E$ and $T : \mathscr{M}_E \to \mathscr{C}$ is determined. Briefly, S preserves injective envelopes (Corollary 3.3), and if $\bar{X}$ denotes the injective envelope of X, then $T(\overline{S(X)}) = \bar{X}$ (Proposition 3.4). It follows that E is right self-injective if and only if U is injective (Theorem 3.5).

Proposition 3.1. *If $A \in \mathscr{C}$ is injective, then $S(A) = \mathrm{Hom}_{\mathscr{C}}(U, A)$ is an injective right E-module.*

Proof. Let

$$\begin{array}{ccc} 0 \to M & \xrightarrow{i} & N \\ \downarrow f & & \\ S(A) & & \end{array}$$

be an exact diagram in $\mathscr{M}_E$. This induces an exact diagram

$$\begin{array}{ccc} 0 \to T(M) & \xrightarrow{T(i)} & T(N) \\ \downarrow T(f) & & \downarrow g \\ TS(A) & \xrightarrow{\Phi_A} & A \end{array}$$

in $\mathscr{C}$, which can be completed since A is injective. Let

$$\Phi(X, Y) : \mathrm{Hom}_{\mathscr{C}}(T(X), Y) \to \mathrm{Hom}_E(X, S(Y))$$

denote the isomorphism associated with the adjoint functors S and T. Then $\Phi : TS \to 1$ by $\Phi_X = \varphi(S(Y), Y)^{-1}(1_{S(Y)})$ for $Y \in \mathscr{C}$. Let $\Psi : 1_{\mathscr{M}_E} \to ST$ be the natural transformation given by

$$\Psi_X = \varphi(X, T(X))(1_{T(X)})$$

for $X \in \mathscr{M}_E$. We have the following diagram in $\mathscr{M}_E$.

$$\begin{array}{ccccc} 0 & \to & M & \xrightarrow{i} & N \\ & & \downarrow f & \swarrow & S(g) \circ \Psi_N \\ & & S(A) & & \end{array}$$

It remains to show that it commutes. We will refer to the following commutative diagram.

$$\begin{array}{ccc}
\mathrm{Hom}_{\mathscr{C}}(TS(A), A) & \xrightarrow{\varphi(S(A), A)} & \mathrm{Hom}_E(S(A), S(A)) \\
\downarrow \mathrm{Hom}_{\mathscr{C}}(T(f), A) & & \downarrow \mathrm{Hom}_E(f, S(A)) \\
\mathrm{Hom}_{\mathscr{C}}(T(M), A) & \xrightarrow{\varphi(M, A)} & \mathrm{Hom}_E(M, S(A)) \\
\uparrow \mathrm{Hom}_{\mathscr{C}}(T(i), A) & & \uparrow \mathrm{Hom}_E(i, S(A)) \\
\mathrm{Hom}_{\mathscr{C}}(T(N), A) & \xrightarrow{\varphi(N, A)} & \mathrm{Hom}_E(N, S(A)) \\
\uparrow \mathrm{Hom}_{\mathscr{C}}(T(N), g) & & \uparrow \mathrm{Hom}_E(N, S(g)) \\
\mathrm{Hom}_{\mathscr{C}}(T(N), T(N)) & \xrightarrow{\varphi(N, T(N))} & \mathrm{Hom}_E(N, ST(N))
\end{array}$$

Now

$$\begin{aligned}
&\mathrm{Hom}_E(f, S(A))\,\varphi(S(A), A)(\Phi_A) = f = \varphi(M, A)\,\mathrm{Hom}_{\mathscr{C}}(T(f), A)(\Phi_A) \\
&\quad = \varphi(M,A)(\Phi_A T(f)) = \varphi(M,A)(g\,T(i)) = \varphi(M,A)\,\mathrm{Hom}_{\mathscr{C}}(T(i),A)(g) \\
&\quad = \mathrm{Hom}_E(i, S(A))\,\varphi(N, A)(g) = \varphi(N, A)(g)\,i = \\
&\quad = \varphi(N, A)(\mathrm{Hom}_{\mathscr{C}}(T(N), g)(1_{T(N)}))\,i \\
&\quad = \mathrm{Hom}_E(N, S(g))(\varphi(N, T(N))(1_{T(N)}))\,i = S(g)\,\Psi_N\,i,
\end{aligned}$$

concluding the proof.

Proposition 3.2. *If A is an essential extension of B in $\mathscr{C}$ then $S(A)$ is an essential extension of $S(B)$ in $\mathscr{M}_E$.*

Proof. Let $f \in S(A) = \mathrm{Hom}_{\mathscr{C}}(U, A)$ with $f \neq 0$. Let $K \xrightarrow{k} U$ be a kernel of the composition $U \xrightarrow{f} A \to A/B$ (i.e., $K = f^{-1}(B)$). Since A is an essential extension of B, $K \neq 0$. Thus there is a map $g \neq 0$, $g : U \to U$

with $\operatorname{Im} g \subset K$. Now $0 \neq fg : U \to B$. Thus $0 \neq f \in S(A)$ implies $0 \neq f(E) \cap S(B)$, and $S(B)$ is essential in $S(A)$.

Corollary 3.3. *The functor S preserves injective envelopes.*

Proposition 3.4. *If $M \in \mathcal{M}_E$ is isomorphic to $S(X)$ for some $X \in \mathcal{C}$, and $\bar{M}$ is an injective envelope of M in $\mathcal{M}_E$, then $T(\bar{M})$ is an injective envelope of $T(M)$ in $\mathcal{C}$.*

Proof. By the previous propositions, if A is an injective envelope of $T(M)$ in $\mathcal{C}$, then $S(A)$ is an injective envelope of $ST(M)$ in $\mathcal{M}_E$. Now since $M \approx S(X)$, $ST(M) \approx STS(X) \approx S(X) \approx M$. Thus $S(A) \approx \bar{M}$, and $T(\bar{M}) \approx TS(A) \approx A$ is an injective envelope of $T(M)$ in $\mathcal{C}$.

If U is a generator in $\mathcal{M}_R$ for some ring R, the previous proposition says that $\bar{E} \otimes_E U$ is an injective envelope of U in $\mathcal{M}_R$.

Theorem 3.5. *The ring E is right self-injective if and only if U is injective in $\mathcal{C}$.*

Proof. If E is right self-injective then $U \approx T(E) = T(\bar{E}) = \bar{U}$ by proposition 3.4.

Assume U is injective in $\mathcal{C}$. Then by Corollary 3.3, $E \approx S(U) = S(\bar{U}) \approx \bar{E}$ so E is right self-injective.

Corollary 3.6. *If U is a generator in $\mathcal{M}_R$, and if E is right self-injective, then the ring R is right self-injective.*

Proof. By the theorem above, U is injective. But R is isomorphic to a summand of a finite sum of copies of U.

4. The Category of Abelian p-groups

The category of Abelian p-groups is Abelian with a generator and exact direct limits. If U is a generator in this category, the trace ideal of U in E is 0 (Theorem 4.2) and U is flat as a module over E (Theorem 4.1). It is not known which generators are proper, but some information about them is provided in 4.4, 4.5 and 4.6. The reader is referred to [2] for the group theoretical facts used in the sequel. The category of Abelian p-groups will be denoted by $\mathcal{C}$.

Theorem 4.1. *Let G be an Abelian p-group with maximum divisible subgroup D. If G/D is unbounded, then G is flat as a module over its endomorphism ring.*

Proof. It is easy to see that the category $\mathcal{C}$ of Abelian p-groups is Abelian with exact direct limits. Also, any $G \in \mathcal{C}$ with G/D unbounded is a generator. Let $E = \mathrm{Hom}_{\mathscr{C}}(G, G)$. It is easy to see that the adjoint of the functor $\mathrm{Hom}_{\mathscr{C}}(G, \cdot): \mathscr{C} \to \mathscr{M}_E$ is just $(\cdot \otimes_E G): \mathscr{M}_E \to \mathscr{C}$. By *Theorem* (*), this functor is exact. In other words, G is flat as a module over E.

A classical result is that any Abelian p-group is determined by its endomorphism ring. The usual construction of a p-group G from its endomorphism ring gives no hint that if G/D is unbounded, then G is E-flat. However 4.1 suggests that if G/D is unbounded, then G may be constructable from E as a direct limit of projectives, showing at once that G is determined by E and is E-flat. This is indeed the case, and that construction has been carried out in another paper [*6*].

Proposition 4.2. *Let $\mathscr{C}$ be the category of Abelian p-groups for some prime p, U a generator in $\mathscr{C}$. Let t be the trace ideal of U in* $E = \mathrm{Hom}_{\mathscr{C}}(U, U)$, *and let* $\Gamma = \cap \{I \mid I$ *is a right ideal of E,* $IU = U\}$. *Then* $t = \Gamma = 0$.

Proof. Let $x \in U$, $x \neq 0$, $o(x) = p^n$ and let m be any positive integer. Since U is a generator, U has an unbounded basic subgroup, so U has a cyclic summand Zy of order $\geqq p^{n+m}$. If $U = Zy \oplus H$, let $\alpha: U \to U$ be the homomorphism defined by $\alpha(ry + h) = rx$ for $r \in Z$, $h \in H$. Write $r = p^k t$ with $(t, p) = 1$. If $rx \neq 0$ then $k < n$ and $o(rx) = p^{n-k}$. Now $o(ry + h) \geqq o(ry) = p^{n+m-k} > p^{n-k} = o(rx) = o(\alpha(ry + h))$. In fact for $u \in U$ with $\alpha(u) \neq 0$, $u = p^k ty + h$ and

$$o(u) - o(\alpha(u)) \geqq p^{n+m-k} - p^{n-k} = p^{n-k}(p^m - 1) \geqq p(p^m - 1).$$

Now let $I(m) = \{e \in E \mid o(u) - o(e(u)) \geqq p(p^m - 1) \text{ if } e(u) \neq 0\}$. Then $I(m)$ is a right ideal of E, for $e \in E$, $\alpha \in I(m)$ and $\alpha e(u) \neq 0$ then $o(u) - o(\alpha e(u)) \geqq o(e(u)) - o(\alpha e(u)) \geqq p(p^m - 1)$. From the first paragraph it is clear that $I(m)U = U$ for any $m > 0$. Thus

$$t \subset \Gamma \subset \bigcap_{m=1}^{\infty} I(m) = 0.$$

Corollary 4.3. $\mathrm{Hom}_E(U, E) = 0$.

The remainder of this section is an attempt to determine the proper generators in the category of Abelian p-groups. First it is shown that not every generator is proper. A p-group is torsion-complete if it is the torsion-subgroup of its closure in the p-adic topology.

Proposition 4.4. *Let G be an unbounded torsion-complete Abelian p-group. Then G is a generator in the category of Abelian p-groups, but is not proper.*

Proof. The group G is a generator since it is unbounded and reduced. Let B be a basic subgroup of G. The pure exact sequence

$$0 \to B \to G \to G/B \to 0$$

yields the exact sequence

$$0 \to \mathrm{Hom}(G/B, G) \to \mathrm{Hom}(G, G) \to \mathrm{Hom}(B, G) \to \mathrm{Pext}(G/B, G).$$

($\mathrm{Pext}(G/B, G)$ is the group of pure extensions of G by G/B. See [5].) Now $\mathrm{Hom}(G/B, G) = 0$ since G/B is divisible and G is reduced. Also $\mathrm{Pext}(G/B, G) = 0$ since G is torsion-complete. Then $\mathrm{Hom}(G, G) \to \mathrm{Hom}(B, G)$ is an isomorphism. But $B \neq G$ since B is a direct sum of cyclic groups and G is not. Hence G is not proper.

Proposition 4.5. *Every non-reduced generator in the category of Abelian p-groups is a proper generator.*

Proof. A non-reduced p-group G has $Z(p^\infty)$ as a summand, and $Z(p^\infty)$ is a co-generator, whence G is proper.

Theorem 4.6. *Let G be a generator in the category of Abelian p-groups. If* $\mathrm{Pext}(G, G) = 0$ *and G is not torsion-complete, then G is proper.*

Proof. By 4.4, G may be assumed to be reduced. Suppose $H \subset G$, $H \neq G$, and $\mathrm{Hom}(G, G) \xrightarrow{\Phi} \mathrm{Hom}(H, G)$ is an isomorphism. Then G/H is divisible, since otherwise $\mathrm{Hom}(G/H, G) \neq 0$ and Φ would not be a monomorphism. Assume that H is not pure in G. Then no basic subgroup of H is pure in G, and so H has a cyclic summand, generated by h, say, that is not a summand of G. If h is of order p^n, then $p^{n-1}h$ is of height at least n in G. That is, there is a $g \in G$ such that $p^n g = p^{n-1}h$. Project H onto the cyclic summand of H generated by h. Suppose this projection can be extended to an endomorphism α of G. Then the order of $\alpha(g)$ is p^{n+1} and so $\alpha(G)/\alpha(H) \neq 0$. However, since G/H is divisible, $\alpha(G)/\alpha(H)$ is divisible. But $\alpha(H)$ is finite and $\alpha(G)$ is reduced. This is an impossibility. Thus H may be assumed to be pure in G, and the pure exact sequence $0 \to H \to G \to G/H \to 0$ yields the exact sequence

$$\mathrm{Hom}(G, G) \xrightarrow{\Phi} \mathrm{Hom}(H, G) \to \mathrm{Pext}(G/H, G) \to \mathrm{Pext}(G, G) = 0.$$

Since G is not torsion-complete, $\mathrm{Pext}(G/H, G) \neq 0$, so Φ is not an isomorphism. This concludes the proof.

One justification for Theorem 4.5 is

Corollary 4.7. *Let G be a generator in the category of Abelian p-groups. If G is a direct sum of cyclic groups, then G is a proper generator.*

Proof. G is not torsion-complete, being an unbounded direct sum of cyclic groups. Since $\mathrm{Pext}(G, G) = 0$, 4.5 applies.

References

[1] Bass, H.: Categories of modules. (Unpublished.)

[2] Fuchs, L.: Abelian groups. Budapest: Publishing House of the Hungarian Academy of Sciences (1958).

[3] GABRIEL, P.: Des catégories abéliennes. Bull. Soc. Math. France **90**, 323—448 (1962).

[4] —, et N. POPESCO: Characterisation des catégories abéliennes avec générateurs et limites inductives exactes. C. R. Acad. Sc. Paris **258**, 4188—4190 (1964).

[5] HARRISON, D. K.: Infinite Abelian groups and homological methods. Annals of Math. **69**, 366—391 (1959).

[6] RICHMAN, F., and E. A. WALKER: Primary Abelian groups as modules over their endomorphism rings. Math. Zeitschr. **89**, 77—81 (1965).

[7] WALKER, C., and E. A. WALKER: Quotient categories and rings of quotients. Submitted for publication.

Institute for Advanced Study
Princeton, New Jersey

New Mexico State University
University Park, New Mexico

Closed Categories *

By

Samuel Eilenberg ** and G. Max Kelly

Introduction

In the usual theory of categories, with any two objects A, B of a category $\mathscr{A}$ there is associated a *set* $\mathscr{A}(A\,B)$ of morphisms of A into B. Frequently the set $\mathscr{A}(A\,B)$ is endowed with an additional structure such as a privileged element or an abelian group structure. It has become clear that as the ramifications of the theory of categories increase, the structures that $\mathscr{A}(A\,B)$ will carry will be richer and more complex. The need for a general theory has been widely felt for some time, and beginnings have been made in various directions and often under restrictive hypotheses; e.g. by MacLane [*15*], Kelly [*10*], Bénabou [*3*], Linton [*12*].

In order to gain sufficient generality one should assume that $\mathscr{A}(A\,B)$ is an object of some category $\mathscr{V}_0$, that this category $\mathscr{V}_0$ is equipped with a functor $V: \mathscr{V}_0 \to \mathscr{S}$ into the category $\mathscr{S}$ of sets, and that $V\mathscr{A}(A\,B)$ is the set of morphisms $A \to B$ in $\mathscr{A}$. One then can write $\mathscr{A}_0(A\,B)$ for $V\mathscr{A}(A\,B)$, and distinguish the "enriched category" $\mathscr{A}$ from the ordinary category $\mathscr{A}_0$ that underlies it. Upon inspection it turned out that the categories $\mathscr{V}_0$ which occur in this connexion are endowed with a structure considerably richer than that of a category. We propose calling these "closed categories", and we may best describe them by citing two examples.

Let $\mathscr{B}$ be the category of real or complex Banach spaces. In order to ensure that an isomorphism is an isometry we take the morphisms $f: A \to B$ to be the linear transformations with norm $\|f\| \leqq 1$; these then form the set $\mathscr{B}(A\,B)$. In addition however we may consider *all* the bounded linear transformations $A \to B$; these form in a natural fashion a Banach space $(A\,B)$. This yields an "internal Hom-functor" $\mathscr{B}^* \times \mathscr{B} \to \mathscr{B}$. The set $\mathscr{B}(A\,B)$ is obtained from the Banach space $(A\,B)$ by applying the functor $\mathscr{B} \to \mathscr{S}$ which to each Banach space assigns its unit ball considered as a set. In addition we have a special "unit"

---

* Received March 7, 1966.

** Supported by Office of Naval Research.

Banach space I, namely $\boldsymbol{R}$ or $\boldsymbol{C}$ as appropriate, and a natural isomorphism $i: A \cong (IA)$. There is also a composition law that will be discussed later.

As a second example consider a topological space X and let $\mathcal{S}h\,X$ be the category of sheaves of sets over X. For any two such sheaves A and B, we then have the set $\mathcal{S}h\,X(A\,B)$. Given any open subset U of X, we may also consider the set $\mathcal{S}h\,U(A\,|\,U,\,B\,|\,U)$ where $A\,|\,U$ is the restriction of the sheaf A to U. These sets form a pre-sheaf on X and define a sheaf that we shall denote by $(A\,B)$. This again yields an "internal Hom-functor". The set $\mathcal{S}h\,X(A\,B)$ is obtained from $(A\,B)$ by applying the functor $\Gamma: \mathcal{S}h\,X \to \mathcal{S}$ which to each sheaf assigns its set of sections. Again there is a privileged unit sheaf I and a natural isomorphism $i: A \cong (IA)$.

The basic elements of the structure of a closed category now become clear. First there is an ordinary category $\mathcal{V}_0$, represented by $\mathcal{B}$ or $\mathcal{S}h\,X$ in the examples above. Next there is a functor $V: \mathcal{V}_0 \to \mathcal{S}$. Then an internal Hom-functor $\mathcal{V}_0^* \times \mathcal{V}_0 \to \mathcal{V}_0$, denoted by (AB), and such that $V(AB)$ is the set $\mathcal{V}_0(AB)$ of morphisms $A \to B$. Further there is a unit object I and a natural isomorphism $i: A \cong (IA)$. What is still lacking is a composition law that generalizes the ordinary composition law in $\mathcal{V}_0$. The notion of composition is usually linked with a notion of "product". However the need of a product for defining composition is only superficial. Indeed if we consider an ordinary category $\mathcal{A}$ and for a fixed $A \in \mathcal{A}$ we wish to consider the left represented functor $L^A = \mathcal{A}(A-): \mathcal{A} \to \mathcal{S}$, then we must indicate the effect of L^A on morphisms; i.e. we must give a morphism

$$L^A_{BC}: \mathcal{A}(BC) \to \mathcal{S}(\mathcal{A}(AB), \mathcal{A}(AC));$$

and this morphism is nothing but the composition law

$$(L^A_{BC}f)g = fg\,.$$

Generalizing this we define the composition law in a closed category to be a morphism

$$L^A_{BC}: (BC) \to ((AB), (AC))\,.$$

This is the last needed primitive term for a closed category, and we denote the whole set of data $(\mathcal{V}_0, V, (AB), I, i, L)$, i.e. the closed category, by $\mathcal{V}$. There are five axioms, but as they involve a term j derived from the other terms, we preferred to include j as a primitive term and add a sixth axiom to fix its value (§ I.2).

In Chapter I we give a precise definition of closed category, and define the corresponding notions of closed functor and closed natural transformation. Then we consider for a closed category $\mathcal{V}$ the notion of a $\mathcal{V}$-category $\mathcal{A}$, i.e. a "category" whose Hom-functor has values $\mathcal{A}(AB)$

in $\mathscr{V}_0$. With such a $\mathscr{V}$-category $\mathscr{A}$ is associated an ordinary underlying category $\mathscr{A}_0$ with $\mathscr{A}_0(A\,B) = V\mathscr{A}(A\,B)$ as indicated above. There is a corresponding notion of $\mathscr{V}$-functor, and also of $\mathscr{V}$-natural transformation. The notations $\mathscr{V}, \mathscr{V}_0$ suggest that $\mathscr{V}$ itself is a $\mathscr{V}$-category with underlying category $\mathscr{V}_0$, which is indeed the case. Each object A of a $\mathscr{V}$-category $\mathscr{A}$ determines a "left represented" $\mathscr{V}$-functor $L^A: \mathscr{A} \to \mathscr{V}$, and this leads to the key representation theorem (Theorem I.8.6.) which is the generalization of the YONEDA theorem for ordinary categories. We here thank JOHN GRAY for impressing upon us the importance of such a theorem; actually the one he wanted was a still higher form which must await a later paper on functor categories. We note throughout the chapter various gross simplifications that ensue when the basic functor $V: \mathscr{V}_0 \to \mathscr{S}$ is faithful.

In Chapter II we consider closed categories which possess a tensor product defined by the adjointness relation

$$(A \otimes B, C) \cong (A(BC)).$$

These considerations lead us to the notion of a monoidal category, which is a *catégorie avec multiplication* in the terminology of BÉNABOU ([*1*], [*2*], [*3*]). A key result here is Theorem II.5.8 which allows us to reconstruct the closed structure on $\mathscr{V}$ from the monoidal structure. A similar result is to be found in [*3*]; cf. also [*10*].

The theory as developed thus far does not allow for a consideration of dual $\mathscr{V}$-categories and therefore all $\mathscr{V}$-functors must remain covariant. In order to introduce contravariance one needs a notion of *symmetry*. In the presence of a tensor product a symmetry takes the form of a natural isomorphism

$$A \otimes B \cong B \otimes A$$

satisfying suitable conditions. This is the subject of Chapter III, where we also show that a symmetry allows us to introduce $\mathscr{V}$-functors of many variables and the appropriate generalized natural transformations. In a separate paper we shall study closed categories with symmetry but without assuming the tensor product. The symmetry then takes the form of an isomorphism

$$(A(BC)) \cong (B(AC)).$$

Chapter IV is devoted to examples. These show the frequency with which closed categories appear in various parts of mathematics. The examples were also chosen to illustrate the various points treated in Chapters I—III. Certain classes of examples will form the subject matter of subsequent papers and such examples have been either completely omitted or treated very sketchily. This in particular applies to the construction of "functor categories" which form an indispensable continuation of the theory presented in this paper.

Chapter I

Closed Categories

1. Notation and Preliminaries

In our notation we use brackets no more than is necessary (logically or psychologically); in particular fx denotes the value of the function f at the argument x. Then $(Kf)x$ denotes the value of Kf at x, $f(xy)$ denotes the value of f at xy, and $f(x,y)$ denotes the value of f at (x,y). We often use dots in place of brackets, as in $Kg.\ Kf.\ x$ for $(Kg)\ (Kf)x$. We are similarly sparing of commas: for a bifunctor T we write $T(AB)$, not $T(A, B)$; but we write $T(f, g)$ since $T(fg)$ would be confusing.

We use $\mathscr{A}^*$ for the dual of a category $\mathscr{A}$; then a functor $T : \mathscr{A} \to \mathscr{B}$ has a dual $T^*: \mathscr{A}^* \to \mathscr{B}^*$, and a natural transformation $\alpha: T \to S : \mathscr{A} \to \mathscr{B}$ has a dual $\alpha^* : S^* \to T^* : \mathscr{A}^* \to \mathscr{B}^*$. We reserve the symbol $\mathscr{S}$ for the category of sets, and we denote by Hom $\mathscr{A}$ the Hom-functor $\mathscr{A}^* \times \mathscr{A} \to \mathscr{S}$; however we abbreviate the values Hom $\mathscr{A}(AB)$ and Hom $\mathscr{A}(f, g)$ of Hom $\mathscr{A}$ to $\mathscr{A}(AB)$ and $\mathscr{A}(f, g)$. Note that we do not require of a category $\mathscr{A}$ that the various $\mathscr{A}(AB)$ be disjoint.

If $\alpha = (\alpha_A)_{A \in \mathscr{A}}$ is a family of morphisms, where say $\alpha_A : TA \to SA$, we often abbreviate to $\alpha : TA \to SA$. Where there are several variables as in $L^A_{BC} : (BC) \to ((AB)\ (AC))$, we may abbreviate L^A_{BC} totally to L, or partially to, say, L^A if we wish to emphasize A. We also use L^A at times to denote the partial family got by fixing A and letting B and C vary.

The reader should note that the criterion for a family of morphisms $\alpha_A : TA \to SA$ to be a natural transformation $\alpha : T \to S$, where $T, S : \mathscr{A} \to \mathscr{B}$, is the commutativity of the diagram

$$\begin{array}{ccc} \mathscr{A}(AB) & \xrightarrow{T_{AB}} & \mathscr{B}(TA, TB) \\ {\scriptstyle S_{AB}}\downarrow & & \downarrow{\scriptstyle \mathscr{B}(1,\alpha_B)} \\ \mathscr{B}(SA, SB) & \xrightarrow[\mathscr{B}(\alpha_A, 1)]{} & \mathscr{B}(TA, SB) \end{array} ; \qquad (1.1)$$

the more usual criterion, got by evaluating (1.1) at $f \in \mathscr{A}(AB)$, is the commutativity of the diagram

$$\begin{array}{ccc} TA & \xrightarrow{Tf} & TB \\ {\scriptstyle \alpha_A}\big\downarrow & & \big\downarrow{\scriptstyle \alpha_B} \\ SA & \xrightarrow[Sf]{} & SB \end{array}$$

As our discourse concerns generalizations of categories, functors, and natural transformations, it will be convenient to use the abstract language of *hypercategories* (the 2-categories of EHRESMANN [*6*]). A hypercategory $\mathfrak{A}$ consists of

(i) a class of *objects* $\mathscr{A}, \mathscr{B}, \ldots$;

(ii) for each pair of objects $\mathscr{A}, \mathscr{B}$ a set of *morphisms*

$$T, S, \ldots : \mathscr{A} \to \mathscr{B};$$

(iii) for each $\mathscr{A}, \mathscr{B}$ and for each pair $T, S : \mathscr{A} \to \mathscr{B}$ a set of *hypermorphisms* $\lambda, \mu, \ldots : T \to S : \mathscr{A} \to \mathscr{B}$;

together with four kinds of composition law:

(i) if $T : \mathscr{A} \to \mathscr{B}$ and $S : \mathscr{B} \to \mathscr{C}$ then $ST : \mathscr{A} \to \mathscr{C}$;

(ii) if $T : \mathscr{A} \to \mathscr{B}$ and $\lambda : S \to R : \mathscr{B} \to \mathscr{C}$ then $\lambda T : ST \to RT : \mathscr{A} \to \mathscr{C}$;

(iii) if $\lambda : S \to R : \mathscr{A} \to \mathscr{B}$ and $T : \mathscr{B} \to \mathscr{C}$ then $T\lambda : TS \to TR : \mathscr{A} \to \mathscr{C}$;

(iv) if $\lambda : T \to S : \mathscr{A} \to \mathscr{B}$ and $\mu : S \to R : \mathscr{A} \to \mathscr{B}$ then $\mu\lambda : T \to R : \mathscr{A} \to \mathscr{B}$;

and two kinds of identity:

(i) $1_{\mathscr{A}} : \mathscr{A} \to \mathscr{A}$;

(ii) $1_T : T \to T : \mathscr{A} \to \mathscr{B}$.

These data are to satisfy the following five axioms:

HC1. The objects and the morphisms form a category $\mathfrak{A}_0$.

HC2. For each $\mathscr{A}, \mathscr{B}$ the morphisms $\mathscr{A} \to \mathscr{B}$ and the hypermorphisms between them form a category $\mathfrak{A}(\mathscr{A}\mathscr{B})$.

HC3. If $\lambda : T \to T' : \mathscr{A} \to \mathscr{B}$, $\mathscr{A}'' \xrightarrow[R']{} \mathscr{A}' \xrightarrow[R]{} \mathscr{A}$, and $\mathscr{B} \xrightarrow[S]{} \mathscr{B}' \xrightarrow[S']{} \mathscr{B}''$, we have

(a) $1_{\mathscr{B}}\lambda = \lambda$, (b) $\lambda 1_{\mathscr{A}} = \lambda$,

(c) $(S'S)\lambda = S'(S\lambda)$, (d) $\lambda(RR') = (\lambda R)R'$,

(e) $(S\lambda)R = S(\lambda R)$.

HC4. If $R : \mathscr{A}' \to \mathscr{A}$ and $S : \mathscr{B} \to \mathscr{B}'$ the assignments $T \mapsto STR$, $\lambda \mapsto S\lambda R$ constitute a functor $\mathfrak{A}(R, S) : \mathfrak{A}(\mathscr{A}\mathscr{B}) \to \mathfrak{A}(\mathscr{A}'\mathscr{B}')$.

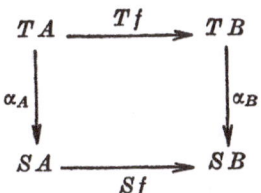

HC5. If $\lambda : T \to S : \mathscr{A} \to \mathscr{B}$ and $\mu : P \to Q : \mathscr{B} \to \mathscr{C}$, the following

diagram commutes:

$$\begin{array}{ccc} PT & \xrightarrow{P\lambda} & PS \\ {\scriptstyle \mu T}\downarrow & & \downarrow{\scriptstyle \mu S} \\ QT & \xrightarrow[Q\lambda]{} & QS \end{array}$$

If $\mathfrak{A}$ and $\mathfrak{B}$ are hypercategories, a *hyperfunctor* $\Phi : \mathfrak{A} \to \mathfrak{B}$ consists of functions assigning

(i) to each object $\mathscr{A}$ of $\mathfrak{A}$ an object $\Phi\mathscr{A}$ of $\mathfrak{B}$;

(ii) to each morphism $T : \mathscr{A} \to \mathscr{B}$ in $\mathfrak{A}$ a morphism $\Phi T : \Phi\mathscr{A} \to \Phi\mathscr{B}$ in $\mathfrak{B}$;

(iii) to each hypermorphism $\lambda : T \to S : \mathscr{A} \to \mathscr{B}$ in $\mathfrak{A}$ a hypermorphism $\Phi\lambda : \Phi T \to \Phi S : \Phi\mathscr{A} \to \Phi\mathscr{B}$ in $\mathfrak{B}$.

These are to satisfy the axioms:

HF1. $\Phi(ST) = \Phi S . \Phi T$ and $\Phi 1 = 1$.
HF2. $\Phi(\lambda T) = \Phi\lambda . \Phi T$.
HF3. $\Phi(T\lambda) = \Phi T . \Phi\lambda$.
HF4. $\Phi(\mu\lambda) = \Phi\mu . \Phi\lambda$ and $\Phi 1 = 1$.

If $\Phi, \Psi : \mathfrak{A} \to \mathfrak{B}$ are hyperfunctors, a *hypernatural transformation* $\eta : \Phi \to \Psi$ is a function assigning to each object $\mathscr{A}$ of $\mathfrak{A}$ a morphism $\eta_{\mathscr{A}} : \Phi\mathscr{A} \to \Psi\mathscr{A}$, and satisfying the axioms:

HN1. If $T : \mathscr{A} \to \mathscr{B}$ in $\mathfrak{A}$, the following diagram commutes:

$$\begin{array}{ccc} \Phi\mathscr{A} & \xrightarrow{\Phi T} & \Phi\mathscr{B} \\ {\scriptstyle \eta_{\mathscr{A}}}\downarrow & & \downarrow{\scriptstyle \eta_{\mathscr{B}}} \\ \Psi\mathscr{A} & \xrightarrow[\Psi T]{} & \Psi\mathscr{B} \end{array}$$

HN2. If $\lambda : T \to S : \mathscr{A} \to \mathscr{B}$ in $\mathfrak{A}$, then

$$\eta_{\mathscr{B}} . \Phi\lambda : \eta_{\mathscr{B}} . \Phi T \to \eta_{\mathscr{B}} . \Phi S : \Phi\mathscr{A} \to \Psi\mathscr{B}$$

coincides with

$$\Psi\lambda . \eta_{\mathscr{A}} : \Psi T . \eta_{\mathscr{A}} \to \Psi S . \eta_{\mathscr{A}} : \Phi\mathscr{A} \to \Psi\mathscr{B}.$$

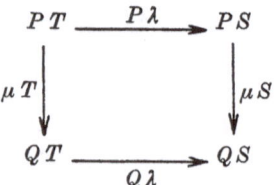

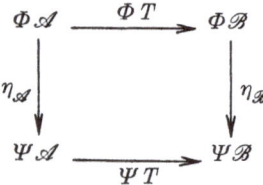

It is clear that small categories, functors, and natural transformations form a hypercategory; so do small hypercategories, hyperfunctors, and hypernatural transformations, if we use the obvious definitions of composition. Since however we shall use hypercategories purely as a convenient language at the formal level, we shall not hesitate to speak of the "hypercategory" $\mathscr{Cat}$ of *all* categories and the "hypercategory" $\mathscr{Hyp}$ of *all* hypercategories, sometimes as here using quotation marks to emphasize this purely formal use. In fact when we speak in this way we suppose $\mathscr{Cat}$ to contain not merely all categories but also all "categories". Note that from any hypercategory $\mathfrak{A}$ we get a category $\mathfrak{A}_0$ by discarding the hypermorphisms; indeed $\mathfrak{A} \mapsto \mathfrak{A}_0$ is clearly the object-function of a forgetful hyperfunctor $\mathscr{Hyp} \to \mathscr{Cat}$.

We recall some special properties of the hypercategory $\mathscr{Cat}$ that provide at once a guideline for our generalizations of categories and a tool for our investigations. The underlying category $\mathscr{Cat}_0$ of $\mathscr{Cat}$ has an initial object, the empty category; and a terminal object, the category $\mathscr{I}$ with a single object and a single morphism. The objects and the morphisms of a category $\mathscr{A}$ may be identified with the functors $\mathscr{I} \to \mathscr{A}$ and the natural transformations between these. The category $\mathscr{S}$ of sets plays a special role in $\mathscr{Cat}$; to each object A in the category $\mathscr{A}$ there is the left represented functor $L^A = \operatorname{Hom} \mathscr{A}(A\ -) : \mathscr{A} \to \mathscr{S}$, and for any functor $T : \mathscr{A} \to \mathscr{S}$ we obtain a bijection between the natural transformations $\alpha : L^A \to T$ and the elements of TA by sending α to $\alpha_A 1_A \in TA$. This representation theorem (Yoneda [17]) occurs most frequently in our applications in the following form, in which we formally state it:

Theorem 1.1. *Let $T : \mathscr{A} \to \mathscr{B}$ be a functor and let $K \in \mathscr{A}$, $M \in \mathscr{B}$. Denote by $\{p\}$ the class of natural transformations*

$$p = p_A : \quad \mathscr{A}(KA) \to \mathscr{B}(M, TA),$$

and define a map $\Gamma : \{p\} \to \mathscr{B}(M, TK)$ by

$$\Gamma p = p_K 1_K . \tag{1.2}$$

Then Γ is a bijection, with inverse $\Omega : \mathscr{B}(M, TK) \to \{p\}$, where $\Omega\theta$ is the composite

$$\Omega\theta : \mathscr{A}(KA) \xrightarrow[T_{KA}]{} \mathscr{B}(TK, TA) \xrightarrow[\mathscr{B}(\theta,1)]{} \mathscr{B}(M, TA). \tag{1.3}$$

Proof. $\Omega\theta$ is indeed natural, since T_{KA} is natural in A and $\mathscr{B}(\theta, 1_B)$ in B. That $\Gamma\Omega = 1$ is obvious, and that $\Omega\Gamma = 1$ follows by a simple naturality argument.

We record the form of (1.3) obtained by evaluating at $f \in \mathscr{A}(KA)$; setting $p = \Omega\theta$ we have the commutative diagram

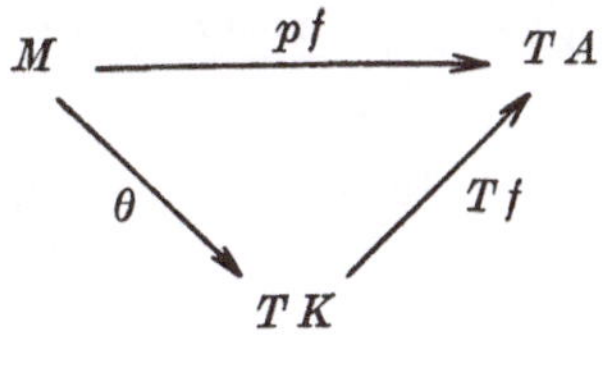

(1.4)

Note that if we take $\mathscr{B} = \mathscr{S}$ and M to be a single point we regain the usual form of the theorem.

The hypercategory $\mathscr{C}at$ is further enriched by its product hyperfunctor $\mathscr{A}, \mathscr{B} \mapsto \mathscr{A} \times \mathscr{B}$ and its duality hyperfunctor $\mathscr{A} \mapsto \mathscr{A}^*$, which allow us to define functors of many variables and both variances. There is a corresponding extension of the concept of natural transformation, which the authors have described in [7], and with which we shall assume familiarity. We record here the appropriate extension to the representation theorem:

Proposition 1.2. *Let $\mathscr{A}, \mathscr{B}, \mathscr{C}, \mathscr{D}$ be categories and let $T : \mathscr{D} \times \mathscr{A} \to \mathscr{B}$, $K : \mathscr{C} \times \mathscr{D}^* \to \mathscr{A}$, $M : \mathscr{C} \to \mathscr{B}$ be functors. Let*

$$p = p_{CDA} : \mathscr{A}(K(C\,D), A) \to \mathscr{B}(M\,C, T(D\,A))$$

be a family of morphisms, natural in A for each fixed C, D; and let

$$\theta = \theta_{CD} : M\,C \to T(D, K(C\,D))$$

be Γp_{CD}. Then p is natural in C (resp. D) if and only if θ is.

Proof. If θ is natural so is p by (1.3). If p is natural so is the composite

$$* \underset{j}{\to} \mathscr{A}(K(C\,D), K(C\,D)) \underset{p}{\to} \mathscr{B}(M\,C, T(D, K(C\,D)))$$

where $*$ is a single point and $j* = 1$; for $j : * \to \mathscr{A}(A\,A)$ is clearly natural in A. It is easy to see that this implies the naturality of $pj*$, which is θ.

2. Closed Categories

We begin by axiomatizing those structures, called *closed categories*, in which the Hom-functors of our generalized categories will take their values.

A closed category $\mathscr{V} = (\mathscr{V}_0, V, \hom \mathscr{V}, I, i, j, L)$ consists of the following seven data:

(i) a category $\mathscr{V}_0$;

(ii) a functor $V : \mathscr{V}_0 \to \mathscr{S}$;

(iii) a functor $\hom \mathscr{V} : \mathscr{V}_0^* \times \mathscr{V}_0 \to \mathscr{V}_0$ (we write $(A\,B)$ for $\hom \mathscr{V}(A\,B)$ and (f, g) for $\hom \mathscr{V}(f, g)$);

(iv) an object I of $\mathscr{V}_0$;

(v) a natural *isomorphism* $i = i_A : A \to (I\,A)$ in $\mathscr{V}_0$;

(vi) a natural transformation $j = j_A : I \to (A\,A)$ in $\mathscr{V}_0$;

(vii) a natural transformation $L = L^A_{BC} : (B\,C) \to ((A\,B)\,(A\,C))$ in $\mathscr{V}_0$.

These data are to satisfy the following six axioms:

CC0. The following diagram of functors commutes:

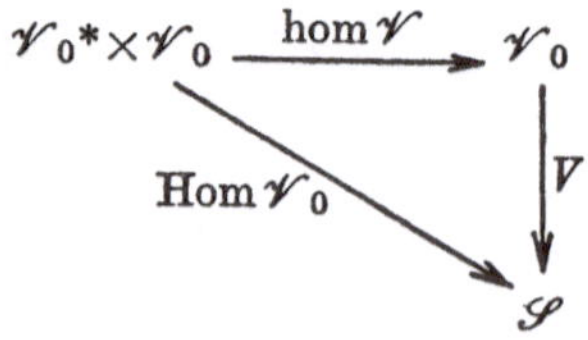

CC1. The following diagram commutes:

$$\begin{array}{ccc} (B\,B) & \xrightarrow{L^A} & ((A\,B)\,(A\,B)) \\ & {\scriptstyle j}\nwarrow \qquad \nearrow{\scriptstyle j} & \\ & I & \end{array}$$

CC2. The following diagram commutes:

$$\begin{array}{ccc} (A\,C) & \xrightarrow{L^A} & ((A\,A)\,(A\,C)) \\ & \searrow{\scriptstyle i} & \downarrow{\scriptstyle (j,1)} \\ & & (I\,(A\,C)) \end{array}$$

CC3. The following diagram commutes:

$$\begin{array}{ccc} (C\,D) & \xrightarrow{\qquad L^B \qquad} & ((B\,C)\,(B\,D)) \\ \downarrow{\scriptstyle L^A} & & \\ ((A\,C)\,(A\,D)) & & \downarrow{\scriptstyle (1,\,L^A)} \\ \downarrow{\scriptstyle L^{(A\,B)}} & & \\ (((A\,B)\,(A\,C)),((A\,B)\,(A\,D))) & \xrightarrow[(L^A,\,1)]{} & ((B\,C),((A\,B)\,(A\,D))) \end{array}$$

CC4. The following diagram commutes:

$$\begin{array}{ccc} (B\,C) & \xrightarrow{L^I} & ((I\,B)\,(I\,C)) \\ & \searrow{\scriptstyle (1,\,i)} & \downarrow{\scriptstyle (i,1)} \\ & & (B\,(I\,C)) \end{array}$$

CC5. The map

$$V\,i_{(AA)} : V(A\,A) \to V(I(A\,A)),$$

which by CC0 may also be written

$$V\,i_{(AA)} : \mathscr{V}_0(AA) \to \mathscr{V}_0(I,(AA)),$$

sends $1_A \in \mathscr{V}_0(AA)$ to $j_A \in \mathscr{V}_0(I,(AA))$.

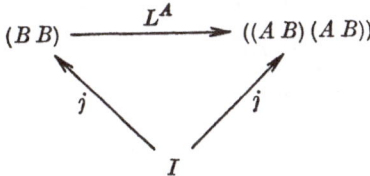

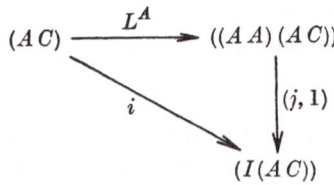

We consider some properties of closed categories that follow directly from the above axioms. Note that by CC0 we have

$$V(A\,B) = \mathscr{V}_0(A\,B)\,, \tag{2.1}$$

$$V(f,g) = \mathscr{V}_0(f,g)\,. \tag{2.2}$$

Define a natural isomorphism

$$\iota = \iota_A : VA \to V(I\,A)$$

by

$$\iota_A = V\,i_A\,. \tag{2.3}$$

Proposition 2.1. *ι provides a representation of the functor $V : \mathscr{V}_0 \to \mathscr{S}$.* Axiom CC5 may be written as

$$j_A = \iota\,1_A; \tag{2.4}$$

of course we could drop j as a primitive term, and drop axiom CC5, using (2.4) as a definition of j; note that j so defined is automatically natural. Any statement about composition with j may be turned into a statement about the image of 1 by means of:

Lemma 2.2. *For any $f : (AA) \to X$ in $\mathscr{V}_0$, the composite*

$$I \xrightarrow[j]{} (AA) \xrightarrow[f]{} X$$

is the image of $1 \in V(AA)$ under the composite map

$$V(AA) \xrightarrow[Vf]{} V\,X \xrightarrow[\iota]{} V(I\,X)\,.$$

Proof. Evaluate at $1 \in V(AA)$ the diagram

$$\begin{array}{ccc} V(A\,A) & \xrightarrow{\;\;Vf\;\;} & V\,X \\ \Big\downarrow{\scriptstyle \iota} & & \Big\downarrow{\scriptstyle \iota} \\ V(I(A\,A)) & \xrightarrow[V(1,f)]{} & V(I\,X), \end{array}$$

which commutes by the naturality of ι.

Proposition 2.3. *In the presence of* CC0 *and* CC5, *the axiom* CC1 *is equivalent to any of the following:*

(a) $(V\,L^A_{BB})\,1_B = 1_{(AB)}\,;$ (2.5)

(b) $(V\,L^A_{BC})\,f = (1,f) \in V((A\,B),(A\,C))$ for $f \in V(BC)$; (2.6)

(c) $V\,L^A_{BC} = (A\,-) : V(BC) \to V((A\,B),(A\,C))\,.$ (2.7)

Proof. Lemma 2.2 shows the equivalence of CC1 with (a), while (b) is merely the evaluated form of (c). The equivalence of (c) with (a) follows by applying the representation theorem, Theorem 1.1, to the natural transformations $V L^A_{BC}$ and $(A -)$ in (2.7), since $(A, 1_B) = 1$.

Proposition 2.4. *For any $f \in V(A B)$ we have a commutative diagram*

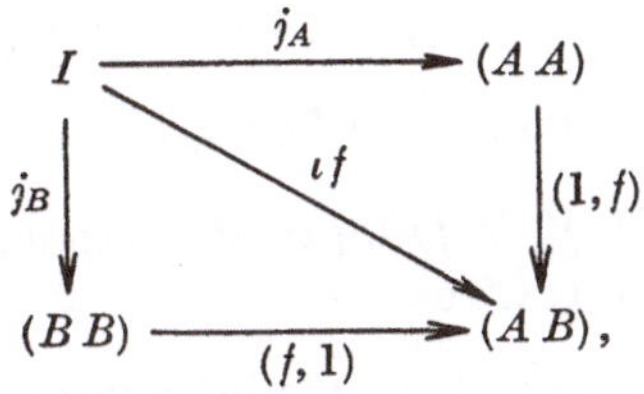

$$(2.8)$$

the diagonal being the image of f under

$$\iota_{(AB)} : V(A B) \to V(I(A B)) .$$

Proof. By Lemma 2.2 we have

$$(1, f) j = \iota . V(1, f) . 1 ,$$

but $V(1, f) 1 = \mathscr{V}_0(1, f) 1 = f$; similarly $(f, 1) j = \iota f$.

Proposition 2.5. $i_{(IA)} = (1, i_A) : (I A) \to (I(I A)) .$

Proof. From the naturality of i we have a commutative diagram

$$
\begin{array}{ccc}
A & \xrightarrow{\ i\ } & (I A) \\
\downarrow{\scriptstyle i} & & \downarrow{\scriptstyle (1, i)} \\
(I A) & \xrightarrow{\ i\ } & (I(I A)) ,
\end{array}
$$

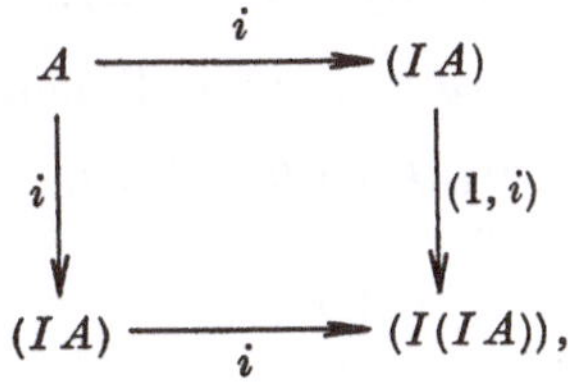

whence the result since i is an isomorphism.

Proposition 2.6. *For any $f \in V(I A)$, the composite*

$$I \xrightarrow{\ j\ } A \xrightarrow{\ i\ } (I A)$$

is the image of f under $\iota : V(I A) \to V(I, (I A))$.

Proof. Apply V to Proposition 2.5 and evaluate at f.

Proposition 2.7. $j_I = i_I : I \to (I I)$.

Proof. Take $A = I$ and $f = 1$ in Proposition 2.6.

Proposition 2.8. *For $f \in V(I I)$ we have $(1, f) = (f, 1) : (I I) \to (I I)$.*

Proof. In (2.8) put $A = B = I$; the result follows because $j_I = i_I$ is an isomorphism.

Proposition 2.9. *The monoid $\mathscr{V}_0(I\,I)$ of endomorphisms of I is commutative.*

Proof. Applying V to Proposition 2.8 gives

$$V(1, f) = V(f, 1) : V(I\,I) \to V(I\,I);$$

evaluating at $g \in V(I\,I)$ now gives $fg = gf$.

Proposition 2.10. *If V is faithful, the axioms* CC2, CC3, CC4 *are consequences of* CC0, CC1, CC5.

Proof. First note that we have made no use of CC2, CC3, CC4 in the deductions above. If V is faithful, a diagram commutes if and only if V of it does so. Applying V to the diagram of CC2 and evaluating at $f \in V(A\,C)$, using (2.6), we get $\iota f = (1, f)\, j$, which is true by Proposition 2.4. Applying V to the diagram of CC3 and using (2.6), we obtain the diagram asserting the naturality in C of L^A_{BC}, which obtains by hypothesis. Similarly V of CC4 is the assertion of the naturality of i.

Proposition 2.11. *Let there be given a category $\mathscr{V}_0$, a faithful functor $V : \mathscr{V}_0 \to \mathscr{S}$, a representation $\iota : VA \cong \mathscr{V}_0(I\,A)$ of V, and, for each A, B in $\mathscr{V}_0$, an object $(A\,B)$ of $\mathscr{V}_0$ with*

$$V(A\,B) = \mathscr{V}_0(A\,B).$$

Then there is a closed category $\mathscr{V} = (\mathscr{V}_0, V, \mathrm{hom}\,\mathscr{V}, I, i, j, L)$ with

$$\mathrm{hom}\,\mathscr{V}(A\,B) = (A\,B)$$

and

$$V i = \iota$$

if and only if

(i) *for each $f : A' \to A$ and $g : B \to B'$, the morphism*

$$\mathscr{V}_0(f, g) : \mathscr{V}_0(A\,B) \to \mathscr{V}_0(A'\,B')$$

is $V(f, g)$ for some morphism $(f, g) : (A\,B) \to (A'\,B')$;

(ii) *for each A, $\iota : VA \to \mathscr{V}_0(I\,A)$ is $V i$ for some isomorphism*

$$i : A \to (I\,A);$$

(iii) *for each A, B, C the map $h \mapsto (1, h) : V(BC) \to V((A\,B)\,(A\,C))$ is $V L^A_{BC}$ for some $L^A_{BC} : (BC) \to ((A\,B)\,(A\,C))$;*

and if these conditions are satisfied $\mathscr{V}$ is unique.

Proof. The conditions are clearly necessary, in view of (2.2) and (2.6). Suppose they are satisfied; then (f, g), i, and L are unique by the faithfulness of V.

For $A'' \underset{f'}{\to} A' \underset{f}{\to} A$ and $B \underset{g}{\to} B' \underset{g'}{\to} B''$ we have

$$\begin{aligned} V(f'f, gg') &= \mathscr{V}_0(f'f, gg') \\ &= \mathscr{V}_0(f, g)\, \mathscr{V}_0(f', g') \\ &= V(f, g)\, V(f', g') \\ &= V((f, g)(f', g')), \end{aligned}$$

whence $(f'f, gg') = (f, g)(f', g')$ by the faithfulness of V. Similarly $(1, 1) = 1$, and thus $(A\,B)$ and (f, g) are the values of a functor hom $\mathscr{V}$ satisfying CC0.

Again since V is faithful the naturality of i follows from that of $Vi = \iota$, and that of L^A follows from that of $VL^A = (A\, -)$. We define j by (2.4), so that CC5 is satisfied; then CC1 is satisfied by Proposition 2.3, and the remaining axioms follow by Proposition 2.10.

Proposition 2.12. *We obtain a closed category, which we denote by* $\mathscr{S}$, *if we set* $\mathscr{V}_0 = \mathscr{S}$, $V = 1$ *and* hom $\mathscr{V}$ = Hom $\mathscr{S}$; *take for* I *a set* $*$, *chosen once for all, consisting of a single point* $*$; *and define* i, j, L *by:*

$$(ia) * = a, \qquad a \in A; \tag{2.9}$$

$$j * = 1; \tag{2.10}$$

$$(Lf)g = fg, \qquad f \in (BC), \quad g \in (A\,B). \tag{2.11}$$

Proof. It is clear that i, j, L are natural. Verification of CC0, CC1 (in the form (2.6)), and CC5 is immediate, and the other axioms follow by Proposition 2.10.

Remark 2.13. For a closed category $\mathscr{V}$ we shall call $\mathscr{V}_0$ the *underlying category*, $V : \mathscr{V}_0 \to \mathscr{S}$ the *basic functor*, and hom $\mathscr{V}$ the *internal* Hom-*functor*.

3. Closed Functors

Let $\mathscr{V} = (\mathscr{V}_0, V, \hom\mathscr{V}, I, i, j, L)$ and $\mathscr{V}' = (\mathscr{V}'_0, V', \hom\mathscr{V}', I', i', j', L')$ be closed categories; we write $(X\,Y)$ for hom $\mathscr{V}'(X\,Y)$ as well as $(A\,B)$ for hom $\mathscr{V}(A\,B)$. A *closed functor* $\Phi = (\phi, \hat{\phi}, \phi^0) : \mathscr{V} \to \mathscr{V}'$ consists of

(i) a functor $\phi : \mathscr{V}_0 \to \mathscr{V}'_0$;
(ii) a natural transformation $\hat{\phi} = \hat{\phi}_{AB} : \phi(A\,B) \to (\phi A, \phi B)$;
(iii) a morphism $\phi^0 : I' \to \phi I$.

These data are to satisfy the following three axioms:

CF 1. The following diagram commutes:

$$\begin{array}{ccc} \phi I & \xrightarrow{\phi j} & \phi(A\,A) \\ {\scriptstyle \phi^0}\big\uparrow & & \big\downarrow{\scriptstyle \hat{\phi}} \\ I' & \xrightarrow[j']{} & (\phi A, \phi A) \end{array}$$

CF 2. The following diagram commutes:

$$\begin{array}{ccc} \phi(I\,A) & \xrightarrow{\hat{\phi}} & (\phi I, \phi A) \\ {\scriptstyle \phi i}\big\uparrow & & \big\downarrow{\scriptstyle (\phi^0, 1)} \\ \phi A & \xrightarrow[i']{} & (I', \phi A) \end{array}$$

CF 3. The following diagram commutes:

$$\begin{array}{ccccc} \phi(B\,C) & \xrightarrow{\phi L} & \phi((A\,B),(A\,C)) & \xrightarrow{\hat{\phi}} & (\phi(A\,B), \phi(A\,C)) \\ {\scriptstyle \hat{\phi}}\big\downarrow & & & & \big\downarrow{\scriptstyle (1, \hat{\phi})} \\ (\phi B, \phi C) & \xrightarrow[L']{} & ((\phi A, \phi B), (\phi A, \phi C)) & \xrightarrow[(\hat{\phi}, 1)]{} & (\phi(A\,B), (\phi A, \phi C)) \end{array}$$

Theorem 3.1. *Closed categories and closed functors form a "category" $\mathscr{C}\ell_0$ if we define the composite of*

$$\Phi = (\phi, \hat{\phi}, \phi^0) : \mathscr{V} \to \mathscr{V}' \quad \textit{and} \quad \Psi = (\psi, \hat{\psi}, \psi^0) : \mathscr{V}' \to \mathscr{V}''$$

to be $X = (\chi, \hat{\chi}, \chi^0) : \mathscr{V} \to \mathscr{V}''$ *where*

(i) χ *is the composite* $\mathscr{V}_0 \xrightarrow[\phi]{} \mathscr{V}'_0 \xrightarrow[\psi]{} \mathscr{V}''_0$; (3.1)

(ii) $\hat{\chi}$ *is the composite* $\psi\phi(A\,B) \xrightarrow[\psi\hat{\phi}]{} \psi(\phi A, \phi B) \xrightarrow[\hat{\psi}]{} (\psi\phi A, \psi\phi B)$; (3.2)

(iii) χ^0 *is the composite* $I'' \xrightarrow[\psi^0]{} \psi I' \xrightarrow[\psi\phi^0]{} \psi\phi I$. (3.3)

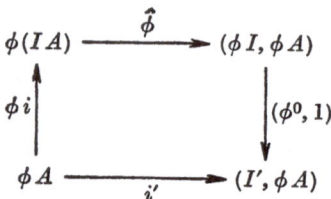

Proof. It is immediate that composition as defined above is associative, with identities $1 = (1, 1, 1)$. What has to be verified is that X satisfies CF 1—CF 3; that is, that the exteriors of the following three diagrams commute:

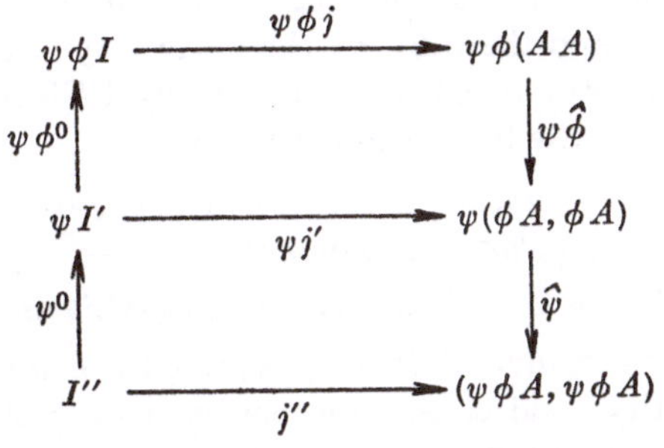

(3.4)

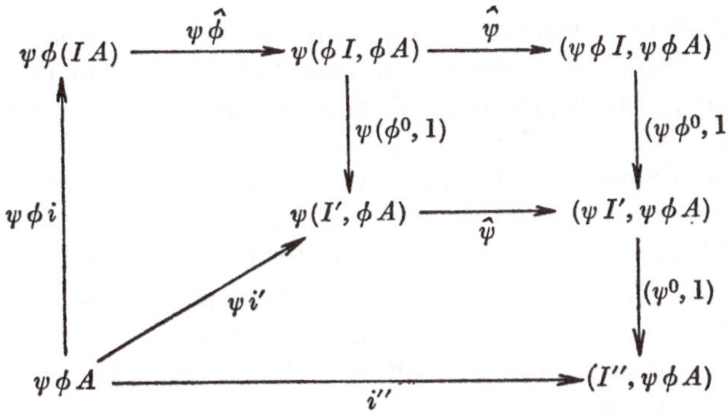

(3.5)

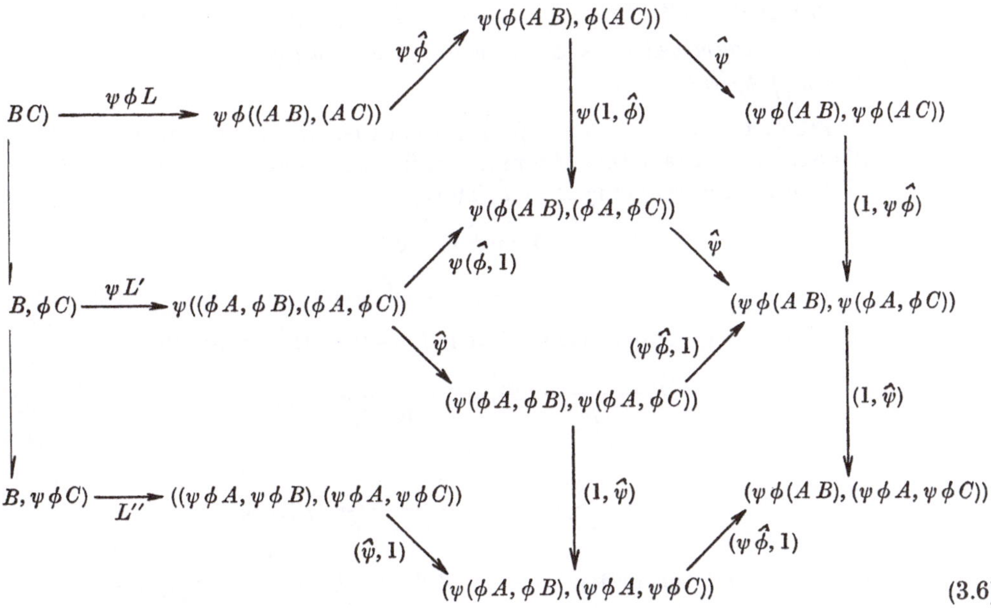

(3.6)

Now in (3.4) one region commutes by ψ of CF1 for Φ, and the other region by CF1 for Ψ; in (3.5) one region commutes by ψ of CF2 for Φ, one by CF2 for Ψ, and the third by the naturality of $\hat{\psi}$; and in (3.6) one region commutes by ψ of CF3 for Φ, one by CF3 for Ψ, two by the naturality of $\hat{\psi}$, and the last region trivially.

Proposition 3.2. *A closed functor $\Phi = (\phi, \hat{\phi}, \phi^0): \mathscr{V} \to \mathscr{V}'_0$ is an isomorphism in the category $\mathscr{C}\ell_0$ if and only if $\phi: \mathscr{V}_0 \to \mathscr{V}'_0$ is an isomorphism of categories, each $\hat{\phi}_{AB}$ is an isomorphism, and ϕ^0 is an isomorphism.*

Proof. If Φ has an inverse $\Psi = (\psi, \hat{\psi}, \psi^0)$, the composites (3.1)—(3.3) are all the identity, and so are the corresponding composites with Φ and Ψ interchanged. It follows at once that ϕ, $\hat{\phi}$, ϕ^0 are all isomorphisms.

If ϕ, $\hat{\phi}$, ϕ^0 are all isomorphisms, define $\psi = \phi^{-1}$ and take for $\hat{\psi}$ and ψ^0 the unique values that render the composites (3.2) and (3.3) equal to the identity; $\hat{\psi}$ is clearly natural, and we have $\Psi\Phi = 1$, but we must show that Ψ satisfies CF1 — CF3.

Consider diagrams (3.4)—(3.6); we know that the exteriors commute and that all the internal regions commute except those that express CF1—CF3 for Ψ. It follows that the latter regions commute also (using, in the case of (3.6), the fact that $\psi\hat{\phi}$ is an isomorphism).

Thus Ψ is a left inverse of Φ. But ψ, $\hat{\psi}$, ψ^0 are all isomorphisms, and so by the same argument Ψ itself has a left inverse. Hence Ψ is a two-sided inverse for Φ.

Proposition 3.3. *In a closed functor $\Phi = (\phi, \hat{\phi}, \phi^0): \mathscr{V} \to \mathscr{V}'$, ϕ^0 is uniquely determined when ϕ and $\hat{\phi}$ are given, and ϕ^0 is an isomorphism if each of ϕ, $\hat{\phi}^0$ is.*

Proof. Let $(\phi, \hat{\phi}, \phi^0)$ and $(\phi, \hat{\phi}, \overline{\phi^0})$ both be closed functors $\mathscr{V} \to \mathscr{V}'$. We express the fact that the first satisfies CF1 and the second satisfies CF2, in each case taking $A = I$. Thus

$$j' = \hat{\phi}.\, \phi j_I.\, \phi^0,$$
$$i' = (\overline{\phi^0}, 1).\, \hat{\phi}.\, \phi i_I.$$

Since $j_I = i_I$ by Proposition 2.7, it follows that the composites

$$I' \xrightarrow{j'} (\phi I, \phi I) \xrightarrow{(\overline{\phi^0}, 1)} (I', \phi I),$$

$$I' \xrightarrow{\phi^0} \phi I \xrightarrow{i'} (I', \phi I),$$

are equal; but the first of these is $\iota' \overline{\phi^0}$ by Proposition 2.4, while the second is $\iota' \phi^0$ by Proposition 2.6. Since ι' is an isomorphism, we have $\overline{\phi^0} = \phi^0$.

If $\hat{\phi}$ is an isomorphism, so too by CF2 is

$$(\phi^0, 1) : (\phi I, \phi A) \to (I', \phi A).$$

If ϕ also is an isomorphism, we can replace ϕA here by any $X \in \mathscr{V}'_0$. Doing this and applying V', we see that

$$V'(\phi^0, 1) : V'(\phi I, X) \to V'(I', X)$$

is an isomorphism for all X; hence ϕ^0 is an isomorphism.

Proposition 3.4. *Let $\mathscr{V}$ and $\mathscr{V}'$ be closed categories and let $\phi : \mathscr{V}_0 \to \mathscr{V}'_0$ be a functor. Then there is a bijection between morphisms*

$$\phi^0 : I' \to \phi I$$

and natural transformations

$$\phi_0 : V \to V'\phi : \mathscr{V}_0 \to \mathscr{S},$$

given by requiring commutativity in the diagram

$$\begin{array}{ccccc} V(IA) & & \xrightarrow{\quad\phi\quad} & & V'(\phi I, \phi A) \\ {\scriptstyle \iota^{-1}}\downarrow & & & & \downarrow{\scriptstyle V'(\phi^0, 1)} \\ VA & \xrightarrow[\phi_0]{} & V'\phi A & \xrightarrow[\iota']{} & V'(I', \phi A) \end{array} \tag{3.7}$$

Proof. Since ι and ι' are natural isomorphisms, this is immediate from the representation theorem.

Thus we can use ϕ, $\hat{\phi}$, ϕ_0 instead of ϕ, $\hat{\phi}$, ϕ^0 as the data for a closed functor. We record the form of (3.7) got by evaluating at $f \in V(IA)$:

$$\phi f . \phi^0 = \iota' \phi_0 \iota^{-1} f. \tag{3.8}$$

Taking in particular $A = I$ and $f = 1_I$, we get

$$\phi^0 = \iota' \phi_{0I} \iota^{-1} 1_I. \tag{3.9}$$

Now replace A by (AB) and let $f = \iota g$ where $g \in V(AB)$; since we also have $f = (1, g) j$ by (2.8), (3.8) becomes

$$\phi(1, g) . \phi j . \phi^0 = \iota' \phi_0 g. \tag{3.10}$$

Taking in particular $B = A$ and $g = 1_A$, we get

$$\phi j_A . \phi^0 = \iota' \phi_0 1_A. \tag{3.11}$$

Proposition 3.5. *Axiom* CF1 *is equivalent to the commutativity of the* *diagram*

$$\begin{array}{ccc} V(A\,B) & \xrightarrow{\phi_0} & V'\phi(A\,B) \\ & \searrow^{\phi} & \downarrow{\scriptstyle V'\hat{\phi}} \\ & & V'(\phi A, \phi B) \end{array} \tag{3.12}$$

Proof. By the representation theorem (3.12) commutes if and only if both legs give the same result when we set $B = A$ and evaluate at 1_A. But $\phi 1 = 1$, and by (3.11)

$$V'\hat{\phi} \,.\, \phi_0 1 = V'\hat{\phi} \,.\, \iota'^{-1}(\phi j \,.\, \phi^0) \,.$$

Since by the naturality of ι' we have

$$V'\hat{\phi} \,.\, \iota'^{-1} = \iota'^{-1} \,.\, V'(1, \hat{\phi}) \,,$$

the commutativity of (3.12) is equivalent to

$$1 = \iota'^{-1} \,.\, V'(1, \hat{\phi})\,(\phi j \,.\, \phi^0) \,,$$

that is,

$$\iota' 1 = V'(1, \hat{\phi})\,(\phi j \,.\, \phi^0)$$

or

$$j' = \hat{\phi} \,.\, \phi j \,.\, \phi^0 \,,$$

which is CF1.

Proposition 3.6. *If* $(\chi, \hat{\chi}, \chi^0)$ *is the composite of the closed functors*

$$(\phi, \hat{\phi}, \phi^0) : \mathscr{V} \to \mathscr{V}' \quad \textit{and} \quad (\psi, \hat{\psi}, \psi^0) : \mathscr{V}' \to \mathscr{V}''$$

then $\chi_0 : V \to V''\psi\phi$ *is the composite*

$$\chi_0 : \qquad V \xrightarrow[\phi_0]{} V'\phi \xrightarrow[\psi_0\phi]{} V''\psi\phi \,. \tag{3.13}$$

Proof. We show that if we *define* χ_0 by (3.13) and then use Proposition 3.4 to define χ^0 we get (3.3). By (3.9) we have

$$\begin{aligned} \chi^0 &= \iota''\, \chi_0 \iota^{-1} 1 \\ &= \iota''\, \psi_0 \phi_0 \iota^{-1} 1 \\ &= \iota''\, \psi_0 \iota'^{-1} \phi^0 \quad \text{by (3.9)} \\ &= \psi \phi^0 \,.\, \psi^0 \quad \text{by (3.8)} \,, \end{aligned}$$

which agrees with (3.3).

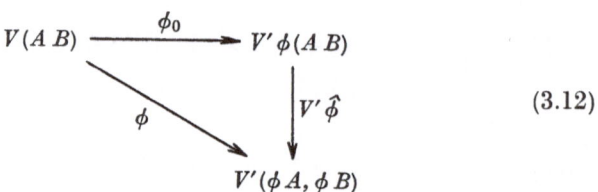

We say that a closed functor $\Phi = (\phi, \hat{\phi}, \phi^0): \mathscr{V} \to \mathscr{V}'$ is *normal* if $V = V'\phi: \mathscr{V}_0 \to \mathscr{S}$ and $\phi_0 = 1: V \to V'\phi$. From Proposition 3.6 we get at once:

Proposition 3.7. *The identity closed functor, and the composite of normal closed functors, are normal; so is the inverse of a normal closed functor that is an isomorphism.*

In view of Proposition 3.5, we may define a normal closed functor $\Phi: \mathscr{V} \to \mathscr{V}'$ directly, as consisting of a functor $\phi: \mathscr{V}_0 \to \mathscr{V}'_0$ and a natural transformation $\hat{\phi}: \phi(AB) \to (\phi A, \phi B)$, satisfying the axioms

NCF0. $$V = V'\phi: \mathscr{V}_0 \to \mathscr{S};$$

NCF1. $$V'\hat{\phi}: V'\phi(AB) \to V'(\phi A, \phi B)$$

coincides with

$$\phi: V(AB) \to V'(\phi A, \phi B);$$

and also the axioms CF2, CF3, in which ϕ^0 is defined by (3.7) with $\phi_0 = 1$.

Proposition 3.8. *The axioms* CF2 *and* CF3 *for a closed functor*

$$\Phi = (\phi, \hat{\phi}, \phi^0): \mathscr{V} \to \mathscr{V}'$$

are consequences of CF1 *if V' is faithful, provided that $\phi_0: VA \to V'\phi A$ is an epimorphism for each A (and so in particular if Φ is normal).*

Proof. For simplicity we shall give the proof only for the case where Φ is normal; the reader will easily provide the proof of the general case, relying on (3.12) instead of its special case NCF1.

Since V' is faithful the diagrams of CF2 and CF3 commute if their images under V' do so. However V' of CF2 coincides, in view of NCF0 and NCF1, with the diagram (3.7) (with $\phi_0 = 1$) which defines ϕ^0. Again V' of CF3 coincides, in view of NCF0, NCF1, and (2.7), with the diagram asserting the naturality in B of $\hat{\phi}_{AB}$.

Proposition 3.9. *Let $\mathscr{V}$ and $\mathscr{V}'$ be closed categories with $V': \mathscr{V}'_0 \to \mathscr{S}$ faithful, and for each $A \in \mathscr{V}_0$ let ϕA be an object of $\mathscr{V}'_0$ with $V'\phi A = VA$. Then there is a normal closed functor $\Phi = (\phi, \hat{\phi}): \mathscr{V} \to \mathscr{V}'$ with the given value on objects if and only if*

(i) *for each $f: A \to B$ in $\mathscr{V}_0$, the morphism $Vf: VA \to VB$ is $V'\phi f$ for some $\phi f: \phi A \to \phi B$;*

(ii) *for each A, B in $\mathscr{V}_0$, the morphism $\phi: V(AB) \to V'(\phi A, \phi B)$ is $V'\hat{\phi}$ for some $\hat{\phi}: \phi(AB) \to (\phi A, \phi B)$;*

and if these conditions are satisfied Φ is unique.

Proof. The conditions are clearly necessary, and if they are satisfied ϕf and $\hat{\phi}$ are unique by the faithfulness of V'. We further conclude from the faithfulness of V' that ϕ is functorial and that $\hat{\phi}$ is natural (because $V'\hat{\phi} = \phi$ is).

Remark 3.10. In concrete cases of closed functors, as in the proposition below, we often by abuse of language denote a closed functor $(\phi, \hat{\phi}, \phi^0)$ by the letter ϕ.

Proposition 3.11. *If $\mathscr{V}$ is a closed category, the functor $V : \mathscr{V}_0 \to \mathscr{S}$ admits a unique extension to a normal closed functor $(V, \hat{V}, V^0) : \mathscr{V} \to \mathscr{S}$, which we still denote by V. We have*

$$\hat{V}_{AB} = V_{AB} : V(AB) \to (VA, VB), \tag{3.14}$$

and $\quad V^0 : * \to VI \quad$ *is given by*

$$V^0 * = \iota_I^{-1} 1 \tag{3.15}$$

where $\iota_I : VI \to V(II)$. Moreover for any $f \in V(IA)$, the image of $*$ under

$$* \xrightarrow[V^0]{} VI \xrightarrow[Vf]{} VA \quad \textit{is} \quad \iota^{-1} f \in VA; \tag{3.16}$$

and in particular the image of $*$ under

$$* \xrightarrow[V^0]{} VI \xrightarrow[Vj]{} V(AA) \quad \textit{is} \quad 1_A. \tag{3.17}$$

Proof. Clearly $\hat{V}$ is unique by NCF1, and NCF0 and NCF1 are in fact satisfied, which suffices by Proposition 3.8. The equations (3.15) and (3.16) are translations of (3.9) and (3.8), and (3.17) is a special case of (3.16).

Proposition 3.12. *A closed functor $\Phi : \mathscr{V} \to \mathscr{V}'$ is normal if and only if the following diagram of closed functors commutes:*

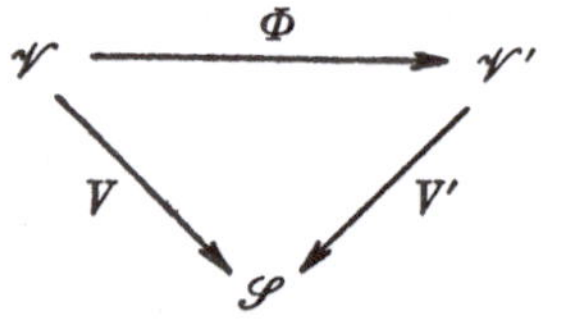

Proof. Set $V'\Phi = X = (\chi, \hat{\chi}, \chi^0)$; then by (3.1), (3.2) and (3.13) we have $\chi = V'\phi$, $\chi = V' \,.\, V'\hat{\phi}$, and $\chi_0 = \phi_0$. So if $X = V$ we have $\phi_0 = V_0 = 1$ and Φ is normal; while if Φ is normal we have $\chi = V'\phi = V$ by NCF0, and $\hat{\chi} = V'.\ V'\hat{\phi} = V'\phi = V$ by NCF0 and NCF1, so that $X = V$.

Remark 3.13. We shall refer to $V : \mathscr{V} \to \mathscr{S}$ as the *basic closed functor* associated with $\mathscr{V}$.

4. Closed Natural Transformations

Let $\Phi = (\phi, \hat{\phi}, \phi^0)$, $\Psi = (\psi, \psi, \psi^0)$ be closed functors $\mathscr{V} \to \mathscr{V}'$. A *closed natural transformation*

$$\eta : \Phi \to \Psi : \mathscr{V} \to \mathscr{V}'$$

consists of a natural transformation

$$\eta : \phi \to \psi : \mathscr{V}_0 \to \mathscr{V}'_0$$

satisfying the following two axioms.

CN 1. The following diagram commutes:

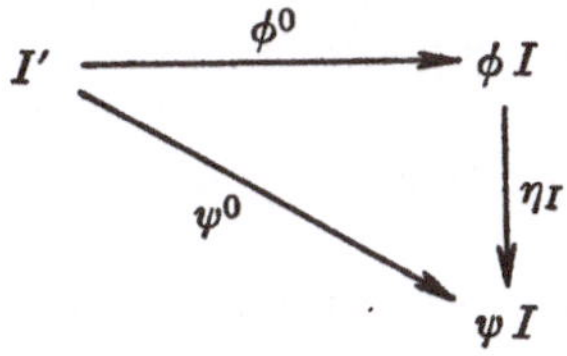

CN 2. The following diagram commutes:

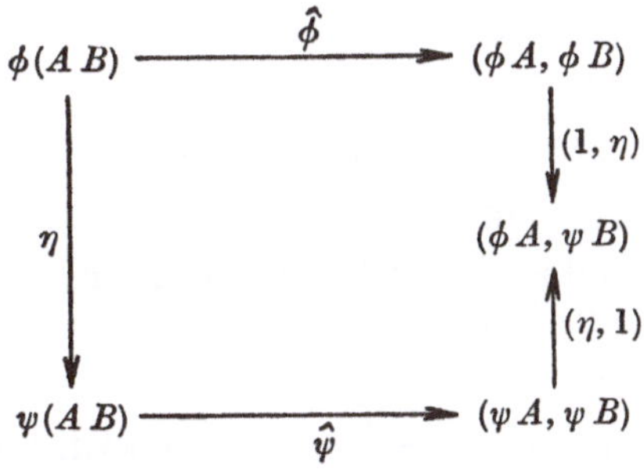

Proposition 4.1. *If we define ϕ_0 and ψ_0 as in Proposition 3.4,* CN 1 *is equivalent to the commutativity of the diagram*

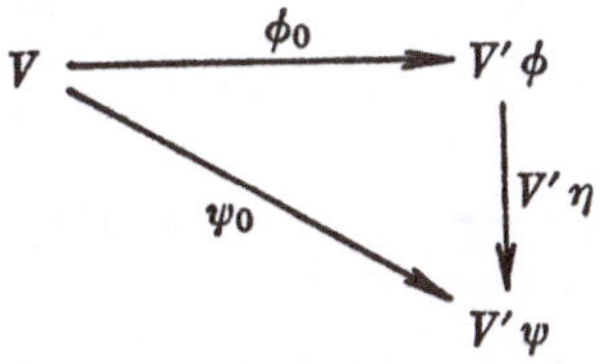

(4.1)

Proof. We show that if we *define* ψ_0 by (4.1), the ψ^0 that corresponds to it by Proposition 3.4 is that given by CN 1. We have by (3.9)

$$\psi^0 = \iota' \psi_0 \iota^{-1} 1 = \iota' \,.\, V'\eta \,.\, \phi_0 \iota^{-1} 1 \,;$$

but by the naturality of ι' we have

$$\iota' \,.\, V'\eta = V'(1, \eta) \,.\, \iota',$$

and thus

$$\begin{aligned} \psi^0 &= V'(1, \eta)\, \iota' \phi_0\, \iota^{-1} 1 \\ &= V'(1, \eta)\, \phi^0 \quad \text{by (3.9)} \\ &= \eta\, \phi^0, \end{aligned}$$

as required.

Theorem 4.2. *Closed categories, closed functors, and closed natural transformations form a hypercategory $\mathscr{Cl}$ if we define the composite of $\eta: \Phi \to \Phi'$ and $\zeta: \Phi' \to \Phi''$ to be the composite $\zeta\eta$ of $\eta: \phi \to \phi'$ and $\zeta: \phi' \to \phi''$, and if for $\Psi: \mathscr{V}' \to \mathscr{V}$, $X: \mathscr{W} \to \mathscr{W}'$, and $\eta: \Phi \to \Phi': \mathscr{V} \to \mathscr{W}$ we define $\eta\Psi$ and $X\eta$ to be $\eta\psi$ and $\chi\eta$. Moreover $\eta: \Phi \to \Phi'$ is an isomorphism if and only if $\eta: \phi \to \phi'$ is.*

The proofs are straightforward, and we leave them to the reader.

Proposition 4.3. *The axiom* CN2 *for a closed natural transformation $\eta: \Phi \to \Psi: \mathscr{V} \to \mathscr{V}'$ is a consequence of* CN1 *if V' is faithful and Φ is normal.*

Proof. If Φ is normal we have $\phi_0 = 1$ and (4.1) gives

$$V'\eta = \psi_0 . \tag{4.2}$$

Since V' is faithful it suffices to show that the image under V' of the diagram CN2 commutes. Using NCF0 and NCF1 for Φ, and (4.2), we may write V' of CN2 in the form

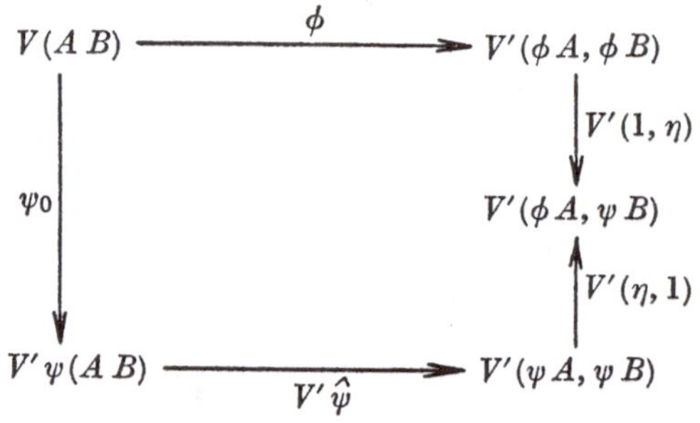

Since $V'\psi \,.\, \psi_0 = \psi$ by (3.12), this is just the diagram that asserts the naturality of η.

Remark 4.4. For the above proposition it would suffice to assume ϕ_0 epimorphic instead of Φ normal.

Proposition 4.5. *If* $\Phi : \mathscr{V} \to \mathscr{V}'$ *is a closed functor, the natural transformation* $\phi_0 : V \to V'\phi : \mathscr{V}_0 \to \mathscr{S}$ *is a closed natural transformation* $\phi_0 : V \to V'\Phi : \mathscr{V} \to \mathscr{S}$. *Moreover if* $\eta : \Phi \to \Psi : \mathscr{V} \to \mathscr{V}'$ *we have the commutative diagram of closed natural transformations*

$$\begin{array}{ccc} V & \xrightarrow{\phi_0} & V'\Phi \\ & \searrow^{\psi_0} & \downarrow{\scriptstyle V'\eta} \\ & & V'\Psi \end{array} \qquad (4.3)$$

Proof. By Proposition 4.3 we need verify only CN 1. If we write this in the equivalent form (4.1) it becomes

$$\begin{array}{ccc} V & \xrightarrow{1} & V \\ & \searrow^{\phi_0} & \downarrow{\scriptstyle \phi_0} \\ & & V'\phi \end{array}$$

which commutes trivially. The diagram (4.3) is just a translation of (4.1).

5. Categories Over a Closed Category

Let $\mathscr{V}$ be a closed category. A *category* $\mathscr{A}$ *over* $\mathscr{V}$, or a $\mathscr{V}$*-category*, consists of the following four data:

(i) a class obj $\mathscr{A}$ of "objects";
(ii) for each $A, B \in \operatorname{obj} \mathscr{A}$, an object $\mathscr{A}(AB)$ of $\mathscr{V}_0$;
(iii) for each $A \in \operatorname{obj} \mathscr{A}$, a morphism

$$j_A : I \to \mathscr{A}(AA)$$

in $\mathscr{V}_0$;
(iv) for each $A, B, C \in \operatorname{obj} \mathscr{A}$, a morphism

$$L^A_{BC} : \mathscr{A}(BC) \to (\mathscr{A}(AB), \mathscr{A}(AC))$$

in $\mathscr{V}_0$.

These data are to satisfy the following three axioms, in which $L^{\mathscr{A}(AB)}$ and $j_{\mathscr{A}(AB)}$ are the L and the j of $\mathscr{V}$, while the other L's and j's are those of $\mathscr{A}$:

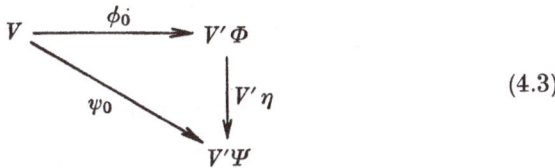

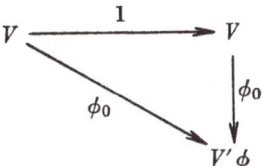

VC1. The following diagram commutes:

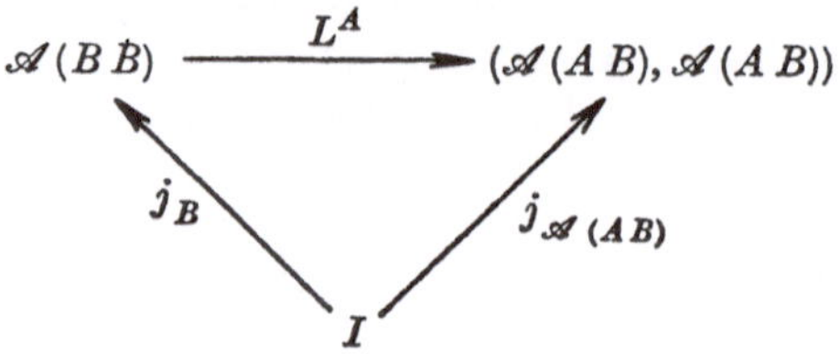

VC2. The following diagram commutes:

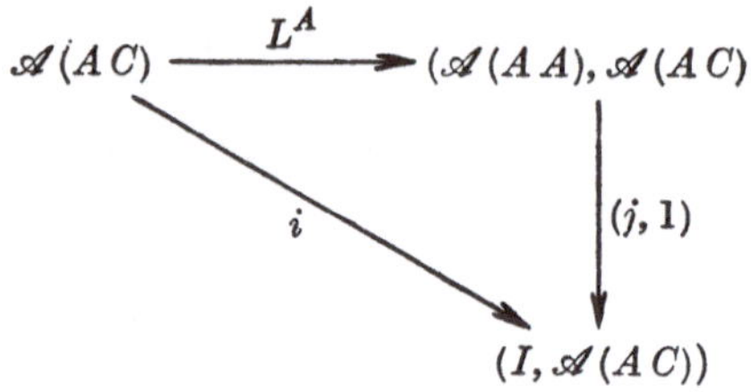

VC3. The following diagram commutes:

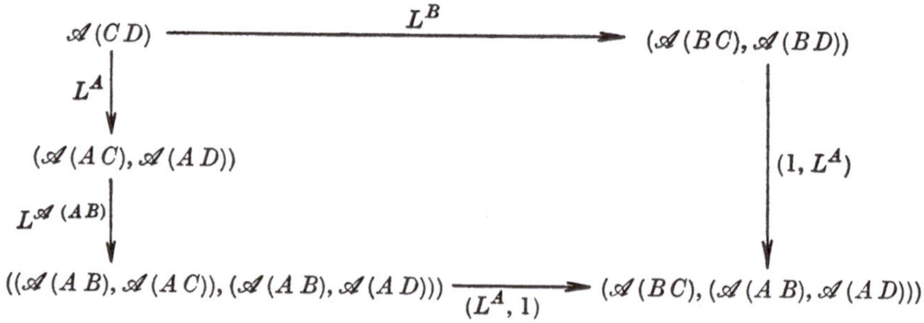

If $\mathscr{A}$ and $\mathscr{B}$ are $\mathscr{V}$-categories, a $\mathscr{V}$-functor $T : \mathscr{A} \to \mathscr{B}$ consists of the following two data:

(i) a function $T : \operatorname{obj} \mathscr{A} \to \operatorname{obj} \mathscr{B}$;

(ii) for each $B, C \in \operatorname{obj} \mathscr{A}$, a morphism

$$T = T_{BC} : \mathscr{A}(BC) \to \mathscr{B}(TB, TC)$$

in $\mathscr{V}_0$.

These data are to satisfy the following two axioms:

VF1. The following diagram commutes:

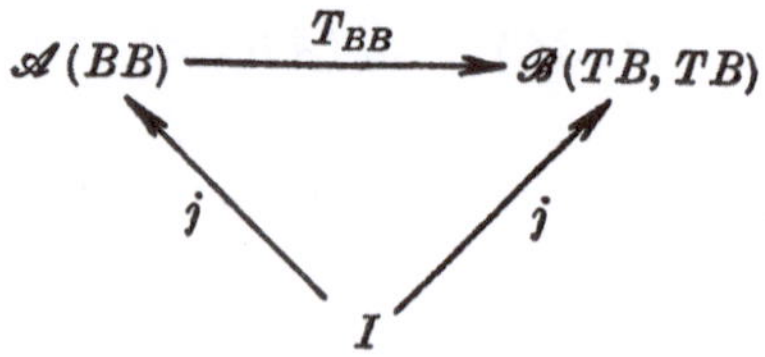

VF2. The following diagram commutes:

$$
\begin{array}{ccc}
\mathscr{A}(CD) & \xrightarrow{\quad L^B \quad} & (\mathscr{A}(BC), \mathscr{A}(BD)) \\
{\scriptstyle T_{CD}}\downarrow & & \\
\mathscr{B}(TC, TD) & & \downarrow {\scriptstyle (1, T_{BD})} \\
{\scriptstyle L^{TB}}\downarrow & & \\
(\mathscr{B}(TB, TC), \mathscr{B}(TB, TD)) & \xrightarrow[(T_{BC}, 1)]{} & (\mathscr{A}(BC), \mathscr{B}(TB, TD))
\end{array}
$$

Theorem 5.1. *$\mathscr{V}$-categories and $\mathscr{V}$-functors form a "category" $\mathscr{V}_*$ if we define the composite of $T : \mathscr{A} \to \mathscr{B}$ and $S : \mathscr{B} \to \mathscr{C}$ to be $P : \mathscr{A} \to \mathscr{C}$ where*

$$PA = STA \tag{5.1}$$

and P_{AB} is the composite

$$P_{AB} : \mathscr{A}(AB) \xrightarrow[T_{AB}]{} \mathscr{B}(TA, TB) \xrightarrow[S_{TA,\, TB}]{} \mathscr{C}(STA, STB) . \tag{5.2}$$

Proof. Clearly composition is associative, with obvious identities $1 : \mathscr{A} \to \mathscr{A}$. We must verify that P satisfies VF1 and VF2, that is, that the exteriors of the following two diagrams commute (See page 446 for diagram (5.4)):

$$
\begin{array}{ccccc}
\mathscr{A}(BB) & \xrightarrow{\quad T \quad} & \mathscr{B}(TB, TB) & \xrightarrow{\quad S \quad} & \mathscr{C}(STB, STB) \\
 & {\scriptstyle j}\nwarrow & \uparrow {\scriptstyle j} & \nearrow {\scriptstyle j} & \\
 & & I & &
\end{array}
\tag{5.3}
$$

In (5.3) one region commutes by VF1 for T and the other by VF1 for S; in (5.4) one region commutes by VF2 for T, one by VF2 for S, and the third trivially.

Theorem 5.2. *If $\mathscr{V}$ is a closed category we get a $\mathscr{V}$-category, also denoted by $\mathscr{V}$, if we take the objects of $\mathscr{V}$ to be those of $\mathscr{V}_0$, take $\mathscr{V}(AB)$ to be (AB), and take for j and L those of the closed category $\mathscr{V}$. Moreover if $\mathscr{A}$ is any $\mathscr{V}$-category and $A \in \mathscr{A}$, we get a $\mathscr{V}$-functor $L^A : \mathscr{A} \to \mathscr{V}$ if we take $L^A B = \mathscr{A}(AB)$ and $(L^A)_{BC} = L^A_{BC}$.*

Proof. The axioms VC1—VC3 for $\mathscr{V}$ reduce to CC1—CC3, and the axioms VF1 and VF2 for L^A reduce to VC1 and VC3.

Remark. In accordance with the above proposition, "object of $\mathscr{V}$" and "object of $\mathscr{V}_0$" are synonyms; we often write $A \in \mathscr{V}$.

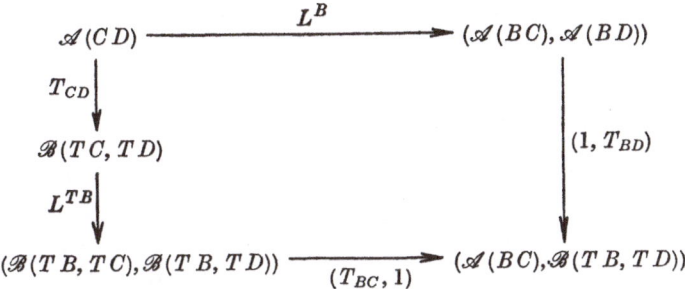

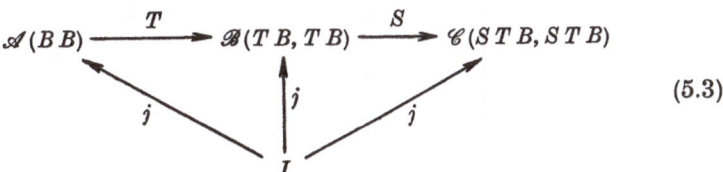

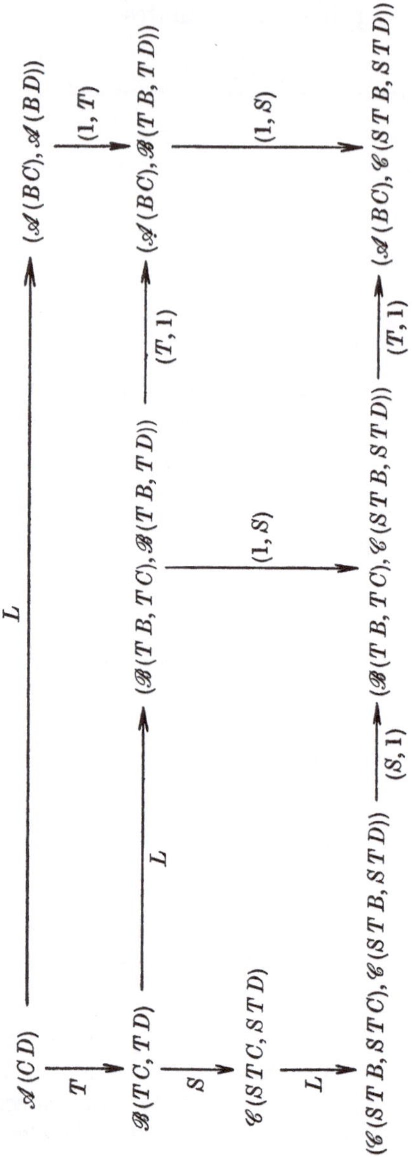

(5.4)

Proposition 5.3. *We obtain a $\mathscr{V}$-category $\mathscr{I}$ with a single object $*$ if we take $\mathscr{I}(**) = I$, take $j: I \to \mathscr{I}(**)$ to be $1: I \to I$, and take $L: \mathscr{I}(**) \to \to (\mathscr{I}(**), \mathscr{I}(**))$ to be $i: I \to (II)$. Moreover if $\mathscr{A}$ is a $\mathscr{V}$-category and $A \in \mathscr{A}$, we get a $\mathscr{V}$-functor $J^A: \mathscr{I} \to \mathscr{A}$ if we set $J^A * = A$ and take*

$J^A : \mathscr{I}(**) \to \mathscr{A}(AA)$ to be $j : I \to \mathscr{A}(AA)$. There are no $\mathscr{V}$-functors $\mathscr{I} \to \mathscr{A}$ other than the J^A.

We leave the verification to the reader.

Proposition 5.4. *An $\mathscr{S}$-category $\mathscr{A}$ may be identified with an ordinary category $\mathscr{A}$ if we identify the image of $j : * \to \mathscr{A}(AA)$ with 1_A and identify $(L^A_{BC} f)g$ with the composite fg, where $f \in \mathscr{A}(BC)$ and $g \in \mathscr{A}(AB)$. An $\mathscr{S}$-functor is then an ordinary functor, and in particular the functor $L^A : \mathscr{A} \to \mathscr{S}$ is then the left represented functor $\mathscr{A}(A\,—)$.*

Proof. The reader will easily verify that VC1—VC3 express the two identity laws and the associative law for composition, while VF1 and VF2 become $T1 = 1$ and $T(fg) = Tf \cdot Tg$.

Remark. By analogy with the above we call the $\mathscr{V}$-functor $L^A : \mathscr{A} \to \mathscr{V}$ a left represented $\mathscr{V}$-functor.

6. The Effect of a Closed Functor

In this section it will be convenient to use j, L for the appropriate data in a $\mathscr{V}$-category $\mathscr{A}$, and j', L' for the corresponding data in *a* $\mathscr{V}'$-category $\mathscr{B}$, etc.

Proposition 6.1. *If $\Phi = (\phi, \hat{\phi}, \phi^0) : \mathscr{V} \to \mathscr{V}'$ is a closed functor and $\mathscr{A}$ is a $\mathscr{V}$-category, the following data define a $\mathscr{V}'$-category $\Phi_* \mathscr{A}$:*

(i) *the objects of $\Phi_* \mathscr{A}$ are those of $\mathscr{A}$;* (6.1)

(ii) $(\Phi_* \mathscr{A})(AB) = \phi\mathscr{A}(AB)$ *(that is, $\phi(\mathscr{A}(AB))$);* (6.2)

(iii) *$j' : I' \to \phi\mathscr{A}(AA)$ is the composite*

$$I' \xrightarrow[\phi^0]{} \phi I \xrightarrow[\phi j]{} \phi\mathscr{A}(AA); \tag{6.3}$$

(iv) *$L' : \phi\mathscr{A}(BC) \to (\phi\mathscr{A}(AB), \phi\mathscr{A}(AC))$ is the composite*

$$\phi\mathscr{A}(BC) \xrightarrow[\phi L]{} \phi(\mathscr{A}(AB), \mathscr{A}(AC)) \xrightarrow[\hat{\phi}]{} (\phi\mathscr{A}(AB), \phi\mathscr{A}(AC)). \tag{6.4}$$

Proof. The axioms VC1—VC3 for $\Phi_* \mathscr{A}$ assert the commutativity of the exteriors of the following three diagrams:

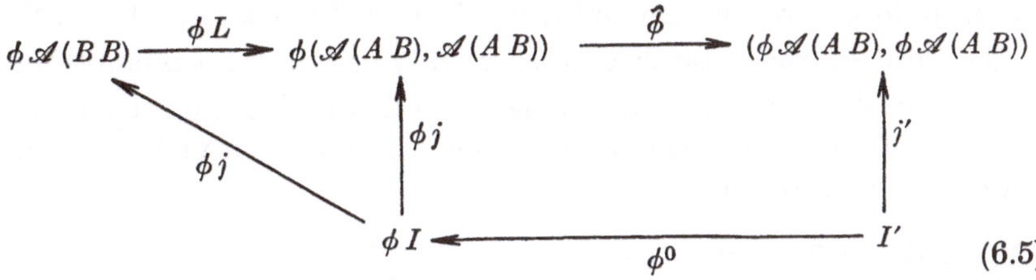

(6.5)

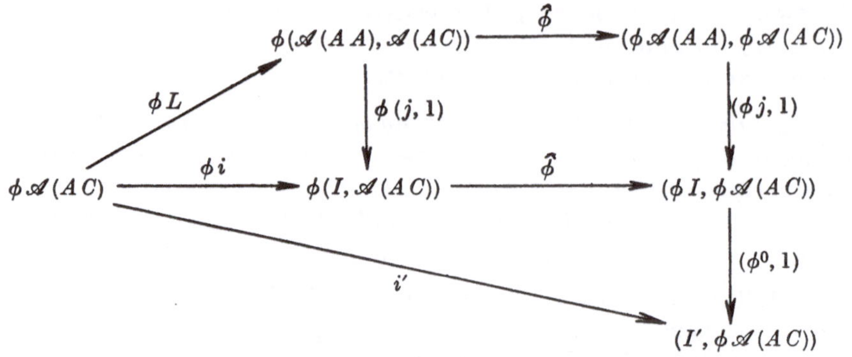

(6.6)

In the next diagram, X, Y, Z stand for $\mathscr{A}(AB)$, $\mathscr{A}(AC)$, $\mathscr{A}(AD)$:

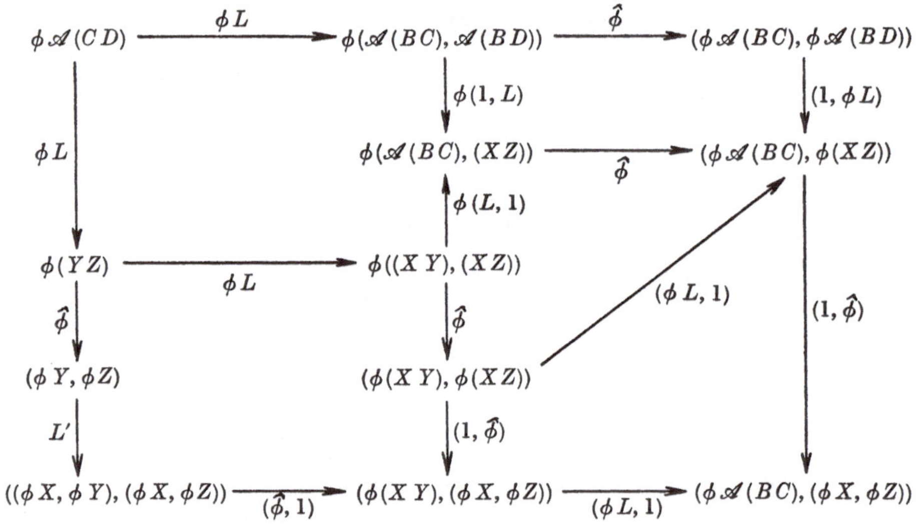

(6.7)

In (6.5) one region commutes by ϕ of VC1 for $\mathscr{A}$, and the other by CF1; in (6.6) one region commutes by ϕ of VC2 for $\mathscr{A}$, one by CF2, and the third by the naturality of $\hat{\phi}$; in (6.7) one region commutes by ϕ of VC3 for $\mathscr{A}$, one by CF3, two by the naturality of $\hat{\phi}$, and one trivially.

The proofs of the following propositions are similar but rather easier: we leave them to the reader.

Proposition 6.2. *If* $\Phi = (\phi, \hat{\phi}, \phi^0) : \mathscr{V} \to \mathscr{V}'$ *is a closed functor and*

$T: \mathcal{A} \to \mathcal{B}$ is a $\mathcal{V}$-functor, the following data define a $\mathcal{V}'$-functor

$$\Phi_* T : \Phi_* \mathcal{A} \to \Phi_* \mathcal{B}:$$

$$\text{(i)} \quad (\Phi_* T) A = TA; \tag{6.8}$$

$$\text{(ii)} \quad (\Phi_* T)_{BC} = \phi T_{BC} : \phi \mathcal{A}(BC) \to \phi \mathcal{B}(TB, TC). \tag{6.9}$$

Proposition 6.3. *The assigments $\mathcal{A} \mapsto \Phi_* \mathcal{A}$, $T \mapsto \Phi_* T$ constitute a functor $\Phi_* : \mathcal{V}_* \to \mathcal{V}'_*$ from the "category" of $\mathcal{V}$-categories and $\mathcal{V}$-functors to that of $\mathcal{V}'$-categories and $\mathcal{V}'$-functors.*

Proposition 6.4. *If $\eta : \Phi \to \Psi : \mathcal{V} \to \mathcal{V}'$ is a closed natural transformation and $\mathcal{A}$ is a $\mathcal{V}$-category, we obtain a $\mathcal{V}'$-functor $\eta_{*\mathcal{A}} : \Phi_* \mathcal{A} \to \Psi_* \mathcal{A}$ if we set*

$$\text{(i)} \quad \eta_{*\mathcal{A}} A = A; \tag{6.10}$$

$$\text{(ii)} \quad (\eta_{*\mathcal{A}})_{BC} = \eta_{\mathcal{A}(BC)} : \phi \mathcal{A}(BC) \to \psi \mathcal{A}(BC). \tag{6.11}$$

Moreover the $\eta_{*\mathcal{A}}$ for $\mathcal{A} \in \mathcal{V}_*$ constitute a natural transformation

$$\eta_* : \Phi_* \to \Psi_* : \mathcal{V}_* \to \mathcal{V}'_*.$$

Theorem 6.5. *The assigments $\mathcal{V} \mapsto \mathcal{V}_*$, $\Phi \mapsto \Phi_*$, $\eta \mapsto \eta_*$ constitute a hyperfunctor $_* : \mathcal{C}\ell \to \mathcal{C}at$ from the "hypercategory" of closed categories to that of categories.*

We now consider the effect of Φ_* on the particular $\mathcal{V}$-category $\mathcal{V}$ and on the particular $\mathcal{V}$-functors $L^A : \mathcal{A} \to \mathcal{V}$.

Theorem 6.6. *If $\Phi = (\phi, \hat{\phi}, \phi^0) : \mathcal{V} \to \mathcal{V}'$ is a closed functor, we obtain a $\mathcal{V}'$-functor $\hat{\Phi} : \Phi_* \mathcal{V} \to \mathcal{V}'$ if we set*

$$\text{(i)} \quad \hat{\Phi} A = \phi A; \tag{6.12}$$

$$\text{(ii)} \quad \hat{\Phi}_{BC} = \hat{\phi}_{BC} : \phi(BC) \to (\phi B, \phi C). \tag{6.13}$$

If $\Psi : \mathcal{V}' \to \mathcal{V}''$ is another closed functor, and if $X = \Psi\Phi : \mathcal{V} \to \mathcal{V}''$, then $\hat{X}$ is the composite

$$\Psi_* \Phi_* \mathcal{V} \xrightarrow[\Psi_* \hat{\Phi}]{} \Psi_* \mathcal{V}' \xrightarrow[\hat{\Psi}]{} \mathcal{V}''. \tag{6.14}$$

If $\mathcal{A}$ is a $\mathcal{V}$-category and $A \in \mathcal{A}$, the following diagram of $\mathcal{V}'$-functors commutes:

$$\begin{array}{ccc} \Phi_* \mathcal{A} & \xrightarrow{\Phi_* L^A} & \Phi_* \mathcal{V} \\ & \searrow^{L'^A} & \downarrow \hat{\Phi} \\ & & \mathcal{V}' \end{array} \tag{6.15}$$

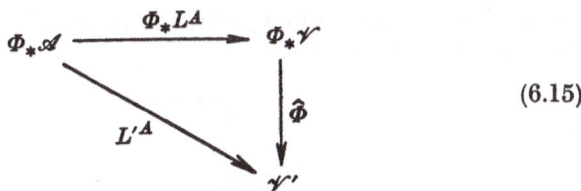

Proof. The axioms VF1 and VF3 for $\hat{\Phi}$ reduce to CF1 and CF3 for Φ. The assertions (6.14) and (6.15) are immediate from (3.1), (3.2), (6.2) and (6.4).

The next proposition does the same for the particular $\mathscr{V}$-category $\mathscr{I}$ and the particular $\mathscr{V}$-functors $J^A : \mathscr{I} \to \mathscr{A}$; we write $\mathscr{I}'$ for the $\mathscr{V}'$-category analogous to $\mathscr{I}$. We leave the proofs to the reader.

Proposition 6.7. *If $\Phi = (\phi, \hat{\phi}, \phi^0) : \mathscr{V} \to \mathscr{V}'$ is a closed functor, we obtain a $\mathscr{V}'$-functor $\Phi^0 : \mathscr{I}' \to \Phi_* \mathscr{I}$ if we set:*

(i) $\Phi^0 * = *;$ (6.16)

(ii) $\Phi^0_{**} : \mathscr{I}'(**) \to \phi \mathscr{I}(**)$ *is* $\phi^0 : I' \to \phi I.$ (6.17)

If $\Psi : \mathscr{V}' \to \mathscr{V}''$ is another closed functor, and if $X = \Psi\Phi : \mathscr{V} \to \mathscr{V}''$, then X^0 is the composite

$$\mathscr{I}'' \xrightarrow[\Psi^0]{} \Psi_* \mathscr{I}' \xrightarrow[\Psi_*\Phi^0]{} \Psi_* \Phi_* \mathscr{I} \tag{6.18}$$

If $\mathscr{A}$ is a $\mathscr{V}$-category and $A \in \mathscr{A}$, the following diagram of $\mathscr{V}'$-functors commutes:

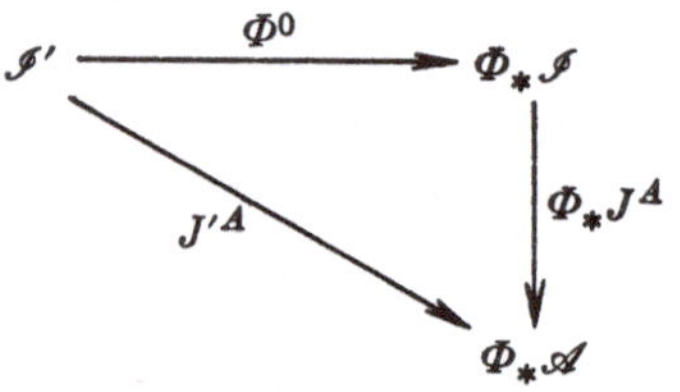

(6.19)

7. The Effect of the Closed Functor $V : \mathscr{V} \to \mathscr{S}$

We apply the results of § 6 to the particular closed functor $V : \mathscr{V} \to \mathscr{S}$. Each $\mathscr{V}$-category $\mathscr{A}$ determines an ordinary category $V_* \mathscr{A}$, with the same objects as $\mathscr{A}$, and with

$$(V_* \mathscr{A})(AB) = V\mathscr{A}(AB). \tag{7.1}$$

The j' of $V_* \mathscr{A}$ is the composite

$$* \xrightarrow[V^0]{} VI \xrightarrow[Vj]{} V\mathscr{A}(AA), \tag{7.2}$$

so that the identity 1_A in $V_* \mathscr{A}$ is the image of $*$ under (7.2). By (3.16) we can express this by:

$$\iota 1_A = j_A \tag{7.3}$$

where

$$\iota : V\mathscr{A}(AA) \to V(I, \mathscr{A}(AA)).$$

Just as (2.4) gives Lemma 2.2, so (7.3) gives:

Lemma 7.1. *For any* $f : \mathscr{A}(AA) \to X$ *in* $\mathscr{V}_0$, *the composite*

$$I \xrightarrow{j} \mathscr{A}(AA) \xrightarrow{f} X$$

is the image of $1 \in V\mathscr{A}(AA)$ *under the composite map*

$$V\mathscr{A}(AA) \xrightarrow{Vf} VX \xrightarrow{i} V(IX).$$

The L' of $V_*\mathscr{A}$ is the composite

$$V\mathscr{A}(BC) \xrightarrow{VL^A} V(\mathscr{A}(AB), \mathscr{A}(AC)) \xrightarrow{V} (V\mathscr{A}(AB), V\mathscr{A}(AC)), \quad (7.4)$$

so that the composite in $V_*\mathscr{A}$ of $g \in V\mathscr{A}(AB)$ and $f \in V\mathscr{A}(BC)$ is

$$fg = (V((VL^A)f))g. \quad (7.5)$$

Taking $\mathscr{A}$ to be the $\mathscr{V}$-category $\mathscr{V}$, we have:

Proposition 7.2. $V_*\mathscr{V} = \mathscr{V}_0$.

Proof. $V_*\mathscr{V}$ has the same objects as $\mathscr{V}$, and so the same objects as $\mathscr{V}_0$; and by (7.1) and (2.1),

$$(V_*\mathscr{V})(AB) = V(AB) = \mathscr{V}_0(AB).$$

By (7.3) and (2.4), $V_*\mathscr{V}$ and $\mathscr{V}_0$ have the same identities; by (7.5), (2.6), and (2.2) the composite in $V_*\mathscr{V}$ is given by

$$\begin{aligned} fg &= (V((VL^A)f))g \\ &= (V(1, f))g \\ &= \mathscr{V}_0(1, f)g, \end{aligned}$$

and this is also the composite fg in $\mathscr{V}_0$.

In accordance with this result we denote $V_*\mathscr{A}$ by $\mathscr{A}_0$, for any $\mathscr{V}$-category $\mathscr{A}$; and we re-write (7.1) for reference as

$$\mathscr{A}_0(AB) = V\mathscr{A}(AB). \quad (7.6)$$

Similarly if $T : \mathscr{A} \to \mathscr{B}$ is a $\mathscr{V}$-functor, we denote V_*T by T_0, so that $T_0 : \mathscr{A}_0 \to \mathscr{B}_0$. The assignment $T \mapsto T_0$ is functorial, and T_0 is given by:

$$T_0 A = TA, \quad (7.7)$$

$$T_0 f = (VT)f \quad \text{for} \quad f \in V\mathscr{A}(BC). \quad (7.8)$$

We call $\mathscr{A}_0$ the underlying category of $\mathscr{A}$, and T_0 the underlying functor of T.

For the underlying functor of $L^A : \mathscr{A} \to \mathscr{V}$ we adopt the special notation $\mathscr{A}(A-) : \mathscr{A}_0 \to \mathscr{V}_0$, so that

$$\mathscr{A}(A-) = V_*L^A. \quad (7.9)$$

The value of $\mathscr{A}(A\,-)$ on the object B is $\mathscr{A}(AB)$, and its value $\mathscr{A}(A\,f)$ on the morphism $f \in V\mathscr{A}(BC)$ is given by

$$\mathscr{A}(A\,f) = (VL^A)\,f. \tag{7.10}$$

Comparing this with (2.6), we see that

$$\mathscr{V}(A\,-) = (A\,-). \tag{7.11}$$

Since the functor $\hat{V} : V_* \mathscr{V} \to \mathscr{S}$ is just $V : \mathscr{V}_0 \to \mathscr{S}$, (6.15) becomes the commutative diagram

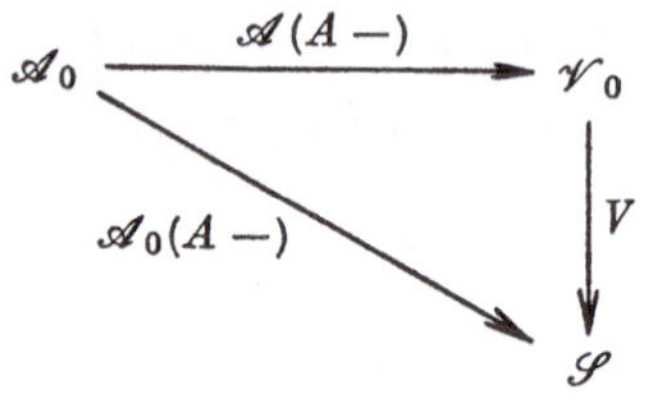

(7.12)

In particular for $f \in \mathscr{A}_0(BC)$ we have

$$V\mathscr{A}(A\,f) = \mathscr{A}_0(A\,f). \tag{7.13}$$

As the reader no doubt suspects, we shall later show the existence of a functor $\mathscr{A}(-\,B) : \mathscr{A}_0^* \to \mathscr{V}_0$, forming together with $\mathscr{A}(A\,-)$ a bifunctor

$$\hom \mathscr{A} : \mathscr{A}_0^* \times \mathscr{A}_0 \to \mathscr{V}_0$$

satisfying the analogue of CC0. Until we do we can state the analogue of only one half of Proposition 2.4, namely the commutativity of

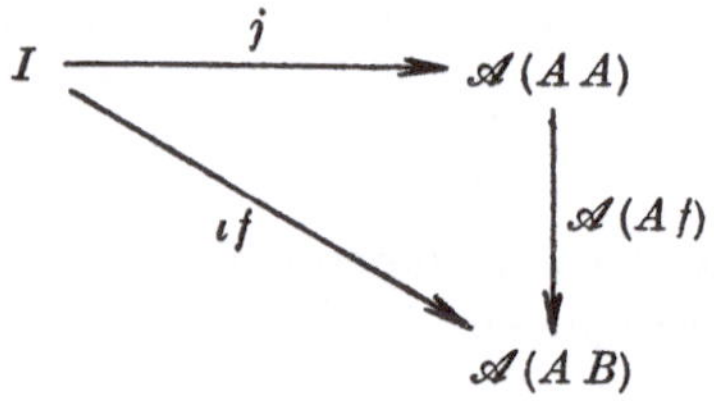

(7.14)

where $f \in V\mathscr{A}(AB)$; this follows at once from Lemma 7.1 and (7.13).

Now let $\Phi = (\phi, \hat{\phi}, \phi^0) : \mathscr{V} \to \mathscr{V}'$ be a closed functor. By Proposition 4.5 the natural transformation $\phi_0 : V \to V'\phi$ is a closed natural transformation $\phi_0 : V \to V'\Phi$. By Proposition 6.4 therefore, it induces a natural transformation $\phi_{0*} : V_* \to V'_* \Phi_* : \mathscr{V}_* \to \mathscr{S}_*$. We shall write Φ_0 for ϕ_{0*}; then Proposition 6.4 may be stated for this special case as:

Proposition 7.3. *If $\Phi : \mathscr{V} \to \mathscr{V}'$ is a closed functor, we have for each*

$\mathscr{V}$-category $\mathscr{A}$ a functor $\Phi_{0\mathscr{A}} : \mathscr{A}_0 \to (\Phi_* \mathscr{A})_0$ given by

(i) $\Phi_{0\mathscr{A}} A = A$; (7.15)

(ii) $\Phi_{0\mathscr{A}} f = \phi_0 f$ *for* $f \in V\mathscr{A}(BC)$, (7.16)

where

$$\phi_0 : V\mathscr{A}(BC) \to V'\phi\mathscr{A}(BC).$$

Moreover if $T : \mathscr{A} \to \mathscr{B}$ is a $\mathscr{V}$-functor, we have a commutative diagram of functors

$$\begin{array}{ccc} \mathscr{A}_0 & \xrightarrow{T_0} & \mathscr{B}_0 \\ {\scriptstyle \Phi_{0\mathscr{A}}}\downarrow & & \downarrow{\scriptstyle \Phi_{0\mathscr{B}}} \\ (\Phi_*\mathscr{A})_0 & \xrightarrow[(\Phi_* T)_0]{} & (\Phi_*\mathscr{B})_0 \end{array} \tag{7.17}$$

If Φ is normal we have $\mathscr{A}_0 = (\Phi_* \mathscr{A})_0$ and $\Phi_{0\mathscr{A}} = 1$.

Consider now the $\mathscr{V}'$-functor $\hat{\Phi} : \Phi_* \mathscr{V} \to \mathscr{V}'$ and its underlying functor $V'_* \hat{\Phi} : (\Phi_* \mathscr{V})_0 \to \mathscr{V}'_0$; the latter is given on objects by $(V'_* \hat{\Phi}) A = \phi A$, and on morphisms by $(V'_* \hat{\Phi}) f = (V'\hat{\phi}) f$, where

$$V'\hat{\phi} : V'\phi(AB) \to V'(\phi A, \phi B).$$

It follows immediately from (3.12) that we have a commutative diagram of functors

$$\begin{array}{ccc} \mathscr{V}_0 & \xrightarrow{\Phi_{0\mathscr{V}}} & (\Phi_*\mathscr{V})_0 \\ & {\scriptstyle \phi}\searrow & \downarrow{\scriptstyle V'_*\hat{\Phi}} \\ & & \mathscr{V}_0' \end{array} \tag{7.18}$$

Proposition 7.4. *If $\Phi : \mathscr{V} \to \mathscr{V}'$ is a closed functor, we have for any $\mathscr{V}$-category $\mathscr{A}$ and any $A \in \mathscr{A}$ a commutative diagram of functors*

$$\begin{array}{ccc} \mathscr{A}_0 & \xrightarrow{\mathscr{A}(A-)} & \mathscr{V}_0 \\ {\scriptstyle \Phi_{0\mathscr{A}}}\downarrow & & \downarrow{\scriptstyle \phi} \\ (\Phi_*\mathscr{A})_0 & \xrightarrow[(\Phi_*\mathscr{A})(A-)]{} & \mathscr{V}_0' \end{array} \tag{7.19}$$

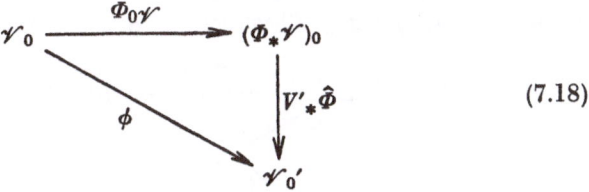

Proof. Consider the diagram:

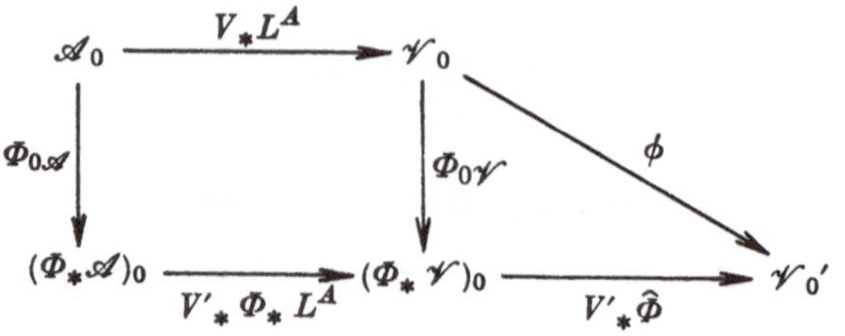

The rectangle commutes by (7.17), and the triangle by (7.18). The top edge is $V_* L^A = \mathscr{A}(A -)$, and the bottom edge is, by (6.15),

$$V'_* L'^A = (\Phi_* \mathscr{A})(A -).$$

Proposition 7.5. *If $\Phi : \mathscr{V} \to \mathscr{V}'$ and $\Psi : \mathscr{V}' \to \mathscr{V}''$ are closed functors with composite $X = \Psi\Phi : \mathscr{V} \to \mathscr{V}''$, we have for any $\mathscr{V}$-category $\mathscr{A}$ a commutative diagram of functors:*

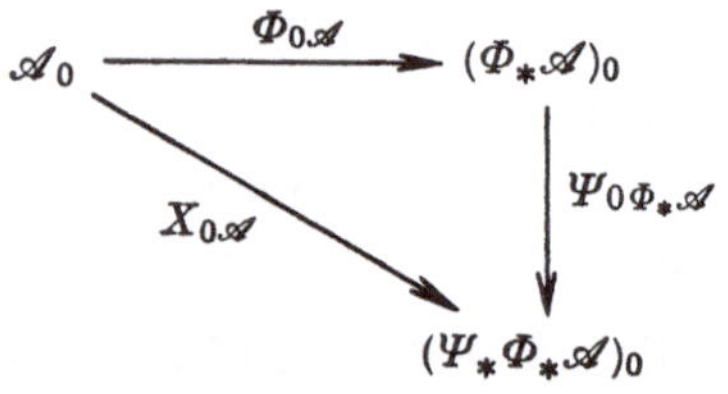

(7.20)

Proof. Immediate from (3.13).

Proposition 7.6. *If $\eta : \Phi \to \Psi : \mathscr{V} \to \mathscr{V}'$ is a closed natural transformation we have for any $\mathscr{V}$-category $\mathscr{A}$ a commutative diagram of functors*

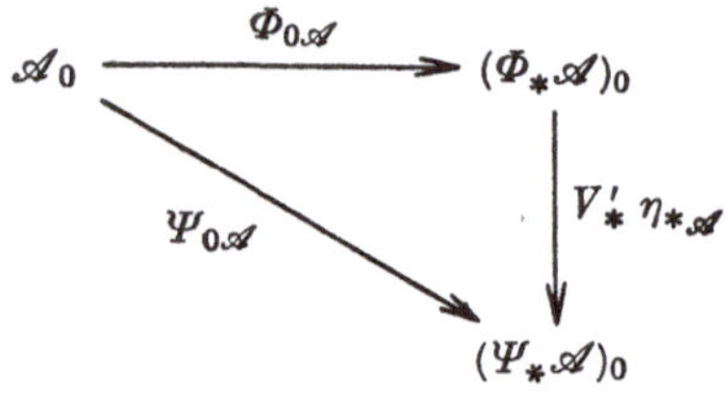

(7.21)

Proof. Immediate from Proposition 4.5.

8. $\mathscr{V}$-functors into $\mathscr{V}$

We shall in § 10 define $\mathscr{V}$-natural transformations between $\mathscr{V}$-functors, turning $\mathscr{V}_*$ into a hypercategory; but we cannot do so until we have defined, for a $\mathscr{V}$-category $\mathscr{A}$, the functor $\mathscr{A}(-B) : \mathscr{A}_0^* \to \mathscr{V}_0$. The easiest way to get this and to establish its properties is by using the representation theorem for $\mathscr{V}$-functors, which we shall prove below. To discuss the representation theorem we need $\mathscr{V}$-natural transformations, but only for functors into $\mathscr{V}$; and here there is no difficulty, for we already have the functor $(-B): \mathscr{V}_0^* \to \mathscr{V}_0$.

For a closed category $\mathscr{V}$ and $\mathscr{V}$-functors $T, S: \mathscr{A} \to \mathscr{V}$, a *$\mathscr{V}$-natural transformation* $\alpha: T \to S$ consists of a family of morphisms $\alpha_A: TA \to$ $\to SA$ in $\mathscr{V}_0$, indexed by the objects of $\mathscr{A}$, satisfying the axiom:

VN($\mathscr{V}$). The following diagram commutes:

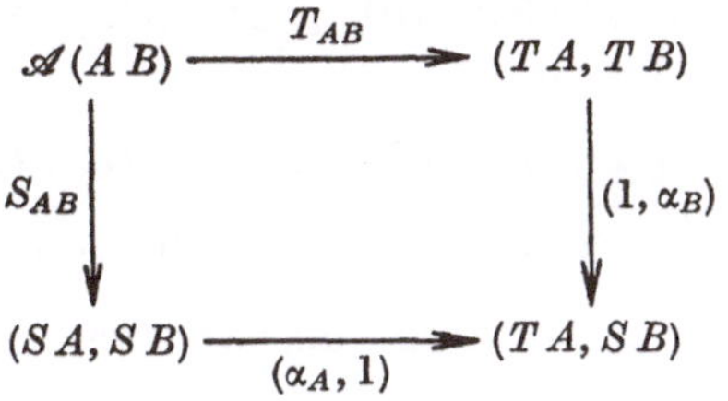

Proposition 8.1. *For a fixed $\mathscr{V}$-category $\mathscr{A}$, the $\mathscr{V}$-functors $\mathscr{A} \to \mathscr{V}$ and the $\mathscr{V}$-natural transformations between them form a "category" if we define the composite $\beta\alpha$ of $\alpha: T \to S$ and $\beta: S \to R$ by*

$$(\beta\alpha)_A = \beta_A \alpha_A . \tag{8.1}$$

Moreover $\alpha: T \to S$ is an isomorphism in this category if and only if each α_A is an isomorphism in $\mathscr{V}_0$.

Proof. The composition law (8.1) is associative, with identities 1_T having components 1_{TA}. The axiom VN for $\beta\alpha$ asserts the commutativity of the exterior of the following diagram:

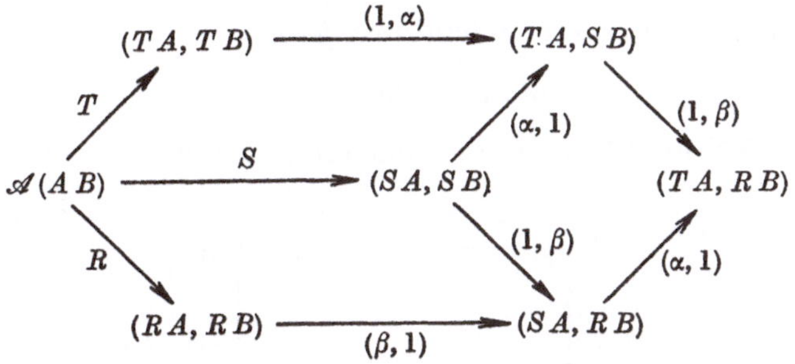

(8.2)

Here one region commutes by VN for α, one by VN for β, and one trivially.

If $\beta = \alpha^{-1}$ then by (8.1) we have $\beta_A = \alpha_A^{-1}$. Conversely if α is $\mathscr{V}$-natural and each α_A is an isomorphism, define β by $\beta_A = \alpha_A^{-1}$. Then in (8.2) the top region, the right region, and the exterior all commute.

Since $(\alpha, 1)$ is an isomorphism, the bottom region commutes also, and this is VN for β.

Proposition 8.2. *Let $Q: \mathscr{C} \to \mathscr{A}$ be a $\mathscr{V}$-functor and $\alpha: T \to S: \mathscr{A} \to \mathscr{V}$ a $\mathscr{V}$-natural transformation. Then there is a $\mathscr{V}$-natural transformation $\alpha Q: TQ \to SQ: \mathscr{C} \to \mathscr{V}$ with components*

$$(\alpha Q)_C = \alpha_{QC}. \tag{8.3}$$

Proof. Write the diagram VN for α, with A and B replaced by QC and QD, and compose both legs with $Q_{CD}: \mathscr{C}(CD) \to \mathscr{A}(QC, QD)$. There results the diagram VN for αQ.

Proposition 8.3. *If $T: \mathscr{A} \to \mathscr{B}$ is a $\mathscr{V}$-functor, the morphisms*

$$T_{BC}: \mathscr{A}(BC) \to \mathscr{B}(TB, TC), \quad C \in \mathscr{A},$$

are the components of a $\mathscr{V}$-natural transformation

$$T_B: L^B \to L^{TB}T: \mathscr{A} \to \mathscr{V}.$$

In particular, taking T to be $L^A: \mathscr{A} \to \mathscr{V}$, the morphisms

$$L^A_{BC}: \mathscr{A}(BC) \to (\mathscr{A}(AB), \mathscr{A}(AC)), \quad C \in \mathscr{A},$$

are the components of a $\mathscr{V}$-natural transformation

$$L^A_B: L^A \to L^{\mathscr{A}(AB)}L^A: \mathscr{A} \to \mathscr{V}.$$

Proof. The axiom VN for T_B reduces to VF2 for T.

Proposition 8.4. *If $f \in \mathscr{V}_0(AB)$, the morphisms*

$$(f, 1): (BC) \to (AC), \quad C \in \mathscr{V},$$

are the components of a $\mathscr{V}$-natural transformation

$$L^f: L^B \to L^A: \mathscr{V} \to \mathscr{V}.$$

Proof. VN for L^f asserts the commutativity of the diagram

$$\begin{array}{ccc} (CD) & \xrightarrow{\quad L^B \quad} & ((BC)(BD)) \\ {\scriptstyle L^A}\downarrow & & \downarrow{\scriptstyle (1,(f,1))} \\ ((AC)(AD)) & \xrightarrow[\quad ((f,1),1) \quad]{} & ((BC)(AD)), \end{array}$$

which is precisely the assertion that L^A_{CD} is natural in A (which it is, by hypothesis).

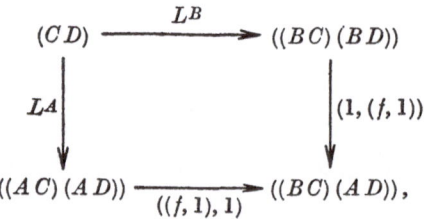

Proposition 8.5. *The morphisms*

$$i_A : A \to (IA)$$

are the components of a $\mathscr{V}$*-natural transformation*

$$i : 1 \to L^I : \mathscr{V} \to \mathscr{V}.$$

Proof. VN for i is CC4 for $\mathscr{V}$.

We now prove the representation theorem for $\mathscr{V}$-functors:

Theorem 8.6. *Let* $T : \mathscr{A} \to \mathscr{V}$ *be a* $\mathscr{V}$*-functor, let* $K \in \mathscr{A}$*, and denote by* $\{p\}$ *the class of* $\mathscr{V}$*-natural transformations*

$$p : L^K \to T : \mathscr{A} \to \mathscr{V},$$

with components

$$p_A : \mathscr{A}(KA) \to TA.$$

Define a map $\Gamma : \{p\} \to V(I, TK)$ *by setting* Γp *equal to the composite*

$$\Gamma p : \ I \xrightarrow[j_K]{} \mathscr{A}(KK) \xrightarrow[p_K]{} TK. \tag{8.4}$$

Then Γ *is a bijection with inverse* $\Omega : V(I, TK) \to \{p\}$ *where* $\Omega\theta$ *is the composite*

$$\Omega\theta: \quad L^K \xrightarrow[T_K]{} L^{TK}T \xrightarrow[L^\theta T]{} L^I T \xrightarrow[i^{-1}T]{} T, \tag{8.5}$$

with components

$$\mathscr{A}(KA) \xrightarrow[T]{} (TK, TA) \xrightarrow[(\theta, 1)]{} (I, TA) \xrightarrow[i^{-1}]{} TA \tag{8.6}$$

Proof. Note that by Propositions 8.1 to 8.5, $\Omega\theta$ defined by (8.5) is indeed a $\mathscr{V}$-natural transformation.

Consider the diagram

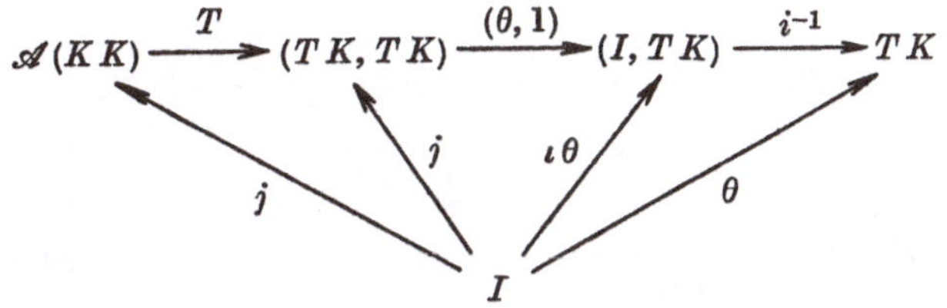

The left region commutes by VF1 for T, the middle region by Proposition 2.4, and the right region by Proposition 2.6. Thus $\Gamma\Omega\theta = \theta$, or $\Gamma\Omega = 1$.

Now let $p \in \{p\}$ and consider the diagram

$$
\begin{array}{ccccccc}
\mathscr{A}(KA) & \xrightarrow{L^K} & (\mathscr{A}(KK), \mathscr{A}(KA)) & \xrightarrow{(j,1)} & (I, \mathscr{A}(KA)) & \xrightarrow{i^{-1}} & \mathscr{A}(KA) \\
\downarrow{T} & & \downarrow{(1,p)} & & \downarrow{(1,p)} & & \downarrow{p} \\
(TK, TA) & \xrightarrow[(p_K,1)]{} & (\mathscr{A}(KK), TA) & \xrightarrow[(j_K,1)]{} & (I, TA) & \xrightarrow[i^{-1}]{} & TA
\end{array}
$$

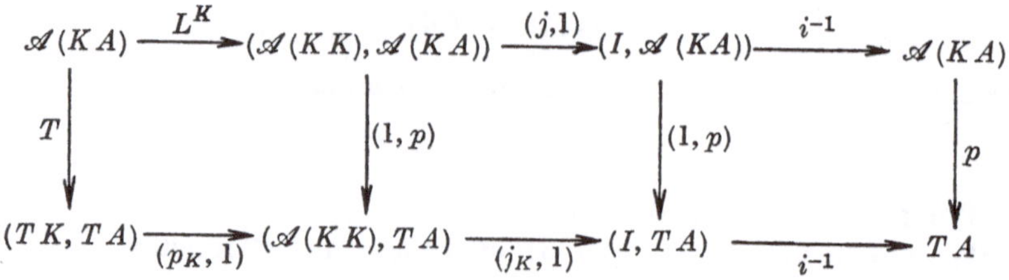

The left region commutes by VN for p, the middle region trivially, and the right region by the naturality of i. The composite $i^{-1}(j, 1)L^K$ along the top edge is 1 by VC2 for $\mathscr{A}$; hence

$$
p = i^{-1}(p_K j_K, 1) T = \Omega \Gamma p,
$$

so that $\Omega \Gamma = 1$.

Corollary 8.7. *In the circumstances of Theorem 8.6 we also have a bijection* $\Gamma' : \{p\} \to VTK$ *given by*

$$
\Gamma' p = (V p_K) 1_K \tag{8.7}
$$

where

$$
V p_K : V\mathscr{A}(KK) \to VTK.
$$

Proof. By Lemma 2.2, $\Gamma' p = \iota^{-1} \Gamma p$. $\tag{8.8}$

We now consider the effect of a closed functor $\mathscr{V} \to \mathscr{V}'$:

Proposition 8.8. *Let* $\Phi = (\phi, \hat{\phi}, \phi^0) : \mathscr{V} \to \mathscr{V}'$ *be a closed functor and let* $\alpha : T \to S : \mathscr{A} \to \mathscr{V}$ *be a* $\mathscr{V}$-*natural transformation. Then the morphisms*

$$
\phi \alpha_A : \phi TA \to \phi SA
$$

are the components of a $\mathscr{V}'$-*natural transformation*

$$
\phi \alpha : \hat{\Phi} . \Phi_* T \to \hat{\Phi} . \Phi_* S : \Phi_* \mathscr{A} \to \mathscr{V}'.
$$

Proof. VN for $\phi \alpha$ asserts the commutativity of the exterior of the diagram:

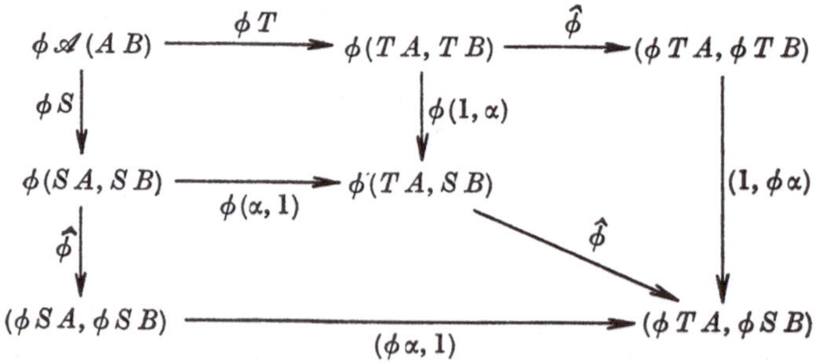

One region commutes by ϕ of VN for α, and two by the naturality of $\hat{\phi}$.

We apply the above proposition to a $\mathscr{V}$-natural transformation

$$p : L^K \to T : \mathscr{A} \to \mathscr{V};$$

note that $\hat{\Phi} \,.\, \Phi_* L^K = L'^K$ by (6.15), so that we have

$$\phi p : L'^K \to \hat{\Phi} \,.\, \Phi_* T : \Phi_* \mathscr{A} \to \mathscr{V}'.$$

Then:

Proposition 8.9. *Let $\Phi : \mathscr{V} \to \mathscr{V}'$ be a closed functor, $T : \mathscr{A} \to \mathscr{V}$ a $\mathscr{V}$-functor and $K \in \mathscr{A}$. Let $\{p\}$ be the class of $\mathscr{V}$-natural transformations*

$$p : L^K \to T : \mathscr{A} \to \mathscr{V}$$

and let $\{q\}$ be the class of $\mathscr{V}'$-natural transformations

$$q : L'^K \to \hat{\Phi} \,.\, \Phi_* T : \Phi_* \mathscr{A} \to \mathscr{V}'.$$

Let $\phi : \{p\} \to \{q\}$ be the map $p \mapsto \phi p$ given by Proposition 8.8, define $\Gamma' : \{p\} \to VTK$ by (8.1), and $\Delta' : \{q\} \to V'\phi TK$ analogously. Then we have a commutative diagram

$$\begin{array}{ccc} \{p\} & \xrightarrow{\ \phi\ } & \{q\} \\ {\scriptstyle \Gamma'}\downarrow & & \downarrow{\scriptstyle \Delta'} \\ VTK & \xrightarrow[\ \phi_{0TK}\ ]{} & V'\phi TK \end{array} \tag{8.9}$$

Proof. Define $\Delta : \{q\} \to V'(I', \phi TK)$ by the analogue of (8.4); it follows at once from (6.3) that $\Delta \phi p$ is the composite

$$\Delta \phi p: \qquad I' \xrightarrow[\ \phi^0\ ]{} \phi I \xrightarrow[\ \phi \Gamma p\ ]{} \phi TK. \tag{8.10}$$

By (3.8) we can write this as $\Delta \phi p = \iota' \phi_0 \iota^{-1} \Gamma p$, which by (8.8) is $\Delta' \phi p = \phi_0 \Gamma' p$; and this is (8.9).

Proposition 8.10. *Let $\eta : \Phi \to \Psi : \mathscr{V} \to \mathscr{V}'$ be a closed natural transformation. Then η is also a $\mathscr{V}'$-natural transformation*

$$\eta : \hat{\Phi} \to \hat{\Psi} \,.\, \eta_{*\mathscr{V}} : \Phi_* \mathscr{V} \to \mathscr{V}'$$

where the second $\mathscr{V}'$-functor here is the composite

$$\Phi_* \mathscr{V} \xrightarrow[\eta_* \mathscr{V}]{} \Psi_* \mathscr{V} \xrightarrow[\hat{\Psi}]{} \mathscr{V}'.$$

Proof. VN for η reduces to CN2 for η.

Note that with η as in Proposition 8.10 each $\mathscr{V}$-functor $T : \mathscr{A} \to \mathscr{V}$ gives rise to a commutative diagram

$$\begin{array}{ccc} \Phi_* \mathscr{A} & \xrightarrow{\Phi_* T} & \Phi_* \mathscr{V} \\ {\scriptstyle \eta_{*\mathscr{A}}} \downarrow & & \downarrow {\scriptstyle \eta_{*\mathscr{V}}} \\ \Psi_* \mathscr{A} & \xrightarrow[\Psi_* T]{} & \Psi_* \mathscr{V} \end{array}$$

so that we have a $\mathscr{V}'$-natural transformation

$$\eta \,.\, \Phi_* T : \hat{\Phi} \,.\, \Phi_* T \to \hat{\Psi} \,.\, \Psi_* T \,.\, \eta_{*\mathscr{A}} .$$

We leave the reader to pursue the relation of this to Proposition 8.9.

Proposition 8.11. *A $\mathscr{V}$-natural transformation $\alpha : T \to S : \mathscr{A} \to \mathscr{V}$ is also a natural transformation $\alpha : T_0 \to S_0 : \mathscr{A}_0 \to \mathscr{V}_0$.*

Proof. Applying V to VN for $\alpha : T \to S$ we get the criterion (1.1) for naturality of $\alpha : T_0 \to S_0$.

9. The Bifunctor hom $\mathscr{A}$

Let $\mathscr{V}$ be a closed category and $\mathscr{A}$ a $\mathscr{V}$-category. For $A, B \in \mathscr{A}$ let $\{p\}$ be the class of $\mathscr{V}$-natural transformations

$$p : L^B \to L^A : \mathscr{A} \to \mathscr{V}$$

with components

$$p_C : \mathscr{A}(BC) \to \mathscr{A}(AC) .$$

Since $L^A B = \mathscr{A}(AB)$, we have by Corollary 8.7 a bijection

$$\Gamma' : \{p\} \to V\mathscr{A}(AB)$$

given by

$$\Gamma' p = (V p_B)\, 1_B . \tag{9.1}$$

For each $f \in V\mathscr{A}(AB)$ $(= \mathscr{A}_0(AB))$, define

$$L^f : L^B \to L^A$$

by

$$\Gamma' L^f = f , \tag{9.2}$$

that is, by

$$(VL^f)\, 1 = f . \tag{9.3}$$

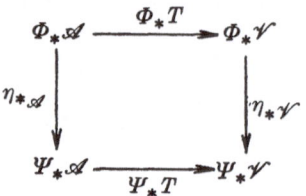

By Proposition 8.11, L^f is also a natural transformation

$$L^f : V_* L^B \to V_* L^A ,$$

that is,

$$L^f : \mathscr{A}(B-) \to \mathscr{A}(A-).$$

Applying V gives another natural transformation

$$VL^f : V\mathscr{A}(B-) \to V\mathscr{A}(A-),$$

or by (7.12)

$$VL^f : \mathscr{A}_0(B-) \to \mathscr{A}_0(A-).$$

Proposition 9.1. $VL^f = \mathscr{A}_0(f, 1) : \mathscr{A}_0(BC) \to \mathscr{A}_0(AC)$.

Proof. By the representation theorem, since VL^f and $\mathscr{A}_0(f, 1)$ are both natural transformations, it suffices to show that, when $C = B$, they have the same value at 1_B. But $\mathscr{A}_0(f, 1)1 = f$, and $(VL^f)1 = f$ by (9.3).

Proposition 9.2. *The assignments $A \mapsto L^A$, $f \mapsto L^f$ constitute a functor from $\mathscr{A}_0^*$ to the "category" of $\mathscr{V}$-functors $\mathscr{A} \to \mathscr{V}$ and $\mathscr{V}$-natural transformations between them.*

Proof. We have to show that

$$L^1 = 1 , \tag{9.4}$$

$$L^{fg} = L^g L^f . \tag{9.5}$$

Since $\Gamma' 1 = (V1)1 = 1$, (9.4) follows since Γ' is a bijection. Now

$$\begin{aligned} \Gamma'(L^g L^f) &= VL^g . VL^f . 1 && \text{by (9.1)} \\ &= VL^g . f && \text{by (9.3)} \\ &= \mathscr{A}_0(g, 1) f && \text{by Proposition 9.1} \\ &= fg ; \end{aligned}$$

so that (9.5) follows since Γ' is a bijection.

Now regarding L^f merely as a natural transformation

$$L^f : \mathscr{A}(B-) \to \mathscr{A}(A-),$$

it follows from Proposition 9.2 that the assigments $A \mapsto \mathscr{A}(A-)$, $f \mapsto L^f$ constitute a functor from $\mathscr{A}_0^*$ to the category of functors $\mathscr{A}_0 \to \mathscr{V}_0$ and natural transformations between them. Since, as is well known, a functor into a functor category corresponds to a bifunctor, we have here a bifunctor

$$\hom \mathscr{A} : \mathscr{A}_0^* \times \mathscr{A}_0 \to \mathscr{V}_0 .$$

Its value $\hom \mathscr{A}(AB)$ on objects is $\mathscr{A}(AB)$, and we agree to write $\mathscr{A}(f, g)$ for $\hom \mathscr{A}(f, g)$. The defining conditions of $\hom \mathscr{A}$ may then be written:

$$\hom \mathscr{A}(A-) = \mathscr{A}(A-)\,(= V_* L^A)\,, \tag{9.6}$$

$$\mathscr{A}(f, 1) = L^f : \mathscr{A}(BC) \to \mathscr{A}(AC)\,. \tag{9.7}$$

Proposition 9.3. *The following diagram of functors commutes:*

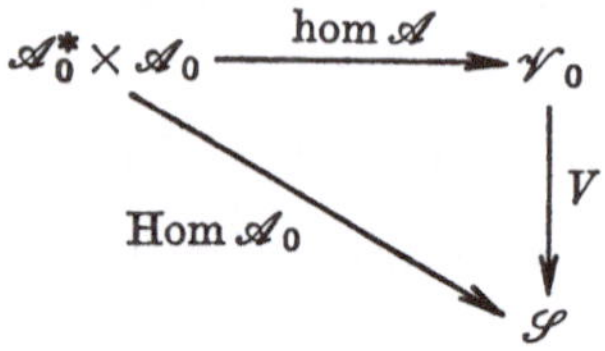

(9.8)

Proof. In view of (7.12), we have only to show that $V\mathscr{A}(f, 1) = \mathscr{A}_0(f, 1)$; this follows from (9.7) and Proposition 9.1.

We record the evaluated form of (9.8):

$$V\mathscr{A}(f, g) = \mathscr{A}_0(f, g)\,. \tag{9.9}$$

Note that when $\mathscr{A} = \mathscr{V}$ the L^f defined in Proposition 8.4 clearly satisfies (9.3), so that the notation L^f is consistent. So is the notation $\hom \mathscr{V}$, as we see by comparing (9.6) and (9.7) with (7.11) and Proposition 8.4.

Now that $\mathscr{A}(AB)$ is the value of a functor $\hom \mathscr{A}$ the question can be raised of the naturality of

$$j_A : I \to \mathscr{A}(AA) \quad \text{and} \quad L^A_{BC} : \mathscr{A}(BC) \to (\mathscr{A}(AB), \mathscr{A}(AC))\,.$$

Similarly if $T : \mathscr{A} \to \mathscr{B}$ is a $\mathscr{V}$-functor, one can discuss the naturality of $T_{BC} : \mathscr{A}(BC) \to \mathscr{B}(TB, TC)$, meaning of course the naturality of $T_{BC} : \mathscr{A}(BC) \to \mathscr{B}(T_0 B, T_0 C)$; recall that $T_0 A = TA$ by (7.7).

Proposition 9.4. *If $\mathscr{A}$ is a $\mathscr{V}$-category the morphisms*

$$j_A : I \to \mathscr{A}(AA) \quad \textit{and} \quad L^A_{BC} : \mathscr{A}(BC) \to (\mathscr{A}(AB), \mathscr{A}(AC))$$

are natural in every variable; and if $T : \mathscr{A} \to \mathscr{B}$ is a $\mathscr{V}$-functor the morphism $T_{BC} : \mathscr{A}(BC) \to \mathscr{B}(TB, TC)$ is natural in both variables.

Proof. For $f \in \mathscr{A}_0(AB) = V\mathscr{A}(AB)$, consider the diagram

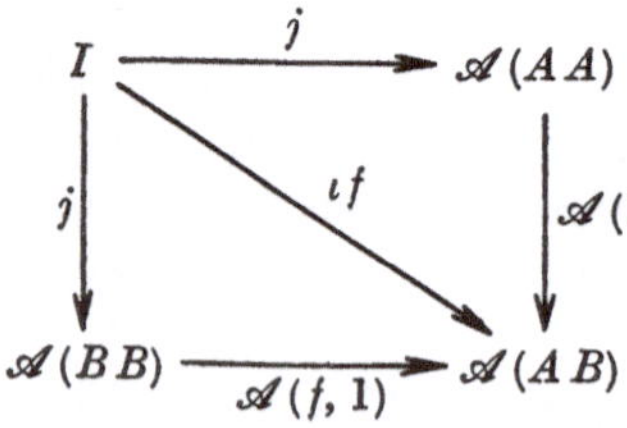

(9.10)

The upper triangle commutes by (7.14); so does the lower one by Lemma 7.1 in view of (9.9). Hence the exterior commutes, so that j is natural.

By Propositions 8.3 and 8.11, T_B is a natural transformation

$$\mathscr{A}(B-) \to \mathscr{B}(T_0 B, T_0 -),$$

i.e. T_{BC} is natural in C. To prove it also natural in B is to show, for each $f \in \mathscr{A}_0(AB)$, the commutativity of

$$\begin{array}{ccc} \mathscr{A}(BC) & \xrightarrow{\quad T_{BC}\quad} & \mathscr{B}(TB, TC) \\ {\scriptstyle \mathscr{A}(f,1)}\downarrow & & \downarrow{\scriptstyle \mathscr{B}(T_0 f, 1)} \\ \mathscr{A}(AC) & \xrightarrow[\quad T_{AC}\quad]{} & \mathscr{B}(TA, TC) \end{array} \tag{9.11}$$

Now (9.11) is the C-component of the following diagram of $\mathscr{V}$-natural transformations:

$$\begin{array}{ccc} L^B & \xrightarrow{\quad T_B\quad} & L^{TB}T \\ {\scriptstyle L^f}\downarrow & & \downarrow{\scriptstyle L^{T_0 f}T} \\ L^A & \xrightarrow[\quad T_A\quad]{} & L^{TA}T \end{array}$$

and so by the representation theorem for $\mathscr{V}$-functors it suffices to apply V to (9.11), put $C = B$, and verify that both legs have the same value at 1_B. Using (7.8) and (9.9), we have

$$V\mathscr{B}(T_0 f, 1) \,.\, VT \,.\, 1 = V\mathscr{B}(T_0 f, 1) \,.\, 1 = T_0 f,$$

$$VT \,.\, V\mathscr{A}(f, 1) \,.\, 1 = VT \,.\, f = T_0 f.$$

Applying the above to the $\mathscr{V}$-functor L^A, we have the naturality in B and in C of L^A_{BC}; it remains to prove its naturality in A, namely the commutativity for $f \in \mathscr{A}_0(DA)$ of

$$\begin{array}{ccc} \mathscr{A}(BC) & \xrightarrow{\quad L^A\quad} & (\mathscr{A}(AB), \mathscr{A}(AC)) \\ {\scriptstyle L^D}\downarrow & & \downarrow{\scriptstyle (1, \mathscr{A}(f,1))} \\ (\mathscr{A}(DB), \mathscr{A}(DC)) & \xrightarrow[\quad (\mathscr{A}(f,1),1)\quad]{} & (\mathscr{A}(AB), \mathscr{A}(DC)) \end{array} \tag{9.12}$$

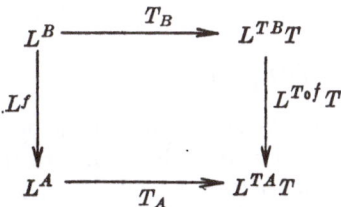

This does indeed commute, being by (9.7) precisely the axiom VN for L^f. This completes the proof.

The category $\mathscr{A}_0$ and the functor hom $\mathscr{A}$ associated with a $\mathscr{V}$-category $\mathscr{A}$, whose existence we have shown constructively, are characterized by the following uniqueness theorem, in which of course $\mathscr{A}_0$ and hom $\mathscr{A}$ cease temporarily to have the meanings we have given them.

Proposition 9.5. *Let $\mathscr{V}$ be a closed category, and suppose we are given a category $\mathscr{A}_0$, a functor* hom $\mathscr{A}: \mathscr{A}_0^* \times \mathscr{A}_0 \to \mathscr{V}_0$ *(whose values on objects and on morphisms we write as $\mathscr{A}(AB)$ and $\mathscr{A}(f, g)$), a natural transformation $j_A: I \to \mathscr{A}(AA)$, and a natural transformation $L_{BC}^A: \mathscr{A}(BC) \to \to (\mathscr{A}(AB), \mathscr{A}(AC))$, satisfying* VC1, VC2, VC3 *and also* (9.8) *and* (7.3). *Then $\mathscr{A}$ is* a fortiori *a $\mathscr{V}$-category, and we necessarily have:*

(i) $\mathscr{A}_0 = V_* \mathscr{A}$;

(ii) $\mathscr{A}(A-) = V_* L^A$;

(iii) $\mathscr{A}(f, 1) = L^f$.

Proof. From (7.3) we get Lemma 7.1, and use it to write VC1 in the form

$$(VL_{BB}^A) 1_B = 1_{\mathscr{A}(AB)}; \tag{9.13}$$

just as in Proposition 2.3 we show that this is equivalent to (7.10). The proof that $V_* \mathscr{A}$ is $\mathscr{A}_0$ is now exactly similar to that of Proposition 7.2, and then (7.10) may be written $V_* L^A = \mathscr{A}(A-)$. The naturality of L_{BC}^A in A gives (9.12), which shows that $\mathscr{A}(f, 1): L^A \to L^D$ is $\mathscr{V}$-natural; since $(V\mathscr{A}(f, 1))1 = \mathscr{A}_0(f, 1)1 = f$, we conclude from (9.3) that

$$\mathscr{A}(f, 1) = L^f.$$

Similarly for $\mathscr{V}$-functors:

Proposition 9.6. *Let $\mathscr{A}$ and $\mathscr{B}$ be $\mathscr{V}$-categories, and suppose we are given a functor $T_0: \mathscr{A}_0 \to \mathscr{B}_0$ and a natural transformation $T_{BC}: \mathscr{A}(BC) \to \to \mathscr{B}(T_0 B, T_0 C)$, such that T satisfies* VF1 *and* VF2 *if we write TA for $T_0 A$. Then T is* a fortiori *a $\mathscr{V}$-functor, and we necessarily have*

$$T_0 = V_* T: \mathscr{A}_0 \to \mathscr{B}_0.$$

Proof. We use Lemma 7.1 to write VF1 for T in the form

$$(VT_{BB}) 1_B = 1_{TB}; \tag{9.14}$$

then by the representation theorem the two natural transformations

$$T_{0BC}, VT_{BC}: V\mathscr{A}(BC) \to V\mathscr{B}(T_0 B, T_0 C)$$

coincide, which completes the proof.

Remark 9.7. It follows from Proposition 9.5 that we could have given an alternative definition of $\mathscr{V}$-category, including the category $\mathscr{A}_0$ and the functor hom $\mathscr{A}$ among the data, insisting on naturality of j and L, and adding (9.8) and (7.3) to VC1—VC3 as axioms. Then j is a superfluous datum, as (7.3) defines it; and VC1 may be expressed without using j in the form (9.13), or equivalently as (7.9) or (7.10). If V is faithful, axioms VC2 and VC3 are then unnecessary, by an argument exactly like that of Proposition 2.10. Indeed when V is faithful the data themselves are somewhat redundant; we leave the reader to formulate an analogue of Proposition 2.11. Similarly we may include T_0 in the definition of a $\mathscr{V}$-functor T, and require T_{BC} to be natural. Then VF1 may be written as (9.14) or (7.8), and VF2 is a consequence if V is faithful.

Proposition 9.8. *If $\Phi : \mathscr{V} \to \mathscr{V}'$ is a closed functor and $\mathscr{A}$ is a $\mathscr{V}$-category, the following diagram of functors commutes:*

$$\begin{array}{ccc}
\mathscr{A}_0^* \times \mathscr{A}_0 & \xrightarrow{\text{hom}\,\mathscr{A}} & \mathscr{V}_0 \\
{\scriptstyle \Phi_{0\mathscr{A}}^* \times \Phi_{0\mathscr{A}}}\downarrow & & \downarrow{\scriptstyle \phi} \\
(\Phi_*\mathscr{A})_0^* \times (\Phi_*\mathscr{A})_0 & \xrightarrow[\text{hom}\,\Phi_*\mathscr{A}]{} & \mathscr{V}_0'
\end{array} \tag{9.15}$$

Proof. As we already have Proposition 7.4, it only remains to show that (9.15) commutes when evaluated at morphisms $f \in \mathscr{A}_0^*(BA)$ and $1 \in \mathscr{A}_0(CC)$; that is, that for $f \in \mathscr{A}_0(AB)$ the morphisms

$$\phi\mathscr{A}(f, 1) : \phi\mathscr{A}(BC) \to \phi\mathscr{A}(AC)\,, \tag{9.16}$$

$$(\Phi_*\mathscr{A})(\phi_0 f, 1) : \phi\mathscr{A}(BC) \to \phi\mathscr{A}(AC)\,, \tag{9.17}$$

coincide.

Now by Proposition 8.8, (9.16) is the C-component of the $\mathscr{V}'$-natural transformation

$$\phi L^f : L'^B \to L'^C : \Phi_*\mathscr{A} \to \mathscr{V}'\,,$$

since $\hat{\Phi} \,.\, \Phi_* L^B = L'^B$ by (6.15); while (9.17) is the C-component of the $\mathscr{V}'$-natural transformation

$$L'^{\phi_0 f} : L'^B \to L'^C : \Phi_*\mathscr{A} \to \mathscr{V}'\,.$$

We have therefore to show that

$$\phi L^f = L'^{\phi_0 f}\,, \tag{9.18}$$

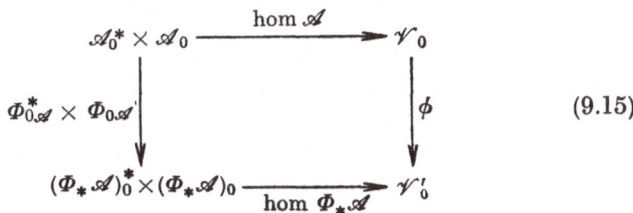

and by (9.3) it suffices to show that

$$(V'\phi L^f)\, 1_B = \phi_0 f. \tag{9.19}$$

By the naturality of ϕ_0 we have a commutative diagram

$$\begin{array}{ccc}
V\mathscr{A}(BB) & \xrightarrow{VL^f} & V\mathscr{A}(AB) \\
\downarrow{\scriptstyle \phi_0} & & \downarrow{\scriptstyle \phi_0} \\
V'\phi\mathscr{A}(BB) & \xrightarrow[V'\phi L^f]{} & V'\phi\mathscr{A}(AB)
\end{array}$$

Evaluating both legs at 1_B gives (9.19), since $(VL^f)1 = f$ by (9.3), and $\phi_0 1 = 1$ since Φ_0 is a functor.

10. $\mathscr{V}$-natural Transformations

We now define $\mathscr{V}$-natural transformations in general. For a closed category $\mathscr{V}$ and $\mathscr{V}$-functors $T, S: \mathscr{A} \to \mathscr{B}$, a *$\mathscr{V}$-natural transformation* $\alpha: T \to S: \mathscr{A} \to \mathscr{B}$ consists of a family of morphisms $\alpha_A: TA \to SA$ in $\mathscr{B}_0$, indexed by the objects of $\mathscr{A}$, satisfying the axiom:

VN. The following diagram commutes:

$$\begin{array}{ccc}
\mathscr{A}(AB) & \xrightarrow{T_{AB}} & \mathscr{B}(TA, TB) \\
\downarrow{\scriptstyle S_{AB}} & & \downarrow{\scriptstyle \mathscr{B}(1, \alpha_B)} \\
\mathscr{B}(SA, SB) & \xrightarrow[\mathscr{B}(\alpha_A, 1)]{} & \mathscr{B}(TA, SB)
\end{array}$$

Proposition 10.1. *A $\mathscr{V}$-natural transformation $\alpha: T \to S: \mathscr{A} \to \mathscr{B}$ is also a natural transformation $\alpha: T_0 \to S_0: \mathscr{A}_0 \to \mathscr{B}_0$. If V is faithful, the naturality of α conversely implies its $\mathscr{V}$-naturality.*

Proof. Both statements follow from the fact that V of the diagram VN is the diagram (1.1) expressing the naturality of α.

Theorem 10.2. *$\mathscr{V}$-categories, $\mathscr{V}$-functors, and $\mathscr{V}$-natural transformations form a "hypercategory" (which we still denote by $\mathscr{V}_*$) if we define the composite of*

(i) $\alpha: T \to S: \mathscr{A} \to \mathscr{B}$ *and* $\beta: S \to R: \mathscr{A} \to \mathscr{B}$ *by*

$$(\beta\alpha)_A = \beta_A \alpha_A; \tag{10.1}$$

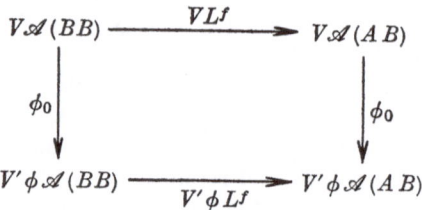

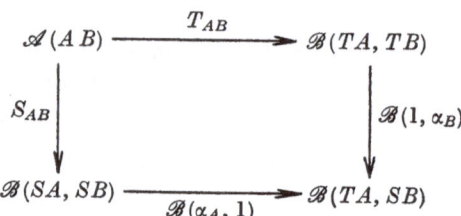

(ii) $Q : \mathscr{C} \to \mathscr{A}$ *and* $\alpha : T \to S : \mathscr{A} \to \mathscr{B}$ *by*

$$(\alpha\, Q)_C = \alpha_{QC}; \tag{10.2}$$

(iii) $\alpha : T \to S : \mathscr{A} \to \mathscr{B}$ *and* $P : \mathscr{B} \to \mathscr{D}$ *by*

$$(P\alpha)_A = P_0 \alpha_A . \tag{10.3}$$

Moreover α *is an isomorphism if and only if each* α_A *is an isomorphism.*

The hypercategory $\mathscr{S}_*$ *is then* $\mathscr{C}at$, *and* $V_* : \mathscr{V}_* \to \mathscr{S}_*$ *is a hyperfunctor if we set*

$$V_* \alpha = \alpha . \tag{10.4}$$

Proof. The proof that $\beta\alpha$ is $\mathscr{V}$-natural, and of the statement about isomorphisms, is a trivial generalization of the proof of Proposition 8.1; similarly the proof of Proposition 8.2 generalizes to show that $\alpha\, Q$ is $\mathscr{V}$-natural. To show that $P\alpha$ is $\mathscr{V}$-natural is to show the commutativity of the exterior of the diagram

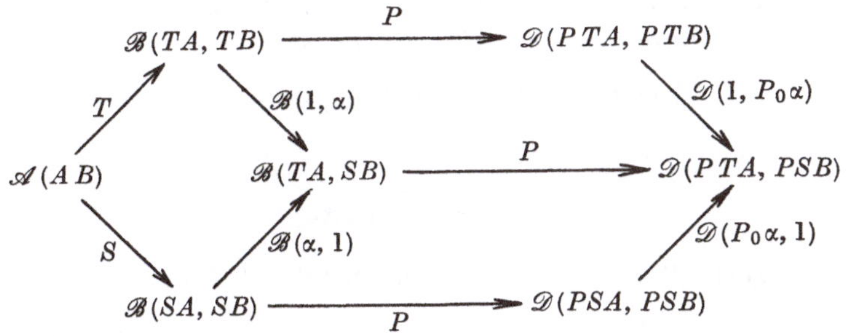

One region commutes by VN for α, and the other two by the naturality of P guaranteed by Proposition 9.4.

V_* clearly preserves all the above laws of composition, so that HC2—HC4 for $\mathscr{V}_*$ follow from HC2—HC4 for $\mathscr{C}at$. As we already have HC1, this completes the proof.

Proposition 10.3. *If* $\Phi : \mathscr{V} \to \mathscr{V}'$ *is a closed functor and*

$$\alpha : T \to S : \mathscr{A} \to \mathscr{B}$$

is a $\mathscr{V}$*-natural transformation, we get a* $\mathscr{V}'$*-natural transformation*

$$\Phi_* \alpha : \Phi_* T \to \Phi_* S : \Phi_* \mathscr{A} \to \Phi_* \mathscr{B}$$

if we set

$$(\Phi_* \alpha)_A = \Phi_{0\mathscr{B}} \alpha_A \tag{10.5}$$

where

$$\Phi_{0\mathscr{B}} : \mathscr{B}_0 \to (\Phi_* \mathscr{B})_0 .$$

In particular if Φ is normal we have (cf. (10.4))

$$\Phi_* \alpha = \alpha . \tag{10.6}$$

Proof. Writing γ for $\Phi_* \alpha$, VN for γ asserts the commutativity of:

$$\begin{array}{ccc} \phi\mathscr{A}(A\,B) & \xrightarrow{\ \phi T\ } & \phi\mathscr{B}(TA, TB) \\ {\scriptstyle \phi S}\downarrow & & \downarrow{\scriptstyle (\Phi_*\mathscr{B})(1,\gamma)} \\ \phi\mathscr{B}(SA, SB) & \xrightarrow[(\Phi_*\mathscr{B})(\gamma, 1)]{} & \phi\mathscr{B}(TA, SB) \end{array} \tag{10.7}$$

Since $\gamma = \Phi_{0\mathscr{B}}\alpha$, it follows from (9.15) that $(\Phi_* \mathscr{B})(1, \gamma) = \phi\mathscr{B}(1, \alpha)$ and $(\Phi_* \mathscr{B})(\gamma, 1) = \phi\mathscr{B}(\alpha, 1)$; thus (10.7) does indeed commute, being ϕ of VN for α.

Proposition 10.4. *If $\Phi : \mathscr{V} \to \mathscr{V}'$ is a closed functor and $\alpha : T \to S : \mathscr{A} \to \mathscr{V}$ is a $\mathscr{V}$-natural transformation, the $\mathscr{V}'$-natural transformation*

$$\phi\alpha : \hat{\Phi}.\Phi_* T \to \hat{\Phi}.\Phi_* S : \Phi_* \mathscr{A} \to \mathscr{V}'$$

of Proposition 8.8 is given by

$$\phi\alpha = \hat{\Phi}.\Phi_* \alpha. \tag{10.8}$$

Proof. By (10.3) and (10.5) the A-component of $\hat{\Phi}.\Phi_* \alpha$ is $V'_* \hat{\Phi}.\Phi_{0\mathscr{V}}.\alpha_A$, and this is $\phi\alpha_A$ by (7.18).

Proposition 10.5. *If $\Phi : \mathscr{V} \to \mathscr{V}'$ is a closed functor, $\Phi_* : \mathscr{V}_* \to \mathscr{V}'_*$ is a hyperfunctor.*

Proof. We have to show that Φ_* respects the composition laws (10.1), (10.2), (10.3). For (10.1) this follows since $\Phi_{0\mathscr{B}}$ is a functor, and for (10.2) it is trivial since $(\Phi_* Q)\, C = QC$ by (6.8). For (10.3), the A-component of $\Phi_*(P\alpha)$ is $\Phi_{0\mathscr{D}} P_0 \alpha_A$, which by (7.17) is equal to $(\Phi_* P)_0 \Phi_{0\mathscr{B}} \alpha_A$, which is the A-component of $\Phi_* P.\Phi_* \alpha$.

Proposition 10.6. *If $\eta : \Phi \to \Psi : \mathscr{V} \to \mathscr{V}'$ is a closed natural transformation, the $\mathscr{V}'$-functors $\eta_{*\mathscr{A}} : \Phi_* \mathscr{A} \to \Psi_* \mathscr{A}$ defined by* (6.10) *and* (6.11) *are the components of a hypernatural transformation $\eta_* : \Phi_* \to \Psi_* : \mathscr{V}_* \to \mathscr{V}'_*$.*

Proof. By Proposition 6.4 η_* is natural, i.e. HN1 is satisfied. We have to verify HN2, namely the coincidence of

$$\eta_{*\mathscr{B}}.\Phi_* \alpha : \eta_{*\mathscr{B}}.\Phi_* T \to \eta_{*\mathscr{B}}.\Phi_* S : \Phi_* \mathscr{A} \to \Psi_* \mathscr{B}$$

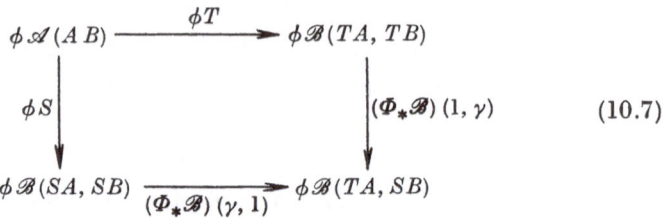

and

$$\Psi_* \alpha . \eta_{*\mathscr{A}} : \Psi_* T . \eta_{*\mathscr{A}} \to \Psi_* S . \eta_{*\mathscr{A}} : \Phi_* \mathscr{A} \to \Psi_* \mathscr{B},$$

where $\alpha : T \to S : \mathscr{A} \to \mathscr{B}$ is a $\mathscr{V}$-natural transformation. By (10.2), (10.3), and (10.5), the A-components of $\eta_{*\mathscr{B}} . \Phi_* \alpha$ and of $\Psi_* \alpha . \eta_{*\mathscr{A}}$ are respectively $V'_* \eta_{*\mathscr{B}} . \Phi_{0\mathscr{B}} . \alpha_A$ and $\Psi_{0\mathscr{B}} \alpha_A$, and these are equal by (7.21).

Theorem 10.7. *The assignments $\mathscr{V} \mapsto \mathscr{V}_*$, $\Phi \mapsto \Phi_*$, $\eta \to \eta_*$ constitute a hyperfunctor $_* : \mathscr{C}\ell \to \mathscr{H}yp$ from the "hypercategory" of closed categories to that of hypercategories.*

Proof. As we already have Theorem 6.5, all that remains to be shown is that, if $\Phi : \mathscr{V} \to \mathscr{V}'$ and $\Psi : \mathscr{V}' \to \mathscr{V}''$ are closed functors and $\alpha : T \to S : \mathscr{A} \to \mathscr{B}$ is a $\mathscr{V}$-natural transformation, we have $(\Psi\Phi)_* \alpha = \Psi_* \Phi_* \alpha$; this is immediate from (7.20).

We now give a form of the representation theorem for $\mathscr{V}$-functors analogous to Theorem 1.1:

Theorem 10.8. *Let $T : \mathscr{A} \to \mathscr{B}$ be a $\mathscr{V}$-functor, let $K \in \mathscr{A}$ and $M \in \mathscr{B}$, and denote by $\{p\}$ the class of $\mathscr{V}$-natural transformations*

$$p : L^K \to L^M T : \mathscr{A} \to \mathscr{V}$$

with components

$$p_A : \mathscr{A}(K A) \to \mathscr{B}(M, TA).$$

Define a map $\Gamma' : \{p\} \to V\mathscr{B}(M, TK)$ by:

$$\Gamma' p = (V p_K) 1_K. \tag{10.9}$$

Then Γ' is a bijection with inverse $\Omega' : V\mathscr{B}(M, TK) \to \{p\}$ where $\Omega' \theta$ is the composite

$$\Omega' \theta : \; L^K \xrightarrow[T_K]{} L^{TK} T \xrightarrow[L^\theta T]{} L^M T \tag{10.10}$$

with components

$$\mathscr{A}(KA) \xrightarrow[T]{} \mathscr{B}(TK, TA) \xrightarrow[\mathscr{B}(\theta, 1)]{} \mathscr{B}(M, TA). \tag{10.11}$$

Proof. Γ' is a bijection by Corollary 8.7, and clearly $\Gamma' \Omega' = 1$.

Remark 10.9. In the circumstances of Theorem 10.8 let $\Phi : \mathscr{V} \to \mathscr{V}'$ be a closed functor and let $\{q\}$ be the class of $\mathscr{V}'$-natural transformations

$$q : L'^K \to L'^M . \Phi_* T : \Phi_* \mathscr{A} \to \mathscr{V}'$$

with components

$$q_A : \phi \mathscr{A}(KA) \to \phi \mathscr{B}(M, TA).$$

Define $\Delta' : \{q\} \to V'\phi\mathscr{B}(M, TK)$ analogously to Γ'; then by Proposition 8.9 we have a commutative diagram

$$\begin{array}{ccc} \{p\} & \xrightarrow{\ \phi\ } & \{q\} \\ {\scriptstyle \Gamma'}\downarrow & & \downarrow{\scriptstyle \Delta'} \\ V\mathscr{B}(M, TK) & \xrightarrow[\phi_0]{} & V'\phi\mathscr{B}(M, TK) \end{array}$$

where $(\phi p)_A = \phi p_A$. In particular if Φ is normal we have $\phi_0 = 1$, and so $\phi : \{p\} \to \{q\}$ is also a bijection. Note especially the case when Φ is $V : \mathscr{V} \to \mathscr{S}$.

Proposition 10.10. *Let $T, S : \mathscr{A} \to \mathscr{B}$ be $\mathscr{V}$-functors and let $\alpha_A : TA \to SA$ be a family of morphisms. Then the following assertions are equivalent:*
(a) *The α_A are the components of a $\mathscr{V}$-natural transformation $\alpha : T \to S$.*
(b) *The morphisms*

$$\mathscr{B}(1, \alpha_B) : \mathscr{B}(TA, TB) \to \mathscr{B}(TA, SB), \quad B \in \mathscr{B},$$

are for each A the components of a $\mathscr{V}$-natural transformation

$$L^{TA}T \to L^{TA}S.$$

(c) *The composite morphisms*

$$\mathscr{A}(AB) \xrightarrow[T]{} \mathscr{B}(TA, TB) \xrightarrow[\mathscr{B}(1,\alpha)]{} \mathscr{B}(TA, SB), \quad B \in \mathscr{B},$$

are for each A the components of a $\mathscr{V}$-natural transformation

$$L^A \to L^{TA}S.$$

Proof. If α is $\mathscr{V}$-natural, $\mathscr{B}(1, \alpha) = L^{TA}\alpha$ is also $\mathscr{V}$-natural, so (a) implies (b). Similarly (b) implies (c) since $T_A : L^A \to L^{TA}T$ is $\mathscr{V}$-natural. We have to show that (c) implies (a).

In the diagram VN for α, namely

$$\begin{array}{ccc} \mathscr{A}(AB) & \xrightarrow{\ T\ } & \mathscr{B}(TA, TB) \\ {\scriptstyle S}\downarrow & & \downarrow{\scriptstyle \mathscr{B}(1,\alpha)} \\ \mathscr{B}(SA, SB) & \xrightarrow[\mathscr{B}(\alpha, 1)]{} & \mathscr{B}(TA, SB) \end{array}$$

the top leg is by hypothesis (c) $\mathscr{V}$-natural for each A; but the bottom leg is also $\mathscr{V}$-natural, being the B-component of $L^{\alpha}S.S_A$. If we put

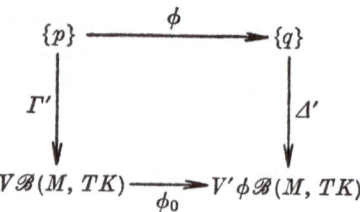

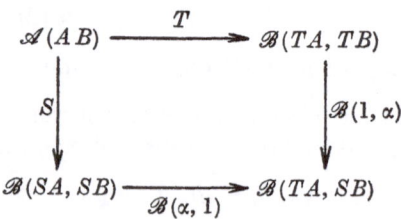

$B = A$, apply V, and evaluate at 1_A, each leg gives α_A; hence by the representation theorem for $\mathscr{V}$-functors the diagram commutes.

Remark 10.11. By a *representation* of a $\mathscr{V}$-functor $T : \mathscr{A} \to \mathscr{V}$ we mean an object K of $\mathscr{A}$ together with a $\mathscr{V}$-natural isomorphism $p : L^K \to T$; if T admits a representation we say it is *representable*. If $p : L^K \to T$ and $q : L^M \to S$ are representations of $T, S : \mathscr{A} \to \mathscr{V}$, it is clear from the representation theorem that a bijection is set up between $\mathscr{V}$-natural transformations $\alpha : T \to S$ and morphisms $f : M \to K$ by requiring commutativity in the diagram

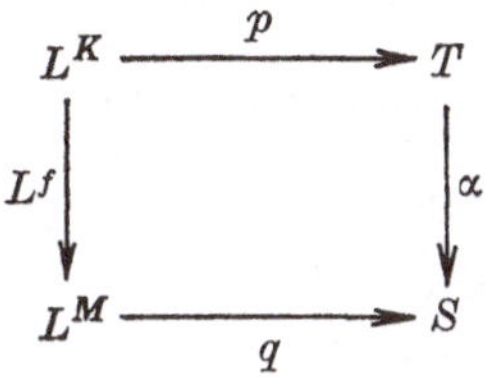

In particular if $p : L^K \to T$ and $q : L^M \to T$ are two representations of T then there is a unique isomorphism $f \in V\mathscr{A}(MK)$ giving a commutative diagram

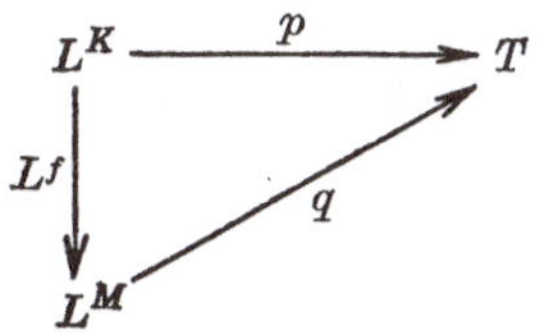

Remark 10.12. If $\mathscr{A}$ is a $\mathscr{V}$-category, any morphism $f \in \mathscr{A}_0(AB)$ is a $\mathscr{V}$-natural transformation $f : J^A \to J^B : \mathscr{I} \to \mathscr{A}$; for VN for f is just (9.10). If $T : \mathscr{A} \to \mathscr{B}$ is a $\mathscr{V}$-functor, it is then consistent with (10.3) to write Tf for $T_0 f$. We shall use both notations, the one for brevity and the other where there is danger of confusion.

Chapter II

Monoidal Closed Categories

1. Monoidal Categories

A *monoidal category* $\mathscr{V} = (\mathscr{V}_0, \otimes, I, r, l, a)$ consists of the following six data:

(i) a category $\mathscr{V}_0$;
(ii) a functor $\otimes : \mathscr{V}_0 \times \mathscr{V}_0 \to \mathscr{V}_0$ (written between its arguments and called the *tensor product* of $\mathscr{V}$);
(iii) an object I of $\mathscr{V}_0$;

(iv) a natural isomorphism $r = r_A : A \otimes I \to A$;
(v) a natural isomorphism $l = l_A : I \otimes A \to A$;
(vi) a natural isomorphism $a = a_{ABC} : (A \otimes B) \otimes C \to A \otimes (B \otimes C)$.

These data are to satisfy the following five axioms:

MC1. The following diagram commutes:

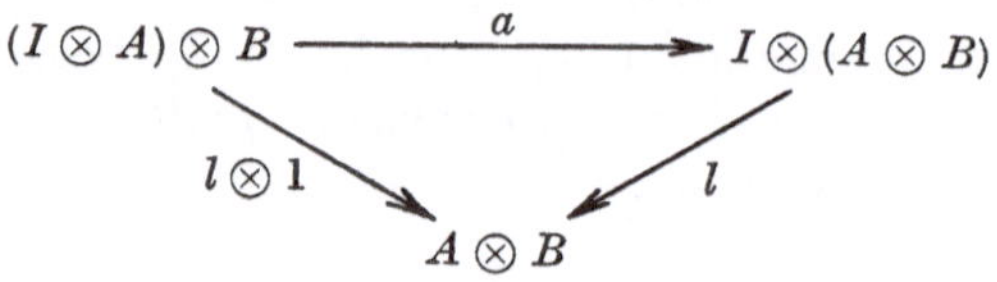

MC2. The following diagram commutes:

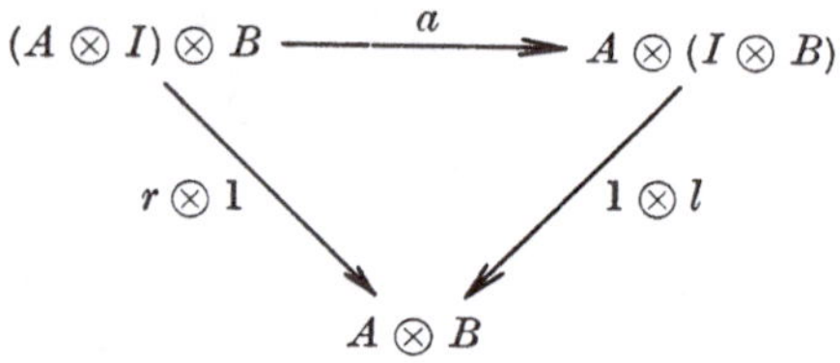

MC3. The following diagram commutes:

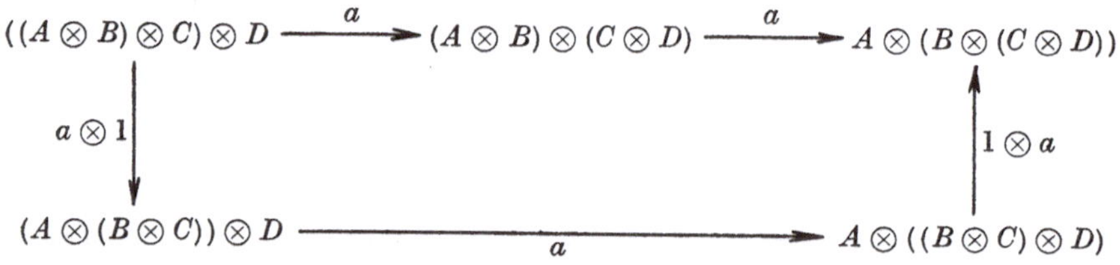

MC4. The following diagram commutes:

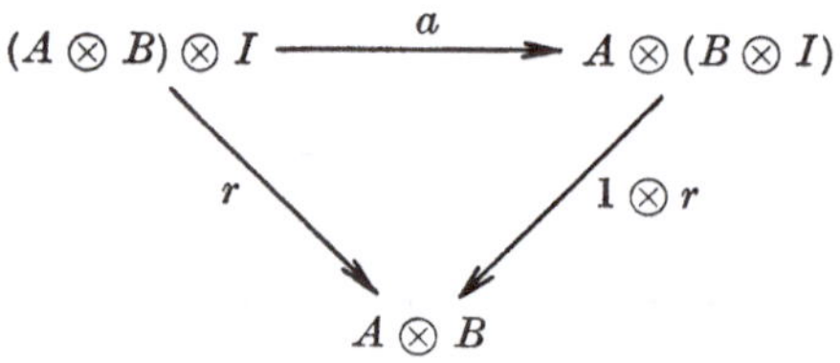

MC5. $$l_I = r_I : I \otimes I \to I.$$

We remark at once that the above axioms are not independent; we have listed them all, and arranged them in the above rather odd order, for later comparison with CC1—CC5. In fact it has been shown (KELLY [9]) that:

Proposition 1.1. MC1, MC4, *and* MC5 *are consequences of* MC2 *and* MC3.

Natural isomorphisms such as a, r, l are said to be *coherent* if, roughly speaking, all diagrams made by their use alone (with their inverses, 1, and $\otimes$), such as the diagrams of MC1—MC5, commute. For an exact description of the meaning of coherence, see MAC LANE [*14*], where it is proved that MC1—MC5 imply:

Proposition 1.2. *The isomorphisms a, r, l are coherent.*

Let $\mathscr{V} = (\mathscr{V}_0, \otimes, I, r, l, a)$ and $\mathscr{V}' = (\mathscr{V}'_0, \otimes', I', r', l', a')$ be monoidal categories; we write $\otimes$ for $\otimes'$ when there is no danger of confusion. A *monoidal functor* $\Phi = (\phi, \tilde{\phi}, \phi^0): \mathscr{V} \to \mathscr{V}'$ consists of

(i) a functor $\phi: \mathscr{V}_0 \to \mathscr{V}'_0$;
(ii) a natural transformation $\tilde{\phi} = \tilde{\phi}_{AB}: \phi A \otimes \phi B \to \phi(A \otimes B)$;
(iii) a morphism $\phi^0: I' \to \phi I$.

These data are to satisfy the following three axioms:

MF1. The following diagram commutes:

$$\begin{array}{ccc} \phi I \otimes \phi A & \xrightarrow{\tilde{\phi}} & \phi(I \otimes A) \\ {\scriptstyle \phi^0 \otimes 1}\uparrow & & \downarrow{\scriptstyle \phi l} \\ I' \otimes \phi A & \xrightarrow[l']{} & \phi A \end{array}$$

MF2. The following diagram commutes:

$$\begin{array}{ccc} \phi A \otimes \phi I & \xrightarrow{\tilde{\phi}} & \phi(A \otimes I) \\ {\scriptstyle 1 \otimes \phi^0}\uparrow & & \downarrow{\scriptstyle \phi r} \\ \phi A \otimes I' & \xrightarrow[r']{} & \phi A \end{array}$$

MF3. The following diagram commutes:

$$\begin{array}{ccc} \phi((A \otimes B) \otimes C) & \xrightarrow{\phi a} & \phi(A \otimes (B \otimes C)) \\ {\scriptstyle \tilde{\phi}}\uparrow & & \uparrow{\scriptstyle \tilde{\phi}} \\ \phi(A \otimes B) \otimes \phi C & & \phi A \otimes \phi(B \otimes C) \\ {\scriptstyle \tilde{\phi} \otimes 1}\uparrow & & \uparrow{\scriptstyle 1 \otimes \tilde{\phi}} \\ (\phi A \otimes \phi B) \otimes \phi C & \xrightarrow[a']{} & \phi A \otimes (\phi B \otimes \phi C) \end{array}$$

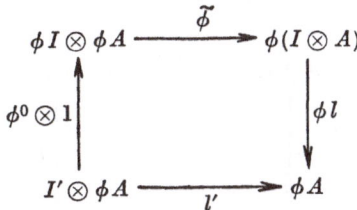

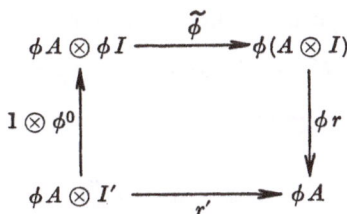

Let $\Phi = (\phi, \tilde{\phi}, \phi^0)$ and $\Psi = (\psi, \tilde{\psi}, \psi^0)$ be monoidal functors $\mathscr{V} \to \mathscr{V}'$. A *monoidal natural transformation*

$$\eta : \Phi \to \Psi : \mathscr{V} \to \mathscr{V}'$$

consists of a natural transformation

$$\eta : \phi \to \psi : \mathscr{V}_0 \to \mathscr{V}'_0$$

satisfying the following two axioms:

MN 1. The following diagram commutes:

$$\begin{array}{ccc} I' & \xrightarrow{\phi^0} & \phi I \\ & \searrow^{\psi^0} & \downarrow{\scriptstyle \eta_I} \\ & & \psi I \end{array}$$

MN 2. The following diagram commutes:

$$\begin{array}{ccc} \phi A \otimes \phi B & \xrightarrow{\tilde{\phi}} & \phi(A \otimes B) \\ {\scriptstyle \eta \otimes \eta}\downarrow & & \downarrow{\scriptstyle \eta} \\ \psi A \otimes \psi B & \xrightarrow[\tilde{\psi}]{} & \psi(A \otimes B) \end{array}$$

We define the composite of monoidal functors

$$\Phi = (\phi, \tilde{\phi}, \phi^0) : \mathscr{V} \to \mathscr{V}' \quad \text{and} \quad \Psi = (\psi, \tilde{\psi}, \psi^0) : \mathscr{V}' \to \mathscr{V}''$$

to be $X = (\chi, \tilde{\chi}, \chi^0) : \mathscr{V} \to \mathscr{V}''$ where

(i) χ is the composite

$$\mathscr{V}_0 \xrightarrow[\varphi]{} \mathscr{V}'_0 \xrightarrow[\psi]{} \mathscr{V}''_0; \tag{1.1}$$

(ii) $\tilde{\chi}$ is the composite

$$\psi\phi A \otimes \psi\phi B \xrightarrow[\tilde{\psi}]{} \psi(\phi A \otimes \phi B) \xrightarrow[\psi\tilde{\varphi}]{} \psi\phi(A \otimes B); \tag{1.2}$$

(iii) χ^0 is the composite

$$I'' \xrightarrow[\psi^0]{} \psi I' \xrightarrow[\psi\varphi^0]{} \psi\phi I. \tag{1.3}$$

We define the composite of monoidal natural transformations $\eta : \Phi \to \Phi' : \mathscr{V} \to \mathscr{W}$ and $\zeta : \Phi' \to \Phi'' : \mathscr{V} \to \mathscr{W}$ to be the composite $\zeta\eta$ of $\eta : \phi \to \phi'$ and $\zeta : \phi' \to \phi''$; and for $\Psi : \mathscr{V}' \to \mathscr{V}$, $\eta : \Phi \to \Phi' : \mathscr{V} \to \mathscr{W}$ and $X : \mathscr{W} \to \mathscr{W}'$ we define $\eta\Psi$ and $X\eta$ to be $\eta\psi$ and $\chi\eta$. We now leave the reader to prove, along the lines of Theorem I.3.1,

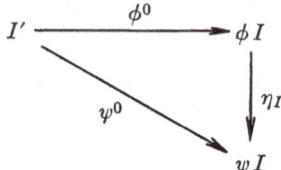

Theorem 1.3. *Monoidal categories, monoidal functors, and monoidal natural transformations form with the above rules of composition a "hyper-category"* $\mathscr{M}on$. *Moreover a monoidal functor* $\Phi = (\phi, \tilde{\phi}, \phi^0)$ *is an isomorphism if and only if each of* $\phi, \tilde{\phi}, \phi^0$ *is an isomorphism, and a monoidal natural transformation* $\eta : \Phi \to \Psi$ *is an isomorphism if and only if* $\eta : \phi \to \psi$ *is.*

2. Monoidal Closed Categories

A *monoidal closed category* (or equally *closed monoidal category*) $\mathscr{V} = ({}^m\mathscr{V}, p, {}^c\mathscr{V})$ consists of the following three data:

(i) a monoidal category ${}^m\mathscr{V} = (\mathscr{V}_0, \otimes, I, r, l, a)$;

(ii) a closed category ${}^c\mathscr{V} = (\mathscr{V}_0, V, \mathrm{hom}\,\mathscr{V}, I, i, j, L)$ with the same $\mathscr{V}_0$ and I as ${}^m\mathscr{V}$;

(iii) a natural isomorphism $p = p_{ABC} : (A \otimes B, C) \to (A(BC))$.

These data are to satisfy the following four axioms, whose bizarre numeration is for later convenience:

MCC2. The following diagram commutes:

$$\begin{array}{ccc} (I \otimes A, B) & \xrightarrow{\;p\;} & (I(AB)) \\ {\scriptstyle (l,1)}\nwarrow & & \nearrow{\scriptstyle i} \\ & (AB) & \end{array}$$

MCC3. The following diagram commutes:

$$\begin{array}{ccccc} ((A \otimes B) \otimes C, D) & \xrightarrow{\;p\;} & (A \otimes B, (CD)) & \xrightarrow{\;p\;} & (A, (B(CD))) \\ \uparrow{\scriptstyle (a,1)} & & & & \uparrow{\scriptstyle (1,p)} \\ (A \otimes (B \otimes C), D) & & \xrightarrow[\;p\;]{} & & (A, (B \otimes C, D)) \end{array}$$

MCC3′. The following diagram commutes:

$$\begin{array}{ccc} (CD) & \xrightarrow{\;L^{A \otimes B}\;} & ((A \otimes B, C), (A \otimes B, D)) \\ \downarrow{\scriptstyle L^B} & & \\ ((BC)(BD)) & & \downarrow{\scriptstyle (1,p)} \\ \downarrow{\scriptstyle L^A} & & \\ ((A(BC)), (A(BD))) & \xrightarrow[\;(p,1)\;]{} & ((A \otimes B, C), (A(BD))) \end{array}$$

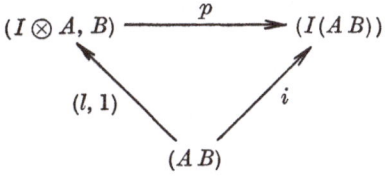

MCC4. The following diagram commutes:

$$\begin{array}{ccc} (A \otimes I, B) & \xrightarrow{\;p\;} & (A(IB)) \\ & {\scriptstyle (r,1)} \nwarrow \quad \nearrow {\scriptstyle (1,i)} & \\ & (AB) & \end{array}$$

Remark 2.1. We shall normally denote the monoidal category ${}^{m}\mathscr{V}$ and the closed category ${}^{c}\mathscr{V}$ by the same symbol $\mathscr{V}$ as the monoidal closed category, except where we wish to distinguish between the three structures.

If $\mathscr{V}$ and $\mathscr{V}'$ are monoidal closed categories, a *monoidal closed functor* $\Phi : \mathscr{V} \to \mathscr{V}'$ is to be a quadruple $(\phi, \tilde{\phi}, \hat{\phi}, \phi^0)$ where ${}^{m}\Phi = (\phi, \tilde{\phi}, \phi^0)$ is a monoidal functor ${}^{m}\mathscr{V} \to {}^{m}\mathscr{V}'$ and ${}^{c}\Phi = (\phi, \hat{\phi}, \phi^0)$ is a closed functor ${}^{c}\mathscr{V} \to {}^{c}\mathscr{V}'$, and where the following axiom is satisfied:

MCF3. The following diagram commutes:

$$\begin{array}{ccc} \phi(A \otimes B, C) & \xrightarrow{\;\phi p\;} & \phi(A(BC)) \\ \Big\downarrow {\scriptstyle \hat{\phi}} & & \Big\downarrow {\scriptstyle \hat{\phi}} \\ (\phi(A \otimes B), \phi C) & & (\phi A, \phi(BC)) \\ \Big\downarrow {\scriptstyle (\tilde{\phi}, 1)} & & \Big\downarrow {\scriptstyle (1, \hat{\phi})} \\ (\phi A \otimes \phi B, \phi C) & \xrightarrow[\;p'\;]{} & (\phi A, (\phi B, \phi C)) \end{array}$$

If $\Phi, \Psi : \mathscr{V} \to \mathscr{V}'$ are monoidal closed functors, a *monoidal closed natural transformation* $\eta : \Phi \to \Psi : \mathscr{V} \to \mathscr{V}'$ is to be a natural transformation $\eta : \phi \to \psi : \mathscr{V}_0 \to \mathscr{V}'_0$ which is both a monoidal natural transformation $\eta : {}^{m}\Phi \to {}^{m}\Psi : {}^{m}\mathscr{V} \to {}^{m}\mathscr{V}'$ and a closed natural transformation $\eta : {}^{c}\Phi \to {}^{c}\Psi : {}^{c}\mathscr{V} \to {}^{c}\mathscr{V}'$.

We define composition of monoidal closed functors by (1.1), (1.2), (1.3), and I(3.2); note that I(3.1) and I(3.3) reproduce (1.1) and (1.3). We define composition of monoidal closed natural transformations, with themselves or with monoidal closed functors, to be their composition as monoidal natural transformations, which is the same as their composition as closed natural transformations. To prove the following theorem requires only the verification that axiom MCF3 survives composition, which we leave to the reader:

Theorem 2.2. *Monoidal closed categories, monoidal closed functors, and monoidal closed natural transformations form a "hypercategory"* $\mathscr{MCl}$.

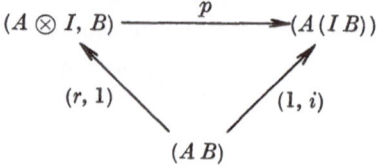

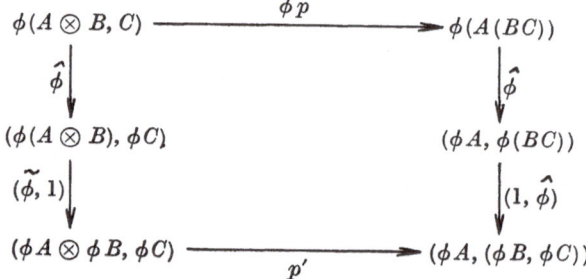

3. Relations Between the Data

Both the data and the axioms for a monoidal closed category are highly redundant (quite apart from the redundancy noted in Proposition 1.1). To examine the interconnexions we place ourselves in the following *basic situation*: we suppose given a category $\mathscr{V}_0$, functors

$$\otimes : \mathscr{V}_0 \times \mathscr{V}_0 \to \mathscr{V}_0 \quad \text{and} \quad \hom\mathscr{V} : \mathscr{V}_0^* \times \mathscr{V}_0 \to \mathscr{V}_0,$$

a natural isomorphism

$$\pi = \pi_{ABC} : \mathscr{V}_0(A \otimes B, C) \to \mathscr{V}_0(A(BC)),$$

and a functor $V : \mathscr{V}_0 \to \mathscr{S}$ satisfying CC0.

The naturality of π gives a commutative diagram

$$\begin{array}{ccc}
\mathscr{V}_0(A \otimes B, C) & \xrightarrow{\quad\pi\quad} & \mathscr{V}_0(A, (BC)) \\
{\scriptstyle \mathscr{V}_0(f \otimes g, h)}\downarrow & & \downarrow{\scriptstyle \mathscr{V}_0(f, (g, h))} \\
\mathscr{V}_0(A' \otimes B', C') & \xrightarrow[\quad\pi\quad]{} & \mathscr{V}_0(A', (B'C'))
\end{array}$$

where $f : A' \to A$, $g : B' \to B$, $h : C \to C'$. Evaluating this at $x \in \mathscr{V}_0(A \otimes \otimes B, C)$ gives a commutative diagram:

$$\begin{array}{ccc}
A' & \xrightarrow{\pi(hx(f \otimes g))} & (B'C') \\
{\scriptstyle f}\downarrow & & \uparrow{\scriptstyle (g, h)} \\
A & \xrightarrow[\pi x]{} & (BC)
\end{array} \tag{3.1}$$

Define natural transformations (natural by Proposition I.1.2)

$$t = t_{BC} : (BC) \otimes B \to C,$$

$$u = u_{AB} : A \to (B, A \otimes B),$$

by

$$t = \pi^{-1} 1_{(BC)}, \tag{3.2}$$

$$u = \pi 1_{A \otimes B}. \tag{3.3}$$

Then for $x : A \otimes B \to C$ and $y : A \to (BC)$ we have commutative diagrams:

$$\begin{array}{ccc}
A & \xrightarrow{\quad\pi x\quad} & (BC) \\
& {\scriptstyle u}\searrow \quad \nearrow{\scriptstyle (1, x)} & \\
& (B, A \otimes B) &
\end{array} \tag{3.4}$$

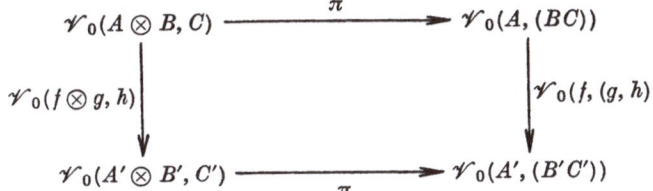

$$\begin{array}{ccc} A \otimes B & \xrightarrow{\pi^{-1} y} & C \\ & {\scriptstyle y \otimes 1} \searrow \quad \nearrow {\scriptstyle t} & \\ & (BC) \otimes B & \end{array} \tag{3.5}$$

These may be regarded as special cases of (3.1), or as coming from the representation theorem (cf. I (1.4)) applied to π and to π^{-1}. Taking in particular $x = t$ and $y = u$ in (3.4) and (3.5) we get commutative diagrams

$$\begin{array}{ccc} (BC) & \xrightarrow{1} & (BC) \\ & {\scriptstyle u} \searrow \quad \nearrow {\scriptstyle (1,\, t)} & \\ & (B, (BC) \otimes B) & \end{array} \tag{3.6}$$

$$\begin{array}{ccc} A \otimes B & \xrightarrow{1} & A \otimes B \\ & {\scriptstyle u \otimes 1} \searrow \quad \nearrow {\scriptstyle t} & \\ & (B, A \otimes B) \otimes B & \end{array} \tag{3.7}$$

The following lemma allows us to use in a systematic way the fact that π is a natural isomorphism.

Lemma 3.1. *With the basic situation as above, let $\mathscr{W}$ be a category, $P, Q : \mathscr{V}_0 \to \mathscr{W}$ functors, and $B \in \mathscr{V}_0$. Then there is a bijection between natural transformations $\alpha = \alpha_A : PA \to Q(A \otimes B)$ and natural transformations $\beta = \beta_A : P(BA) \to QA$, given by*

(i) *β_A is the composite*

$$P(BA) \xrightarrow[\alpha_{(BA)}]{} Q((BA) \otimes B) \xrightarrow[Qt]{} QA; \tag{3.8}$$

(ii) *α_A is the composite*

$$PA \xrightarrow[Pu]{} P(B, A \otimes B) \xrightarrow[\beta_{A \otimes B}]{} Q(A \otimes B). \tag{3.9}$$

Proof. Consider the diagram

$$\begin{array}{ccc} \mathscr{V}_0(A \otimes B, C) & \xrightarrow{\pi} & \mathscr{V}_0(A(BC)) \\ \Big\downarrow {\scriptstyle Q} & & \Big\downarrow {\scriptstyle P} \\ \mathscr{W}(Q(A \otimes B), QC) & & \mathscr{W}(PA, P(BC)) \\ & {\scriptstyle \mathscr{W}(\alpha,\, 1)} \searrow \quad \swarrow {\scriptstyle \mathscr{W}(1,\, \beta)} & \\ & \mathscr{W}(PA, QC) & \end{array} \tag{3.10}$$

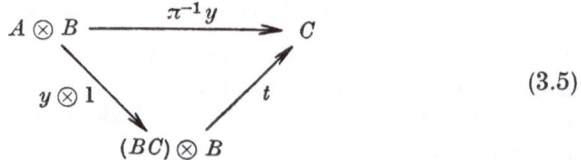

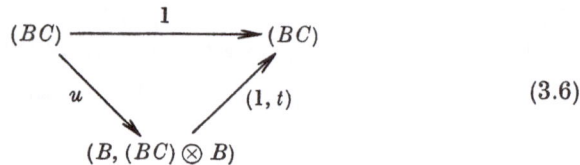

In the language of Theorem I.1.1, the left edge of (3.10) is $\Omega\alpha$ and the right edge is $\Omega\beta$ (different Ω's, the second with reversed variance). It follows, since π is an isomorphism, that the commutativity of (3.10) sets up a bijection between α's and β's. Putting $C = A \otimes B$ and evaluating at 1 gives (3.9); putting $A = (BC)$ and evaluating at t gives (3.8).

Note that by evaluating (3.10) at $x \in \mathscr{V}_0(A \otimes B, C)$ we get

$$Qx.\alpha = \beta.P\pi x : PA \to QC. \tag{3.11}$$

Again, since π is a natural isomorphism, the representation theorem shows that the commutativity of the diagram

$$\begin{array}{ccc} \mathscr{V}_0(I \otimes A, B) & \xrightarrow{\pi} & \mathscr{V}_0(I(AB)) \\ & \nwarrow_{\mathscr{V}_0(l,1)} \quad {}^{v}\nearrow & \\ & \mathscr{V}_0(AB) & \end{array} \tag{3.12}$$

sets up a bijection between natural isomorphisms $l : I \otimes A \to A$ and natural isomorphisms

$$v = v_{AB} : \mathscr{V}_0(AB) \to \mathscr{V}_0(I(AB)).$$

Further the representation theorem applied to v shows that there is a bijection between natural transformations v and natural transformations $j = j_A : I \to (AA)$, given by

$$j_A = v_{AA} 1_A. \tag{3.13}$$

Then if $f \in \mathscr{V}_0(AB)$ we have by I(1.4) the commutative diagram (cf. I(2.8))

$$\begin{array}{ccc} I & \xrightarrow{j} & (AA) \\ {\scriptstyle j}\downarrow & \searrow^{vf} & \downarrow{\scriptstyle (1,f)} \\ (BB) & \xrightarrow[(f,1)]{} & (AB) \end{array} \tag{3.14}$$

Evaluating (3.12) at 1_A after putting $B = A$ now gives:

$$\pi l = j. \tag{3.15}$$

In the same way commutativity of the diagram

$$\begin{array}{ccc} \mathscr{V}_0(A \otimes I, B) & \xrightarrow{\pi} & \mathscr{V}_0(A(IB)) \\ & \nwarrow_{\mathscr{V}_0(r,1)} \quad {}^{\mathscr{V}_0(1,i)}\nearrow & \\ & \mathscr{V}_0(AB) & \end{array} \tag{3.16}$$

sets up a bijection between natural isomorphisms $r: A \otimes I \to A$ and natural isomorphisms $i: A \to (IA)$. Putting $B = A$ and evaluating at 1 gives:

$$\pi r = i. \tag{3.17}$$

Again, commutativity of the diagram

$$\begin{array}{ccccc} \mathscr{V}_0((A\otimes B)\otimes C, D) & \xrightarrow{\pi} & \mathscr{V}_0(A\otimes B, (CD)) & \xrightarrow{\pi} & \mathscr{V}_0(A, (B(CD))) \\ \Big\uparrow {\scriptstyle \mathscr{V}_0(a,1)} & & & & \Big\uparrow {\scriptstyle \mathscr{V}_0(1,p)} \\ \mathscr{V}_0(A\otimes(B\otimes C), D) & & \xrightarrow[\pi]{\qquad\qquad} & & \mathscr{V}_0(A, (B\otimes C, D)) \end{array} \tag{3.18}$$

sets up a bijection between natural isomorphisms

$$a: (A \otimes B) \otimes C \to A \otimes (B \otimes C)$$

and natural isomorphisms $p: (B \otimes C, D) \to (B(CD))$. Evaluating (3.18) at $x \in \mathscr{V}_0(A \otimes (B \otimes C), D)$ gives a commutative diagram

$$\begin{array}{ccc} A & \xrightarrow{\pi\pi(xa)} & (B(CD)) \\ & {\scriptstyle \pi x}\searrow \quad \nearrow{\scriptstyle p} & \\ & (B\otimes C, D) & \end{array} \tag{3.19}$$

Finally, an application of Lemma 3.1 with $\mathscr{W} = \mathscr{V}_0^*$ (watch the variances!), $P = (-, (BC))$, and $Q = (-\, C)$, gives a bijection between natural transformations $p: (A \otimes B, C) \to (A(BC))$ and natural transformations $L: (AC) \to ((BA)(BC))$, determined by either of the commutative diagrams

$$\begin{array}{ccc} (AC) & \xrightarrow{L} & ((BA)(BC)) \\ & {\scriptstyle (t,1)}\searrow \quad \nearrow{\scriptstyle p} & \\ & ((BA)\otimes B, C) & \end{array} \tag{3.20}$$

$$\begin{array}{ccc} (A\otimes B, C) & \xrightarrow{p} & (A(BC)) \\ & {\scriptstyle L}\searrow \quad \nearrow{\scriptstyle (u,1)} & \\ & ((B, A\otimes B), (BC)) & \end{array} \tag{3.21}$$

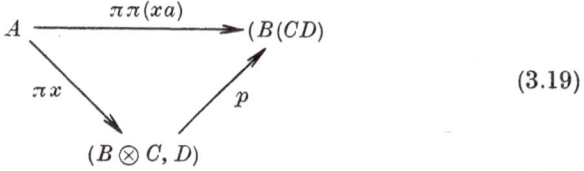

If we write (3.11) for this special case, we get for $x : A \otimes B \to C$ a commutative diagram

$$\begin{array}{ccc} (CD) & \xrightarrow{(x,\,1)} & (A \otimes B,\, D) \\ {\scriptstyle L}\downarrow & & \downarrow{\scriptstyle p} \\ ((BC)(BD)) & \xrightarrow[(\pi x,\,1)]{} & (A(BD)) \end{array} \tag{3.22}$$

Now suppose that we have besides the basic situation $\mathscr{V}_0$ etc., a second one $\mathscr{V}'_0$ etc.; and that we have a functor $\phi : \mathscr{V}_0 \to \mathscr{V}'_0$. Then since π and π' are both natural isomorphisms, a trivial generalization of the argument of Lemma 3.1 shows that the commutativity of the diagram

$$\begin{array}{ccc} \mathscr{V}_0(A \otimes B, C) & \xrightarrow{\pi} & \mathscr{V}_0(A(BC)) \\ {\scriptstyle \phi}\downarrow & & \downarrow{\scriptstyle \phi} \\ \mathscr{V}'_0(\phi(A \otimes B), \phi C) & & \mathscr{V}'_0(\phi A, \phi(BC)) \\ {\scriptstyle \mathscr{V}'_0(\tilde{\phi},\,1)}\downarrow & & \downarrow{\scriptstyle \mathscr{V}'_0(1,\,\hat{\phi})} \\ \mathscr{V}'_0(\phi A \otimes \phi B, \phi C) & \xrightarrow[\pi']{} & \mathscr{V}'_0(\phi A, (\phi B, \phi C)) \end{array} \tag{3.23}$$

sets up a bijection between natural transformations

$$\tilde{\phi} : \phi A \otimes \phi B \to \phi(A \otimes B)$$

and natural transformations

$$\hat{\phi} : \phi(BC) \to (\phi B, \phi C).$$

If we evaluate (3.23) at $x \in \mathscr{V}_0(A \otimes B, C)$, we get a commutative diagram

$$\begin{array}{ccc} \phi A & \xrightarrow{\phi \pi x} & \phi(BC) \\ & {\scriptstyle \pi'(\phi x.\,\tilde{\phi})}\searrow & \downarrow{\scriptstyle \hat{\phi}} \\ & & (\phi B, \phi C) \end{array} \tag{3.24}$$

Proposition 3.2. *If $\mathscr{V}$ is a monoidal closed category, define π and v by*

$$\pi_{ABC} = V p_{ABC}, \tag{3.25}$$

$$v_{AB} = V i_{(AB)} \; (= \iota_{(AB)}); \tag{3.26}$$

then the basic situation obtains and we have (3.12), (3.13), (3.16), (3.18), *and* (3.20). *Moreover if* $\Phi : \mathscr{V} \to \mathscr{V}'$ *is a monoidal closed functor we have* (3.23).

Proof. Applying V to MCC2 and using (3.26) gives (3.12); while (3.13), in view of (3.26), is CC5. Similarly applying V to MCC4 and to MCC3 gives (3.16) and (3.18).

Axiom MCC3′ may be interpreted as an instance of VN, stating: p_{ABC} is the C-component of a $\mathscr{V}$-natural transformation

$$p_{AB} : L^{A\otimes B} \to L^A L^B. \tag{3.27}$$

It follows that (3.20) is a diagram of $\mathscr{V}$-natural transformations

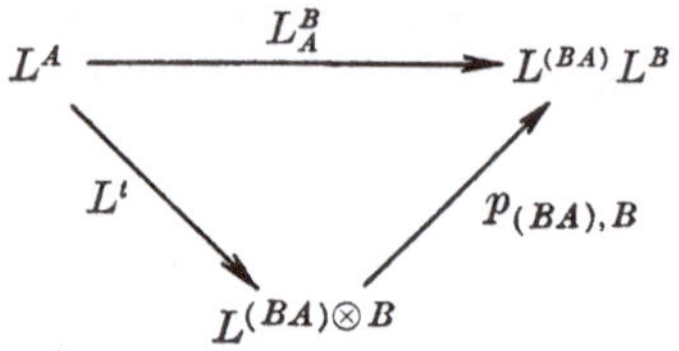

so that its commutativity will follow if, after putting $C = A$ and applying V, both legs have the same value at 1. But $(VL)\ 1 = 1$ by I (2.5), and $Vp.\ V(t, 1).\ 1 = Vp.\ t = \pi t = 1$.

If Φ is a monoidal closed functor, diagram (3.23) is the exterior of:

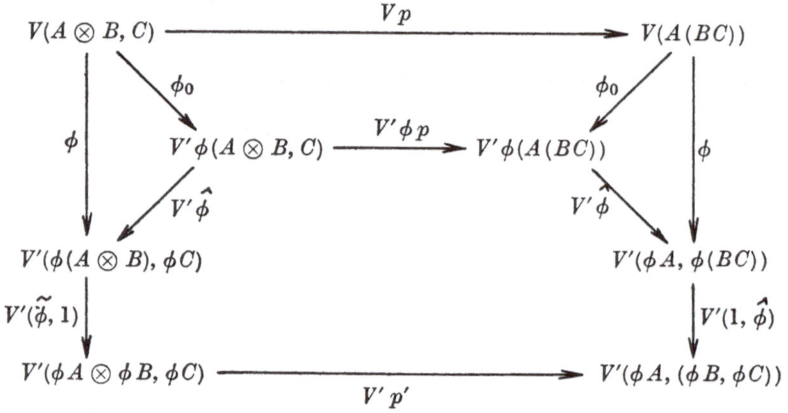

Here one region commutes by V' of MCF3, two regions by I(3.11), and one by the naturality of ϕ_0.

4. Relations Between the Axioms

Proposition 4.1. *Suppose that in the basic situation of* § 3 *we have natural isomorphisms* a, l, r, p, v, i *and natural transformations* j, L, *con-*

nected by (3.12), (3.13), (3.16), (3.18), *and* (3.20). *Then the following implications hold between the axioms* MC, MCC, *and* CC:

(i) *in the presence of* CC5, *we have* MC1 $\Leftrightarrow$ $Vp = \pi$ $\Leftrightarrow$ CC1;

(ii) MC2 $\Leftrightarrow$ MCC2 $\Leftrightarrow$ CC2;

(iii) MC3 $\Leftrightarrow$ MCC3 $\Leftrightarrow$ MCC3′ $\Leftrightarrow$ CC3;

(iv) MC4 $\Leftrightarrow$ MCC4 $\Leftrightarrow$ CC4;

(v) CC5 $\Rightarrow$ MC5 *(one way only!)*.

Proof of (i). Note first that Lemma I.2.2, Proposition I.2.3, and Proposition I.2.4 use only CC0 and CC5, and so are available here.

Since π is an isomorphism we get from MC1 an equivalent diagram by applying π twice to each leg. Now

$$\begin{aligned} \pi\pi(la) &= p.\pi l && \text{by (3.19)} \\ &= pj && \text{by (3.15)}; \end{aligned}$$

and

$$\begin{aligned} \pi\pi(l \otimes 1) &= \pi\pi(1(l \otimes 1)) \\ &= \pi(\pi 1 . l) && \text{by (3.1)} \\ &= \pi(ul) && \text{by (3.3)} \\ &= (1, u)\pi l && \text{by (3.1)} \\ &= (1, u)j && \text{by (3.15)}. \end{aligned}$$

Thus our equivalent diagram to MC1 is

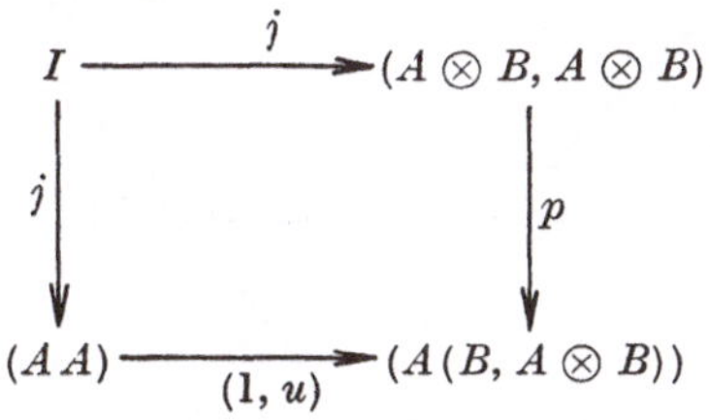

and by Lemma I.2.2 this may be expressed as $(Vp)1 = V(1, u)1$; that is, $(Vp)1 = u$. Since we also have $\pi 1 = u$, and since Vp and π are both natural, the statement $(Vp)1 = u$ is further equivalent to $Vp = \pi$, by the representation theorem.

If $Vp = \pi$, we get by applying V to (3.20), putting $C = A$, and evaluating at 1,

$$\begin{aligned} (VL)1 &= \pi . V(t, 1) . 1 \\ &= \pi t \\ &= 1. \end{aligned}$$

Conversely if $(VL)1 = 1$, (3.21) gives

$$\begin{aligned}(Vp)1 &= V(u, 1) . VL . 1\\ &= V(u, 1) . 1\\ &= u,\end{aligned}$$

and hence $Vp = \pi$.

Thus $Vp = \pi$ is equivalent to $(VL)1 = 1$, and this is, in the presence of CC0 and CC5, equivalent to CC1 by Proposition I.2.3.

Proof of (ii). Applying π twice to each leg of MC2, we get

$$\begin{aligned}\pi\pi((1 \otimes l)a) &= p.\pi(1 \otimes l) && \text{by (3.19)}\\ &= p(l, 1).\pi 1 && \text{by (3.1)}\\ &= p(l, 1)u,\end{aligned}$$

and

$$\begin{aligned}\pi\pi(r \otimes 1) &= \pi(\pi 1.r) && \text{by (3.1)}\\ &= \pi(ur)\\ &= (1, u).\pi r && \text{by (3.1)}\\ &= (1, u)i && \text{by (3.17)}\\ &= iu && \text{by the naturality of } i.\end{aligned}$$

Thus the equivalent diagram to MC2 is

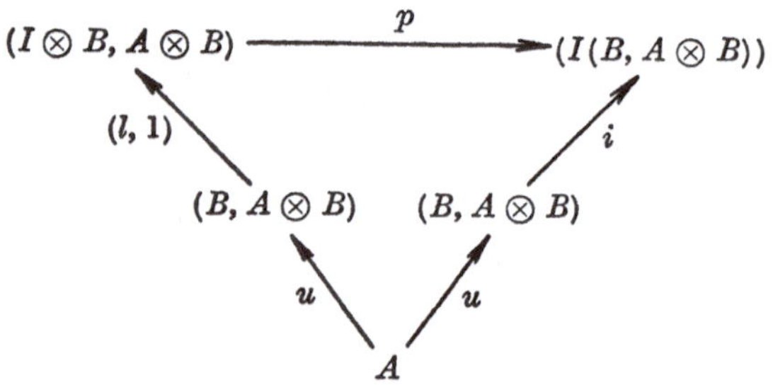

and this is equivalent to MCC2 by an application of Lemma 3.1.

Now use (3.22) to replace $p(l, 1)$ in MCC2 by $(\pi l, 1)L$, which by (3.15) is $(j, 1)L$; the result is precisely CC2.

Proof of (iii). First note that if we put $D = (A \otimes B) \otimes C$ and $x = a^{-1}$ in (3.19) we get

$$\begin{aligned}p.\pi a^{-1} &= \pi\pi 1\\ &= \pi u\\ &= (1, u)u && \text{by (3.4);}\end{aligned}$$

thus we have

$$\pi a^{-1} = p^{-1}(1, u)\, u\,. \tag{4.1}$$

Now write MC3 in the form

$a^{-1}(1 \otimes a)\, a = a(a^{-1} \otimes 1) : (A \otimes (B \otimes C)) \otimes D \to (A \otimes B) \otimes (C \otimes D)$
and apply π twice to each term. We have

$$\begin{aligned}
\pi\pi(a^{-1}(1 \otimes a)\, a) &= p.\, \pi(a^{-1}(1 \otimes a)) && \text{by (3.19)}\\
&= p(a, 1).\, \pi a^{-1} && \text{by (3.1)}\\
&= p(a, 1)\, p^{-1}(1, u)\, u && \text{by (4.1);}
\end{aligned}$$

and

$$\begin{aligned}
\pi\pi(a(a^{-1} \otimes 1)) &= \pi(\pi a.\, a^{-1}) && \text{by (3.1)}\\
&= (1, \pi a).\, \pi a^{-1} && \text{by (3.1)}\\
&= (1, \pi a)\, p^{-1}(1, u)\, u && \text{by (4.1)}\\
&= p^{-1}(1, (1, \pi a))\, (1, u)\, u && \\
& && \text{by the naturality of } p^{-1}\\
&= p^{-1}(1, (1, \pi a)\, u)\, u && \\
&= p^{-1}(1, \pi\pi a)\, u && \text{by (3.4)}\\
&= p^{-1}(1, p.\, \pi 1)\, u && \text{by (3.19)}\\
&= p^{-1}(1, p)\, (1, u)\, u\,. &&
\end{aligned}$$

The equivalent diagram to which we have now reduced MC3 is:

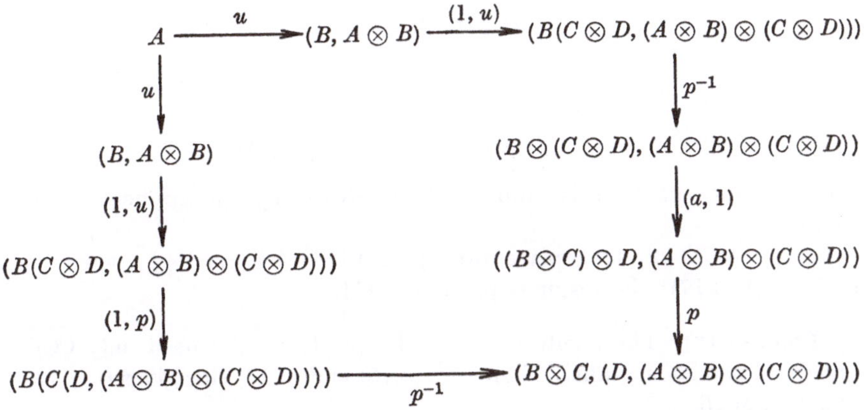

and this is equivalent to MCC3 by two applications of Lemma 3.1.

To show that MCC3 $\Leftrightarrow$ MCC3′, we transform both legs of MCC3;

we have

$$\begin{aligned} pp(a,1) &= p(\pi a, 1)L && \text{by (3.22)} \\ &= (\pi\pi a, 1)LL && \text{by (3.22)} \\ &= (p.\pi 1, 1)LL && \text{by (3.19)} \\ &= (u,1)(p,1)LL; \end{aligned}$$

and

$$\begin{aligned} (1,p)p &= (1,p)(u,1)L && \text{by (3.21)} \\ &= (u,1)(1,p)L. \end{aligned}$$

The diagram we now have is equivalent to MCC3′ by an application of Lemma 3.1.

To show that MCC3′ ⇔ CC3, consider the proof of Proposition I.8.1; it makes no use of the properties of closed categories, beyond the fact that hom $\mathscr{V}$ is a bifunctor, and shows purely formally that if two families of morphisms satisfy diagrams of the form VN, so does their composite. We can interpret MCC3′ as VN for p_{AB} and CC3 as VN for L^A_B; and diagrams (3.20) and (3.21) allow us to deduce each of these from the other, if we know that $(t, 1)$ and $(u, 1)$ satisfy VN. This is indeed the case, for the proof in Proposition I.8.4 that $(f, 1)$ satisfies VN uses only the naturality of L.

Proof of (iv). Applying π twice to each leg of MC4 we get

$$\begin{aligned} \pi\pi((1\otimes r)a) &= p.\pi(1\otimes r) && \text{by (3.19)} \\ &= p(r,1).\pi 1 && \text{by (3.1)} \\ &= p(r,1)u, \end{aligned}$$

and

$$\begin{aligned} \pi\pi r &= \pi i && \text{by (3.17)} \\ &= (1,i)u && \text{by (3.4)}; \end{aligned}$$

the resulting diagram is equivalent to MCC4 by an application of Lemma 3.1.

Now use (3.22) to replace $p(r, 1)$ in MCC4 by $(\pi r, 1)L$, which is $(i, 1)L$ by (3.17); the result is precisely CC4.

Proof of (v). The result $j_I = i_I$ of Proposition I.2.7 used only CC0 and CC5, and so is available here. By (3.15) and (3.17) this gives $l_I = r_I$, which is MC5.

Proposition 4.2. *Under the conditions of Proposition 4.1,* MC2 *is a consequence of* MC1, MC3, MC4, *and* MC5.

Proof. From the naturality of l we have a commutative diagram

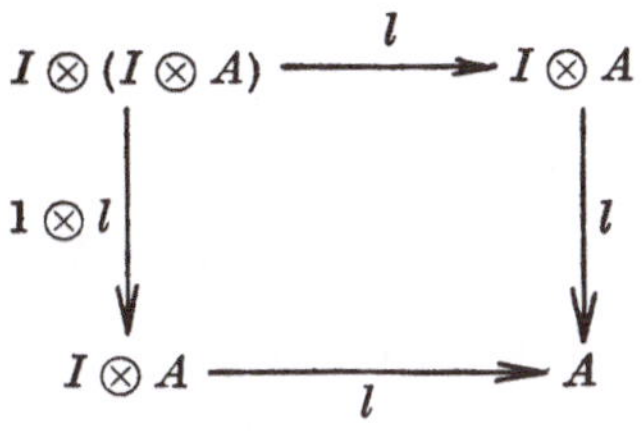

and since l is an isomorphism we have

$$l = 1 \otimes l : I \otimes (I \otimes A) \to I \otimes A\,. \tag{4.2}$$

Consider the special case of MC2 when A is put equal to I; (4.2) enables us to replace therein $1 \otimes l$ by l, and MC5 to replace $r \otimes 1$ by $l \otimes 1$; thus this special case becomes an instance of MC1, and is therefore available under the present hypotheses.

Now consider the diagram

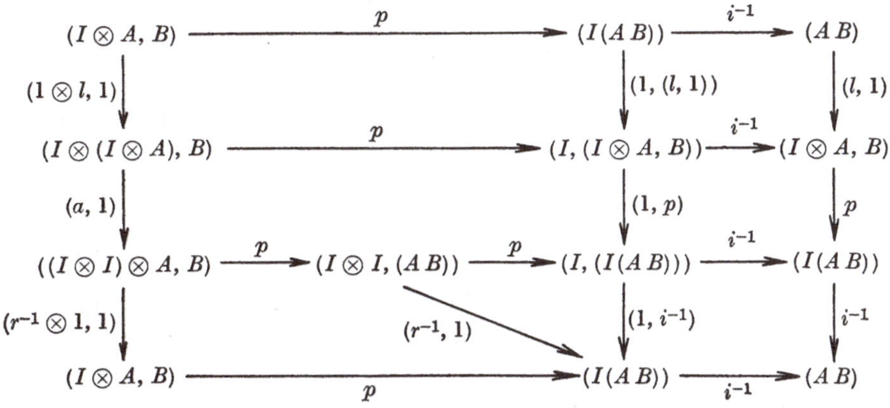

The diagram commutes by MCC3 and MCC4 (which are available here by Proposition 4.1) together with the naturality of p and of i^{-1}. The composite left edge is 1 by the above special case of MC2, so that the composite right edge is also 1. This is MCC2, which is equivalent to MC2 by Proposition 4.1.

Proposition 4.3. *Let $\mathscr{V}$ and $\mathscr{V}'$ be monoidal closed categories, $\phi : \mathscr{V}_0 \to$ $\to \mathscr{V}'_0$ a functor, $\phi^0 : I' \to \phi I$ a morphism, and $\tilde{\phi}$ and $\hat{\phi}$ natural transformations connected by* (3.23). *Then the axioms* CF, MF, *and* MCF *are* *related by:*

(i) MF 1 $\Leftrightarrow$ CF 1;

(ii) MF 2 $\Leftrightarrow$ CF 2;

(iii) MF 3 $\Leftrightarrow$ MCF 3 $\Leftrightarrow$ CF 3.

Proof of (i). Apply π' to both legs of MF 1. We have

$$\begin{aligned} \pi'(\phi l \,.\, \tilde{\phi}(\phi^0 \otimes 1)) &= \pi'(\phi l \,.\, \tilde{\phi}) .\, \phi^0 && \text{by} \quad (3.1) \\ &= \hat{\phi} \,.\, \phi \pi l \,.\, \phi^0 && \text{by} \quad (3.24) \\ &= \hat{\phi} \,.\, \phi j \,.\, \phi^0 && \text{by} \quad (3.15), \end{aligned}$$

and

$$\pi' l' = j' \qquad \text{by} \quad (3.15);$$

the equivalent diagram to which we have reduced MF 1 is now precisely CF 1.

Proof of (ii). Applying π' to MF 2 we get

$$\begin{aligned} \pi'(\phi r \,.\, \tilde{\phi}(1 \otimes \phi^0)) &= (\phi^0, 1)\, \pi'(\phi r \,.\, \tilde{\phi}) && \text{by} \quad (3.1) \\ &= (\phi^0, 1)\, \hat{\phi} \,.\, \phi \pi r && \text{by} \quad (3.24) \\ &= (\phi^0, 1)\, \hat{\phi} \,.\, \phi i && \text{by} \quad (3.17), \end{aligned}$$

and

$$\pi' r' = i' \qquad \text{by} \quad (3.17);$$

the resulting diagram is precisely CF 2.

Proof of (iii). Apply π' twice to each leg of MF 3. We get

$$\begin{aligned} \pi'\pi'(\tilde{\phi}(1 \otimes \tilde{\phi})\, a') &= p' \,.\, \pi'(\tilde{\phi}(1 \otimes \tilde{\phi})) && \text{by} \quad (3.19) \\ &= p'(\tilde{\phi}, 1) \,.\, \pi' \tilde{\phi} && \text{by} \quad (3.1) \\ &= p'(\tilde{\phi}, 1)\, \hat{\phi} \,.\, \phi u && \text{by} \quad (3.24) \text{ with } x = 1; \end{aligned}$$

and

$$\begin{aligned} \pi'\pi'(\phi a \,.\, \tilde{\phi}(\tilde{\phi} \otimes 1)) &= \pi'(\pi'(\phi a \,.\, \tilde{\phi}) \,.\, \tilde{\phi}) && \text{by} \quad (3.1) \\ &= \pi'(\hat{\phi} \,.\, \phi \pi a \,.\, \tilde{\phi}) && \text{by} \quad (3.24) \\ &= (1, \hat{\phi}) \,.\, \pi'(\phi \pi a \,.\, \tilde{\phi}) && \text{by} \quad (3.1) \\ &= (1, \hat{\phi})\, \hat{\phi} \,.\, \phi \pi \pi a && \text{by} \quad (3.24) \\ &= (1, \hat{\phi})\, \hat{\phi} \,.\, \phi(p u) && \text{by} \quad (3.19) \text{ with } x = 1 \\ &= (1, \hat{\phi})\, \hat{\phi} \,.\, \phi p \,.\, \phi u; \end{aligned}$$

giving a diagram equivalent to MCF 3 by an application of Lemma 3.1.

To show MCF3 $\Leftrightarrow$ CF3, we use (3.22) to transform the legs of the former. We have

$$\begin{aligned} p'(\tilde{\phi},1)\,\hat{\phi} &= (\pi'\tilde{\phi},1)\,L\hat{\phi} && \text{by (3.22)} \\ &= (\hat{\phi}\,.\,\phi u,1)\,L\,\hat{\phi} && \text{by (3.24) with } x=1 \\ &= (\phi u,1)\,(\hat{\phi},1)\,L\hat{\phi}; \end{aligned}$$

and

$$\begin{aligned} (1,\hat{\phi})\,\hat{\phi}\,.\,\phi p &= (1,\hat{\phi})\,\hat{\phi}\,.\,\phi(u,1)\,.\,\phi L && \text{by (3.21)} \\ &= (1,\hat{\phi})\,(\phi u,1)\,\hat{\phi}\,.\,\phi L && \text{by the naturality of } \hat{\phi} \\ &= (\phi u,1)\,(1,\hat{\phi})\,\hat{\phi}\,.\,\phi L; \end{aligned}$$

giving a diagram equivalent to CC3 by an application of Lemma 3.1.

Proposition 4.4. *Let $\Phi, \Psi: \mathscr{V} \to \mathscr{V}'$ be monoidal closed functors and $\eta: \phi \to \psi$ a natural transformation. Then* CN1 *is identical with* MN1, *and* CN2 *is equivalent to* MN2.

Proof. We apply π' to both legs of MN2, getting

$$\begin{aligned} \pi'(\eta\tilde{\phi}) &= (1,\eta)\,.\,\pi'\tilde{\phi} && \text{by (3.1)} \\ &= (1,\eta)\,\hat{\phi}\,.\,\phi u && \text{by (3.24) with } x=1; \end{aligned}$$

and

$$\begin{aligned} \pi'(\tilde{\psi}(\eta\otimes\eta)) &= (\eta,1)\,.\,\pi'\tilde{\psi}\,.\,\eta && \text{by (3.1)} \\ &= (\eta,1)\,\hat{\psi}\,.\,\psi u\,.\,\eta && \text{by (3.24) for } \Psi \text{ with } x=1 \\ &= (\eta,1)\,\hat{\psi}\eta\,.\,\phi u && \text{by the naturality of } \eta; \end{aligned}$$

giving a diagram equivalent to CN2 by an application of Lemma 3.1.

5. The Forgetful Hyperfunctors

We say that a hyperfunctor $\Phi: \mathfrak{A} \to \mathfrak{B}$ is *locally isomorphic* if for each pair of objects $\mathscr{A}, \mathscr{B}$ in $\mathfrak{A}$ the functor $\mathfrak{A}(\mathscr{A}\mathscr{B}) \to \mathfrak{B}(\Phi\mathscr{A}, \Phi\mathscr{B})$ determined by Φ is an isomorphism of categories.

We have forgetful hyperfunctors $\mathscr{MC\ell} \to \mathscr{M}on$ and $\mathscr{MC\ell} \to \mathscr{C\ell}$ given by $\mathscr{V} \mapsto {}^m\mathscr{V}$, $\Phi \mapsto {}^m\Phi$, $\eta \mapsto \eta$ and $\mathscr{V} \mapsto {}^c\mathscr{V}$, $\Phi \mapsto {}^c\Phi$, $\eta \mapsto \eta$. From Propositions 3.2, 4.3 and 4.4 we have at once:

Theorem 5.1. *The forgetful hyperfunctors $\mathscr{MC\ell} \to \mathscr{M}on$ and $\mathscr{MC\ell} \to \mathscr{C\ell}$ are locally isomorphic.*

It remains to examine which monoidal categories and which closed categories admit enrichment to a monoidal closed category. From Theorem 5.1 we have:

Corollary 5.2. *If a monoidal category* ${}^m\mathscr{V}$ *(resp. a closed category* ${}^c\mathscr{V}$*) admits enrichment to a monoidal closed category* $\mathscr{V}$, *then* $\mathscr{V}$ *is unique to within an isomorphism of the form* $(1, 1, \hat{\phi}, 1)$ *(resp.* $(1, \tilde{\phi}, 1, 1)$*).*

We shall say that a monoidal category (resp. a closed category) is *closed* (resp. *monoidal*) if it admits enrichment to a monoidal closed category.

Theorem 5.3. *A closed category* $\mathscr{V}$ *is monoidal if and only if the* $\mathscr{V}$*-functor* $L^A L^B : \mathscr{V} \to \mathscr{V}$ *is representable for each* $A, B \in \mathscr{V}$. *If representations*

$$p_{AB} : L^{A \otimes B} \to L^A L^B$$

with components

$$p_{ABC} : (A \otimes B, C) \to (A(BC))$$

are chosen, there is exactly one monoidal closed structure with the given p.

Proof. The necessity is clear from (3.27). If representations as above are given, there is a unique functor $\otimes$, with the given values $A \otimes B$ on objects, rendering p natural in A and B — it is already natural in C by Proposition 8.11. For the naturality of p means the commutativity of

$$\begin{array}{ccc} L^{A' \otimes B'} & \xrightarrow{p_{A'B'}} & L^{A'} L^{B'} \\ & & \downarrow{\scriptstyle L^f L^{B'}} \\ \downarrow{\scriptstyle L^{f \otimes g}} & & L^A L^{B'} \\ & & \downarrow{\scriptstyle L^A L^g} \\ L^{A \otimes B} & \xrightarrow[p_{AB}]{} & L^A L^B \end{array} \tag{5.1}$$

with components

$$\begin{array}{ccc} (A' \otimes B', C) & \xrightarrow{p} & (A'(B'C)) \\ \downarrow{\scriptstyle (f \otimes g, 1)} & & \downarrow{\scriptstyle (f, (g, 1))} \\ (A \otimes B, C) & \xrightarrow[p]{} & (A(BC) \end{array}$$

and by Remark I.10.11, $f \otimes g$ is uniquely determined by (5.1); the functoriality of $\otimes$ is then clear.

Similarly the representation theorem for $\mathscr{V}$-functors gives the existence of unique isomorphisms l, a, r satisfying MCC2, MCC3, MCC4. Defining

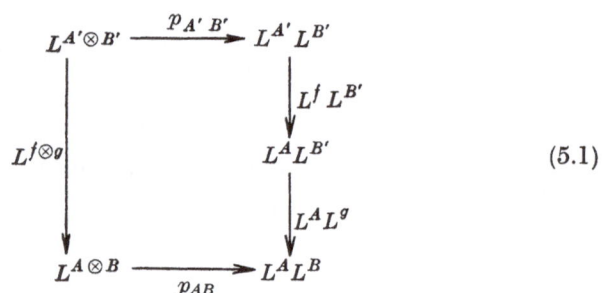

π and v by (3.25) and (3.26), we obtain (3.12), (3.16), and (3.18) by applying V to MCC2, MCC3, and MCC4; the naturality of l, a, r now follows by Proposition I.1.2. Also (3.13) is just CC5, and (3.20) follows exactly as in the proof of Proposition 3.2. Proposition 4.1 now ensures the validity of the remaining axioms.

Corollary 5.4. *If a closed category $\mathscr{V}$ is monoidal, so is any closed category isomorphic to $\mathscr{V}$.*

Proof. The representability of $L^A L^B$ is easily seen to survive passage to an isomorph.

The question of which monoidal categories are closed is somewhat more complicated due to the necessity of constructing the functor $V : \mathscr{V}_0 \to \mathscr{S}$; we shall deal first with the case where V is given. By a *normalization* of a monoidal category $\mathscr{V}$ we shall mean a functor $V : \mathscr{V}_0 \to \mathscr{S}$ together with a natural isomorphism $\iota = \iota_A : VA \to \mathscr{V}_0(IA)$; a monoidal category with a given normalization is said to be *normalized.* A monoidal closed category has a canonical normalization given by the V and $\iota = Vi$ it already possesses; any monoidal category admits a normalization, namely $V = \mathscr{V}_0(I\, -)$ and $\iota = 1$, but if $\mathscr{V}$ is also closed this differs in general from the canonical one. A normalized monoidal category shall be said to be closed only if it admits enrichment to a monoidal closed category with the given V and with Vi equal to the given ι.

Theorem 5.5. *A normalized monoidal category $\mathscr{V}$ is closed if and only if the following two conditions are satisfied:*

(i) *the functor $\mathscr{V}_0(- \otimes B, C) : \mathscr{V}_0^* \to \mathscr{S}$ is representable for each $B, C \in \mathscr{V}_0$;*

(ii) *representing objects (BC) and representations*

$$\pi = \pi_{ABC} : \mathscr{V}_0(A \otimes B, C) \to \mathscr{V}_0(A\,(BC))$$

of the above functors may be so chosen that

$$V(BC) = \mathscr{V}_0(BC) \tag{5.2}$$

and

$$\text{the composite } \mathscr{V}_0(BC) \xrightarrow[\mathscr{V}_0(l,1)]{} \mathscr{V}_0(I \otimes B, C) \xrightarrow[\pi]{} \mathscr{V}_0(I\,(BC)) \text{ is } \iota_{(BC)}\,. \tag{5.3}$$

$\mathscr{V}$ then admits a unique monoidal closed structure with Vp equal to the given π.

Proof. The conditions are necessary by Proposition 3.2; suppose they are satisfied.

By the representation theorem, there is a unique way of extending

(BC) to a functor hom $\mathscr{V}$ with respect to which π_{ABC} is natural in B and in C as well as in A.

For $f: B' \to B$ and $g: C \to C'$ we have by the naturality of π and of l a commutative diagram

$$\begin{array}{ccccc}
\mathscr{V}_0(BC) & \xrightarrow{\mathscr{V}_0(l,1)} & \mathscr{V}_0(I \otimes B, C) & \xrightarrow{\pi} & \mathscr{V}_0(I(BC)) \\
\downarrow{\scriptstyle \mathscr{V}_0(f,g)} & & \downarrow{\scriptstyle \mathscr{V}_0(1 \otimes f, g)} & & \downarrow{\scriptstyle \mathscr{V}_0(1,(f,g))} \\
\mathscr{V}_0(B'C') & \xrightarrow[\mathscr{V}_0(l,1)]{} & \mathscr{V}_0(I \otimes B', C') & \xrightarrow[\pi]{} & \mathscr{V}_0(I(B'C'))
\end{array}$$

Now by (5.3) and the naturality of ι the left edge of this diagram must also be $V(f, g)$; thus CC0 is satisfied.

Next define v, j, i, p, L by (3.12), (3.13), (3.16), (3.18), and (3.20); these definitions are all forced by Proposition 3.2. By (3.12) and (5.3) we have $v_{BC} = \iota_{(BC)}$; axiom CC5 will follow if we prove $Vi = \iota$.

Since $A \cong (IA)$, to show $Vi_A = \iota_A$ it suffices to show $Vi_{(IA)} = \iota_{(IA)}$. By Proposition I.2.5, which uses only the fact that i is a natural isomorphism, we have $i_{(IA)} = (1, i_A)$. Since $\iota_{(IA)} = v_{IA}$, we have to prove that

$$V(1, i_A) = v_{IA} : \mathscr{V}_0(IA) \to \mathscr{V}_0(I(IA)).$$

By the representation theorem, it suffices to put $A = I$ and evaluate at 1_I; since $V(1, i_I)\,1 = i_I$, and $v_{II}\,1 = j_I$ by (3.13), we need $i_I = j_I$. This follows at once from MC5, namely $r_I = l_I$, by (3.15) and (3.17).

From Proposition 4.1 it now follows that the remaining axioms are satisfied and that $Vp = \pi$.

Remark 5.6. A normalized monoidal category may satisfy condition (i) of Theorem 5.5 but admit no π satisfying condition (ii). For instance if $\mathscr{V}_0$ has only a finite number of objects and the $\mathscr{V}_0(BC)$ are all different, it is clearly impossible to satisfy (5.2). However if $\mathscr{V}_0$ is so large that V *admits transport of structure*, which is the case in most of the large categories that occur in nature, condition (ii) can always be satisfied. We say that a functor $V: \mathscr{V}_0 \to \mathscr{S}$ admits transport of structure if, for any $A \in \mathscr{V}_0$, $X \in \mathscr{S}$, and isomorphism $f: VA \to X$, there is a $B \in \mathscr{V}_0$ with $VB = X$ and an isomorphism $g: A \to B$ with $Vg = f$.

Proposition 5.7. *If $\mathscr{V}$ is a normalized monoidal category and $V: \mathscr{V}_0 \to \mathscr{S}$ admits transport of structure, $\mathscr{V}$ is closed if and only if condition* (i) *of Theorem 5.5 is satisfied.*

Proof. First choose representing objects $(BC)'$ and representations

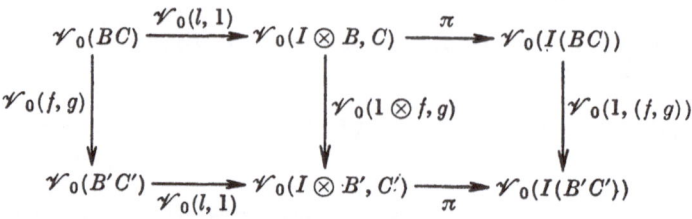

$\pi' : \mathscr{V}_0(A \otimes B, C) \to \mathscr{V}_0(A, (BC)')$. Then we have an isomorphism

$$\mathscr{V}_0(BC) \xrightarrow[\mathscr{V}_0(l,1)]{} \mathscr{V}_0(I \otimes B, C) \xrightarrow[\pi']{} \mathscr{V}_0(I, (BC)') \xrightarrow[\iota^{-1}]{} V(BC)' . \quad (5.4)$$

Now choose for each B, C an object (BC) with $V(BC) = \mathscr{V}_0(BC)$ and an isomorphism $k_{BC} : (BC) \to (BC)'$ with Vk_{BC} equal to (5.4). Define a new representation π as the composite

$$\pi: \mathscr{V}_0(A \otimes B, C) \xrightarrow[\pi']{} \mathscr{V}_0(A, (BC)') \xrightarrow[\mathscr{V}_0(1,k^{-1})]{} \mathscr{V}_0(A(BC)) . \quad (5.5)$$

We already have (5.2) by our choice of (BC); we show that π satisfies (5.3). Consider the diagram

$$\begin{array}{ccc}
\mathscr{V}_0(BC) & \xrightarrow{\mathscr{V}_0(l,1)} & \mathscr{V}_0(I \otimes B, C) \\
{\scriptstyle Vk}\downarrow & & \downarrow{\scriptstyle \pi'} \\
V(BC)' & \xrightarrow{\iota} & \mathscr{V}_0(I, (BC)') \\
{\scriptstyle Vk^{-1}}\downarrow & & \downarrow{\scriptstyle \mathscr{V}_0(1,k^{-1})} \\
V(BC) & \xrightarrow[\iota]{} & \mathscr{V}_0(I(BC))
\end{array}$$

The top region commutes by our choice of k, and the bottom region by the naturality of ι. In view of (5.5) this gives (5.3).

We now consider criteria for an unnormalized monoidal category to be closed.

Theorem 5.8. *A monoidal category $\mathscr{V}$ is closed if and only if the following two conditions are satisfied:*

(i) *the functor $\mathscr{V}_0(- \otimes B, C) : \mathscr{V}_0^* \to \mathscr{S}$ is representable for each $B, C \in \mathscr{V}_0$;*

(ii) *representing objects (BC) and representations*

$$\pi = \pi_{ABC} : \mathscr{V}_0(A \otimes B, C) \to \mathscr{V}_0(A(BC))$$

of the above functors may be so chosen that $\mathscr{V}_0(BC)$ and the composite

$$\mathscr{V}_0(BC) \xrightarrow[\mathscr{V}_0(l,1)]{} \mathscr{V}_0(I \otimes B, C) \xrightarrow[\pi]{} \mathscr{V}_0(I(BC)) \quad (*)$$

depend only on (BC).

With such representations chosen, a monoidal closed structure with Vp equal to the given π is unique except for some indeterminacy in the definition of V.

Proof. We define V and ι on the full subcategory of $\mathscr{V}_0$ determined by objects of the form (BC). We define $V(BC)$ to be $\mathscr{V}_0(BC)$ and $\iota_{(BC)}$ to be (*); these definitions are consistent by condition (ii). Moreover they are forced by Proposition 3.2, which shows that condition (ii) is necessary; we already know that condition (i) is necessary. We take the value of Vf for $f: (BC) \to (DE)$ to be $\iota^{-1}\mathscr{V}_0(1, f)\,\iota$; this is forced if ι is to be natural, and does make V a functor and ι natural.

We have considerable liberty in completing the definitions of V and of ι. For definiteness let us define VA, where A is not of the form (BC), to be $\mathscr{V}_0(IA)$, and define $\iota: VA \to \mathscr{V}_0(IA)$ to be 1; then so define V on morphisms that ι is natural.

We now have (5.2) and (5.3), and Theorem 5.5 gives the desired result. We have only to note that our manner of completing the definitions of V and of ι makes no difference to the forced definitions of j, i, L, p.

Theorem 5.9. *A monoidal category $\mathscr{V}$ possesses an isomorph $\mathscr{V}'$ which is closed if and only if it satisfies condition* (i) *of Theorem 5.8. Moreover if representations $\pi: \mathscr{V}_0(A \otimes B, C) \to \mathscr{V}_0(A(BC))$ are given, there is a canonical way of constructing the monoidal closed category $\mathscr{V}'$.*

Proof. Condition (i) is necessary because it clearly survives passage to an isomorph. Suppose representations π as above are given.

We define a new category $\mathscr{V}'_0$ and an isomorphism $\phi: \mathscr{V}_0 \to \mathscr{V}'_0$. The objects of $\mathscr{V}'_0$ are those of $\mathscr{V}_0$, and ϕ is the identity on objects. We set

$$\mathscr{V}'_0(BC) = \mathscr{V}_0(I, (BC)) \tag{5.6}$$

and define $\phi_{BC}: \mathscr{V}_0(BC) \to \mathscr{V}'_0(BC)$ to be the composite

$$\phi_{BC}: \mathscr{V}_0(BC) \xrightarrow[\mathscr{V}_0(l,1)]{} \mathscr{V}_0(I \otimes B, C) \xrightarrow[\pi]{} \mathscr{V}_0(I(BC)). \tag{5.7}$$

Finally we define composition in $\mathscr{V}'_0$ so that ϕ becomes a functor.

We now use the isomorphism ϕ to transfer to $\mathscr{V}'_0$ the structure of monoidal category on $\mathscr{V}_0$, getting a monoidal category $\mathscr{V}'$ and an isomorphic monoidal functor $\Phi = (\phi, \tilde{\phi}, \phi^0): \mathscr{V} \to \mathscr{V}'$. To be precise, we have $A \otimes' B = A \otimes B, f \otimes' g = \phi(\phi^{-1} f \otimes \phi^{-1} g), I' = I, a' = \phi a$, $l' = \phi l,\ r' = \phi r,\ \tilde{\phi} = 1,\ \phi^0 = 1$.

We next define a normalization of $\mathscr{V}'$. We define

$$V'A = \mathscr{V}_0(IA) \tag{5.8}$$

and define $\iota': V'A \to \mathscr{V}'_0(IA)$ to be

$$\iota' = \phi: \mathscr{V}_0(IA) \to \mathscr{V}'_0(IA); \tag{5.9}$$

then define V' on morphisms so that ι' is natural.

Finally we define representations $\pi': \mathscr{V}'_0(A \otimes B, C) \to \mathscr{V}'_0(A(BC))$ by setting π' equal to the composite

$$\pi': \mathscr{V}'_0(A \otimes B, C) \xrightarrow[\phi^{-1}]{} \mathscr{V}_0(A \otimes B, C) \xrightarrow[\pi]{} \mathscr{V}_0(A(BC)) \xrightarrow[\phi]{} \mathscr{V}'_0(A(BC)); \tag{5.10}$$

this is clearly natural in A.

$\mathscr{V}'$ satisfies (5.2) by (5.6) and (5.8); we show that it also satisfies (5.3). Consider the diagram

$$\begin{array}{ccccc}
\mathscr{V}_0(BC) & \xrightarrow{\mathscr{V}_0(l,\,1)} & \mathscr{V}_0(I \otimes B,\, C) & \xrightarrow{\pi} & \mathscr{V}_0(I(BC)) \\
\Big\downarrow{\scriptstyle \phi} & & \Big\downarrow{\scriptstyle \phi} & & \Big\downarrow{\scriptstyle \phi} \\
\mathscr{V}'_0(BC) & \xrightarrow[\mathscr{V}'_0(l',\,1)]{} & \mathscr{V}'_0(I \otimes B,\, C) & \xrightarrow[\pi']{} & \mathscr{V}'_0(I,\, (BC))
\end{array}$$

The left region commutes by the naturality of ϕ, since $l' = \phi l$, and the right region by (5.10). The left edge and the top edge are equal isomorphisms by (5.7); hence the bottom edge equals the right edge, which is ι' by (5.9). This is (5.3) for $\mathscr{V}'$, and we now appeal to Theorem 5.5.

We complete this section by describing an economical way of giving a monoidal closed category:

Theorem 5.10. *Suppose given a category $\mathscr{V}_0$, functors $\otimes: \mathscr{V}_0 \times \mathscr{V}_0 \to \mathscr{V}_0$ and* hom $\mathscr{V}: \mathscr{V}_0^* \times \mathscr{V}_0 \to \mathscr{V}_0$, *and a functor $V: \mathscr{V}_0 \to \mathscr{S}$ satisfying* CC0. *Suppose further given an object I of $\mathscr{V}_0$, a natural isomorphism $i: A \to (IA)$, and a natural isomorphism $p: (A \otimes B, C) \to (A(BC))$. Then these data can be completed to give a monoidal closed category if and only if the r and the a defined by* (3.16) *and* (3.18), *with $\pi = Vp$, satisfy* MCC4 *and* MCC3; *and $\mathscr{V}$ is then unique. Moreover if V is faithful the satisfaction of* MCC4 *and* MCC3 *is automatic.*

Proof. The necessity of the conditions follows from Proposition 3.2; moreover if V is faithful MCC4 and MCC3 follow from their images under V, which are (3.16) and (3.18).

If the conditions are satisfied we define v by (3.26), j by (3.13), l by (3.12), and L by (3.20). Since we have forced CC5 by our definition of v, and since we have $Vp = \pi$, it follows from Propositions 4.1 and 4.2 that all the axioms are satisfied.

6. Categories over a Monoidal Closed Category

If $\mathscr{V}$ is a monoidal category, we define a $\mathscr{V}$-category $\mathscr{A}$ to consist of the following four data:

(i) a class obj $\mathscr{A}$ of "objects";

(ii) for each $A, B \in$ obj $\mathscr{A}$, an object $\mathscr{A}(AB)$ of $\mathscr{V}_0$;

(iii) for each $A \in$ obj $\mathscr{A}$, a morphism

$$j_A : I \to \mathscr{A}(AA)$$

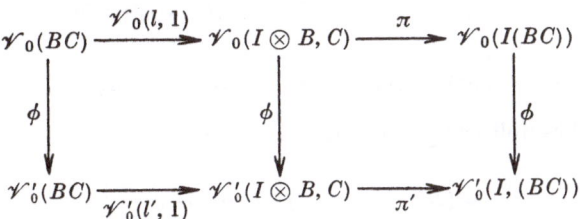

in $\mathscr{V}_0$;

(iv) for each $A, B, C \in \operatorname{obj} \mathscr{A}$, a morphism

$$M^B_{AC} : \mathscr{A}(BC) \otimes \mathscr{A}(AB) \to \mathscr{A}(AC)$$

in $\mathscr{V}_0$.

These data are to satisfy the following three axioms:

VC1′. The following diagram commutes:

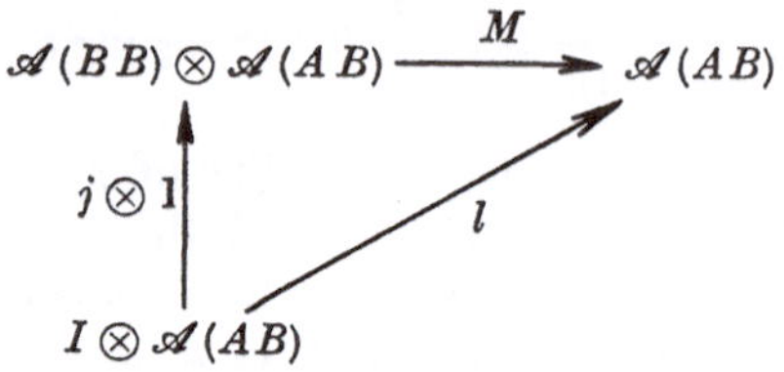

VC2′. The following diagram commutes:

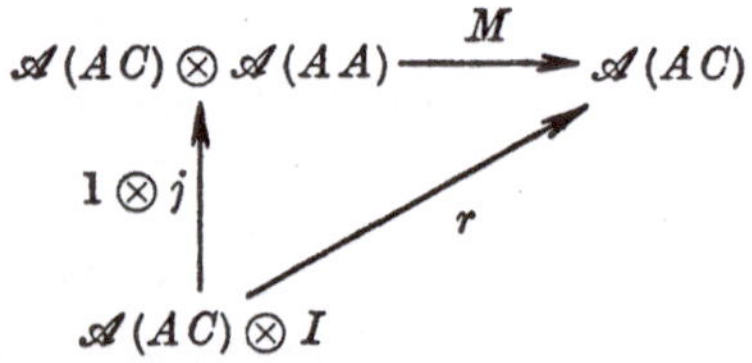

VC3′. The following diagram commutes:

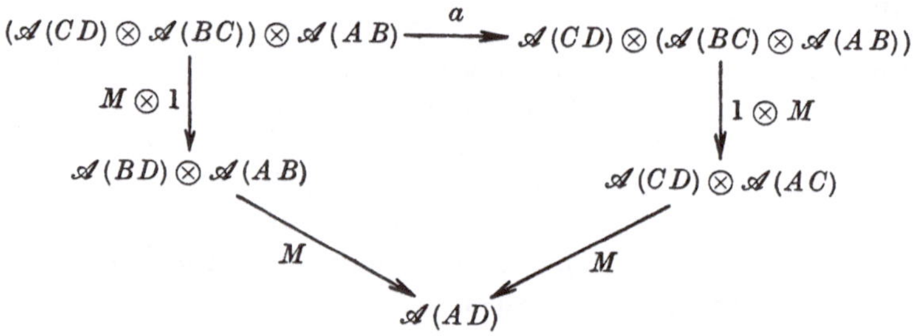

If $\mathscr{A}$ and $\mathscr{B}$ are $\mathscr{V}$-categories where $\mathscr{V}$ is a monoidal category, a $\mathscr{V}$-functor $T : \mathscr{A} \to \mathscr{B}$ is to consist of the following two data:

(i) a function $T : \operatorname{obj} \mathscr{A} \to \operatorname{obj} \mathscr{B}$;

(ii) for each $B, C \in \operatorname{obj} \mathscr{A}$, a morphism

$$T = T_{BC} : \mathscr{A}(BC) \to \mathscr{B}(TB, TC)$$

in $\mathscr{V}_0$.

These data are to satisfy the following two axioms:

VF 1′. The following diagram commutes:

$$\begin{array}{ccc} \mathscr{A}(B\,B) & \xrightarrow{T_{BB}} & \mathscr{B}(T\,B, T\,B) \\ & {\scriptstyle j}\nwarrow \quad \nearrow{\scriptstyle j} & \\ & I & \end{array}$$

VF 2′. The following diagram commutes:

$$\begin{array}{ccc} \mathscr{A}(C\,D) \otimes \mathscr{A}(B\,C) & \xrightarrow{M} & \mathscr{A}(B\,D) \\ {\scriptstyle T \otimes T}\downarrow & & \downarrow{\scriptstyle T} \\ \mathscr{B}(T\,C, T\,D) \otimes \mathscr{B}(T\,B, T\,C) & \xrightarrow[M]{} & \mathscr{B}(T\,B, T\,D) \end{array}$$

We leave the reader to verify:

Proposition 6.1. *If $\mathscr{V}$ is a monoidal category, $\mathscr{V}$-categories and $\mathscr{V}$-functors form a "category" $\mathscr{V}_*$ if we define composition by* I (5.1) *and* I (5.2).

Remark 6.2. If $\mathscr{V}$ is monoidal without being closed, $\mathscr{V}$ itself does not in general have the structure of a $\mathscr{V}$-category. However there is always a $\mathscr{V}$-category $\mathscr{I}$, defined as in Proposition I.5.3 except that in place of L we give $M : \mathscr{I}(**) \otimes \mathscr{I}(**) \to \mathscr{I}(**)$, defining it to be $l_I : I \otimes I \to I$. The rest of Proposition I.5.3 then applies word for word.

We also leave the reader to verify:

Proposition 6.3. *From a monoidal functor $\Phi = (\phi, \tilde{\phi}, \phi^0) : \mathscr{V} \to \mathscr{V}'$ we get a functor $\Phi_* : \mathscr{V}_* \to \mathscr{V}'_*$ if we define $\Phi_* \mathscr{A}$ by* I (6.1), I (6.2), I (6.3), *and:*

$$M' : \phi\,\mathscr{A}(B\,C) \otimes \phi\,\mathscr{A}(A\,B) \to \phi\,\mathscr{A}(A\,C)$$

is the composite

$$\phi\,\mathscr{A}(\,BC) \otimes \phi\,\mathscr{A}(A\,B) \xrightarrow[\tilde{\phi}]{} \phi(\mathscr{A}(B\,C) \otimes \mathscr{A}(A\,B)) \xrightarrow[\phi M]{} \phi\,\mathscr{A}(A\,C); \quad (6.1)$$

and define $\Phi_* T$ by I (6.8) *and* I (6.9). *Further if $\eta : \Phi \to \Psi : \mathscr{V} \to \mathscr{V}'$ is a monoidal natural transformation we get a natural transformation*

$$\eta_* : \Phi_* \to \Psi_* : \mathscr{V}_* \to \mathscr{V}'_*$$

if we define $\eta_{*\mathscr{A}}$ by I (6.10) *and* I (6.11). *Finally the assignments $\mathscr{V} \mapsto \mathscr{V}_*$, $\Phi \mapsto \Phi_*$, $\eta \mapsto \eta_*$ constitute a hyperfunctor $_* : \mathscr{M}on \to \mathscr{C}at$.*

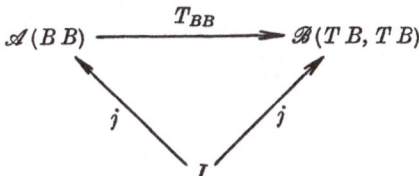

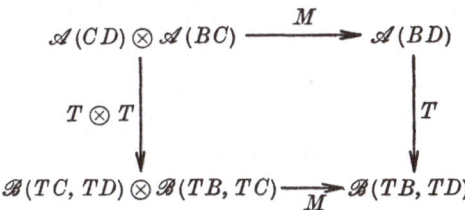

Theorem 6.4. *Let $\mathscr{V}$ be a monoidal closed category. Then the "categories" ${}^m\mathscr{V}_*$ and ${}^c\mathscr{V}_*$ coincide if we identify the ${}^m\mathscr{V}$-category $\mathscr{A} = (\text{obj}\,\mathscr{A}, \mathscr{A}(A\,B), j, M)$ with the ${}^c\mathscr{V}$-category $\mathscr{A} = (\text{obj}\,\mathscr{A}, \mathscr{A}(A\,B), j, L)$ where*

$$L^A_{BC} = \pi\, M^B_{AC}; \tag{6.2}$$

here π is

$$\pi = V p : V(\mathscr{A}(B\,C) \otimes \mathscr{A}(A\,B), \mathscr{A}(A\,C)) \to$$
$$\to V(\mathscr{A}(B\,C), (\mathscr{A}(A\,B), \mathscr{A}(A\,C)))\,.$$

Moreover if $\Phi : \mathscr{V} \to \mathscr{V}'$ is a monoidal closed functor, ${}^m\Phi_* : {}^m\mathscr{V}_* \to {}^m\mathscr{V}'_*$ coincides with ${}^c\Phi_* : {}^c\mathscr{V}_* \to {}^c\mathscr{V}'_*$, and if $\eta : \Phi \to \Psi$ is a monoidal closed natural transformation, $\eta_* : {}^m\Phi_* \to {}^m\Psi_*$ coincides with $\eta_* : {}^c\Phi_* \to {}^c\Psi_*$.

Proof. We shall prove (i) VC1 $\Leftrightarrow$ VC1$'$; (ii) VC2 $\Leftrightarrow$ VC2$'$; (iii) VC3 $\Leftrightarrow$ VC3$'$; (iv) VF2 $\Leftrightarrow$ VF2$'$; (v) I(6.4) and (6.1) are related by (6.2). The other matters to be verified are trivial.

Proof of (i). Apply π to both legs of VC1$'$; we get

$$\begin{aligned} \pi(M(j \otimes 1)) &= \pi M\,.\,j && \text{by} \quad (3.1)\\ &= L j && \text{by} \quad (6.2);\end{aligned}$$

and

$$\pi l = j \qquad \text{by } (3.15);$$

the resulting diagram equivalent to VC1$'$ is precisely VC1.

Proof of (ii). Applying π to both legs of VC2$'$ we get

$$\begin{aligned} \pi(M(1 \otimes j)) &= (j, 1)\,.\,\pi M && \text{by} \quad (3.1)\\ &= (j, 1)\, L && \text{by} \quad (6.2);\end{aligned}$$

and

$$\pi r = i \qquad \text{by } (3.17);$$

the resulting diagram is VC2.

Proof of (iii). Applying π twice to each leg of VC3$'$ we get

$$\begin{aligned} \pi\pi(M(1 \otimes M)\,a) &= p\,.\,\pi(M(1 \otimes M)) && \text{by} \quad (3.19)\\ &= p(M, 1)\,.\,\pi M && \text{by} \quad (3.1)\\ &= p(M, 1)\, L && \text{by} \quad (6.2)\\ &= (\pi M, 1)\, L L && \text{by} \quad (3.22)\\ &= (L, 1)\, L L && \text{by} \quad (6.2);\end{aligned}$$

and

$$\begin{aligned} \pi\pi(M(M \otimes 1)) &= \pi(\pi M\,.\,M) && \text{by} \quad (3.1)\\ &= \pi(L M) && \text{by} \quad (6.2)\end{aligned}$$

$$= (1, L) \,.\, \pi M \qquad \text{by} \quad (3.1)$$
$$= (1, L)\, L \qquad \text{by} \quad (6.2);$$

the resulting diagram is precisely VC3.

Proof of (iv). Applying π to each leg of VF2′ we get

$$\pi(T M) = (1, T) \,.\, \pi M \qquad \text{by} \quad (3.1)$$
$$= (1, T)\, L \qquad \text{by} \quad (6.2);$$

and

$$\pi(M(T \otimes T)) = (T, 1) \,.\, \pi M \,.\, T \qquad \text{by} \quad (3.1)$$
$$= (T, 1)\, L T \qquad \text{by} \quad (6.2);$$

the resulting diagram is VF2.

Proof of (v) We have, with M' defined by (6.1),

$$\pi' M' = \pi'(\phi M \,.\, \tilde{\phi})$$
$$= \hat{\phi} \,.\, \phi \pi M \qquad \text{by} \quad (3.24)$$
$$= \hat{\phi} \,.\, \phi L \qquad \text{by} \quad (6.2)$$
$$= L' \qquad \text{by} \quad \text{I}(6.4).$$

Remark 6.5. If $\mathscr{V}$ and $\mathscr{V}'$ are monoidal closed categories we shall, in view of Theorems 5.1 and 6.4, identify a monoidal closed functor $\Phi = (\phi, \tilde{\phi}, \hat{\phi}, \phi^0) : \mathscr{V} \to \mathscr{V}'$ with the monoidal functor $(\phi, \tilde{\phi}, \phi^0)$ and the closed functor $(\phi, \hat{\phi}, \phi^0)$.

7. The $\mathscr{V}$-functor K^B

Let $\mathscr{V}$ be a monoidal closed category. By Lemma 3.1 the natural transformation

$$p^{-1} : (A(BC)) \to (A \otimes B, C)$$

determines a natural transformation

$$K^B_{AC} : (AC) \to (A \otimes B, C \otimes B)$$

connected with p^{-1} by the diagrams

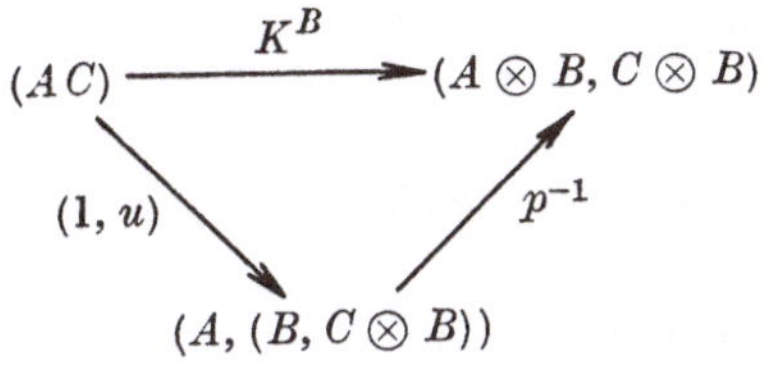

(7.1)

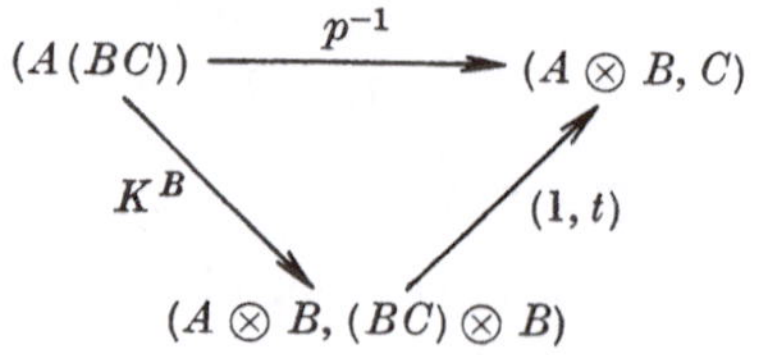

(7.2)

Theorem 7.1. *Let $\mathscr{V}$ be a monoidal closed category. For each $B \in \mathscr{V}$ we obtain a $\mathscr{V}$-functor $K^B : \mathscr{V} \to \mathscr{V}$ if we set $K^B A = A \otimes B$ and $(K^B)_{AC} = K^B_{AC}$, and the underlying functor $V_* K^B : \mathscr{V}_0 \to \mathscr{V}_0$ is $- \otimes B$. Moreover the morphisms $t_{BC} : (BC) \otimes B \to C$ and $u_{CB} : C \to (B, C \otimes B)$ are the C-components of $\mathscr{V}$-natural transformations $t_B : K^B L^B \to 1 : \mathscr{V} \to \mathscr{V}$ and $u_{\cdot B} : 1 \to L^B K^B : \mathscr{V} \to \mathscr{V}$.*

Proof. Applying V to (7.1) and evaluating at $f \in V(AC)$ gives

$$\begin{aligned}(VK^B) f &= \pi^{-1} V(1, u) f \\ &= \pi^{-1}(u f) \\ &= \pi^{-1} u \,.\, (1 \otimes f) \quad \text{by (3.1)} \\ &= 1 \otimes f \qquad\qquad \text{by (3.3)}.\end{aligned}$$

Thus $V_* K^B = - \otimes B$, and we have VF1 for K^B in the form $(VK^B) 1 = 1$ (cf. Remark I.9.7). Leave aside for the moment the question of VF2 for K_B.

From (3.20) and (7.2) we get a commutative diagram

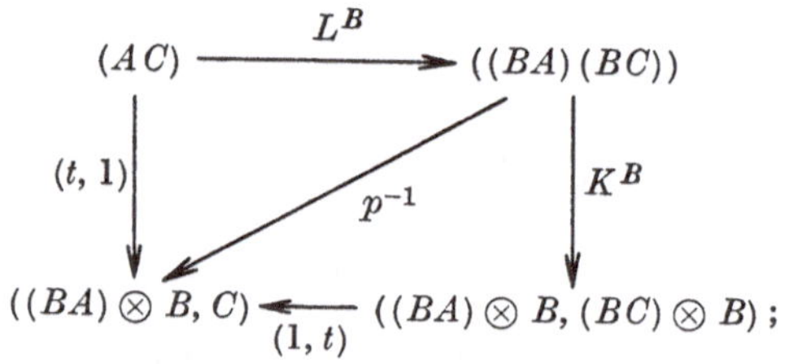

the exterior of this is VN for t_B. Similarly from (3.21) and (7.1) we get a commutative diagram

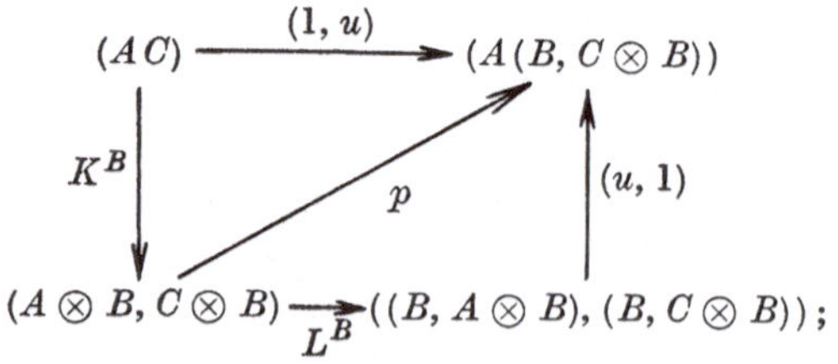

the exterior of this is VN for $u_{\cdot B}$.

Since the proofs of the assertions of Theorem I.10.2 make no use of VF2 for the $\mathscr{V}$-functors involved, we can use them here before we have VF2 for K^B. The composite $p^{-1}(1, u)$ in (7.1) is therefore the C-component of a $\mathscr{V}$-natural transformation

$$L^A \xrightarrow[L^A u_{\cdot B}]{} L^A L^B K^B \xrightarrow[p^{-1}_{AB} K^B]{} L^{A \otimes B} K^B ,$$

and so this composite, which by (7.1) is K^B_{AC}, satisfies VN; and this is VF2 for K^B.

Proposition 7.2. *If $\mathscr{V}$ is a monoidal closed category and $f \in \mathscr{V}_0(AB)$, the morphisms*

$$1 \otimes f : C \otimes A \to C \otimes B$$

are the C-components of a $\mathscr{V}$-natural transformation

$$K^f : K^A \to K^B .$$

Proof. VN for K^f asserts the commutativity of

$$\begin{array}{ccc}
(CD) & \xrightarrow{\quad K^A \quad} & (C \otimes A, D \otimes A) \\
{\scriptstyle K^B}\downarrow & & \downarrow{\scriptstyle (1, 1 \otimes f)} \\
(C \otimes B, D \otimes B) & \xrightarrow[(1 \otimes f, 1)]{} & (C \otimes A, D \otimes B)
\end{array}$$

and this is just the assertion that K^B_{CD} is natural in B, which it is by its definition (7.1).

The following result shows the relation of t and u to special cases of M (i.e. the M of $\mathscr{V}$ itself) and K:

Proposition 7.3. *In a monoidal closed category we have commutative diagrams:*

$$\begin{array}{ccc}
(BC) \otimes B & \xrightarrow{\quad t \quad} & C \\
{\scriptstyle 1 \otimes i}\downarrow & & \downarrow{\scriptstyle i} \\
(BC) \otimes (IB) & \xrightarrow[M]{} & (IC)
\end{array} \tag{7.3}$$

$$\begin{array}{ccc}
C & \xrightarrow{\quad u \quad} & (B, C \otimes B) \\
{\scriptstyle i}\downarrow & & \downarrow{\scriptstyle (l, 1)} \\
(IC) & \xrightarrow[K]{} & (I \otimes B, C \otimes B)
\end{array} \tag{7.4}$$

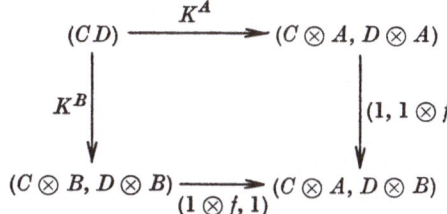

Proof. Applying π to both legs of (7.3) we get

$$\begin{aligned}\pi(it) &= (1, i) \,.\, \pi t && \text{by (3.1)}\\ &= (1, i) && \text{by (3.2)},\end{aligned}$$

and

$$\begin{aligned}\pi(M(1 \otimes i)) &= (i, 1) \,.\, \pi M && \text{by (3.1)}\\ &= (i, 1) L && \text{by (6.2)};\end{aligned}$$

thus (7.3) reduces to CC4.

We compose each leg of (7.4) with the isomorphism

$$p : (I \otimes B, C \otimes B) \to (I(B, C \otimes B)),$$

getting

$$\begin{aligned}pKi &= (1, u)i && \text{by (7.1)}\\ &= iu && \text{by the naturality of } i,\end{aligned}$$

and

$$p(l, 1)u = iu \qquad \text{by MCC2};$$

thus (7.4) commutes.

The following is an adjoint form, as it were, of the representation theorem for $\mathscr{V}$-functors:

Proposition 7.4. *Let $\mathscr{V}$ be a monoidal closed category, $\mathscr{A}$ a $\mathscr{V}$-category, $T : \mathscr{A} \to \mathscr{V}$ a $\mathscr{V}$-functor, $A \in \mathscr{A}$ and $B \in \mathscr{V}$. Denote by $\{q\}$ the class of $\mathscr{V}$-natural transformations*

$$q : K^B L^A \to T : \mathscr{A} \to \mathscr{V}$$

with components

$$q_C : \mathscr{A}(AC) \otimes B \to TC.$$

Define a map $\Delta : \{q\} \to V(B, TA)$ by setting Δq equal to the composite

$$\Delta q: \qquad B \xrightarrow[l^{-1}]{} I \otimes B \xrightarrow[j \otimes 1]{} \mathscr{A}(AA) \otimes B \xrightarrow[q_A]{} TA. \tag{7.5}$$

Then Δ is a bijection with inverse Π, where $\Pi\theta$ for $\theta : B \to TA$ is the composite

$$\Pi\theta: \qquad K^B L^A \xrightarrow[K^B T_A]{} K^B L^{TA}\, T \xrightarrow[K^B L^\theta T]{} K^B L^B\, T \xrightarrow[i_B T]{} T, \tag{7.6}$$

with components

$$\mathscr{A}(AC) \otimes B \xrightarrow[T \otimes 1]{} (TA, TC) \otimes B \xrightarrow[(\theta, 1) \otimes 1]{} (B, TC) \otimes B \xrightarrow[t]{} TC. \tag{7.7}$$

Proof. By (3.4), πq_C is the composite

$$\mathscr{A}(AC) \xrightarrow[u]{} (B, \mathscr{A}(AC) \otimes B) \xrightarrow[(1, q)]{} (B, TC)$$

which is the component of a $\mathscr{V}$-natural transformation

$$\pi q: \qquad L^A \xrightarrow[u_{\cdot B}L^A]{} L^B K^B L^A \xrightarrow[L^B q]{} L^B T.$$

Similarly, if $\bar{q}_C : \mathscr{A}(AC) \to (B, TC)$ are the components of a $\mathscr{V}$-natural transformation $\bar{q} : L^A \to L^B T$, then $\pi^{-1}\bar{q}_C$, which by (3.5) is the composite

$$\mathscr{A}(AC) \otimes B \xrightarrow[\bar{q}\otimes 1]{} (B, TC) \otimes B \xrightarrow[t]{} TC,$$

is the component of a $\mathscr{V}$-natural transformation

$$\pi^{-1}\bar{q}: \qquad K^B L^A \xrightarrow[K^B\bar{q}]{} K^B L^B T \xrightarrow[t_B T]{} T.$$

Thus we have a bijection $\pi : \{q\} \to \{\bar{q}\}$, and so by Theorem I.10.8 we have a bijection

$$\{q\} \xrightarrow[\pi]{} \{\bar{q}\} \xrightarrow[\Gamma']{} V(B, TA),$$

with inverse

$$V(B, TA) \xrightarrow[\Omega']{} \{\bar{q}\} \xrightarrow[\pi^{-1}]{} \{q\}.$$

Comparison of I (10.11), (3.4), and (7.6) shows at once that $\pi^{-1}\Omega' = \Pi$. It remains to show that $\Gamma'\pi = \Delta$.

Evaluating (3.12) at $x \in \mathscr{V}_0(AB)$ and using (3.26) gives $\iota x = \pi(xl)$, and (3.4) then gives

$$\iota x = (1, xl)\, u. \tag{7.8}$$

Applying this to Δq gives

$$\begin{aligned} \iota\Delta q &= (1, q)\,(1, j \otimes 1)\, u \\ &= (1, q)\, u j \quad \text{by the naturality of } u \\ &= \pi q . j \qquad \text{by (3.4)} \\ &= \iota(\Gamma'\pi q) \quad \text{by I(10.9) and Lemma I.2.2;} \end{aligned}$$

thus $\Delta q = \Gamma'\pi q$, as required.

If in the above proposition we take $\mathscr{A} = \mathscr{V}$ and $A = I$, and use the isomorphism $i : 1 \to L^I$, we get:

Corollary 7.5. *If $\mathscr{V}$ is a monoidal closed category and $T : \mathscr{V} \to \mathscr{V}$ is a $\mathscr{V}$-functor, there is a bijection between $\mathscr{V}$-natural transformations $q : K^B \to T$ and morphisms $\theta : B \to TI$, where θ is the composite*

$$B \xrightarrow[l^{-1}]{} I \otimes B \xrightarrow[q_I]{} TI. \tag{7.9}$$

In particular, q is determined by q_I.

8. The Underlying Category of a $\mathscr{V}$-category

The closed category $\mathscr{S}$ is monoidal, with the cartesian product $A \times B$ for $A \otimes B$, since $L^A L^B$ admits the representation

$$p : (A \times B, C) \to (A(BC))$$

where

$$((pf)x)y = f(x,y), \quad f \in (A \times B, C), \quad x \in A, \quad y \in B. \tag{8.1}$$

We verify at once that a, r, l have their expected values and that M given by (6.2) corresponds to the usual composition law in categories.

Let $\mathscr{V}$ be a normalized monoidal category, and define $V^0 : * \to VI$ by

$$V^0 * = \iota_I^{-1} 1_I; \tag{8.2}$$

then by the naturality of ι (cf. I(3.16)) the image of $*$ under the composite $* \xrightarrow[V^0]{} VI \xrightarrow[Vf]{} VA$ is given by

$$Vf . V^0 * = \iota^{-1} f. \tag{8.3}$$

Now define a natural transformation

$$\tilde{V} : VA \times VB \to V(A \otimes B)$$

by the commutative diagram

$$\begin{array}{ccccc} VA \times VB & & \xrightarrow{\quad\tilde{V}\quad} & & V(A \otimes B) \\ \Big\downarrow{\scriptstyle \iota \times \iota} & & & & \Big\uparrow{\scriptstyle \iota^{-1}} \\ \mathscr{V}_0(IA) \times \mathscr{V}_0(IB) & \xrightarrow[\otimes]{} & \mathscr{V}_0(I \otimes I, A \otimes B) & \xrightarrow[\mathscr{V}_0(l^{-1}, 1)]{} & \mathscr{V}_0(I, A \otimes B) \end{array} \tag{8.4}$$

where $\otimes$ is the map sending (f, g) to $f \otimes g$; we record the evaluated form of (8.4) as

$$\iota \tilde{V}(x, y) = (\iota x \otimes \iota y) l_I^{-1}. \tag{8.5}$$

Proposition 8.1. *If* $\mathscr{V}$ *is a normalized monoidal category, and* V^0 *and* $\tilde{V}$ *are defined by* (8.2) *and* (8.4), *the triple* $(V, \tilde{V}, V^0)$ *is a monoidal functor* $V : \mathscr{V} \to \mathscr{S}$.

Proof. Consider for example axiom MF3 for V, which reads:

$$\begin{array}{ccc} V((A \otimes B) \otimes C) & \xrightarrow{\quad Va\quad} & V(A \otimes (B \otimes C)) \\ \Big\uparrow{\scriptstyle \tilde{V}} & & \Big\uparrow{\scriptstyle \tilde{V}} \\ V(A \otimes B) \times VC & & VA \times V(B \otimes C) \\ \Big\uparrow{\scriptstyle \tilde{V} \times 1} & & \Big\uparrow{\scriptstyle 1 \times \tilde{V}} \\ (VA \times VB) \times VC & \xrightarrow[a']{} & VA \times (VB \times VC) \end{array}$$

Since both legs are natural and since $V \cong \mathscr{V}_0(I-)$, repeated application of Proposition 7.4 shows that it suffices to put $A = B = C = I$ and verify that both legs have the same value at $((\iota^{-1}1, \iota^{-1}1), \iota^{-1}1)$; this verification is immediate. Similarly we verify MF1 and MF2.

Essentially the following result is given by Bénabou [3]:

Theorem 8.2. *Let $\mathscr{V}$ be a normalized monoidal category and let $\mathscr{A}$ be a $\mathscr{V}$-category. Then we can find in exactly one way a category $\mathscr{A}_0$ and a functor* hom $\mathscr{A}$: $\mathscr{A}_0^* \times \mathscr{A}_0 \to \mathscr{V}_0$ *such that:*

(i) *$\mathscr{A}_0$ has the same objects as $\mathscr{A}$;*

(ii) $\hom \mathscr{A}(AB) = \mathscr{A}(AB)$; (8.6)

(iii) *the following diagram commutes:*

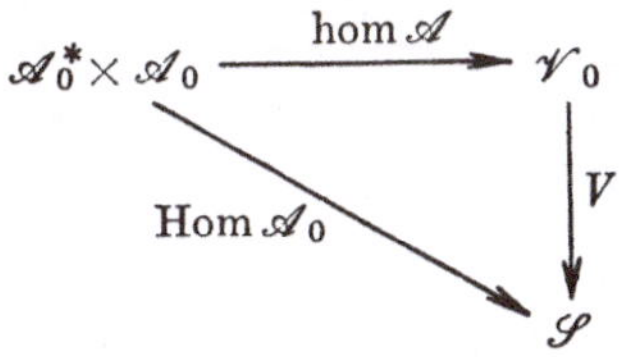

(8.7)

(iv) $j_A = \iota 1_A$, (8.8)

where $\iota: V\mathscr{A}(AA) \to \mathscr{V}_0(I, \mathscr{A}(AA))$;

(v) *M^A_{BC} is natural in A, B, and C, and j_A in A.*

Proof. We first prove the uniqueness. The objects and the morphisms of $\mathscr{A}_0$ are fixed by (8.6) and (8.7); we must have

$$\mathscr{A}_0(AB) = V\mathscr{A}(AB). \tag{8.9}$$

Next, $\iota_{\mathscr{A}(AB)}$ gives a natural isomorphism (using (8.9))

$$\iota: \mathscr{A}_0(AB) \to \mathscr{V}_0(I, \mathscr{A}(AB)).$$

Since $\iota 1 = j$ by (8.8), the representation theorem (applied both to $\mathscr{A}_0(A-)$ and $\mathscr{A}_0(-B)$) shows that for $f \in \mathscr{A}_0(AB)$ we have a commutative diagram

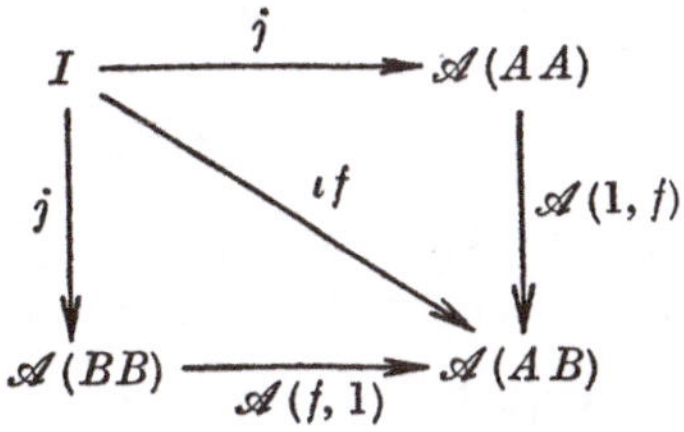

(8.10)

For $f \in \mathscr{A}_0(BC)$ the following diagram commutes by the naturality of M and by (8.10):

$$\begin{array}{ccccc}
I \otimes \mathscr{A}(AB) & \xrightarrow{j \otimes 1} & \mathscr{A}(BB) \otimes \mathscr{A}(AB) & \xrightarrow{M} & \mathscr{A}(AB) \\
& \searrow^{\iota f \otimes 1} & \downarrow{\scriptstyle \mathscr{A}(1,f) \otimes 1} & & \downarrow{\scriptstyle \mathscr{A}(1,f)} \\
& & \mathscr{A}(BC) \otimes \mathscr{A}(AB) & \xrightarrow[M]{} & \mathscr{A}(AC)
\end{array}$$

Composing both legs of this with $l^{-1}: \mathscr{A}(AB) \to I \otimes \mathscr{A}(AB)$, and using VC1′, we get a commutative diagram

$$\begin{array}{ccc}
\mathscr{A}(AB) & \xrightarrow{\mathscr{A}(1,f)} & \mathscr{A}(AC) \\
{\scriptstyle l^{-1}}\downarrow & & \uparrow{\scriptstyle M} \\
I \otimes \mathscr{A}(AB) & \xrightarrow[\iota f \otimes 1]{} & \mathscr{A}(BC) \otimes \mathscr{A}(AB)
\end{array} \tag{8.11}$$

This fixes the value of $\mathscr{A}(1, f)$. Similarly we get the following diagram, which fixes the value of $\mathscr{A}(f, 1)$:

$$\begin{array}{ccc}
\mathscr{A}(CD) & \xrightarrow{\mathscr{A}(f,1)} & \mathscr{A}(BD) \\
{\scriptstyle r^{-1}}\downarrow & & \uparrow{\scriptstyle M} \\
\mathscr{A}(CD) \otimes I & \xrightarrow[1 \otimes \iota f]{} & \mathscr{A}(CD) \otimes \mathscr{A}(BC)
\end{array} \tag{8.12}$$

Thus the functor hom $\mathscr{A}$ is unique. Finally (8.7) gives $\mathscr{A}_0(1, f) = V\mathscr{A}(1, f)$; so that $\mathscr{A}_0(1, f)$ is determined, and with it the law of composition in $\mathscr{A}_0$; for $\mathscr{A}_0(1, f)g = fg$.

We now prove the existence. We take the monoidal functor V of Proposition 8.1 and define

$$\mathscr{A}_0 = V_* \mathscr{A}. \tag{8.13}$$

We then have (i) and (8.9). By the definition I(6.3) of j' (cf. Proposition 6.3), the identity of $\mathscr{A}_0(AA)$ is the image of $*$ under

$$* \xrightarrow[V^0]{} VI \xrightarrow[Vj]{} V\mathscr{A}(AA),$$

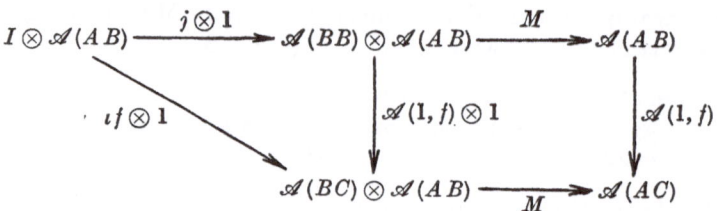

so that by (8.3) we have (8.8). The naturality of j now follows from that of ι.

By (6.1) the M of $\mathscr{A}_0$ is the composite

$$V\mathscr{A}(BC) \times V\mathscr{A}(AB) \xrightarrow[\tilde{V}]{} V(\mathscr{A}(BC) \otimes \mathscr{A}(AB)) \xrightarrow[VM]{} V\mathscr{A}(AC);$$

evaluating this at $f \in V\mathscr{A}(BC)$ and $g \in V\mathscr{A}(AB)$, and using (8.5) and the naturality of ι, we find that $\iota(fg) \in \mathscr{V}_0(I, \mathscr{A}(AC))$ is the composite

$$I \xrightarrow[l_I^{-1}]{} I \otimes I \xrightarrow[\iota f \otimes \iota g]{} \mathscr{A}(BC) \otimes \mathscr{A}(AB) \xrightarrow[M]{} \mathscr{A}(AC). \tag{8.14}$$

We now define $\mathscr{A}(1, f)$ by (8.11), and observe at once that by (8.8) and VC1′, $\mathscr{A}(1, 1) = 1$. Similarly we define $\mathscr{A}(f, 1)$ by (8.12). We have yet to prove that these definitions, together with (8.6), give a bifunctor hom $\mathscr{A}$, that (8.7) is satisfied, and that M is natural.

Since we have (8.9), (8.7) will follow if we prove $V\mathscr{A}(1, f) = \mathscr{A}_0(1, f)$ and $V\mathscr{A}(f, 1) = \mathscr{A}_0(f, 1)$; by symmetry we need only prove one of these. We take the first and express it in the evaluated form

$$(V\mathscr{A}(1, f))g = fg, \quad \text{where} \quad g \in \mathscr{A}_0(AB) \quad \text{and} \quad f \in \mathscr{A}_0(BC).$$

By the naturality of ι, the same statement may be expressed:

$$\text{the composite} \quad I \xrightarrow[\iota g]{} \mathscr{A}(AB) \xrightarrow[\mathscr{A}(1,f)]{} \mathscr{A}(AC) \quad \text{is} \quad \iota(fg). \tag{8.15}$$

Consider the diagram

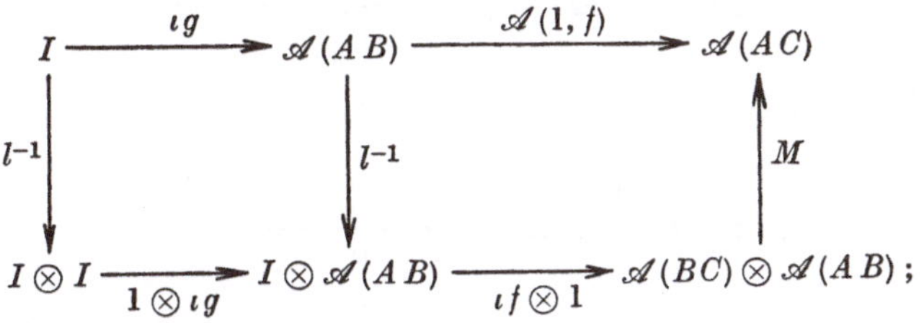

the left region commutes by the naturality of l, and the right region by (8.11); since by (8.14) the long leg is $\iota(fg)$, we have (8.15).

We now prove part of the naturality of M, namely the commutativity of

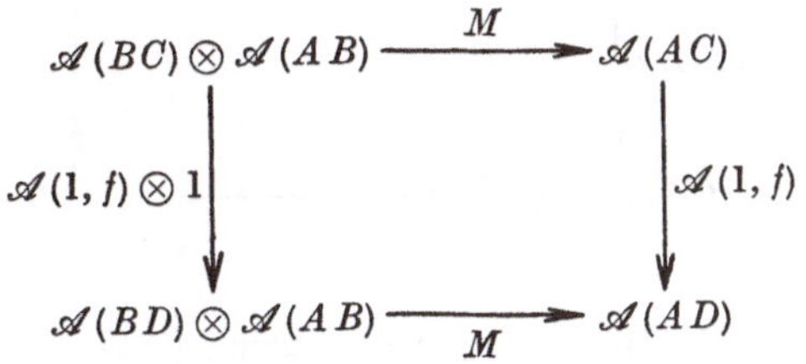

(8.16)

where $f: C \to D$. Writing x for ιf and using (8.11), (8.16) is the exterior of the following diagram:

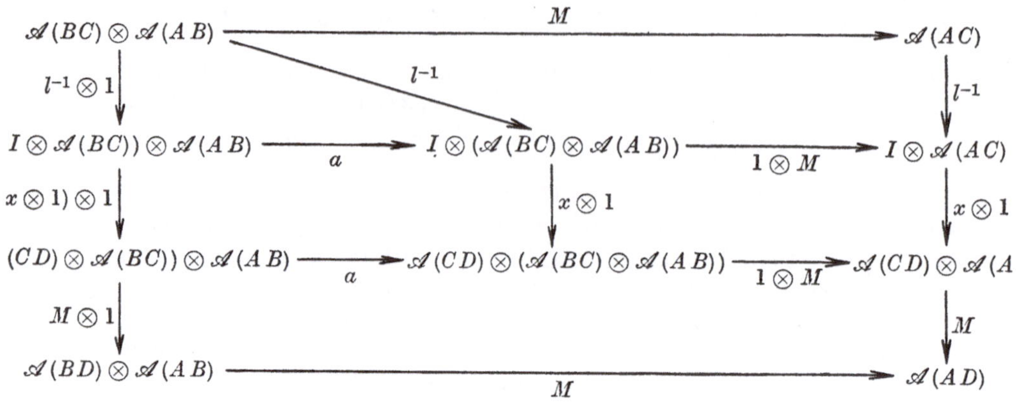

One region commutes by VC3′, one by MC1, one by the naturality of l, one by the naturality of a, and one trivially.

Another part of the naturality of M, namely the commutativity of

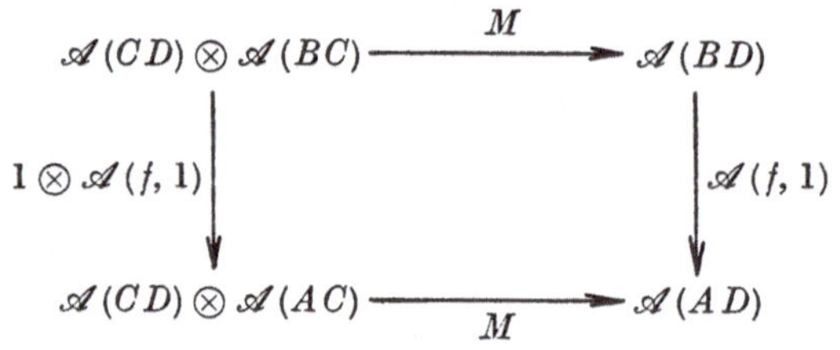

(8.17)

where $f: A \to B$, follows by symmetry. The final part is the commutativity of

$$
\begin{array}{ccc}
\mathscr{A}(CD) \otimes \mathscr{A}(AB) & \xrightarrow{\mathscr{A}(f,1)\otimes 1} & \mathscr{A}(BD) \otimes \mathscr{A}(AB) \\
{\scriptstyle 1 \otimes \mathscr{A}(1,f)} \downarrow & & \downarrow {\scriptstyle M} \\
\mathscr{A}(CD) \otimes \mathscr{A}(AC) & \xrightarrow{M} & \mathscr{A}(AD)
\end{array}
$$

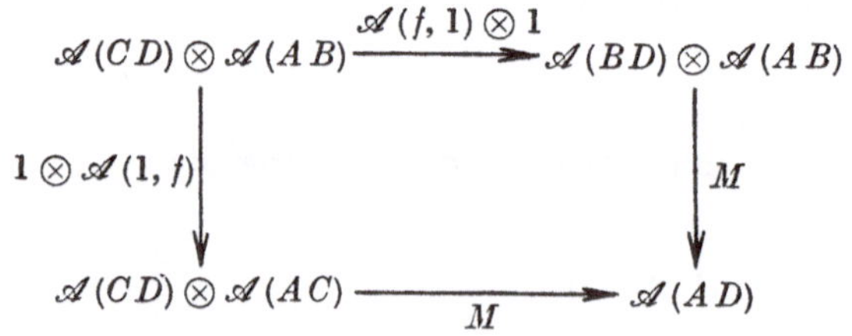

(8.18)

where $f: B \to C$. Writing x for ιf and using (8.11) and (8.12), (8.18) is the exterior of the following diagram:

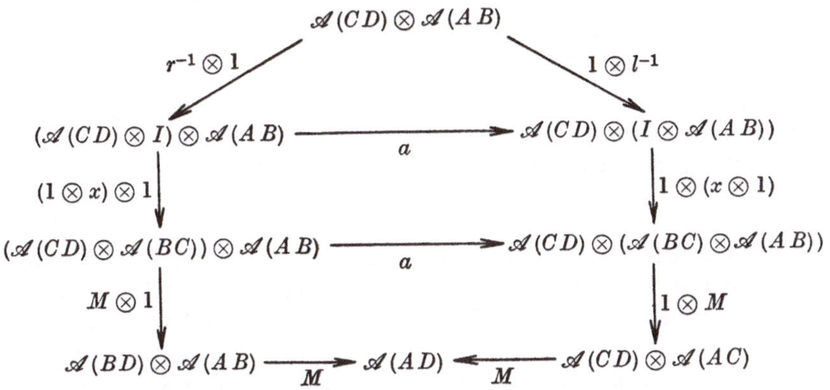

This commutes by MC2, the naturality of a, and VC3′.

It now remains to prove that hom $\mathscr{A}$ is a functor; we need to show that $\mathscr{A}(1, f)\mathscr{A}(1, g) = \mathscr{A}(1, fg)$, that $\mathscr{A}(g, 1)\mathscr{A}(f, 1) = \mathscr{A}(fg, 1)$, and that $\mathscr{A}(1, f)\, \mathscr{A}(g, 1) = \mathscr{A}(g, 1)\mathscr{A}(1, f)$. We need not prove the second of these, for it will follow by symmetry when we have proved the first. Consider the diagram, where $g: B \to C$ and $f: C \to D$:

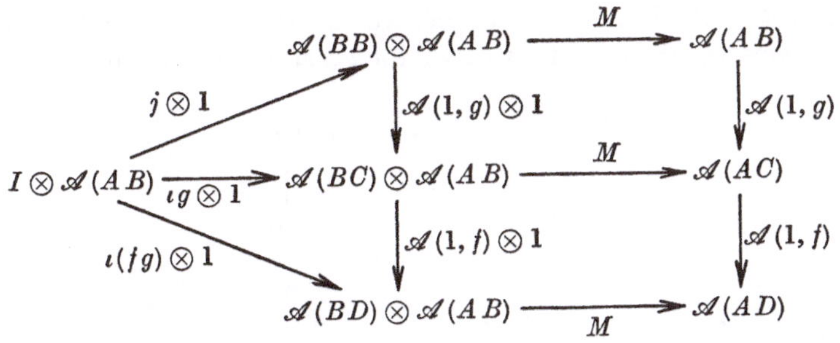

The triangles commute by (8.15) (using $j = \iota 1$) and the rectangles by (8.16). Since $M(j \otimes 1) = l$ by VC1′, one leg is $\mathscr{A}(1, f)\mathscr{A}(1, g)l$; the other leg is $\mathscr{A}(1, fg)l$ by (8.11). Thus $\mathscr{A}(1, f)\mathscr{A}(1, g) = \mathscr{A}(1, fg)$.

Now let $g: A \to B$ and $f: C \to D$, and consider the diagram

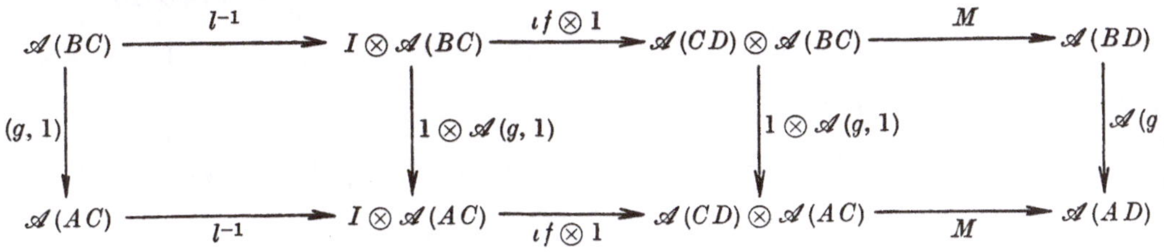

One region commutes by the naturality of l, one trivially, and one by (8.17). The top and the bottom edge are each $\mathscr{A}(1, f)$, by (8.11); hence $\mathscr{A}(g, 1)\mathscr{A}(1, f) = \mathscr{A}(1, f)\mathscr{A}(g, 1)$. This completes the proof.

We leave to the reader the proof of:

Proposition 8.3. *If $\mathscr{V}$ is a normalized monoidal category and $T: \mathscr{A} \to \mathscr{B}$ is a $\mathscr{V}$-functor, there is exactly one functor $T_0: \mathscr{A}_0 \to \mathscr{B}_0$, with $T_0 A = TA$, such that $T_{BC}: \mathscr{A}(BC) \to \mathscr{B}(T_0 B, T_0 C)$ is natural; namely*

$$T_0 = V_* T. \tag{8.19}$$

Remark 8.4. If $\mathscr{V}$ is a normalized monoidal category, it is now possible to define $\mathscr{V}$-natural transformations exactly as in § I.10; the proofs of Proposition I.10.1 and Theorem I.10.2 remain valid word for word.

To discuss the effect of a monoidal functor $\Phi: \mathscr{V} \to \mathscr{V}'$, where $\mathscr{V}$ and $\mathscr{V}'$ are both normalized, we define $\phi_0: V \to V'\phi$ exactly as in Proposition I.3.4, except that we must now write $\mathscr{V}_0(AB)$ and not $V(AB)$, etc. Note that equations I(3.8) and I(3.9) are still valid, and these give I(3.13) and I(4.1).

The analogue of Proposition I.4.5, namely that ϕ_0 is a monoidal natural transformation

$$\phi_0: V \to V'\Phi: \mathscr{V} \to \mathscr{S}, \tag{8.20}$$

is still valid but needs a new proof; we can then write I(4.1) as I(4.3).

To prove (8.20), note that MN 1 for ϕ_0 states the commutativity of

$$\begin{array}{ccc} * & \xrightarrow{V^0} & VI \\ {\scriptstyle V'^0}\downarrow & & \downarrow{\scriptstyle \phi_0} \\ V'I' & \xrightarrow[V'\phi_0]{} & V'\phi I \end{array}$$

which follows immediately from I(3.9) and (8.3). MN 2 for ϕ_0 states the commutativity of

$$\begin{array}{ccccc} VA \times VB & & \xrightarrow{\widetilde{V}} & & V(A \otimes B) \\ {\scriptstyle \phi_0 \times \phi_0}\downarrow & & & & \downarrow{\scriptstyle \phi_0} \\ V'\phi A \times V'\phi B & \xrightarrow[\widetilde{V}']{} & V'(\phi A \otimes \phi B) & \xrightarrow[V'\widetilde{\phi}]{} & V'\phi(A \otimes B) \end{array} \tag{8.21}$$

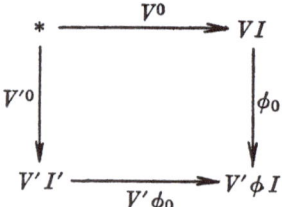

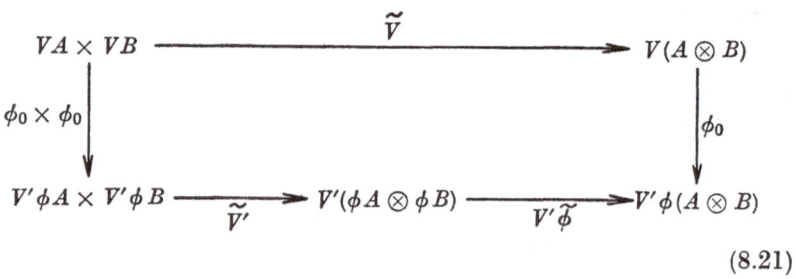

Since $V \cong \mathscr{V}_0(I\,-)$, it follows by repeated application of Proposition 7.4 that it suffices to put $A = B = I$ and verify that both legs of (8.21) have the same value at $(\iota^{-1}1, \iota^{-1}1)$. Using (8.5) and I(3.9), the resulting assertion is the commutativity of

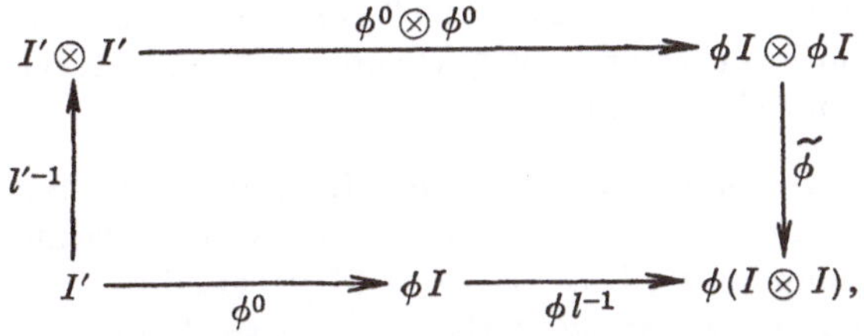

which is immediate from MF1 for Φ and the naturality of l'.

We then have at once the analogue of Propositions I.7.3, I.7.5, and I.7.6; we also have that of Proposition I.9.8, but we need a new proof for this. By symmetry it suffices to prove that

$$\phi\mathscr{A}(1, f) = (\Phi_*\mathscr{A})(1, \phi_0 f)$$

where $f \in \mathscr{A}_0(BC)$; since by I(3.8) (with ιf in place of f) we have $\phi\iota f.\,\phi^0 = \iota'\phi_0 f$, we are led by the definition (8.11) to proving the commutativity of the exterior of:

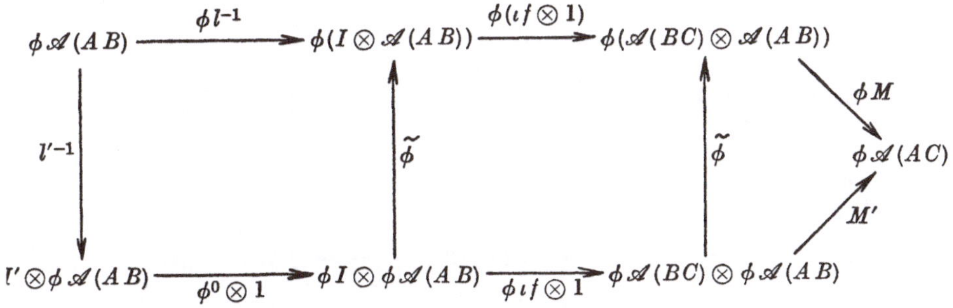

The left region commutes by MF1, the middle region by the naturality of $\tilde{\phi}$, and the right region by (6.1).

We now have at once the analogues of Propositions I.10.3, I.10.5, and I.10.6, and Theorem I.10.7. What we have no analogue of, if $\mathscr{V}$ is not closed, is propositions referring to the $\mathscr{V}$-category $\mathscr{V}$.

Theorem 8.5. *If $\mathscr{V}$ is a monoidal closed category, the monoidal functor $V : \mathscr{V} \to \mathscr{S}$ of Proposition 8.1 coincides with the closed functor $V : \mathscr{V} \to \mathscr{S}$ of Proposition I.3.11. Moreover the category $\mathscr{A}_0$ and the functors* hom $\mathscr{A}$, *T_0, $\Phi_{0\mathscr{A}}$ defined above then coincide with those of Chapter I.*

Proof. Let the unique monoidal closed functor extending $(V, \tilde{V}, V^0)$ be $(V, \tilde{V}, \hat{V}, V^0)$. Since we have the same V^0 as in Proposition I.3.11, and since V_0 depends only on V and V^0, it follows that $(V, \hat{V}, V^0)$ is normal. Hence by NCF1 we have $\hat{V} = V$, as required.

The assertions about $\mathscr{A}_0$, hom $\mathscr{A}$, and T_0 are clear from the uniqueness clauses of Propositions I.9.5 and I.9.6, Theorem 8.2, and Proposition 8.3. The assertion about $\Phi_{0\mathscr{A}}$ is obvious.

Remark 8.6. We have included a definite functor $V: \mathscr{V}_0 \to \mathscr{S}$ as part of the definition of a closed category $\mathscr{V}$, while treating it as an "extra" for a monoidal $\mathscr{V}$, because when $\mathscr{V}$ is closed $\mathscr{V}$ itself is a $\mathscr{V}$-category, and it is most tedious if the $\mathscr{V}_0$ and the hom $\mathscr{V}$ constructed in § I.7 and § I.9 differ from those given as part of the data of $\mathscr{V}$.

Chapter III

Symmetric Monoidal Closed Categories

1. Symmetric Monoidal Categories

A *symmetry* for a monoidal category $\mathscr{V}$ consists of a natural isomorphism $c = c_{AB}: A \otimes B \to B \otimes A$ in $\mathscr{V}_0$, satisfying the following two axioms:

MC6. $c_{BA}\, c_{AB} = 1 : A \otimes B \to A \otimes B$.

MC7. The following diagram commutes:

$$\begin{array}{ccccc} (A \otimes B) \otimes C & \xrightarrow{a} & A \otimes (B \otimes C) & \xrightarrow{c} & (B \otimes C) \otimes A \\ \downarrow{\scriptstyle c \otimes 1} & & & & \downarrow{\scriptstyle a} \\ (B \otimes A) \otimes C & \xrightarrow[a]{} & B \otimes (A \otimes C) & \xrightarrow[1 \otimes c]{} & B \otimes (C \otimes A) \end{array}$$

A monoidal category $\mathscr{V}$ together with a symmetry c for $\mathscr{V}$ is called a *symmetric monoidal category*. Note that a monoidal category $\mathscr{V}$, even a closed one, may admit several distinct symmetries; an example of this is given in § IV.6 below.

We have from MACLANE [*14*] and KELLY [*9*]:

Proposition 1.1. *In a symmetric monoidal category $\mathscr{V}$ the natural isomorphisms a, r, l, c are coherent.*

If $\mathscr{V}$ and $\mathscr{V}'$ are symmetric monoidal categories, a monoidal functor $\Phi = (\phi, \tilde{\phi}, \phi^0): \mathscr{V} \to \mathscr{V}'$ is said to be *symmetric* if the following axiom is satisfied:

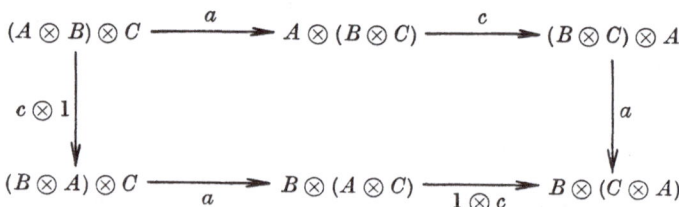

MF4. The following diagram commutes:

$$\begin{array}{ccc} \phi(A\otimes B) & \xrightarrow{\ \phi c\ } & \phi(B\otimes A)\\ \Big\uparrow{\scriptstyle\tilde\phi} & & \Big\uparrow{\scriptstyle\tilde\phi}\\ \phi A\otimes\phi B & \xrightarrow[\ c'\ ]{} & \phi B\otimes \phi A\end{array}$$

One easily verifies:

Proposition 1.2. *Composites and inverses of symmetric monoidal functors are symmetric.*

Thus symmetric monoidal categories, symmetric monoidal functors, and monoidal natural transformations (no change in the definition of these last) form a sub-hypercategory $\mathscr{S}\mathscr{M}on$ of $\mathscr{M}on$.

The monoidal closed category $\mathscr{S}$ admits an obvious symmetry $c: A \times B \to B \times A$ given by $c[x, y] = [y, x]$. (In this chapter we shall use square brackets to denote ordered pairs to avoid confusion with our use of (— —) in a closed category.)

Proposition 1.3. *If the symmetric monoidal category $\mathscr{V}$ has a normalization V, ι, the monoidal functor $V: \mathscr{V} \to \mathscr{S}$ is symmetric.*

Proof. To verify the commutativity of

$$\begin{array}{ccc} V(A\otimes B) & \xrightarrow{\ Vc\ } & V(B\otimes A)\\ \Big\uparrow{\scriptstyle\tilde V} & & \Big\uparrow{\scriptstyle\tilde V}\\ VA\times VB & \xrightarrow[\ c'\ ]{} & VB\times VA\end{array} \tag{1.1}$$

it suffices by repeated application of Proposition II. 7.4, since $V \cong \cong \mathscr{V}_0(I-)$, to put $A = B = I$ and show that both legs have the same value at $[\iota^{-1}1, \iota^{-1}1]$. This reduces to showing the commutativity of

$$\begin{array}{ccc} I\otimes I & \xrightarrow{\ c\ } & I\otimes I\\ {\scriptstyle\iota^{-1}}\nwarrow & & \nearrow{\scriptstyle\iota^{-1}}\\ & I & \end{array}$$

which we have by coherence.

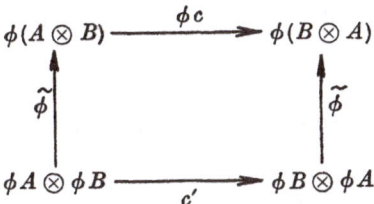

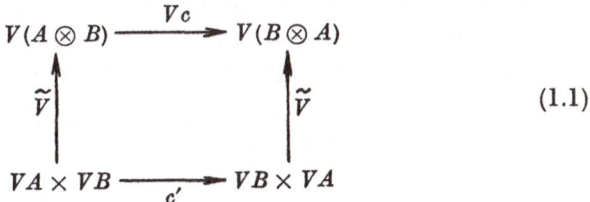

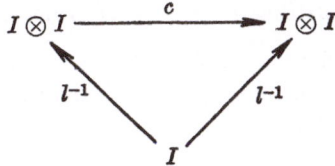

Remark 1.4. In future we shall suppose without explicit mention, wherever the context requires it, that a normalization is chosen for the monoidal category $\mathscr{V}$; if $\mathscr{V}$ is closed the normalization is of course to be the canonical one.

2. Duality for $\mathscr{V}$-categories

If $\mathscr{V}$ is a symmetric monoidal category, $\mathscr{V}$-categories and $\mathscr{V}$-functors are defined as in Chapter II, the symmetry playing no part in these definitions; similarly, given a normalization of $\mathscr{V}$, $\mathscr{V}$-natural transformations are defined as in Chapter II.

Proposition 2.1. *If $\mathscr{V}$ is a symmetric monoidal category and $\mathscr{A}$ is a $\mathscr{V}$-category, the following data define a $\mathscr{V}$-category $\mathscr{A}^*$ called the dual of $\mathscr{A}$:*

(i) $\operatorname{obj} \mathscr{A}^* = \operatorname{obj} \mathscr{A}$; (2.1)

(ii) $\mathscr{A}^*(AB) = \mathscr{A}(BA)$; (2.2)

(iii) $j: I \to \mathscr{A}^*(AA)$ *is* $j: I \to \mathscr{A}(AA)$; (2.3)

(iv) $M: \mathscr{A}^*(BC) \otimes \mathscr{A}^*(AB) \to \mathscr{A}^*(AC)$ *is the composite*

$$\mathscr{A}(CB) \otimes \mathscr{A}(BA) \xrightarrow[c]{} \mathscr{A}(BA) \otimes \mathscr{A}(CB) \xrightarrow[M]{} \mathscr{A}(CA). \quad (2.4)$$

Proof. We verify VC3′ for $\mathscr{A}^*$, leaving the reader to verify VC1′ and VC2′. We need the commutativity of the exterior of (see page 515):

The hexagon commutes by coherence, the pentagon by VC3′ for $\mathscr{A}$, and the two quadrangles by the naturality of c.

In the following propositions the absence of a proof indicates that they are straightforward and that their verification is left to the reader.

Proposition 2.2. *If $\mathscr{V}$ is a symmetric monoidal category and $T: \mathscr{A} \to \mathscr{B}$ is a $\mathscr{V}$-functor, the following data define a $\mathscr{V}$-functor $T^*: \mathscr{A}^* \to \mathscr{B}^*$:*

(i) $T^*A = TA$; (2.5)

(ii) $T^*_{BC}: \mathscr{A}^*(BC) \to \mathscr{B}^*(TB, TC)$ *is* $T_{CB}: \mathscr{A}(CB) \to$

$$\to \mathscr{B}(TC, TB). \quad (2.6)$$

Proposition 2.3. *If $\mathscr{V}$ is a symmetric monoidal category, the assignments $\mathscr{A} \mapsto \mathscr{A}^*$, $T \mapsto T^*$ constitute an involutory functor $D: \mathscr{V}_* \to \mathscr{V}_*$.*

Remark 2.4. It will be clear from the context whether $\mathscr{V}_*$ denotes the hypercategory or the underlying category; a notational distinction here would be cumbersome.

Proposition 2.5. *Let $\mathscr{V}$ be a symmetric monoidal category and let $\mathscr{I}$ be the $\mathscr{V}$-category of Remark II.6.2; then $\mathscr{I}^* = \mathscr{I}$ and $J^{A*} = J^A$.*

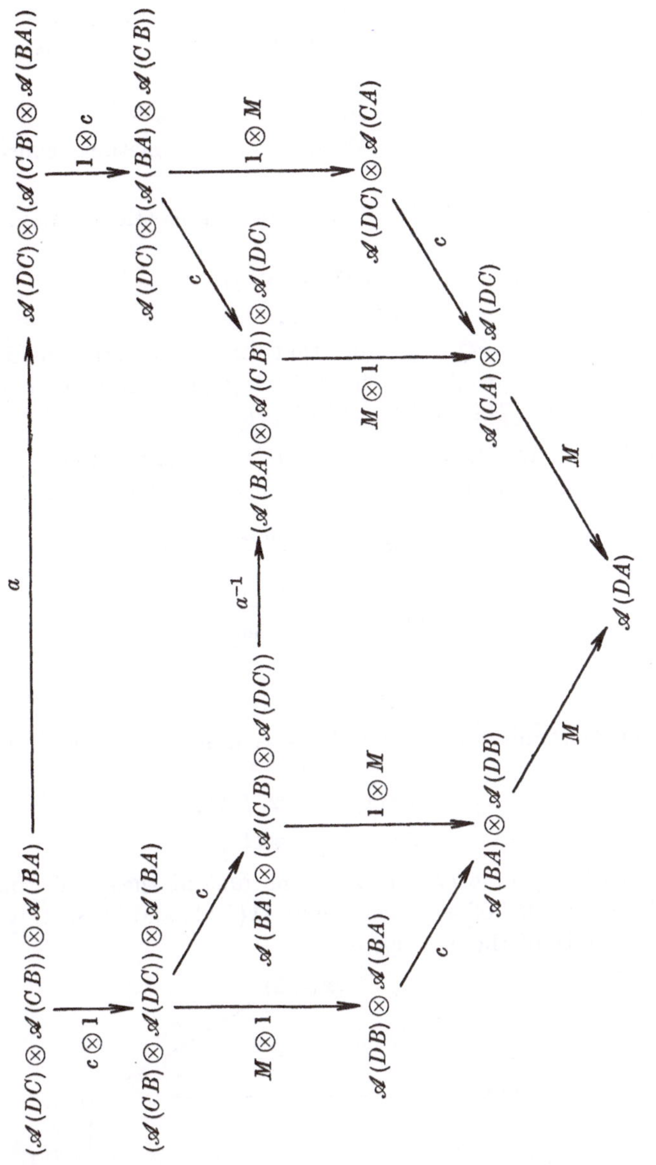

Proposition 2.6. *If $\Phi : \mathscr{V} \to \mathscr{V}'$ is a symmetric monoidal functor, the functor $\Phi_* : \mathscr{V}_* \to \mathscr{V}'_*$ commutes with D; that is,*

$$(\Phi_* \mathscr{A})^* = \Phi_* \mathscr{A}^*, \tag{2.7}$$

$$(\Phi_* T)^* = \Phi_* T^*. \tag{2.8}$$

Proposition 2.7. *Let $\eta : \Phi \to \Psi : \mathscr{V} \to \mathscr{V}'$ be a monoidal natural transformation between symmetric monoidal functors. Then if $\mathscr{A}$ is a $\mathscr{V}$-category we have*

$$\eta_{*\mathscr{A}^*} = (\eta_{*\mathscr{A}})^* : \Phi_* \mathscr{A}^* \to \Psi_* \mathscr{A}^*. \tag{2.9}$$

Remark 2.8. A $\mathscr{V}$-functor $T : \mathscr{A}^* \to \mathscr{B}$ is sometimes called a *contravariant $\mathscr{V}$-functor $T : \mathscr{A} \to \mathscr{B}$.*

Applying Proposition 2.6 to the symmetric monoidal functor $V : \mathscr{V} \to \mathscr{S}$ gives:

$$(\mathscr{A}_0)^* = (\mathscr{A}^*)_0, \tag{2.10}$$

$$(T_0)^* = (T^*)_0; \tag{2.11}$$

we therefore write $\mathscr{A}_0^*$, T_0^*; note that for $\mathscr{S}$-categories duality reduces to the classical concept. In the following proposition $\mathfrak{c}$ is the functor sending $[AB]$ to $[BA]$ and $[f, g]$ to $[g, f]$.

Proposition 2.9. *If $\mathscr{V}$ is a symmetric monoidal category and $\mathscr{A}$ is a $\mathscr{V}$-category, the following diagram of functors commutes:*

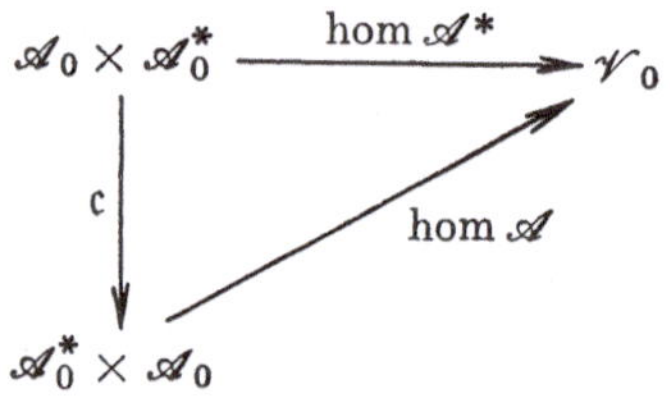

(2.12)

Proof. Commutativity on objects is immediate, and it remains to show that

$$\mathscr{A}^*(1, f) = \mathscr{A}(f, 1), \tag{2.13}$$

$$\mathscr{A}^*(f, 1) = \mathscr{A}(1, f). \tag{2.14}$$

It suffices by symmetry to prove the first of these. Writing x for ιf and $\overline{M}$ for the "M" of $\mathscr{A}^*$, we have by II (8.11) and II (8.12) to prove the commutativity of the exterior of:

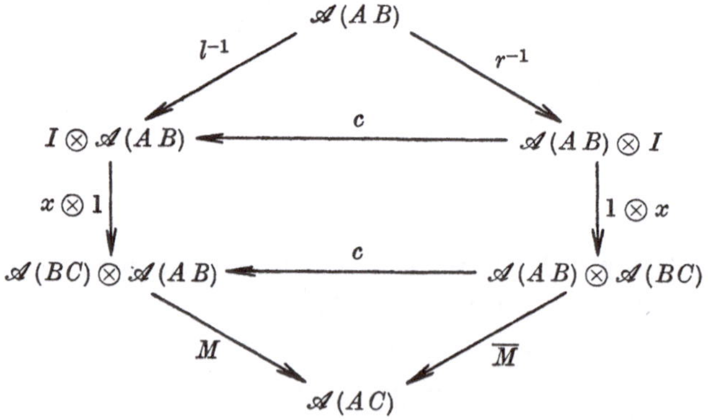

The top region commutes by coherence, the middle one by the naturality of c, and the bottom one by (2.4).

Proposition 2.10. *Let $\mathscr{V}$ be a symmetric monoidal category and $\alpha : T \to S : \mathscr{A} \to \mathscr{B}$ a $\mathscr{V}$-natural transformation. Then the $\alpha_A : TA \to SA$ are also the components of a $\mathscr{V}$-natural transformation*

$$\alpha^* : S^* \to T^* : \mathscr{A}^* \to \mathscr{B}^*.$$

Proof. In view of Propositions 2.2 and 2.9, VN for α^* is identical with VN for α.

Proposition 2.11. *If $\mathscr{V}$ is a symmetric monoidal category, $D : \mathscr{V}_* \to \mathscr{V}_*$ becomes an involutory hyperfunctor if we set $D\alpha = \alpha^*$; and if $\Phi : \mathscr{V} \to \mathscr{V}'$ is a symmetric monoidal functor, D and Φ_* commute as hyperfunctors; that is,*

$$(\Phi_* \alpha)^* = \Phi_* \alpha^* . \tag{2.15}$$

Proof. Clearly D respects, in a contravariant way, composition of $\mathscr{V}$-natural transformations, with themselves and with $\mathscr{V}$-functors; to be precise we have

$$(\beta \alpha)^* = \alpha^* \beta^* , \tag{2.16}$$

$$(T \alpha)^* = T^* \alpha^* , \tag{2.17}$$

$$(\alpha S)^* = \alpha^* S^* . \tag{2.18}$$

Proposition 2.7 applied to $\phi_0 : V \to V'\Phi : \mathscr{V} \to \mathscr{S}$ gives

$$\Phi_{0\mathscr{B}^*} = (\Phi_{0\mathscr{B}})^* , \tag{2.19}$$

whence (2.15) from I(10.5).

Remark 2.12. Hypercategories will be shown in § IV.2 to be $\mathscr{V}$-categories for a suitable $\mathscr{V}$, namely the category of small categories with an appropriate symmetric monoidal closed structure. The dual of a hypercategory $\mathfrak{A}$ is therefore given by $\mathfrak{A}^*(\mathscr{A}\mathscr{B}) = \mathfrak{A}(\mathscr{B}\mathscr{A})$. However there is another kind of dual given by $\mathfrak{A}^\dagger(\mathscr{A}\mathscr{B}) = \mathfrak{A}(\mathscr{A}\mathscr{B})^*$, the dual of the category $\mathfrak{A}(\mathscr{A}\mathscr{B})$. The type of contravariance exhibited by the hyperfunctor $D : \mathscr{V}_* \to \mathscr{V}_*$ in Proposition 2.11 is that appropriate to this second kind of duality.

3. Tensor Products of $\mathscr{V}$-categories

For a symmetric monoidal category $\mathscr{V}$ we construct by suitable combinations of a, a^{-1}, and c, (the details being irrelevant by coherence), a natural isomorphism

$$m : (A \otimes B) \otimes (C \otimes D) \to (A \otimes C) \otimes (B \otimes D) ,$$

called the *middle-four interchange.*

Proposition 3.1. *If $\mathscr{V}$ is a symmetric monoidal category and $\mathscr{A}$, $\mathscr{B}$ are $\mathscr{V}$-categories, the following data define a $\mathscr{V}$-category $\mathscr{A} \otimes \mathscr{B}$:*

(i) *the objects of $\mathscr{A} \otimes \mathscr{B}$ are the ordered pairs $[AB]$, $A \in \mathscr{A}$, $B \in \mathscr{B}$;* (3.1)

(ii) $(\mathscr{A} \otimes \mathscr{B})([AB], [A'B']) = \mathscr{A}(AA') \otimes \mathscr{B}(BB')$; (3.2)

(iii) $j : I \to (\mathscr{A} \otimes \mathscr{B})([AB], [AB])$ *is the composite*

$$I \xrightarrow{l^{-1}} I \otimes I \xrightarrow{j \otimes j} \mathscr{A}(AA) \otimes \mathscr{B}(BB); \tag{3.3}$$

(iv) $M : (\mathscr{A} \otimes \mathscr{B})([A'B'], [A''B'']) \otimes (\mathscr{A} \otimes \mathscr{B})([AB], [A'B']) \to$
$\to (\mathscr{A} \otimes \mathscr{B})([AB], [A''B''])$

is given by the commutative diagram

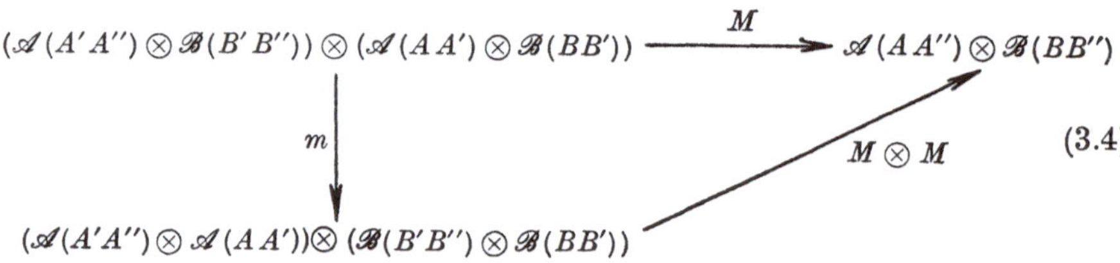

(3.4)

The proof is straightforward, although the diagrams are hard to fit on a page. The diagram proving VC3′ for $\mathscr{A} \otimes \mathscr{B}$ looks like this:

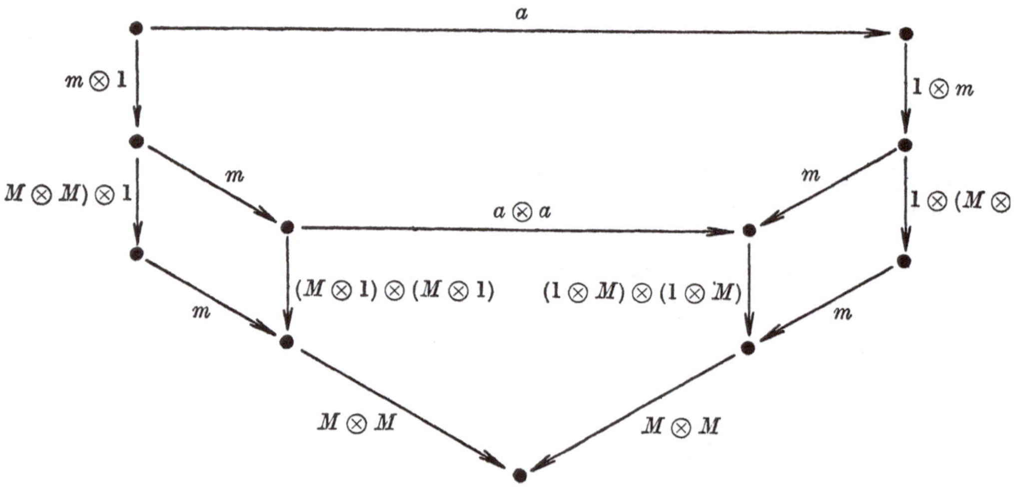

The hexagon commutes by coherence, the two quadrangles by the naturality of m, and the pentagon by VC3′ for $\mathscr{A}$ and for $\mathscr{B}$. The reader may verify VC1′ and VC2′. Similarly we easily verify:

Proposition 3.2. *If $\mathscr{V}$ is a symmetric monoidal category and $T:\mathscr{A}\to\mathscr{C}$ and $S:\mathscr{B}\to\mathscr{D}$ are $\mathscr{V}$-functors, the following data define a $\mathscr{V}$-functor $T\otimes S:\mathscr{A}\otimes\mathscr{B}\to\mathscr{C}\otimes\mathscr{D}$:*

(i) $(T\otimes S)[A\,B]=[T\,A,S\,B]$; (3.5)

(ii) $(T\otimes S)_{[AB],[A'B']}:(\mathscr{A}\otimes\mathscr{B})([A\,B],[A'\,B'])\to$
$\to(\mathscr{C}\otimes\mathscr{D})([T\,A,S\,B],[T\,A',S\,B'])$ *is*

$$\mathscr{A}(A\,A')\otimes\mathscr{B}(B\,B')\xrightarrow[T_{AA'}\otimes S_{BB'}]{}\mathscr{C}(TA,TA')\otimes\mathscr{D}(S\,B,S\,B'). \tag{3.6}$$

Proposition 3.3. *If $\mathscr{V}$ is a symmetric monoidal category the assignments $\mathscr{A},\mathscr{B}\mapsto\mathscr{A}\otimes\mathscr{B}$ and $T,S\mapsto T\otimes S$ constitute a functor $\otimes:\mathscr{V}_*\times\mathscr{V}_*\to\to\mathscr{V}_*$.*

We now define $\mathscr{V}$-functors

$$\begin{aligned}
\mathfrak{a}&:(\mathscr{A}\otimes\mathscr{B})\otimes\mathscr{C}\to\mathscr{A}\otimes(\mathscr{B}\otimes\mathscr{C}),\\
\mathfrak{r}&:\mathscr{A}\otimes\mathscr{I}\to\mathscr{A},\\
\mathfrak{l}&:\mathscr{I}\otimes\mathscr{A}\to\mathscr{A},\\
\mathfrak{c}&:\mathscr{A}\otimes\mathscr{B}\to\mathscr{B}\otimes\mathscr{A}.
\end{aligned}$$

For instance, $\mathfrak{a}[[AB]\,C]=[A\,[B\,C]]$, and
$\mathfrak{a}:((\mathscr{A}\otimes\mathscr{B})\otimes\mathscr{C})([[A\,B]\,C],[[A'\,B']\,C'])\to$
$\to(\mathscr{A}\otimes(\mathscr{B}\otimes\mathscr{C}))([A\,[B\,C]],[A'\,[B'\,C']])$

is

$$a:(\mathscr{A}(AA')\otimes\mathscr{B}(B\,B'))\otimes\mathscr{C}(C\,C')\to\mathscr{A}(AA')\otimes(\mathscr{B}(B\,B')\otimes\mathscr{C}(C\,C')).$$

We then easily verify:

Proposition 3.4. *If $\mathscr{V}$ is a symmetric monoidal category, then $\mathfrak{a}$, $\mathfrak{r}$, $\mathfrak{l}$, $\mathfrak{c}$ are coherent natural isomorphisms in $\mathscr{V}_*$, defining on $\mathscr{V}_*$ the structure of a symmetric monoidal "category" $\mathscr{V}_\#$.*

To discuss the effect of a symmetric monoidal functor we need the following lemma, which is an easy consequence of MF3 and MF4:

Lemma 3.5. *For a symmetric monoidal functor $\Phi:\mathscr{V}\to\mathscr{V}'$ we have a commutative diagram:*

$$\begin{array}{ccc}
\phi((A\otimes B)\otimes(C\otimes D)) & \xrightarrow{\ \phi m\ } & \phi((A\otimes C)\otimes(B\otimes D))\\
\uparrow{\scriptstyle\tilde\phi} & & \uparrow{\scriptstyle\tilde\phi}\\
\phi(A\otimes B)\otimes\phi(C\otimes D) & & \phi(A\otimes C)\otimes\phi(B\otimes D)\\
\uparrow{\scriptstyle\tilde\phi\otimes\tilde\phi} & & \uparrow{\scriptstyle\tilde\phi\otimes\tilde\phi}\\
(\phi A\otimes\phi B)\otimes(\phi C\otimes\phi D) & \xrightarrow[m']{} & (\phi A\otimes\phi C)\otimes(\phi B\otimes\phi D)
\end{array} \tag{3.7}$$

Now if $\Phi : \mathscr{V} \to \mathscr{V}'$ is a symmetric monoidal functor and $\mathscr{A}, \mathscr{B}$ are $\mathscr{V}$-categories, we define a $\mathscr{V}'$-functor

$$\tilde{\phi}_{\#} : \Phi_* \mathscr{A} \otimes \Phi_* \mathscr{B} \to \Phi_* (\mathscr{A} \otimes \mathscr{B});$$

$\tilde{\phi}_{\#}$ is the identity on objects, and $\tilde{\phi}_{\#[AB][A'B']}$ is

$$\tilde{\phi} : \phi \mathscr{A}(AA') \otimes \phi \mathscr{B}(BB') \to \phi(\mathscr{A}(AA') \otimes \mathscr{B}(BB')) . \tag{3.8}$$

Verification that $\tilde{\phi}_{\#}$ is indeed a $\mathscr{V}'$-functor is easy using (3.7). Similarly we define a $\mathscr{V}'$-functor

$$\phi^0_{\#} : \mathscr{I}' \to \Phi_* \mathscr{I};$$

$\phi^0_{\#} * = *$, and $\phi^0_{\#} : \mathscr{I}'(**) \to \phi \mathscr{I}(**)$ is $\phi^0 : I' \to \phi I$.

Proposition 3.6. *If $\Phi : \mathscr{V} \to \mathscr{V}'$ is a symmetric monoidal functor, the triple $(\Phi_*, \tilde{\phi}_{\#}, \phi^0_{\#})$ is a symmetric monoidal functor $\Phi_{\#} : \mathscr{V}_{\#} \to \mathscr{V}'_{\#}$.*

Proposition 3.7. *If $\eta : \Phi \to \Psi : \mathscr{V} \to \mathscr{V}'$ is a monoidal natural transformation where Φ and Ψ are symmetric monoidal functors, then*

$$\eta_* : \Phi_* \to \Psi_* : \mathscr{V}_* \to \mathscr{V}'_*$$

is a monoidal natural transformation

$$\eta_* : \Phi_{\#} \to \Psi_{\#} : \mathscr{V}_{\#} \to \mathscr{V}'_{\#}.$$

Proposition 3.8. *The assignments $\mathscr{V} \mapsto \mathscr{V}_{\#}$, $\Phi \mapsto \Phi_{\#}$, $\eta \mapsto \eta_*$, constitute a hyperfunctor from $\mathscr{S}\mathscr{M}on$ to itself.*

Proposition 3.9. *If $\mathscr{V}$ is a symmetric monoidal category we have*

$$(\mathscr{A} \otimes \mathscr{B})^* = \mathscr{A}^* \otimes \mathscr{B}^* , \tag{3.9}$$

$$(T \otimes S)^* = T^* \otimes S^* . \tag{3.10}$$

Proposition 3.10. *If $\mathscr{V}$ is a symmetric monoidal category, D becomes a symmetric monoidal functor $D : \mathscr{V}_{\#} \to \mathscr{V}_{\#}$ if we set $\tilde{D} = 1$, $D^0 = 1$. Moreover if $\Phi : \mathscr{V} \to \mathscr{V}'$ is a symmetric monoidal functor, D commutes with $\Phi_{\#}$.*

We now consider underlying categories. Note that $(\mathscr{A} \otimes \mathscr{B})_0$ is not expressible in terms of $\mathscr{A}_0$ and $\mathscr{B}_0$. Applying Proposition 3.6 to the symmetric monoidal functor $V : \mathscr{V} \to \mathscr{S}$ gives a symmetric monoidal functor $V_{\#} : \mathscr{V}_{\#} \to \mathscr{S}_{\#}$. In particular we have the functor

$$\tilde{V}_{\#} : \mathscr{A}_0 \times \mathscr{B}_0 \to (\mathscr{A} \otimes \mathscr{B})_0;$$

it is the identity on objects and is given on morphisms by

$$\tilde{V}_{\#}[f, g] = \tilde{V}[f, g] . \tag{3.11}$$

Let us introduce for this last morphism the notation

$$f \otimes g = \tilde{V}[f, g]; \tag{3.12}$$

thus if $f: A \to A'$ in $\mathscr{A}_0$ and $g: B \to B'$ in $\mathscr{B}_0$ we have $f \otimes g: [AB] \to \to [A' B']$ in $(\mathscr{A} \otimes \mathscr{B})_0$. (If $\mathscr{V}$ happens to be closed and if $\mathscr{A} = \mathscr{B} = \mathscr{V}$, one must be careful to distinguish this from $f \otimes g: A \otimes B \to A' \otimes B'$ in $\mathscr{V}_0$.)

Since $\tilde{V}_{\#}$ is a functor we have

$$hf \otimes kg = (h \otimes k)(f \otimes g), \tag{3.13}$$

$$1 \otimes 1 = 1. \tag{3.14}$$

From the naturality of $\tilde{V}_{\#}$ we get, for $\mathscr{V}$-functors $T: \mathscr{A} \to \mathscr{C}$ and $S: \mathscr{B} \to \mathscr{D}$,

$$(T \otimes S)_0 (f \otimes g) = T_0 f \otimes S_0 g. \tag{3.15}$$

The fact that $V_{\#}$ satisfies MF1—MF4 can be expressed in terms of the functors $\mathfrak{a}_0: ((\mathscr{A} \otimes \mathscr{B}) \otimes \mathscr{C})_0 \to (\mathscr{A} \otimes (\mathscr{B} \otimes \mathscr{C}))_0$, etc., as follows:

$$\mathfrak{l}_0(1 \otimes f) = f; \tag{3.16}$$

$$\mathfrak{r}_0(f \otimes 1) = f; \tag{3.17}$$

$$\mathfrak{a}_0((f \otimes g) \otimes h) = f \otimes (g \otimes h); \tag{3.18}$$

$$\mathfrak{c}_0(f \otimes g) = g \otimes f. \tag{3.19}$$

Note that the 1 of (3.16) and (3.17) is the 1 of $\mathscr{I}_0$, and is the element $V^0* = \iota^{-1} 1$ of VI.

Lemma 3.11. Let $f \in \mathscr{A}_0(BC)$ and $g \in \mathscr{B}_0(YZ)$, so that

$$f \otimes g \in (\mathscr{A} \otimes \mathscr{B})_0 ([B\,Y], [C\,Z]).$$

Then $\iota f: I \to \mathscr{A}(BC)$, $\iota g: I \to \mathscr{B}(YZ)$, and $\iota(f \otimes g): I \to \mathscr{A}(BC) \otimes \mathscr{B}(YZ)$ are connected by the commutative diagram:

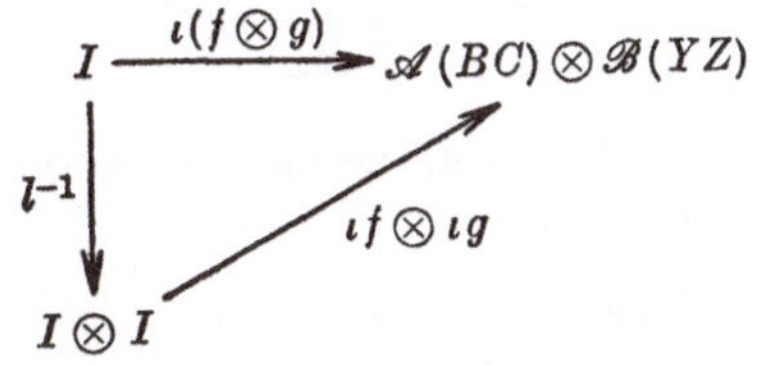

$$\begin{array}{ccc} I & \xrightarrow{\iota(f \otimes g)} & \mathscr{A}(BC) \otimes \mathscr{B}(YZ) \\ {\scriptstyle \mathfrak{l}^{-1}} \downarrow & \nearrow & \\ & {\scriptstyle \iota f \otimes \iota g} & \\ I \otimes I & & \end{array} \tag{3.20}$$

Proof. Immediate from II (8.5).

In the following proposition $\mathfrak{m}$ is the middle-four interchange for categories:

Proposition 3.12. If $\mathscr{V}$ is a symmetric monoidal category and $\mathscr{A}$, $\mathscr{B}$

are $\mathscr{V}$-categories, the following diagram of functors commutes:

$$
\begin{array}{ccc}
(\mathscr{A} \otimes \mathscr{B})_0^* \times (\mathscr{A} \otimes \mathscr{B})_0 & \xrightarrow{\text{hom } (\mathscr{A} \otimes \mathscr{B})} & \mathscr{V}_0 \\
\uparrow \tilde{V}_\#^* \times \tilde{V}_\# & & \uparrow \\
(\mathscr{A}_0^* \times \mathscr{B}_0^*) \times (\mathscr{A}_0 \times \mathscr{B}_0) & & \otimes \\
\downarrow \mathfrak{m} & & \\
(\mathscr{A}_0^* \times \mathscr{A}_0) \times (\mathscr{B}_0^* \times \mathscr{B}_0) & \xrightarrow{\text{hom } \mathscr{A} \times \text{hom } \mathscr{B}} & \mathscr{V}_0 \times \mathscr{V}_0
\end{array}
$$

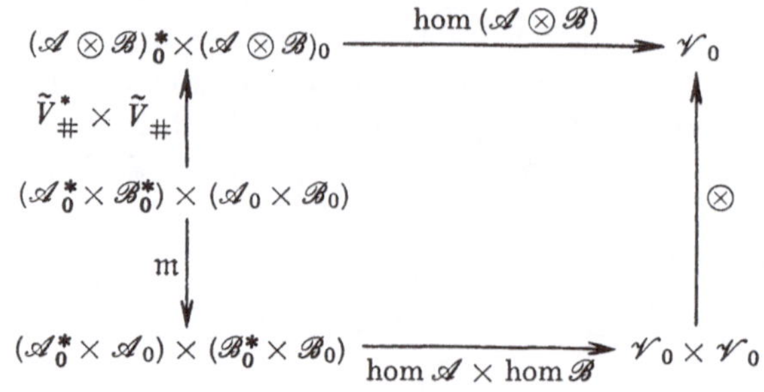

(3.21)

Proof. Commutativity on objects is immediate, and commutativity on morphisms states:

$$(\mathscr{A} \otimes \mathscr{B})(h \otimes k, f \otimes g) = \mathscr{A}(h, f) \otimes \mathscr{B}(k, g). \qquad (3.22)$$

It suffices by symmetry to prove:

$$(\mathscr{A} \otimes \mathscr{B})(1, f \otimes g) = \mathscr{A}(1, f) \otimes \mathscr{B}(1, g). \qquad (3.23)$$

Writing x and y for ιf and ιg, and $\bar{M}$ for the "M" of $\mathscr{A} \otimes \mathscr{B}$, we need by II(8.11) and by (3.20) the commutativity of the exterior of:

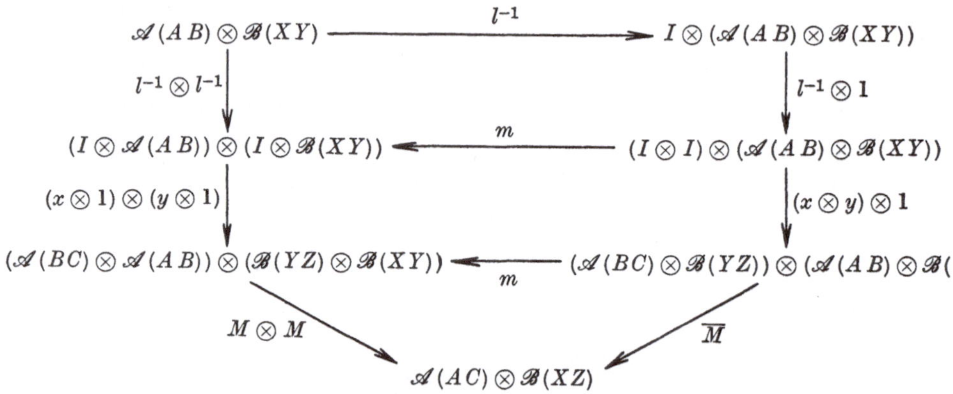

The top region commutes by coherence, the middle region by the naturality of m, and the bottom region by (3.4).

Proposition 3.13. *Let $\mathscr{V}$ be a symmetric monoidal category and let $\alpha : T \to S : \mathscr{A} \to \mathscr{C}$ and $\beta : P \to Q : \mathscr{B} \to \mathscr{D}$ be $\mathscr{V}$-natural transformations. Then the $\alpha_A \otimes \beta_B : [TA, PB] \to [SA, QB]$ are the components of a $\mathscr{V}$-natural transformation $\alpha \otimes \beta : T \otimes P \to S \otimes Q : \mathscr{A} \otimes \mathscr{B} \to \mathscr{C} \otimes \mathscr{D}$.*

Proof. In view of (3.22), VN for $\alpha \otimes \beta$ is the tensor product of VN for α and VN for β.

Proposition 3.14. *If $\mathscr{V}$ is a symmetric monoidal category, $\otimes : \mathscr{V}_* \times \times \mathscr{V}_* \to \mathscr{V}_*$ is in fact a hyperfunctor and the natural isomorphisms $\mathfrak{a}, \mathfrak{r}, \mathfrak{l}, \mathfrak{c}$ are in fact hypernatural.*

Proof. To show that $\otimes$ is a hyperfunctor we need to show

$$(\alpha \otimes \beta)(P \otimes Q) = \alpha P \otimes \beta Q, \tag{3.24}$$

$$(T \otimes S)(\alpha \otimes \beta) = T\alpha \otimes S\beta, \tag{3.25}$$

$$(\gamma \otimes \delta)(\alpha \otimes \beta) = \gamma\alpha \otimes \delta\beta, \text{ and } 1 \otimes 1 = 1. \tag{3.26}$$

Of these, (3.24) is trivial, (3.25) is immediate from (3.15), and (3.26) is immediate from (3.13) and (3.14).

The hypernaturality of $\mathfrak{a}, \mathfrak{r}, \mathfrak{l}, \mathfrak{c}$ is immediate from (3.16)—(3.19).

Proposition 3.15. *If $\Phi : \mathscr{V} \to \mathscr{V}'$ is a symmetric monoidal functor, the natural transformation $\tilde{\phi}_{\#} : \Phi_* \mathscr{A} \otimes \Phi_* \mathscr{B} \to \Phi_* (\mathscr{A} \otimes \mathscr{B})$ is in fact hypernatural.*

Proof. Let $\alpha : T \to T' : \mathscr{A} \to \mathscr{C}$ and $\beta : S \to S' : \mathscr{B} \to \mathscr{D}$ be $\mathscr{V}$-natural transformations. By the naturality of $\tilde{\phi}_{\#}$ we have a commutative diagram

$$\begin{array}{ccc}
\Phi_*\mathscr{A} \otimes \Phi_*\mathscr{B} & \xrightarrow{\tilde{\phi}_{\#}} & \Phi_*(\mathscr{A} \otimes \mathscr{B}) \\
{\scriptstyle \Phi_* T \otimes \Phi_* S}\big\downarrow & & \big\downarrow{\scriptstyle \Phi_*(T \otimes S)} \\
\Phi_*\mathscr{C} \otimes \Phi_*\mathscr{D} & \xrightarrow[\tilde{\phi}_{\#}]{} & \Phi_*(\mathscr{C} \otimes \mathscr{D})
\end{array} \tag{3.27}$$

and a similar diagram for T', S'. The hypernaturality of $\tilde{\phi}_{\#}$ means that we also have

$$\tilde{\phi}_{\#} \cdot (\Phi_* \alpha \otimes \Phi_* \beta) = \Phi_* (\alpha \otimes \beta) \cdot \tilde{\phi}_{\#}; \tag{3.28}$$

in view of the various definitions this follows from II (8.21).

Proposition 3.16. *If $\mathscr{V}$ is a symmetric monoidal category and α and β are $\mathscr{V}$-natural transformations, we have*

$$(\alpha \otimes \beta)^* = \alpha^* \otimes \beta^*. \tag{3.29}$$

4. $\mathscr{V}$-bifunctors

If $\mathscr{V}$ is a symmetric monoidal category, a $\mathscr{V}$-functor $T : \mathscr{A} \otimes \mathscr{B} \to \mathscr{C}$ is often called a *$\mathscr{V}$-bifunctor*. For its value on objects we use the usual notation $T(AB)$ instead of $T[AB]$. Given such a $\mathscr{V}$-bifunctor we define for each $A \in \mathscr{A}$ a $\mathscr{V}$-functor $T(A-) : \mathscr{B} \to \mathscr{C}$ as the composite

$$T(A-): \qquad \mathscr{B} \xrightarrow[l^{-1}]{} \mathscr{I} \otimes \mathscr{B} \xrightarrow[J^A \otimes 1]{} \mathscr{A} \otimes \mathscr{B} \xrightarrow[T]{} \mathscr{C}. \tag{4.1}$$

Similarly for each $B \in \mathscr{B}$ we define a $\mathscr{V}$-functor $T(-B) : \mathscr{A} \to \mathscr{C}$ as the composite

$$T(-B): \qquad \mathscr{A} \xrightarrow[r^{-1}]{} \mathscr{A} \otimes \mathscr{I} \xrightarrow[1 \otimes J^B]{} \mathscr{A} \otimes \mathscr{B} \xrightarrow[T]{} \mathscr{C}. \tag{4.2}$$

We call $T(A-)$ and $T(-B)$ the *partial functors* of T. For their values on objects we have

$$T(A-)\,B = T(-B)\,A = T(A\,B). \tag{4.3}$$

Proposition 4.1. *If $\mathscr{V}$ is a symmetric monoidal category and $T : \mathscr{A} \otimes \mathscr{B} \to \mathscr{C}$ is a $\mathscr{V}$-functor, the following diagram commutes, and each leg is equal to* $T_{[AB][A'B']} : \mathscr{A}(AA') \otimes \mathscr{B}(BB') \to \mathscr{C}(T(AB), T(A'B'))$:

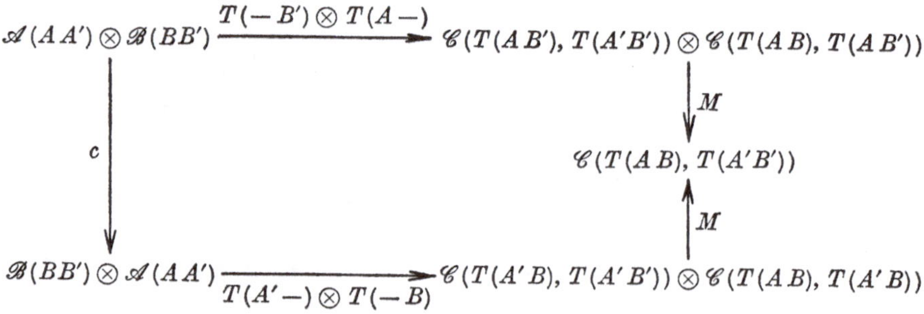

(4.4)

Proof. Consider the diagram

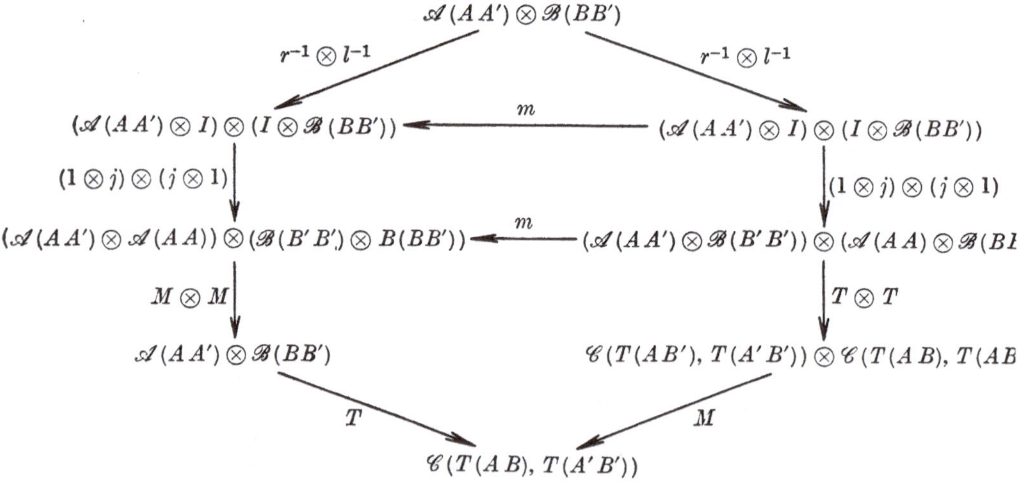

The top region commutes by coherence, the middle region by the naturality of m, and the bottom region by VF 2′ for T (in view of (3.4)). Thus the exterior commutes. By VC 2′ for $\mathscr{A}$ and VC 1′ for $\mathscr{B}$, the left leg is just T; by (4.1) and (4.2) the right leg is the upper leg of (4.4).

The proof that the lower leg of (4.4) is also T is entirely similar, using the diagram

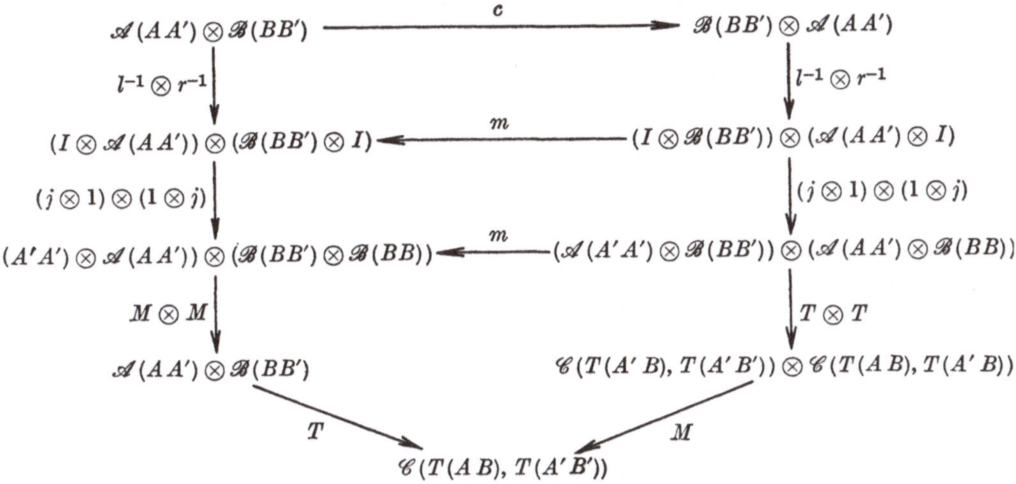

Proposition 4.2. *Let $\mathscr{V}$ be a symmetric monoidal category and let $T(A\,-): \mathscr{B} \to \mathscr{C}$ and $T(-\,B): \mathscr{A} \to \mathscr{C}$ be families of $\mathscr{V}$-functors indexed by $A \in \mathscr{A}$ and $B \in \mathscr{B}$ respectively. Suppose that $T(A\,-)\,B = T(-\,B)\,A$, and write $T(A\,B)$ for their common value; and suppose that (4.4) commutes. Then there is a unique $\mathscr{V}$-functor $T: \mathscr{A} \otimes \mathscr{B} \to \mathscr{C}$ of which $T(A\;\;-)$ and $T(-\,B)$ are the partial functors.*

Proof. Define $T: \mathscr{A}(AA') \otimes \mathscr{B}(BB') \to \mathscr{C}(T(A\,B), T(A'\,B'))$ to be the top leg of (4.4); this is forced by Proposition 4.1, which proves the uniqueness.

VF 1′ for T requires the commutativity of the exterior of the diagram

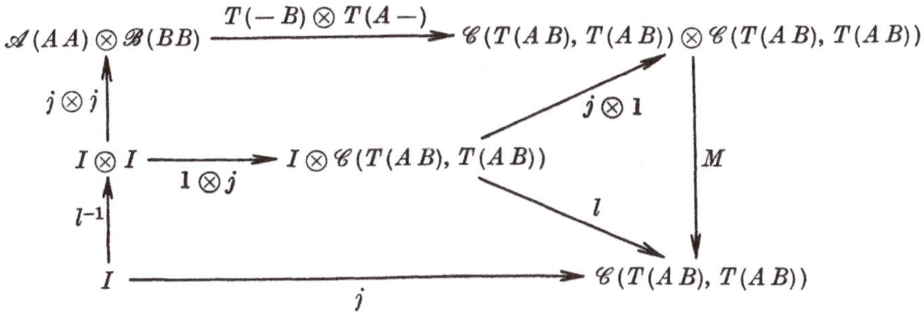

The top region commutes by VF1′ for $T(-B)$ and for $T(A-)$, the bottom region by the naturality of l, and the right region by VC1′ for $\mathscr{C}$.

VF2′ for T requires the commutativity of the following diagram, in which $\alpha, \alpha', \beta', \beta''$ stand for $T(A-)$, $T(A'-)$, $T(-B')$, $T(-B'')$, and in which $(A'\,B', A''\,B'')$ stands for $\mathscr{C}(T(A'\,B'), T(A''\,B''))$, etc.:

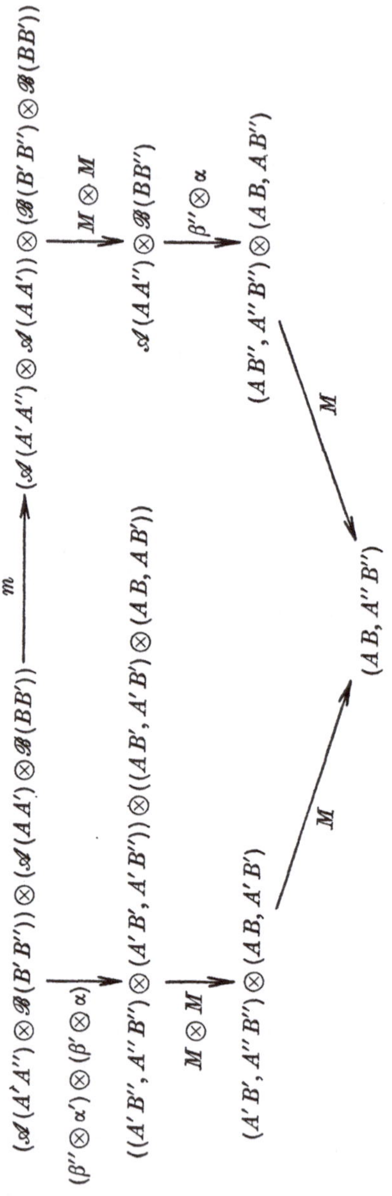

Writing m out in terms of a and c, this becomes the exterior of the following diagram, in which we leave the reader to fill in the objects:

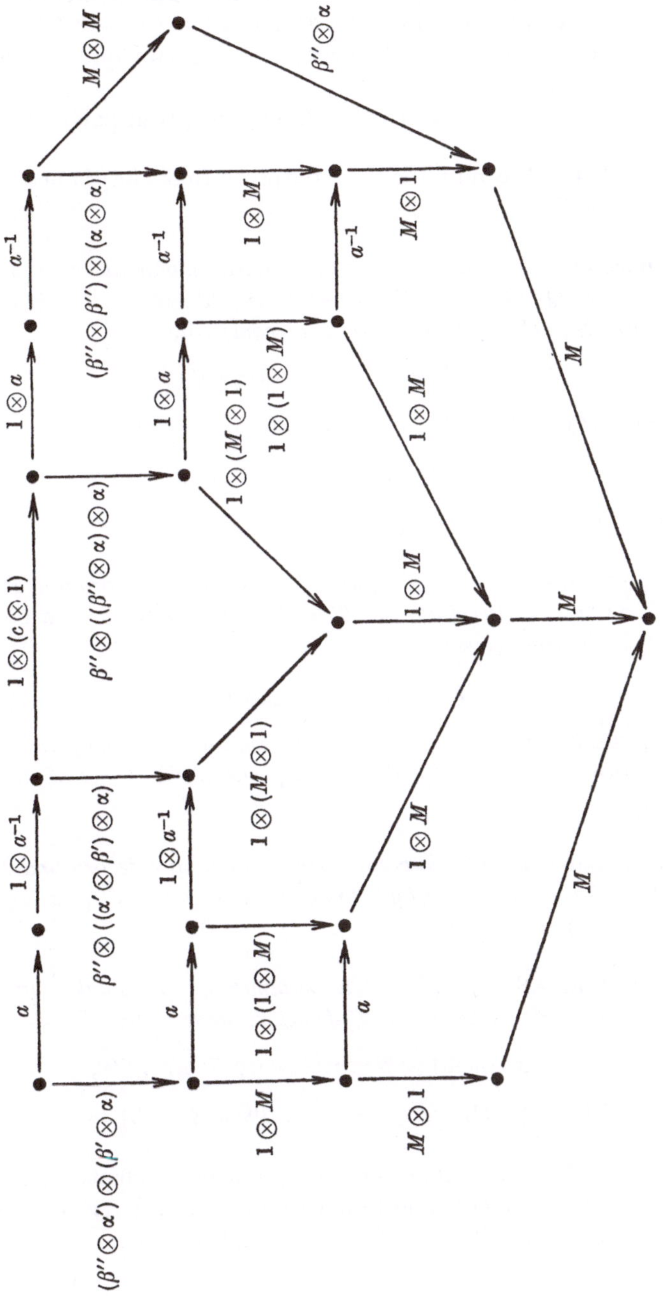

The two hexagons at the top commute by the naturality of a. The pentagon in the middle of the top commutes by (4.4) (tensored with two other diagrams that commute trivially). The pentagon at the far right commutes by VF2′ for $T(-B'')$ and for $T(A-)$. The four other pentagons commute by VC3′ for $\mathscr{C}$, and the two quadrangles by the naturality of a.

Finally it is easily verified that T has the given partial functors.

Remark 4.3. We note for later purposes that (4.4) is needed only to get VF2′ for T.

Proposition 4.4. *Let $\mathscr{V}$ be a symmetric monoidal category and let $P : \mathscr{A}' \to \mathscr{A}$, $Q : \mathscr{B}' \to \mathscr{B}$, $N : \mathscr{C} \to \mathscr{C}'$, and $T : \mathscr{A} \otimes \mathscr{B} \to \mathscr{C}$ be $\mathscr{V}$-functors. Define $S : \mathscr{A}' \otimes \mathscr{B}' \to \mathscr{C}'$ to be the composite*

$$\mathscr{A}' \otimes \mathscr{B}' \xrightarrow[P \otimes Q]{} \mathscr{A} \otimes \mathscr{B} \xrightarrow[T]{} \mathscr{C} \xrightarrow[N]{} \mathscr{C}'.$$

Then $S(-B') : \mathscr{A}' \to \mathscr{C}'$ is the composite

$$\mathscr{A}' \xrightarrow[P]{} \mathscr{A} \xrightarrow[T(-,QB')]{} \mathscr{C} \xrightarrow[N]{} \mathscr{C}',$$

with a similar formula for $S(A'-)$.

Proposition 4.5. *If $\mathscr{V}$ is a symmetric monoidal category and $P : \mathscr{A} \to \mathscr{C}$, $Q : \mathscr{B} \to \mathscr{D}$ are $\mathscr{V}$-functors, set $T = P \otimes Q : \mathscr{A} \otimes \mathscr{B} \to \mathscr{C} \otimes \mathscr{D}$. Then $T(-B)$ is the composite*

$$\mathscr{A} \xrightarrow[\mathrm{r}^{-1}]{} \mathscr{A} \otimes \mathscr{I} \xrightarrow[P \otimes JQB]{} \mathscr{C} \otimes \mathscr{D}.$$

Proposition 4.6. *If $\mathscr{V}$ is a symmetric monoidal category and $P : \mathscr{A} \to \mathscr{B}$ is a $\mathscr{V}$-functor, denote by T the composite $\mathscr{I} \otimes \mathscr{A} \xrightarrow[\mathrm{l}]{} \mathscr{A} \xrightarrow[P]{} \mathscr{B}$. Then $T(*-) = P$.*

Proposition 4.7. *If $\mathscr{V}$ is a symmetric monoidal category and $P : \mathscr{A} \otimes \mathscr{B} \to \mathscr{C}$ is a $\mathscr{V}$-functor, denote by T the composite $\mathscr{B} \otimes \mathscr{A} \xrightarrow[\mathrm{c}]{} \mathscr{A} \otimes \mathscr{B} \xrightarrow[P]{} \mathscr{C}$. Then $T(-A) = P(A-)$.*

Proposition 4.8. *If $\mathscr{V}$ is a symmetric monoidal category and $P : \mathscr{A} \otimes (\mathscr{B} \otimes \mathscr{C}) \to \mathscr{D}$ is a $\mathscr{V}$-functor, denote by T the composite*

$$(\mathscr{A} \otimes \mathscr{B}) \otimes \mathscr{C} \xrightarrow[\mathrm{a}]{} \mathscr{A} \otimes (\mathscr{B} \otimes \mathscr{C}) \xrightarrow[P]{} \mathscr{D},$$

and write S for $T(-C) : \mathscr{A} \otimes \mathscr{B} \to \mathscr{D}$. Then $S(-B) = P(-[BC])$.

Remark 4.9. Propositions 4.6—4.8 allow us to identify $(\mathscr{A} \otimes \mathscr{B}) \otimes \mathscr{C}$ with $\mathscr{A} \otimes (\mathscr{B} \otimes \mathscr{C})$, etc., and to write for example $T : \mathscr{A} \otimes \mathscr{B} \otimes \mathscr{C} \to \mathscr{D}$. We then write $T(-BC)$ etc. for the partial functors.

If $\Phi : \mathscr{V} \to \mathscr{V}'$ is a symmetric monoidal functor, and $T : \mathscr{A} \otimes \mathscr{B} \to \mathscr{C}$ is a $\mathscr{V}$-functor, let us write $\Phi_{**} T : \Phi_* \mathscr{A} \otimes \Phi_* \mathscr{B} \to \Phi_* \mathscr{C}$ for the composite

$$\Phi_{**} T: \qquad \Phi_* \mathscr{A} \otimes \Phi_* \mathscr{B} \xrightarrow[\tilde{\phi}_{\#}]{} \Phi_* (\mathscr{A} \otimes \mathscr{B}) \xrightarrow[\Phi_* T]{} \Phi_* \mathscr{C}. \tag{4.5}$$

Proposition 4.10. *If $\Phi : \mathscr{V} \to \mathscr{V}'$ is a symmetric monoidal functor and $T : \mathscr{A} \otimes \mathscr{B} \to \mathscr{C}$ is a $\mathscr{V}$-functor, we have*

$$(\Phi_{**} T)(A\, -) = \Phi_* (T(A\, -)) : \Phi_* \mathscr{A} \to \Phi_* \mathscr{C}. \tag{4.6}$$

Proposition 4.11. *If $\eta : \Phi \to \Psi : \mathscr{V} \to \mathscr{V}'$ is a monoidal natural transformation where Φ, Ψ are symmetric monoidal functors, and if $T : \mathscr{A} \otimes \mathscr{B} \to \mathscr{C}$ is a $\mathscr{V}$-functor, we have a commutative diagram:*

$$\begin{array}{ccc} \Phi_* \mathscr{A} \otimes \Phi_* \mathscr{B} & \xrightarrow{\eta_{*\mathscr{A}} \otimes \eta_{*\mathscr{B}}} & \Psi_* \mathscr{A} \otimes \Psi_* \mathscr{B} \\ {\scriptstyle \Phi_{**}T}\downarrow & & \downarrow{\scriptstyle \Psi_{**}T} \\ \Phi_* \mathscr{C} & \xrightarrow[\eta_{*\mathscr{C}}]{} & \Psi_* \mathscr{C} \end{array} \tag{4.7}$$

Proof. This follows easily from the naturality of η_* together with Proposition 3.7.

Proposition 4.12. *Let $\mathscr{V}$ be a symmetric monoidal category and T, S : $\mathscr{A} \otimes \mathscr{B} \to \mathscr{C}$ be $\mathscr{V}$-functors. Let*

$$\alpha_{AB} : T(A\,B) \to S(A\,B), \quad A \in \mathscr{A},\ B \in \mathscr{B},$$

be a family of morphisms in $\mathscr{C}_0$. Then the α_{AB} are the components of a $\mathscr{V}$-natural transformation $\alpha : T \to S$ if and only if, for each A, α_{AB} is the B-component of a $\mathscr{V}$-natural transformation $\alpha_A : T(A\, -) \to S(A\, -)$ and further, for each B, α_{AB} is the A-component of a $\mathscr{V}$-natural transformation $\alpha_{\cdot B} : T(-\, B) \to S(-\, B)$.

Proof. Suppose that $\alpha : T \to S : \mathscr{A} \otimes \mathscr{B} \to \mathscr{C}$ is $\mathscr{V}$-natural; then so is $\alpha(J^A \otimes 1)\, l^{-1} : T(A\, -) \to S(A\, -) : \mathscr{B} \to \mathscr{C}$; and the B-component of $\alpha(J^A \otimes 1)\, l^{-1}$ is α_{AB}. Thus α_A is $\mathscr{V}$-natural, and a similar argument shows that $\alpha_{\cdot B}$ is $\mathscr{V}$-natural.

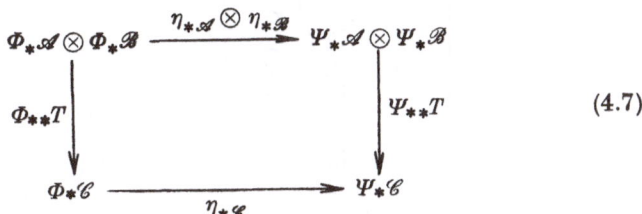

Now suppose that α_A and $\alpha_{\cdot B}$ are $\mathscr{V}$-natural. Using the top leg of (4.4) to express T and S, VN for α becomes the exterior of the following diagram:

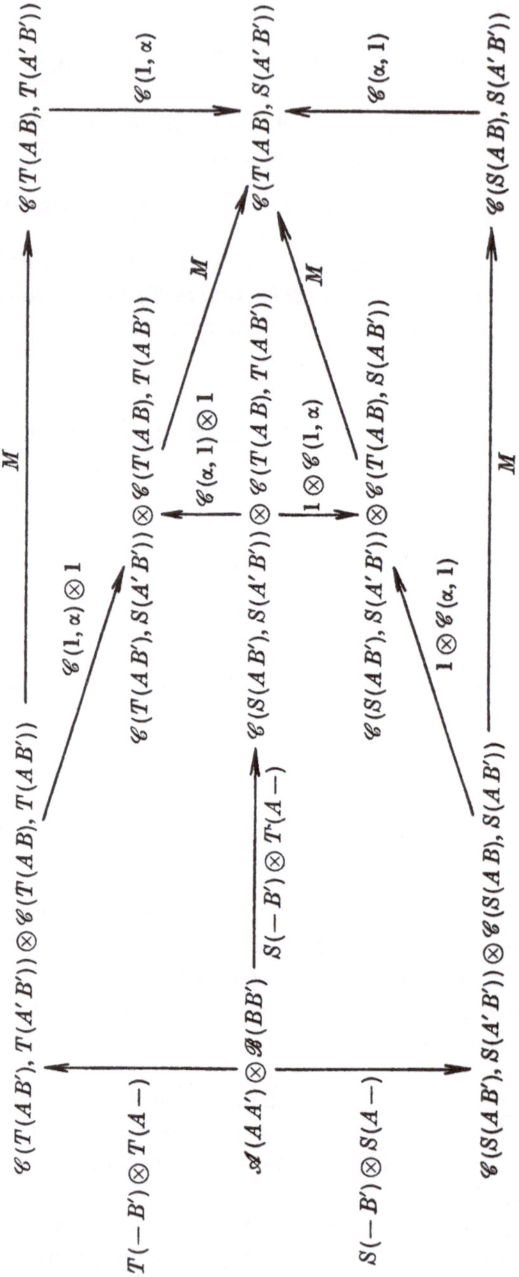

𝒞(T(AB′), T(A′B′)) ⊗ 𝒞(T(AB), T(AB′))
M
𝒞(T(AB), T(A′B′))
𝒞(1, α)
𝒞(T(AB), S(A′B′))
𝒞(α, 1)
𝒞(S(AB), S(A′B′))
T(−B′) ⊗ T(A−)
𝒞(1, α) ⊗ 1
𝒞(T(AB′), S(A′B′)) ⊗ 𝒞(T(AB), T(AB′))
M
𝒜(AA′) ⊗ ℬ(BB′)
S(−B′) ⊗ T(A−)
𝒞(S(AB′), S(A′B′)) ⊗ 𝒞(T(AB), T(AB′))
𝒞(α, 1) ⊗ 1
1 ⊗ 𝒞(1, α)
S(−B′) ⊗ S(A−)
1 ⊗ 𝒞(α, 1)
𝒞(S(AB′), S(A′B′)) ⊗ 𝒞(T(AB), S(AB′))
M
𝒞(S(AB′), S(A′B′)) ⊗ 𝒞(S(AB), S(AB′))
M

The left regions commute by VN for $\alpha_{.B'}$ and α_A, and the other regions by the naturality of M.

Proposition 4.13. *The map $\alpha \mapsto \alpha_A$ of Proposition 4.12 satisfies:*

(i) $(N\alpha)_A = N\alpha_A$ *where* $N : \mathscr{C} \to \mathscr{C}'$; (4.8)

(ii) $(\alpha(P \otimes Q))_A = \alpha_{PA} Q$ *where* $P : \mathscr{A}' \to \mathscr{A}$ *and* $Q : \mathscr{B}' \to \mathscr{B}$; (4.9)

(iii) $(\beta\alpha)_A = \beta_A \alpha_A$ *where* $\beta : S \to U : \mathscr{A} \otimes \mathscr{B} \to \mathscr{C}$. (4.10)

If $\Phi : \mathscr{V} \to \mathscr{V}'$ is a symmetric monoidal functor and $\alpha : T \to S : \mathscr{A} \otimes \otimes \mathscr{B} \to \mathscr{C}$ is a $\mathscr{V}$-natural transformation, define

$$\Phi_{**}\alpha : \Phi_{**}T \to \Phi_{**}S : \Phi_*\mathscr{A} \otimes \Phi_*\mathscr{B} \to \Phi_*\mathscr{C}$$

by

$$\Phi_{**}\alpha = \Phi_*\alpha . \tilde{\phi}_{\#} . \tag{4.11}$$

Then, since $\tilde{\phi}_{\#}$ is the identity on objects, we have

$$(\Phi_{**}\alpha)_{AB} = (\Phi_*\alpha)_{AB}, \tag{4.12}$$

which is $\Phi_{0\mathscr{C}}(\alpha_{AB})$ by I(10.5). Thus we have trivially:

Proposition 4.14. *If $\Phi : \mathscr{V} \to \mathscr{V}'$ is a symmetric monoidal functor and $\alpha : T \to S : \mathscr{A} \otimes \mathscr{B} \to \mathscr{C}$ is a $\mathscr{V}$-natural transformation, we have*

$$(\Phi_{**}\alpha)_A = \Phi_*\alpha_A . \tag{4.13}$$

From the hypernaturality of η_* together with Proposition 3.7, we get:

Proposition 4.15. *If $\eta : \Phi \to \Psi : \mathscr{V} \to \mathscr{V}'$ is a monoidal natural transformation where Φ and Ψ are symmetric monoidal functors, and if $\alpha : T \to \to S : \mathscr{A} \otimes \mathscr{B} \to \mathscr{C}$ is a $\mathscr{V}$-natural transformation, then*

$$\eta_{*\mathscr{C}} . \Phi_{**}\alpha : \eta_{*\mathscr{C}} . \Phi_{**}T \to \eta_{*\mathscr{C}} . \Phi_{**}S : \Phi_*\mathscr{A} \otimes \Phi_*\mathscr{B} \to \Psi_*\mathscr{C}$$

coincides with

$$\Psi_{**}\alpha . (\eta_{*\mathscr{A}} \otimes \eta_{*\mathscr{B}}) : \Psi_{**}T . (\eta_{*\mathscr{A}} \otimes \eta_{*\mathscr{B}}) \to$$
$$\to \Psi_{**}S . (\eta_{*\mathscr{A}} \otimes \eta_{*\mathscr{B}}) : \Phi_*\mathscr{A} \otimes \Phi_*\mathscr{B} \to \Psi_*\mathscr{C} .$$

Proposition 4.16. *Let $\Phi : \mathscr{V} \to \mathscr{V}'$ be a symmetric monoidal functor, and $\alpha : T \to S : \mathscr{A} \otimes \mathscr{B} \to \mathscr{C}$ a $\mathscr{V}$-natural transformation. We have*

$$T^*(A-) = (T(A-))^* , \tag{4.14}$$

$$\Phi_{**}T^* = (\Phi_{**}T)^* , \tag{4.15}$$

$$\Phi_{**}\alpha^* = (\Phi_{**}\alpha)^* . \tag{4.16}$$

Proposition 4.17. *Let $\Phi : \mathscr{V} \to \mathscr{V}'$ be a symmetric monoidal functor, let $T : \mathscr{A} \otimes \mathscr{B} \to \mathscr{C}$ be a $\mathscr{V}$-functor, and let $\alpha : P \to P' : \mathscr{A}' \to \mathscr{A}$ and $\beta : Q \to Q' : \mathscr{B}' \to \mathscr{B}$ be $\mathscr{V}$-natural transformations. Then*

$$\Phi_{**}T.(\Phi_* P \otimes \Phi_* Q) = \Phi_* T.\Phi_{**}(P \otimes Q) : \Phi_* \mathscr{A}' \otimes \Phi_* \mathscr{B}' \to \Phi_* \mathscr{C}, \tag{4.17}$$

and

$$\Phi_{**}T.(\Phi_* \alpha \otimes \Phi_* \beta) = \Phi_* T.\Phi_{**}(\alpha \otimes \beta). \tag{4.18}$$

Proof. We have a commutative diagram

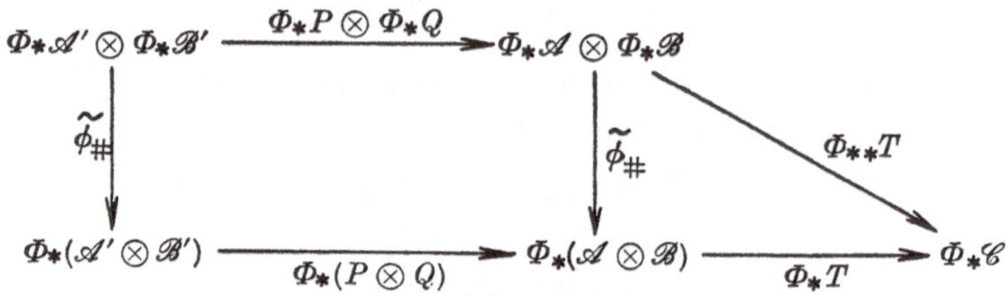

by the naturality of $\tilde{\phi}_{\#}$ and the definition of $\Phi_{**}T$; and by the hyper-naturality of $\tilde{\phi}_{\#}$ we have a similar diagram with α, β in place of P, Q.

Taking the symmetric monoidal functor $V : \mathscr{V} \to \mathscr{S}$ and a $\mathscr{V}$-functor $T : \mathscr{A} \otimes \mathscr{B} \to \mathscr{C}$, we write

$$T_{00} = V_{**} T, \tag{4.19}$$

so that $T_{00} : \mathscr{A}_0 \times \mathscr{B}_0 \to \mathscr{C}_0$ is the composite

$$\mathscr{A}_0 \times \mathscr{B}_0 \xrightarrow[\tilde{V}_{\#}]{} (\mathscr{A} \otimes \mathscr{B})_0 \xrightarrow[T_0]{} \mathscr{C}_0. \tag{4.20}$$

If $\alpha : T \to S : \mathscr{A} \otimes \mathscr{B} \to \mathscr{C}$ is a $\mathscr{V}$-natural transformation, we have by (4.12) since V is normal

$$(V_{**}\alpha)_{AB} = \alpha_{AB}; \tag{4.21}$$

we shall in practice identify $V_{**}\alpha$ with α, so that we also have $\alpha : T_{00} \to$ $\to S_{00} : \mathscr{A}_0 \times \mathscr{B}_0 \to \mathscr{C}_0$. Proposition 4.17 gives in this case

$$T_{00}(\alpha_A, \beta_B) = T_0(\alpha_A \otimes \beta_B), \tag{4.22}$$

so that by I (10.3) we have

$$(T(\alpha \otimes \beta))_{AB} = T_{00}(\alpha_A, \beta_B). \tag{4.23}$$

Proposition 4.18. *Let $\mathscr{V}$ be a symmetric monoidal category and let $T : \mathscr{A} \otimes \mathscr{B} \to \mathscr{C}$ be a $\mathscr{V}$-functor. Then*

$$T(A\,-) : \mathscr{B}(BB') \to \mathscr{C}(T_{00}(AB), T_{00}(AB'))$$

is natural in A; with a similar result for $T(-\,B)$.

Proof. Let $f \in \mathscr{A}_0(AA')$ and compose both legs of (4.4) with

$$\mathscr{B}(BB') \xrightarrow[l^{-1}]{} I \otimes \mathscr{B}(BB') \xrightarrow[if \otimes 1]{} \mathscr{A}(AA') \otimes \mathscr{B}(BB').$$

By the naturality of ι, the composite

$$I \xrightarrow[\iota f]{} \mathscr{A}(AA') \xrightarrow[T(-B')]{} \mathscr{C}(T(AB'), T(A'B'))$$

is $\iota((VT(-B'))f)$, that is, $\iota((T(-B'))_0 f)$, or $\iota T_{00}(1, f)$ by (4.6). Using the naturality of c and II (8.11) and II (8.12), we find we have the diagram

$$\begin{array}{ccc} \mathscr{B}(BB') & \xrightarrow{T(A-)} & \mathscr{C}(T(AB), T(AB')) \\ {\scriptstyle T(A'-)}\downarrow & & \downarrow{\scriptstyle \mathscr{C}(1, T_{00}(f,1))} \\ \mathscr{C}(T(A'B), T(A'B')) & \xrightarrow[\mathscr{C}(T_{00}(f,1),1)]{} & \mathscr{C}(T(AB), T(A'B')). \end{array} \qquad (4.24)$$

which expresses the naturality in A of $T(A-)$.

Proposition 4.19. *Let $\mathscr{V}$ be a symmetric monoidal category and let $T: \mathscr{A} \otimes \mathscr{B} \to \mathscr{C}$ be a $\mathscr{V}$-functor. Then for each $f \in \mathscr{A}_0(AA')$, the morphism*

$$T_{00}(f, 1): T(AB) \to T(A'B)$$

is the B-component of a $\mathscr{V}$-natural transformation

$$T(f, 1): T(A-) \to T(A'-);$$

with a similar result for $T(1, g)$.

Proof. VN for $T(f, 1)$ is (4.24).

5. Extraordinary $\mathscr{V}$-natural Transformations

Let $\mathscr{V}$ be a symmetric monoidal category. We now introduce for $\mathscr{V}$-categories the extraordinary kinds of natural transformation introduced for ordinary categories in [7].

Let $\mathscr{A}$ and $\mathscr{B}$ be $\mathscr{V}$-categories, $T: \mathscr{A}^* \otimes \mathscr{A} \to \mathscr{B}$ a $\mathscr{V}$-functor, and B a fixed object of $\mathscr{B}$. A family of morphisms in $\mathscr{B}_0$,

$$\gamma_A : B \to T(AA), \quad A \in \mathscr{A},$$

is said to be $\mathscr{V}$-natural if the following axiom is satisfied:

VN′. The following diagram commutes:

$$\begin{array}{ccc} \mathscr{A}(AA') & \xrightarrow{T(A-)} & \mathscr{B}(T(AA), T(AA')) \\ {\scriptstyle T(-A')}\downarrow & & \downarrow{\scriptstyle \mathscr{B}(\gamma_A, 1)} \\ \mathscr{B}(T(A'A'), T(AA')) & \xrightarrow[\mathscr{B}(\gamma_{A'}, 1)]{} & \mathscr{B}(B, T(AA')) \end{array}$$

Similarly a family of morphisms in $\mathscr{B}_0$,

$$\delta_A : T(A\,A) \to B\,, \quad A \in \mathscr{A}\,,$$

is said to be $\mathscr{V}$-natural if the following axiom is satisfied:

VN″. The following diagram commutes:

$$\begin{array}{ccc} \mathscr{A}(A\,A') & \xrightarrow{\;T(-A)\;} & \mathscr{B}(T(A'A), T(A\,A)) \\ {\scriptstyle T(A'-)}\downarrow & & \downarrow{\scriptstyle \mathscr{B}(1,\delta_A)} \\ \mathscr{B}(T(A'A), T(A'A')) & \xrightarrow[\;\mathscr{B}(1,\delta_{A'})\;]{} & \mathscr{B}(T(A'A), B) \end{array}$$

Now if we have $\mathscr{V}$-functors

$$T : \mathscr{A}_1^* \otimes \mathscr{A}_1 \otimes \cdots \otimes \mathscr{A}_p^* \otimes \mathscr{A}_p \otimes \mathscr{C}_1 \otimes \cdots \otimes \mathscr{C}_r \to \mathscr{D}\,,$$

$$S : \mathscr{B}_1^* \otimes \mathscr{B}_1 \otimes \cdots \otimes \mathscr{B}_q^* \otimes \mathscr{B}_q \otimes \mathscr{C}_1 \otimes \cdots \otimes \mathscr{C}_r \to \mathscr{D}\,,$$

and morphisms

$$\alpha_{A_1 \cdots A_p B_1 \cdots B_q C_1 \cdots C_r} : T(A_1 A_1 \cdots A_p A_p C_1 \cdots C_r) \to S(B_1 B_1 \cdots B_q B_q C_1 \cdots C_r)\,,$$

we define α to be $\mathscr{V}$-natural if it is so in each variable $A_1 \cdots C_r$ separately when the others are held fixed. Proposition 4.12 shows that we may group the variables $C_1, \ldots, C_r$ at pleasure; the situation is entirely similar for the other variables, and we leave the reader to adapt the proof of Proposition 4.12 to prove:

Proposition 5.1. *If $\mathscr{V}$ is a symmetric monoidal category, if*

$$T : \mathscr{A}^* \otimes \mathscr{B}^* \otimes \mathscr{A} \otimes \mathscr{B} \to \mathscr{C}$$

is a $\mathscr{V}$-functor, and if $C \in \mathscr{C}$, a family of morphisms

$$\alpha_{AB} : T(ABAB) \to C$$

is $\mathscr{V}$-natural in $[AB]$ if and only if it is so in each of A, B separately.

Proposition 5.2. *The rules for composition of extraordinary $\mathscr{V}$-natural transformations are formally identical with those of* [7].

Proof. The considerations of [7] use only the formal properties of diagrams VN, VN′, VN″.

For composition of $\mathscr{V}$-natural transformations with $\mathscr{V}$-functors, we have:

Proposition 5.3. *Let $\mathscr{V}$ be a symmetric monoidal category, let*

$$T : \mathscr{A}^* \otimes \mathscr{A} \to \mathscr{B}, \quad P : \mathscr{C} \to \mathscr{A}, \quad Q : \mathscr{B} \to \mathscr{D}$$

be $\mathscr{V}$-functors, and let $\gamma_A : B \to T(A\,A)$ be $\mathscr{V}$-natural. Then the family of morphisms

$$Q_0\,\gamma_{PC} : Q\,B \to Q\,T(PC, PC)$$

is also $\mathscr{V}$-natural, the relevant bifunctor now being

$$Q\,T(P^* \otimes P) : \mathscr{C}^* \otimes \mathscr{C} \to \mathscr{D}\,.$$

We leave the reader to adapt the proof from that of Theorem I.10.2. There is of course a corresponding result for $\mathscr{V}$-natural transformations of the VN″ type, but it is only the dual of the above and needs no separate proof. Note that by Remark I.10.12 we may write $Q\gamma_{PC}$ instead of $Q_0\gamma_{PC}$; for the name of the family we shall use $Q\gamma P$.

The proof of the following proposition is exactly like that of Proposition I.10.3:

Proposition 5.4. *Let $\Phi : \mathscr{V} \to \mathscr{V}'$ be a symmetric monoidal functor, $T : \mathscr{A}^* \otimes \mathscr{A} \to \mathscr{B}$ a $\mathscr{V}$-functor, and $\gamma_A : B \to T(A\,A)$ a $\mathscr{V}$-natural transformation. Then $\Phi_{0\mathscr{B}}\gamma_A : B \to (\Phi_{**}\,T)(A\,A)$ is a $\mathscr{V}'$-natural transformation, which we shall write $\Phi_*\gamma$.*

Analogously to Proposition I.10.6 we have (cf. also Proposition 4.11):

Proposition 5.5. *Let $\eta : \Phi \to \Psi : \mathscr{V} \to \mathscr{V}'$ be a monoidal natural transformation where Φ, Ψ are symmetric monoidal functors, and let $\gamma_A : B \to$ $\to T(A\,A)$ be a $\mathscr{V}$-natural transformation, where $T : \mathscr{A}^* \otimes \mathscr{A} \to \mathscr{B}$. Then the $\mathscr{V}'$-natural transformations*

$$\eta_{*\mathscr{B}} \cdot \Phi_*\,\gamma : B \to \eta_{*\mathscr{B}} \cdot (\Phi_{**}\,T)\,(A\,A)$$

and

$$\Psi_*(\gamma\,\eta_{*\mathscr{A}}) : B \to \Psi_{**}\,T\,(\eta_{*\mathscr{A}}A, \eta_{*\mathscr{A}}A)$$

coincide.

6. Symmetric Monoidal Closed Categories

A *symmetric monoidal closed category* $\mathscr{V}$ shall mean a monoidal closed category $\mathscr{V}$ with a symmetry as in § 1. If $\mathscr{V}$ and $\mathscr{V}'$ are symmetric monoidal closed categories, a symmetric monoidal functor $\Phi = (\phi, \tilde{\phi}, \phi^0) : \mathscr{V} \to \mathscr{V}'$, identified with the monoidal closed functor $\Phi = (\phi, \tilde{\phi}, \hat{\phi}, \phi^0) : \mathscr{V} \to \mathscr{V}'$, shall be called a *symmetric closed functor*. Then symmetric monoidal closed categories, symmetric closed functors, and closed natural transformations form a sub-hypercategory $\mathscr{SMCl}$ of $\mathscr{MCl}$.

Proposition 6.1. *If V is faithful, the monoidal closed category $\mathscr{V}$ admits at most one symmetry.*

Proof. Let c and $\bar{c}$ be two symmetries. From (1.1) we get

$$Vc.\tilde{V} = V\bar{c}.\tilde{V}.$$

Now apply II(3.24) with $\Phi = V$ and $x = c$; we get, since $\hat{V} = V$,

$$V.V\pi c = V.V\pi\bar{c}.$$

Since V is faithful and π is an isomorphism, we deduce $c = \bar{c}$.

If $\mathscr{V}$ is a symmetric monoidal closed category and $\mathscr{A}$ is a $\mathscr{V}$-category, we have for each $A \in \mathscr{A}$ the $\mathscr{V}$-functor $L^A : \mathscr{A}^* \to \mathscr{V}$. To distinguish this from the L of $\mathscr{A}$ we write it as $R^A : \mathscr{A}^* \to \mathscr{V}$, and treat it as an attribute of $\mathscr{A}$ rather than $\mathscr{A}^*$. We have

$$R^A B = \mathscr{A}(BA), \tag{6.1}$$

and

$$R^A_{BC} : \mathscr{A}^*(BC) \to (\mathscr{A}^*(AB), \mathscr{A}^*(AC));$$

that is,

$$R^A_{BC} : \mathscr{A}(CB) \to (\mathscr{A}(BA), \mathscr{A}(CA)). \tag{6.2}$$

Since the M of $\mathscr{A}^*$ is given by (2.4), we have by II(6.2)

$$R = \pi(Mc). \tag{6.3}$$

It follows that R^A_{BC} in (6.2) is natural in all variables, since π, M, c are. We call R^A the *right represented $\mathscr{V}$-functor*.

The underlying functor $V_* R^A : \mathscr{A}^*_0 \to \mathscr{V}_0$ is given by

$$V_* R^A = \mathscr{A}(-A) : \mathscr{A}^*_0 \to \mathscr{V}_0, \tag{6.4}$$

in view of Proposition 2.9. Translating Proposition I.8.3 we get:

Proposition 6.2. *If $\mathscr{V}$ is a symmetric monoidal closed category and $T : \mathscr{A} \to \mathscr{B}$ is a $\mathscr{V}$-functor, the morphisms*

$$T_{BC} : \mathscr{A}(BC) \to \mathscr{B}(TB, TC), \quad B \in \mathscr{A},$$

are the components of a $\mathscr{V}$-natural transformation

$$T_{\cdot C} : R^C \to R^{TC} T^* : \mathscr{A}^* \to \mathscr{V}.$$

As in Proposition I.8.4, the naturality of R^A in A gives:

Proposition 6.3. *If $\mathscr{V}$ is a symmetric monoidal closed category and $\mathscr{A}$ is a $\mathscr{V}$-category, let $f \in \mathscr{A}_0(AB)$. Then the morphisms*

$$\mathscr{A}(1, f) : \mathscr{A}(CA) \to \mathscr{A}(CB), \quad C \in \mathscr{A},$$

are the components of a $\mathscr{V}$-natural transformation

$$R^f : R^A \to R^B : \mathscr{A}^* \to \mathscr{V}.$$

Theorem 6.4. *If $\mathscr{V}$ is a symmetric monoidal closed category and $\mathscr{A}$ is a $\mathscr{V}$-category, the $\mathscr{V}$-functors $R^B : \mathscr{A}^* \to \mathscr{V}$ and $L^A : \mathscr{A} \to \mathscr{V}$ are the partial functors of a $\mathscr{V}$-functor*

$$\operatorname{Hom} \mathscr{A} : \mathscr{A}^* \otimes \mathscr{A} \to \mathscr{V}.$$

We defer the proof of this, which consists in verifying (4.4), to § 7. By I(7.9) and (6.4) we have:

$$(\operatorname{Hom} \mathscr{A})_{00} = \operatorname{hom} \mathscr{A} : \mathscr{A}_0^* \times \mathscr{A}_0 \to \mathscr{V}_0. \tag{6.5}$$

Proposition 6.5. *If $\mathscr{V}$ is a symmetric monoidal closed category and $\mathscr{A}$ is a $\mathscr{V}$-category, the following diagram of $\mathscr{V}$-functors commutes:*

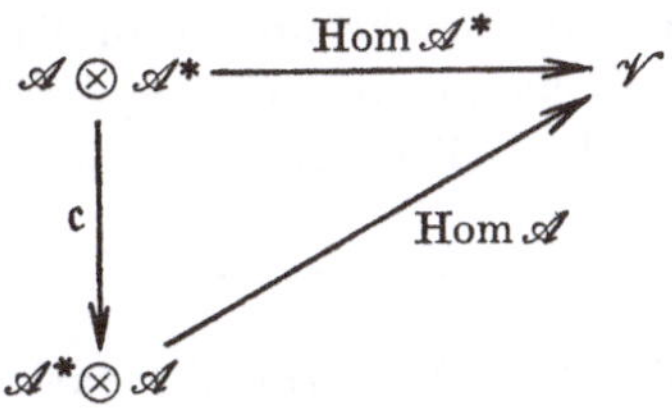

(6.6)

Proof. By the uniqueness in Proposition 4.2, it suffices to check that the partial functors agree. This is obvious from Proposition 4.7 and the definition of R^A.

Proposition 6.6. *Let $\Phi : \mathscr{V} \to \mathscr{V}'$ be a symmetric closed functor and $\mathscr{A}$ a $\mathscr{V}$-category. Then*

$$\operatorname{Hom} \Phi_* \mathscr{A} : \Phi_* \mathscr{A}^* \otimes \Phi_* \mathscr{A} \to \mathscr{V}'$$

is the composite

$$\Phi_* \mathscr{A}^* \otimes \Phi_* \mathscr{A} \xrightarrow[\Phi_{**}\operatorname{Hom}\mathscr{A}]{} \Phi_* \mathscr{V} \xrightarrow[\hat{\Phi}]{} \mathscr{V}'.$$

Proof. It suffices to check the partial functors; in view of Proposition 4.10 the result follows from I(6.15) for $\mathscr{A}$ and for $\mathscr{A}^*$.

Now for a symmetric monoidal closed category $\mathscr{V}$ define a natural transformation

$$H^B_{AC} : (A\,C) \to (B \otimes A, B \otimes C)$$

as the composite

$$(A\,C) \xrightarrow[K^B_{AC}]{} (A \otimes B, C \otimes B) \xrightarrow[(c,c)]{} (B \otimes A, B \otimes C); \tag{6.7}$$

see § II.7 for the definition of K.

Proposition 6.7. *For each B in the symmetric monoidal closed category $\mathscr{V}$ we obtain a $\mathscr{V}$-functor $H^B : \mathscr{V} \to \mathscr{V}$ if we set $H^B A = B \otimes A$ and $(H^B)_{AC} = H^B_{AC}$, and the underlying functor $V_* H^B : \mathscr{V}_0 \to \mathscr{V}_0$ is $B \otimes -$.*

Proof. We have, for $f \in \mathscr{V}_0(AC)$,

$$\begin{aligned}(VH^B)f &= V(c,c).\,VK^B.f && \text{by (6.7)}\\ &= V(c,c)(1 \otimes f) && \text{by Theorem II.7.1}\\ &= c(1 \otimes f)c \\ &= f \otimes 1 && \text{by the naturality of } c.\end{aligned}$$

Thus we have $V_* H^B = B \otimes -$, and we have VF1 for H^B in the form $(VH^B)1 = 1$ (cf. Remark I.9.7).

Axiom VF2 for H^B is axiom VN for H^B_A. Since the proofs of the assertions of Theorem I.10.2 make no use of VF2 for the $\mathscr{V}$-functors involved, we can use them here before we have VF2 for H^B. Since $c^2 = 1$, the definition (6.7) may be written in the form VN to show that

$$c_{CB} : C \otimes B \to B \otimes C$$

is the C-component of a $\mathscr{V}$-natural transformation

$$c_{\cdot B} : K^B \to H^B .$$

The composite (6.7) is therefore the C-component of a $\mathscr{V}$-natural transformation

$$L^A \xrightarrow[K^B_A]{} L^{A \otimes B} K^B \xrightarrow[L^c K^B]{} L^{B \otimes A} K^B \xrightarrow[L^{B \otimes A} c_{\cdot B}]{} L^{B \otimes A} H^B ;$$

thus H^B_{AC} is $\mathscr{V}$-natural in C, which is VF2 for H^B.

From the naturality in B of H^B_{AC}, which is immediate from its definition (6.7), we get just as in Proposition II.7.2:

Proposition 6.8. *If $\mathscr{V}$ is a symmetric monoidal closed category and $f \in \mathscr{V}_0(AB)$, the morphisms*

$$f \otimes 1 : A \otimes C \to B \otimes C, \quad C \in \mathscr{V},$$

are the components of a $\mathscr{V}$-natural transformation

$$H^f : H^A \to H^B : \mathscr{V} \to \mathscr{V} .$$

We defer to § 7 the proof of:

Theorem 6.9. *If $\mathscr{V}$ is a symmetric monoidal closed category, the $\mathscr{V}$-functors $K^B : \mathscr{V} \to \mathscr{V}$ and $H^A : \mathscr{V} \to \mathscr{V}$ are the partial functors of a $\mathscr{V}$-*

functor

$$\mathrm{Ten} : \mathscr{V} \otimes \mathscr{V} \to \mathscr{V} .$$

Since $V_* K^B = - \otimes B$ and $V_* H^A = A \otimes -$, we have

$$(\mathrm{Ten})_{00} = \otimes : \mathscr{V}_0 \times \mathscr{V}_0 \to \mathscr{V}_0 . \tag{6.8}$$

Proposition 6.10. *If $\mathscr{V}$ is a symmetric monoidal closed category and $\mathscr{A}$ and $\mathscr{B}$ are $\mathscr{V}$-categories, the following diagram of $\mathscr{V}$-functors commutes:*

$$\begin{array}{ccc} (\mathscr{A}^* \otimes \mathscr{B}^*) \otimes (\mathscr{A} \otimes \mathscr{B}) & \xrightarrow{\mathrm{Hom}(\mathscr{A} \otimes \mathscr{B})} & \mathscr{V} \\ \Big\downarrow \mathfrak{m} & & \Big\uparrow \mathrm{Ten} \\ (\mathscr{A}^* \otimes \mathscr{A}) \otimes (\mathscr{B}^* \otimes \mathscr{B}) & \xrightarrow[\mathrm{Hom}\,\mathscr{A} \otimes \mathrm{Hom}\,\mathscr{B}]{} & \mathscr{V} \otimes \mathscr{V} \end{array} \tag{6.9}$$

Proof. It suffices to check the partial functors; we must therefore show that the $L^{[AB]}$ of $\mathscr{A} \otimes \mathscr{B}$ is given by the composite

$$\mathscr{A} \otimes \mathscr{B} \xrightarrow[L^A \otimes L^B]{} \mathscr{V} \otimes \mathscr{V} \xrightarrow[\mathrm{Ten}]{} \mathscr{V} ; \tag{6.10}$$

the same result applied to $\mathscr{A}^*$ and $\mathscr{B}^*$ then gives equality of the other partial functors.

Let the partial functors of $L^{[AB]} : \mathscr{A} \otimes \mathscr{B} \to \mathscr{V}$ be $L^{[AB]}(-D) = P : \mathscr{A} \to \mathscr{V}$ and $L^{[AB]}(C-) = Q : \mathscr{B} \to \mathscr{V}$. Then to show that $L^{[AB]}$ is (6.10) is to show that P is the composite

$$\mathscr{A} \xrightarrow[L^A]{} \mathscr{V} \xrightarrow[K\mathscr{B}(BD)]{} \mathscr{V} \tag{6.11}$$

and that Q is the composite

$$\mathscr{B} \xrightarrow[L^B]{} \mathscr{V} \xrightarrow[H\mathscr{A}(AC)]{} \mathscr{V} , \tag{6.12}$$

as we see from Proposition 4.4.

We immediately verify that P agrees with (6.11) and Q with (6.12) on objects. By (4.2), P_{XY} is given by:

$$\begin{aligned} &\mathscr{A}(XY) \xrightarrow[r^{-1}]{} \mathscr{A}(XY) \otimes I \xrightarrow[1 \otimes j]{} \mathscr{A}(XY) \otimes \mathscr{B}(DD) \xrightarrow[L^{[AB]}]{} \\ &\qquad \to (\mathscr{A}(AX) \otimes \mathscr{B}(BD), \mathscr{A}(AY) \otimes \mathscr{B}(BD)) . \end{aligned} \tag{6.13}$$

Taking π^{-1} of (6.13), using II (3.1), II (6.2), and (3.4), we get the upper leg of the following diagram:

$$
\begin{array}{ccccc}
\mathscr{A}(XY)\otimes(\mathscr{A}(AX)\otimes\mathscr{B}(BD)) & \xrightarrow{r^{-1}\otimes 1} & (\mathscr{A}(XY)\otimes I)\otimes(\mathscr{A}(AX)\otimes\mathscr{B}(BD)) & \xrightarrow{(1\otimes j)\otimes 1} & (\mathscr{A}(XY)\otimes\mathscr{B}(DD))\otimes(\mathscr{A}(AX)\otimes\mathscr{B}(BD)) \\
\downarrow{\scriptstyle a^{-1}} & & \downarrow{\scriptstyle m} & & \downarrow{\scriptstyle m} \\
(\mathscr{A}(XY)\otimes\mathscr{A}(AX))\otimes\mathscr{B}(BD) & \xrightarrow{1\otimes l^{-1}} & (\mathscr{A}(XY)\otimes\mathscr{A}(AX))\otimes(I\otimes\mathscr{B}(BD)) & \xrightarrow{1\otimes(j\otimes 1)} & (\mathscr{A}(XY)\otimes\mathscr{A}(AX))\otimes(\mathscr{B}(DD)\otimes\mathscr{B}(BD)) \\
 & \searrow{\scriptstyle M\otimes 1} & & & \downarrow{\scriptstyle M\otimes M} \\
 & & & & \mathscr{A}(AY)\otimes\mathscr{B}(BD)
\end{array}
\qquad (6.14)
$$

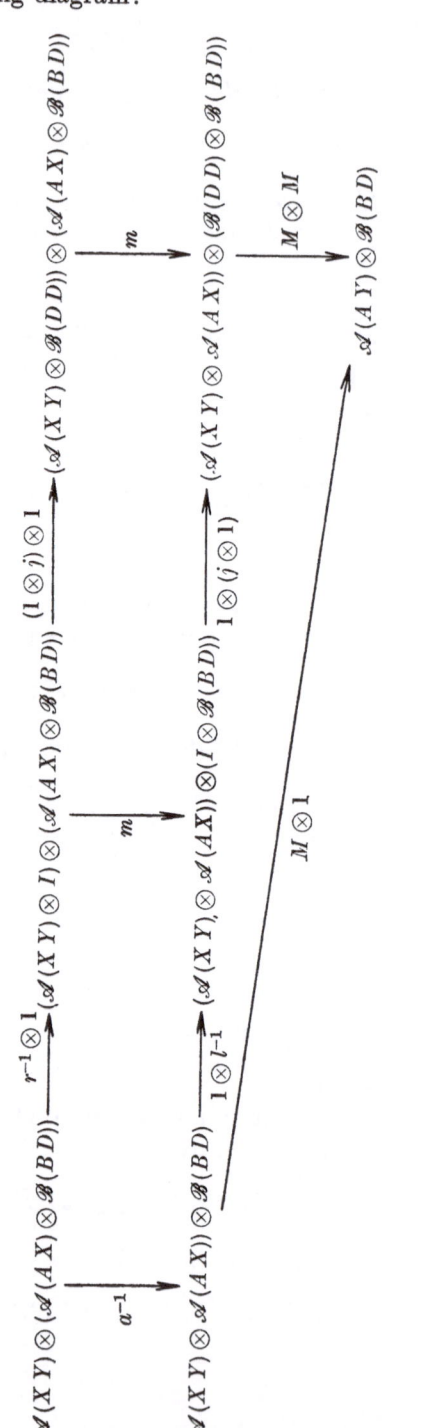

Now in (6.14) the top left region commutes by coherence, the top right region by the naturality of m, and the bottom region by VC1′ for $\mathscr{B}$. It follows that we have

$$\pi^{-1} P = (M \otimes 1)\, a^{-1}. \tag{6.15}$$

Thus

$$\begin{aligned} P &= \pi((M \otimes 1)\, a^{-1}) & & \\ &= p^{-1} . \pi\pi(M \otimes 1) & &\text{by II}(3.19) \\ &= p^{-1} . \pi(u M) & &\text{by II}(3.1) \text{ with } x = 1 \\ &= p^{-1}(1, u) . \pi M & &\text{by II}(3.1) \\ &= K L & &\text{by II}(7.1) \text{ and II}(6.2); \end{aligned}$$

which proves that P is (6.11).

In an exactly similar way one proves that

$$\pi^{-1} Q = (1 \otimes M)\, c\, a^{-1}(1 \otimes c); \tag{6.16}$$

by the naturality of c we also have

$$\pi^{-1} Q = c(M \otimes 1)\, a^{-1}(1 \otimes c);$$

II(3.1) then gives

$$\begin{aligned} Q &= (c, c)\, \pi((M \otimes 1)\, a^{-1}) & & \\ &= (c, c)\, K L & &\text{by the calculation above} \\ &= H L & &\text{by } (6.7). \end{aligned}$$

This completes the proof.

7. The $\mathscr{V}$-naturality of the Canonical Morphisms

Proposition 7.1. *Let $\mathscr{V}$ be a symmetric monoidal closed category and let $T(A\,-) : \mathscr{B} \to \mathscr{C}$ and $T(-\,B) : \mathscr{A} \to \mathscr{C}$ be families of $\mathscr{V}$-functors indexed by $A \in \mathscr{A}$ and $B \in \mathscr{B}$, with $T(A\,-)B = T(-\,B)A = T(AB)$. Then these are the partial functors of a bifunctor $T : \mathscr{A} \otimes \mathscr{B} \to \mathscr{C}$ if and only if either of the following diagrams commutes:*

$$\begin{array}{ccccc} \mathscr{A}(AA') & \xrightarrow{T(-B')} & \mathscr{C}(T(AB'), T(A'B')) & \xrightarrow{L} & (\mathscr{C}(T(AB), T(AB')), \mathscr{C}(T(AB), T(A'B'))) \\ {\scriptstyle T(-B)}\downarrow & & & & \downarrow{\scriptstyle (T(A-),1)} \\ \mathscr{C}(T(AB), T(A'B)) & & & & \\ {\scriptstyle R}\downarrow & & & & \\ (\mathscr{C}(T(A'B), T(A'B')), \mathscr{C}(T(AB), T(A'B'))) & & \xrightarrow[(T(A'-),1)]{} & & (\mathscr{B}(BB'), \mathscr{C}(T(AB), T(A'B'))) \end{array} \tag{7.1}$$

$$\begin{array}{ccccc}
\mathscr{B}(B\,B') & \xrightarrow{T(A'\,-)} & \mathscr{C}(T(A'\,B),T(A'\,B')) & \xrightarrow{L} & (\mathscr{C}(T(A\,B),T(A'\,B)),\mathscr{C}(T(A\,B),T(A'\,B'))) \\
{\scriptstyle T(A\,-)}\downarrow & & & & \\
\mathscr{C}(T(A\,B),T(A\,B')) & & & & \downarrow{\scriptstyle (T(-\,B),1)} \\
{\scriptstyle R}\downarrow & & & & \\
(\mathscr{C}(T(A\,B'),T(A'B')),\mathscr{C}(T(A\,B),T(A'\,B'))) & & \xrightarrow[(T(-\,B'),1)]{} & & (\mathscr{A}(A\,A'),\mathscr{C}(T(A\,B),T(A'\,B')))
\end{array} \tag{7.2}$$

Proof. By Propositions 4.1 and 4.2, it suffices to show that each of (7.1), (7.2) is equivalent to (4.4). Applying π to both legs of (4.4), we get

$$\begin{aligned}
\pi(M(T(-\,B')\otimes T(A\,-))) &= (T(A\,-),1).\pi M.T(-\,B') && \text{by II}(3.1)\\
&= (T(A\,-),1)\,LT(-\,B') && \text{by II}(6.2);
\end{aligned}$$

and

$$\begin{aligned}
\pi(M(T(A'-)\otimes T(-\,B))\,c) &= \pi(M\,c(T(-\,B)\otimes T(A'-))) && \text{by the naturality of } c,\\
&= (T(A'-),1)\,RT(-\,B) && \text{by II}(3.1) \text{ and } (6.3);
\end{aligned}$$

thus (4.4) is equivalent to (7.1).

Similarly we get (7.2) if we reverse the direction of the arrow c in (4.4) before applying π.

Corollary 7.2. *If $\mathscr{V}$ is a symmetric monoidal closed category and $T:\mathscr{A}\otimes\mathscr{B}\to\mathscr{C}$ is a $\mathscr{V}$-functor, then*

$$T(A\,-):\mathscr{B}(BB')\to\mathscr{C}(T(AB),T(AB'))$$

and

$$T(-\,B):\mathscr{A}(AA')\to\mathscr{C}(T(AB),T(A'B))$$

are $\mathscr{V}$-natural in A and B respectively, with respect to the bifunctors

$$\mathscr{A}^*\otimes\mathscr{A}\xrightarrow[T(-\,B)^*\otimes T(-\,B')]{}\mathscr{C}^*\otimes\mathscr{C}\xrightarrow[\mathrm{Hom}\,\mathscr{C}]{}\mathscr{V},$$

$$\mathscr{B}^*\otimes\mathscr{B}\xrightarrow[T(A\,-)^*\otimes T(A'\,-)]{}\mathscr{C}^*\otimes\mathscr{C}\xrightarrow[\mathrm{Hom}\,\mathscr{C}]{}\mathscr{V}.$$

Proof. VN′ for $T(A\,-)$ is (7.1), and for $T(-\,B)$ is (7.2).

Remark 7.3. As we noted in Remark 4.3, we needed (4.4), or equivalently (7.1) or (7.2), only to get VF2 for T. Now VF2 for T is not involved in the definitions of extraordinary $\mathscr{V}$-natural transformations in § 5, nor in Propositions 5.2, 5.3, 5.4. We can therefore use all of these before proving Theorem 6.4 and 6.9. Indeed, we shall prove these precisely by appealing to (7.1) and (7.2), stated in terms of $\mathscr{V}$-naturality; the proofs are contained in the following theorem. The bifunctors, or in the first instance their partial functors, with respect to which the stated $\mathscr{V}$-naturality obtains, are obvious compositions of H, K, L, and R.

Theorem 7.4. *If $\mathscr{V}$ is a symmetric monoidal closed category, the morphisms a, r, l, c, p, t, u, H, K are $\mathscr{V}$-natural in every variable. Moreover if $\mathscr{A}$ is a $\mathscr{V}$-category, the morphisms M, L, R, j are $\mathscr{V}$-natural in every variable.*

Remark 7.5. The $\mathscr{V}$-naturality in A of H^A and of L^A establishes Theorems 6.4 and 6.9 by means of Proposition 7.1.

Proof of Theorem 7.4. We first observe that c_{AB} is $\mathscr{V}$-natural in both variables, for since $c^2 = 1$ the definition (6.7) of H may be interpreted as VN for c in either variable. We next prove the $\mathscr{V}$-naturality of t_{BC}; it is $\mathscr{V}$-natural in C by Theorem II.7.1. Consider the diagram:

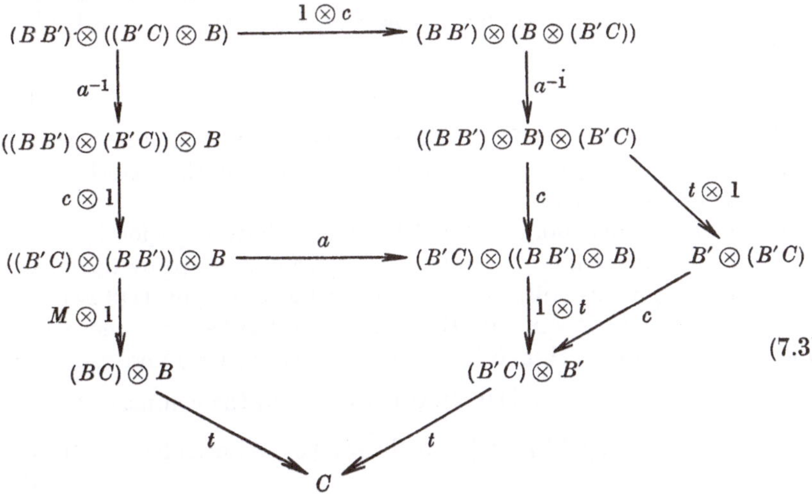

(7.3)

The top region commutes by coherence, and the right region by the naturality of c. The bottom region would commute by VC3′ for $\mathscr{V}$ if we had (IB), (IB'), (IC) in place of B, B', C at the extreme right of each object, and M in place of t; it therefore commutes by II(7.3).

Now apply π to each leg of (7.3). We get

$$\begin{aligned}
\pi(tc(t\otimes 1)a^{-1}(1\otimes c)) &= (1,t)(c,c)\pi((t\otimes 1)a^{-1}) && \text{by II(3.1)}\\
&= (1,t)(c,c)p^{-1}\pi\pi(t\otimes 1) && \text{by II(3.19)}\\
&= (1,t)(c,c)p^{-1}\pi(ut) && \text{by II(3.1)}\\
& && \text{with } x=1\\
&= (1,t)(c,c)p^{-1}(1,u) && \text{by II(3.1)}\\
& && \text{since } \pi t = 1\\
&= (1,t)(c,c)K && \text{by II(7.1)}\\
&= (1,t)H && \text{by (6.7);}
\end{aligned}$$

and

$$\begin{aligned}
\pi(t(M\otimes 1)(c\otimes 1)a^{-1}) &= \pi(\pi^{-1}(Mc).a^{-1}) && \text{by II(3.5)}\\
&= p^{-1}\pi\pi(\pi^{-1}(Mc)) && \text{by II(3.19)}\\
&= p^{-1}\pi(Mc) && \\
&= p^{-1}R && \text{by (6.3)}\\
&= (1,t)KR && \text{by II(7.2);}
\end{aligned}$$

the resulting diagram is precisely VN'' for the $\mathscr{V}$-naturality of t_{BC} in B. Note that by stopping one line before the end of the above calculation we have

$$(1,t)H = p^{-1}R. \tag{7.4}$$

We now prove the $\mathscr{V}$-naturality of u_{AB}; it is $\mathscr{V}$-natural in A by Theorem II.7.1. For its $\mathscr{V}$-naturality in B, VN' is the exterior of the diagram: (see page 545):

The top left region commutes by (7.4), and the bottom region by II(3.7). The other regions, reading from left to right, commute (i) by the naturality of H, (ii) trivially, (iii) by the naturality of p, (iv) by II(3.21).

We turn to L, R, and M. Because L^A and R^A are $\mathscr{V}$-functors, L^A_{BC} and R^A_{BC} are $\mathscr{V}$-natural in B and C by Proposition I.8.3 and Proposition 6.2. Since $M = \pi^{-1} L$ by II(6.2), it is by II(3.5) the composite:

$$M^B_{AC} : \mathscr{A}(BC) \otimes \mathscr{A}(AB) \xrightarrow[L\otimes 1]{} (\mathscr{A}(AB), \mathscr{A}(AC)) \otimes \mathscr{A}(AB) \xrightarrow[t]{} \mathscr{A}(AC). \tag{7.5}$$

Since t is $\mathscr{V}$-natural in everything and $L\otimes 1$ is $\mathscr{V}$-natural in everything except A, M^B_{AC} is $\mathscr{V}$-natural in B and C. (We are implicitly using Propositions 5.2 and 5.3, and Proposition 6.8. We continue to use these implicitly, as well as I(9.7), Proposition II.7.2, and Proposition 6.3; note that these last and Proposition 6.8 become subsumed under Proposition

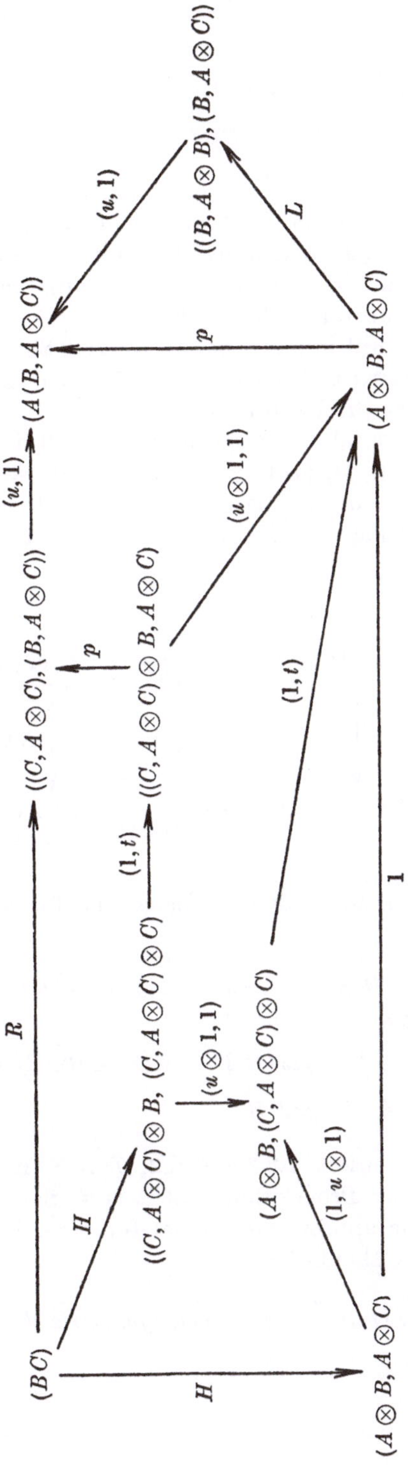
(BC)
H
R
H
(A ⊗ B, A ⊗ C)
((C, A ⊗ C) ⊗ B, (C, A ⊗ C) ⊗ C)
(u ⊗ 1, 1)
(A ⊗ B, (C, A ⊗ C) ⊗ C)
(1, u ⊗ 1)
1
(1, t)
((C, A ⊗ C), (B, A ⊗ C))
((C, A ⊗ C) ⊗ B, A ⊗ C)
p
(1, t)
(u ⊗ 1, 1)
(u, 1)
(A(B, A ⊗ C))
p
(A ⊗ B, A ⊗ C)
(u, 1)
L
((B, A ⊗ B), (B, A ⊗ C))

4.19 only *after* we know that we have bifunctors Hom $\mathscr{A}$ and Ten.) We also have $M = \pi^{-1} R \,.\, c$ by (6.3), so that M is the composite:

$$M^B_{AC} : \mathscr{A}(BC) \otimes \mathscr{A}(AB) \xrightarrow[c]{} \mathscr{A}(AB) \otimes \mathscr{A}(BC) \to$$
$$\xrightarrow[R \otimes 1]{} (\mathscr{A}(BC), \mathscr{A}(AC)) \otimes \mathscr{A}(BC) \xrightarrow[t]{} \mathscr{A}(AC)\,. \tag{7.6}$$

Since t and c are $\mathscr{V}$-natural in everything, and $R \otimes 1$ in everything except C, M^B_{AC} is $\mathscr{V}$-natural in B and A. Thus M is $\mathscr{V}$-natural in all variables. Now $L = \pi M$, and so by II(3.4) we have $L = (1, M)\, u$; thus L is $\mathscr{V}$-natural in all variables since M and u are. Similarly $R = \pi(Mc) = (1, Mc)\, u$ is $\mathscr{V}$-natural in all variables.

p is now $\mathscr{V}$-natural in all variables by II(3.21), then K by II(7.1) (the inverses of *ordinary* $\mathscr{V}$-natural transformations are $\mathscr{V}$-natural by Theorem I.10.2), then H by (6.7).

Since i is $\mathscr{V}$-natural by Proposition I.8.5, it follows from MCC2, MCC3, MCC4 that $(l, 1)$, $(a, 1)$, $(r, 1)$ are $\mathscr{V}$-natural in every variable. The $\mathscr{V}$-naturality of l, a, r themselves now follows by Proposition I.10.10.

Finally we consider j. VN′ for j is the exterior of the following diagram:

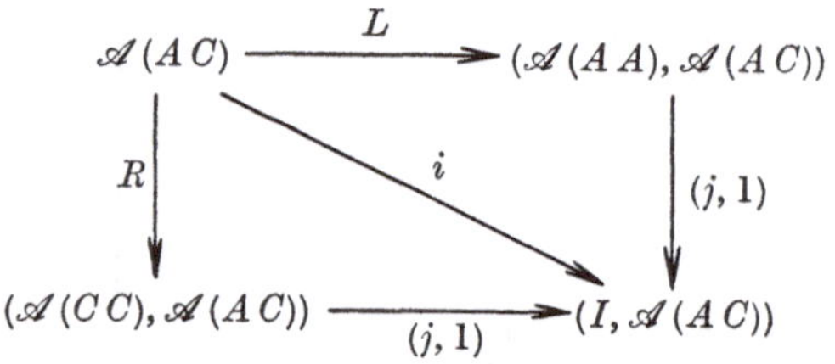

The two regions commute by VC2 for $\mathscr{A}$ and for $\mathscr{A}^*$. This completes the proof.

Proposition 7.6. *If* $\mathscr{V}$ *is a symmetric monoidal closed category and* $T : \mathscr{A} \otimes \mathscr{B} \to \mathscr{C}$ *is a* $\mathscr{V}$*-functor, then*

$$T : \mathscr{A}(AA') \otimes \mathscr{B}(BB') \to \mathscr{C}(T(AB), T(A'B'))$$

is $\mathscr{V}$*-natural in* A, A', B, *and* B'.

Proof. By Propositions I.8.3 and 6.2, T is $\mathscr{V}$-natural in $[AB]$ and $[A'B']$. Therefore by Proposition 4.12 it is $\mathscr{V}$-natural in A, B, A', B' with respect to the appropriate partial functors. By Proposition 6.10 these are what they should be.

Proposition 7.7. *If* $\Phi : \mathscr{V} \to \mathscr{V}'$ *is a symmetric closed functor,* $\hat{\phi}_{AB}$ *and*

$\tilde{\phi}_{AB}$ *are* $\mathscr{V}'$*-natural in all variables. More precisely we have for fixed* A*:*

$$\hat{\phi}_A : \hat{\Phi} \,.\, \Phi_* L^A \to L'^{\Phi A} \,.\, \hat{\Phi} : \Phi_* \mathscr{V} \to \mathscr{V}',$$

$$\tilde{\phi}_A : H'^{\Phi A} \,.\, \hat{\Phi} \to \hat{\Phi} \,.\, \Phi_* H^A : \Phi_* \mathscr{V} \to \mathscr{V}',$$

with similar results for fixed B.

Proof. Since $\hat{\phi}_{AB}$ are the components of the $\mathscr{V}'$-functor $\hat{\Phi} : \Phi_* \mathscr{V} \to \mathscr{V}'$, the result for $\hat{\phi}$ follows by Propositions I.8.3 and 6.2, in view of I(6.15).

Now consider the diagram of MCF3; p', $(1, \hat{\phi})$, and $\hat{\phi}$ are $\mathscr{V}'$-natural in all variables, and so is ϕp by Proposition I.8.8. We deduce the $\mathscr{V}'$-naturality of $(\tilde{\phi}, 1)\, \hat{\phi}$, and so by composition that of $(\tilde{\phi}, 1)\, \hat{\phi} \,.\, \phi H$. Since $\hat{\phi} \,.\, \phi H$ are the components of $\hat{\Phi} \,.\, \Phi_* H^A$, the $\mathscr{V}'$-naturality of $\tilde{\phi}$ follows from Proposition I.10.10.

Lemma 7.8. *Let* $\mathscr{V}$ *be a symmetric monoidal closed category and let* $P, Q : \mathscr{A} \to \mathscr{B}$ *and* $T : \mathscr{A}^* \otimes \mathscr{A} \to \mathscr{B}$ *be* $\mathscr{V}$*-functors. Consider families of morphisms*

$$\alpha_A : PA \to QA,$$

$$\gamma_A : B \to T(AA),$$

$$\delta_A : T(AA) \to B.$$

Then the $\mathscr{V}$*-naturality of* α, γ, δ *is equivalent to that of the families*

$$\iota\alpha_A : I \to \mathscr{B}(PA, QA),$$

$$\iota\gamma_A : I \to \mathscr{B}(B, T(AA)),$$

$$\iota\delta_A : I \to \mathscr{B}(T(AA), B).$$

Proof. We give the proof for α, leaving the others to the reader. Since by I(7.14) $\iota\alpha$ is the composite

$$I \xrightarrow{j} \mathscr{B}(PA, PA) \xrightarrow{\mathscr{B}(1,\alpha)} \mathscr{B}(PA, QA),$$

the $\mathscr{V}$-naturality of $\iota\alpha$ follows from that of α in view of the $\mathscr{V}$-naturality of j.

Now suppose that $\iota\alpha$ is $\mathscr{V}$-natural. By Proposition I.10.10, to prove the $\mathscr{V}$-naturality of α it suffices to prove the $\mathscr{V}$-naturality in C of the composite

$$\mathscr{A}(AC) \xrightarrow{P} \mathscr{B}(PA, PC) \xrightarrow{\mathscr{B}(1,\alpha)} \mathscr{B}(PA, QC),$$

and since i is $\mathscr{V}$-natural, it suffices to prove the $\mathscr{V}$-naturality in C of the composite

$$\mathscr{A}(AC) \xrightarrow{P} \mathscr{B}(PA, PC) \xrightarrow{\mathscr{B}(1,\alpha)} \mathscr{B}(PA, QC) \xrightarrow{i} (I, \mathscr{B}(PA, QC)). \quad (7.7)$$

Now (7.7) is certainly $\mathscr{V}$-natural in A. If we put $A = C$, apply V, and evaluate at 1_C we get $\iota\alpha_C$. Thus by Theorem I.10.8 (with $\mathscr{A}^*$ in place of $\mathscr{A}$!), (7.7) is the composite

$$\mathscr{A}(A\,C) \xrightarrow[P]{} \mathscr{B}(PA, PC) \xrightarrow[R]{} (\mathscr{B}(PC, QC), \mathscr{B}(PA, QC)) \rightarrow \xrightarrow[(\iota\alpha, 1)]{} (I, \mathscr{B}(PA, QC)). \tag{7.8}$$

By the $\mathscr{V}$-naturality of $\iota\alpha$, this is $\mathscr{V}$-natural in C, as required.

We now prove the $\mathscr{V}$-analogue of Proposition I.1.2:

Proposition 7.9. *Let $\mathscr{V}$ be a symmetric monoidal closed category and let $T: \mathscr{D} \otimes \mathscr{A} \to \mathscr{B}$, $P: \mathscr{C} \otimes \mathscr{D}^* \to \mathscr{A}$, $Q: \mathscr{C} \to \mathscr{B}$ be $\mathscr{V}$-functors. Let*

$$q_{CDA}: \mathscr{A}(P(C\,D), A) \to \mathscr{B}(QC, T(DA))$$

be a family of morphisms, $\mathscr{V}$-natural in A for each C, D; and in the language of Theorem I.10.8 let $\Gamma' q_{CD}$ be

$$\theta_{CD}: QC \to T(D, P(C\,D)).$$

Then q is $\mathscr{V}$-natural in C (resp. D) if and only if θ is.

Proof. q as $\Omega'\theta$ is the composite

$$\mathscr{A}(P(C\,D), A) \xrightarrow[T(D-)]{} \mathscr{B}(T(D, P(C\,D)), T(D\,A)) \xrightarrow[\mathscr{B}(\theta, 1)]{} \mathscr{B}(QC, T(D\,A))$$

and so is $\mathscr{V}$-natural if θ is. Similarly $\iota\theta$ is the composite

$$I \xrightarrow[j]{} \mathscr{A}(P(C\,D), P(C\,D)) \xrightarrow[q]{} \mathscr{B}(QC, T(D, P(C\,D)))$$

and is $\mathscr{V}$-natural if q is. Lemma 7.8 then gives the $\mathscr{V}$-naturality of θ.

Chapter IV

Examples

1. Elementary Examples

We have seen that the category $\mathscr{S}$ of sets admits an obvious structure of symmetric monoidal closed category, and that $\mathscr{S}$-categories are ordinary categories, etc.

The category $\mathscr{P}_0$ of pointed sets has as objects sets A with a distinguished element a_0 and as morphisms maps $f: A \to B$ with $fa_0 = b_0$. If we take for the tensor product the "smash" product $A \times B$, consisting of the cartesian product $A \times B$ with $A \times b_0 \cup a_0 \times B$ shrunk to a single point, $\mathscr{P}_0$ becomes a symmetric monoidal category $\mathscr{P}$. It is closed, (AB) being $\mathscr{P}_0(AB)$ with the distinguished element $f_0: A \to B$ where $f_0 A = b_0$; the basic functor $P: \mathscr{P}_0 \to \mathscr{S}$ is the forgetful functor assigning to each pointed set its underlying set, and I is a set with two points, one of them distinguished. A $\mathscr{P}$-category $\mathscr{A}$ is a pointed category, i.e. one in which each $\mathscr{A}(A\,B)$ has a distinguished element 0 such that $f0g = 0$; and a $\mathscr{P}$-functor $T : \mathscr{A} \to \mathscr{B}$ is one such that $T\,0 = 0$. Since P is faithful, a $\mathscr{P}$-natural transformation is just a natural transformation.

The category $\mathscr{G}_0$ of abelian groups admits a symmetric monoidal closed structure $\mathscr{G}$ that is too familiar to need description; indeed it is by analogy with the situation here that we use the name "tensor product" for $\otimes$ in any monoidal category. $\mathscr{G}$-categories are just pre-additive categories, and $\mathscr{G}$-functors are additive functors, i.e. those for which $T(f+g) = Tf + Tg$.

In the same way we get the symmetric monoidal closed category $\mathscr{M}K$ of modules over a commutative ring K; $\otimes$ is now $\otimes_K$, I is K, and the morphisms $f : A \to B$ form a K-module (AB). A ring-morphism $L \to K$ induces a symmetric closed functor $\mathscr{M}K \to \mathscr{M}L$, and in particular we have the forgetful closed functor $\mathscr{M}K \to \mathscr{M}\mathbf{Z} = \mathscr{G}$. (In future, when only symmetric monoidal closed categories are in question, "closed functor" shall mean "symmetric closed functor" unless the contrary is stated.) We also have forgetful closed functors $\mathscr{G} \to \mathscr{P} \to \mathscr{S}$; all of these are normal, so that the composite $\mathscr{M}K \to \mathscr{G} \to \mathscr{P} \to \mathscr{S}$ is the basic closed functor $\mathscr{M}K \to \mathscr{S}$.

In our definition of closed category we began with an ordinary category $\mathscr{V}_0$ and a functor $V : \mathscr{V}_0 \to \mathscr{S}$. The reader will guess that we might instead lay down a basic symmetric monoidal closed category $\mathscr{W}$ in place of $\mathscr{S}$, and then define a *closed category over* $\mathscr{W}$, starting with a $\mathscr{W}$-category $\mathscr{V}_0$ and a $\mathscr{W}$-functor $V : \mathscr{V}_0 \to \mathscr{W}$, and taking all the data to be $\mathscr{W}$-functors and $\mathscr{W}$-natural transformations. This is the case, and we shall show in a later paper that to give a symmetric monoidal closed category $\mathscr{V}$ *over* $\mathscr{W}$ is the same thing as to give a symmetric monoidal closed category $\overline{\mathscr{V}}$ and a *normal* closed functor $\Phi : \overline{\mathscr{V}} \to \mathscr{W}$; then $\mathscr{V}_0$ is the $\mathscr{W}$-category $\Phi_* \overline{\mathscr{V}}$, and V is $\hat{\Phi}$. Thus $\mathscr{M}K$ may be considered as a closed category over $\mathscr{G}$, or for that matter over $\mathscr{P}$; and many other examples will appear below.

All of the above examples fit into the class of "algebraic categories" in the sense of LAWVERE [*11*], also known as "varieties" or as "equational categories". The following considerations are due to LINTON [*13*]. (Cf. also FREYD [*8*].)

An algebraic category $\mathscr{K}_0$ comes equipped with a faithful forgetful functor K into $\mathscr{S}$, which has an adjoint F, where FX is the free algebra on the set X. Thus K admits a representation $\iota : KA \to \mathscr{K}_0(IA)$ where I is the free algebra on one generator. Let us call $\mathscr{K}_0$ *commutative* if, for each n-ary operation t of the algebraic theory (we allow n to be any cardinal) and each algebra A, the map $t : (KA)^n \to KA$ is a morphism

$A^n \to A$ in $\mathscr{K}_0$; this means that if s is any m-ary operation of the theory we have, with an obvious notation, commutativity in the diagram

$$\begin{array}{ccc} (KA)^{mn} & \xrightarrow{\;s^n\;} & (KA)^n \\ \Big\downarrow{\scriptstyle t^m} & & \Big\downarrow{\scriptstyle t} \\ (KA)^m & \xrightarrow[\;s\;]{} & KA \end{array} \tag{1.1}$$

It is easy to see that the set $\mathscr{K}_0(AB)$ forms a subalgebra (AB) of the cartesian power algebra B^{KA}, for all A and B, if and only if $\mathscr{K}_0$ is commutative.

Supposing henceforth $\mathscr{K}_0$ commutative, define (AB) as above; it is clearly a functor, and satisfies CC0. Moreover ι is now a natural isomorphism $i: A \to (IA)$. The set $\mathscr{K}_0(A, (BC))$ may be identified with the set of *bimorphisms* $f: KA \times KB \to KC$, a bimorphism being a map f for which the two partial maps $f(a-)$ and $f(-b)$ are, for each $a \in KA$ and $b \in KB$, morphisms in $\mathscr{K}_0$. It is further clear that there is a bijection between bimorphisms $f: KA \times KB \to KC$ and morphisms $g: A \otimes B \to C$, where $A \otimes B$ is a suitable quotient algebra of $F(KA \times KB)$ (impose upon the latter the relations ensuring "bilinearity"). There results an isomorphism

$$\pi: \mathscr{K}_0(A \otimes B, C) \to \mathscr{K}_0(A(BC))$$

which is at once seen to be a natural isomorphism of algebras

$$p: (A \otimes B, C) \to (A(BC)).$$

Since K is faithful, we have by Theorem II.5.10 a monoidal closed structure $\mathscr{K}$ on $\mathscr{K}_0$, which is moreover clearly symmetric by the definition of $\otimes$. Such a $\mathscr{K}$ will be called an *algebraic closed category.* A less elementary example is given by Mac Lane's theory of affine modules ([*4*], Chapter XII).

Note that it is not true in a general symmetric monoidal closed category $\mathscr{V}$ that the tensor product is the universal object for bimorphisms; this cannot be the case unless V is faithful, and need not be the case then, as is shown by the example of quasi-topological spaces in § 2 below.

2. Cartesian Closed Categories

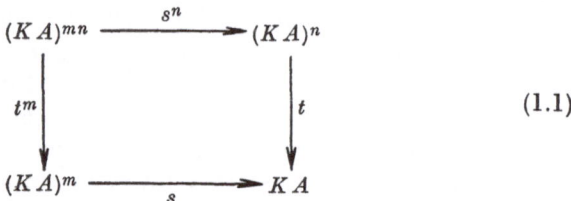

Any category $\mathscr{V}_0$ that admits finite products (including the product of *no* objects, i.e. a terminal object I) admits a structure of symmetric monoidal category $\mathscr{V}$ in which $A \otimes B$ is taken to be $A \times B$; for the canonical isomorphisms

$$(A\times B)\times C\cong A\times(B\times C),\ I\times A\cong A,\ A\times I\cong A,\ A\times B\cong B\times A$$

are easily seen to be coherent. Such a symmetric monoidal category is said to be *cartesian*. We are of course supposing that for each A, B a definite product $A\times B$ with its projections is chosen; different choices would replace $\mathscr{V}$ by an isomorph.

If a symmetric monoidal category is given, it is easily seen that its monoidal structure coincides with some cartesian structure if and only if the following two conditions are satisfied:

(i) I is terminal, so that for each A there is a unique $\theta: A\to I$;

(ii) the morphisms $A\otimes B\xrightarrow[1\otimes\theta]{} A\otimes I\xrightarrow[r]{} A$ and $A\otimes B\xrightarrow[\theta\otimes 1]{} I\otimes B\xrightarrow[l]{} B$ are the projections of a product.

In particular a and c are then uniquely determined.

By Theorem II.5.9, a cartesian monoidal category $\mathscr{V}$ is closed (or is so after replacing $\mathscr{V}_0$ by an isomorph) if and only if the functor $-\times B$ has a coadjoint. Thus for $\mathscr{V}$ to be closed it is necessary that $-\times B$ preserve colimits, which places severe restrictions on $\mathscr{V}_0$.

In particular, if $\mathscr{V}_0$ has an initial object O, it is a colimit, and so we must have $O\times B=O$. If $\mathscr{V}_0$ is pointed, i.e. if $O\cong I$, we then have

$$B\cong I\times B\cong O\times B\cong O,$$

so that every object is initial. Thus the only cartesian closed categories that are pointed are those equivalent to the unit category $\mathscr{I}$ with a single object and a single morphism. The closed category $\mathscr{S}$ is cartesian, but the closed categories $\mathscr{P}$, $\mathscr{G}$, $\mathscr{M}K$ of § 1, being pointed, are not.

A prime example of a cartesian closed category is the category of small categories. Let $\mathscr{C}_0$ be the category with small categories A as its objects and functors $T: A\to B$ as its morphisms, and give it the cartesian monoidal structure $\mathscr{C}$. Then $\mathscr{C}$ is closed, for we get an adjunction $\pi:\mathscr{C}_0(A\times B, C)\cong\mathscr{C}_0(A(BC))$ if we take (BC) to be the functor category whose objects are functors $T: B\to C$ and whose morphisms are natural transformations $\alpha: T\to S$. The basic functor $C:\mathscr{C}_0\to\mathscr{S}$ sends the category A to its set of objects, and I is the category with one object and one morphism.

One easily verifies that $\mathscr{C}$-categories, $\mathscr{C}$-functors, and $\mathscr{C}$-natural transformations are precisely hypercategories, hyperfunctors, and hypernatural transformations. $\mathscr{C}$ itself is a $\mathscr{C}$-category, and hence a hypercategory.

If we ignore considerations of smallness and legitimacy, we can identify $\mathscr{C}$ with $\mathscr{C}at$, which is now a closed "category". The hypercategory $\mathscr{S}_*$ is $\mathscr{C}at$ *qua* hypercategory, while the symmetric monoidal category $\mathscr{S}_{\#}$ (cf. Proposition III.3.4) is $\mathscr{C}at$ *qua* closed category. Then $\mathscr{C}at_*$ is the hypercategory $\mathscr{H}yp$, while the symmetric monoidal category $\mathscr{C}at_{\#}$ is in fact a cartesian closed category which we shall still call $\mathscr{H}yp$. The basic closed functor $C : \mathscr{C}at \to \mathscr{S}$ induces a hyperfunctor $C_* : \mathscr{H}yp \to \mathscr{C}at$ whose effect is to ignore the hypermorphisms; it is in fact a normal closed functor (Proposition III.3.6) and exhibits $\mathscr{H}yp$ as a closed category *over* $\mathscr{C}at$.

If in a hypercategory the morphisms are regarded as objects and the hypermorphisms as morphisms there results a category. In this way we get a hyperfunctor $\mathscr{H}yp \to \mathscr{C}at$, and it is easy to see that it is induced by the closed functor $M = (M, \tilde{M}, M^0) : \mathscr{C}at \to \mathscr{S}$ defined as follows. For any category A, MA is the set of all morphisms $f : X \to X'$ in A. If $g : Y \to Y'$ is an element of MB then $\tilde{M} : MA \times MB \to M(A \times B)$ maps the pair (f, g) to the morphism $(f, g) : (X, X') \to (Y, Y')$ in $A \times B$. $M^0 : * \to MI$ is uniquely defined since I has only one morphism.

Any category can be made into a hypercategory by giving it identities as its only hypermorphisms. The resulting hyperfunctor $\mathscr{C}at \to \mathscr{H}yp$ is induced by a closed functor $D : \mathscr{S} \to \mathscr{C}at$, where DX is the discrete category based on X and $\tilde{D}$, D^0 are suitably defined.

Just as a class has only objects; a category has also morphisms between objects; and a hypercategory has also hypermorphisms between morphisms; so one may define an n-category with morphisms of every type i, $1 \leqq i \leqq n$, a morphism of type i connecting two of type $i - 1$ (cf. Ehresmann [6]). Then n-categories form a cartesian closed category $\mathscr{C}^n$, and $\mathscr{C}^n_{\#} = \mathscr{C}^{n+1}$; in particular $\mathscr{C}^0 = \mathscr{S}$, $\mathscr{C}^1 = \mathscr{C}at$, $\mathscr{C}^2 = \mathscr{H}yp$. Forgetting the morphisms of type n gives a normal closed functor $\mathscr{C}^n \to \mathscr{C}^{n-1}$, so that $\mathscr{C}^n$ may be regarded as a closed category over $\mathscr{C}^{n-1}$. Note that there are many kinds of contravariance for $\mathscr{C}^n$-functors; for $\mathscr{C}^n$, besides its duality involution D, inherits involutions from the D's of the $\mathscr{C}^i$ with $i < n$ (cf. Remark III.2.12).

Another interesting example of a cartesian closed category is that of simplicial sets (i.e. complete semi-simplicial complexes). This is best viewed as a functor category, and as such will be treated in a later paper.

If $\mathscr{W}_0$ is any category admitting finite products then, although the cartesian monoidal structure on $\mathscr{W}_0$ may not be closed, it may be possible to find a full product-preserving embedding of $\mathscr{W}_0$ into a category $\mathscr{V}_0$ whose cartesian monoidal structure is closed. An example of this is Spanier's [16] embedding of the category of topological spaces in that of quasi-topological spaces, which is a cartesian closed category. The "compactly defined" hausdorff spaces (sometimes called k-spaces; cf. Ronald Brown [5]) form a full closed cartesian subcategory. A detailed examination and generalization of Spanier's construction will appear elsewhere.

3. Closed Categories with one Object

Let M be an abelian monoid, written multiplicatively, and let $\mathscr{V}_0$ be the category with a single object I and with M for the monoid $\mathscr{V}_0(I\,I)$ of endomorphisms of I. Define $I \otimes I = I$, and $f \otimes g = fg$ for $f, g \in M$; then $\otimes$ is a functor, and gives $\mathscr{V}_0$ the structure of a symmetric monoidal category $\mathscr{V}$ if we take a, l, r, c all to be 1. $\mathscr{V}$ is in fact closed, with $(I\,I) = I$ and $(f, g) = fg$; it suffices to take 1 for the adjunction π. Then i, L, p all turn out to be 1, and $V : \mathscr{V}_0 \to \mathscr{S}$ is given by $VI = M$ (regarded as a set) and $(Vf)\, g = fg$.

It is an easy exercise to show that any closed category $\mathscr{V}$ with a single object must be isomorphic to that constructed above for some M. If M consists of the identity alone, we obtain the closed category $\mathscr{I}$ with one object and one morphism.

With $\mathscr{V}$ as above let $\mathscr{G}$ be the closed category of abelian groups and consider the (not necessarily symmetric) closed functors

$$\Phi = (\phi, \tilde{\phi}, \phi^0) : \mathscr{V} \to \mathscr{G}.$$

First, ϕI is to be some abelian group A; and to be a functor, ϕ must map the monoid M into the monoid of endomorphisms of A; let us write fa for $(\phi f)a$, where $f \in M$ and $a \in A$. Next we have $\tilde{\phi} : \phi I \otimes \phi I \to \to \phi(I \otimes I)$, that is, $\tilde{\phi} : A \otimes A \to A$; write ab for $\tilde{\phi}(a \otimes b)$. The naturality of $\tilde{\phi}$ is expressed by: $(fa)\,(gb) = (fg)\,(ab)$, for $f, g \in M$ and $a, b \in A$. Finally we have $\phi^0 \colon \mathbf{Z} \to \phi I$; write 1 for $\phi^0 1 \in A$. The axioms MF1—MF3 give $1a = a$, $a1 = a$, and $(ab)\,c = a(bc)$. Thus a closed functor $\Phi : \mathscr{V} \to \mathscr{G}$ is just an algebra over the monoid ring $\mathbf{Z}(M)$ of M, and the closed functor Φ is symmetric if and only if this algebra is commutative. One easily verifies that a closed natural transformation $\Phi \to \Psi$ corresponds to a morphism of $\mathbf{Z}(M)$-algebras.

One can generalize by considering closed functors $\mathscr{V} \to \mathscr{W}$, where $\mathscr{W}$ is any (not necessarily symmetric monoidal) closed category, and so obtain what we might call a $\mathscr{W}$-algebra over $\mathbf{Z}(M)$. Again we may suppose that M is itself a ring, so that $\mathscr{V}_0$ is pre-additive; if we restrict $\Phi : \mathscr{V} \to \mathscr{G}$ to be additive, it corresponds to an M-algebra.

In particular, closed functors $\mathscr{I} \to \mathscr{G}$ correspond to rings, and closed functors $\mathscr{I} \to \mathscr{S}$ to monoids.

4. Ordered Sets

Any full subcategory of $\mathscr{S}$ that is closed under $A \times B$ and (AB) has a cartesian closed structure consistent with that of $\mathscr{S}$. Thus we get the closed category of finite sets; and if n is any integer > 1 or is ∞, we get a still smaller closed category by excluding all sets of cardinal c with $1 < c < n$.

Taking $n = \infty$ gives the closed category of sets with at most one element, a closed category that is at once cartesian and algebraic. For our purposes it is more convenient to replace this category by a skeleton, namely the full subcategory of $\mathscr{S}$ determined by the empty set $\emptyset$ and a fixed one-element set $*$ whose only member is also called $*$. This category admits the cartesian monoidal structure given by $* \times * = *$, $A \times B = \emptyset$ otherwise; but to make it closed we must replace it by an isomorph, which we do by relabelling each of its three morphisms $\emptyset \to \emptyset \to * \to *$ by the same symbol, namely $*$. Then we can take $(* \, \emptyset) = \emptyset$ and $(A\,B) = *$ otherwise, and we get a closed category $\mathscr{T}$ (for "tiny"). $T: \mathscr{T}_0 \to \mathscr{S}$ is given by $T\,\emptyset = \emptyset$ and $T\,* = *$, and thanks to our relabelling we have $T(A\,B) = \mathscr{T}_0(AB)$ as required.

It is clear that a $\mathscr{T}$-category is a category $\mathscr{A}$ in which each $\mathscr{A}(AB) = \emptyset$ or $*$, and that a $\mathscr{T}$-functor is just a functor. In such a category all diagrams commute, and any category in which all diagrams commute becomes a $\mathscr{T}$-category when we relabel all its morphisms with the same symbol $*$. If we write $A < B$ whenever $\mathscr{A}(AB) = *$, we see that a small $\mathscr{T}$-category $\mathscr{A}$ is the same thing as a pre-ordered set, i.e. a set $\mathscr{A}$ with a binary relation $A < B$ satisfying

$$A < B \quad \text{and} \quad B < C \quad \text{imply} \quad A < C,$$
$$A < A;$$

while a $\mathscr{T}$-functor is an order-preserving map.

If $A < B$ and $B < A$ then A and B are isomorphic and we write $A \sim B$; if $A \sim B$ implies $A = B$, the pre-order is an order and the category $\mathscr{A}$ is skeletal. The passage from a pre-ordered set $\mathscr{A}$ to the associated ordered set $\overline{\mathscr{A}}$, consisting in factoring out the equivalence relation $A \sim B$, corresponds to the passage from the category $\mathscr{A}$ to a skeleton $\overline{\mathscr{A}}$.

A monoidal structure on a given pre-ordered set (i.e. small $\mathscr{T}$-category) $\mathscr{V}_0$ is determined by a function $A \otimes B$ and an object I of $\mathscr{V}_0$; the fact that $\otimes$ is a functor is expressed by the condition

$$A < B \quad \text{implies} \quad A \otimes C < B \otimes C \quad \text{and} \quad C \otimes A < C \otimes B, \tag{4.1}$$

while the existence of a, r, l (which are then unique and coherent) is expressed by

$$(A \otimes B) \otimes C \sim A \otimes (B \otimes C), \quad A \otimes I \sim A, \quad I \otimes A \sim A. \tag{4.2}$$

This monoidal category has a normalization given by

$$VA = * \quad \text{if} \quad I < A, \quad VA = \emptyset \quad \text{otherwise}. \tag{4.3}$$

For it to be closed we need a function (BC) satisfying

$$A \otimes B < C \quad \text{if and only if} \quad A < (BC)\,; \tag{4.4}$$

condition (ii) of Theorem II.5.5 is automatically satisfied. The existence of i, L, p now implies

$$A \sim (IA)\,, \tag{4.5}$$

$$(BC) < ((AB)(AC))\,, \tag{4.6}$$

$$(A \otimes B, C) \sim (A(BC))\,. \tag{4.7}$$

Clearly a monoidal structure $\mathscr{V}$ on $\mathscr{V}_0$ induces under passage to the quotient a monoidal structure $\overline{\mathscr{V}}$ on the skeleton $\overline{\mathscr{V}}_0$, which is closed if $\mathscr{V}$ is; we have only to replace $\sim$ by $=$ in the above.

As an example let $\mathscr{V}_0$ be both an ordered set and a group, the two structures being related by

$$A < B \quad \text{implies} \quad AC < BC \quad \text{and} \quad CA < CB\,. \tag{4.8}$$

Then if we take $A \otimes B$ to be AB and I to be 1, (4.1) and (4.2) are satisfied, and we have a monoidal structure, which is symmetric if the group is abelian. This monoidal structure is closed, for (4.4) is satisfied with $(BC) = CB^{-1}$. Note the special case when $\mathscr{V}_0$ is given the trivial (i.e. discrete) order.

A further example, suggested by Lawvere, is the following. Let $\mathscr{V}_0$ be a pre-ordered set with finite products; thus there is a greatest element 1 and there is a greatest lower bound $A \wedge B$ of any two elements A, B. Now suppose that the cartesian monoidal structure $\mathscr{V}$ is closed, and write $B \Rightarrow C$ instead of (BC). Then (4.4) becomes

$$A \wedge B < C \quad \text{if and only if} \quad A < B \Rightarrow C\,, \tag{4.9}$$

while its consequences (4.5)—(4.7) become

$$A \sim 1 \Rightarrow A\,, \tag{4.10}$$

$$B \Rightarrow C < (A \Rightarrow B) \Rightarrow (A \Rightarrow C)\,, \tag{4.11}$$

$$(A \wedge B) \Rightarrow C \sim A \Rightarrow (B \Rightarrow C)\,. \tag{4.12}$$

A pre-ordered set with the above properties is called a *Brouwerian logic*, the motivation being as follows. Let $\mathscr{V}_0$ be the set of all sentences in some given first-order theory or some propositional calculus, classical or intuitionistic, and interpret "$A < B$" as "A entails B", "$A \wedge B$" as "A and B", and "$A \Rightarrow B$" as "A implies B".

Assume now that $\mathscr{V}_0$ has a least element 0, and define *negation* as

$$A^{\#} = A \Rightarrow 0\,. \tag{4.13}$$

Then (4.9) with $C = 0$ gives

$$A \wedge B \sim 0 \quad \text{if and only if} \quad A < B^{\#} \tag{4.14}$$

which gives an alternative definition of $B^{\#}$. Since the functor $A \Rightarrow B$ is contravariant in A, we have

$$A < B \quad \text{implies} \quad B^{\#} < A^{\#}. \tag{4.15}$$

The Brouwerian logic is said to be *classical* if it has a least element 0 and if the negation satisfies

$$A^{\#\#} \sim A. \tag{4.16}$$

Theorem 4.1. (LAWVERE). *For a Brouwerian logic $\mathscr{V}_0$ the following three conditions are equivalent:*

(i) *$\mathscr{V}_0$ is classical.*

(ii) *The dual $\mathscr{V}_0^*$ (i.e. the set $\mathscr{V}_0$ with the order reversed) is also a Brouwerian logic and the two negations are isomorphic.*

(iii) *The ordered set $\overline{\mathscr{V}}_0$ associated to the pre-ordered set $\mathscr{V}_0$ is a Boolean algebra.*

Proof. (i) *implies* **(ii).** By (4.15) and (4.16), $\#$ is an order-reversing involution and so $\mathscr{V}_0$ has least upper bounds given by de Morgan's law

$$A \vee B = (A^{\#} \wedge B^{\#})^{\#}$$

and $\mathscr{V}_0^*$ is also a Brouwerian logic. The dual of (4.14) shows that the negation in $\mathscr{V}_0^*$ is again $\#$.

(ii) *implies* **(iii).** By (4.14) applied to $\mathscr{V}_0^*$, which has the same negation as $\mathscr{V}_0$, we have

$$A \vee B \sim 1 \quad \text{if and only if} \quad B^{\#} < A.$$

Combining this with (4.14) we have

$$A \wedge B \sim 0 \quad \text{and} \quad A \vee B \sim 1 \quad \text{if and only if} \quad B^{\#} \sim A.$$

Since the relation on the left between A and B is symmetric, we have

$$B^{\#} \sim A \quad \text{if and only if} \quad A^{\#} \sim B,$$

that is, $A^{\#\#} \sim A$. From this and (4.15), $\#$ is an order-reversing involution, and so we have de Morgan's law

$$(A \vee B)^{\#} = A^{\#} \wedge B^{\#}.$$

From (4.14) with $A = B^{\#}$ we get

$$B \wedge B^{\#} \sim 0.$$

Because $- \wedge B$ has a coadjoint it commutes with coproducts, giving

$$(A \vee C) \wedge B = (A \wedge B) \vee (C \wedge B).$$

Thus we have a Boolean algebra.
(iii) *implies* (i): trivial.

Corollary 4.2. *In a classical logic we have*

$$B \Rightarrow C \sim B^{\#} \vee C.$$

Proof. In a Boolean algebra, $A \wedge B < C$ if and only if $A < B^{\#} \vee C$.

The following is an example of a non-classical Brouwerian logic. Let X be a topological space, and let $\mathscr{V}_0$ be the set of open subsets of X ordered by

$$A < B \quad \text{if and only if} \quad \bar{A} \supset B$$

where $\bar{A}$ denotes the closure of A. Then $A \wedge B$ is $A \cup B$, and

$$A \wedge B < C \quad \text{iff} \quad \overline{A \cup B} \supset C \quad \text{iff} \quad \bar{A} \cup \bar{B} \supset C \quad \text{iff} \quad \bar{A} \supset C - \bar{B};$$

thus (4.9) is satisfied with $C - \bar{B}$ for $B \Rightarrow C$. The greatest element 1 is $\emptyset$ and the least element 0 is X; $A^{\#}$ being $A \Rightarrow 0$ is $X - \bar{A}$. Thus $A^{\#\#}$ is the interior of $\bar{A}$, and is in general different from A.

Finally let $\mathscr{V}_0$ be a linearly ordered set with a greatest element 1. Then it is a Brouwerian logic, for $A \wedge B = \min(A, B)$ and we obtain (4.9) if we set $B \Rightarrow C$ equal to 1 if $B < C$ and equal to C otherwise. If $\mathscr{V}_0$ has a least element 0 we find that $A^{\#} = 0$ if $A \neq 0$ while $0^{\#} = 1$. We have $A^{\#\#} = 0$ or 1, so that $\mathscr{V}_0$ is classical if and only if it has either one or two elements, i.e., if and only if it is either $\mathscr{I}$ or $\mathscr{T}$.

5. Modules over Algebras

Let Λ be an algebra over the commutative ring K and let $\mathscr{V}_0$ be the category of two-sided Λ-modules. For $A, B \in \mathscr{V}_0$ define $A \otimes B$ to be $A \otimes_\Lambda B$, made into a two-sided Λ-module by using the left Λ-operation on A and the right Λ-operation on B. With Λ itself as I and the obvious definitions of a, r, l we obtain a monoidal category $\mathscr{V}$ (not in general symmetric) over $\mathscr{M}K$.

This monoidal structure is closed, for we have $\pi : \mathscr{V}_0(A \otimes B, C) \cong \cong \mathscr{V}_0(A(BC))$, where (BC) is the K-module of those K-morphisms $f : B \to C$ satisfying $f(b\lambda) = (fb)\lambda$ for $b \in B$ and $\lambda \in \Lambda$, made into a two-sided Λ-module by setting

$$(\gamma f \lambda) b = \gamma(f(\lambda b)), \quad b \in B, \quad \gamma, \lambda \in \Lambda.$$

Then i and p turn out to have their expected values, and L to correspond to the usual composition. The basic functor $V : \mathscr{V}_0 \to \mathscr{M}K$ takes $A \in \mathscr{V}_0$ to the K-module $\{a \in A \mid \lambda a = a\lambda \text{ for all } \lambda \in \Lambda\}$.

Now suppose that Λ is a Hopf algebra over K, with co-algebra structure given by algebra-morphisms $\varepsilon : \Lambda \to K$, $\eta : \Lambda \to \Lambda \otimes \Lambda$. Let

$\mathscr{W}_0$ be the category of left Λ-modules, and for $A, B \in \mathscr{W}_0$ define $A \otimes B$ to be $A \otimes_K B$, which is at first a $(\Lambda \otimes \Lambda)$-module, and which we make into a Λ-module by pull-back along η. Similarly make the K-module K into a Λ-module by pull-back along ε. Then the a, r, l of $\mathscr{M}K$ are easily verified to be Λ-morphisms, and define on $\mathscr{W}_0$ a monoidal structure $\mathscr{W}$ with K as I. $\mathscr{W}$ is symmetric if and only if the co-algebra structure of Λ is commutative. Define a normalization $W: \mathscr{W}_0 \to \mathscr{M}K$ of $\mathscr{W}$ by setting

$$WA = \{a \in A \mid \lambda a = (\varepsilon \lambda) a \quad \text{for all} \quad \lambda \in \Lambda\};$$

then we have $\iota: WA \cong \mathscr{W}_0(KA)$ where $(\iota a) k = k a$.

Now the right operation of Λ on itself gives to $\Lambda \otimes B$ the structure of a right Λ-module, and this in turn gives to $\mathscr{W}_0(\Lambda \otimes B, C)$ the structure of a left Λ-module which we call (BC). We have an isomorphism of K-modules

$$\mathscr{W}_0((\Lambda \otimes B) \otimes_\Lambda A, C) \cong \mathscr{W}_0(A, \mathscr{W}_0(\Lambda \otimes B, C)) \tag{5.1}$$

where $(\Lambda \otimes B) \otimes_\Lambda A$ gets its Λ-module structure from the left Λ-module structure of $\Lambda \otimes B$. The right member of (5.1) is $\mathscr{W}_0(A(BC))$; we assert that the left member is isomorphic to $\mathscr{W}_0(A \otimes B, C)$. Indeed the isomorphism

$$(\Lambda \otimes_K B) \otimes_\Lambda A \cong (\Lambda \otimes_\Lambda A) \otimes_K B \cong A \otimes B$$

is an isomorphism of $(\Lambda \otimes \Lambda)$-modules, and so *a fortiori* of Λ-modules. Thus, since $\mathscr{W}_0$ admits transport of structure, $\mathscr{W}$ is closed. (It is in fact necessary to replace the above (BC) by an isomorph in order to get actual equality $W(BC) = \mathscr{W}_0(BC)$.)

6. Complexes and Graded Modules

Let K be a commutative ring. A complex over K is a diagram

$$A: \quad \cdots \longrightarrow A_n \xrightarrow{\ d\ } A_{n-1} \longrightarrow \cdots$$

in the category of K-modules, satisfying $dd = 0$; and a morphism of complexes is a morphism of diagrams, so that we have a category $\mathscr{C}_0 K$. The category $\mathscr{G}_0 K$ of graded K-modules is the full subcategory determined by the complexes A in which $d = 0$.

In $\mathscr{C}_0 K$ we define a tensor product $A \otimes B$ by

$$(A \otimes B)_n = \sum A_p \otimes B_q, \quad p + q = n;$$

$$d(a \otimes b) = da \otimes b + (-1)^p a \otimes db, \quad a \in A_p, \quad b \in B_q.$$

For I we take the complex, denoted by K, which has $K_0 = K$ and $K_n = 0$ for $n \neq 0$. With the obvious definitions of a, r, l we obtain a monoidal category $\mathscr{C}K$ over $\mathscr{M}K$. We easily verify that $\mathscr{C}K$ is closed,

(BC) being given by

$$(BC)_n = \prod (A_p, B_{n+p}), \quad p \in \mathbf{Z};$$
$$(df)_p a = d(f_p a) + (-1)^{p+1} f_{p-1} da, \quad f \in (AB)_n, \quad a \in A_p.$$

Then p and i turn out to have their expected values, and L and M correspond to the usual composition: for instance the n-component of M, mapping $\sum_{p+q=n} (BC)_p \otimes (AB)_q$ into $(AC)_n$, takes $g \otimes f$ to gf, where $(gf)_r = g_{r+p} f_r$ for $g \in (BC)_q$ and $f \in (AB)_p$. The basic functor $\mathscr{C}_0 K \to \mathscr{M} K$ turns out to be Z_0, the functor sending each complex over K to its K-module of 0-cycles; the morphisms $A \to B$ are clearly the 0-cycles of (AB).

The subcategory $\mathscr{G}_0 K$ is closed under $A \otimes B$ and (AB), and so inherits from $\mathscr{C} K$ the structure of a monoidal closed category $\mathscr{G} K$ over $\mathscr{M} K$; the functor $Z_0 : \mathscr{C}_0 K \to \mathscr{M} K$ when restricted to $\mathscr{G}_0 K$ merely sends each graded K-module A to its 0-component.

We now discuss possible symmetries for $\mathscr{C} K$ and $\mathscr{G} K$. If c is such a symmetry, consider the object $K^p \in \mathscr{G} K$ satisfying $K^p_p = K$, $K^p_q = 0$ if $q \neq p$, and let $1^p \in K^p_p$ be the identity of K. Since

$$K^p \otimes K^q \cong K^{p+q} \cong K^q \otimes K^p,$$

it follows that

$$c(1^p \otimes 1^q) = \varepsilon(p,q)\, 1^q \otimes 1^p \quad \text{where} \quad \varepsilon(p,q) \in K.$$

By naturality it follows that $c : A \otimes B \to B \otimes A$ must be given by

$$c(a \otimes b) = \varepsilon(p,q)\, b \otimes a, \quad a \in A_p, \quad b \in B_q.$$

The conditions MC6, MC7 on c now become

$$\varepsilon(p,q)\,\varepsilon(q,p) = 1,$$
$$\varepsilon(p, q+r) = \varepsilon(p,q)\,\varepsilon(p,r).$$

If we set

$$k = \varepsilon(1,1)$$

we have

$$k^2 = 1$$

and

$$c(a \otimes b) = k^{pq}(b \otimes a), \quad a \in A_p, \quad b \in B_q.$$

For the category $\mathscr{G} K$ there are no further conditions, and thus we have one symmetry for every $k \in K$ with $k^2 = 1$; in particular we can take k to be 1 or -1, getting in one case $c(a \otimes b) = b \otimes a$ and in the other $c(a \otimes b) = (-1)^{pq}(b \otimes a)$; if $K = \mathbf{Z}$ these are the only symmetries for $\mathscr{G} K$. For the category $\mathscr{C} K$ however we still must ensure that c commutes with the differentiation d. Taking $p = q = 1$ we have

$$dc(a \otimes b) = kd(b \otimes a) = kdb \otimes a - kb \otimes da;$$
$$cd(a \otimes b) = c(da \otimes b) - c(a \otimes db) = b \otimes da - db \otimes a.$$

Since this is to hold for all a and b we must have $k = -1$; and for $k = -1$ we do in fact have $dc = cd$. Thus $\mathscr{C}K$ has a unique symmetry $c(a \otimes b) = (-1)^{pq} b \otimes a$.

We define three closed functors $\Phi, Z, H : \mathscr{C}K \to \mathscr{G}K$, which are symmetric if $\mathscr{G}K$ is given the symmetry with $k = -1$. The functor

$$\Phi : \mathscr{C}_0 K \to \mathscr{G}_0 K$$

forgets the differential structure, so that ΦA is A considered merely as a graded module; $\tilde{\Phi}$ and Φ^0 are the identity; clearly Φ is not normal. The functor Z (resp. H) assigns to each complex A its cycles ZA (resp. its homology HA) regarded as an object of $\mathscr{G}K$. $\tilde{Z}$ and $\hat{Z}$ are the usual natural transformations

$$ZA \otimes ZB \to Z(A \otimes B), \quad Z(AB) \to (ZA, ZB),$$

and similarly for H; Z^0 and H^0 are the identity. It is clear that Z is normal while H is not.

There is a completely different monoidal closed structure $\mathscr{G}'K$ on $\mathscr{G}_0 K$ given by

$$(A \otimes' B)_n = A_n \otimes B_n,$$
$$(AB)'_n = (A_n, B_n),$$
$$I'_n = K \quad \text{for all } n;$$

the basic functor $\mathscr{G}_0 K \to \mathscr{M} K$ in this case sends A to $\prod A_n$, and is faithful.

7. Simplicial Complexes

A simplicial complex A is a set together with a family of finite subsets of A called the spanning subsets of A; there are two axioms, namely that every subset of a spanning subset spans, and that every single-point subset spans. A morphism or simplicial map $f : A \to B$ is a map of the set A into the set B such that if T spans in A then fT spans in B. There results a category $\mathscr{K}_0$; there is a faithful forgetful functor $K : \mathscr{K}_0 \to \mathscr{S}$ sending the simplicial complex A to A regarded merely as a set, and K is represented by the simplicial complex I consisting of a single point, which is also the terminal object of $\mathscr{K}_0$.

We shall describe three different structures of closed category on $\mathscr{K}_0$, in each of which K is the basic functor. Two of these are symmetric monoidal; the third is an example of a non-monoidal closed category.

For the first, give $\mathscr{K}_0$ the cartesian monoidal structure $\mathscr{K}$; the product $A \times B$ is the simplicial product, i.e. the product of the underlying sets with $T \subset A \times B$ spanning if and only if its projections in A and in B both span. This monoidal structure is closed, with (AB) consisting of all simplicial maps $f: A \to B$ with a subset $\{f_1, \ldots, f_n\}$ spanning if and only if, for each spanning subset T in A, the set $f_1 T \cup \cdots \cup f_n T$ spans in B.

For the second monoidal structure $\mathscr{K}'$ on $\mathscr{K}_0$, we define $A \otimes' B$ so that it solves the problem of bimorphisms; we take $A \otimes' B$ to be the product $A \times B$ of the underlying sets with, for its spanning sets, those of the forms

$$a \times S, \quad a \in A, \quad S \text{ spans in } B,$$
$$T \times b, \quad T \text{ spans in } A, \quad b \in B.$$

Again I is the identity for $\otimes'$, and $\mathscr{K}'$ is symmetric and closed; $(AB)'$ consists of the simplicial maps $f: A \to B$, with $\{f_1, \ldots, f_n\}$ spanning if and only if, for each $a \in A$, the set $\{f_1 a, \ldots, f_n a\}$ spans in B.

The third closed structure $\mathscr{K}''$ is not monoidal, and we start by defining $(AB)''$ to consist of the simplicial maps $f: A \to B$, with $\{f_1, \ldots, f_n\}$ spanning if and only if either $n = 1$ or $f_1 A \cup \cdots \cup f_n A$ spans in B. We easily verify conditions (i)—(iii) of Proposition I.2.11, so that since K is faithful we have a closed category.

To see that $\mathscr{K}''$ is not monoidal, it suffices to show that the functor $(A\,-)''$ does not preserve products and so cannot have an adjoint. Let s^n denote the complex consisting of $n+1$ points with all subsets spanning, and let $+$ denote the coproduct in $\mathscr{K}_0$, i.e., the disjoint union; we write ps^n for the p-fold coproduct $s^n + \cdots + s^n$. Then

$$(2s^0, 2s^0)'' = 4s_0,$$
$$(2s^0, s^1)'' = s^3,$$
$$(2s^0, 2s^0 \times s^1)'' = (2s^0, 2s^1)'' = 2s^3 + 8s^0.$$

References

[1] Bénabou, J., Catégories avec multiplication. C. R. Acad. Sci. Paris **256** 1887—1890, (1963)

[2] —, Algèbre élémentaire dans les catégories avec multiplication. C. R. Acad. Sci. Paris **258** 771—774, (1964)

[3] —, Catégories relatives. C. R. Acad. Sci. Paris **260** 3824—3827, (1965)

[4] Birkhoff, G., and Mac Lane, S., Algebra. Macmillan (1967) (to appear).

[5] Brown, R., Function spaces and product topologies. Quart. J. Math. Oxford Ser. II **15** 238—250, (1964)

[6] Ehresmann, C., Catégories structurées. Ann. Sci. École Norm. Sup. **80** 349—425, (1963)

[7] Eilenberg, S., and Kelly, G. M., A generalization of the functorial calculus. J. Algebra **3** 366—375, (1966)
[8] Freyd, P., Algebra-valued functors in general and tensor products in particular. Colloq. Math. **14** 89—106, (1966)
[9] Kelly, G. M., On Mac Lane's conditions for coherence of natural associativities, commutativities, etc. J. Algebra **1** 397—402, (1964)
[10] — Tensor products in categories. J. Algebra **2** 15—37, (1965)
[11] Lawvere, F. W., Functorial semantics of algebraic theories. (Dissertation, Columbia Univ., 1963).
[12] Linton, F. E. J., Autonomous categories and duality of functors. J. Algebra **2** 315—349, (1965)
[13] — Autonomous equational categories. J. Math. Mech. (to appear).
[14] MacLane, S., Natural associativity and commutativity. Rice Univ. Studies **49** 28—46, (1963)
[15] — Categorical algebra. Bull. Amer. Math. Soc. **71** 40—106, (1965)
[16] Spanier, E., Quasi-topologies. Duke Math. J. **30** 1—14, (1963)
[17] Yoneda, N., On the homology theory of modules. J. Fac. Sci. Univ. Tokyo Sect. I. **7** 193—221, (1954)

Department of Mathematics
Columbia University
New York, N.Y. 10027

Pure Mathematics Department
University of Sydney
Sydney, N. S. W.

Herstellung: Konrad Triltsch, Graphischer Betrieb, Würzburg